Commission of the European Communities

Information Symposium

ROADWAY DRIVAGE TECHNIQUES

IN THE COAL MINES OF THE EUROPEAN COMMUNITY

Luxembourg, 9 and 10 November 1983

The three language versions of these proceedings are being published in partnership by:
— Verlag Glückauf GmbH, Essen (Federal Republic of Germany), German language edition
— Graham and Trotman Limited, London (United Kingdom), English language edition
— Société de l'industrie minérale, Saint-Etienne (France), French language edition

GRAHAM & TROTMAN LIMITED

EUR 9126 d, e, f

Organisation of the Information Symposium
by
The Commission of the European Communities
Directorate-General Energy
and Directorate-General for Information Market and Innovation

Publication arrangements:
P.P. ROTONDO'

Cover by
T. ROTA

LEGAL NOTICE

Published by:
Graham & Trotman Limited, Sterling House
66 Wilton Road London, SW1V 1DE

Printed in Belgium

ISBN 0-86010-575-X

CONTENTS

SESSION II - MECHANIZED DRIVAGE IN COAL

Chairman : A. WHEELER, Director of the Scottish Area, National Coal Board (United Kingdom)

SESSION III - MECHANIZATION OF SUPPORTS

Chairman : J. VANDERPUTTE, Director-general NV Kempense Steenkolenmijnen, Houthalen (België)

SESSION IV - SHOTFIRING AND LOADING, SHAFTS

Chairman : F. HEISING, Geschäftsführer, Steinkohlenbergbauverein, Essen (Deutschland)

ROUND TABLE DISCUSSIONS

Chairman : P. PHILIPPRON, Ingénieur principal, siège Zolder, N.V. Kempense Steenkolenmijnen, Zolder (Belgique)

Chairman : P.M. DAVIES, Senior Mining Engineer, National Coal Board, North Derby Area (United Kingdom)
M. Phil., F.I.Min.E., C. Eng.

OPENING SESSION

- Coal policy in the european Community
- Roadway drivage - The changing scene

COAL POLICY IN THE EUROPEAN COMMUNITY

Dr. K. REICHERT

Director for Coal, Directorate-General for Energy
Commission of the European Communities, Brussels

Over the last ten years there have been drastic changes in the energy situation in the European Community. In 1973 oil accounted for over 63% of our total energy supplies, and almost all of the oil required had to be imported. In the ten-year period leading up to the oil crisis, consumption of coal fell by one-third, and nuclear energy was of comparatively minor importance. Ten years later the Community is a region with a lower level of oil consumption, and one in which there is a respectable level of oil production. Nuclear energy's contribution has quadrupled. Consumption of hard coal and other solid fuel for electricity generation has increased perceptibly, and energy efficiency has improved considerably.

Whenever people talk about energy nowadays they always allude to the lessons of the past. However, we must look to the future, for the energy situation in the European Community is such that the risks ahead must not be under-estimated. Consequently, in June the Commission of the European Communities proposed a five-year programme for energy and energy research in the Community which includes, among other things, a balanced solid fuels policy.

The main question for the Community in the energy sector is whether the process of adjustment will continue. The following problems in particular are involved: the reduction of dependence on imported oil, the diversification of energy supplies, greater energy efficiency, and restructuring both on the demand side and on the supply side. Energy politicians are afraid that the present situation on the oil market could result in a relaxation of efforts and a slowing down of the process of adjustment, or even a return to the previous patterns of behaviour and consumption.

Market forces have been very effective. Since 1979 there has been a break in the hitherto upward trend in energy consumption and a dissociation in the hitherto close relationship between economic growth and energy consumption (sometimes referred to as the energy coefficient). However, a feature of the present conditions is the continuing uncertainty as to which way things will go: will we get over the recession, and will we be spared a third oil price shock which could send us into another international economic crisis?

In 1980 the Community agreed on a number of energy-policy objectives for 1990, e.g.:

(a) to reduce oil consumption to 40% of total primary energy consumption,
(b) to reduce the energy coefficient to 0.7 or less,
(c) to cover the primary energy input for electricity generation by solid fuel and nuclear energy to the extent of 70 to 75%.

The prime responsibility for attaining these objectives lies with the Governments of the Member States. I am pleased to be able to say that the statistics indicate that progress has been made in this direction, and that further progress is on the cards in the individual countries between now and 1990. However, these forecasts are based on assumptions which are no longer valid. For example, some Governments are reckoning on substantial investment in the power station sector, whereas the number of planned power stations and power stations under construction is at the lowest level since 1977. Some Governments are reckoning on substantial investment in indigenous energy sources, whereas investment capital is in short supply and expensive. Others are reckoning on the penetration of coal and other solid fuel in industry, whereas conversion from oil to coal has been a slow process and is now virtually at a standstill. Most Governments are assuming a level of economic growth which is unelastically high. And all these hypotheses were made before the price of oil fell by 15%.

In view of this state of affairs, the Commission considers that the Governments must maintain the process of restructuring, and that some of the efforts must be made at Community level and must be financed by the Community. The Commission has therefore proposed a five-year programme (1984-88) the funding of which would entail a seven-fold increase in the previous budget appropriations for energy: 770 MioECU. Expenditure would rise to 1 500 MioECU to 2 000 MioECU by 1986/87 and would then fall. This may be regarded as an ambitious programme, but we shouldnot lose sight of the fact that this year the Community will spend over 4 000 MioECU on milk products and 400 MioECU on tobacco alone.

In addition, the European Council meeting in Stuttgart instructed the Council and the Commission to work out new Community policies to give renewed vigour to the European Community. Energy policy could perhaps be developed as a counterbalance to agricultural policy.

The question of a Community policy for solid fuel is of course bound up in this, since solid fuel can make a contribution towards the diversification of the pattern of supply and a change in the pattern of consumption. It is a fact that consumption of hard coal in the European Community has not increased significantly since the first oil price crisis of 1973/74. It then totalled 310 million tonnes, rising to all of 314 million tonnes by 1980. Since then it has fallen again and this year will total only 292 million tonnes. The steel crisis and the low level of economic growth are responsible.

Coking coal is the biggest loser at the moment, and nobody expects an increase in activity in the steel industry in Europe or in the rest of the world. In 1983 the steel industry will require only 42 million tonnes of coke; that is one-third less than in 1977. This decline has been only partially offset by the slight increase in specific coke consumption in blast furnaces. The coke rate seems to have stabilized at 530 kg.

On the other hand, the prospects for steam coal are better. Although power stations are not producing substantially more electricity than last year,better results are in prospect as regards coal consumption in the electricity generating sector. However, there are considerable differences in the situation from one country to another. By way of an example, I would mention France where this year only 12 million tonnes instead of the 17 million tonnes originally planned will be needed by Electricité de France. In addition, the power stations pursue a wide variety of stockpiling policies. All in all, a coal input of 185 million tonnes is likely, corresponding to around 43% of the primary energy used for electricity generation.

In recent years impressive rates of increase in coal consumption have been recorded in the industrial heat market.

There are two reasons for this: firstly, the cement industry has now fully converted back to coal, as it has no environmental problems to contend with. Secondly, this market is so small - 17.5 million tonnes of coal and 4 million tonnes of coke - that even small tonnages have a large impact in terms of percentages. Industry in the European Community still consumes 75 million tonnes of fuel oil and considerable quantities of natural gas. The potential for converting to coal is considerable, but there is clearly little inclination to do so - under present conditions at any rate.

On the supply side, Community coal has been unable to maintain its market share, despite all the hopes immediately after the first oil crisis. In 1973 production totalled 270 million tonnes. In 1983 it will amount to only 233 million tonnes. On the other hand, imports have steadily increased, rising from under 30 million tonnes to 74 million tonnes in 1980, and this year they will amount to around 62 million tonnes. This trend clearly reflects the competitive situation on the market.

The energy-policy debate at Community level has placed considerable emphasis on coal and the other solid fuels in the last 20 months, and the discussions are continuing, as the Council meeting in Brussels on 4 November has shown. It would be an exaggeration to speak of clear results and policy guidelines. This is attributable to the fact that coal issues are viewed and assessed differently by the individual Member states depending on whether they themselves produce coal or are dependent on imports. In addition, there are now the special interests of Ireland and Greece as peat and brown coal (lignite) producers are to be taken into account.

In the spring of 1983 the Energy Ministers agreed that the discussions should focus on the following four matters:

(a) obstacles to greater use of solid fuel;
(b) ways and means of promoting greater use of solid fuel through research, development and demonstration;

(c) the influence of Community production of solid fuel (under satisfactory economic conditions) on the security of energy supplies;
(d) what the Community can do to resolve the social and regional problems closely linked to the future of the coal industry.

Where the first point - greater use of solid fuel - is concerned, the Community already has a number of instruments, and further measures have been proposed.

Two Council Directives adopted in 1975 prohibit the construction of new (single-fuel) oil or gas power stations. Exceptions must be justified at Community level.

In the spring of 1983 the Council addressed two recommendations to the Governments of the Member States calling upon them to promote investment in the conversion of oil-fired industrial boilers to solid fuel and to support the use of coal for the heating of public buildings and in district heating systems.

In 1982 alone, loans totalling 832 MioECU were granted for investments in the use of coal,notably in power stations. Some of these loans were granted at reduced rates of interest.

A proposal designed to enable such subsidized loans to be granted for investment projects, including conversion to coal in industry, has been before the Council for a year. Unfortunately, it seems to have little chance of being adopted.

The second point - research, development and demonstration - is particularly important in the context of greater use of solid fuel in industry. The conversion costs are often regarded as being excessively high in comparison to the proportion which energy costs represent for a company overall, and solid fuels are still regarded as being old-fashioned, dirty and inconvenient.

However, there are new technologies and developments which give the lie to this. As it can be demonstrated that such plants constitute a genuine alternative to the more expensive hydrocarbons oil and gas, a large market could emerge here. New technologies entail risks, however, and so the decision to employ them is deferred or only taken if there is sufficient financial incentive.

For this reason, the Commission has embarked upon efforts to support demonstration projcets in the various energy sectors, one of which is the combustion, transportation, transfer and storage of solid fuels and the use of waste fuels. The emphasis is of course on techniques such as fluidized-bed combustion, coal/liquid suspensions, and combined cycles etc. The Commission will shortly be allocating 20 MioECU to 16 projects selected from 68 proposals. Basically, an operation of this kind can only be effective if it is multiannual in nature. The Commission had therefore also proposed a five-year demonstration programme. However, the discussions in the Council have proved to be particularly difficult and complicated, not least because of the financial questions. No final decision for the period 1984-87 has yet been taken.

In parallel with this programme in the field of combustion technology, the Commission has also proposed a research programme which is part of a multiannual non-nuclear energy reserach programme and will, we hope, be adopted shortly.

For the sake of completeness, however, mention should also be made of the long-running demonstration programme for gasification and liquefaction, which has also been extended for 1983, but no further so far. The Commission regards it as being particularly important not to relax efforts in this area despite the declining interest worldwide, since it could be of considerable importance for energy supplies, and also from the point of view of exports of equipment and technology.

The focal point of the solid fuels policy which the Commission has proposed to the Council is the question as to how the production of hard coal in theCommunity can be made more competitive. Only a small proportion of the existing production capacity can compete with imported coal, and only an industry operating on a sound economic basis can contribute towards the security of energy supplies.

For this reason, the Commission has proposed that 300 MioECU should be made available from the Community Budget each year for 5 years in order to pormote investment designed to increase productivity. A non-repayable subsidy of up to 25% of the investment costs is envisaged.

In addition, investments designed to increase capacity in the peat and brown coal (lignite) industries in Ireland and Greece should also receive a Community subsidy of this kind in order to enable these two Member States, which are among the least prosperous, to reduce more quickly their dependence on oil and the related costs.

The comparatively high proportion of coal mining capacity in the Community which is completely or marginally uneconomic or uncompetitive, and the resulting lack of adaptability to changing market conditions have presented the Community with the problem of large stocks of unsaleable coal. On the other hand, these stocks constitute a financial burden for the firms in question in two ways: they reduce the urgently-needed investment funds and increase the total costs of production. In addition one should not under-estimate the adverse psychological effects of large stocks of coal where the miners and coal users are concerned.

For this reason, the Commission proposed a three-year action for the reduction of stocks, whereby the undertakings would receive a payment of 10 ECU per tonne of reduction. However, the Member States raised fundamental objections to this proposal and, moreover, non-Member countries have protested vociferously that such a procedure appears to go against the International Energy Agency's principles relating to world coal trade.

Alongside the question of investment aid for the coal industry, the extension of the coking coal arrangement is also being discussed. This system, which has operated for17 years, combining national subsidies and Community sales aid and thus securing a coking coal basis for the steel industry in the European Community, is to be adapted once again to the economic conditions, before it finally expires in 1988. The Commission is of the opinion that there is no longer a serious problem where security of supply is concerned, and that the steel industry itself should bear the costs of its supplies. The expiry date should enable suppliers and customers to adapt to the new conditions.

As the last component in the series of policy proposals, the Commission has proposed to the Council that additional funds totalling 60 MioECU for each of the next three years should be transferred to the ECSC budget. It is likely that the restructuring of the coal industry, and in particular the closure of unprofitable pits, will result in increased spending on social measures. Article 56 of the ECSC Treaty, which provides for various types of aid for the workers affected, will continue to play a very important role. In addition, the scheme for creating new jobs for ECSC workers through the granting of subsidized loans to industrial firms, which has also proved its worth, will be continued.

Last Friday the energy ministers discussed the whole package of measures at length, but without reaching any definite decisions or conclusions. The next Council meeting for energy questions, which is planned to take place in mid-December, could produce a decision on the coking coal system which will involve, however, a shorter duration and different - more limited - financial arrangements than originally proposed.

All other questions will be considered in the light of the outcome of the European Council that is to take place in Athens at the beginning of December. The Commission was asked toputforward a new analysis of the market for solid fuels and the competitiveness of the Community's industry with regard to improted coal and other energy vectors. In addition, the questions of solid fuel use, including problems of environmental protection and the socio-economic repercussions of the restructuring of the coal industry will be examined afresh.

In conclusion, I would repeat that the Commission is determined to do all that it can to support and facilitate the switch away from oil, and that increased use of coal and other solid fuel would certainly be one of the most successful ways of achieving this. At present, and in the short term, a rapid increase in coal consumption is unlikely. It would seem that the general economic trend, which in the final analysis is decisive for the position of Community coal mining, has now bottomed out. A revival in the world economy will bring about stability on the world energy market and coal must take a substantial share of this market. Indigenous coal must maintain its position vis-à-vis competing imported coal, since the coal market must be as free as possible if diversification and cheap supplies are to be guaranteed.

I now declare the Information Symposium on Roadway Drivage Techniques in the Coal Mines of the European Community open.

ROADWAY DRIVAGE - THE CHANGING SCENE

L J MILLS, CBE

Deputy Chairman, National Coal Board, United Kingdom

Summary

This technical introductory paper sets the theme for the Symposium and touches broadly on the individual aspects of roadway drivage and shaft sinking that will be covered by subsequent technical papers. The paper indicates the continued requirement for productive and financial improvements in the context of the market for coal and its competition. It touches on the progress made in all operational spheres, but with particular emphasis on those areas of research and development which have potential for further improvement and application, and which are considered to be essential contributors to the better health of the mining industry. Particular attention is paid to new machine and systems control techniques which, with the advent of the microprocessor, provide means of obtaining precise engineering and management information and of controlling, in a consistent and automatic way, operations which previously relied on a high degree of manual skill. A general concern with the reliability of equipment and systems is expressed, and the question of devising industry procedures for obtaining proper planning information and the means of acquiring proper financial outcome via life cycle costing is also discussed. The message of the paper is that, while recognising past and current achievements, we should nevertheless regard them mainly as stepping stones towards realising the potential for improvement that exists, through the application of new technologies and new techniques to production, control and financial and operational planning.

1. INTRODUCTION

The coal industry of the European Community is the largest tunnelling contractor in Europe. Each year, in Europe, hundreds of kilometres of tunnel are driven (more than 1500 kilometres). Roadways range from small in-seam drivages to the very large cross-measure tunnels, which combine mines or open new coalfields. Thus, the mining engineer has a rich and varied experience. He will work in widely varying strata, from hard, solid rock to the incompetent rocks encountered in faulted ground. He will cope with water, dust and gas or rock outbursts.

Against this huge and complex background of challenge, problem solving and arduous experience, machines and techniques have been developed, which are second to none, and which have been exported to all other coalmining areas of the world. Strenuous efforts have been made to strengthen and develop the efficiency of our operation, through exchange with other industries, notably in the civil engineering field, and more recently in that of control technology.

Any coal drivage can now be mechanised, from thin coal headings of less than 1 metre, using the smaller In-seam Miners, up to seams of more than 3 m, employing continuous Miners, Dintheaders, etc, and small roadheaders. Machines are available for the smallest access roadways, eg from the steep seam workings in France to the large circular roadways in excess of 6 m diameter used in colliery combination work in Germany.

Since the last Community sponsored symposium on Roadway Drivage ten years ago, immense changes have occurred, both in equipment and in the structure of production capacity. Concentration of production has been planned and achieved, new coalfields have been opened, many thousands of kilometres of roadway have been driven and many shafts have been sunk.

Several papers will outline this history and the current situation. They will show that, often, excellent operational results are being obtained. We will see, however, that both development engineers and operators must use improved techniques, new materials and control technology to achieve even better results. This is an appropriate time for practitioners of the art of tunnelling to meet and exchange information and experience and to indicate the future direction. The opportunity of reflecting on past experience and of using it in the interests of efficiency and competence will not be lost. Every effort must now be directed towards developing the next stages of progress, as a contribution to the better economic outcome which the coal industry needs in order to survive at an appropriate level within the energy scene.

I do not propose in this paper to review in any detail the experience of the last ten years. Other papers will do that. Nor do I intend to provide tables of performance statistics, rates of advance, costs, etc, except for one: that concerning the rate of productivity increase over the last 10 years. The example taken is from the NCB and it illustrates the rapid rise of coalface productivity and the much flatter curve for elsewhere below ground and its effect on overall oms (Figure 1). Tunnelling's contribution can improve this vital area.

2. BROAD PROGRESS AND CURRENT SITUATION

2.1 Rock Removal

An engineer choosing a machine for cutting rock has a choice of several basic methods. He can drill the rock ready for shotfiring; currently available for this purpose are hydraulically powered rotary and rotary percussive machines, besides the normal heavy-duty compressed-air drifter. He can use one of the many roadheaders, ranging from the light-duty 20 tonners up to the heavy-duty machines, weighing 100 tonnes or more. He can deploy a variety of cutter picks, from the conventional radial and point-attack types, right up to the very strongly built rotating point-attack picks for the hardest rocks, or alternatively, the very large radial picks as used on the two heaviest NCB machines. The thrust forces he can use to urge the cutting head into the rocks can be as low as 2 tons or as high as 50 tonnes.

Given the requisite scale of operation, it may be sensible to use discs and roller cutters as applied on full-face circular machines. The heaviest of these machines weigh around 300 tons and are designed to provide massive thrust at the cutting head, which can be the equivalent of more than 20 tonnes per tool. They are capable of cutting any rock likely to be encountered in coalmining.

Figure 2 demonstrates this wide choice of machines by identifying cutting head torque, cutting head power, and slewing and elevating forces for roadheaders or thrust forces for full-face circular machines; it also shows the weight. These selected examples, which range from the lighter roadheaders up to full-face tunnelling machines, do not provide a comprehensive listing.

2.1.1 Drilling

Even now, light or medium-duty equipment can be brought to an abrupt halt by unexpected geological conditions, e.g. faulting, sandstone intrusions or other anomalies, causing mechanical difficulties and rendering further progress uneconomic. The substitution of equipment powerful enough to deal with these circumstances is a costly nuisance. However, alternatives are on the horizon. The polycrystalline (Stratapax or Syndite) tipped drill bit, properly applied to pure rotary machines, has already shown that it can provide - in certain circumstances - rapid drilling, consistency and bit life many times that of the conventional tungsten carbide tool. Work done in France, Germany and Great Britain has shown that similar results can be obtained by using high-pressure water jets to assist conventional drill bits. These are just two, quickly applied, 'turn on, turn off' solutions, which await refinement of the production engineering to promote wider use.

Drilling with jumbos is well established and offers many advantages, but is very dependent on the skill of the operator. Automation offers a solution to this problem and already (in Norway for example) automated drilling has reached the stage of production demonstration. Microprocessors analyse data collected from transducers which are positioned to measure boom element orientation, and from attitude instruments on the machine itself, so as to assess and determine the commands that must then be given via another microprocessor to the electro-hydraulic control system. This ensures the correct programmed drilling of a number of positioned holes with maximum efficiency by two-boom or three-boom machines. The NCB is currently arranging with a manufacturer for the certification and trial of such equipment on coalmining machines. No doubt similar work is going on in other parts of Europe and I look forward to hearing more of this in the relevant papers.

2.1.2 Cutting (Picks)

Much more cross-measure drifting is being carried out by roadheaders. Following papers will show the extent of this, the results achieved and problems exposed. Essentially, machine development has been in providing greater power at the cutting head, greater slewing and elevating forces and greater rock clearance capacity. The consequence has been an increase in machine size and weight, in order to provide the necessary constructional strength to resist the very high reactive forces induced during operation, and also to keep the machine stable in the roadway by virtue of its weight, rather than anchorage (Figure 2). Larger and more powerful machines need larger, stronger and more expensive cutter picks. It has not been conclusively proved that the increase in power, weight and cost has proportionally raised the degree of hardness of rock they can cut satisfactorily and economically. The increase in rate of excavation has probably been all that was hoped for.

We will see from a British paper that augmentation of strength, weight and force is not the only approach to making available at the cutting head that extra power which circumstances may demand. In theory, this can be achieved (Figure 2) without increasing the reactive force, by raising the pressure of water through pumps or intensifiers and transmitting it to the cutting head and a nozzle discharge system. This extra power can be efficiently utilised as an aid to rock cutting. There is an NCB/US Department of Energy project on the use of high-pressure water jets to assist mechanical cutting. This work has reached the point of demonstration, where up to 112 kW equivalent has been applied at a roadheader cutting head via nozzles positioned near cutting picks and emitting water jets at pressures of around 700 bar. The jet impinges on the rock immediately in front of the cutting pick. With attention to selection of equipment and detail design this practice raises the cutting capability of the machine and, in the most abrasive conditions, prolongs the life of the cutting picks, thereby affecting the economy of the operation.

Production engineering improvement is required to ensure safety in operation, as well as reliability. Perhaps automatically controlled phasing devices are desirable to limit the amount of water used, particularly on ripping-type heads. Perhaps much more fundamental research is required. But clearly, here is an opportunity to try to break away from the costly process of making machines bigger, heavier and more expensive.

This subject is of considerable interest in Europe and with the goodwill of our project partners in the US Department of Energy, we practise interchange of information with France, Germany and Belgium. Let us look forward to rapid, practical progress.

Apart from the high capital cost of some roadheaders, there is the continuing cost of maintenance, repair and component replacement. Some of this may be regarded as a natural consequence of designing, for employment in arduous conditions, a machine that is made to do its work via a cutting head at the end of a long boom, itself connected to a machine chassis which is generally not stabilised in the roadway except through its own weight. The advantages of such a machine are its flexibility in operation, its ability to cut almost any shape of roadway, and sometimes its quality of manoeuvrability for loading, etc. The mechanical disadvantages are fairly obvious.

In the large cross-measure drivages with perhaps, more standard requirements, it might now be possible to agree on certain limiting parameters of roadway size and shape. This should permit machines to be designed with a certain loss of flexibility in operation, offset by the application of better design principles: slower linear cutting speeds, more rigidity, consistent application of tool to rock (an approach to the principle of machining of metals). It would lead to the accommodation of gearboxes, bearings, cylinders, etc with reliability factors well in excess of design requirement. Centre-line operation of the machine, via full width loading aprons, or some form of loading independent of machine position would offer benefits in the design and operation of traction elements.

The benefits arising from such a specification and design approach are already evidenced by the rock cutting ability of the MRDE circular tunnelling machine at Cadley Hill Colliery. A number of the principles mentioned above are also incorporated in the Paurat E200 machine, and other German machines, the Anglo/Soviet Roadheader and the NCB/Eimco machine. We look forward to examining the wider experience arising from their operation.

2.1.3 Cutting (Discs)

The papers from Germany and Great Britain on full-face tunnelling machines will show the benefit of operation with discs and roller cutters. Papers from France and Belgium will discuss the mining applications of these tools with a different method of application.

Extensive experience in Germany, and two installations in Great Britain, have indicated the production capabilities. Robbins, Wirth and Demag machines have the ability to cut through both hard and abrasive rock, which would probably not be amenable to even the most powerful of the existing heavy duty roadheaders. Drivage rates of up to 150 m per week in circular roadways of diameters between 5 and 6 m have been demonstrated. Wherever a major roadway has to be driven, over some minimum - but still debated - economical length, where large roadway cross-sections for ventilation and/or transport are required, then this type of machine must be a candidate for selection.

There are economic limitations to its use, and these generally derive from the nature of its operation, which requires to cater for the very high thrust necessary to provide the individual loading on more than 30 rolling cutters or discs, and the high torque necessary to provide maximum cutting efficiency. This leads to massive machine design for high anchorage forces and resistance to the generated reactions.

If the thrust and torque requirements of the tool itself can be changed, then perhaps machine design, weight, cost and therefore range of application can be changed. In Germany, much effort has gone into effecting this and we now have the vibrated roller disc ready for production trials. In a paper from the same country, we will hear of surface trials of the application of high-pressure water via jets disposed to assist the discs, which reduce torque requirements by significant amounts.

The French and Belgian papers deal with the 'Bouygue' machine and illustrate a principle of discs attacking a circular face, but mounted on rangeable arms. By virtue of a considerably reduced number of tools, the machine thrust requirement is reduced, hopefully in the same proportion.

We do not yet know much about the mechanical reliability of the vibrating discs, or the equation - extra installed horsepower versus improved cutting - on the high-pressure water project. We are not sure if we can eventually cope with the higher tool speeds that the Bouygue machine principle will probably induce. But these are valuable and worthwhile undertakings, which we need to see successfully developed in order to widen the scope for full face operation in our coalmines.

2.2 Roadway Supports

Perhaps the aspect of roadway drivage which has changed least, is the roadway lining. Generally D-shaped steel roadway supports have been used, sometimes including yielding elements and sometimes with different lagging or strutting. It is true that conditions in Belgian coalfields have caused them for many years to use circular roadways lined with concrete blocks or segments, but until recently, this practice has not spread widely in the rest of Europe. In Germany and in France, many kilometres of roadway have been supported during the last two years, solely by roof bolts or by a combination of roof bolts and shotcreting. In Great Britain we have experimented with roof bolting as an aid to conventional supports and the value of shotcreting has been demonstrated in difficult conditions. However, I would suggest that in Europe, 90% of roadways are still supported by the steel arch erected in the old labour intensive manner.

I therefore particularly welcome the papers showing new approaches.

The opportunities for change are clearly available. The relatively widespread use of circular full-face machines in Germany has indicated the superiority of the full circular support over the arch. Certainly at Selby where we have a circular roadway running parallel to the conventional D-shaped one, direct comparisons tend to confirm the better resistive capability of the circular roadway.

The most adventurous experiment being carried out in Great Britain is the use of full concrete circular supports in a major roadway at Cadley Hill Colliery, in conjunction with the MRDE tunnelling machine. Much of the technology here has been taken from civil engineering, with the segment design based, to a considerable extent, on Belgian experience. This is a totally new experience and the development of the operation and management control has been intensive and sometimes difficult. However, these supports, set in 1 m increments, weighing 10 tonnes per metre, requiring the inclusion of proper yielding elements between segments, as well as efficient packing and grouting behind the supports, now line more than 700 metres of roadway and by February/March next year, will hopefully line the complete 1500 m length.

In spite of the weight and size of the segments - the roadway is 5.5 m diameter - it has been convincingly demonstrated that up to 5 m of advance and support setting can be made in a shift and, since the roadway passes through notoriously difficult strata, the suitability of the support will be rigorously tested.

The opportunity has been taken to expand this work into a piece of long-term strata control research and at the same time provide an early warning system. At approximately every 50 metres, one set of supports is equipped with strain gauges on the inner and outer skin to measure and indicate the load-induced strain in the support. Data is collected and analysed by microcomputer. The results obtained so far show the distribution of strain around supports is by no means consistent. This probably indicates point loading arising from the difficulty of executing ideal grouting behind the support. In some places, loads are exhibited which are well within the designed resistive range (1 MPa) of the supports, but which would already have started to deform steel circular supports. Measurement of the effects of remedial measures taken in one particularly bad spot has been extremely beneficial, and emphasises the indispensible value to the engineer of instrumentation in an otherwise 'blind' situation. Figure 3 shows strain data recorded and analysed at one instrument station and indicates the 'point' loadings, and also the trend of strain with time.

Where conventional supports are used there has been a persistent problem of erecting them quickly enough to avoid lost machine time. Many attempts have been made to solve this problem and many pieces of equipment have been tried, particularly in conjunction with roadheaders: walking shields, solid shields, monorail mounted erection devices. By and large, no approach has achieved widespread application.

The German paper on mechanical aids in support setting will show us where success has been obtained, and which of it can be built upon with further development. But such a well known problem, which has remained largely unsolved over so many years, requires a more radical rethink, involving a completely new look at the design, disposition and combination of the basic equipment.

In a small number of cases, support setting does not represent a complete loss of machine time. The full-face circular tunnelling machines, for example, are designed and operated so that all or most of the support setting can take place during cutting and loading. In the case of the MRDE tunnelling machine, the supports are completely erected while cutting and loading continue. The Westphalia Shield machine is similar. With the NCB/Dosco In-seam Miner, the design of the machine is such that support erection - in this case props and bars - can be effected while the machine is advancing into the coal.

In these cases, the records of machine utilisation show that the machine advance is constrained mainly by its ability to cut through the strata. Support erection is carried out under cover and in comparative safety. These are tremendous advantages over the use of, say, conventional roadheaders or continuous miners, in which cutting and loading, followed by support erection, are cyclic operations and 50% machine utilisation represents the best normally achievable. Figure 4 shows actual power recordings on an In-seam Miner and a Mk 2 Dintheader illustrating the difference in machine-utilisation. These examples give us cause to reflect whether we need, not additions to and around our machines, but rather, a fundamental change in machine design which will give us an inherent facility for continuous operation.

2.3. Coal Heading Machines

There is, within the older coalfields of Europe, an increasing interest in the ability of coal heading machines to operate as coal production units, or as very rapid face opening installations.

The instantaneous capacity of our coal heading machines has never been in doubt and in well planned installations, with properly provisioned and organised services, excellent results may be obtained. Nevertheless, some of the deficiencies of system operation, which were apparent a long time ago, have persisted. A mobile machine does not respond well to being serviced by immobile ancillaries.

Temporary and less efficient conveying facilities have to be provided when a machine is turning off a roadway, or making a fresh entry. This is perfectly illustrated by the partial extraction system at Whitwick Colliery where an in-seam miner on its production run of 700 m makes an advance of 25-30 m per shift - on one occasion up to 50 m - but at the end may require one working week for re-establishment prior to the next coalwinning advance.

At Ensdorf Colliery in Saarland, excellent advance rates have been obtained with a shearer type machine operating on a short curved conveyor, again after the initial installation and start up procedures.

We have tried to obtain the necessary mobility of back-up equipment with ideas mainly from the USA, (tracked conveyor tail ends, etc.), but their use is not widespread. Experiments recently carried out, using load/haul/dump trucks behind large heading machines has indicated a very flexible system and one which might be developed further in conjunction with some form of mobile bunkerage. Perhaps our concentration on production of coal from longwall faces has deflected our attention from solving this problem, but with a new interest in the extraction of coal from remnant pillars, under urban areas etc, better answers will evolve from our greater dedication.

2.4 Dust

A priority matter in tunnelling, and one of constant concern for reasons of health hazard and loss of operating efficiency, is dust. The problems of dust control change with the degree of power and complexity of machines, but the change takes different forms. For example, full-face circular tunnelling machines, with very high extraction rates, may make a good deal of dust, but the machine is so designed that the cutting area can be enclosed behind a flexible bulkhead and the dust laden air extracted for filtration, without any real effect on other operations. It is also possible that, even though subject to heavy operating forces, the rolling discs will not make as much dust as a drag bit and so, to some extent, the problem here is less acute.

The same is not true of heavy-duty roadheaders. Here, cutter picks are working hard, excavation rates are high, and, for operational reasons, it is not possible to enclose the worked area. Exhaust ventilation applied in these situations (if the extraction ducts can be positioned close to the cutting head) can work efficiently enough to keep the operator in fresh air and meet statutory requirements for dust control. That does not mean to say that the area around the cutting head and part of the boom is free from dust. Visibility is severely restricted and, consequently, so is operational efficiency.

Dealing with the collected dust also presents problems. The quantity is considerable and in Great Britain it has been found that our heavy-duty wet type filter units have needed pre-cleaning facilities to remove the bulk of coarser particles. Even so, when 5 or 6 cubic metres per second of dust laden air is entrained, the filters are on their limit.

In Germany, the problem has been dealt with by using the highly efficient dry type filter, whose disadvantage is its much greater size.

Our research and development work on dust indicates the benefits of high-torque machines with low pick speeds, and better and more reliable pick face flushing systems, which not only assist with dust at the cutting head, but also have an effect on dust made during the loading process and at transfer points.

2.5 Shafts And Surface Drifts

It is unfortunate that, the cycles of economic rise and decline in the coal industries, have made shaft sinking and surface drifting - themselves dependent upon new investment - periodic activities. In Europe the development of new shaft sinking methods, shaft lining and associated work, has therefore not been as consistent and enduring as other production activities. However, our contractors have, by dint of obtaining work overseas, kept in touch with the general trend of development, and practised this wherever possible.

We shall find from the papers presented, that up to the present almost all the shafts in Europe, have been sunk by conventional methods and that the linings, although much improved in design, have been placed by conventional methods.

We know there are alternatives to the conventional methods of shaft sinking through wet ground, with its requirement for high cost processes such as freezing. For example, a surface shaft has been drilled in Germany using the well drilling techniques developed for shaft sinking in the United States, and associated with lowering of a prefabricated steel lining into the completed shaft. There has been at least one other experiment with a shaft sinking machine in the United States.

Shaft drilling machines have recently been used for deepening shafts in Germany. We must accept that total 'shaft drilling' has so far been a production method only on smaller diameter shafts than we normally require. We do accept that recent experience with conventional operations has shown that tightly managed, tightly controlled and well organised operations can produce remarkable results.

However, I would hope that we, and the contractors who serve us, will not relax our efforts in the development of the newer, more radical approaches, which could, if successful, make a considerable contribution towards reducing the very high cost of new colliery development.

2.6 Loading

The papers on loading and haulage reflect two aspects of development, which are of growing interest. The use of loaders in drilled and shotfired circular roadways presents unusual problems, but problems which have been dealt with, in an evolutionary manner, as experience has built up in Belgium.

The use of free-steered, vehicles for loading out, particularly in shotfired headings, is becoming more and more established in France and in Germany, and also has a foothold in Great Britain. The ready discharge of mineral from the working place and the prompt supply of men and materials to that area are key factors in the efficiency of the operation, and the rubber-tyred vehicle with its great flexibility, assists in this.

The development of the quick detachable system (QDS) for use on such vehicles adds to this flexibility. This equipment allows the same vehicle to be used in its normal loading mode, as a materials carrier or for riding a small team of men into the heading (Figure 5). The idea is being developed to a point where the QDS gear is used to allow the machine to carry and operate drill booms and later, small light-duty cutting heads.

For those operations, where the length of the roadway or other circumstances would render a more complex system unviable, a workhorse to perform a wide range of tasks will be a tremendous advantage.

3. SYSTEMS CONTROL AND AUTOMATION

3.1 Machine Condition Monitoring

We look forward to hearing presentations on the application of the range of new technology, which can be used in the testing and the permanent monitoring of health and condition of equipment.

Most machines can now be specified for purchase with some element of built-in instrumentation, which will assist during the commissioning of the machine and in fault diagnosis. Measurements are usually of hydraulic pressure and flow levels, oil temperature and electrical power level. Where instruments are not built in they can be added. For some time now it has also been possible to use certified equipment for data collection onto chart recorders, data loggers and other tape recorders, so as to study the technical operation of a machine over periods ranging from a few minutes - where fast response is required to catch the transients - to several days, when scanning at 4 to 5 second intervals is acceptable.

Where machines and systems are large and complex, it is worthwhile considering a more permanent monitoring system. The advent of the microcomputer and better facilities for data transmission and analysis make this feasible. The first examples are appearing. The MRDE tunnelling machine is equipped with a prototype system, based on the NCB's General Purpose MINOS Package. The instrumentation covers hydraulic and electrical levels and status, on the machine, on the supporting power packs and system conveyors, and on the ventilation. Other specific information on machine orientation can be displayed.

The information is transmitted by hard wire to a surface outstation in the colliery control room, where a dynamic display of machine condition can be obtained. Historical data can be called up in the form of 8-hour, 32-hour and 7-day histograms. These can be used to provide trend information for the engineer as well as operational information for the colliery management.

In Germany, production monitoring equipment is available for the heavier machines, where similarly, a large number of selected technical measurements can be made. The data can also transmitted to a surface outstation for ready display and recording of information. An interesting development is the interpretation of data by the computer, based on evolving experience, which enables it to suggest the location of problems within a hydraulic circuit or other machine element.

This practice will undoubtedly spread. Its impact on production costs and performance will be considerable. The next steps in Britain in this direction will be the coverage of the Robbins tunnelling machine at Selby and a 'next generation' monitoring system for the NCB/EIMCO machine will be applied. We will be looking for sensible application to our general production machines, and I have no doubt that the same process will apply in Germany, France and Belgium.

3.2 Automation

This is a priority area for attention.

The tunnel is a neglected area of opportunity for automation. Compared with a coalface, a tunnel presents a limited geographical area with a convenient concentration of machine and ancillary equipment. The machine operations can be made to originate from a stable base - better, say, than a shearer operating on a flexible snaking panzer - and the consequences of a tunnelling machine action are more predictable.

Clearly, a machine which is not only expensive, but also an essential tool for the prompt performance of work necessary to our coalfield development and pit extensions, should normally be in the hands of skilled operators and management. Nevertheless, there are many operator-dependent functions which would be better done in some more consistent, automatic manner, eg the cutting of the correct roadway shape with a large roadheader. Operator visibility is restricted by machine bulk, and sometimes by large amounts of dust due to the rapid excavation rate, so there is an incentive to provide an automatic means of making the roadway shape. Over the years, there have been demonstrations of intention to do this, but there has been no large commercial application. Fortunately, with the advent of powerful microprocessors and considerable instrument development, the process is again underway. Germany has already applied an operator aid in the form of an indication system with signalled profile limits, which is used in conjunction with their machine condition monitoring system. Within the NCB, we have arranged to purchase and test two systems of positive profile cutting control, linked with devices to ensure that the shape is cut on correct roadway alignment. Other work in Europe is on the same lines.

Can we look further ahead? Once cutting control has been established on a roadheader, which then must perforce operate continuously on the centre-line of the roadway, is there any great difficulty in extending the automation process to proper sequencing of machine advance and other operations?

All of the operations on a full-face circular machine are relatively easily measured, sequenced and controlled. The machine lends itself to a much greater degree of automation, and hence consistency of operation, and should be the subject of study with this in mind.

3.3 Computer-Aided Design For Cutting Heads

The crucial matter of pick and cutting tool design will, no doubt, have been well discussed by the end of the Symposium. Research work carried out in all coalmining countries of the European Community has led to a more exact knowledge of the capability of individual cutting picks, the required angle of attack, lacing and loading patterns and techniques such as the rotation of point-attack picks to retain a practical level of sharpness. More difficult to pin down precisely have been the overall design of the cutting head and the marriage of ideal pick lacing with practical considerations of engineering and production design. With the aid of the ubiquitous computer, it is possible to improve this process considerably.

Computer programs are now available to examine the design of cutting heads for a variety of strata. One such NCB program contains individual programs, for specific aspects of design, all of them inter-linked by a master control program to provide them in the correct logical sequence. Designs can be permanently stored or retrieved for amendment. Pick boxes, scrolls, etc may be moved, deleted or placed in original positions, to achieve the optimum design or engineering compromise. Dynamic display of the pick breakout/cutting pattern, progressing across a screen, can be obtained.

This is an illustration of the way in which elements of machine and system design can be made more certain and interrelated with other interface factors to improve operation and reliability, to a previously unforeseen extent.

4. THE WAY AHEAD

4.1 Guided Longhole Drilling

We have no papers dealing specifically with the classification of strata and determination of cuttability of rock, but we are all aware that this is a vital adjunct to any planning we do.

Significant scientific advance has been made in determining suitability for rock cutting with, of course, the intention of determining the capability of machine and method selected for particular drivages.

There is no doubt that experts using the full range of tests developed in Europe (a characteristic range of compressive strength, tensile strength, abrasivity, quartz content, specific energy and even fracture count) can successfully assess the machine requirement to break down that strata. The problem is to know the incidence of strata along the line of a tunnel, particularly cross-measure drivages into new areas of exploitation. Normally, there will be vertical borehole information, but it is generally true that the information is precise at the point of drilling and the geology between boreholes has to be assumed.

We have had a revival of guided longhole drilling for exploration in the coal seam. The consequent development of natural gamma sensors, tilt and attitude instruments, suitable for incorporation in a probe of small enough diameter to be accommodated in the conventional drill string, improves the possibility of making our knowledge of strata along the tunnel route positive. We now have available an intrinsically safe system whereby an instrument probe can be pumped along whatever length of hole has been drilled - up to 1000 m. It can transmit signals relative to strata density and geometric position of the drill head through a trailing wire to a small computer situated near the drill operator, with data analysis and immediate read-out facilities. We have also developed machines of the correct physical size, capable of drilling more than 1000 m through cross-measure strata, and have devised and developed techniques for steering the drill and drill string in a desired direction.

The total package will enable us to drill and survey the complete line of most of our proposed tunnels, assessing the strata along the way by preliminary information from the instrument probe and by taking core samples at selected points for detailed examination in the laboratory. Figure 6 illustrates the overall procedure.

This process should remove uncertainties from the operational planning, known difficulties and anomalies being easier to deal with than surprises.

4.2 Economics

In the discussions that follow, we shall all be making our judgements on the financial benefit of using one machine and/or system as opposed to another, and our interest in this aspect of operations is sharpened according to the availability of money for capital investment and reserve extension schemes.

Once more, the computer comes to our aid in making relevant calculations and, because of its ability to assess and arrange information and its capacity for storage, filing and retrieval of large quantities of data, the use of comprehensive life cycle costing exercises becomes a less labour intensive and tedious exercise. Providing we arrange to set up an appropriate financial model, it is quite feasible for the larger centralised organisations to handle a very comprehensive input,e.g. 1. The capital cost of a machine and its ancillary equipment. 2. Design and development costs. 3. All that is involved in surface and underground testing with consequent modifications leading to acceptance trials. 4. Spares replacement costs and associated maintenance and repairs. 5. Costs associated with its operation, eg operator training and operator related facilities, its demand on mineral and material transport, etc. 6. Cost of installation, and later of the withdrawal of the machine and transport to the surface or to another operating site.

For an absolutely comprehensive exercise, it would be necessary to take into account such factors as the financial effect of reaching and exploiting reserves in a specified period, and the cost of any particular and peculiar demands its operation might have on the working of the rest of the colliery and the specialist facilities of the production area.

To put the results obtained in perspective, one must also find a way of taking into account physical factors, such as the amenability or otherwise of the strata and general conditions, the intensity of operation and the limits of back-up service which can be provided.

From exercises such as this, it should be possible to assess and compare equipments and methods on several useful financial bases, eg cost of unit of work done, (cubic metre extracted or metres advanced) or an industry cost per machine, etc.

This precise and detailed cost information should be developed to become part of the planning information we need when setting out on our larger scale ventures.

4.3 Management Information

Associated with machine condition monitoring and automation is monitoring for management information and control.

A major tunnelling installation can be as expensive in capital cost and as complex a matter for organisation and management as a coalface. It is important that management on site and at Colliery Office level are provided with good information, (historical and in real time) in order to react to topical situations and to have available, material for analysis and study as an input to planning subsequent installations.

Papers from Germany will indicate interest in applying this philosophy to full-face tunnelling machines and the paper from Selby will show in more detail what is planned for the Robbins machine at that site.

At Selby the first stage is intended to cover production activity.

For this, the pit-wide MINOS system will be employed, with FIDO software (using part of the coal clearance standby system for data transmission), control room VDU displays and printer, with a link to a secondary computer for long term analysis/storage and a remote terminal in the Manager's Office. There will be an outstation for data transmission on the pantechnicon from sufficient transducers to indicate the ON/OFF status of cutter head power - with distinction between idling and cutting - the first machine conveyor, the bridge conveyor, and all conveying systems out to the main trunk belt.

The output is similar to that of our coalface FIDO system, giving shift and weekly running delay times with appropriate percentages, and a chronological log of delays (with reasons for delay being input via a keyboard), with trend analysis on a shift, weekly or other selected basis via a secondary computer.

The second stage of the installation will add dynamic and historical information on relative machine element movement, and this will complete the management information coverage. Since a parallel road is being driven at Selby, utilising a heavy-duty roadheader, a second sub-station will be added which will provide comparative information on that machine and installation.

4.4 Reliability Improvement

While the size of roadways may well, in general terms, have increased in cross section, we still have an abnormal constriction on the permitted physical size of mining machinery, as compared with machines for similar duties on the surface. In terms of trying to accommodate the necessary power and strength in the machine volume available, we are always trying to get 'a quart into a pint pot'. Our standards of engineering design and manufacture must always be high and our means of checking that we are achieving and maintaining these standards must be constantly reviewed.

In Germany and in Great Britain we have a considerable surface facility for the development and the acceptance testing of complete machines and systems. We all have laboratory facilities, either within our industries or at the premises of our machine suppliers, for comprehensive testing of machine elements, particularly gear transmissions and hydraulic circuit components. Structural testing facilities are readily available and can be disposed to carry out almost any combination of tests.

Nowhere is the organised use of these facilities and developed procedures more necessary than in the field of rock cutting and loading, where operating conditions can never be guaranteed to remain stable, and can be very demanding.

Within the NCB a 'Reliability' system and procedure has evolved. All proposals for new equipment are assessed for likely operational satisfaction by the relevant specialist department at Headquarters, and safety aspects are rigorously examined, against a set of guidelines and codes of practice in addition to the statutory regulations. Critical design features can also be assessed through Headquarters Engineering Department and the Design Office at MRDE, with particular reference to geared and hydrostatic transmissions.

It is often necessary to run laboratory tests on, for example, a new type or a new power level of gearbox. When the machine is complete, surface trials will be carried out at Swadlincote Test Site to examine the operational facets and its suitability within a system. An earlier check on system compatability may have been carried out via a computer aided graphics system, which is capable of a three-dimensional view and examination at any selected cycle stage.

The machine will go into a colliery for underground trials, sponsored and monitored by the Headquarters Field Trials procedure. This monitoring will include instrumented commissioning trials. The longer term monitoring will be carried out using the same transducers, but logging or recording data over a wider scanning interval and at regular periods over some months. Technical information can be stored on desk top computers and analysed to highlight areas of unreliability and indicate necessary changes in design. An understanding of the trend information which may show decline in component efficiency can be used to reduce the incidence of breakdowns by preventative maintenance.

All of the information obtained is reviewed on a national basis via our Major Development Committee on Reliability and by other specialist departments such as the Defects and Modification Committee. The information is used to assist manufacturers in improving reliability locally and on a longer-term national basis.

Apart from any 'breakthrough' advances in design or significant improvements in manufacturing or materials technology, it is only by this rational, logical process of obtaining precise information and analysing it to promote a competent understanding of reliability trends, that we can make continuous and certain progress.

5. CONCLUSIONS

I said at the beginning of this paper that I would not attempt to detail progress in the art and science of roadway drivage over the last ten years. This information is available in other places and should certainly not be neglected. However, I know you will agree that at this time, when coal is once again fighting for its share in the marketplace, in competition with the now apparently plentiful oil, natural gas, imported coal and, of course, nuclear power, we should be totally concerned with the future and the steps we have to take, the opportunities we have to grasp, to ensure that technological advance assists in keeping our performance at a level which retains our competitive ability. Therefore I suggest that we listen with great interest to the very competent and enlightening technical papers which are to follow, congratulating ourselves on our past achievements, using them as a spring-board to exploit the undoubted potential for further improvement.

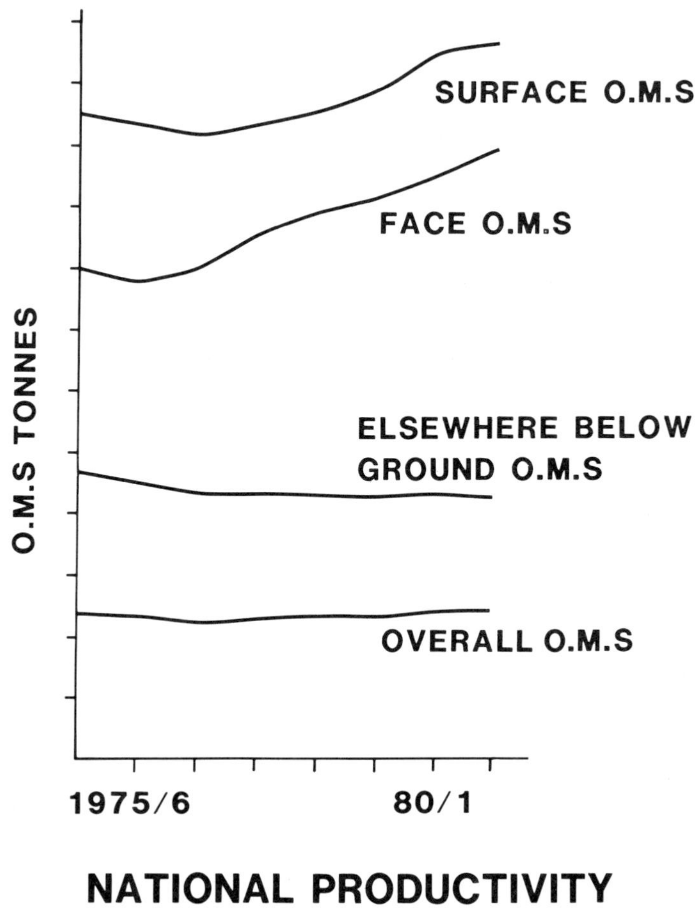

FIGURE 1.

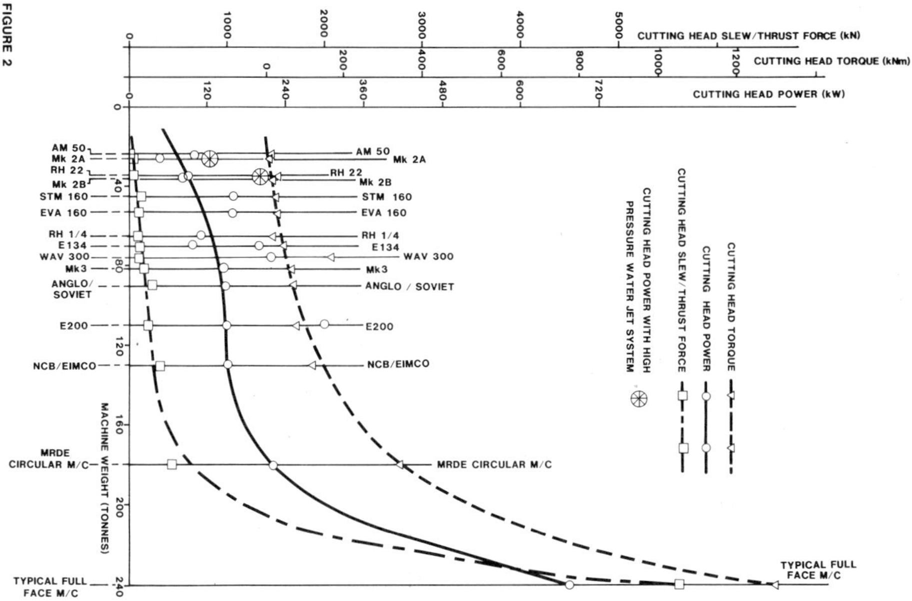

FIGURE 2

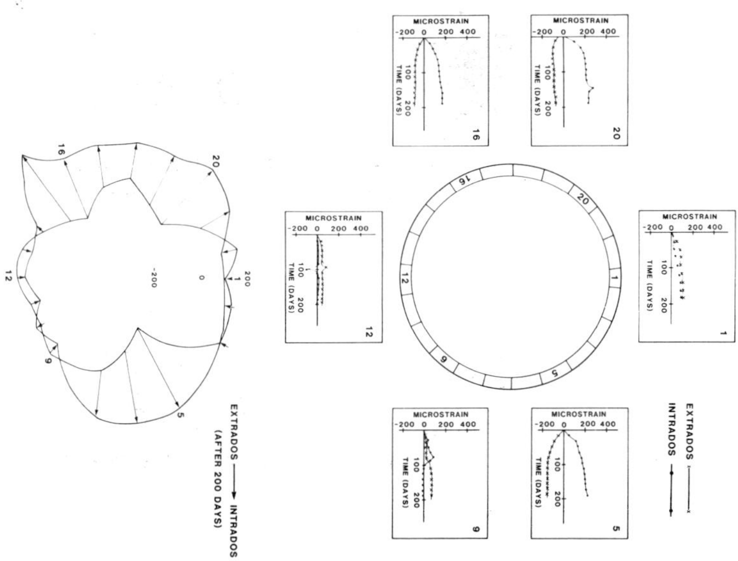

FIGURE 3.

SURFACE STRAINS FOR RING No 161

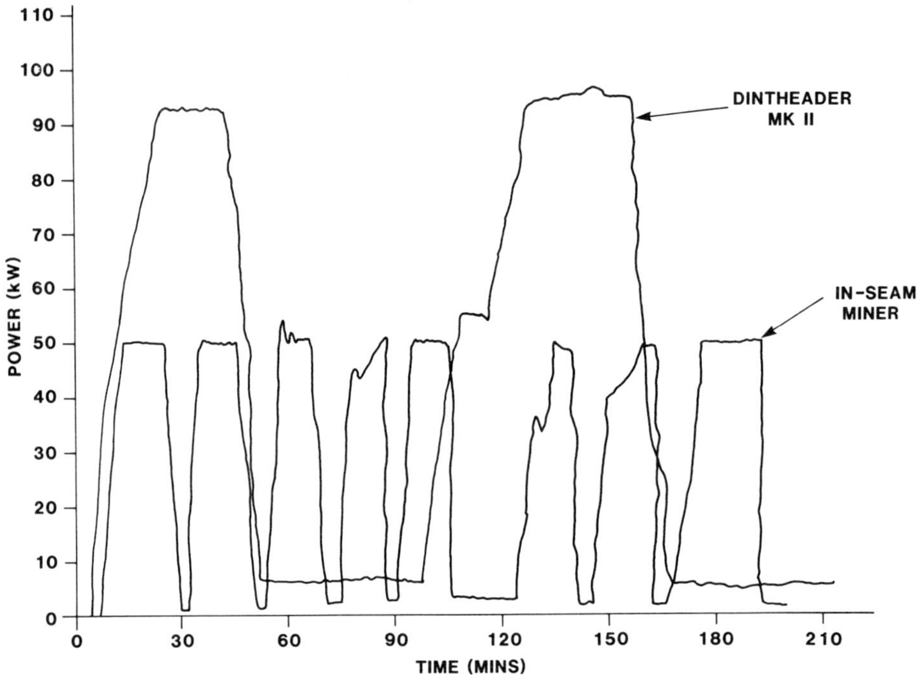

POWER UTILIZATION - REPRESENTS M/C ULTILIZATION

FIGURE 4.

QUICK DETACHABLE
SYSTEM EQUIPMENT

FREE STEERED VEHICLE
WITH LOADING BUCKET

F.S.V. FITTED WITH
MANRIDING MODULE

FORK LIFT ATTACHMENT

SPECIAL MATERIALS CARRIER

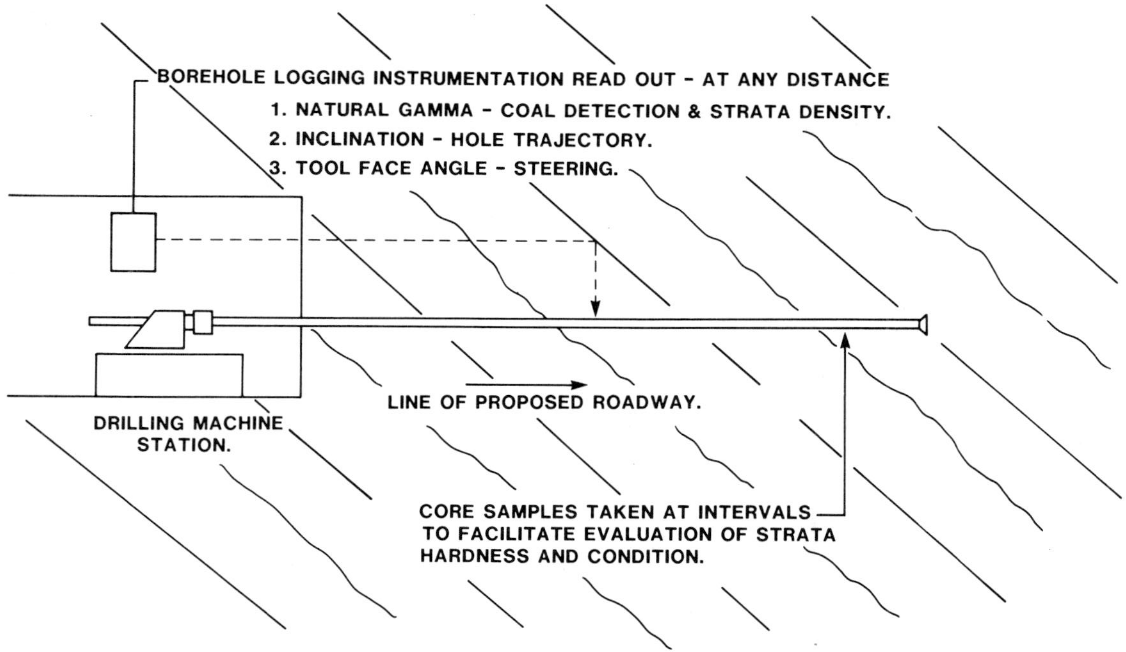

INVESTIGATION IN ADVANCE OF PROPOSED ROADWAY BY DRILLING THROUGH CROSS MEASURE STRATA.

FIGURE 6.

EIMCO
EIMCO

EIMCO

EIMCO

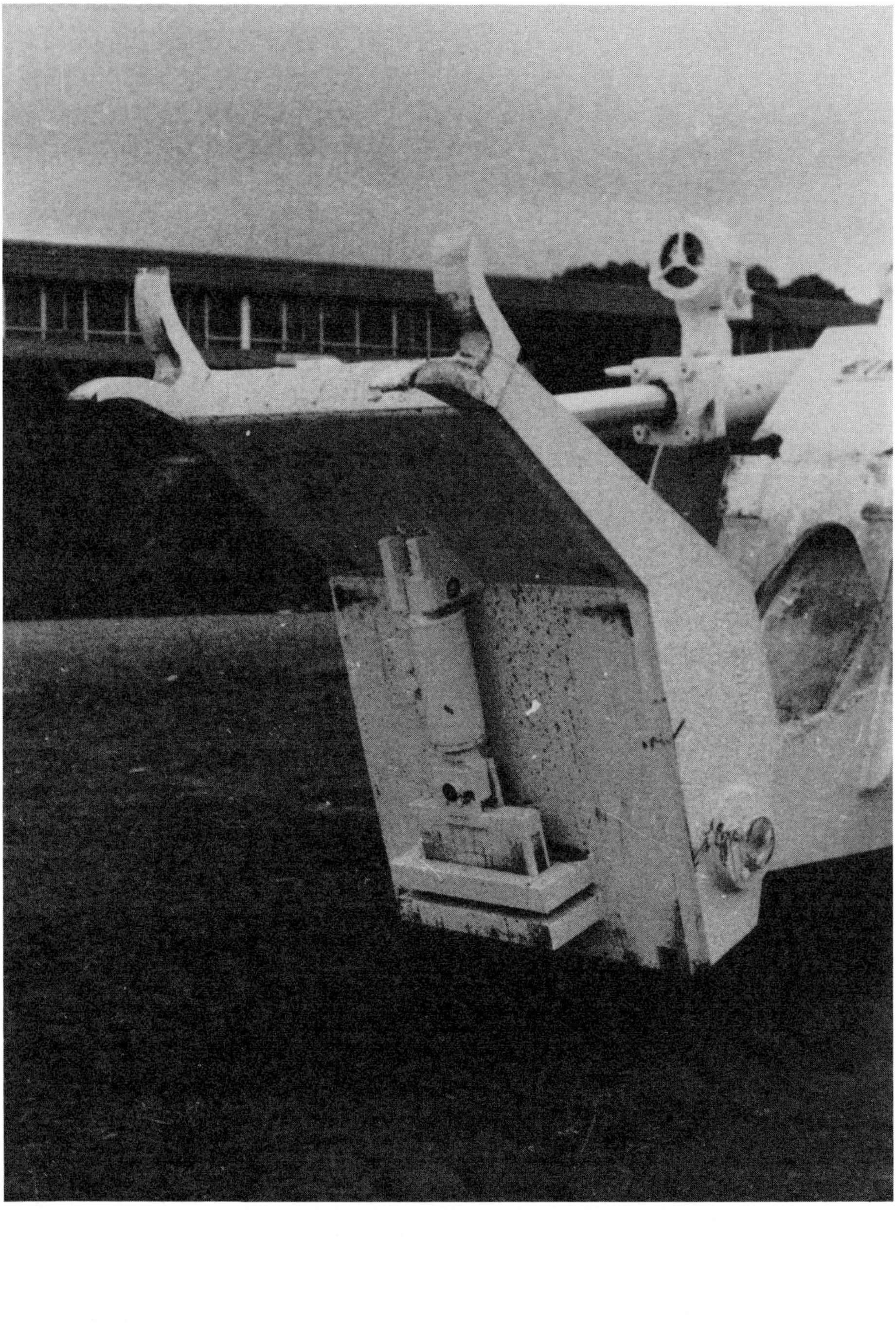

SESSION I - MECHANIZED DRIVAGE IN STONE

- Drivages with full-face roadheaders in the coalmines of the federal republic of Germany
- The Robbins tunnel boring machine at Selby
- Deployment and technical specification of a Bouygues tunneling machine at the Nord et Pas-de-Calais Coalfield
- Improvement in roadheaders
- First operational experiences with the E 200 selective roadheader in stonedrifting

DRIVAGES WITH FULL-FACE ROADHEADERS IN THE COALMINES OF THE FEDERAL REPUBLIC OF GERMANY

H.W. Hövelhaus

Director of Operations, Nordstern/Zollverein
Combined Mine, Bergbau AG Lippe

Summary

The utilisation of full-face roadheaders is becoming steadily more significant in line with the underground accessing of large adjacent takes from existing collieries. The first successful application of a full-facer in the coal industry of the FRG started in 1971. A marked upturn began after 1976, both as regards the annual distances driven and the number of full-face machines in service in the collieries of the Ruhr and Saar. A total of 7 full-facers were in service in 1982 and 1983, 3 of them with a borehead diameter of 6.5 m. However, it has not proved possible to obtain any further significant increase in the net daily drivage rate over the past 6 years. Machine types of similar design from 3 manufacturers are available; in every case they have provided convincing proof of their capacity. The first 12 years of full-face drivages have been 12 years of constant improvement in small and large steps. It was, in particular, the improvements made to points of detail in the outbye systems which brought about the breakthrough to today's big drivage performances. The focus of development over the last 4 years has been on safety improvements to the borehead in terms of maintenance and repairs, enhanced flameproofing and more effective prevention of potential gas fires. Also, crucial improvements have been made in terms of machine steering by means of laser beams and of a rapid and uninterrupted extension of ducting strings by means of cassette-type magazines. Particular attention has been paid to capacious and high-performing supply and clearance systems for the drivages. The costs as identified indicate that the method is very cost effective, provided that the rapid drivage of long sections of road is accepted as an indispensable component of the objective in hand.

1. THE GENERAL BACKGROUND, AND THE STATISTICAL COURSE OF EVENTS

Full-facer drivages started in the early Seventies, and during the first 6 years this revolutionary technique made only gradual headway in the German coal industry. It is true that the first application of a Robbins machine in 1971 was able to draw on a drivage technology which had been thoroughly tried and tested in civil engineering and tunnelling over the previous 10 years, but the specific demands and objectives of roadway drivages in the Ruhr Carboniferous created additional problems, compared to tunnelling, which necessitated new and specifically orientated research and development. In particular, the constant changes in rock types in the Carboniferous, ranging from soft shale to ultra-hard sandstone and conglomerate, put the boring tools of full-face roadheaders under extremely high stress. Then there was the problem of transporting the machines underground and of assembling them in special chambers, which urgently required solution, and also problems related to the rapid and trouble-free and (so far as at all possible) fully mechanised setting of roadway supports, to dust prevention and air conditioning, to machine steering and to flameproofing. In retrospect it can be said that most of these problems were solved in good time and in a satisfactory manner.

From the start of full-face drivages in 1971 to the end of 1982, a total of 73 km of stoneroads were driven by that method in German coalmines; 28% of that distance was driven during the first 6 years, and 72% in the other six (Figure 1). The yardage, therefore, more than doubled during the second half of the period under review.

However, this big increase, particularly during the last 6 years, cannot be attributed solely to improved problem-solving in the matter of full-face drivages; it also reflects changes in the development factor, i.e. the ratio between the quantity of rock excavated and the production of saleable coal. This factor has increased from 17.7 cu.m. per 1000 tonnes of saleable output at the start of the oil crisis in

1973 to 34.5 cu.m. in 1982; in other words, it has nearly doubled. However, the increase in length of road, in metres per 1000 tonnes, did not keep pace with the development factor during that period, since the section in level stoneroads also increased, from 21.4 to 26 sq.m.

7 full-face roadheaders are currently in service, 5 at Ruhrkohle AG and one each at Eschweiler Bergwerksverein and Saarbergwerke AG (Table 1). Full-face drivage projects have already been completed at another 5 collieries. A length of road of 88 km has so far been planned for the machines which are now in service. The longest individual section is 10 300 m, and the shortest is 1300 m. The greatest depth of operation is at 1270 m b.s.l., the shallowest at 400 m, and the average at the other collieries is about 1000 m. 3 of the full-facers now in service were manufactured by the Robbins Company, 3 were designed and fabricated by Mannesmann-Demag AG, and one by Wirth GmbH. In the course of time the diameter of the borehead has settled at between 6 to 6.5 m.

There has also been an upsurge in the percentage of full-face stone drivages, in terms of the total, in the German coal industry (from 2% in 1976 to 15% in 1982).

The explanation for this upsurge lies not only in the improved technology and enhanced economic viability of this drivage method; overlying aspects and developments have also contributed. In particular, there has been the need to find a rapid connection to adjacent virgin takes for collieries along the edge of the Ruhr whose reserves are in depletion. In this connection there have been several reasons why it has proved more economic and advantageous to ensure the winding capacity of the old take for several more decades by establishing an extensive underground connection between the old and the new take, rather than to sink a new colliery.

At all German collieries where full-facers are currently used for stone drivages and objective is to establish this connection, and to access the new take, as quickly as possible. To mention just one example, production from the workable seam areas in the old take of General Blumenthal colliery can only be maintained for a limited period because of depletion. The colliery, therefore, has to speed-up the development of the adjacent Haltern take, so that workable coal reserves can be accessed and prepared in good time. Among other things, 10.5 km of large-section stoneroads have to be driven for this purpose; only a powerful full-face roadheader will complete the job on schedule. Speaking in general terms, with major projects of this kind time is often such a dominating factor that full-facers are the only answer, in view of their high performance in comparison to conventional drivages where the rate of advance is relatively slow.

However, there is a further argument in favour of more full-face drivages. The connection of an old and a new take usually also necessitates long underground haulage routes of up to 30 km (as in the case of Haltern), for the 'backwards' mass transport of run-of-mine to the old winding shaft. This is where unmanned locomotives with automatic train sequence control is a proved system, and also the most economic. A big-train system with, say, sixteen 28-cu.m. hopper cars in each train and with 300-kW twin-unit trolley wire locomotives which can handle an attached load of 800 tonnes and travel at speeds of up to 40 km p.h. requires a permanent way, for a track gauge of 1000 mm, of a design and quality which cannot yet be found in coalmines. It is only the interaction of large circular roads and carefully laid and compacted ballast which permits this new dimension of high-performance underground transport media.

2. TECHNICAL DEVELOPMENT TO THE PRESENT STATE OF THE ART

2.1 The General Background

The first attempts at fully-mechanised stone drivages were made in the early Sixties with purpose-designed machines. The Wohlmeyer machine should be mentioned in this connection (although its method of cutting was different from that of present-day full-facers) and also the Bade & Co. stonedrifter. The latter machine was brought into service at Prosper colliery; it cut a diameter of 4 metres, and its construction and method of working already comprised features such as roller bits and staking systems which were perfectly comparable to modern machines.

These first applications of roadheaders in the coal industry did not lead to a breakthrough, because the technical designs and the various components which were used at that time did not permit a continual operation on a technically sound and economically viable basis. The world's tunnelling industry, therefore, devoted its attention to the technical improvement of boring tools and their bearings, and of the actual machine bodies.

The breakthrough then finally came in the mid-Sixties in the drivage of galleries in highly abrasive and strong sandstone and greywacke. These successes were repeated in machine drivages for numerous road tunnels in Switzerland and Austria, and a comparison between the rock strata in question and those in the Ruhr Carboniferous made it likely that a similar success would be obtained in deep mines. Early in 1971 a consortium of specialist mining contractors plucked up courage to drive the first 7 km in stone on the 7th level at Minister Stein colliery with a Robbins machine with a diameter of 4.80 to 5.20 m. The operation was successful, and the machine fully met the design requirements which the coal industry expected from a roadheader. Also, it was found that the correct method was to drive the deep-lying road to its final section from the start, and not to apply a two-stage reaming method. Another thing which was already found at that time was that the road had to be fully completed (including the permanent way, the track and the pipelines) behind the machine or the backtrain, without having to require any further work.

The next 12 years of full-face drivages which followed this first application in a coalmine in 1971 have been 12 years of steady improvement in small and large steps, not only improvements to the actual fullfacers but also to all the back equipment of the machines. It was, in particular, the improvements made to points of detail in the outbye systems which brought about the breakthrough to today's big drivage performances.

The total system of a modern full-face unit can be seen in Figure 2. Behind the machine body and tender, which is roughly 25-metres long, there is the outbye system, starting with the floor chute. This is followed by the bridge conveyor, with a big clearance for the transloading of materials and supports, and the backtrain with the equipment for power supply, dedusting, ventilation, airconditioning, spoil clearance and loading facilities for the large-capacity spoil cars and the hose and cable magazine. The overall length of the drivage system illustrated, from the borehead to the end of the hose and cable magazine, varies quite considerably for the 7 full-face roadheaders which are currently in service, from 377 metres at Monopol to 655 metres at Westfalen.

2.2 Full-face Roadheaders

As already mentioned, when a road which has been driven by the full-face method comes into service, its purpose is to provide a long-distance connection between takes, which has to handle an exchange of large volumes of materials and products or a substantial airflow per unit of time, and even both in some special cases. However, large airflows and transport volumes call for large roadway cross-sections. There has, therefore, been a leap forward, from 4.8 or 5.20 metres to the present maximum diameter of 6.5 metres, which can currently be handled by 3 machines.

This increase in the borehead diameter also produced benefits in terms of conditions at the working place and safety during operations. Support setters, transport workers, craftsmen, maintenance men and, in particular, special mining personnel engaged in supporting work, can move more freely and work more safely in the roadhead when the latter's cross-section is wider. It is not without reason that full-face drivages are now much safer than conventional drivages.

The type of horizontal staking (Figure 3) and the shape of the borehead are the salient distinguishing features of all machines currently in service. The Mannesmann-Demag machine at Prosper-Haniel has a twin static horizontal anchoring at both ends; this type of staking has done well in stable and unstable ground, but particularly in the former. In the course of development, Mannesmann-Demag have modified this twin staking; their machine which has been in service at Haus Aden colliery since 1981 is equipped with single staking at the back and with a friction-type anchor shield at the front.

The Robbins machines, too, have a single anchoring system and a forward friction-type shield (Figure 4). Such machines are in service at Monopol colliery and, in a more recent design, at General Blumenthal and Westfalen. They are easy to steer by way of the staker system which acts in a transverse plane.

Another type of anchoring system is represented by the Wirth machine (Figure 5) which is now in service at Lohberg colliery. This also has a twin staker system, but of a cruciform design, which has the advantage of releasing the anchoring force into those wall sections of the circular road which, as experience has shown, are under less strata pressure.

The three companies which have been mentioned - Mannesmann-Demag AG and Wirth GmbH (Germany) and the Robbins Company (United States) - are, broadly speaking, the leading manufacturers of full-face roadheaders on the international market. Their products, which are also in use in the coal industry of the FRG at this time, are technically comparable in terms of machine types and have done well.

2.2.1 Boreheads and bits, securing the boring area, support technology

Over recent years there have been changes in the principles which govern the shape of the borehead or, more precisely, the surface design of the frontal side of the borehead. The intention was to have the actual head and its dust screen as narrow as possible, so as to be able to set the supports as early as possible. This was made feasible by the change from conical heads, whose feature was that they had a sort of self-centering facility, to the almost flat heads which were used with newly-designed and constructed full-facers. In terms of a stable heading face the flat borehead had the advantage of greater stability against unexpected falls of rock and coal. The flat design also met the demand for a shorter borehead so as to be able and set the first supporting ring, or back-up strata bolting if required, behind the protective canopy as soon as at all possible. The Robbins Company took the lead in flattening the borehead, and the two German manufacturers have followed this lead. Such a flat Demag head is shown in Figure 6 as an example of the line which is being pursued by all three manufacturers.

So far as boring tools are concerned, single-ring discs with a usual diameter of 350 mm, with 75 to 100 mm between the cutting teeth and with a contact pressure of 15 to 20 tonnes now hold the stage. Button roller tools are no longer talked about. The discs have also been greatly improved in terms of bearing and permanent lubrication in connection with a tight seal against the penetration of drilling mud. It was the reliability and inherent output of roller boring tools which were the main factors of greater drilling speeds, and thus also for the general progress of full-face driving as a whole.

However, the shaping and design of the borehead also has to be considered from another aspect. During the daily maintenance and checking routine it is still essential to systematically examine the borehead with a view to the state of its roller tools and their bearings and the state of the scoop lips, and to deal immediately with any deficiencies. In order to do this, the machine has to be withdrawn from the heading face, and the space thus created has to be entered. No irresponsible risks may be taken in this connection, whether or not the face is stable. However, the risk cannot be avoided unless the heading face is additionally secured. After a fatal accident had occurred because a rock face had not been secured properly, a great deal of thought, backed up by tests, was given to reaching a sensible and effective immediate measure by means of sprags and airbags. However, practical experience has shown that these immediate measures can only be an interim or ad-hoc solution if the actual borehead is not modified. Really suitable measures, to secure not only the heading face and roof but also part of the walls, meant changes in the design of the boreheads, and this called for quite a lot of development time and work.

A technically thought-out solution of this type is shown in Figure 7. The system which secures the boring area can be advanced from inside the borehead and keyed against the heading face. But even this design was unable to meet all requirements for a comprehensive securing of the boring area during maintenance. So that, at least, not all work on the boring tools would have to be done in the cutting area, the manufacturers devoted increased effort to finding a way whereby these jobs, and the changing of the rollers, could be performed from a safe place behind the borehead (Figure 8). The first application of such a borehead is proceeding at General Blumenthal colliery, but the design still has to prove itself in practice.

The dependence of the full-face drivage rate on support-setting performance has always been regarded as the major problem. The top performance of the whole drivage system is still determined by the time required to set the support rings, but the remarkable improvement in mechanised setting has at least eased this bottleneck to a considerable extent.

Rigid or yielding steel rings in 4 to 6 sections are used to support the roads; depending on the nature of the ground, the profile weight varies from 36 to more than 40 kg per m, with a support density of not less than 0.75 to at most 1.5 metres. The paletised support materials are brought directly to the machine along overhead monorails, from where they go along a permanent conveyor belt to the support-setting aid directly behind the borehead. Annular racks, into which the various sections are located in the correct sequence and are connected together, have become widely adopted as support-setting aids. The setting process is largely mechanised, and takes between 15 and 20 minutes with rigid supports, with an extra 5 minutes or so in the case of yielding supports.

If it becomes necessary to strengthen the support system without increasing its weight, this can be done by increasing the support density, but for technical reasons the distance between supports cannot be reduced to less than 0.75 m. Another option is to set rock bolts according to a pre-determined pattern; here again, the bolts should be set as soon as possible. Either option involves extra work around the road-header, and will thus impede the drivage rate to some degree.

2.2.2 Machine operation and monitoring

Only one man is needed to operate a full-face roadheader (Figure 9). The operator's stand contains regulator and setting valves for the machine's hydraulics, i.e. for, basically, the thrust and staker rams and the steering controls.

The machine is steered according to the directional control data which is displayed to the operator. The direction to be followed is indicated by a laser which is suspended in the roof. The laser beam hits a receiver which is fitted to the dust shield and is equipped with photo-electric cells. The line of the laser beam is displayed by means of an LED monitor in the operator's stand. There is also an instrument panel which gives a visual status display for a very wide range of appliances in the machine and on the backtrain. There are various safety monitoring system which largely prevent operator error. Wired logic systems are used, and also individually programmable micro-processors.

It is precisely this successful application of micro-processors, and R & D based on measurements taken on full-face roadheaders, that is the launching pad for expanded monitoring of steering and regulating systems. It is thought that this will allow a substantial reduction in the weight of machines, for instance by way of a rock-specific operating mode. A step in this direction would have a beneficial effect on the size of assembly chambers, on the transport outlay and on underground machine mobility.

2.3 Backtrain and outbye systems, ventilation, airconditioning, dedusting

Nowadays the most difficult item in the planning of a full-face drivage is to correctly size the backtrain of the roadheader. Not only does the backtrain have to be synchronised with the transport medium in the roadway, but the system must be able to negotiate curves without the slightest difficulty, the floor has to be kept very clean and all controls and other appliances must be easily accessible. The general trend with platforms has been that they have become steadily longer in recent years (up to 153 m) and that there has been a disproportionate increase in their surface area (up to 774 m^2). Considering the large size of the dirt cars (e.g. a capacity of 10 m^3) and the need to relocate the travelway from the floor to the platform, the decision will necessarily have to be in favour of suspended platforms. There are cases when a floor-mounted portal design will offer advantages, because this eliminates labour-intensive overhead monorails. So far as size is concerned, the present trend is most unlikely to continue; rather, the optimisation of the backtrain will lead to compact designs. The cost-effective utilisation of full-face drivage systems for shorter roads (which remain to be developed) will depend basically on how the complete backtrain is designed.

So far as the auxiliary ventilation of long roadways is concerned, the main demands were for low air losses and big airflows. The former demand was met by using long individual sections of smooth and tightly connected ducting to make up the strings (usually two), and the latter by big duct cross-sections and high pressure. Substantial benefits have also accrued from cassette-type ducting magazines (Figure 10). The 1.5-metre long cassettes can accommodate some 100 metres of ducting, which is pulled out of the magazine automatically in line with the advance of the drivage. The ducting in the cassettes is specially folded into a twin shell, and it only takes a few minutes to replace a cassette. To prevent a total cut of ventilation in the drivage road during the exchange of cassettes, two strings of 1200-mm ducting are used instead of one string with a bigger diameter. The two cassettes are synchronised in such a manner that they never have to be replaced at the same time. The auxiliary airflow is of the order of 1600 m^3 per minute.

Preference has been given of late to smaller types of dry dust extractors, for an airflow of 2 x 150 m^3 per minute, at the borehead and the many transfer points. The make of heading dust is continually evacuated by means of an appliance which was developed by Bergbau-Forschung; the dust is not dry bagged, which means that potential fly dust cannot be created even if the procedure is not applied correctly. Two dedusters, working independently, are usually operated at any one time.

The technical development of hose and cable stores has been completely satisfactory in recent years (Figure 11). With the lengths of road and fast drivage rates which are involved, they cover a great deal of space.

2.4 Special equipment

Unfortunately there have been some recent gas fires in the area of the borehead, with all the consequences which this entails, and all the specialist establishments, authorities, collieries and operators have, therefore, concerned themselves with early warning systems and preventive measures, among which there is a water spraying system whose primary purpose is to cool the front parts of the machine, and a high-output dry extinguisher for the cutting area.

Special measures have to be taken if the permissible escape time is reduced. In particular, a 12-man mobile escape chamber has been developed, which is pressurised by means of on-board compressed air cylinders (Figure 12). Mobile water trough barriers on the backtrain are aimed at always having the statutory immediate protection against explosions as close-up as possible.

3. SPECIAL FEATURES OF DRIVAGES

3.1 Preparatory work for machine drivages

Creating a suitable assembly chamber and transporting the machine down the mine accounts for a great deal of technical and economic outlay during the planning and execution of a machine drivage. The machine manufacturers have largely met the demand for roadheaders which can be completely broken down into component parts which, for transport purposes, are not greater than 5 x 1.5 x 3 metres (length x height x width) and do not weigh more than 16 tonnes. These limits are set by the available inside diameters from the shaft mouth to the sumping-in point of the roadheader, and by the maximum permissible suspended weight down the shaft (Figure 14).

Full-facers to bore a diameter of between 6 and 6.5 metres require an assembly chamber with a clear height of at least 8 or 9 metres and a floor width of not less than 7 to 8 metres. The length of the chamber will depend on the type of machine and, if the case arises on what is needed for a further utilisation after assembly, such as a pump pack or a conveyor drivehead. Purely in terms of a machine assembly chamber, the typical length is between 20 and 25 metres. The preparations for machine assembly also include a crone of appropriate lifting power, and electrical equipment (Figure 15).

It is only very seldom that the backtrain system can be completely built up during the start-up phase. Stage haulage to evacuate the spoil is, therefore, usually needed until the completed backtrain allows rail haulage.

3.2 Working through faults

The control of difficult geological conditions is one of the special problems in full-face drivages. It has to be borne in mind that, in coalmining the presence of friable, unstable strata and of a high cover load has to be expected outside the actual main area of faulting much more frequently than in other branches of mining.

In the course of the machine drivage of (by now) more than 70 km of road, various methods of working through faults have been applied. They have ranged from strata consolidation by fan-shaped preliminary injection (Figure 16) to temporary manual drivage and hand-set supports in the faulted zone (Figure 17). The range of measures naturally depends on the extent of the geological fault; a complete listing of the many measures which have been applied so far would exceed the framework of this paper. All of them reduce output, but they do make it possible to continue the machine drivage when stable ground is re-encountered.

Let us merely point out that it is not only the fact of highly developed machine technology which demands correspondingly high professional qualifications from the operators and supervisors, but that, in particular, working through difficult sections of rock and seams which tend to fall and spall often demands the last ounce of mining and engineering know-how. Once a team, and its leaders, has learned how to handle a particular type of machine it should, therefore, be kept together for deployment to other sites.

4. DRIVAGE PERFORMANCES

There has been no marked change in the average net drivage rate in 1982 compared to the period from 1977 to 1981, and thus no discernible improvement. The average for all machines in service in 1982 was about 12 metres per day; the maximum of 14.49 was achieved at General Blumenthal, and the biggest monthly drivage (573 metres = 27 metres per day) was reached at Neu-Monopol. The average excavation output, in terms of all full-face drivages, in 1982 was 4.5 m^3 per MS (manshift), related to all shifts in the roadhead and outbye. In terms of roadhead machine shifts, the average excavation output was between 10 and 12 m^3 per MS. Compared to 1973/79 (when that output ranged from 5.3 to 9.3 m^3 per MS) there has been a marked improvement in at least that area.

A work and downtime analysis is available for the period prior to 1979 for machine-running and machine-shifting time of full-face roadheaders. This shows that boring and shifting accounted for 30% of total time. A more recent analysis shows that this figure has now risen to 35%; this could explain at least part of the improvement in roadhead OMS (output per man-shift). However, these figures have to be further improved, so as to get a further increase in the drivage performance and thus in the cost effectiveness of full-face drivages. The difficulties which are encountered in improving matters in precisely this area indicate that, in order to achieve this, the whole drivage system, including the backtrain, including the materials supply system, including the spoil evacuation system, and many more areas, have to be better synchronised.

5. COSTS AND COST EFFECTIVENESS

Comparable full-face drivage costs are of value only if they are precisely demarcated, one against the other. This includes the identification of all upstream and downstream costs for transport, assembly, dismantling etc., and it also includes the identification of machine depreciation rates, repair costs, the risk supplement, the outlay for working through faults and full details of the chargeable first capital equipment. According to these criteria, the average costs of all full-face drivages currently proceeding are not comparable. The only thing that is possible is to quote the following figures by way of a rough guide: The average costs for a full-face drivage, excluding preparatory and post-production outlay, are likely to be between 5300 and about 6500 DM per metre.

In the final analysis, the per-metre drivage cost for a total development project in stone is affected by the depreciation rate for the drivage system and by the number of assembly and dismantling processes, plus the creation of the infrastructure which is required. 1500 MS are needed to create the assembly chamber and the start pipe, and often more than 2000 MS for transport and system assembly. Underground transfers of the system create additional outlay of roughly the same order of magnitude; for instance, one of the most recent transfers cost 4.5 m DM.

The minimum length of drivage is a much-discussed subject so far as cost effectiveness is concerned. Apart from the aforementioned factors it depends, in particular, on the conditions pertaining to the site of the drivage. A drivage which starts near pit bottom, with all the advantages which this entails in terms of transport and assembly, is obviously a more economic proposition than a drivage which starts well away from the shaft. However, the minimum length of a section of road should involve a drivage of about 3 km. A total drivage of that length is normally not enough to apply a full-face drivage economically. Against this, according to extracts from the Ruhrkohle AG data bank the m^3 cost of blasting a level stoneroad with an excavated section of between 20 and 25 m^2 is not all that much less than the comparable cost for a full-face machine drivage, and with the latter the average drivage rate is a good 4 times greater than in the case of blasting.

An appraisal of all the advantages and drawbacks of the various drivage methods in level roads shows that nowadays there is no longer any efficient alternative to full-facing if a major project is involved. Then there is the cost effectiveness of stone drivages with full-face machines, which is basically due to the high rate of drivage, particularly bearing in mind that a conventional drivage for development roads of that length would have to start years sooner than a machine drivage.

6. FURTHER TECHNICAL DEVELOPMENT

The machines now in service have a high technological level, in terms of their present area of application, and, broadly speaking, are not really in need of further improvement; there is, therefore, no reason to expect any more breakthroughs. The main points to consider for the future is to improve the cost effectiveness and to widen the scope of application. A start should be made by reducing the preliminary and post-production outlay, by reducing the size of the assembly and dismantling chambers, by making the transport of the machines and backtrains simpler and faster by means of lighter and compact machine systems, by introducing new rock-breaking methods and by applying modern measurement and regulating technology.

A new Mannesmann-Demag concept represents a first step in that direction. The purpose of the short machine which is illustrated in Figure 18 is, inter alia, to reduce the weight of the individual machine components, so that assembly is not only made easier but can actually be carried out within the cross-section of the boring profile. Also, the actual machine is to have the same length as its boring diameter. With an appropriately flexible coupled-on backtrain system it will then be possible to negotiate curve radii of 50 m in all directions, as against the 150 to 170 m with the current machines. If this machine really makes it possible to dispense with assembly and start chambers, it would open new dimensions in terms of a cost-effective drivage of shorter stone roads, or in-seam roads with solid surrounding rock.

Another interesting option is to combine high-pressure water jets and conventional roller boring tools. Following some promising preliminary trials on the surface, Bergbau-Forschung is now field-testing high-pressure jets fitted on an instrumented Demag machine at Göttelborn colliery. These tests are expected to confirm the possibility of reducing static thrust, of benefits in terms of dust prevention and of a reduction in the wear on the roller boring tools.

R & D on measurements conducted on full-face roadheaders, also by Bergbau-Forschung, is the basis for an expansion of electronic monitoring equipment for steering and regulating systems. It is precisely the successful application of microprocessors which is expected to allow the optimal utilisation of the machine, for instance by a rock-specific operating mode. Overall, this is, therefore, a programme which holds out the expectation of continued successful development of the full-face method.

The Key to illustrations is at the end of the report.

Table 1 - Drivages with full-face roadheaders in 1982

Colliery Company		Neu-Monopol RAG	Westfalen EVV	General Blumenthal RAG	Göttelborn Saarberg	Prosper -Haniel RAG	Haus Aden RAG	Lohberg RAG
Manufacturer		Robbins	Robbins	Robbins	Demag	Demag	Demag	Wirth
Borehead diameter	m	5,4	6,1	6,5	6,0	6,0	6,5	6,5
Start of drivage		Jan. 77	Okt. 79	Dez. 79	Jan. 80	Sept.80	Mai 81	Jan. 82
Depth	m b.s.l.	960	1.035 1.250	1.025	400	932	940	1.270
Total driven length	m	11.732	5.850	8.232	7.951	5.385	4.638	2.056
Average drivage rate	m per day	10,86	10,05	14,49	13,09	11,68	10,74	10,88
Maximum monthly drivage	m per month	573	314	459	330	427	333	328
Excavation performance	cu.m. per MS	4,2	-	4,4	8,0	3,3	4,8	3,9

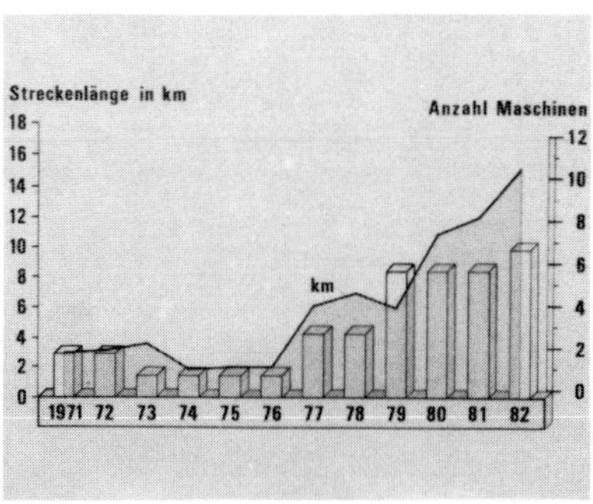

Figure 1

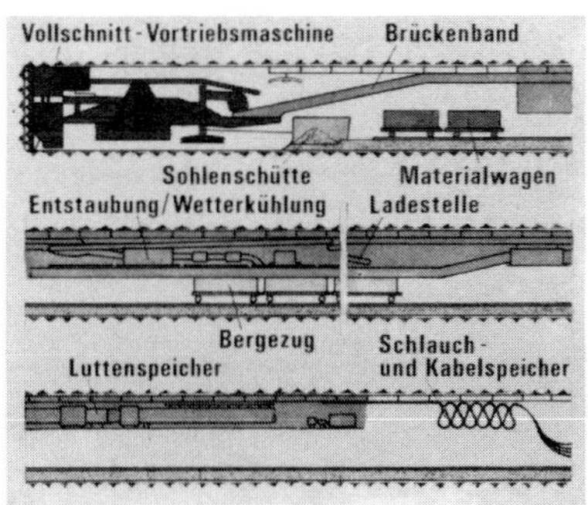

Figure 2

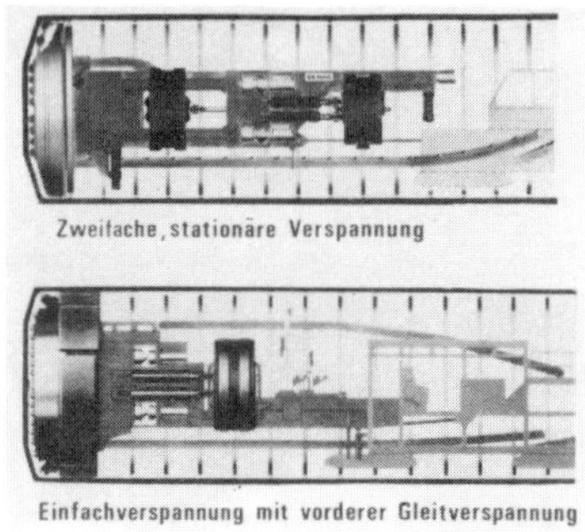

Figure 3

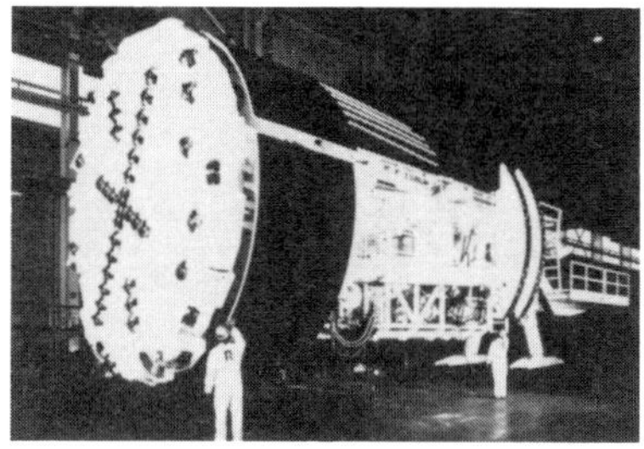

Figure 4

Figure 5

Figure 6

Figure 7

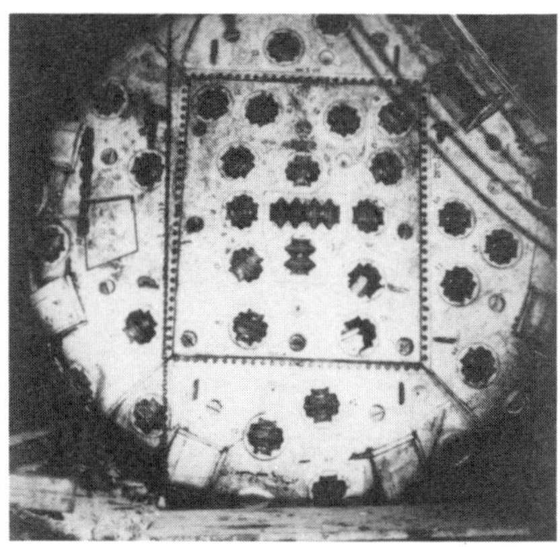

Figure 8

Figure 9

Figure 10

Figure 11

Figure 12

Figure 13

Figure 14

Figure 15

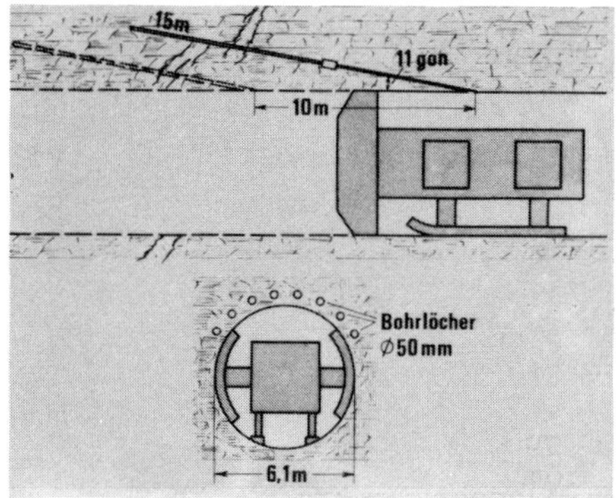

Figure 16

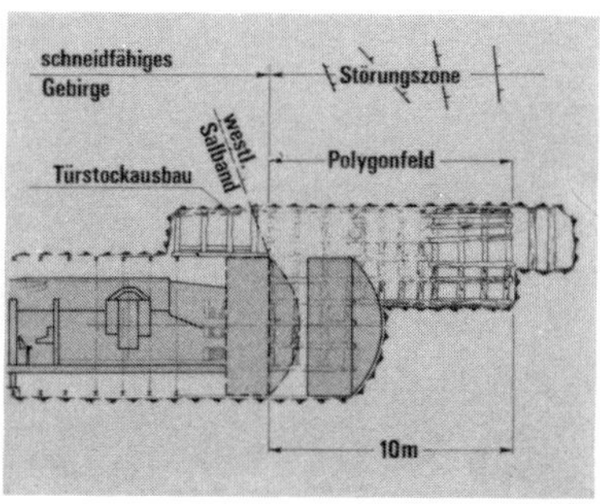

Fiaure 17

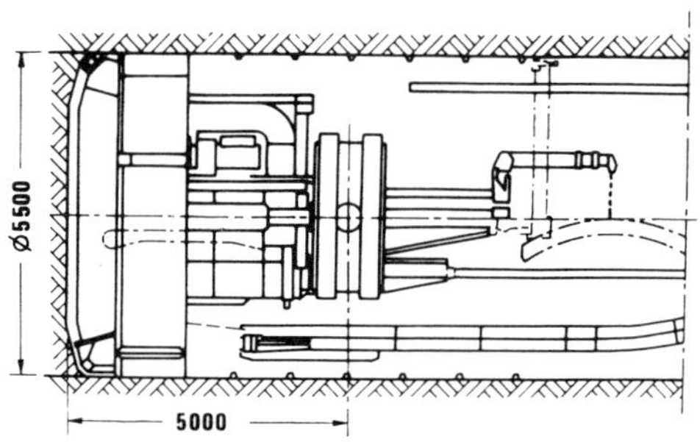

Figure 18

<u>KEY TO ILLUSTRATIONS</u>

Streckenlänge	length of road
Anzahl Maschinen	number of machines

Figure 1. Full-face roadheaders in stoneroads.

Vollschnitt-Vortriebsmaschine	full-face roadheader
Brückenband	bridge conveyor
Sohlenschütte	floor chute
Materialwagen	materials car
Entstaubung/Wetterkühlung	de-dusting and ventilation
Ladestelle	loading point
Bergezug	spoil train
Luttenspeicher	ducting magazine
Schlauch- und Kabelspeicher	hose and cable store

Figure 2. Roadheader with its backtrain system.

Zweifache, stationäre Verspannung	stationary double staking
Einfachverspannung mit vorderer Gleitverspannung	single staking with forward friction-type anchoring

Figure 3. The main DEMAG staking systems.

Figure 4. Robbins full-face roadheader.

Figure 5. Wirth TBM 650 Sche E full-face roadheader.

Figure 6. Short and flat borehead of a DEMAG roadheader.

Figure 7. System to secure the immediate roadhead, fitted to a Wirth roadheader.

Figure 8. Robbins borehead, on which the picks can be changed from behind.

Figure 9. Control stand of a DEMAG roadheader.

Figure 10. Magazine with a 100-m section of folding duct.

Figure 11. Hose and cable store.

Figure 12. Mobile escape chamber.

Figure 13. Mobile water trough barrier on the backtrain.

Figure 14. A Wirth roadheader in the assembly chamber.

Figure 15. Roadheader assembly chamber at Lohberg colliery.

Bohrlöcher	boreholes

Figure 16. Injection into the area ahead of the rockface in a machine drivage.

schneidfähiges Gebirge	cuttable rock
Störungszone	faulted zone
Türstockausbau	porch-type supports
Polygonfeld	polygonal section
westl. Salband	western claycourse

Figure 17. Working through a fault with a roadheader.

Figure 18. DEMAG roadheader for medium-hard rock.

THE ROBBINS TUNNEL BORING MACHINE AT SELBY

C.T. MASSEY, B.Sc., C.Eng., F.I.Min.E.

DEPUTY DIRECTOR (MINING) - SELBY PROJECT

National Coal Board, North Yorkshire Area

Synopsis

The Paper describes the large Coalfield found in the Vale of York in the 1960's and illustrates the concept of working this coal with the minimum effect on the environment. The concept requires the drivage of two fifteen kilometre spine roadways across the Coalfield, which are critical drivages. To improve the speed of drivage of the South Spine Roadway, and to give a better quality roadway, a Robbins Tunnel Boring Machine was installed. The special features of the machine are described and experience up to the end of May 1983 are reported in the paper.

1. INTRODUCTION

The National Coal Board was in the fortunate position of having found a large new Coalfield in the Vale of York in England just before the oil crisis of the 1970's which signalled a revival of the coal industry.

Fig. 1 shows that the find had additional advantages. The market place, in the form of three large power stations, is adjacent to the Coalfield and within twenty miles is an experienced labour force working in mines which are nearing exhaustion.

No mineral find is ideal in all its specifications. In the case of the Selby Coalfield, it lies under very rich agricultural land generally only a few metres above sea level, and is traversed by rivers which are tidal and subject to flooding. The working of the Coalfield had to recognise these factors by limiting subsidence and the amount of land taken from agriculture for the Coalfield operations.

The size of the Coalfield, around 2,000 million tons, would clearly sustain an annual output of 10 million tons per year but it was obviously not possible to impose five large independent coalmines around the coalfield with all the services they need without a major impact on the environment. Some new concept had to be applied.

2. CONCEPT

The concept shown on Fig. 2 consists of a site (Gascoigne Wood) outside the area of the coalfield to act as the receiving point for the coal from the underground workings and despatch it in a suitable homogenous quality to the market by rail. To receive the coal from underground, two drifts are driven 15 kilometres and it is along these that the coal will be transported from the underground mine workings. Five satellite mine sites are spaced around the Coalfield, each operating independently, providing services for manpower, equipment and ventilation.

Fig. 3, showing an isometric projection of the Coalfield, illustrates the interaction between the mine developments and the drivage of the main spine tunnels.

Clearly, the drivage of the spine tunnels and the installation of the conveyor in each tunnel is the critical factor in the success of the project. The position of the spine tunnels is approximately 70 metres below the Barnsley Seam and the connection between them and the workings in the seam is by large diameter boreholes or staple bunkers.

It was in these circumstances that a Tunnel Boring Machine was considered.

3. THE ROBBINS TUNNEL BORING MACHINE

The specification of the Selby machine is included in Appendix 1 of the Paper.

The Selby machine does have certain special features which are worthy of additional comment.

The Cutterhead and Cutters

The cutterhead assembly consists of an inner cutterhead and four outer cutterhead segments and it is on to these segments that the bucket scoops are constructed. These buckets have replaceable hard face metal tips. The flat face of the cutterhead is designed to be within a close distance of the tunnel face to attempt to prevent large lumps of stone getting wedged in the buckets. The aim is to grind lumps into smaller pieces before they pass to the bucket openings. Normally the segments making up the cutterhead are welded together underground but the Selby machine has bolted joints. The bolts used were 25mm diameter Grade 8 set to a torque of 124.47 kg-m. and have proved a secure joint. The cutterhead also utilises a new technique for changing the cutters.

With earlier Robbins full face tunnelling machines the normal procedure for cutter changing is for it to be undertaken by men working at the face side of the cutting head.

It was decided that any working at the face side should be minimised and could not be part of normal procedure for the operation of the machine.

Changes in the head design of the Selby Machine and the method of fitting of the cutter assemblies has resulted in the possibility of replacing cutters from the rear of the head. The single disc cutter housings are welded directly to the cutterhead. The cutter assemblies are saddle mounted and bolted into the cutter housings.

The cutter assemblies consist of shaft, hub disc type cutter, tapered roller bearing and caterpillar type face seals. When cutters are changed on the machine the whole cutter assembly is replaced. The used assemblies are refurbished with a new disc cutter and returned to service.

Support Setting

For setting supports the machine uses a Ring Beam Erector which will place rings of 100mm, 125mm or 150mm cross section as near as 3 metres back from the tunnel face. The rings are backed with weld mesh and are initially set within the tail skin fingers and are expanded as they pass out of the fingers. To allow this expansion a wrap around 3 bolt fishplate is used for normal joints but for the centre joint fishplate a Holywell friction type is used which allow 200mm of closure. The rings are fastened together with square tubular tie bars.

Filling the Invert

It is necessary to fill the invert to a depth of 1.2 metres to form a composite floor on which to install the rail track for the diesel locomotive haulage and the support girders for the trunk conveyor. A swivel conveyor is used at the junction of the machine conveyor with the pantechnican conveyor. This swivel conveyor is rammed hydraulically beneath the transfer point whenever infill dirt is required and takes infill from the main flow of dirt. This method fills the invert twentyfive metres back from the face of the tunnel and has proved good for compaction and consolidation and all that is needed to give a good rail track is a layer of 20mm dolomitic limestone ballast.

Machine Steering Equipment

The steering of the machine is controlled by a laser target installation, a bubble level indicator, and a side support indicator.

The laser equipment utilises two targets mounted on the machine and is a Vickers FLP 2 which produces light intensity of 1 milliwatt. It is kept within 300 metres of the tunnel face.

Ventilation of the Heading

The heading is ventilated with a forcing system through 1,200mm flexible ducting complimented by an exhaust overlap filter fan on the pantechnican. The flexible ducting is fed out of a cassette 100 metres at a time which is mounted on the pantechnican. The cassette carrier hold a spare cassette which can be changed in ten minutes for the empty one. The system has worked satisfactorily for an advance of over 3,000 metres from the site of the 2 x 120 h.p. forcing fans. The system is illustrated in Figure 3A.

Supplying the Tunnel Drivage

Bail track for the supply is installed to the machine head and is laid in 6 metre lenghts of 60lb rails supplied from the surface in complete 6 metres unit

The supplies system operated is currently Rack/Adhesion Locomotives propelling four car GMT trains, each train loaded with materials for 10 metres of advance, i.e. rings, tie bars, fishplates, mesh, rails, pipes and Anderson Strathclyde conveyor and base girders.

The train is propelled into the heading with the ocomotive at the rear and the guard seated at the front end of the leading car equipment so that he can apply the brakes to the locomotive and he can also communicate with the locomotive driver.

Ballast and other materials are on separate trains and e ballast trains are required per day to accomodate an advance rate of 24 metres.

The system of operation demands that once the material train arrives at the heading delivery point it is unloaded by the machine team and returns outbye.

This supply system has proved very effective and is capable of maintaining a high performance from the T.B.M. for the whole of its 15 kilometres of travel.

Operating Experience

The tunnel boring machine was installed underground in a special chamber constructed 1,700 metres from the mouth of the South Drift. Build up of the machine and associated equipment was carried out during July, August and September 1981. Cutting commenced in October under supervision of Robbins staff and training of the contractors' (Thyssen) workforce was undertaken. The machine was staffed throughout the twentyfour hours on four working shifts but one shift is retained for maintenance and back up work. A Method Study carried out by the N.C.B. indicated that the machine was capable of 164 metres over a five day week and during week ending 19th March, 1982 it achieved 151 metres. The total drivage done to May 1983 by the machine is 4,784 metres and the finished product is a roadway larger and of better quality than is usual in Coal Mines see Fig.4. The rate of progress of the tunnel allowed a rephasing of the drivages from the satellite mines to give earlier connections to the spine tunnels and the Gascoigne Wood surface.

The application of the Robbins T.B.M. in the South Tunnel on a circular roadway, in parallel with the drivage of an arch shape roadway by a boom type heading machine, in the North Tunnel, soes allow a comparision of the two types of roadway and the two methods of drivage.

From December, 1980 to the end of July, 1983, the North Tunnel has been driven a total of 5,014 metres at an average of 44,3 metres per week. In the same period, the South Tunnel has advanced a total of 4,859 metres. The average advance was 84.9 metres per week during the time that the Robbins T.B.M. was working. Excluded from the time, is the period to construct the erection chamber and install the machine, and the period when the machine has been sttod due to bad conditions.

The circular South Spine roadway has needed negligbl maintenance work done on it, and is a larger roadway than the arch type roadway. The arch roadway has required a total of 2,600 metres of dinting at a cost of £ 120 per metre. While the roadway is a good mine roadway, it does appear likely that it will require substantially more one-going maintenance than the circular roadway.

Integrating the Tunnel Boring Machine operations with Coal Production

Coal production from the first Mine Wistow started passing down a 1.8 metre diameter lined borehole between the Barnsley Seam and the Spine Tunnels in January 1982. Before the South Spine tunnel is finished around November, 1984 several million tonnes of run of mine coal will have been produced at Wistow. As this coal is sold direct to the Generating Board without treatment, it is essential that no dirt from the T.B.M. is mixed with it. To overcome this problem some form of bunkerage for the dirt from the T. B. M. is required . It is also necessary to ensure that the installation of the Anderson Strathclyde steel cord conveyor proceeds at a similar rate as the T.B.M..

At the end of May 1982 the steel cord conveyor started running for the first 5,000 metres of the South Tunnel.

Both these problems have been tackled by the installation of a purpose made bunker conveyor manufactured by Dowty Meco.

The conveyor consists of two 1,000 metre long overlapping conveyors - one 750 mm wide and the other 900 mm wide. They are hydraulically driven at varying speeds, and the speeds of the conveyors can be linked automatically, or operate independently when emptying the outbye conveyor. At full length they will bunker 200 tonnes of mineral from the heading machine.

The overlap will be used to allow the extension of the steel cord conveyor belt by 200 metres over a long weekend. The bunker has proved effective at bunkering dirt and allowing coal to be run out of the Spine Tunnel from Wistow workings.

<u>Experience with difficult Geology</u>

At the beginning of 1983 for the Selby Project as a whole, the Robbins machine and the South Tunnel drivage was well clear of the critical path and the only concern was the uncertainty of the detailed geology at the shaft sites and its possible impact on roadway drivage there.

Faulting was found of a complex nature in drivages to the North East at Wistow in roadways slightly adjacent and 70 metres above the position of the Robbins Tunnel. The ground was very weak and cavities over the roadway supports had to be infilled with fly ash and cement. The roadway, was however maintained with the original supports and progress, while slow, was not halted. Other roadways at Wistow adjacent to the first one found the faulting system different and variable. I was not possible to predict how the faulting would show itself 70 metres below in the Robbins tunnel, but as a precaution, the section of the supports was increased to 150 mm x 125 mm and the spacing was reduced to 650 mm.

In fact the faulting was very different from the projection with the strata highly disturbed by a fault running parallel with the tunnel as well as faults crossing the tunnel as it advanced. See Fig. 5. It became apparent that the horizon of the tunnel could not have been worse, with very hard strata in the bottom of the invert and a very strong sandstone about 4 ft. above the top of the tunnel. In between was a very weak and shattered mudstone which was being squeezed into the excavation.

The approach to this problem was to drive a pilot heading over the top of the tunnel and forward of the cutting head. From this the strata was reinforced by roof bolts and then the heading and its associated cavity was filled with cement and fly ash.

Progress was resumed and appeared successful except that the fingers on the machine did distort the arches as the machine advanced. However, when the operations stopped over a Saturday night shift during the weekend of 19th/20th March, 1983 the machine became fast due to pressure from the strata on the cutting head and the shield and a cavity developed forward of the head. The whole area of the machine was subject to severe pressure with distortion of the supports.

It was thus necessary to repair along the length of the machine with supports set 325 mm apart and the supports have been encased in concrete to allow the grippers to grip onto the supports rather than the strata. The weak strata in front of the cutterhead is being removed and replaced by a competent cementatious infill which can be cut and profiled by the cutterhead as the machine advances. Two important lessons have been learned in the last three months up to May 1983. Firstly, where the machine is cutting in weak shales or mudstone and there are beds of sandstone near which are likely to put pressure on the heading, the machine must operate continuously to avoid the head and shield getting fast. Secondly, as arches are set over the fingers they must be restrained so that the fingers do not pull them out of position as the machine advances. The arches also need to be quickly strengthened after they are set either with additional struts or by a concrete infill.

4. CONCLUSIONS

The Selby Project is an adventurous concept to interpose a large mining operation into a rural area with minimum effect. Indeed, on completion, the whole of the sites will only take a little over three hundred acres of land from agriculture. To be effective, clearly it is essential that the main spine roadways from Gascoigne Wood should be driven quickly and to a high standard. Despite the recent delays to May 1983 this has been the result of the application of the Robbins T.B.M..

The success of this full face tunnel boring machine should lead to further applications in British Coal Mines. It does offer rates of drivage which make possible much larger areas of exploitation from one mine site and this gives great cost savings by reducing the number of shaft sinkings. The quality of a roadway profiled in this way does raise the question of considering some form of T.B.M. for driving in-seam roadways for the working faces.

ACKNOWLEDGEMENTS

The author wishes to thank Mr. Mills, Deputy Chairman of the N.C.B. and Mr. Eaton, Director of the North Yorkshire Area for permission to produce this Paper. He would like to record his appreciation of the help given by Mr. J. Stenton and members of the Selby Team in the preparation of the Paper. The views expressed are those of the author, and not necessarily those of the N.C.B..

APPENDIX 1

GENERAL SPECIFICATIONS

ROBBINS T.B.M.

MACHINE DIAMETER	19 FT. (5.8M)
MAIN BEARING	TAPERED ROLLER TYPE
CUTTERS:	TYPE: ROBBINS DISC QUANTITY: 35, 14" DISC PLUS 6, 13" DISC
MAXIMUM RECOMMENDED AVERAGE CUTTER LOAD:	30,000 POUNDS (13.600 KG)
MAXIMUM RECOMMENDED TOTAL THRUST:	578 M TONS
CUTTERHEAD HORSEPOWER:	900 HP (6-112 KW MOTORS)
CUTTERHEAD DRIVE:	AC MOTORS/GEAR REDUCERS MOTORS 6-112 KW, 1500 RPM 1100V, 3-PHASE, 50 HZ, SIZE 150 HP
BORING STROKE:	1.6 METRES
HYDRAULIC SYSTEM:	
PUMP & MOTORS:	2400 PSI MAX. (165 BARS)
CYLINDERS:	3000 PSI MAX. (208 BARS)
HYDRAULIC FLUID:	WATER/OIL EMULSION
ELECTRICAL SYSTEM:	
MOTOR CIRCUIT:	1100V - 3 PHASE - 50 HZ
CONTROLS & LIGHTING:	240V - 1 PHASE - 50 HZ
POWER SUPPLY:	6600 - 3 PHASE - 50 HZ
TRANSFORMER:	1500 KVA - 2 EACH 750 KVA
WEIGHT:	240 METRIC TONS (APPROXIMATE)

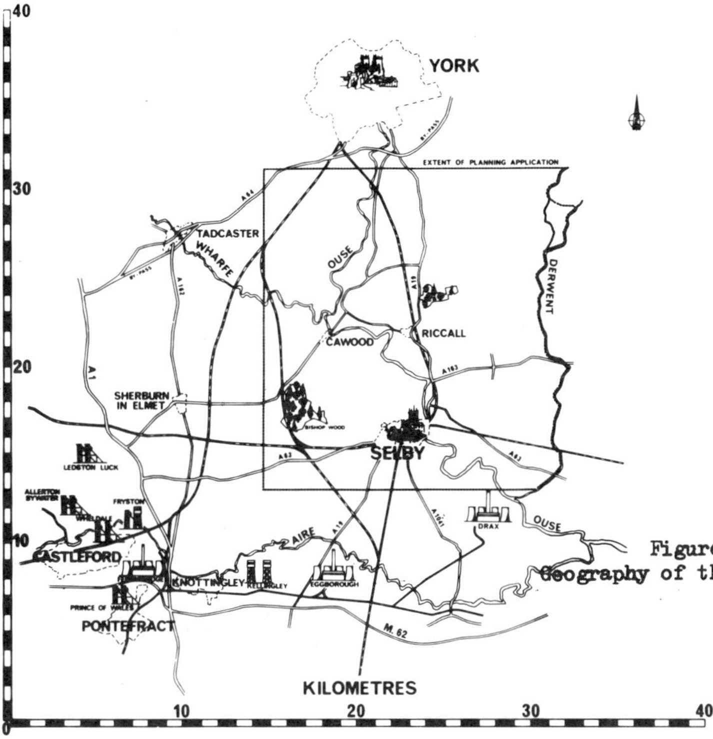

Figure 1:
Geography of the Selby Area

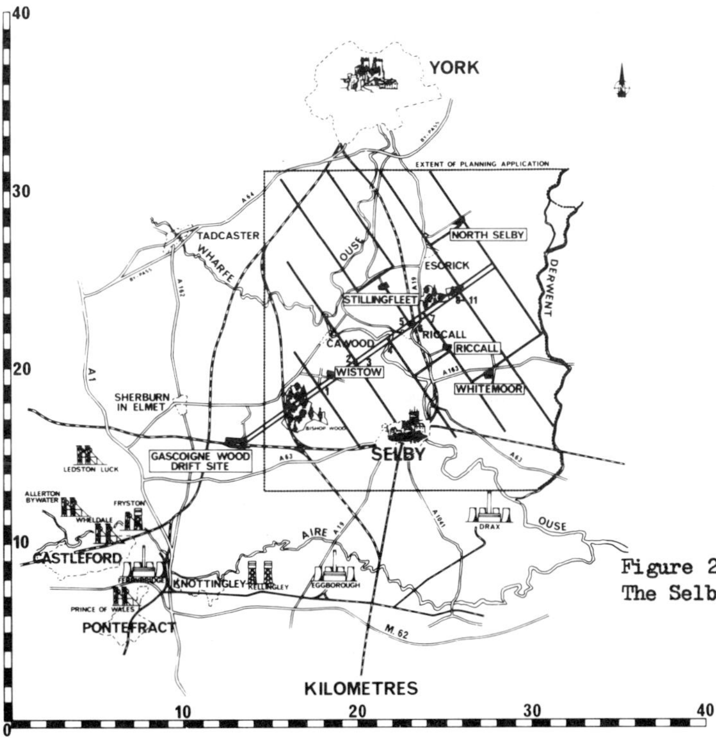

Figure 2 :
The Selby Concept

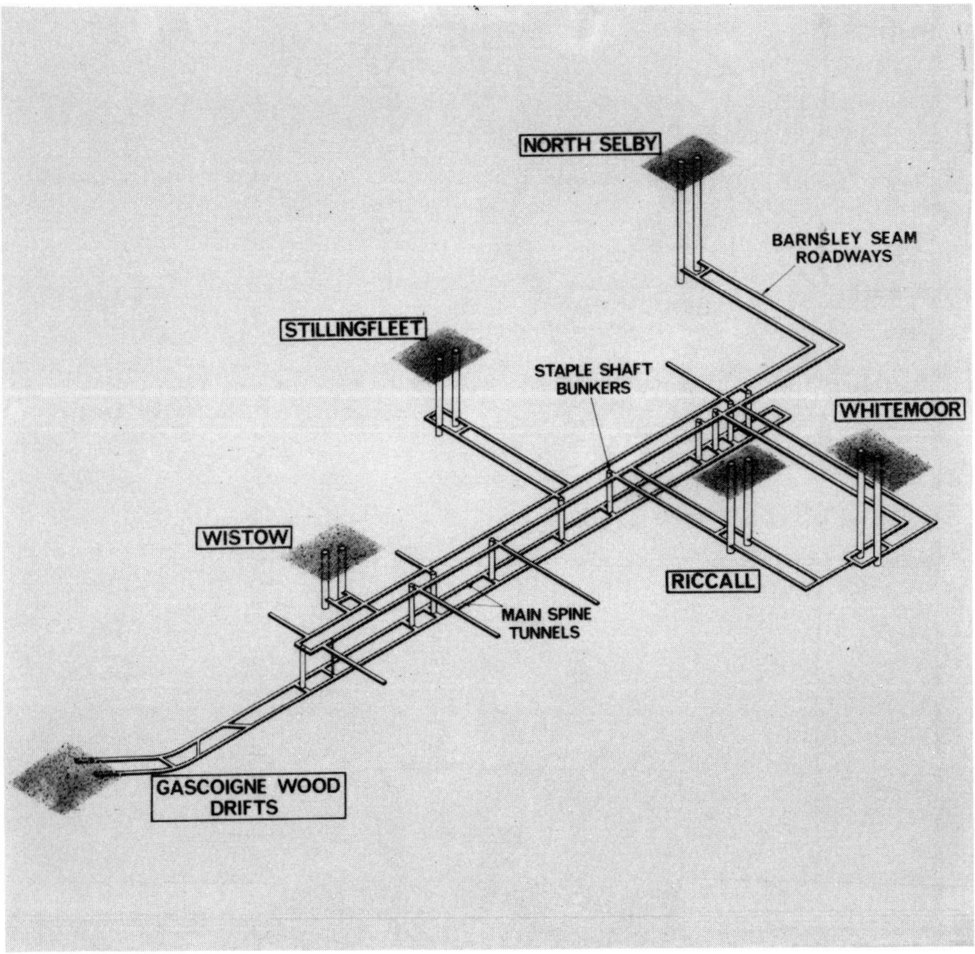

Figure 3 : Isometric of the Selby Concept

Fig. 3A

PLAN SHOWING LAYOUT OF FAN DUCTS AND CASSETTE ON ROBBINS MACHINE

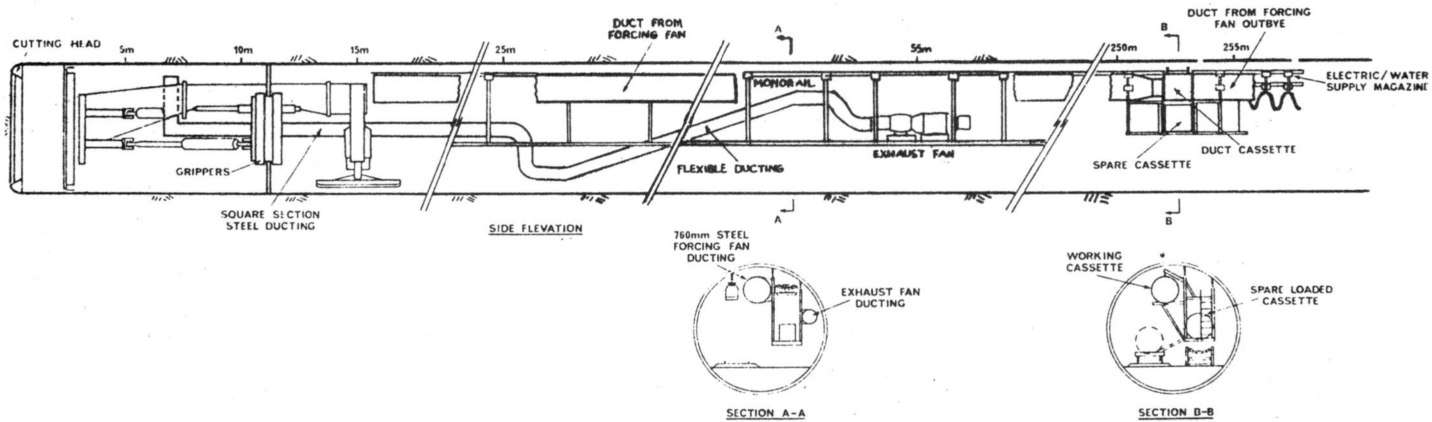

Figure 4 : South Tunnel Drivage

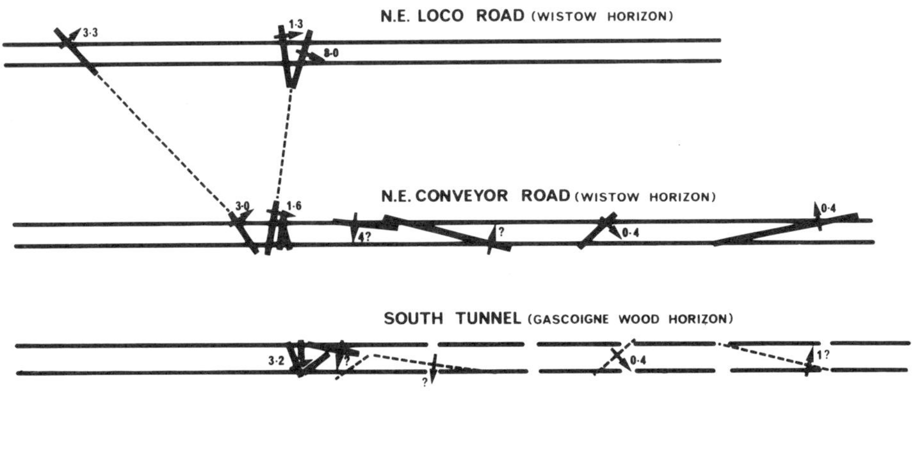

MAY 1983

Figure 5 : Faulting in Barnsley Seam and at Tunnel Horizon

DEPLOYMENT AND TECHNICAL SPECIFICATION OF A BOUYGUES TUNNELLING MACHINE AT THE NORD ET PAS-DE-CALAIS COALFIELD

A. Blanc

Chief Engineer, Nord et Pas-de-Calais Coalfield

Summary

The management of the Nord et Pas-de-Calais coalfield have signed a contract with the Bouygues Company to drive roadways in rock to open up a new level at Arenberg colliery. The road will be driven with a 5 m diameter tunnelling machine designed and developed by Bouygues.

This tunnelling machine is an original design and mounts only four disc cutters on 4 oscillating jibs. This mode of boring provides a number of advantages over its competitors with full-face configuration: lightness, flexibility, ease of installation and transfer.

Roadway drivage is to commence early in 1984 and will last for about 3 years. Envisaged daily advance rate is 10 m.

1. DESCRIPTION OF PROJECT

1.1 DESCRIPTION OF THE WORKING (Photo 1)

Arenberg colliery, close to Valenciennes is working a very undulating measure of lean coal consisting of thin and inclined seams. In order to maintain continuity of extraction within the production programme it is necessary to create a new level which has to come into service in mid-1985 at the latest.

The roadway spines of this new level at -670 m will consist of two main drifts which will intersect the take from East to West for a distance of some 6 km. Access to the panels to be won will be by North-South cross-cuts every 500 m. The roadways will have a cross-section of 19.6 m^2. They will be supported by TH section units weighing 29 kg at 1 m spacing.

Since May 1983 the -670 level has been approached by a 13° inclined shaft some 400 m long. It is from this working that the main drivage of the new level will start because the main shaft, currently being deepened, will not be commissioned until the end of 1984.

1.2 PLANNING THE OPERATIONS (Photo 2)

In order to extract the measure in an orderly way and especially to take proper account of strata pressure on the infrastructures, it is desirable to practise retreat working towards the shaft and divide the workings in the different districts and networks. A conventional drivage of the level roadways by two heading teams working simultaneously towards the East and West would not provide access to the most distant panels from the shaft before the end of 1986, yet extraction is scheduled to begin in the middle of 1985.

To prevent starting the level by panels the extraction of which would rapidly endanger the infrastructure, it was necessary to introduce a rapid drivage system with advance rates of over 200 m a month. We believed that a tunnelling machine would enable this problem to be solved.

The Bouygues Company offered us a contract after due consultation for drivage which would meet the set timetable by means of a tunnelling machine of its own design.

The machine will be finally installed at the end of 1983. It should thus be possible to reach the Eastern extremity of the measure in the middle of 1985. Subsequently, and after transfer to the West, the main roadway system could be entirely finished at the end of 1986.

1.3 MAIN FEATURES OF BOUYGUES-HBNPC AGREEMENT

This is a conventional provision-of-service agreement. The company provides the equipment and the manpower needed for the drivage. The contracting side provides the supports and other equipment and covers outbye requirements.

The prices agreed are comparable to those of similar agreements awarded to companies employing conventional drivage techniques.

The agreement relates to normal conditions in the deposit (well-established hardness, regularity of strata, absence of water and firedamp). A price table is laid down for ancillary work and for exceptional conditions.

The rate of advance envisaged is 200 m/month after a period of running-in and testing of the machine of 4 months. Any benefits that may be gained by improved productivity during the life of the working will be shared in equal parts between the contractor and the sub-contractor.

2. THE BOUYGUES TECHNOLOGY

2.1 DESCRIPTION OF TUNNELLING MACHINE (Photos 3,4,5,6,7,8,9)

2.1.1 Rotary head

The boring diameter is 5 m. The rotary head mounts four sweeping jibs, each with a 400 mm diameter steel cutting tool.

The sweeping motion is achieved by a single jack fitted forward of the head. Transmission is via a gearbox with 4 linked shafts. The rated power for rotation is 800 kW. Speed is 25 rpm. A swivel joint provides oil feed to the sweeper jack and water feed to the cutter discs.

2.1.2 Machine body

Its main components are:

- a fixed head providing mounting for bearings and arrestor,
- a three-part skid for forward floor support,
- two chromed tubes to guide the runner,
- an underframe for 4 gearboxes and motors.

2.1.3 Runner and gripper shoes

The sliding tube runner carries a set of clamps linked to height-adjusting jacks which control the height of the tunnelling machine and enclose the torque reaction.

Two thrust cylinders engage with the gripper plates to move the tunnelling machine forward.

Two jacks on the rear skids stabilise the machine body while the grippers are repositioned.

Two forward jacks under the haulage will lift the tunnelling machine to allow it to travel back.

2.1.4 Debris conveyor

A vibrating chain conveyor is installed under the fixed head. In sequence behind it, a belt conveyor moves the debris to the rear of the machine.

2.1.5 Canopy

The three-piece canopy protects the roadhead area. It is retractable and enables the tunnelling machine to be raised and retracted inside the already-set arches during tool-changing.

2.1.6 Arch-setting equipment

A carriage moves the arch segments immediately behind the canopy and onto a setter, consisting of two lateral guides and a haulage chain.

Once the segments have been set and the mesh placed, the arch is taken out of the setter and locked in position.

2.1.7 Equipment train

A linking beam drawn by the tunnelling machine covers the

zone of road support setting and runs behind on rails.

Six conveyors are suspended from two monorails and make up a flexible debris clearance chain.

The linking beam draws a power pack train consisting of the control cabinets, units providing for automation of machine functions; the electro-hydraulic pack, the dust-extractor and fans, a supply line of 150 m effective length strung under a monorail to link the equipment train to the stationery transformer station. The tunnelling machine is fed electricity at 1000 V, with water, compressed air and telecommunications.

2.2 PRINCIPLES OF OPERATION

2.2.1 Boring

The Bouygues tunnelling machine as designed by the French company mounts four articulated sweeping jibs on its rotating head, which oscillate in the planes passing through the road axis. The oscillating motion is produced by a single central hydraulic cylinder linked to the jibs by means of rods.

The tools at the end of the jibs describe increasing or decreasing spirals of anything between 20 and 100 mm, depending on the hardness of the strata. Each tool is allocated to a different zone of the roadhead. These zones are concentric. The tools are synchronised and return to their zone at the same time at the end of the cycle.

After each cycle of working inwards or outwards the machine advances automatically on programme to give an advance of 0.50 m. At the end of the cycle the gripper shoes have to be repositioned.

2.2.2 Guide system

The front end of the machine rests on a skid which slides on the rock floor. The extensible part consists of two tubes, which slide inside an articulated runner, and is braced at the rear by means of the gripper shoes.

2.2.3 Alignment

The optical alignment system used consists of a gas laser, a twin mirror arrangement and one leveller bar.

2.2.4 Debris clearance

The deflector paddles which are part of the moving head of the machine scoop the debris towards the vibrating chain conveyor which sweeps the roadhead along the forward skid. A flexible run of belt conveyors capable of negotiating curves of up to 75 m radius moves the debris further outbye.

2.2.5 Ventilation and dust-extraction

Air is extracted at the roadhead behind the canopy, passed through a dry de-duster at a rate of 7 m^3/s, blowing out the cleaned air rearwards.

2.3 ORGANISATION OF HEADING

2.3.1 Assembly or dismantling of tunnelling machine

The tunnelling machine and its back-up train can be dismantled in the heading into small-sized components capable of being accommodated in a shaft cage 3.10 m long.

The room required for assembly can be a normal roadway with vertical side walls 4 m high and 100 m long.

The end of the roadway which is to accommodate the cutting head has to be circular and concreted for a length of 5 m and of a 5 m diameter.

2.3.2 Tunnelling

The organisation needed for the tunnelling phase is as follows:

- boring proceeds simultaneously with arch setting. The actual speed planned is around 2 m/h in shales and 1.30 m/h in sandstones, with one arch being set each meter.
- the tunnelling team consists, apart from the section and personnel heads, of some ten men (haulage excluded). Half the personnel is allocated to clearance and arch setting, the other half to loading mine cars, operating the tunnelling machine, laying track and monorails.
- two traction locos pull 12 mine car train units, each corresponding to one machine cycle. The train units are changed while the gripping shoes are repositioned.
- at the weekend an extra team completes outbye fixtures:

positioning ducts, moving transformers and cables forward, laying rail points.

3. REASONS FOR SELECTING BOUYGUES TUNNELLING MACHINE

The Bouygues tunnelling machine, designed from an idea of M. Montacie, an engineer with the firm, attacks the strata with 4 disc cutters mounted on the boom ends which sweep the roadhead while turning.

As a result, the thrust to be applied for advance is proportional to the number of disc cutters (10 to 40 t per disc cutter, depending on strata) and so the force of gripper shoes and pusher rams is small.

This characteristic distinguishes the machine from the competitor full-facers and gives it these advantages:

3.1 Advantages from jib sweeping

The operator can readily see the roadhead and can thus stop his machine as soon as an anomaly occurs: water ingress, malfunction, wear of cutter discs, etc.

It is easy to reach the roadhead face to make exploratory drillings or infusions because one of the jibs has a drill rest.

The jibs can be retracted. This means that the machine can be travelled back several metres in a few minutes. It is then possible to proceed entirely safely to change the disc cutters, check the roadhead, repair a component or set a support rapidly, all under the protection of the already-set arches.

Once the drivage is completed the machine can be moved away by retracting it and travelling it along the rails in the road it has just driven.

3.2 Advantages from the low thrust needed on the head

The machine weighs 90 t, or three times less than a full-face tunnelling machine of the same diameter, hence its compactness. Its length is less than 7 m. No major component weighs more than 6 t and each of these will fit within the dimensions of a mine car (width 0.85 m, length 3.30 m, height 1.70 m). As a result:

- component transport and machine installation are simple.
- it is not necessary to fashion an assembly room prior to installation. This is a very valuable saving.
- it is easy to work around the machine, which helps to do repairs, support setting and related work, such as dealing with any water inrushes.

The thrust of the gripper shoes, being proportional to that of the tunneller head, is also very low. It disturbs the strata only slightly. This makes it possible to advance the machine inside concrete block panels placed 1 m from the face without risking damage to them. This important innovation has been developed in conjunction with engineers of the Campine coalfield and allows the tunnelling machine to be used in the most fragile strata.

4. CONCLUSION

For the first time the Nord et Pas-de-Calais coalfield has chosen to drive the main roadways of a new level of a colliery with a tunnelling machine for reasons which have been set out in this paper.

The need to keep pace with planning led us to go for completely mechanised drivage in order to maintain high speeds. The solution selected will enable us to prepare the opening-up of the measure more effectively. The increase in drivage speed and the concentration on a single working instead of two or three should bring down the total cost of the operation. The mechanised drivage and circular supports should make for excellent roadway stability. The use of a sophisticated technology and the hoped-for improvement in working conditions (dust, noise, mechanised handling) will favour the personnel working on the project. All these factors combine to give a considerable new impetus to drivage in rock, a technique which had not developed for many years.

The choice of the coalfield fell on a French system which

seemed to meet the needs in a very exact way.

The Bouygues tunnelling machine is powerful yet of reasonable dimensions. It is the only one commercially available to be adaptable to the access requirements of our mines.

We hope that the results we will be able to log during 1984 will match our expectations.

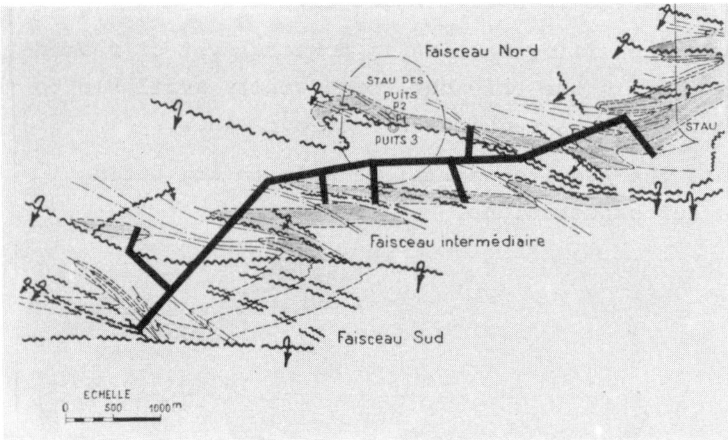

FIG. 1

FIG. 2

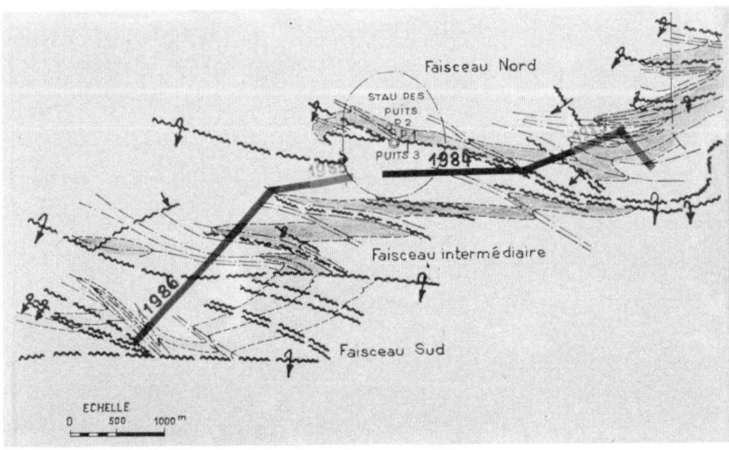

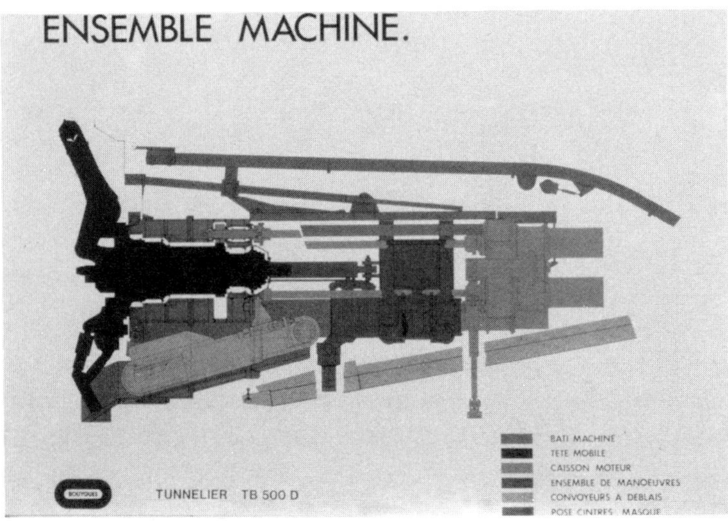

FIG. 3

FIG. 4

FIG. 5

FIG. 6

FIG. 7

FIG. 8

FIG. 9

FIG. 10

FIG. 11

FIG. 12

FIG. 14

FIG. 15

FIG. 13

IMPROVEMENT IN ROADHEADERS

P G TREGELLES and A H MORRIS

National Coal Board, Mining Research and Development Establishment
United Kingdom

Summary

The paper briefly describes the evolutionary progress of roadheaders in Europe from the early 1960s to the present date, in terms of development from light duty machines to a heavier generation of improved capacity and reliability.

Means of improving the cutting capability of machines are discussed with particular emphasis on the application of high pressure water jets as an aid to mechanical cutting, and the provision of a stable anchored base from which boom and turret can work to provide more positive application of cutting head to the rock.

A section of the paper is devoted to control technology with descriptions of the computer aided design systems for cutting heads and two of the technical information monitoring systems which are now being developed and applied.

Systems design and management is touched upon, as is automation, with examples of application of these techniques to remove operator dependency and provide more consistent operation.

The paper concludes with a discussion on aspects of design and new methods where a potential exists for achieving a more continuous non-cyclic operation and data acquisition for analysis and use in subsequent system and machine design.

1. INTRODUCTION

Although there may be some historical debate concerning the date when the roadheader first appeared in Western Europe, for all practical purposes it was in 1961, when the NCB purchased a PK3 machine from Russia, modified it to carry out experiments and production trials, and then used the principle as a base for the development of the roadheader industry. An evolutionary process has resulted in the current availability of a wide variety of machines from a relatively large number of manufacturers: some from Great Britain and some from Germany and Austria, where development also began in the 1960s. Figure 1 lists some examples of the roadheaders currently in Europe today, with their salient characteristics. The table includes two machines, which are not yet established in series production, but will be in the near future and which, in any case, provide interesting statistics.

A review of European roadheaders carried out in 1976 showed that the heaviest machines at that time were the 60-65 t Paurat Roboter and the 70 t Westfalia WAV200. Figure 1 shows that since then the machines have become heavier: 110 t for the Paurat E200, around 120 for the NCB/EIMCO machine, and 90 for the Anglo/Soviet Roadheader. The Voerst Alpine AM 100 is around 90 t. If the MRDE Circular Tunnelling Machine is included as a roadheader type by virtue of operation through a boom and turret, its weight is about 170 tons; that of the Westphalia Messerschildt is nearer 200. Thus, the incremental improvement of the roadheader is along the classical development lines of shearers and powered face supports into the heavy-duty category, with assumptions of better capability and greater reliability as the end product. The results obtained from this application of heavier equipment have not yet been fully assessed. In normal strata, the excavation rate has risen progressively, since the increase in size affords greater physical space and facility for larger, higher powered loading aprons and conveyors. The power applied at individual picks has increased, but not at the same rate. The - cautious - estimate of ability to cut harder rock has no apparent straightline relationship to increase in size and weight. In fact, it is only recently, with the application of the E200, the MRDE tunnelling machine, the Messerschildt, and the Anglo/Soviet collaboration machine, that a marked upturn in the ability to cut harder rock can be claimed with a degree of certainty.

2. IMPROVEMENT IN CUTTABILITY

Several distinct and not necessarily exclusive approaches to the improvement of cutting capability are now discernible. One relates to the vastly improved knowledge and ability to apply research findings in the design of the cutting tools and their disposition on cutting heads, which in turn, are themselves more scientifically designed for attack on the rock and loading out of debris. Another approach is represented by the evolutionary development into heavier and more powerful machines which still retain flexibility in operation. A third follows civil engineering practice, where anchorage in the roadway via a shield or some other method is used to provide the stable base from which more efficient cutting can be practised and a fourth is the application of fluid force, ie high pressure water jets.

2.1 Water Jet Assistance

If we can find an acceptable means of applying significant extra power at the cutter head without recourse to greater stabilising weight of machine, or anchorage in the roadway, then the whole course of roadheader development might be changed. The prizes to be won by a successful development of this nature are obvious in terms of reduction in machine cost, without sacrifice of cutting ability, and facility for handling the consequently smaller machines into and out of coalmines.

One such experiment is contained in the NCB/US Department of Energy Collaboration Project, which has the objective of applying high-pressure water jets on the cutter head of a roadheader to assist the mechanical cutting process.

This work has been reported elsewhere in technical papers and need not be repeated in detail. Suffice it say that in the course of the project, fundamental work in MRDE laboratories and at the HSE Laboratories in Buxton, showed that the use of water pressures of up to 700 bar could significantly reduce the lateral and normal forces required for a pick to cut a predetermined groove through rock. Engineering development then established the application of the technique to a Dosco Mk IIA roadheader in such a way that water could be released at these pressures on a number of picks in the cutting head. Combinations of 6, 9 or 15 picks in strategic positions were covered. Figure 2.

Trials of the machine at our Middleton Limestone Mine Test Site indicated that in rock (which averages 103 MPa in compressive strength and is fairly homogeneous) much better extraction rates could be obtained, the life of the picks was considerably extended and perhaps equally significantly, vibration of the machine was reduced. This reduction in vibration is well illustrated by power recordings taken on the cutter motor and hydraulic services of the machine in the 'jets on' and 'jets off' condition. The relatively steady power requirement in the 'jets on' condition contrasts with the fluctuation occasioned by cutter head bounce in the 'jets off' condition.

Of some importance also is the evidence that dust suppression is very much improved and the incidence of frictional sparking very much reduced.

Underground production trials have been delayed because of the amount of engineering development required on the pressure intensification equipment, the water transmission and release systems, and particularly the rotating seal where water is passed from a static pipe to a rotating head. These problems have been overcome by change in design of booms and gearboxes to allow the use of small diameter, easily changed seals.

A position has now been reached where both Anderson Strathclyde and Dosco have made prototype machines based on technical information provided by MRDE. MRDE have also modified two machines for underground trials and experience should be obtained during 1983.

With the agreement of the US Department of Energy we have practised information exchange with our European partners. Development is also proceeding in France and Germany, notably with the requirement for phased release of water where ripping type heads are used.

2.2 Machines With Anchorage

A major improvement in cutting capability, or at least in the ability to apply much higher unit pickloads, cutting head torque, and traversing forces, arises if a machine is designed to contain the reacted forces, not by virtue of physical machine weight but by anchoring in the roadway. This is common in civil engineering, where booms are used within shields as part of the conventional tunnel drivage and roadway lining process. MRDE, in conjunction with Cadley Hill Colliery, is currently conducting trials with a machine which takes advantage of this design approach. A very rigid boom secured to a turret, mounted in slides within a machine, is powered by a 225 kW motor through a gearbox of such a reduction, (210-1) that the cutting head speed is very low and the consequent torque very high, (312 kNm). The unit pickforce is approx 1 ton and the forces which can be generated to push the cutting head into the rock are up to 50 t. The sum of these forces is of course reacted back through the boom, turret and slides to the shield, which is split horizontally so that it can be firmly anchored in the roadway.

Experience already obtained has indicated that this high-force, slow-speed approach to rock cutting is a correct one and even though, initially, several problems with cutting head design and such mundane matters as ability to rapidly change worn picks, severely hampered the operation, rock of more than 206 MPa in compressive strength has been successfully negotiated and rock at more than 103 Mpa can be dealt with routinely.

The advantages arising from anchorage in the roadway, and a general increase in machine rigidity are confirmed by experience from Germany with the Westphalia Messerschildt, and also from the initial trials of the NCB/Eimco machine.

One problem, which arises from the rigid anchoring of the boom and turret in combination with high output torque, concerns the cutter gearbox. Because every reaction, every transient force and every inducement to vibration is rigidly resisted, the reflected and alternating loads on the cutting head are absorbed in the gearbox and transmission and mechanical breakdowns have been experienced with the MRDE machine.

Advantages of using a roadheader-type boom in a shield are apparent. A circular shape roadway can be produced, which, in the conditions of high strata pressures experienced at Cadley Hill, is beneficial, and it is possible to place a high-strength concrete panel lining inside the tailskin of the shield in a completely mechanised manner and with complete safety to the operators.

Although the overall machine and system is more complex than a standard roadheader mounted on caterpillar tracks, it is less so than the total machine system used in disc and roller cutter machines, and may well provide a satisfactory application in the gap between full-face machines and heavy-duty roadheaders.

An interesting feature of this machine is the vertical mounting of the boom on the turret, the traverse motion being driven via motors on an annular gear. This method ensures that, when the boom is set at the required angle and radius, simple rotation of the turret provides accurate profiling of the roadway with a measured amount of overcut.

The instrumentation and monitoring facilities of the machine are described in a separate section of this paper, but in addition it has been found necessary to provide automatic control of traversing speeds according to the measured load on the cutter motor, and a read-out panel for the operator showing him clearly the angular position of his boom and the status of conveyors.

A similar machine, though it cuts an arch shape rather than a circle and uses a ripping head on the cutting boom, is the Westphalia Messerschildt. Although not practising the same low speed, high torque approach as on the MRDE machine - output speed 59 rev/min as against 7 rev/min - the rigid anchoring of the machine and shield in the roadway has indicated better cutting capability. The facility for setting supports safely, at the time cutting and loading are carried out, has provided the same continuous operational characteristic.

3. CONTROL TECHNOLOGY

3.1 Cutting Heads - CAD

A most significant, but less remarked factor affecting the performance of machines, is the detail design of the cutting head. Over the past years, a much greater understanding of the design requirement for a properly balanced, vibration-free cutting head, with picks placed so as to ensure maximum break-out efficiency and maximum durability of the pick has been achieved. The biggest step in this direction has been made possible by the provision of computer programs, which were originally developed to check existing cutter head designs, and which have since been improved to allow initiation and rationalisation of design.

Tied into the development and use of the computer programs is the fundamental and practical information obtained from extensive research programmes in all countries of the Community. These culminate in practical development trials in Germany and Great Britain and via the full-scale cutting facilities at Steinkohlenbergbauverein at MRDE. The optimum attack (and other) angles for both axial and for the radial picks, has been closely determined, and combined with other research work which has indicated force requirement, pick lacing and line frequency for different types of strata. This wealth of data forms the input to the computer programs.

The programs are written for desk-top computers and can therefore be widely distributed at little cost. Several programs are used to look at different aspects of the design sequence. Also available is a library file facility, whereby designs may be permanently stored, or retrieved for review and/or amendment at will.

The system allows the introduction of design engineering competence and the requirements of production engineering, ie the inevitable compromise between the ideal and the practical. At every design stage, the pick boxes can be moved, deleted, or added to the cutting head as required, to achieve an optimum layout. Data on the three-dimensional geometry of various pick boxes can be retrieved from memory in order to ensure that they do not clash and that water nozzle placement, loading vanes, etc can be engineered successfully.

During the design process, schematic representations of mineral removed by each pick are available for checking the pick breakout/cutting pattern for all lines of picks. The VDU screen display for this is dynamic, and shows the cutting process progressing across the screen with each pick appearing in sequence as defined by the lacing diagram. Figure 3. This shows where picks have unequal duty and allows such errors to be corrected by manipulation of pick box positions. Expert interpretation of this breakout pattern will indicate where picks are prone to side rubbing and when grooving is probable. It is also possible to check the shaft torque, traversing and vertical force components for a cutting head operating through one revolution. Figure 4. This search for out-of-balance forces is very important when cutter heads are operating in tough and difficult strata, since any induced vibrations are reflected through the boom and gear transmission and can be a contributory cause of machine damage.

3.2 Technical Information Monitoring

When machines dispose larger power and weight, it is important to know in what way forces are acting and combining to affect the physical condition of the machine, and equally important for management to be able to examine the system operation, analyse the activity sequence and plan and test modifications. Significant steps have been taken in this direction. An example of a machine condition monitoring system is available for the E200. A variety of analogue pressure transducers are used in selected positions to pick up critical pressures on elements within the hydraulic circuits, with instruments for measuring electrical power level, and temperature measuring devices in the hydraulic oil and in the vicinity of the cutter gearbox. These instruments feed data through a microprocessor and then to a visual display unit positioned close to the machine driver. The data displayed can be arranged in a number of helpful ways. For example he is able to continually operate the machine to maximum load levels without actually going into electrical stall or hydraulic oil dump condition.

In addition, the data can be transmitted and displayed at another static station, either below ground or on the surface. Here the information can be viewed or recalled on a printer to be examined for change of relative values and therefore machine element condition. It is claimed that an experienced engineer can use the recorded pressure, temperature, and electrical power levels to check the condition of the cutter gearbox, the efficiency of hydraulic components, the degree of pick wear and even such things as chain tension and condition of flights, etc in the machine conveyor. Precise interpretation of data is said to indicate which element within a hydraulic circuit has failed or is failing, thus reducing diagnostic time and component replacement to a minimum.

MRDE has devised and installed a technical and management information monitoring system for the Circular Boom Tunnelling Machine at Cadley Hill Colliery. Measurement is carried out by four electric power transducers, eight hydraulic pressure transducers, two linear position transducers and eight digital transducers. The information is collected underground and transmitted to the surface for display on a VDU in the Colliery Control Centre and another one in the office of the Development Engineer at MRDE, some 3 miles away. Figure 5.

The VDUs show (a) a dynamic display of analogue and digital transducers readings and (b) histograms of all analogues over 8 hour, 32 hour, and 7 day periods. The histograms do not show the same precision as the dynamic display, but average values taken over a period. The information can be used to examine the instantaneous machine state, ie power levels, hydraulic pressure levels etc, detecting any abnormal operating conditions. The same information - recorded - can be analysed for trends which will indicate the condition of mechanical components, and rate and incidence of deterioration. The histograms can also be used to detect operational trends, pinpoint machine utilisation problems and plan improvements.

The equipment has already proved extremely beneficial in the commissioning period and in helping to establish the causes of particular failures. It has been instrumental in indicating a proper matching of operations in order to ensure maximum machine utilisation.

These are only two examples of an increasing application. Other manufacturers are providing similar systems with the same objectives.

3.3 System Management

Following from 3.2 above it is clear that an area of work which must develop from its present embryonic form is that of management monitoring of the system as it operates. The aim must be to get a continuous measurement of the running time of the machine and record of the reasons for the lost time. A basic system for the measurement of lost time on coalfaces has been adapted to the measurement of lost time in drivages. The evidence from the coalface work is that it will take a considerable time for data to be collected before there is sufficient to produce valuable analyses from the computer for the engineers and managers to have total system information.

Already it is evident that a key element in continuity of operation must be to make the setting of supports a function independent of the cutting operation. There is growing evidence that this may be achieved if the machine is designed with the roof setting mechanisation integral with the cutting task.

3.4 Automation

A recognised requirement is control of the boom for cutting the varied shapes that a roadheader is required to produce. The advent of the bigger machines with poorer operator visibility, with aggravated make of dust around the cutting head - has accentuated this.

Previously, the process of development and application has been sporadic, and although experiments have been made there has been little application until recently. But this is now underway. Systems to visually aid the operator are in use in Germany. The same system which houses the machine condition monitoring on the E200 includes linear potentiometers, which continually measure the position of the boom as moved by the vertical and the horizontal slewing systems. A proximity switch also checks the position of the boom relative to the loading apron. The information is processed and fed back to a screen near the operator, so that the position of the cutting head within a programmed shape is displayed. As the operator moves the cutting head within the overall envelope a light and/or noise warning is produced when he touches the limit at any point.

An interesting feature of the system is that the roadway form can be programmed into the microprocessor via the operator cutting a model shape and imprinting this on the system memory. The warning system will keep the operator within the 'taught' envelope until he decides to change it.

In Britain we are going straight for a completely automatic profile guidance and roadway alignment device.

A specification was written at MRDE, which calls for a system whereby information on the machine position and attitude within the roadway are established by the use of tilt instruments on the body of the machine and a measurement of the direction and manner in which a laser beam registers on a light sensitive receptive target. This information is to be analysed by a small microprocessor, capable of storing and transmitting information as required.

Another microprocessor is programmed to accept signals from linear or rotary transducers which constantly measure the position and angles of the cutting boom, analyse them, and command the cutting head to follow a pattern of extraction within a programmed arch shape. Alternatively, while the boom can be operated manually for the major part of the extraction, the computer will control cutting on the profile.

This second microprocessor is constantly updated by the first as to machine position and attitude, commanding it to adjust the position of the cut arch shape in space to ensure it is on correct roadway alignment. The size and shape of the roadway is programmed into the computer via a keyboard. Figure 6.

Quotations obtained against this specification have resulted in orders being placed with ZED and with PTT for competitive systems. These have commenced surface trials on a Dosco Mk III and providing the project goes as planned, further experimental equipments will be applied to the Anglo/Soviet Roadheader and other large machines.

The use of systems such as this throws into more direct focus the need for constant cutter power measurement and more sensitive control to match hydraulic flow and traversing force for optimum cutting and control purposes.

Again, these are only examples of a general advance. Similar work is well advanced with other manufacturers.

4. RELIABILITY

To assess improvement in reliability is difficult without access to a large amount of detailed information. It can be expressed in terms of reduced machine down time due to breakdown, or in financial terms, which relate to the cost of spares and repairs, or in terms of cost per cubic metre extracted, but unless comprehensive information on a large number of installations in a variety of conditions is available, any comparison may be misleading.

However one indication of the improved reliability of the heavier duty machines is the increased length of time they spend underground and operating between major overhauls. Although the heavy duty machines have now been in use in British coalmines for 5 years and now number more than 30, it has not yet been thought necessary to allocate one of the Regional Workshops to the overhaul and repair of this category of machine as is the case with others.

Generally, improving reliability depends on specific investigatory and design improvement exercises, using all the modern diagnostic, design and testing aids now available. It also depends on establishing regular measurement routines providing precise information to display efficiency trends, patterns of spares usage and relevant conditions.

5. CONCLUSIONS

The roadheader is as fundamental to the drivage of roadways in the relatively soft rock conditions of Europe as is the shearer or the plough to coal getting. The machine is maturing from its relatively light-duty antecedents to a heavier generation, capable of greater capacities and of dealing with a greater range of rocks. The technical advances that will take this further are those of staking the machine, to ensure a firm presentation of the cutting tool to the rock at slow speeds and high torque, and the use of high-pressure water jets to assist the pick in its cutting function.

Progress is being made in monitoring the condition of the machinery from electrical, mechanical and hydraulic points of view. The introduction of appropriate test routines with a strong discipline of regularity, will improve the maintenance of the machine and result in reduced lost time. Storage of the information on a computer will ensure that the maintenance schedule can become much more anticipatory than at present. The long-term storage of the data, together with record of cost of replacement, lubrication and so forth, will ensure that life cycle costing will lead to improvements in the design.

This advance in monitoring will be extended to recording the system behaviour. It is certain that improved operation will result from better visibility. In addition, there will be much better information on which engineers can work to improve the system design.

Automation of the machine will inevitably accompany the improved reliability of the equipment and of the system. The advantages of the automation will be not a reduction in the manpower but an improvement of the operation as a result of reducing the element of chance. Performance will be much more consistent, in rate of advance, loading of the machine and line of cutting.

The overall aim of the engineer is to maximise the running time of the available machines. Their effort will be directed towards improvement of the reliability of the machine and of the systems in which it works.

This progress will be achieved by an input of high technical competence both by the traditional engineer and by the new generation of control engineers with their instrumentation and computer capabilities. Their role will be to support the traditional rugged operator attitudes, which have been responsible for the high rates of drivage achieved with present-day equipment.

	WESTFALIA MESSER-SCHILDT WAV 200	MRDE C.T.M.	NCB/ EIMCO	PAURAT E200	ALPINE AM100	ANGLO/ SOVIET	DOSCO Mk3	PAURAT E134	DEMAG VS3	EICKHOFF EVA160	ANDERSON STRATH-CLYDE RH 22	DOSCO Mk2A
MACHINE WEIGHT - tonnes	190	170	120	110	84	90	81	70	63	53	35	26
TOTAL INSTALLED POWER - kW	310	435	300	462	456	300	300	300	264	310	160	105
CUTTING MOTOR POWER (MAX) - kW	200	225	150	300	228	150	142	200	160	160	90	48
CUTTING HEAD SPEED (MAX) - rpm	59.0	6.4	6.7	36.3	59.0	29.0	32.0	40.3	53.0	48.0	50.0	60.0
CUTTING HEAD DIAMETER (MAX) - m	1.00	2.00	2.00	1.21	1.20	1.00	1.04	1.21	0.90	1.00	0.81	0.53
PICK SPEED (MAX) - m/s	3.10	0.67	0.70	2.30	3.71	1.52	1.74	2.60	2.50	2.50	2.10	1.66
PICK FORCE (AT MAX PICK SPEED)- kN	65	336	212	130	62	99	81	77	64	64	43	29
ARCING FORCE - kN	98	540	334	196	88	235	157	98	74	98	54	34
ELEVATING FORCE - kN	98	490	255	196	88	235	157	98	83	98	51	32

FIG. 1

FIGURE 2 - WATER JET ROADHEADER

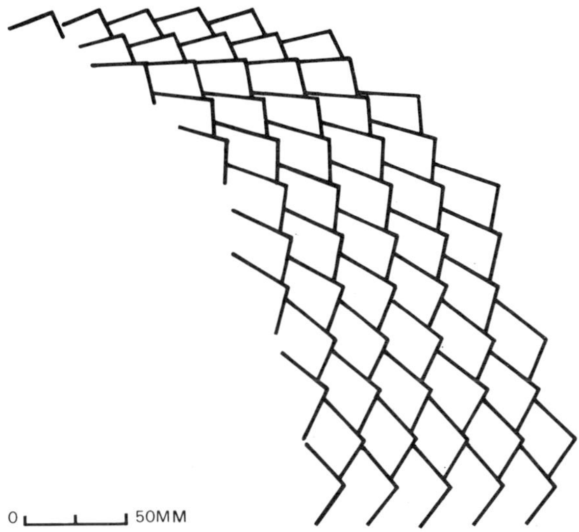

YOU CAN NOW REDRAW THE BREAKOUT PATTERN FOR SEVERAL REVOLUTIONS OF THE HEAD. ALSO YOU CAN ENLARGE A SECTION OF THE PATTERN BY DRAWING IT ONLY FOR THE PICKS OF INTEREST. DO YOU WANT EITHER OF THESE OPTIONS (Y/N) ? *

FIG.3 PICK BREAKOUT PATTERN – ROADHEADER

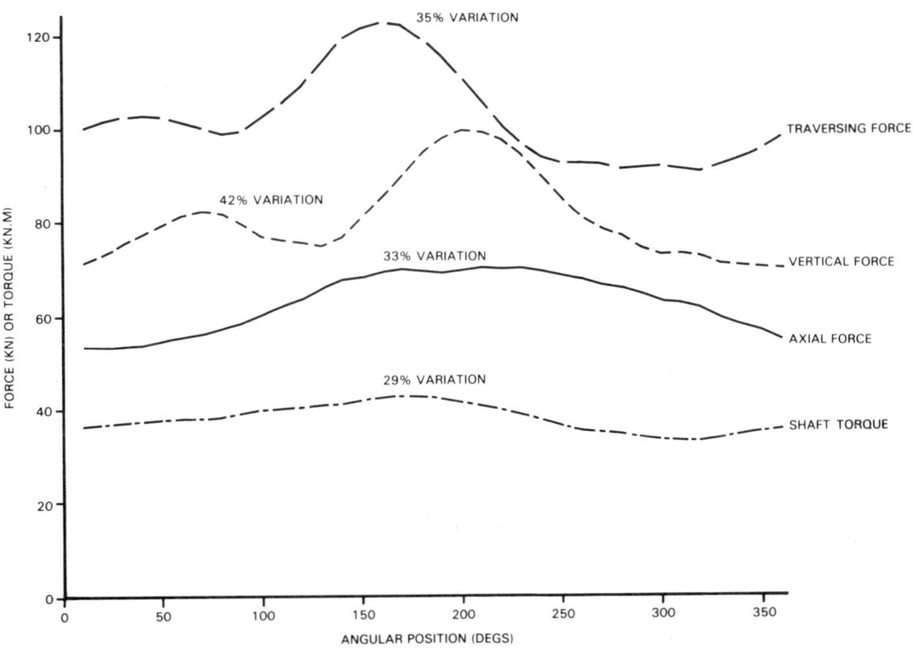

FIG.4 STANDARD HEAD – 180 DEG CUT SECTOR

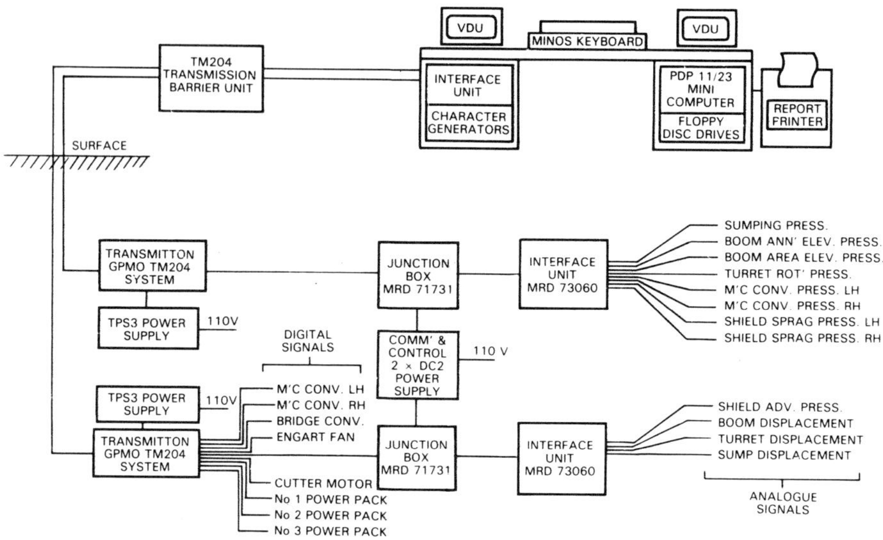

FIG. 5 CTM MONITORING SYSTEM

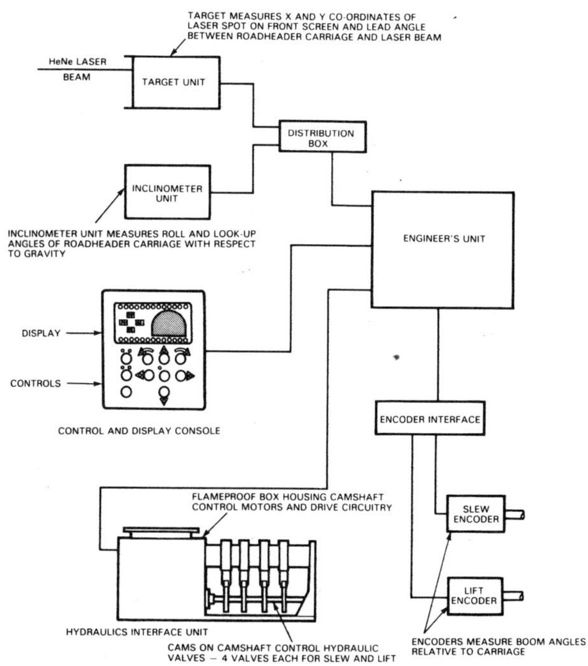

FIG. 6 SYSTEM INTERCONNECTION

FIRST OPERATIONAL EXPERIENCES WITH THE E 200 SELECTIVE ROADHEADER IN STONEDRIFTING

RUDOLF SANDER

Director, Bergbau AG Niederrhein,
Rheinland Combined Mine

Summary

In the West German coal industry 100 km of stonedrifts and 420 km of in-seam roads were driven in 1982. Whilst only limited use of full-face roadheaders is possible, increasing numbers of selective-cut roadheaders are being used to head in-seam roads. Since the proportion of rock being cut in drivage is increasing, the trend towards heavy types of machine is continuing. Good results achieved with selective-cut roadheaders led to the decision to extend their use to stonedrifting. Taking as a basis the successful E 134 Roboter built by Paurat GmbH, the E 200, weighing 115 tonnes, was developed in the context of a research project. In early 1981 the first E 200 came into service at the Rheinland Combined Mine at Bergbau AG Niederrhein and shortly thereafter another began to be used at Saarbergwerke AG's Ensdorf pit. Since then these roadheaders have driven, respectively, 1642 and 2069 metres of stonedrifts. Even though the advance rates of 4 to 8 metres per day are not entirely satisfactory, it has been demonstrated that even sandy mudstone can be cut at not unreasonable cost. Successful practical experience has led to the delivery of two further E 200s. Development work is being concentrated mainly on improving the cutting heads and raising the slewing forces when the machine is anchored. And application of metering and monitoring systems like the "automatic pilot" successfully installed at the Rheinland Combined Mine.

1. REASONS FOR THE DEVELOPMENT OF HEAVY SELECTIVE-CUT ROADHEADERS

In 1982 about 100 km of stonedrifts and 420 km of in-seam roads were driven in coal-mines in the Federal Republic of Germany.

Although the number of full-face roadheaders used for stone-drifting has increased in recent years, only 15% of roads are machine-headed. This proportion is likely to increase further in the next few years, but the conditions required for full-face roadheaders - roadway sections that are at least 3 km long, sufficient space for repositioning and adequate dirt removal capacity - are only present to a limited extent. It looks as if 70 - 80% of stonedrifts will continue to be driven by the conventional, comparatively slow and costly blasting method.

In-seam heading has developed along very different lines over the past few years (Figure 1). In 1982 150 km or 36% of in-seam roads were headed with selective-cut roadheaders. Since it is now well established that machine-driven in-seam roads independent of a face are significantly cheaper than blasted roads given the appropriate conditions, it is likely that there will be a further increase in the use of selective-cut roadheaders (Figure 2). At the end of 1982 the German coal industry had 110 machines in use. A breakdown of the individual manufacturers shows that the main machines used are those made by Paurat GmbH and Voest Alpine AG (Figure 3).

The steadily growing number of selective-cut roadheaders was only made possible by the extension of their range of application. The good results achieved by the machines encouraged their transfer to seams which at first sight seemed unsuited to machine drivage, such as thin seams with hard surrounding rock and large roadway cross-sections. The proportion of rock being cut in drivage increased steadily (Figure 4). Attempts were also made to increase natural drivage lengths by working systematically through faults or by interposing inclines, in order to save on the number of machine transfers required. The increased demands thus made on selective-cut roadheaders are conditioning the developments made by the manufacturers over recent years (Figure 5), the main feature emerging being the trend towards heavy machines. The E 134 and AM 100 Roboters yielded outstanding results in hard surrounding rock or in sections drifted entirely in stone, but the limits of these selective-cut roadheaders also became apparent, particularly with regard to thrust and stability.

2. DEVELOPMENT OF A SELECTIVE-CUT ROADHEADER FOR STONE-DRIFTS

The limited applications of full-face roadheaders and the unsatisfactory drivage results achieved by blasting led to a decision, taken in the light of the good results achieved in in-seam heading, to develop a heavy selective-cut roadheader for stone-drifting. In the context of a project subsidised by the Federal Ministry for Research and Technology, detailed specifications for such a machine were worked out in the years 1978/79. The key requirements were as follows:-

- Raising cutting performance and exploiting it better to produce higher breaking performance
- Making the cutting heads run more smoothly and optimising machine stability
- Development of cutting heads with internal feed spraying
- Integrated dust-extractor ducts
- Development of a reinforced conveyor system.

As had happened almost 10 years previously with the Roboter, responsibility for managing this project was once again given to the Rheinland combined mine, Bergbau AG Niederrhein. Building on the successful design of the E 134 Roboter the working in close co-operation with Bergbau-Forschung GmbH, the supplier - Paurat GmbH, Voerde - developed the E 200 in the years 1979 to 1981.

3. DESCRIPTION OF AN E 200 SELECTIVE-CUT ROADHEADER

The machine has a service weight of 115 t and is 13.6 m long (Figure 6). Its height is 2.7 m and its width, apart from the loading apron, is 4.8 - 6.0 m. With a sumping depth of max. 1.2 m, the machine cuts a road width of 5.5 to 7.0 m and a height of 3.5 - 6.0 m.

The installed rating is 473 or 523 kW with a mains voltage of 1000 V. The forward-rotating cutting head is driven by 175/350 kW pole-reversible motors. The triple-spiral cutting head cuts at an average peripheral speed of 1.05 or 2.1 m/sec. The raising and slewing jacks produce a thrust of 20 to 28 Mp. The crawler transmits a thrust of 60 Mp for sumping. 16 jets with a diameter of 0.8 mm are used for spraying the picks. Up to 100 l/min of water at max. 100 bar can be supplied by a booster pump. The hydraulic power unit has a rating of 90 kW. All cutting and manoeuvring functions are supplied either by a pressure-balanced radial-piston two-stage pump with an output of in total 180 l/min at max. 250 bar or by 2 two-stage pumps of the same design with a total throughput of 360 l/min (Figure 7). The debris is taken through the machine by two reinforced EKF roller curve conveyors and goes via a stage loader to the outbye conveyor. The machine's portal has a cross-section of 1.0 x 0.65 m. The loading apron and the rear vertical stabilising jacks can be hydraulically pressed, at a pre-set thrust, onto the roadway floor. Extending the tipping edges and increasing the area of ground contact in this way improve the machine's standing moment.

The dust make is extracted in the roadhead by two 600 ducts and taken via extractor pipes integrated into the machine frame to dry and wet dedusters with a throughput of 600 m^3/min.

4. PRELIMINARY TESTS ON THE SURFACE

In 1979/80 cutting heads made by Paurat GmbH were tested on Bergbau-Forschung GmbH's test site. This series of tests was designed to improve the cutting heads' performance and to determine pick stability.

The prototype of the E 200 stone-drifting Roboter was tested in early 1981 by Paurat GmbH at their own surface test facility. The tests were carried out on a concrete test block. The series of measurements were monitored by the Mechanical Engineering Establishment of the WBK. Machine vibration measurements were taken and the specific extraction rate in KWh/m^3 was established.

5. EARLY APPLICATIONS OF THE E 200 STONE-DRIFTING ROBOTER

Once the surface tests had been concluded and had been found to be very promising, the first E 200 stone-drifting Roboter was delivered at the end of March 1981 to Bergbau AG Niederrhein's Rheinland combined mine (Figure 8). Shortly thereafter a second machine was supplied to Saarbergwerke AG's Ensdorf mine.

At the Rheinland combined mine, the E 200 is being used to drive the main northern lateral road on the 703 m level of the Rossenray colliery. The ground dips at 5 - 6 gon and is made up, over a 3000 m stretch, of 81% mudstone, 11% sandy shale, 6% sandstone. Numerous geological faults with considerable throws have to be worked through. The road is supported by 4-section BnC 22.8 yielding arches with an excavated cross-section of 26 m^2. Fine-meshed matting is used as lagging. The roof is filled directly in the roadhead using ready-mix concrete made by Elskes KG, whilst the walls are shotcreted 6 m behind the roadhead.

Following the assembly of the machine and the training of the team the machine has been in operation since October 1981. By May 1983 1642 m had been driven. The average rate of advance is about 4 m/d. This admittedly unsatisfactory rate is mainly due to geological difficulties, but also to mechanical adjustments and problems with backfilling the supports.

It has so far been possible to far outstrip the specified cutting speed of 0.45 m^3/min with an average speed of 1.1 m^3/min. The average rate of pick consumption is 4 per metre. The main feature is that, since the beginning of the year, when the sandy shale and sandstone strata were reached, cutting performance sank to about 0.4 m^3/min (Figure 9). In pick consumption there were peaks of up to 50 picks per metre.

Since May 1981 the second E 200 has been in service in Ensdorf colliery. Up to and including February 1983, 2039 m lateral roads have been driven with an excavated section of 21.2 m^2 (Figure 10). In the first lateral road, 1277 m were driven with rates of advance of 4.4 - 8 m/d. Average pick consumption in sandy mudstone was 0.5 per metre. It rose to 4.2 per metre in the harder sandy shale and sandstone. The second drivage section with the E 200 has been in progress since August 1982 and by February 1983 762 m of roadway had been driven. The average rate of advance so far has been of the order of 6.7 m/d.

One of the special features of this roadheader is that the cutting power rating has been raised to 350 kW. A telescopic advancing mechanism has also been fitted.

6. CONCLUSIONS AND FURTHER DEVELOPMENT OF THE E 200

Practical experience so far gained with this selective-cut roadheader designed for stone-drifting provides important guidelines for further development of the E 200.

It has been demonstrated that even sandy mudstone can be excavated with reasonable pick costs and good cutting times, but - the hard sandstone strata that have been encountered represent the limits on the use of selective-cut roadheaders.

From the aspect of cutting technique, improvements are envisaged with regard to raising slewing forces as this would enable better utilisation to be made of installed cutting ratings and torques at the cutting head.

It has also been shown to be necessary to improve the cutting heads themselves. There is a particular danger of wear in the forward zone of the cutting head, as this is what must effect the sumping into the solid material in stonedrifts.

The higher exploitation of cutting potential demanded by mine operators has led to a further development in the direction of higher thrust forces and an increase in travel speeds.

More metering and monitoring systems than previously will be needed for the optimal use of the machines' performance. They should not only register information but should give the machine operator real-time facilities to achieve the optimal cutting orientation together with data on machine status from moment to moment. Paurat GmbH's master "automatic pilot", which is already in service at the Rheinland combined mine, is a promising system of this type. The automatic pilot consists of a microcomputer system, which processes three system packages: malfunction diagnosis, cutting profile control and performance optimisation.

Appropriate support techniques should be adopted to improve the E 200's utilisation factor. The latest developments in rock bolting techniques would appear to be well suited to this.

The water supply for the internal ducting of the picks is technically solved by a rotary feedthrough system, but the service lives of the seals need to be increased; a segment regulator is to be developed to reduce the volume of water for the same working pressure.

The layout and design of the E 200, together with the successful drivage results so far achieved, make it appear worthwhile to develop this roadheader so that it can also be used for small roadway sections of about 16 m^2. One possibility being considered is the reduction of the dimensions of the machine by transferring the fittings from the machine chassis to the equipment train.

Two more E 200's have come into service since the first two were used. At Walsum colliery the machine is used for driving a stonedrift. At Ensdorf colliery another E 200 is working a seam which is difficult to dint. This machine was equipped with the boosted hydraulic unit mentioned above and had increased thrust at the cutting head in order to achieve better excavation rates with greater radial forces and higher travel speeds. The objective is to reach rates of advance averaging 22 m/d.

KEY TO ILLUSTRATIONS

Roadway length in km

Blasting

+ 76%

Machine drivage

Increase

Figure 1. Proportion of machine-driven roadways.

Number of machines

Figure 2. Development of cutting-head machines.

Number of machines

Total machines: 110

Situation: Dec. 1982

Figure 3. Manufacturers of selective-cut roadheaders.

Sandstone
Sandy shale
Mudstone
Coal

Excavated cross-section

Figure 4. Proportion of surrounding rock in road drivage with selective-cut roadheaders.

Number of machines

Weight category

Figure 5. Weight categories of selective-cut roadheaders.

Length	13.6 m
Width	4.8 - 6.0 m
Height	2.7 m
Weight	115 t
Total rating	512 or 562 kW
Rating (cutting head)	300 or 350 kW
Slewing force (cutting head)	196 or 280 kN
Road width	5.5 - 7.0 m
Road height	3.5 - 6.0 m

Figure 6. Technical data for the E 200.

Figure 7. E 200, Paurat GmbH.

Figure 8. E 200, Paurat GmbH.

SAMPLE 1
Compressive strength
Tensile strength
Wear factor
Toughness

SAMPLE 2
Compressive strength
Tensile strength
Wear factor
Toughness

Figure 9. Parameters for rock mechanics measured during operating runs made by the E 200, Paurat GmbH.

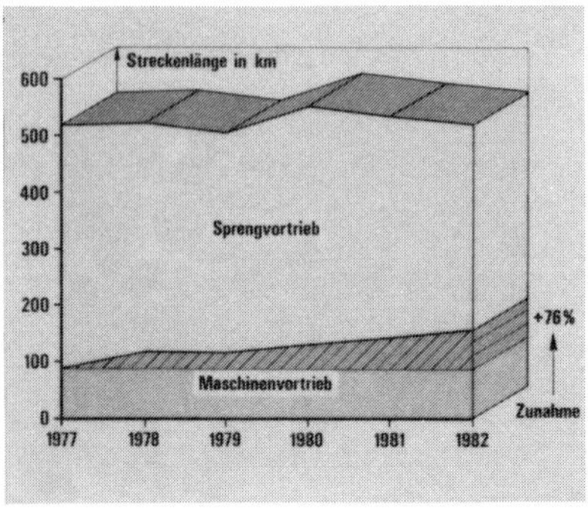

Bild 1: Anteil maschinell aufgefahrener Strecken

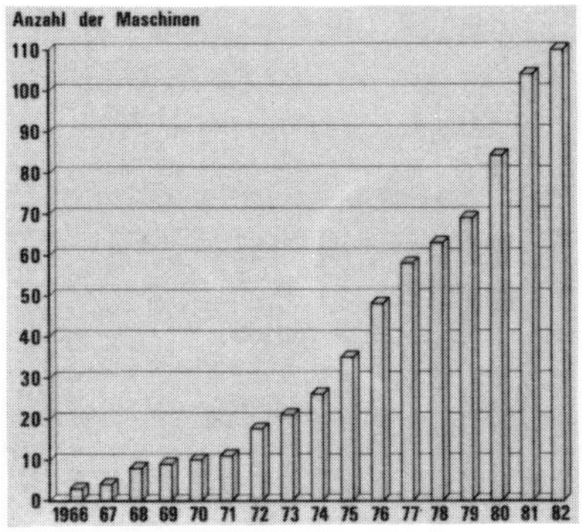

Bild 2: Entwicklung von Schneidkopfmaschinen

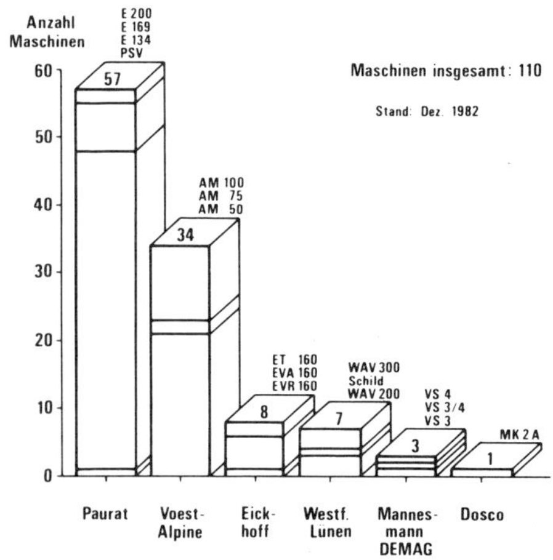

Bild 3: Hersteller von TVM

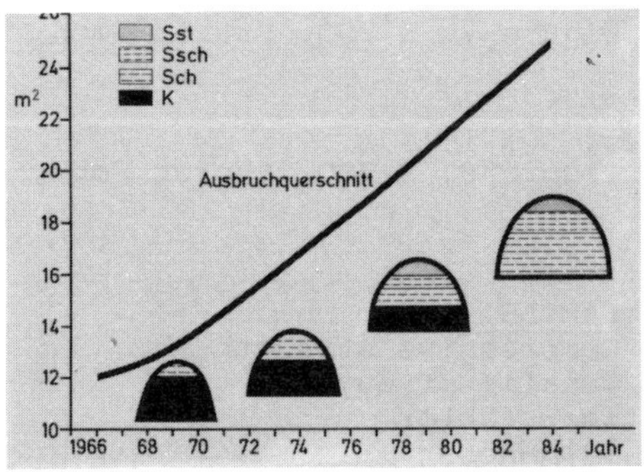

Bild 4: Nebengesteinsanteil im Streckenvortrieb mit Teilschnitt-Vortriebsmaschinen

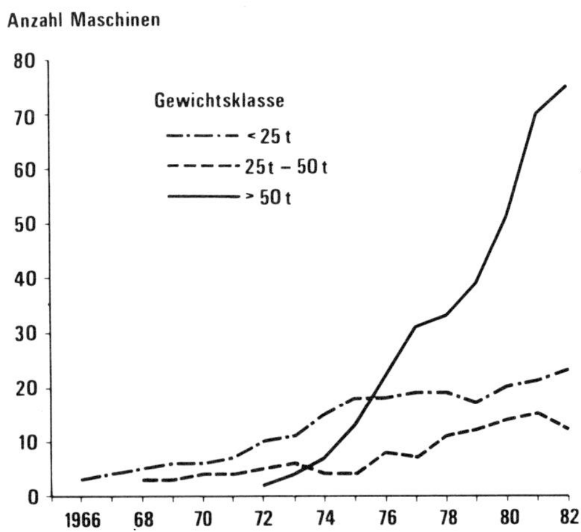

Bild 5: Gewichtsklassen von TVM

Lange	13,6 m
Breite	4,8 - 6,0 m
Höhe	2,7 m
Gewicht	115 t
Leistung (ges.)	512 oder 562 kW
Leistung (Schneidkopf)	300 oder 350 kW
Schwenkkraft (Schneidkopf)	196 oder 280 kN
Streckenbreite	5,5 - 7,0 m
Streckenhöhe	3,5 - 6,0 m

Bild 6: Technische Daten der E 200

Bild 7: E 200, Paurat GmbH

Bild 8: E 200, Paurat GmbH

Kennwert		Mittelwert
PROBE 1		
Druckfestigkeit	N/mm²	124.3
Zugfestigkeit	N/mm²	13.8
Verschleißkoeffizient	N/mm	0.342
Zähigkeit	$\sigma_D : \sigma_Z$	9.0
PROBE 2		
Druckfestigkeit	N/mm²	70.1
Zugfestigkeit	N/mm²	10.9
Verschleißkoeffizient	N/mm	0.632
Zähigkeit	$\sigma_D : \sigma_Z$	6.4

Bild 9: Gesteinsmechanische Kennwerte aus Einsätzen der E 200, Paurat GmbH

Colliery	Ensdorf		Rheinland
Operating area	Lateral	Lateral	Lateral
Commencement of drivage	May 1981	Aug. 1982	June 1981
Excavated section m^2	21	21	25-26
Distance so far driven m	1277	762 (Feb. 83)	1642 (May 83)
Rate of advance			
Average m/d		6.7	4.0
Best month m/d	7.4	8.0	6.4
Monthly drivage			
Average m/month	85	113	76
Maximum m/month	148	160	103

Figure 10. Operating data for the E 200, Paurat GmbH.

SESSION II - MECHANIZED DRIVAGE IN COAL

- Drivage in the Seam (Light duty roadheaders, dintheaders, and in-seam miners)
- Driving trapezoidal roadways with the SSM machine
- The use of impact rippers in the german coalmining industry
- Mechanization of shortwall heading at Zolder Colliery in the Kempen coalfield

DRIVAGE IN THE SEAM

(Light Duty Roadheaders, Dintheaders and In-Seam Miners)

by L. HARRIS

Deputy Director (Mining) South Midlands Area
National Coal Board

Summary

The National Coal Board's annual programme of mining development work which last year exceeded 570 kilometres is aimed at ensuring continuity of working for the right number of longwall, shortwall, bord and pillar and single entry workings.

Mechanisation of in-seam development drivage continues and there is now available machines to suit almost any roadway size or profile, in differing strata conditions.

This paper describes the machines and systems currently in use.

1. INTRODUCTION

Underground drivages within the National Coal Board collieries exceeds 550 Km per year. This level of activity is necessary in order to sustain the annual changeover of 380 - 400 coalfaces, and is approximately equivalent to the construction at each mine of three kilometres of new drivage each year.

The annual mining development drivage distance nationally is divided more or less evenly between face access drivage and drivage for new districts and seams. Face access drivages can be further sub-divided: 40% for advance faces and 60% for retreat faces.

In common with other coal industries timing and rate of advance are of critical importance in mining development operations. Slippage in a face changeover time, whether from deficiencies in rates of drivage, installation or salvage, can result in substantial shortfalls in output and proceeds. Consequently getting faces ready on time is still the main purpose of mining development systems.

In addition to drivage work for longwall faces, more than 90 Km of additional coal heading work is done on a production basis to provide about two million tonnes of coal from bord and pillar, and single entry retreat mining methods.

2. DRIVAGES IN THE SEAM

2.1 Drivages All Coal (At least 90% coal)

In 1982/83 66 Km were driven of which 59 Km was face line drivage. Heading machines account for 78% of this drivage, predominantly Dosco Dintheaders and Roadheaders. Hand-loading at 10% is mainly in face headings in thin seams for which there is no suitable type of machine available.

(See Figure 1 for drivages by method of work and Table I for average performance levels.)

2.2 Drivages Mainly Coal (50% to 90% coal)

In 1982/83 225 Km were driven and the proportion extracted by heading machines was higher at 87%. Here again Dosco Dintheaders and Roadheaders are most prominent.

(See Figure 1 for drivages by method of work and Table I for average performance levels.)

2.3 Drivages Mainly Stone (10% to 50% coal)

In 1982/83 200 Km were driven and the proportion extracted by heading machines again in excess of 80%. In this strata Roadheaders take the ascendency over Dintheaders with more than 50% of the total drivage. (See Figure 1 for drivages by method of work and Table I for average performance levels.)

2.4 Comparison of Heading Machines in Different Strata

Continuous Miners obtain by far the highest results in coal and soft stone. Dintheaders can maintain their performance for square work roadways into slightly harder strata, although In-seam Miners matched them for performance in face line drivages last year. (See Figure 2 for Average Rate of Drivage by Method and Strata.)

3. TYPES OF HEADING MACHINE IN USE

3.1 Dosco Mk.2A Roadheader

This light duty single boom roadheader is the most popular type in use in British coal mines. More than 600 have been purchased by the National Coal Board and the utilisation factor is high with about 540 machines in regular daily use. It is used extensively for all types of drivage work, particularly in seam drivages of D shape in conjunction with steel arch girders.

Good results have been obtained in coal, mainly coal, and mainly stone, in all the British Coalfields. Performances in excess of 10 metres per day advance have been achieved at Frickley Colliery in the Doncaster Area, at Lea Hall Colliery in the Western Area, and in my own Area of South Midlands at Donisthorpe Colliery.

The machine dimensions are:-

Length	7.2 metres.
Width	2.9 metres.
Height	1.6 metres.

The total weight and installed power are respectively :

24 tonnes and 105 Kw.

3.2 Dosco Dintheader

This is the most popular machine in British mines for the drivage of rectangular profiled roadways. Here again the utilisation is high with about 315 machines in regular use out of a holding of 400 machines.

It is used extensively for roadway drivage for retreat mining access work, face headings, scours in soft stone and other in seam drivage work. Good results have been obtained in many installations throughout N.C.B. Mines, in particular 18 metres per day has been exceeded at Woolley Colliery in the Barnsley Area, and 16 metres per day exceeded at Trelewis Drift and Betws in South Wales.

The machine has a working range as follows :-

Height - Minimum 1.47 metres.

- Maximum 2.59 metres.

Minimum Width 1.83 metres.

The total weight and installed power are respectively :-

14 tonnes and 90 Kw.

A new uprated version is now on trial with 4 machines in use. The objective being to have a higher powered machine (232 Kw) capable of cutting in harder conditions with a height range up to 3.31 metres. The best performance so far has been 75 metres advance in one week (5 day working).

3.3 Dosco In Seam Miner (I.S.M.)

Designed at the Mining Research and Development Establishment (M.R.D.E.) and later developed by Dosco Overseas Engineering Ltd. The main use is for winning out new face lines and there have been successful installations in most Areas of the N.C.B. One of the main advantages is its ability to allow installation of powered supports and face drivage to take place concurrently. Range of application is 0.9 metres to 1.7 metres section. Width can be varied by increasing the length of the cutting frame starting at 5.33 metres and varying up to 13.41 metres. The original types were hydraulic powered, whilst the later versions have electric motor drive to the cutting frame, with hydraulics retained to advancing and setting roof supports. In addition to winning out new coal faces the machine has been successful in the following applications :-

Stable Elimination

Partial Extraction

Single Entry Face System

Dirt Absorbing headings.
About 40 machines are in use currently with good results being obtained in a variety of places.
In April 1982 a record performance of 367 metres advance in one week (5 day working) was achieved at Whitwick in my own Area of South Midlands, in a coal heading 6.00 m wide by 1.4 m high.
The machine weight depends on width of the frame, a typical one being 14 tonnes, with the cutter chain powered by two 37.5 Kw motors.

3.4 Joy 1CM4 Continuous Miner

These machines have been used successfully over many years for all types of drivage work with a rectangular profile, in coal and mainly coal strata. Also used extensively for bord and pillar mining in the North East Area of the National Coal Board. Most of these machines are in the age group 15 - 20 years and are gradually being phased out and replaced by the newer heavier duty designs of continuous miner. They have given good service over the years and most of the 20 machines still operating are still giving satisfactory performance, particularly at Ellington Colliery where good rates of advance and high productivity levels are still being obtained.
It has a loading capacity of 3½ tonnes per minute in the height range of 1.5 metres to 3 metres, and is capable of negotiating 90° turns in a 4.3 metre roadway.
Total weight and installed power are respectively :-
23 tonnes and 180 Kw.

3.5 Anderson Strathclyde Boom Miner

This is regarded as a soft strata light duty roadheader, hence used for coal or mainly coal drivages.
It is capable of operating in roadways as small as 1.7 metres high and 2.56 metres wide. The cutting head can be axial or forward rotating type, and the combination of these features make it suitable for both arched roadway work as well as lower in seam mining work. In suitable strata some good performance levels have been achieved, particularly in my own Area at Rawdon Colliery.
Total weight and installed power are respectively :-
19.5 tonnes and 120 Kw.

10 machines are currently in use.

3.6 Anderson Strathclyde RH 22 Roadheader

This machine is used for in seam drivages where harder strata is in close proximity to the seam. About 28 machines are currently in use and good performance levels have been obtained at Solsgirth in Scotland, Ollerton in North Nottinghamshire, and Rawdon in South Midlands.

The machine dimensions are :-

Length	8.69 metres	
Height	2.05 metres	
Width	2.75 metres	(Frame)

Total weight and installed power are respectively :-
35 tonnes and either 138 Kw or 180 Kw.

3.7 Dosco Mk.2B Roadheader

This machine is also used for in-seam drivages where harder strata is in close proximity to the seam.

This machine was a logical addition to the range of Dosco Roadheading machinery, and takes its place between the highly successful Mk.2A and the heavy duty Mk.3.

One of the first machines to go underground was at Donisthorpe Colliery in the South Midlands Area where 1500 metres have been driven through fairly hard strata over an 18 month period in a roadway 4.9 m x 3.7 m setting heavy duty steel arches.

The machine dimensions are :-

Length	9.7 metres	
Height	1.9 metres	
Width	2.7 metres	(Frame)

Total weight and installed power are respectively :-
36 tonnes and 165 Kw.

3.8 Joy 12CM5 Continuous Miner

This is the newer heavier version of the successful continuous miner.

Four machines have been purchased all in the North East Area for coal production in bord and pillar workings. Capable of high production rates up to 800 tonnes per day and 40 tonnes per manshift productivity.

3.9 B.J.D./Dresser Heliminer 122 M

This is another type of the newer designs of continuous miner that are now being purchased to replace the old machines. Three machines were originally purchased and two put to use at Ellington in the North East Area for coal production in bord and pillar workings. The high daily rate of advance achieved has led to the purchase of two more machines to block out for retreat mining at Wistow Colliery in the Selby Complex, to match the high development rates obtained by the other types of heading machines used in the access roadway drivages. Good results have been obtained, with weekly performances of up to 130 m being achieved.

Machine specification is as follows :-

Length 10.13 metres
Width 2.34 metres (Frame)
Height 0.99 metres
Mining Height - 1.47 to 2.82 metres
Auger cutting elements variable 2.89 - 4.72 metres
weight 42.4 tonnes

3.10 Dosco L.H.1300 Roadheader

This roadheader is of medium duty designed to operate in low height conditions (1.3 m) having the ability to cut out square or D shaped roadway profiles. It is a very powerful machine having some 239 Kw installed electrical power and is extremely stable due to it's low centre of gravity and overall weight of 35 tonnes. The main attractions are flexibility of application and ability to cut hard strata whilst matching the rates of advance achieved with Dintheaders. From a centre line position it is capable of extracting a section 5.5 metres wide by 3.5 metres high.

To date four machines have been purchased and three put to use to develop roadways for retreat face access. The initial results have been very promising with 77 metres achieved in one week (5 day working) and 275 metres over a four week period.

Machine dimensions are :-

Length 9.7 metres
Width 3.0 metres
Height 1.3 metres

3.11 Demag Unicorn Miner

Six machines were purchased and 5 put to use in Scotland and 1 in Western Area.

The main advantage is the ability to drive and support on steep gradients. In Scotland face gradients of 1 in 1 and steeper have been driven. Their use has been confined mainly to Seafield and Frances Collieries where rates of 60 metres per week have been achieved.

Excavation height - 1.2 metres up to 2.8 metres.
width - 3.5 metres up to 5.0 metres.

3.12 Lee Norse CM 48 Continuous Miner

Many of these old machines have now been phased out, although a number are still in use for development work and bord and pillar extraction. Creditable performance levels are still being obtained.

4. BORD AND PILLAR PRODUCTION SYSTEMS

Bord and Pillar production systems occur largely in the North East Area of the N.C.B., where highly mechanised systems of partial extraction operate at high levels of production and productivity.

A good example is at Ellington Colliery in the Yard seam, where the bord and pillar layout consists of driving roadways 2.54 metres high by 6.10 metres wide on a square pillar system of 28.35 metre centres.

The main conveyor and supplies roadways are taken ahead of the panel being extracted, then the panel is extracted taking a system of five headings being driven by the continuous miner, which is backed up by two shuttle cars, one gathering arm loader and usually one Joy mobile tail end delivery conveyor. Supports are 150 mm x 125 mm R.S.J. girders set on wooden legs of 1 m to 1.2 m centres with punch props (wood) set in the web of the R.S.J.'s to act as struts.

In March this year total bord and pillar output at Ellington Colliery from ten sites totalled 120,643 tonnes for a total advance of 8,574 metres. Three types of the Continuous Miners described are in use : Joy 1CM4, Joy 12CM5, and Dresser Heliminer 122M.

5. SINGLE ENTRY PRODUCTION SYSTEMS

5.1 Coal Pillars

Coal pillars are left for a variety of reasons. Over the years multi-seam working has created problems associated with seam

interaction as well as the usual obstacles such as faults and washouts. Many coalfields have reserves which lie beneath extensive residential areas and industrial activities, in addition to buildings which have protection orders. Consequently considerable reserves can be sterilised by the need to leave protective coal pillars for surface features and for surface drifts. Add to these the coal left alongside and between faults, plus pillar protection required to assist roadway stability and minimise interaction problems between seams underground. In this way the total sterilised reserves within a given area can be substantial. Within one N.C.B. Area it was estimated at close to 100 million tonnes.
This is not a new problem and over the years many pillars have been extracted by conventional longwall methods. With the need to drive two roadways, however, there are obvious limits on viability.

5.2 Single Entry Concept

The concept of a single entry face is to drive a gate entry into an area of coal to be worked and subsequently retreat a shortwall face back alongside the initial entry road.
Two collieries having the benefit of wide experience in longwall retreat mining pioneered the early work and have now designed and operated a number of successful single entry faces. Following the initial success in the 1.5 m thick Newhill seam at Wath Colliery in the South Yorkshire Area on a 7 metre long face the system was further exploited by increasing the face length to 35 metres in the same seam at South Kirkby Colliery in the N.C.B. Barnsley Area.
The coal area to be worked at South Kirkby forms part of a protective pillar for an adjoining drift mine and the adjacent South Kirkby town centre, and is large enough to allow a number of faces to be worked with runs of 600 metres.
The gate entries are driven by Dintheaders and supported by Hollybank (4.88 m x 2 m) rectangular supports with splayed legs. Face equipment has been kept as much as possible to standard versions, although new designs offering improved production and productivity have resulted from the lessons learned on the early faces. Manpower on the face normally consists of a

machine operator and three others to deal with supports and the gate entry junction work. One mechanic, one electrician and the Deputy in charge complete the team.

5.3 Results Obtained and Prospects

The following results for a 600 metre face run on the second face to be worked gave a contribution to the colliery in excess of £800,000.

	Average	Best
Strips per week	64	140
Tonnes per manshift	16.96	78.63
Tonnes per day	278	990
Tonnes per week	1332	3117

Other Areas and Collieries are undergoing trials or are planning to do so, including my Area of South Midlands at Measham Colliery.

In order to facilitate the rapid retreat mentioned it is necessary that the single entry access is purpose driven. There are examples of using old existing roadways, although speed of movement is generally limited. This in turn means that the real success of the system for sustained good performance is the ability to drive rapidly in the gate entry. To date the Dintheader has proved to be the best machine for this purpose.

6. PARTIAL EXTRACTION BY IN-SEAM MINER UNDER THE TOWN OF COALVILLE

6.1 Background Information

Whitwick Colliery sunk in 1820 is one of the North West Leicestershire collieries, which form part of the South Midlands Area of the N.C.B. This coalfield which will be exhausted by the end of this decade is confined to a geographical area approximately 15 Km long (North to South) and 3 Km wide, is bounded on the eastern side by the large Thringstone fault and on the western side by the incrop of seams. Figure 3 shows the proximity of the town of Coalville and the cover line to the triassic strata. In total 14 different seams have been worked in the coalfield, 12 of them having been worked at Whitwick contained in a vertical section of strata some 200 m thick. The Five Feet seam represents more than 50% of the remaining reserves, although unfortunately a substantial part of this is underneath the town of Coalville at a depth of 76 m ranging to 120 m. Obviously any longwall working at 1.5 m section would have had serious subsidence effects on the town centre, which includes a new Shopping Precinct. It was decided to examine the position with the view to partial extraction working.

6.2 Planning the Layout (Figure 4 refers)

With subsidence protection being of paramount importance it was decided to extract on the basis of 6 m headings and 15 m pillars giving approximately 30% extraction whilst ensuring surface stability. The original layout (Block A) was designed to drive two parallel roadways (Main laterals) 200 m apart and 600 m long using the Dosco Dintheader, supported by Hollybank flat top arches, and work the In-Seam Miner at right angles to these laterals. This first extraction was also planned to enable surface subsidence monitoring to be undertaken, so that subsequent amendments to the layout could be made if necessary.

6.3 Revision of Layout (Figure 4 refers)

Results from the first trial heading indicated that with some technical and layout modifications the machine potential was much higher than originally envisaged.

Consequently the following changes were made :

i) It was decided to run the In-seam Miner headings parallel to the main laterals giving a potential run of 570 m, in order to reduce the proportion of total time in turning the machine at the end of each run.

ii) A new 660 mm low type conveyor was installed.

iii) The In-seam Miner discharge conveyor was extended to provide 27 metres overlap (previously 10 m).

iv) Steel straps for roof support were replaced by two 2.74 m wood bars and four wood props at 1.1 m centres.

The conditions with a strong roof and hard floor remained good throughout.

Block B shows the revised layout.

6.4 Technical Details of I.S.M. and System of Work

Machine extraction - overall width 6 m (See Figure 5)
height 1.37 m

Overall weight	-	14 tonnes
Cutter chain drive	-	Two 37.5 Kw Motors
Cutter chain speed	-	1.35 m/second
Cutter picks	-	75 mm gauge HB 300 type
Power pack	-	22.5 Kw motor
Pump capacity	-	80 litres/minute
Coal Thickness Indicator	-	Gamma Ray
Pushing rams	-	2 metre stroke
Conveyor	-	22.5 Kw electric drive, 0.46 m wide overlap 27 m.

As previously stated - Supports are wood bars and props with coal clearance onto 0.66 m low type belt conveyor.

Ventilation: Exhausting system using 0.61 m flexiduct.

Supplies transport: Double Drum hydraulic haulage and special colliery designed transporter. Mini winch fitted to In-seam miner.

Turning the machine at the end of a run and manoeuvring into position for the next extraction required a special technique and is shown in Figure 6.

6.5 Results Obtained

During the first full year of operation 1981/82 some 2,751 m of Dintheader drivage was completed to develop the two 600 m x 200 m blocks (Block A and Block B - Figure 4) known as 50's area. The in-seam miner extracted 2,724 metres of advance during the same period, producing a total saleable output from the area of 54,698 tonnes. This included the record performance for week ending 10th April 1982 when 367 metres were driven in 10 machine shifts to produce 4,512 tonnes of output - (See Table II).

The net contribution for the first year of operation was £443,711.

6.6 Future Prospects

Since the first year of operation a second In-seam miner has been installed in another block and with the two machines working concurrently, 30%/40% of the colliery's total output of around 10,000 tonnes saleable per week has been mined in this way. It is planned to step this level up to 50% in order to balance out the remaining reserves between partial extraction and conventional longwall working.

7. CONCLUSIONS

The main conclusions which I draw from the information given are :-

i) Whilst the total development drivage in N.C.B. mines for 1982/83 of 573 Km was a new high, it is unlikely that the future drivage programme will increase significantly beyond this level. The move towards more retreat mining and the reduction in total number of faces required are likely to be self cancelling.

ii) The growth rate in the use of heading machines for in-seam drivages will slow down as the plateau of about 85% of total drivage distance has been reached.

iii) A number of excellent well proven machines are now available for a variety of seam sections and strata conditions. It is likely that improvements to existing designs, particularly increased reliability and better cutting tools will take precedence over the development of new types.

iv) Production from heading machines at high productivity levels is making an improved contribution to overall efficiency.

8. ACKNOWLEDGEMENTS

I would first acknowledge the invitation to make a contribution to this session on mechanised drivage in coal which has provided a valuable opportunity to review and discuss our systems.

I am most grateful for the assistance received from colleagues in Headquarters Mining Development Branch, Doncaster and in my own Area at Coleorton.

Finally I would like to thank Mr. R. Rawlinson, O.B.E., Area Director, South Midlands Area, N.C.B. for permission to present this paper, and emphasise that expressions of opinion given are my own, and not necessarily those of the National Coal Board.

NATIONAL COAL BOARD

DEVELOPMENT DRIVAGE 1982/83 (IN SEAM)

TABLE I - DISTANCE DRIVEN AND RATES OF DRIVAGE BY STRATA CATEGORIES

METHOD OF WORKING	ALL COAL		MAINLY COAL		MAINLY STONE	
	Distance (Km)	Adv/Day (m)	Distance (Km)	Adv/Day (m)	Distance (Km)	Adv/Day (m)
Drill and Fire - Hand Load	6.39	3.00	5.11	2.74	4.60	1.89
Drill and Fire - Mech Load						
Bucket Loader	3.08	3.38	14.51	2.14	26.65	2.10
Gathering Arm Loader	0.83	3.26	2.73	2.48	2.88	2.22
Slusher	1.70	3.20	2.37	2.76	1.13	2.51
Flight Loader	2.18	4.69	3.21	4.28	1.17	2.77
Others	-	-	0.30	6.18	0.33	2.05
Total Mech Load	7.79	3.69	23.12	2.42	32.16	2.18
Heading Machines						
Roadheaders	11.19	3.70	80.19	3.80	105.30	3.75
Dintheaders	12.82	5.35	102.25	6.45	57.74	5.65
In Seam Miners	4.09	4.77	2.51	6.56	0.21	2.88
Continuous Miners	23.46	24.28	11.33	8.50	-	-
Total Heading Machines	51.56	6.99	196.28	4.95	163.25	4.19
GRAND TOTAL	65.74	5.66	224.51	4.40	200.01	3.56

Please Note: In addition 82.61 Km were driven in all stone conditions making a combined total of 572.87 Km

NATIONAL COAL BOARD
SOUTH MIDLANDS AREA

WHITWICK COLLIERY

RECORD PERFORMANCE FOR IN SEAM DRIVAGE

TABLE II - SHOWING DAILY/SHIFT PERFORMANCES

DAY	SHIFT	ADVANCE METRES	CUMULATIVE ADVANCE METRES	TONNES PER SHIFT	CUMULATIVE TONNES
Monday	Days	39	39	479	479
	Afts.	43	82	529	1008
Tuesday	Days	32	114	393	1401
	Afts.	49	163	602	2003
Wednesday	Days	50	213	615	2618
	Afts.	29	242	357	2975
Thursday	Days	50	292	615	3590
	Afts.	53	345	652	4242
Friday	Days	10	355	123	4365
	Afts.	12	367	147	4512

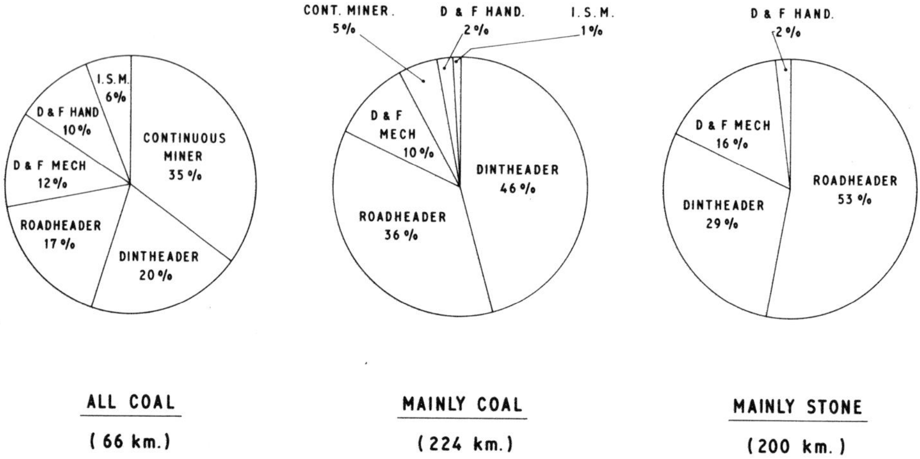

FIGURE 1. DRIVAGES BY METHOD OF WORK AND STRATA 1982/83.

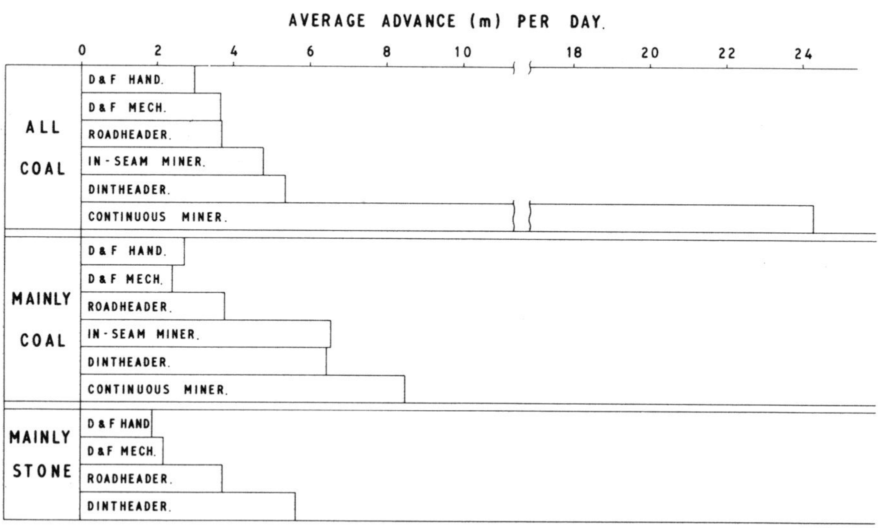

FIGURE 2. AVERAGE RATE OF DRIVAGE BY MACHINE AND STRATA 1982/83.

FIGURE 3.

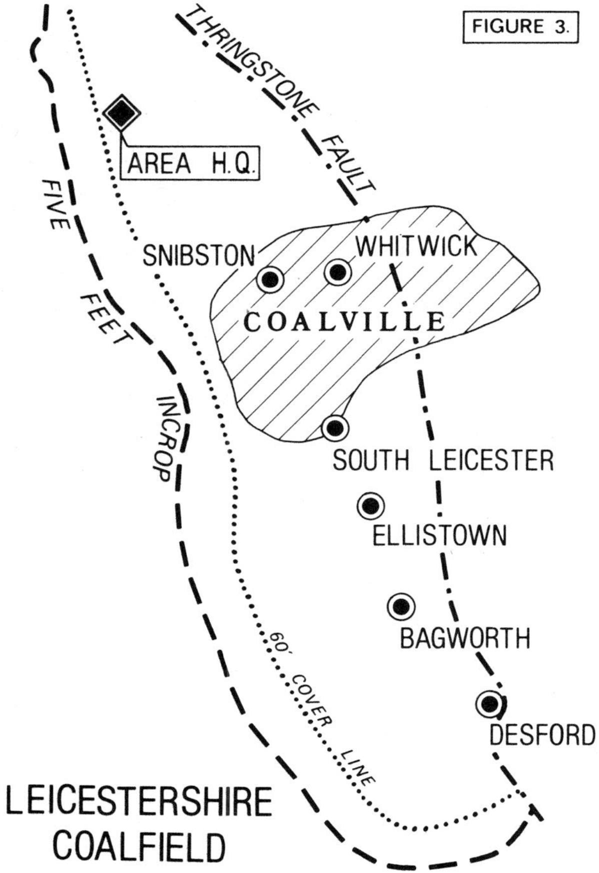

LEICESTERSHIRE COALFIELD

FIGURE 4.

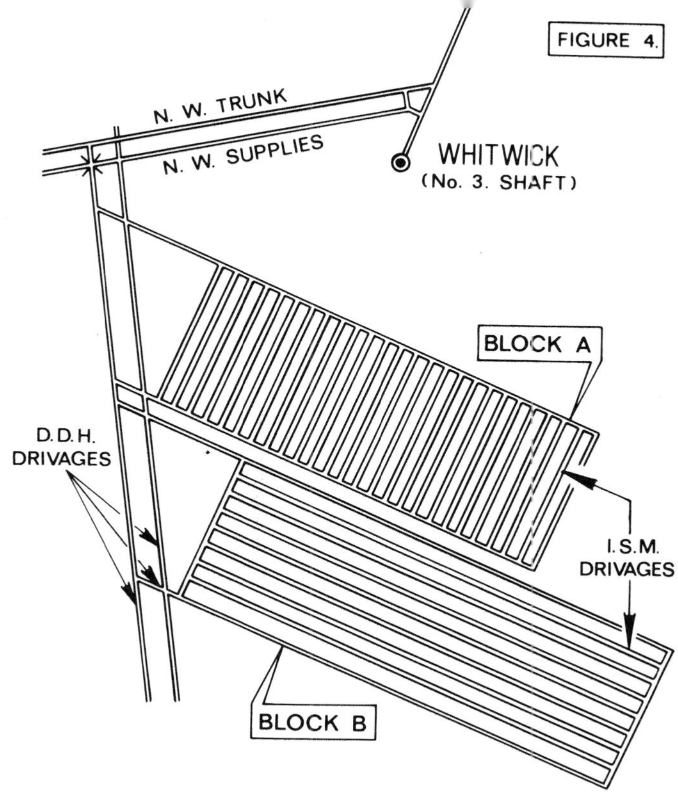

WHITWICK COLLIERY

ORIGINAL & REVISED LAYOUT OF FIVE FEET PARTIAL EXTRACTION – 50's AREA

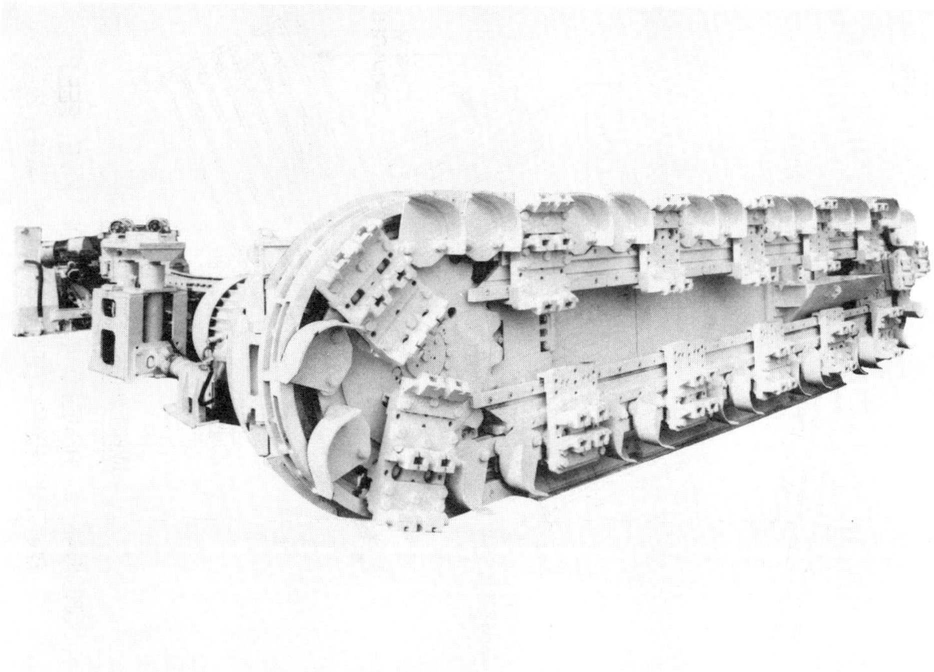

Figure 5 : Face Side View of machine

General view of machine

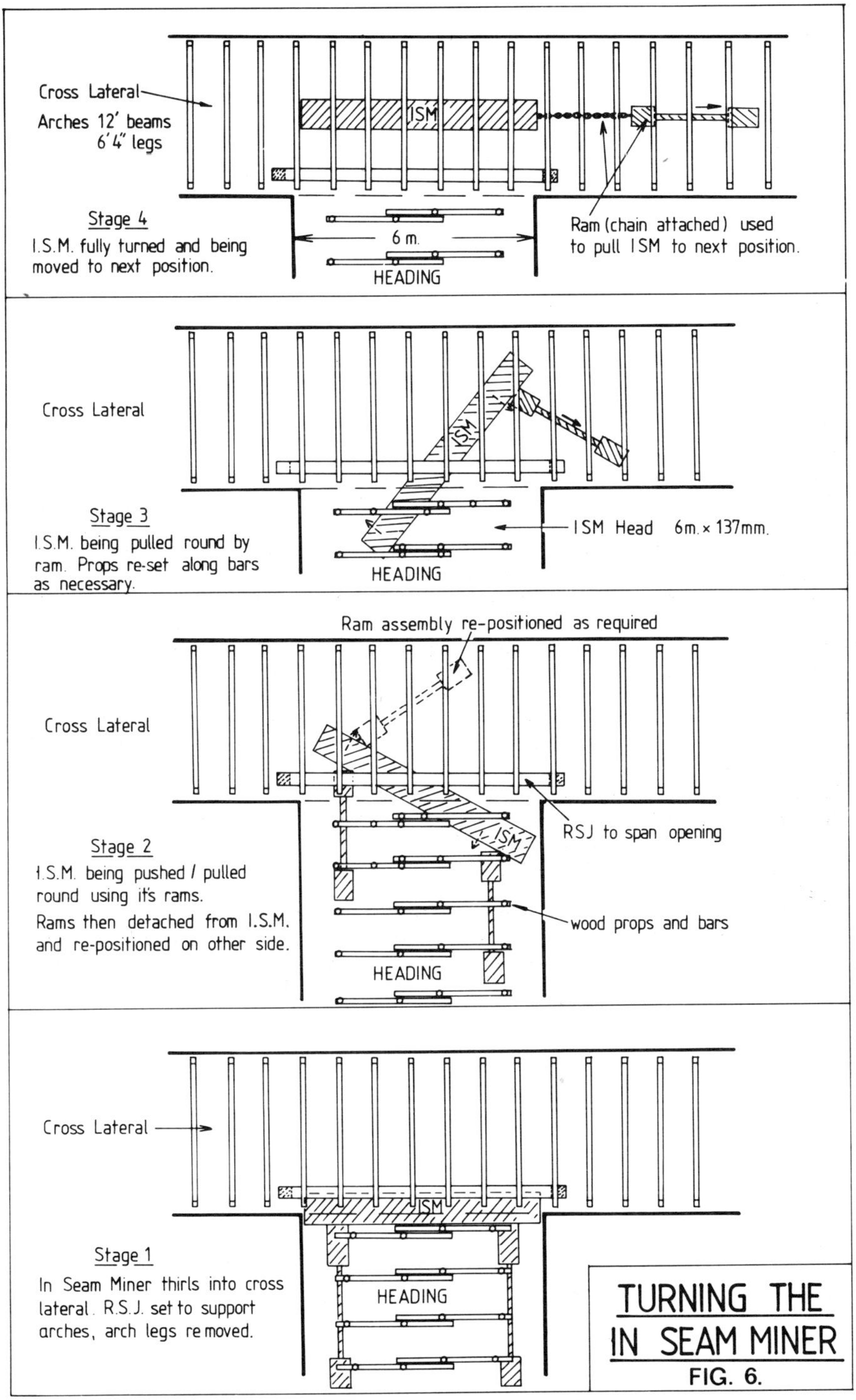

TURNING THE IN SEAM MINER
FIG. 6.

DRIVING TRAPEZOIDAL ROADWAYS WITH THE SSCM MACHINE

C. Geraud

Chief Engineer, Head of Testing,
Houillères du Bassin de Lorraine,
Charbonnages de France
MERLEBACH

Summary

Heading operations are of paramount importance in the Mines de Potasse d'Alsace (MDPA). The main characteristics of the roadways are their reduced cross-section, their high temperature and the fact that only a small amount of water can be used for dust suppression, given the nature of the deposits. Despite the advent in 1950 of continuous miners, as a result of the drop in output with conventional heading machines, new machines have had to be developed in recent years which are especially suited to headings with substantial lateral inclines. More particularly, in conjunction with the SSCM, the MDPA undertook to design and test the MTP 52. This original design of machine features inter alia a ravageur-type cutting head with twin drums and a variable tilt device enabling it to drive roadways with a roof slope of up to 27°. In addition the machine has a sound operating temperature and on account of its ancillary equipment it can travel long distances independently of the power supply.

In conclusion, the performance level achieved by this machine is in line with that forecast, which justified the purchase of a second machine and has proved that it would be suitable for use in other types of deposit.

1. INTRODUCTION

In the Mines de Potasse d'Alsace it is necessary to drive 6.1 m of roadway in order to extract every 1000 t of mineral - hence the importance of heading operations.

This figure can be explained by:

- the choice of a retreating system of mining, sometimes covering long distances.
- the need to double the total number of roadways for mining purposes on account of the reduced roadway cross-section.

In addition, the development of as yet unexplored working areas in order to replace areas which are currently being exhausted further increases the current outlay on drivage.

Finally, the increasing number of difficulties encountered in the workings and the corresponding substantial drop in performance in the case of the conventional heading machines, despite attempts at modernisation, have led the MDPA to introduce new heading machines which are better adapted, more efficient and cheaper to run.

2. CHARACTERISTICS OF HEADING AREAS

2.1. Cross-section extracted

In order to minimise support setting operations in the workings and underground maintenance work during the life of the roadway, we limit the width of the gate roads to 3.90 m (or 5.50 m in the case of main access roads).

2.2. Type of deposits

The traditional workings which are uniform and subhorizontal are now being exhausted and consequently new headings are being driven into zones which are clearly more inclined, more irregular and faulted in places.

2.3. Climate and environment

The temperature is about 43°C at the current average depth of 700 m.

The deposits are classified as gassy and so flameproof materials and equipment have to be used.

Finally rock salt and sylvinite are both water-soluble and so water sprays cannot effectively be used for the purpose of dust control.

3. DEVELOPMENTS IN HEADING TECHNIQUES

3.1. Conventional method

The conventional heading method is by drilling and blasting along the kerf; loaders and shuttle cars are used for loading out and two adjacent working zones linked by an intersection are worked simultaneously by the heading team. These machines have proved unsuitable for operations in workings which show a pronounced lateral incline, in irregular workings and where the strata is faulted. Various attempts to improve the machines have not been successful enough to compensate for the increased stress.

3.2. Initial experiences with continuous-miner-type machines

The first continuous miners were introduced underground towards 1950. These machines yielded good results in conventional workings (subhorizontal, at shallow depth and uniform) and gave relatively unsatisfactory performances where the workings were laterally inclined, in irregular and faulted ground. Whilst testing the Paurat E 195-type selective cut machine (of which there are at present five) the MDPA in conjunction with the SSCM undertook to design and test the SSCM MTP 52 machine, an original design of continuous miner.

3.3. Testing the 'Ravageur-type' SSCM MTP 52 continuous miner

This machine was introduced into service at the beginning of 1981 (see Figures 1 & 2; Photos 1, 2 and 3) and has a 'ravageur'-type cutting head with twin drums mounted on a transverse drive shaft, which penetrates 0.50 m per unit of advance. This is the basic difference between the 'ravageur'-type machine and transverse cutting head machines of comparable output.

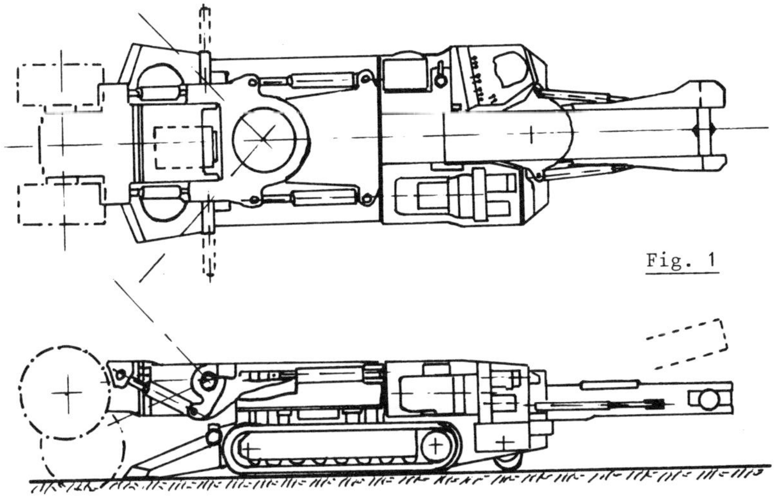

Fig. 1

Figure 1

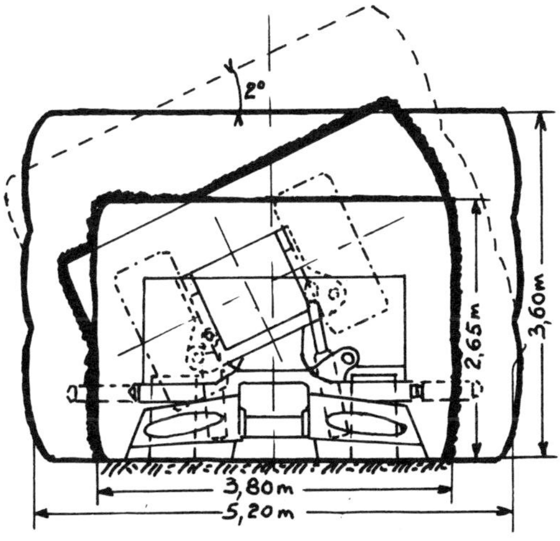

Fig. 2

Figure 2

Its main features are set out below in Table 1.

Overall ground space	(length	11.20 m
	(width	3.0 m
	(height	1.75 m
	(machine weight	54 t
Power rating	(cutting	170 kW
	(total rating	280 kW
Cutting head	(diameter	1.50 m
	(width	2 x 0.63 m
	(No. of picks	214
	(Max. pick speed	3.15 m/s
Loading & Clearance	(width of front loading blade	3.0 m
	(rotation speed of gathering-arm plates	42 r.p.m.
	(width of conveyor	0.76 m
	(speed of conveyor	1.80 m/s
	(height of conveyor	0.40 m
	(Max. discharge height	1.97 m
Heading profile	(max. width	5.20 m
	(max. height	3.60 m
	(min. width	3.80 m
	(min. height	2.20 m
	(max. unit of advance	0.50 m
	(max. tilt at roof-cutting drums ...	27°

The following points should be noted:

- Drum geometry and layout are designed for a minimum heading width of 3.80 m. Variations in cutter head level and the inclusion of a variable tilt device make it possible to drive a trapezoidal roadway with a deviation from the horizontal of up to 27°.
- Output at cutting head = 170 kW.
- High-speed articulated conveyor. Its wide section, high performance design allows it to start up under full load at each loading cycle.

- Total weight of the installation plus its forward centre of gravity provide maximum stabilisation when cutting into the roof and when making vertical cuts (on descent). A stabilizer roller located at the rear of the machine frame further improves the stability of the machine during these two stages of operation and also in the case of convex floor conditions.
- The front loading blade features heavy duty transverse stabilisers.
- Water cooling of the cutting drum motor and hydraulic fluid.
- The machine can move over long distances independently of the power supply by means of a trailer-mounted diesel hydraulic power pack driving the hydraulic unit of the continuous miner.

Development of this prototype resulted in:
- Modification of the design of the lower part of the front loading blade to facilitate floor-cleaning operations. This was essential if the operator was to be able to control the direction of the machine when moving and stationary.
- Suppression of clouds of dust created during heading operations, which travelled the length of the conveyor, by fitting a dust-proof cover along the central section of the conveyor.
- Replacement of certain hydraulic valves and modification of the driver's cab.
- Full mechanisation of the tilt control at the cutting head, which was originally achieved by manually operating a hydraulic pump.
- Original gathering arms replaced by rotary plates fitted with three radial arms and relocation of the cutting head lifting jacks in order to improve loading.

The modified machine was then introduced into service in October 1981, and the following points were noted:
- The teething troubles disappeared.
- The machine showed good dynamic stability, even under difficult heading conditions (height 3.40 m - roadway gradient up to 10°) where conventional continuous miners could not be used.
- Its suitability for heading normal cross-sections in an incline of up to 14° and its sound loading performance on down-

hill slopes make it suitable for driving rise headings for thick-seam strike faces.

- Capable of driving roadways with a roof slope of up to 27° with no appreciable drop in performance.
- It proved superior to the Paurat E195-type machine in terms of operating temperature and on account of its improved hydraulic efficiency.
- The sturdy design of its basic mechanical, electrical and hydraulic components.
- The advantage of travel over long distances independently of the power supply.

4. CONCLUSION

The performance level achieved by this machine (20 - 25 m per day) was in line with that forecast and this in turn justified the purchase of a second machine (introduced into service in March, 1983) which retained all the main sub-assemblies - the reliability of which had already been tried and tested - and in addition incorporated further improvements (especially regarding the supply system and ease of access in the event of a breakdown).

It constitutes an original mechanised heading method which presents a neat solution to the problem of driving access roads in level and inclined deposits: a study is currently under way to adapt the machine for heading operations at La Houve colliery in the HBL.

MACHINE SSCM MTP 52

Photo 1

MACHINE SSCM MTP 52

Photo 2

Photo 3

THE USE OF IMPACT RIPPERS IN THE GERMAN COALMINING INDUSTRY

M.R. Liesner

Dipl.-Ing. and Assessor
Ruhrkohle AG Holding Company, Essen

Summary

The introduction of impact rippers in the West German coalmining industry some ten years ago marked the debut of a new heading technique, mainly for driving roadways in line with and behind the face. During the intervening ten years, some 50 of these machines have been used to drive 110 km of gateroads. Although the use of impact rippers was initially more common in roadways driven behind the face, the emphasis gradually shifted to roadways driven in line with the face. Roadways driven in advance of the face were always the exception. At one point, a comparative standstill appeared to have been reached in the application of this technique, but new developments may extend the impact ripper's field of application in the near future. The limits within which it is feasible to use impact rippers are relatively narrow. These machines have gained an accepted place for themselves in situations where retreat mining of roadways driven in advance of the face is impossible, and where their breaking capacity is sufficient to keep pace with the rate of face advance. First and foremost they require thinly-bedded stratified surrounding rock in the roof, and the possibility of dispensing with dinting of the floor. In roadways driven behind the face, a seam thickness of at least 1.80 m is required, whereas in roadways driven in line with the face the coal/stone ratio must be relatively favourable. Provided these requirements are met, impact rippers are a cost-effective means of mechanisation in the gateroads. Additionally, in roadways driven in line with and behind the face, the impact ripper has the advantage of causing less convergence, which means that the roadway cross-section can be smaller, thus reducing heading costs.

1. INTRODUCTION

Almost exactly ten years have passed since the first impact rippers were introduced in the West German coalmining industry, for enlarging stonedrifts and driving the gateroads at Ruhrkohle AG. This marked the introduction of a method which had already gained widespread acceptance in the UK and France. This step had been preceded by an in-depth study of the advantages and disadvantages of this new method in the, to some extent, rather different conditions in which it would be applied in the German deposits.

Following the introduction of the first of these machines which were used to drive approx. 400 m of gateroads during 1973, a further fifty or so have since been introduced and have driven a total of approx. 110 km of roadway to date. With a total annual heading rate of 18 km, this method now accounts for just under 6% of the total in-seam roadways driven (Figure 1). Insufficient statistical data is available with regard to their numerous other uses, in terms of roadway enlargement and maintenance. This paper, therefore, deals exclusively with gateroads.

It would seem like a good idea to start by listing the requirements and expectations upon which the decision to use impact rippers in this type of roadway depended.

2. ROADWAY ADVANCE SEQUENCE AND INFLUENCE OF CONVERGENCE

By the early Seventies, face technology had reached a level which enabled a high face output and a correspondingly high rate of face advance to be achieved. Wherever conditions permitted, retreat mining had already largely replaced other methods, which meant that heading operations presented no obstacle in terms of face advance.

At some pits, however, the advantages of retreat mining could not be fully turned to account. These included pits at which the rate of face advance was restricted due to the high concentrations of firedamp found in the deposit, and pits at which there was so much convergence in roadways driven in advance of the face that the resultant high costs made retreat mining uneconomical.

Possible methods of driving the gateroads during coal-winning are:

- roadways driven in advance of the face, i.e. approx. 40 to 70 m ahead of the face
- roadways driven in line with the face, i.e. heading and face at the same level, and
- roadways driven behind the face, i.e. the ripping following 10 to 15 m behind the face.

In gateroads driven in advance - as indeed in retreating gateroads - the entire front abutment pressure of the face has an adverse effect on convergence behaviour. Convergence is at its worst here, and necessitates extensive maintenance work. Roadways driven in line with the face are somewhat of an improvement in geomechanical terms, as the front abutment pressure produces less of an effect. Gateroads driven behind the face are the most stable, as they are relatively unaffected by the rear abutment pressure. Figure 2 illustrates the different types of roadway convergence, depending on the roadway advance sequence.

One method of reducing convergence is to opt for smaller roadway cross-sections, and in deep workings this can reduce roadway costs considerably.

3. USE OF IMPACT RIPPERS IN ROADWAYS DRIVEN BEHIND THE FACE

In principle, the above information shows that mining engineers must be highly in favour of roadways driven behind the face, provided that no other considerations take precedence. Formerly, this method of drivage used to be quite common in the West German coalmining industry, and constituted perhaps the most common method in the UK. But during the Sixties, this method was gradually replaced in Germany by roadways driven in advance of the face, as it suffered a number of drawbacks due to the preponderance of drilling and blasting work involved:

- After blasting, the face ventilation was often interrupted or at least restricted, which proved an increasingly serious safety hazard, due to the increasing amount of gas produced as production increased;

- when the CH_4 concentration exceeded 1%, blasting operations had to be discontinued;
- the increasing size of roadway cross-sections, resulting in more need for supports (closer spacing of support units, heavier support profiles), also affected the rate of roadway advance, so that the ripping could no longer keep up with the face and the distance between the two increased steadily;
- this, in turn, made manriding, coal and materials transport conditions considerably more difficult, particularly in narrow seams;
- cables and pipelines were frequently damaged, despite all safety precautions.

This meant that the arguments in favour of using impact rippers applied more to locations in which the above disadvantages and restrictions were less important.

The first time one of these machines was used by the West German coalmining industry in an in-seam roadway was in March 1973. This was in the loader gate in the approx. 2 m-thick seam B at Hugo colliery. The equipment consisted of a Gullick Dobson Impact Ripper mounted on a walking frame, distributed in Germany by Becorit - today Klöckner-Becorit in Recklinghausen - plus a side-discharge loader manufactured by Salzgitter Maschinen- und Anlagen-AG (Figure 3). To give an excavated cross-section of 16 m^2, some 6 m^3 of roof had to be brought down per m of roadway, the roof consisting of thinly-bedded shale with a compressive strength of 30 to 40 N/mm^2. The floor of the seam did not need to be dinted, as the roadway was driven on the floor of the seam. Thus, in terms of breaking work, the demands made on the machine, which was fitted with a relatively lightweight Mk I impact head, were not too extreme.

The roadway driven behind the face, shown again in Figure 4, achieved a rate of advance amounting to between 3.20 and

3.70 m/d, corresponding to the rate of face advance. Two teams of two men were employed at the ripping, and the rate of advance was greatly dependent on the rate of face advance. On the basis of method studies and performance calculations, it would have been entirely feasible to reckon on a rate of advance of 6 to 7 m/d, if all available reserve capacities and manpower reserves had been mobilised.

In April 1973, soon after the first machine, a second manufactured by Hausherr and Söhne, Sprockhövel, was introduced in a gateroad at Prosper-Haniel colliery. This same manufacturer had also been responsible for supplying the German coalmining industry's very first machine, for enlarging stonedrifts at Radbod colliery.

The model chosen for Prosper-Haniel colliery was unfortunately not such a success, as soon became clear. The transverse-mounted front conveyor was unable to load and evacuate the excavated material in the required fashion, snaking of the conveyor was often difficult, and loading out thus proved the most critical aspect of heading operations. As a result, heading operations were discontinued after a short experimental period, when a roadway driven from the opposite direction was about to thirl.

On the whole, the drawbacks and difficulties were not of a serious nature - or at least they were no real fault of the machine, but rather of the system as a whole, which was not always able to cope with the impact ripper's requirements. In addition to the firms of Hausherr and Klöckner-Becorit mentioned above, another manufacturer, Salzgitter Maschinen- und Anlagen-AG, based in Salzgitter-Bad, soon joined the circle of competitors with a model of their own, a portal-type impact ripper mounted on a walking frame.

The steady, unhurried pace at which impact rippers were gradually introduced led to a considerable variety of models being available on the market, each tailored to the customer's needs. This meant that, from the outset, impact rippers had a sound basis for later extensions to their range.

4. ROADWAYS DRIVEN IN LINE WITH THE FACE

The impact rippers' initial success in roadways driven behind the face soon encouraged us to employ them in roadways driven in line with the face. Whereas until 1978 they predominated in the former type of roadway, after that date they became more common in the latter, and have retained this position to the present day, gradually increasing the margin until in 1982 the total in-line drivages amounted to over 13 km (Figure 1).

In roadways driven in line with the face, these machines have retained the characteristics of a ripping machine. In this type of roadway, the coal is won by the winning machine used on the face, the impact ripper's only task being to bring down the surrounding rock (Figures 5 and 6).

In roadways driven in line with the face, the selection criteria are somewhat simpler than in those driven behind the face. Whereas a minimum seam thickness of at least 1.80 m is required in the latter, the former can be driven in lesser thicknesses, provided that the surrounding rock can be brought down in the time available, which is dependent on the rate of face advance. Additionally, in centre and tail gates, a method of haulage not otherwise required can be dispensed with. On the other hand, the surrounding rock has not been loosened to the same extent in roadways driven in line with the face, so that the impact ripper has more difficulty in breaking the ground. Furthermore, the impact ripper must always remain fairly exactly in line with the face, whereas the roadways driven behind the face, the distance can afford to vary to some degree.

In gateroads driven in line with the face, the impact rippers are encountering a certain amount of competition from the new generation of drum shearer loaders. These machines, possessing longer booms and a high installed capacity of at least 150 kW, enable the roadway profile to be cut in line with the face provided that the surrounding rock is not too hard, and thus save having an extra machine in the face end region. In contrast to the impact ripper, they are not so much dependent on thinly-bedded stratified rock, but do require softer conditions. They also require considerably more in the way of dust suppression. In the West German coalmining industry, a total of approx. 10 km of roadway per annum are now cut simply by using the shearer loader.

5. GATEROADS DRIVEN IN ADVANCE OF THE FACE

The very first impact ripper introduced at Hugo colliery started off by driving several metres of roadway in the solid, with relative success. Many subsequent attempts have been made to use impact rippers as fully-fledged heading machines in gateroads driven in advance of the face. These attempts have not, however, proved entirely successful, apart from several isolated cases in which the particularly favourable nature of the seam and the surrounding rock contributed to the success of the operation (Figure 1).

The main reasons for the impact rippers' on the whole unsatisfactory results to date in roadways driven in advance of the face are:

- insufficient breaking capacity of the impact head, even when cutting the coal of the heading face in the solid, thus resulting in too much time being wasted in producing a second free face, and
- the fact that the strata formations in the surrounding rock were still undisturbed, due to insufficient abutment pressure, thus making ripping more difficult.

A widespread effort has undoubtedly been made to increase mechanisation in roadways driven in advance of the face, but precisely this roadway advance sequence has largely defeated all such attempts, in terms of financial viability. Thus our hope was that we could make up for this deficiency with the aid of the impact ripper.

6. EXPERIENCE TO DATE

After ten years of using impact rippers in the West German coalmining industry, these machines have won for themselves an accepted position in the driving of roadways. This method is universally so familiar that its advantages and disadvantages can be weighed up and included in calculations at the planning stage.

I should like to conclude by saying a few words about our considerable experience to date, and the accredited need for future improvements, based in part on this experience.

6.1 Mechanical aspects

The walking frames used on the first models have largely been replaced by crawler assemblies. Crawlers provide greater mobility and are also more stable in the case of transverse gradients. Where the impact ripper straddles the scraper conveyor, the cheaper walking frame will, as a rule, still be preferred.

Following initial difficulties, machine-available-time has improved considerably, and mechanical failures seldom occur. The constantly updated impact rippers are manoeuvrable, versatile, and are accepted by the workforce.

The prospect of improving breaking performance, even in stronger, less obviously stratified surrounding rock, has led to the use of increasingly heavier impact heads. Whereas the first heads were hydraulic hammers weighing 250 to 400 kg, as in the Gullick Dobson Mk I model or the Krupp HM 200 mentioned previously, today the impact hammers weigh a standard 700 to 800 kg, and still heavier models are currently being tested. Figure 7 shows a selection of the impact heads now available.

Heavier heads, in for example the 1000 kg range, will, however, necessitate strengthening the boom and the entire machine. The resultant increase in cost, which may well increase the cost of their hire, will of course have to be offset by improved performance, if they are to remain cost-effective.

This also applies to the concept of mounting two separate booms on one chassis, the idea being that these booms would then provide mutual back-up during heading and difficult breaking operations.

For in-depth testing of further development possibilities, Bergbau-Forschung GmbH in Essen have constructed a test rig, whose principal aim is to solve the problems listed in Figure 8. The answer to the first of these has, in practice, so far been that a slightly truncated pyramid-shaped tip has proved the most effective pick design, next to the chisel pick. Improvements to this design could contribute towards increasing the impact ripper's breaking capacity and extending its range of application.

Dust suppression has not, in general, caused any major problems, but an effort should still be made to improve dust control by means of spray jet design and location, in order to keep dust exposure levels to a minimum.

More stringent noise control would also serve to further improve working conditions underground, and to make this a more pleasant place to work.

6.2 Operational and geological aspects

Despite the heavier impact heads now available, the main requirement for success in using impact rippers is still thinly-bedded stratified surrounding rock. Individual thin bands of relatively hard rock have less effect on breaking performance and are easier to cope with than cohesive non-stratified areas, even if these are softer. Provided that the more difficult strata conditions do not occur too frequently, a temporary switch can be made to drilling and blasting operations, where possible simply by fitting a high-performance hammer drill to the hammer mounting to facilitate drilling operations.

The critical factors in breaking operations are, in particular, the roadway circumference and in some cases a floor wedge. The excavation of these often minimal volumes of rock frequently takes up so much time that the impact ripper's very effectiveness is brought into question. From our experience so far, we will, therefore, be extremely careful to avoid having to dint a wedge of the floor, by using impact rippers exclusively in roadways where the transverse inclination of the roadway floor does not exceed the critical 10 gon.

For roadways driven behind the face, a certain minimum seam thickness should be postulated, because of the relevant ventilation, manriding and transport requirements. Experiences to-date indicate a minimum thickness of 1.80 m as the desirable aim. In roadways driven in line with the face, the limit should be determined in another way, the decisive factor being the volume of surrounding rock to be brought down and temporarily supported within the duration of the face winning cycle. This has currently led to the necessity of dispensing with the use of impact rippers in thicknesses of less than approx. 1.50 m. At Nordstern/Zollverein colliery, however, a research and development project is underway in which an exceedingly heavy-duty impact hammer is being used in a roadhead driven in line with the face in a thin seam, in order to ensure that heading operations do not slow down the rate of face advance.

One encouraging result which says a lot for the impact ripper's flexibility is the relatively low transport and assembly time of these machines, which makes them viable propositions for use in short sections of roadway.

Performance and costs in roadways driven by means of impact rippers are something of which I intend to mention only one example here, as comparability can only be guaranteed by uniform accounting methods and predefined limitations. Thus, to drive approx. 4.2 km of roadway by means of blasting would yield an average rate of advance of 4.6 fm^3/MS and would cost approx. 150 DM/m^3 in specific production costs. To drive approx. 1.6 km of roadway in line with the face in the same length of time, using impact rippers, would produce corresponding values of 6.4 fm^3/MS and 116 DM/m^3, which amounts to an improvement of almost 40% and 23% respectively.

I should like to conclude by paying a brief tribute to the men who operate these machines. For with no other heading technique does success depend so greatly on the skill of the operator as in this case. An inexperienced operator may take an entire shift in order to cut just one metre of rock, whereas an experienced operator can possibly achieve this - perhaps even more expertly - in under an hour. With our industry's recognizedly high level of technology, this shows, in no uncertain terms, to what extent some of our engineering success depends on the mining experience and know-how of the men who have to deal with the technology at close quarters in their everyday working lives.

USE OF IMPACT RIPPERS	1974	1976	1978	1980	1982	TOTAL DRIVAGES SINCE 1974
Total drivages (in km)	2.2	6.0	10.5	17.8	17.6	107.1
Roadways driven in advance of the face (in km)	0.6	0.4	1.0	1.1	0.5	8.5
Roadways driven in line with the face (in km)	0.3	1.6	5.3	11.8	13.3	63.5
Roadways driven behind the face (in km)	1.3	4.9	4.2	4.8	3.6	36.1
No. of machines used	13	24	33	36	48	50

Figure 1 : Use of the impact rippers in gateroads

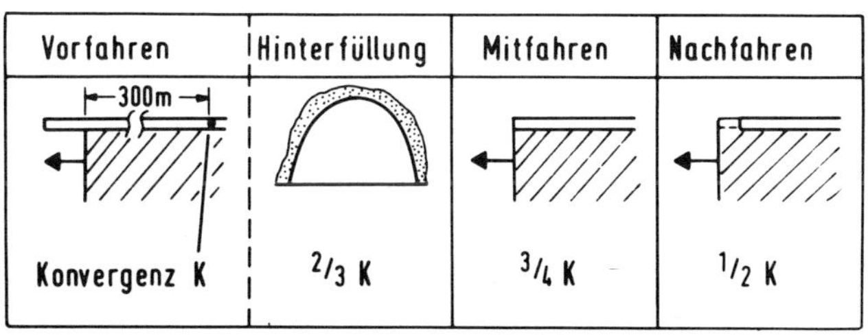

Figure 2

Figure 3

Figure 4

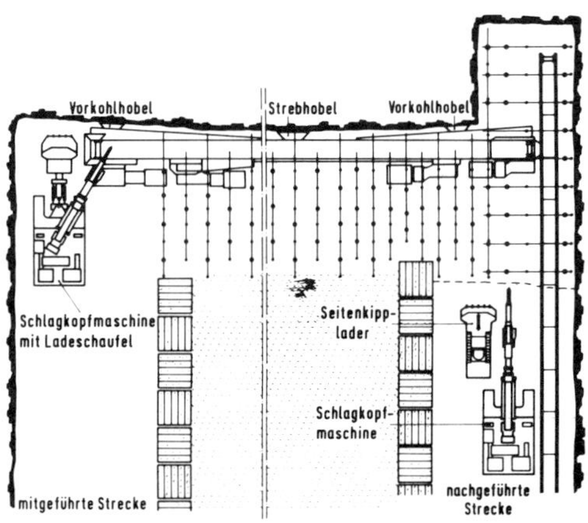

Figure 5

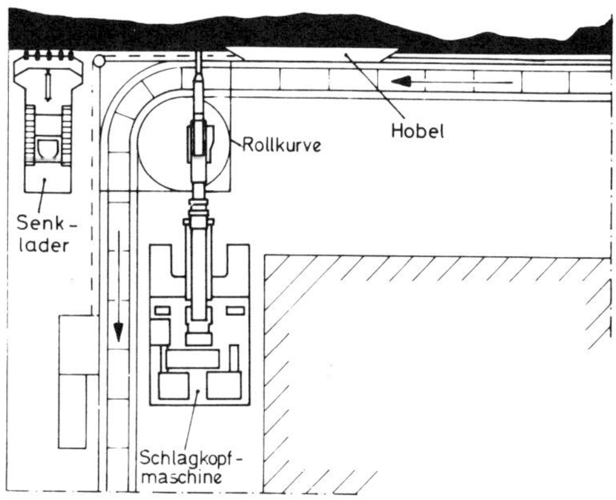

Figure 6

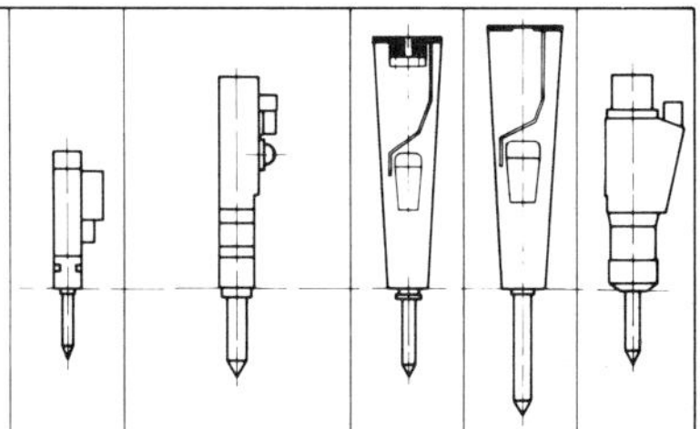

Technische Daten	Krupp HM 200	Krupp HM 800	Krupp HM 900	Montabert BRH 501	Montabert BRH 750	Gullick Dobson GD 3000
Hammergewicht ohne Halterung (kg)	255	712	712	800	895	760
Schlagzahl (min^{-1})	480/650	420/750	420/750	500	300-700	1-600
Einzelschlagarbeit max (Nm)	700	3020	3600	3000	3690	2950

Figure 7

- **Optimierung der Meißelform**
- **Ermittlung der Schlagenergie verschiedener Schlagköpfe**
- **Staubbekämpfung**
- **Geräuschdämpfung**

Figure 8

KEY TO ILLUSTRATIONS

Vorfahren	Roadways driven in advance of the face
Hinterfüllung	Backfilling
Mitfahren	Roadways driven in line with the face
Nachfahren	Roadways driven behind the face
Konvergenz	Convergence

Figure 2. Influence of roadway advance sequence on convergence.

Figure 3. First use of Gullick-Dobson impact ripper and SMAG HL 80 side-discharge loader.

Figure 4. Model of roadway driven behind the face.

Vorkohlhobel	pre-coaling plough
Strebhobel	main plough
Seitenkipplader	side-discharge loader
Schlagkopfmaschine	impact ripper
mit Ladeschaufel	with shovel attachment
mitgeführte Strecke	roadway driven in line with the face
nachgeführte Strecke	roadway driven behind the face

Figure 5. Diagram showing use of impact rippers in roadways driven in line with and behind the face.

Hobel	plough
Rollkurve	roller curve
Senklader	dint header
Schlagkopfmaschine	impact ripper

Figure 6. Diagram showing use of impact ripper in a roadway driven in line with the face.

Technische Daten	technical data
Hammergewicht ohne Halterung	weight of hammer without mounting
Schlagzahl	blow frequency
Einzelschlagarbeit	striking energy per blow

Figure 7. Relative size and characteristics of various impact heads.

Optimierung der Meißelform	optimisation of pick design
Ermittlung der Schlagenergie verschiedener Schlagköpfe	determination of striking energy of various impact heads
Staubbekämpfung	dust suppression
Geräuschdämpfung	noise attenuation

Figure 8. Aims of Bergbau-Forschung's test rig for impact heads.

MECHANISATION OF SHORTWALL HEADING AT ZOLDER COLLIERY IN THE KEMPEN COALFIELD

J. Legrand and R. Resseler

Underground Manager - Zolder Colliery
Head of Planning and Services Underground - Zolder Colliery

Summary

Driving gate-roads behind the coal face involves the extraction of short faces at the main-gate and tail-gate ends. At our colliery we have been striving to mechanize these operations since 1967. In recent years we have had at our disposal a compact and reliable machine which is capable of extracting both the coal and the surrounding rock in these short faces.

In the Kempen coalfield this machine is used for cutting main-gate and tail-gate stables on both plough and shearer faces. The same machine can even be employed for driving rise headings and arch-supported roadways. While in all cases the coal and rock can be cut by the same machine, the installation as a whole tends to vary enormously according to the type of short face being extracted.

It is certain that this machine will continue to be used for the extraction of other types of shortwall faces. One such application may be the heading of circular-section roadways where such a profile would serve to improve roadway stability and thereby allow a retreating system of working. An increase in machine output would also help extend the operating scope of this type of equipment.

1. INTRODUCTION

High strata pressure in the Kempen coalfield has meant that many gate-roads are driven behind the face which is extracted by advancing working. At Zolder and Beringen collieries all gate-roads are now driven behind the face line. This method of drivage means that main-gate and tail-gate stables have to be cut, an operation which proves time-consuming and arduous under difficult conditions. For this reason Zolder colliery commenced work on the development of machines suitable for shortwall extraction. The first such machine was introduced in 1967, initially in a rise heading and later in a main-gate stable. Since then their design has been continuously revised, their output increased and their operating reliability improved. Their scope of operation is now so diversified that the same machine can be employed in totally different areas of application.

In this report we shall restrict ourselves to an examination of the current situation as regards the mechanisation of different types of face operations at Zolder colliery. It will be necessary, on occasions, to give brief descriptions of various stages in the development of the mechanisation process owing to the fact that these correspond with changes in the design of heading and face-mechanisation procedures. We shall conclude by describing future development possibilities.

2. TYPES OF SHORTWALL

In coal mines one finds different types of shortwall workings. We shall not attempt to define these here. However, it is important to draw attention to the fundamental differences between "shortwall" and "longwall" workings.

On longwall faces the various operations, i.e. winning, product loading and clearance, support functions and conveyor advance, are carried out on an identical basis throughout the entire length of the face. The mechanization of these operations has reached an amazing level of technical perfection. However, the techniques applied on the coal face cannot be used at the face ends and it is here that one finds the most typical types of "shortwall". As on the coal face, the coal has to be won, loaded and cleared, the roof supported and the various installations advanced; this must likewise be done over the entire length of the shortwall. In addition, these working places are areas where the face installations must regularly be shortened or extended. As a result, shortwall operations are directly linked to the advance of the main coal face.

We shall also see that various underground workings not associated with the coal face can also be treated as shortwalls. We shall begin by examining two types of shortwall which are associated with the coal face. Figure 1 represents a working face with the main-gate and tail-gate being driven behind the face line. We shall describe in sequence the mechanized systems used for extracting the coal face A in the main-gate ahead of the main face line and for cutting the tail-gate stable B which is in line with the coal face itself.

Finally we shall describe two cases of mechanisation on "shortwalls" which are not associated with the coal face, namely a roadway drivage and a face development heading.

That one can deal with such a variety of workings in a single report is due to the fact that the same machine, or rather a machine composed of the same components, can cut both coal and stone under very different profile conditions. The basic components are those of the Eickhoff ESA 60 L stable-hole machine. In each case all the work associated with the winning operation, i.e. product loading and clearance, roof support and advance of various installations, requires particular care and attention. In fact the success of the main operation, in this case coal getting, depends on the degree of efficiency achieved by all the face operations as a whole.

We shall now proceed to describe different instances where such a technique has been employed at Zolder colliery and shall use the first of these to explain in greater detail the actual cutting machine and its operating potential.

3. MECHANIZED CUTTING OF MAIN-GATE STABLES

Figure 2 shows a typical layout at the main-gate end. The diagram features the main face conveyor, the stage loader and the support system comprising link bars supported on single hydraulic props. The cutting machine travels to and fro along the face and runs on top of a curving conveyor. The latter features two successive curves of 90 and 45 degrees which allow it to discharge directly on to the stage loader. The entire installation is snaked forward by hydraulic rams.

The winning machine is composed of the following main components (Figure 3):

1. Cutting drum (A) having a maximum diameter of 1.80 m and a width of 0.40 to 0.65 m.
2. Ranging arm (B) which carries the drum and can rotate continuously about its axis (G). For main-gate stables a 250-mm-long arm is used. The arm incorporates the planetary gearing for the cutting drum drive.
3. Electric motor (D), rated output 60 kW, whose axis is situated directly in line with the axis of rotation of the ranging arm.
4. A variable-flow hydraulic pump (F) supplying the hydraulic motors. This pump is driven via a reduction gear (G) coupled directly to the other end of the electric-motor shaft.
5. Hydraulic motor (E) with planetary reduction gear to drive the ranging arm.
6. Second hydraulic motor (H), identical to (E), also provided with planetary gearing to drive the haulage box. The latter features a toothed wheel which engages in the haulage rack.
7. All the above components are mounted on a machine frame (I) which also includes the oil reservoir (K).
8. The machine travels from one side of the face to the other on a conveyor which features a tubular guide on the face side and spill-plates on the goaf side, these being superposed by the rack which also acts as a guide.
9. The machine also carries the electric-motor control buttons, the hydraulic control valves for the ranging arm and haulage

box, devices for monitoring the liquid level and temperature of the hydraulic circuits, oil cooler and water-flow and pressure monitors.

10. Hydraulic jacks (L) located on the goaf side of the conveyor provide continuous control of the cutting level and so prevent 'stepping'.

11. Lateral shields (Figure 2-G) are mounted at each end of the face. By means of hydraulic jacks these shields can be adjusted so as to provide control of the machine direction.

12. The cooling water for the electric motor is used to provide dust control; this water passes through the oil cooler and then along the ranging arm to reach the spray jets mounted on the cutting drum.

The main technical features of the ESA 60 L machine are listed in the Table below. The figures given in the first column are for machines generally used in main-gate stables while the second column gives the maximum variations which this type of machine can achieve.

	Main-gate stables	Max. variations
Motor output	60 kW	60 kW
Drum diameter	1.70 m	1.30-1.80 m
Drum width	0.40 m	0.40-0.65 m
Drum speed	32 r.p.m.	32 r.p.m.
Peripheral speed at picks	2.85 m/sec	2.18-3.02 m/sec
Length of ranging arm	250 mm	250-500-750 mm
Maximum travel of drum in relation to machine axis	500 mm	500-1000-1500 mm
Machine travel speed	0 - 3 m/min	0 - 4.5 m/min
Effort delivered by haulage box	110 kN	110 kN
Total machine weight	5500 kg	4000-7000 kg

Method of operation in a main-gate stable is as follows:

The machine is brought to the left-hand end of the face. The face conveyor is then snaked forward at the right-hand end of the face by about 0.40 m. The machine then travels to the right-hand end, cutting a wedge of coal some 0 to 40 cm thick. When the drum reaches the end of its run the left-hand end of the conveyor is snaked forward by about 0.40 m and the cycle begins again. If coal is left hanging above the drum the ranging arm is raised accordingly and a second cutting run is made. A machine fitted with a 250 mm arm and a 1.70 m-diameter drum can cut a face 1.70 to 2.20 m in height. The EKFOV conveyor incorporates a roller-disc curve which creates a conveyor angle of 90 degrees. Several short-length line pans are mounted behind the curve to give the conveyor sufficient flexibility to absorb the alternating movement of the snaking action. When the face has been cut to a depth of 60 cm the machine is stopped to give time to install the cruciform roof bars and set the hydraulic props.

While the work of winning, loading and clearing the coal is fully mechanized, the setting of this type of support system, comprising straight and cruciform link bars, remains an arduous task which requires skilled manpower. Nevertheless, this type of support method is still generally used at the main-gate end since no other system is better suited for providing sufficient roof support above the drive unit of the face conveyor.

The curving conveyor, with its 90 and 45 degree curves, has the advantage of being able to serve the entire stable area with a single machine; however, such an arrangement means that the supports must be withdrawn and re-set each time the conveyor is to be snaked over. A T-shaped conveyor of suitable size and performance and tested to a high degree of operational reliability would eliminate this drawback; unfortunately, such a piece of equipment is not yet available on the market.

Since 1979 ESA 60 L machines using this working method have driven a total of 3750 m of stable at Zolder colliery; this represents an average output of 44 cm/MS. These shortwall installations are only put into action once per day whenever the machine can cut in one shift the average advance of the main coal face. When the stable is properly organized an advance of 3 m per shift can be regularly achieved on a 9 m-long face; to date the maximum advance achieved in a single shift has been 4.5 m.

4. TAIL-GATE STABLE AS AN EXTENSION OF THE COAL FACE

In this case the ESA 60 L machine is mounted on the main face conveyor. This method can be employed equally well on shearer and on plough faces. Figure 4 shows such a layout used on a coal-plough face.

The shortwall face is free of props throughout its entire length. The machine is moved by an Eicotrack haulage system. The drive motors for the conveyor and coal plough are separated from each other so that an anchor chock can be located between the two units. A hydraulic support system, which can be advanced as the conveyor is snaked forward, is installed above the AFC and plough drive motors. By causing the machine to deviate by 20 cm towards the face and by the same amount towards the roof it can clear the AFC drive head. In this position the machine can still cut low enough to permit corrections to its cutting level. Here the ESA 60 L operates exactly like a coal-face shearer.

The stone face follows the coal face at a distance of about 5 m. The dirt is loaded out by an auxiliary conveyor which discharges on to the AFC.

The shortwall machine uses the basic elements of the ESA 60 L though these are arranged differently from those which go to make up the main-gate stable machine. Figure 5 clearly shows the cutting drum of 1.60 m diameter, the 250 mm-long ranging arm (as used on the main-gate machine), the electric motor, the hydraulic pump with drive reduction gear, the hydraulic drive motor for the ranging arm and the haulage box. All these components are mounted on a low machine frame which rests on two rollers on the face side. Such an arrangement increases machine stability and improves product loading and clearance. The oil reservoir is no longer located in the frame but on it, on the face side and in front of the haulage box. The machine naturally retains the same control and monitoring facilities.

The machine can be fitted with a ranging arm 250, 500 or 750 mm in length. Figure 6 shows the machine positioned above a UH8 drive unit on a Halbach & Braun EKF2 face conveyor. The stable is cut to a height of at least 2 m throughout its length irrespective of the seam thickness. At the changeover point between the plough and the stable-hole machine the latter no longer runs on its rollers in front of the AFC but rests directly on the conveyor thanks to steps in the guide rollers. This arrangement is necessary to eliminate the use of ramp plates on which the machine runs when working the high face, but which would impede the passage of the coal plough. On the face the machine cuts to the same height as the plough, or at least equal to the diameter of the drum - in this case 1.60 m.

The work of winning, loading and clearing the coal etc. is, therefore, fully mechanized while the anchorage and support work is only partly mechanized. In fact, link bars supported by single hydraulic props are still used to support the zone in the roadway axis on account of the fact that modern methods have not been adapted to these local conditions. The stable-hole machine can cut up to 20 m into the face.

The same system is employed with an identical machine on shearer faces. In this case the face shearer is completely independent of the snaking operations at the tail-gate end and once the shearer reaches the end of its run it can return immediately without wasting any time.

This method of heading with the ESA 60 L in the tail-gate is a relatively recent development. It was first used in April 1981 and since then has been applied on one plough face and on three shearer faces. A total of 2000 m have been driven at an average advance of 2.00 m per shift and an output of about 40 cm/MS (including those engaged in shifting the AFC drive unit).

The disadvantage of this system is that the advance of the tail-gate end is totally dependent on the advance of the coal face. The stable-hole machine cannot operate as an independent unit.

5. <u>ROADHEADING</u>

As regards the extraction of short faces not associated

with coal-winning areas we shall begin by examining roadway drivages of 11 m^2 section. In this case a roadheading operation can be compared with the extraction of a shortwall face. Figure 7 shows the layout of the installation. An EKFOV conveyor is installed right across the width of the heading face. At one of the face ends the conveyor turns to run parallel to the roadway axis by means of a roller-curve section. Its telescopic arrangement thus allows it to overlap the roadway conveyor by about 4.5 m.

The cutting machine also comprises the components of the ESA 60 L and travels to and fro across the heading face on that section of the conveyor which is parallel to the face. The diagram clearly shows the cutting drum, the 750 mm ranging arm, the electric motor, the drive reduction gear for the arm and the haulage box complete with rack. These components are mounted on the machine frame at a height which allows the drum, when set low, to cut to the floor of the roadway. Height control is achieved by raising the AFC by means of hydraulic jacks positioned both at the front and at the rear of the conveyor. The machine is operated by remote control via electrovalves.

The heading sequence is carried out in the following manner. When the machine is in the right-hand corner with its cutting drum resting on the floor, the left-hand end of the installation is snaked forward by about 40 cm. The machine then travels to the left-hand corner of the face, with its drum set low, cutting as it goes. The right-hand side is then advanced. The drum is then brought back to the right-hand end, this time with a series of machine flits being made in a clearly-defined sequence followed by partial rotations of the ranging arm towards the roof. In this manner the arm describes a complete circle of 360 degrees across the heading face, the drum ending at floor level in the right-hand corner. In doing this it traces a kind of spiral while cutting a strip of constant thickness. A very simple technique is used to achieve a satisfactory degree of profile accuracy. This is done by painting the various zones different colours, i.e. the travel rack and a disc fitted to the ranging arm. The cutting drum, starting from its low position in the left-hand corner of the face, travels the entire length of the white zone. A rotating action, corresponding to the entire white area of the disc, is then undertaken by the ranging arm, and the process is repeated for each of the colours in succession: the machine moves to the red zone on the rack and the ranging arm is rotated to correspond with the red portion of the disc and so on. In this way an accurate profile is cut, provided that at the outset the conveyor is in a horizontal position and correctly aligned in the direction of the roadway. By cutting the profile circumference in 24 stages a cutting accuracy of 3 cm is achieved.

During the mechanized roadheading trials special attention was continuously focussed on the support setting system. For the first tests the support system was composed of a double overlapping chock comprising 2 x 4 props each having a load-bearing capacity of 70 tonnes. The two chocks were fitted with two roof bars having articulated, telescopic extensions. The double chock also featured an articulating arm capable of picking up an arch crown at floor level and setting it against the roof. Subsequent tests involved modifications to the chock structure, this being simplified to 6 and then to 4 props, each leg retaining a load capacity of 70 tonnes. In practice it was learnt that the arch-setting performance could be improved by concentrating more on the handling of the arches than on the load-bearing capacity of the temporary support system. Our development work, which was based on the concept that to devise an effective mechanisation process one had to have a very reliable support system, led to the construction of chocks which were very heavy and much too difficult to manipulate. On account of this, the final series of tests saw the construction of a working table equipped with a hoist. This was designed basically as a support-setting aid rather than a support structure, though the hoist is still capable of supporting 25 000 daN above the machine. The arches are pre-assembled on the table and the crowns can be positioned at the roof in sets of 2 or 3 at one time. Such a method means that 6 arches can regularly be installed in a single shift.

To date the ESA 60 L has driven a total of 680 m of roadway with an average advance of 7.50 m per day. During the heading operation the thickness of the seam has varied from 1.54 to 1.96 m. The hardness of the roof strata has also varied during the tests, reaching a maximum of 880 kg/cm^2. Even such a very compact rock was cut by the machine at a sufficient rate.

The main drawback to this method is the fact that a particular machine can only cut the roadway to one height, this being determined by the length of the arm and by the diameter of the drum. When fitted with a 750 mm ranging arm and a 1.80 m-diameter drum the machine at Zolder can cut to a height of 3.30 m, which is currently the maximum possible level. The minimum height which can be cut with the machine is about 1.90 m. Obviously, the machine is not restricted to any particular roadway width.

A further drawback to the machine is that the cutting drum always leaves a wedge of material at the foot of the arch supports; this must continually be removed by hand. It is vital that a solution be found to this problem so as to improve the heading performance per shift in cases where the rock in the roadway floor is very hard.

6. RISE HEADINGS

When driving rise headings it is possible to use the machine arrangement employed for cutting main-gate stables, as shown in Figure 1. However, the special problems associated with rise headings have resulted in the following modifications:

The heading debris is loaded out by an EKFOV conveyor. The first section of this conveyor runs parallel to the heading face while a roller-curve is used to divert the flow in the axis of the heading and so discharge on to the standard AFC. The roadhead conveyor has a telescopic action to overlap the roadway conveyor in the same way as that described in the previous section. In this case, it is vital that the extension of the main roadway conveyor can be undertaken very quickly.

The winning machine could be kept absolutely identical to that which is used in main-gate stables. This allows the extraction of seams 1.70 to 2.20 m thick. In order to create a mechanized system for driving rise headings in thinner seams we designed a fixed-arm machine whose more compact construction enables it to operate in seam thicknesses down to 1.30 m. This particular machine is shown in Figure 9. The diagram shows the electric motor, the haulage box, the hydraulic pump and the arm of the ESA 60 L. The arm is 250 mm in length and does not turn, being fixed in a specific position. The rotary gear for turning the arm has also been removed. The machine can be fitted with a choice of 3 cutting drums with diameters of 1.30, 1.50 and 1.70 m. Each drum diameter corresponds with a different rack height, the latter being adapted by means of a support piece which is inserted between the rack and the spill-plates.

Cutting height is adjusted by hydraulic jacks located behind the spill-plates. Alignment is also corrected by means of hydraulic rams.

The type of support system selected is of vital importance since it is this which often determines the rate of advance. At Zolder colliery the method is to use Van Wersch link bars set parallel to the direction of the heading. Such a system is not an inconvenience since these temporary supports are replaced by powered supports prior to the start-up of the face.

We shall continue to stress that the winning capacity of the machine is limited by the support setting rate, irrespective of the type of system employed.

7. FURTHER DEVELOPMENT PROSPECTS

It is feasible to expect an increase in the operating range of machines of this nature.

In the first place one could increase the machine output. This would allow the machine to cut harder rock, to operate on longer faces and to achieve higher cutting speeds. Work is currently under way on a 150 kW ESA machine which may be fitted with a ranging arm 1.60 m in length.

A more fundamental improvement to the machine involves mounting the support housing of the cutting arm on a rotating shaft so as to raise the axis of rotation of the arm. Figure 10 shows an ESA machine which has been modified in this manner. The support housing contains the electric motor, the hydraulic pump and the arm rotating mechanism. The housing is raised and brought into a horizontal position by means of a hydraulic jack. The axis of rotation of the arm can thus be brought to such a height that the cutting drum can be positioned exactly at floor level with the ranging arm facing vertically downwards. With the housing in a low position the drum can clear a passage for the machine in both directions.

This arrangement has the following advantages over the conventional fixed-housing machine:

- increase in the cutting height of the drum
- extra-low machine height
- a wider passage can be cut to the right and left of the machine
- the same machine can cut shortwall faces of different height and width
- by making suitable modifications to the conveyor on which the machine runs it becomes possible to cut a wide variety of profiles.

An actual working site has already been found for the machine at Zolder colliery. This will involve the simultaneous heading of a roadway and the tail-gate stable of a production face some 6 to 9 m in front of the face conveyor (Figure 11). By keeping this distance as short as possible it is hoped to minimize the level of convergence due to the passage of the face. Material is discharged on to the face conveyor via an intermediate conveyor which is overlapped by the telescopic-action EKFOV conveyor; such an arrangement makes for a conveying flexibility of about 3 m. The intermediate AFC is located in the tail-gate stable.

It is also possible to drive both the stable and the roadway in line with the coal face by mounting the machine directly on the face AFC. Figure 12 shows such an installation on a plough face. The advance of the stable-hole machine is directly dependent on the face advance and there is no longer any flexibility. Prototype trials under such conditions would not be advisable because of the risk of production losses. While a greater output could be expected with the machine mounted on the face AFC, we opted to test the prototype at a point some 9 m ahead of the face in order to ensure a certain degree of freedom due to the flexibility provided by such an arrangement.

Another possible application lies in the heading of circular-section roadways. Up until now there have been very few retreat faces worked in the Kempen coalfield. One reason for this is the amount of maintenance work needed in the gate-roads. As a result of experiences acquired in circular stone-drifts lined with concrete panels it is logical to believe that circular gate-roads would offer better resistance to the very high strata pressures found in our coalfield. When supported with circular or even elliptical rings such roadways should, in theory, stand better and require less maintenance than roads lined with conventional supports. The machine described above is highly capable of cutting both circular and elliptical roadways. Before undertaking such a project, however, it would be necessary to conduct tests in order to determine the behaviour of circular or elliptical roadways serving a retreat face.

8. CONCLUSIONS

Shortwall faces are to be found underground in a wide variety of forms. The type of work required at such points is identical to that employed on the coal face, namely winning, support setting, product loading and clearance. The development of the ESA 60 L means that a highly-reliable machine is now available which can extract shortwall faces of widely-varying types. Product loading and clearance, and more especially support setting, may differ from one kind of shortwall to the next. Given the current state of technology in this field it is impossible to avoid the multiplicity of work needed in these areas of operation.

Such a wide variety of applications is clearly established by examining those instances where ESA 60 L machines have been employed to extract shortwall faces at collieries both in the Kempen coalfield and abroad. The area of application of the ESA machine is certainly not restricted to those operations described in this paper and a major extension to its sphere of operation is expected in the years to come.

Figure 1

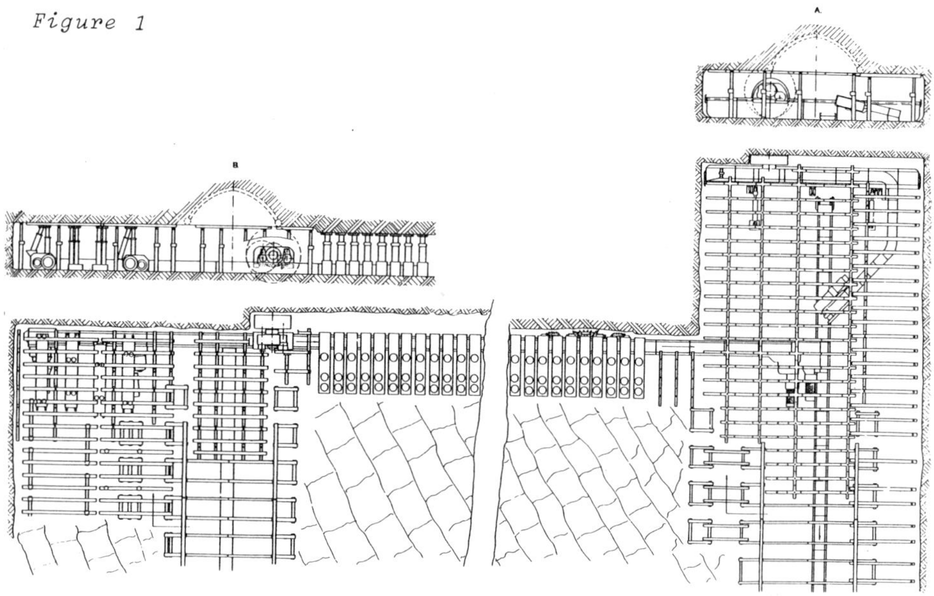

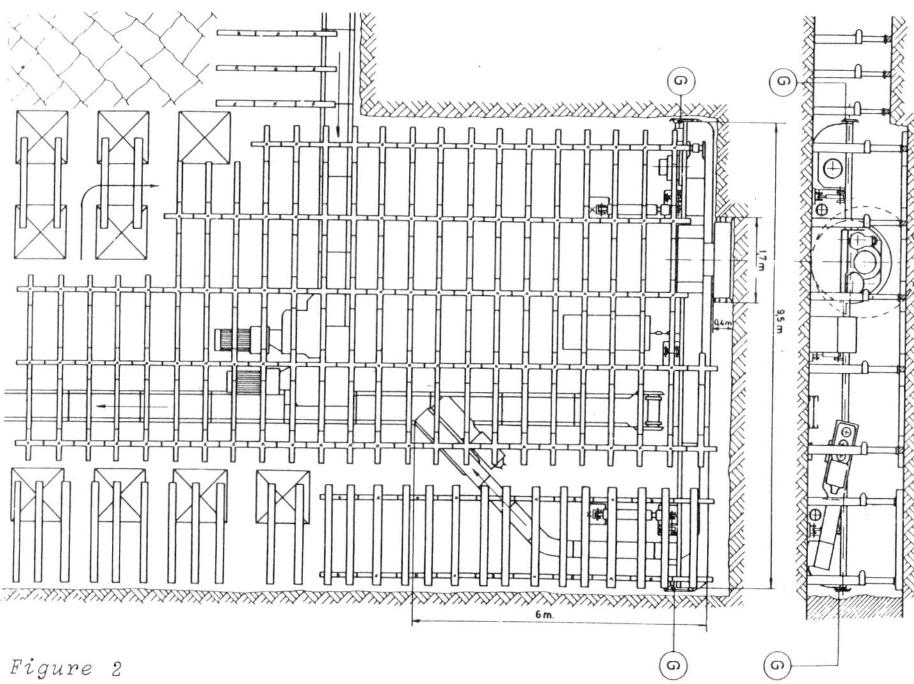

Figure 2

Figure 3

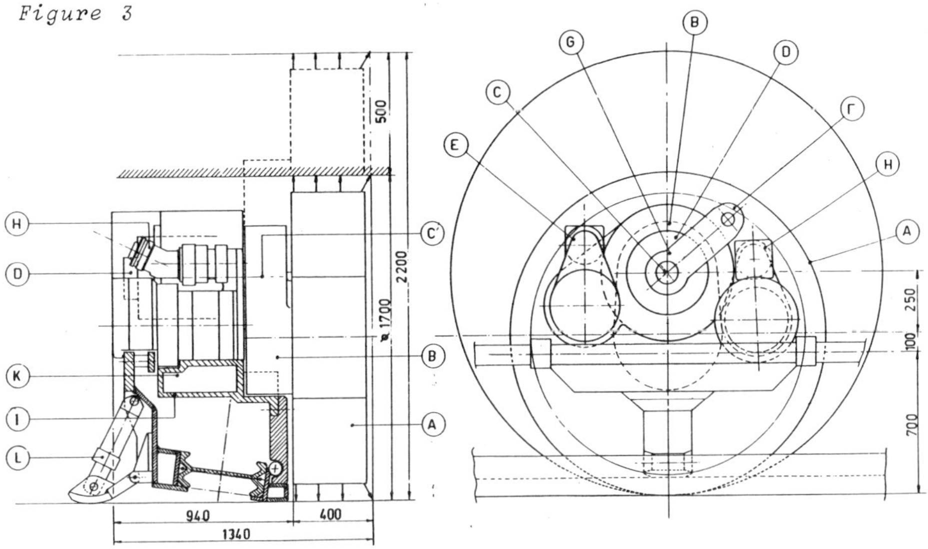

Figure 3 A

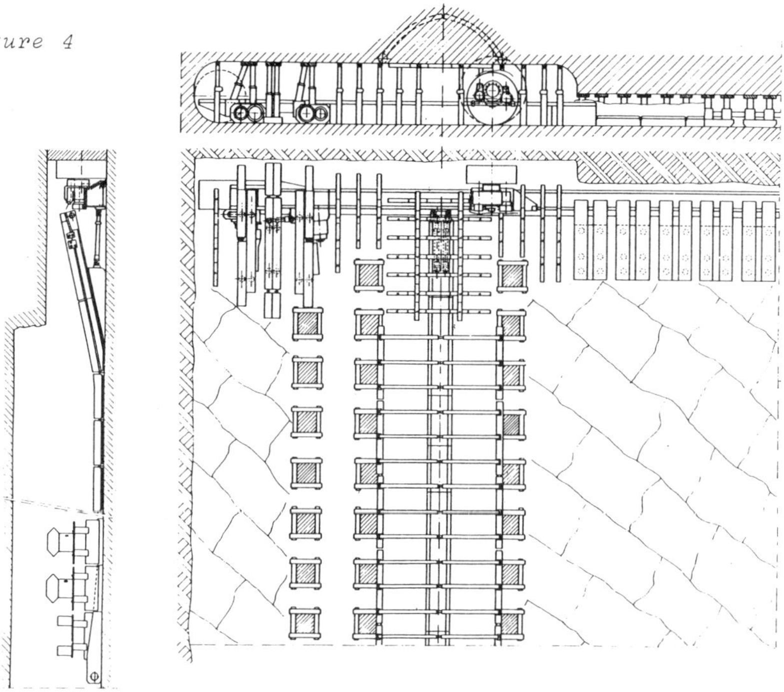

Figure 4

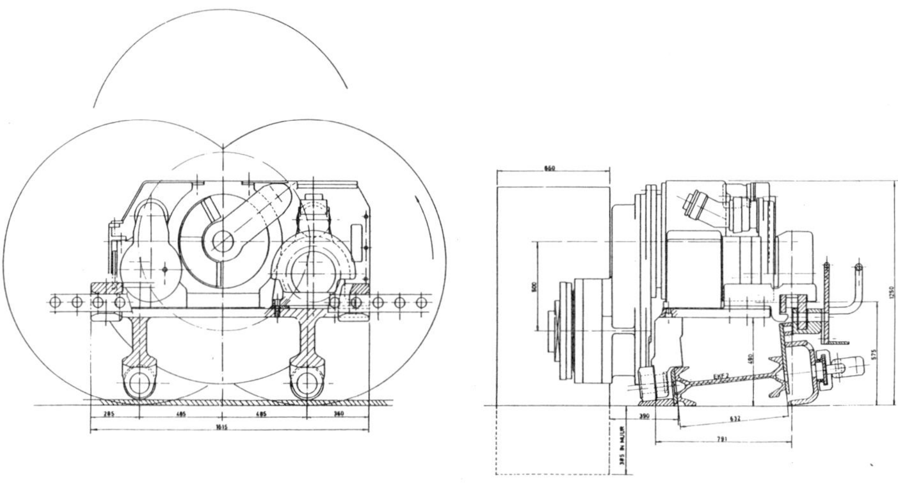

Figure 5

Figure 5 A

Figure 5 B

Figure 6

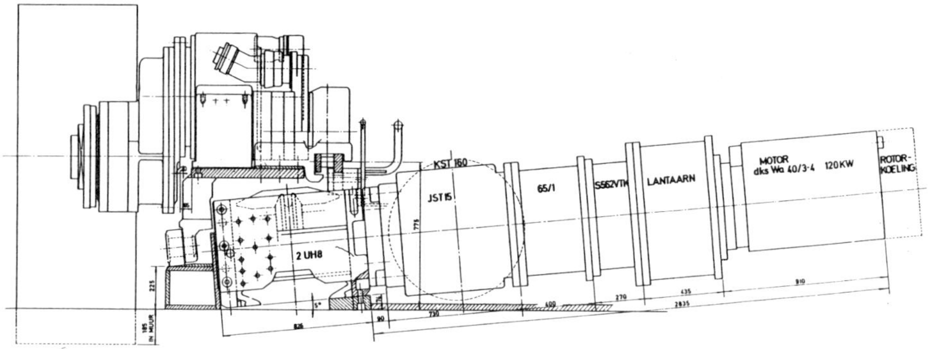

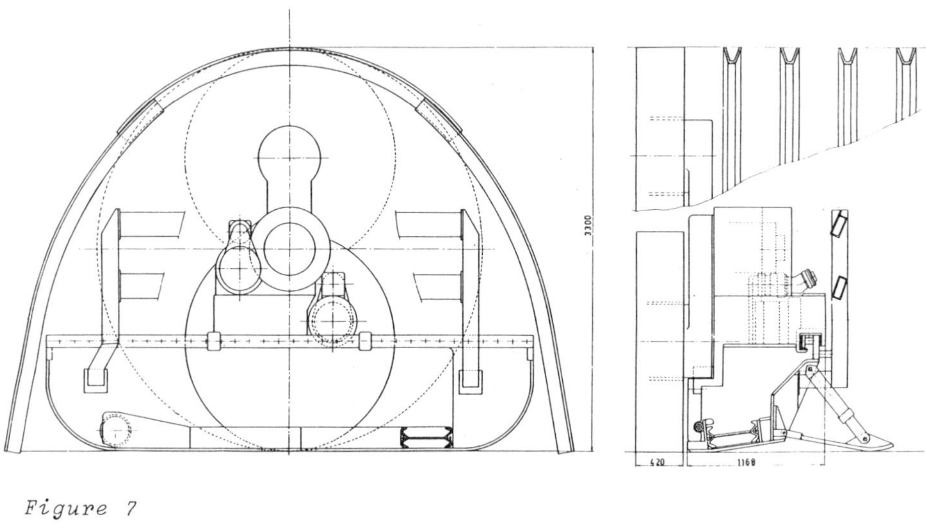

Figure 7

Figure 7 A

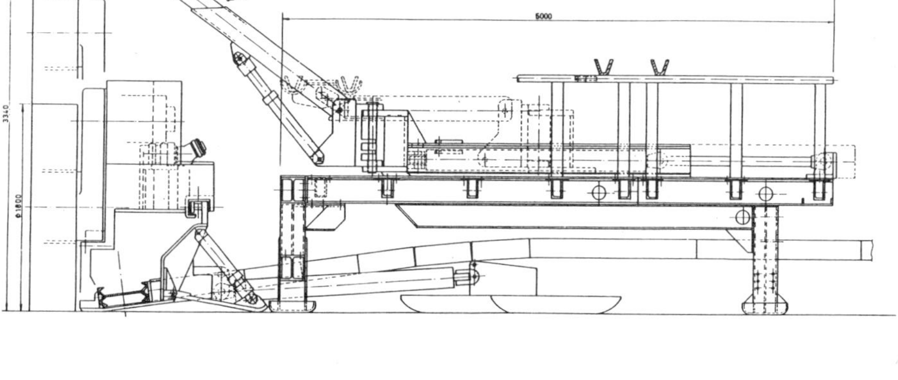

Figure 8

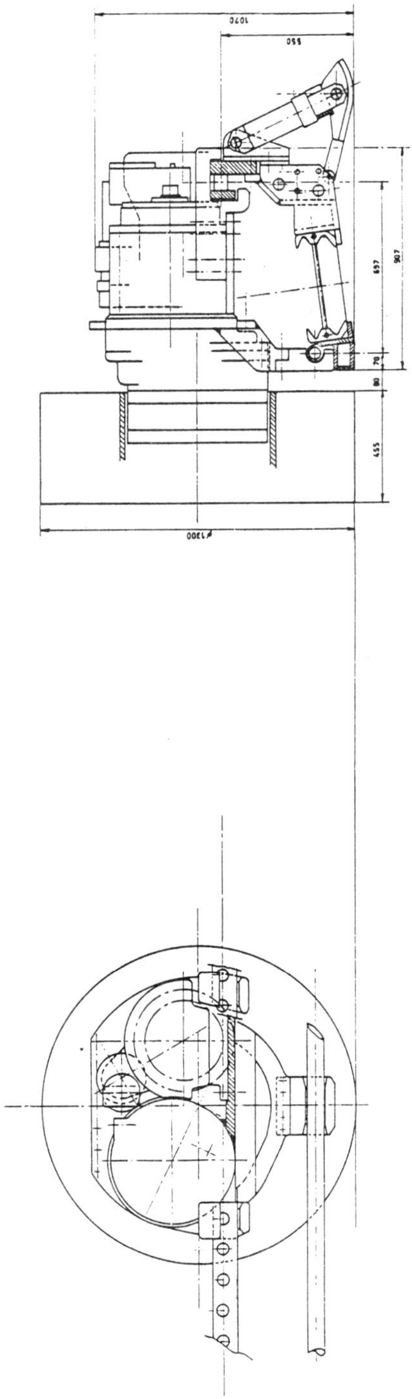

Figure 9

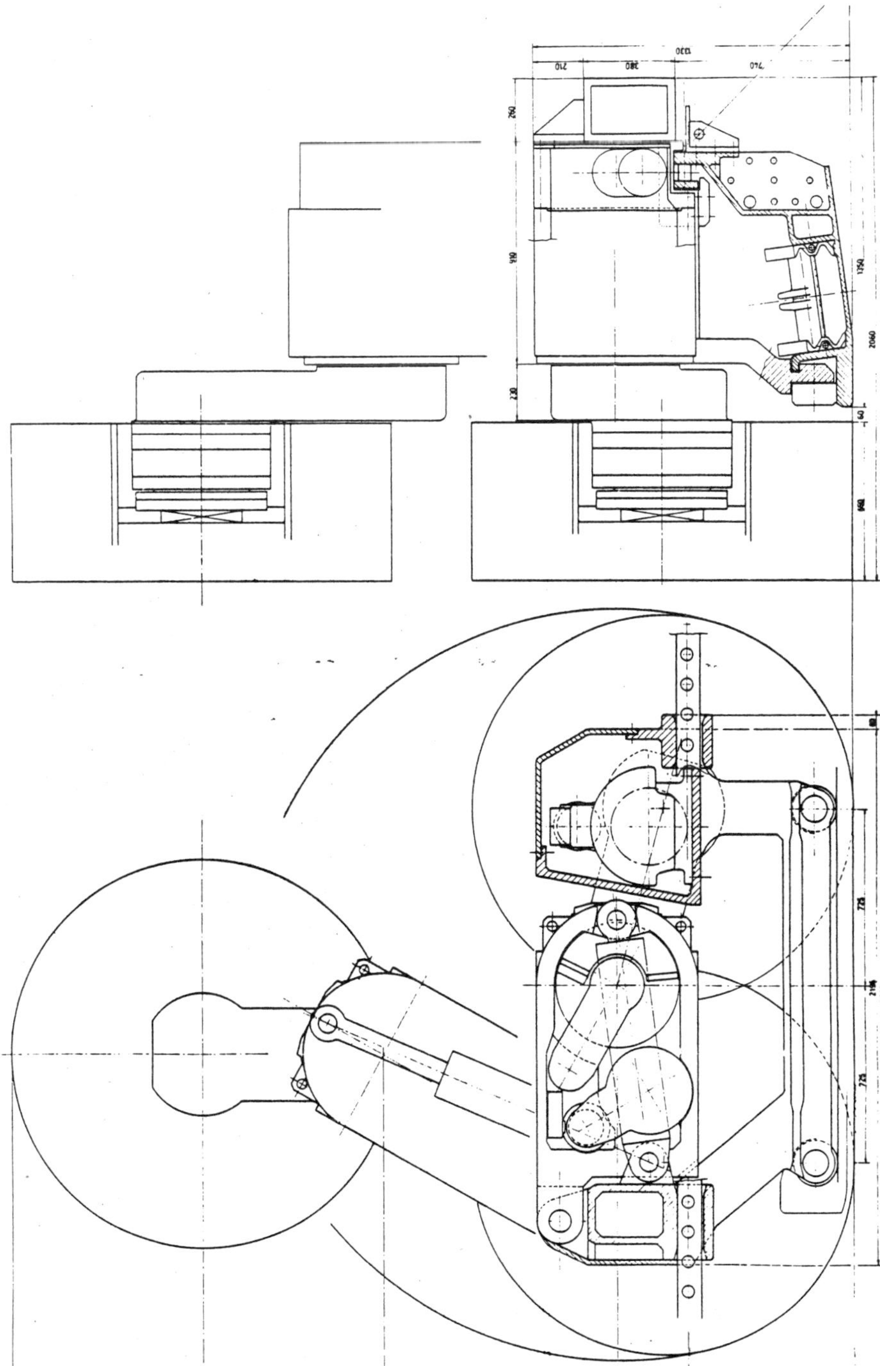

Fig. 10

Figure 11

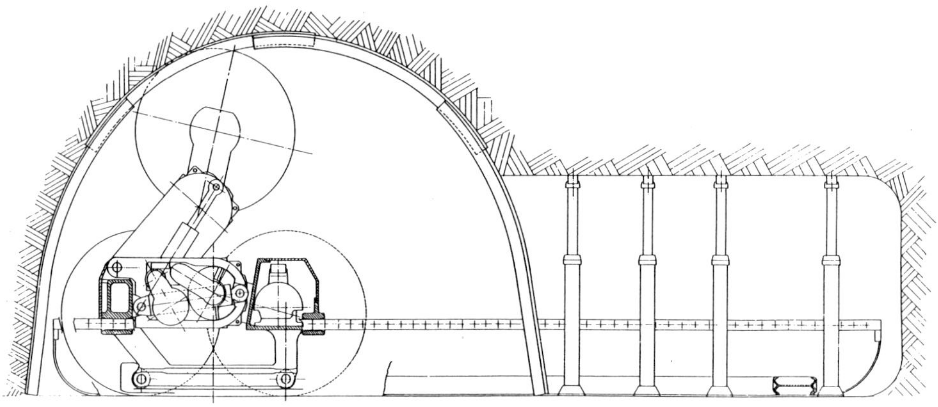

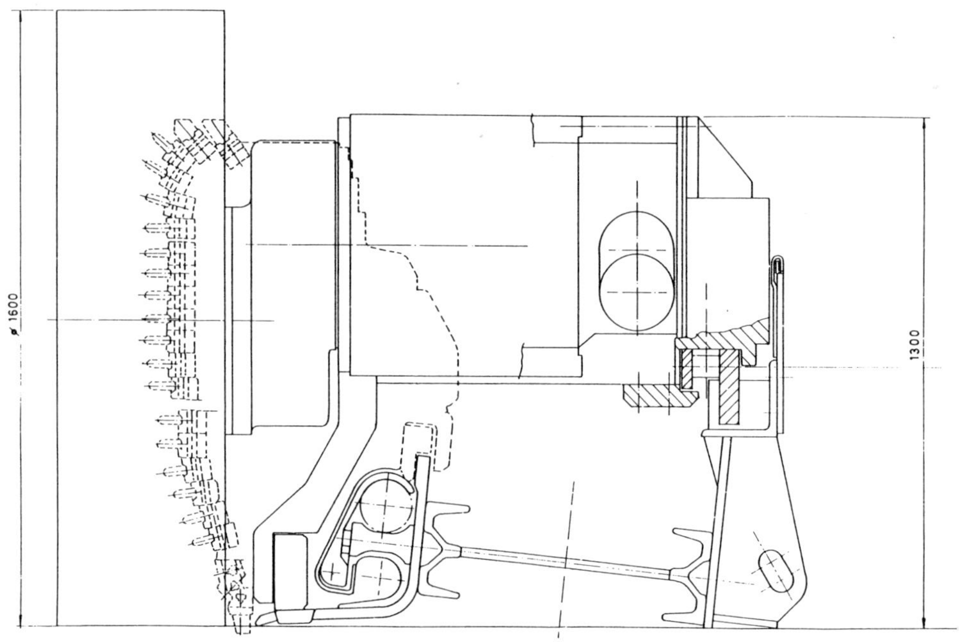

Figure 12

SESSION III - MECHANIZATION OF SUPPORTS

- New developments in rock bolting
- Use of cement products underground
- Linings for circular roadways
- Mechanized setting of roadway supports

NEW DEVELOPMENTS IN ROCK BOLTING

L. Tinchon
Divisional Engineer
Lorraine Coalfield
MERLEBACH

M. Dejean
Head of Strata Group
Centre d'Etudes et de
Recherches des Charbonnages
de France
VERNEUIL

Summary

After a brief historical survey of the introduction of rock bolting into French coalmines and of the conditions under which research was undertaken, the authors discuss the improvements made to bolts and the new types of bolt on the market which are anchored along their entire length either by friction or grouting. Virtually all the techniques currently used by the mining profession are thus covered.

The authors go on to consider the development of techniques for drilling and bolt setting, laying emphasis on the research already conducted or in progress to extend the field of application of rotary drilling, which has great advantages in terms of machine dimensions, working conditions, space and noise.

Finally they study the practical uses of bolting; they put forward the rules for determining bolting parameters, taking into account the nature of the strata and their classification together with the type of anchorage and its characteristics, and then they cover the ways in which bolting was actually used in the coalfields in 1982, whether as the sole means of support or as a supplement to a conventional support system.

1. INTRODUCTION

Up to 1947, the use of bolting as the sole means of roof support was very limited. At that date in the USA, the US Bureau of Mines became interested in this method of treatment with a view to reducing the frequency of accidents caused by rock falls. It recommended its use in coal mines and in less than two years, bolting had found full acceptance in the American mining industry.

This rapid growth can be explained by very favourable strata conditions and by the fact that bord and pillar working was well suited to the use of bolting on a large scale.

Following these spectacular results, the technique spread throughout the world where it often had to be adapted to less favourable conditions.

In France, bolting was introduced in 1948 in the Lorraine iron mines, whilst in the coal mines it was only used as a provisional support system or in reinforcement work.

Trials of support systems relying solely on rock bolting (hereinafter referred to as integral bolting), carried out in the various coalfields between 1950 and 1963, met with only limited success and then only in good strata conditions.

The diversity of the results obtained was doubtless due to inadequate preparation, insufficient strata movement studies and lack of monitoring, but in particular to the limited potential for point anchorage in coal-mining conditions.

The emergence of full-column resin-bonded bolts modified these results and brought about a resurgence in the popularity of bolting. Also, more thorough investigations of rock mechanics including measurements in situ led to the establishment of rules and monitoring methods which made bolting more reliable.

Integral bolting in in-seam trapezoidal and semi-circular roads grew very rapidly from 1964 onwards, then stabilised and even declined in in-seam roadways from 1977. The reasons for this slowdown were:

- the too rapid spread of the method, which meant that bolting was applied to all types of strata with the disadvantages that entails
- the need for regular monitoring
- drilling equipment which was not always suitable
- a longer setting time than required by arches.

The growing trend towards mechanised roadheading is currently associated with arch supports, mainly due to the limited space available around the machine and the rapidity with which arches can be positioned.

The absence of lagging in this type of working soon showed up the weakness of this support method, as the lagging needed for the normal functioning of the arch could only be put in place at considerable expense and with a considerably increased cycle time. Only the Bullflex method can currently meet the criteria for a load-bearing support system in the machine zone which can be integrated with the normal operation of the machine.

The profession, therefore, renewed its interest in bolting as an effective method, while trying to find rapid setting techniques.

Trials conducted at the Niederberg pit in the Federal Republic of Germany, where semicircular-section integral bolting in a thin seam has yielded spectacular results, led to the reconsideration of basic principles such as not cutting into the roof and reinforcement as the face passes. Research has thus been resumed in the field of integral bolting with the objectives of:

- finding a drilling technique which is compact, effective and silent irrespective of rock characteristics
- using the optimal bolts to improve cycle time
- providing support as near to the face as possible
- encompassing by this method a high percentage of roadways, whatever the characteristics of the surrounding rock, with the addition, if necessary, of supplementary treatment (guniting).

In what follows we shall describe the improvements made to bolts and the new models on the market, the bolts in question being all of the full-column anchorage type.

We shall then consider the trials carried out or in progress on the drilling and setting of these bolts.

We shall conclude by considering the practical applications for bolting in France, discussing first the rules determining bolting parameters, and the actual methods of utilisation.

2. BOLTS (Figure 1)

This heading covers:

- resin-bonded bolts
- cement-bonded bolts
- full-column friction-bonded bolts
- sliding bolts

2.1 Full-column bonded bolts

2.11 Rods (Figure 2)

In the French coalfields, the rods used are concrete-ribbed cylinders (nominal diameter 20 mm), with a high capacity for elongation and a breaking load of 20 tonnes. For practical purposes, the capacity for elongation is proportional to the initial length and is of the order of 20%. The steel rod is threaded at one end over a length of 12 to 13 cm, the other end being bevelled so that it can pierce the cartridge.

Four types of bolt are used:

- the conventional bolt
- the Artop bolt
- the top bolt
- the bolt with a shearable pin (Figure 3).

The Artop and top bolts have been in use for many years and have proved highly efficient in mechanised setting.

The bolt with a shearable pin, the latest in the range, combines ease of setting with a high degree of effectiveness.

A calibrated pin crosses the nut and the bolt at the end of the rod. As the grout sets rapidly, the increase in torque is sufficient to cause the pin to break and the end-plate tightens automatically with the aid of a driving helve. This bolt is easier to install than the Artop bolt as it only needs to be turned in one direction. The gain in setting time is approximately 1 minute per bolt.

Other types of full-column bonded bolt are:

- the fibre-glass rod, reserved specially for coal walls where the shearer drum has to penetrate
- the mixed rod made of steel and fibreglass, used as above. It has the advantage that the surrounding rock sets better around it and it also withstands the shock waves caused by shotfiring
- the cable bolt (Figure 4)

This is a 19 mm diameter cable with 4 strands of 4-core 3.3 mm diameter hard-steel wire. Its strength is 180 to 200 kg/mm^2 and its breaking load 20 tonnes. One end is slant cut and the other end has an 80 mm long steel tube (3/4 inch, HP 60 bars) fitted with a nut to form an anchorage point.

This bolt is mainly used as an anchorage point but it can also be used in thin seams to provide a sound support system at the desired depth.

Two variants of these cable bolts have been made:

- the 22 mm diameter cable, with a 30 tonne breaking load
- the 25 mm diameter cable, with a breaking load of 40 tonnes.

2.12 Grouts

(a) Resin (Figure 5)

There are many types of resin grout on the market, all of which have much the same components. They are polyester resins, a mixture of unsaturated ether salts, also containing styrene to control their fluidity. The action of an organic peroxide as a catalyst generates free radicals which bring about polymerisation. This reaction changes the resin from a viscous liquid state to the solid state.

The various types of grout differ according to the positioning of the catalyst (central, peripheral or off-centre) and according to the catalyst's consistency (pasty or solid). Pastiness gives the grouts better durability and year-old grouts are often found which have retained all their properties.

Setting speeds depend on whether insertion is manual or mechanised. The percentage of catalyst used has to be varied accordingly.

(b) Cement grouts (Figure 6)

Cement grouts are enjoying renewed popularity due to the trends in raw material prices and the improvements made to this type of grout. They have been in use for a long time, their main disadvantage being the time they take to set, but they have now been improved and setting times of between one and two minutes are now possible.

Two types of grout are available on the market:

- grouts which need to be soaked for about 3 minutes before use
- grout with water already incorporated.

The various tests carried out on these grouts have not so far shown any material advantages over resin grouts, but they do provide improved bonding in fissured strata.

2.2 The full-column friction-bonded bolt

Two types of bolt are currently used in France, mainly in ore mines:

- the Ingersoll Rand split-set bolt
- the Swellex bolt manufactured by Atlas Copco.

2.21 The split-set bolt (Figure 7)

This is a device measuring 38.1 mm in diameter, which is forced into holes which are 35 mm in diameter. Depending on the hole's length, diameter and straightness, and the hardness and friction coefficient, an anchorage of 2250 to 4500 kg per metre of contact can be obtained on insertion.

These bolts give good results in ore mines in weak ground.

Trials are under way in coal mines, but some reservations are being expressed as regards their behaviour over the long term in humid and abrasive conditions.

2.22 The Swellex bolt (Figure 8)

This bolt is made up of a tube with an external diameter of 41 mm and a wall thickness of 2 mm, which is subjected to mechanical deformation to reduce its external diameter to a mere 25.5 mm. The ends are reinforced by a pressed and welded sleeve. The lower, stronger sleeve has fastenings to keep the end-plate in place. A hole through this sleeve and the tube wall allows water to be injected under pressure into the bolt. On installation, the bolt swells under the water pressure (300 bars) supplied by a pneumatic pump, and meshes with the irregularities inside the borehole. Once the swelling has taken place, the bolt becomes shorter and the end-plate becomes wedged against the rock.

These bolts are inserted in boreholes with a 38 mm diameter and have a rapid installation rate (50 per hour). They are currently being tested at Dauphiné in the Centre-Midi coalfield. During these trials there have been many cases when the adaptor on the water-injection tube has broken.

2.3 Sliding bolts (Figure 9)

For a long time bolt users have wished to increase bolts' capacity for elongation, with a view to enabling bolting to be used in strata with major deformations. Several systems designed to bring about a significant increase in elongation capacity have been tested, particularly in the United States. Over the last few years, our German colleagues have developed an interesting original system.

The Bochumer Eisenhütte system comprises a standard 24 mm diameter rod, which is resin-bonded over 1.80 m. At its base there is a threaded sleeve, which is 35 mm in diameter and 30 cm long. This sleeve is separated from the rod by a substance composed of resin and plastic nodules which allows the internal rod to slide inside the sleeve under a constant pressure of approximately 10 tonnes. This bolt calls for a 35 mm diameter reaming of the standard 28 mm diameter hole. It costs 50 DM per 2.00 m length.

These bolts are attractive in principle and, as other types of sliding bolt have been tested by Bergbauforschung, it will be interesting to know their behaviour and effect on the strata before considering their introduction to our mines, in view of their cost.

3. DRILLING MACHINES FOR ROCK BOLTING

In French coalmines, the drilling diameter used is 28 mm for rods which are 20 mm in diameter.

We shall not deal here with manual drilling, which is a long and costly procedure, but shall concentrate instead on current trends in the mechanised installation of bolts.

There are many types of drilling jumbo, with various French firms (Secoma, Montabert and CMM) manufacturing to order equipment which is adapted to the working whatever its height and width. Each of these firms supplies both rotary drilling and electrohydraulic rotary-percussive drilling (Figures 10, 11 and 12).

In conventional headings using drilling and shotfiring, tyred or crawler-mounted jumbos with 1 or 2 booms are perfectly satisfactory. For hydraulic machines, the average setting rate for a bolt is 3 minutes or 20 bolts per hour per boom.

In mechanised headings it is not possible to bring a jumbo up to the roadhead, even if suspended on a monorail, and research has been conducted into fixing 2 booms on to the body of the machine to allow bolts to be inserted as near the roadhead as possible.

To facilitate the fitting of these booms the objective has been to give them the minimal dimensions compatible with reliability; this leads on to the problem of the type of drilling.

Rotary-percussive drilling is cumbersome and noisy but can be used anywhere.

Rotary drilling, which is quiet, fast and more compact, has limits which are often reached in the rock surrounding our seams. These limits are mainly due to the cutting tool which is incapable of performing effectively beyond 30 points on the "wear index". This term, also known as "drillability", is the product of the hardness in Cerchar points multiplied by abrasiveness also in Cerchar points. In conjunction with the fixing of the booms, therefore, attempts have also been made to push back the limits on the potential for rotary drilling, thus permitting a form of mechanisation which is better adapted to the lack of space in the machine workings.

Two courses have been followed with very promising results:

- the use of high-strength cutting tools with diamond tips
- the use of high-pressure auxiliary water jets (1000 - 1200 bars).

3.1 Statapax bits with "diamond" tips (Figure 13)

The initial bench tests were carried out by Cerchar.

A second test carried out at HBL confirmed the bench test results. In this second test the following parameters were varied:

- rotation speed 750 - 900 - 1250 rpm
- thrust 400 - 750 - 1300 kg
- cutting angle of the bit - 6° - 15°
- flow rate of cooling water 9 - 15 - 20 - 40 litres per minute.

The results obtained show that the "diamond" bit requires a high speed of rotation (1250 rpm), a low thrust (400 kg), a negative cutting angle of - 15° and a high flow rate for the cooling water (40 litres per minute). In ground with an abrasive index of 40, it has a service life which is 60 times longer than that of a standard bit. On the other hand, the tips have been found to be somewhat fragile on collaring and on encountering ground which is fractured or non-homogeneous.

The cost of these bits is very high (2300 to 2700 F each) and their use is only justified when the standard bit has reached its limit. It has proved reliable, however, not only in view of the improvements in working conditions, but also because of the reduction in cost of the drill rig.

3.2 High-pressure water jets known as "auxiliary" jets (Figure 14)

In addition to high-pressure jets (3600 bars) used on their own, tests have been carried out at Cerchar on standard rotary drilling supplemented by water jets of 1000 to 1300 bars.

After a phase of preliminary tests carried out in various types of rock to check the feasibility of this new technique, systematic experiments were conducted in an artificial rock mass of average hardness and abrasiveness.

The experimental schedule was so designed that for each of the parameters studied, it was possible to measure the contribution made by the addition of the water jets.

This test comprised two phases:

- the development of the bits and the positioning of the jets
- the drilling of the solid with variation of the parameters.

The main results were as follows:

- the performance of conventional rotary drilling can be substantially improved if the design of the bits is modified
- auxiliary jets of water increase drilling speed
- the contribution made by these water jets is greater when the water outlets are in the tips
- it is preferable to have two jets rather than one, the first directed towards the periphery of the bit, the other towards the central core
- there is a marked decrease in the power required for cutting if these jets are used.

The study is continuing in highly abrasive and resistant strata. Tests in a working will take place at the end of 1983 in a HBNPC mine.

4. RULES FOR DETERMINING BOLTING PARAMETERS

Many parameters can be varied to ensure that bolting allows the ground to find its proper state of equilibrium: type and diameter of rod, type of end-plate, bolt density. Their choice must comply with both of the following fundamental requirements:

- tolerate rock deformations without breakage or too marked a drop in resistance
- exert on the ground, while this deformation is taking place, the maximum possible constraint, but one which is compatible with the resistance of the rock to shearing along the bond.

It is, therefore, necessary to adapt the bolting pattern to the lithological features of the ground and to its previous degree of fissuring, and to the amount of stress to which the working will be subjected.

This analysis has been made in previous papers.

Table I summarises the bolting parameters which have been found to be relevant (Figure 15).

4.1 Strata

This Table is based on a simple classification of the type and behaviour of strata:

- under the term "type of strata" have been grouped their structural characteristics, in other words their density of fracturation, whether the ground to be bolted is stratified or homogeneous, its mechanical strength and weatherability. The strata have been classified into 3 categories:

1 - Ground which is homogeneous and only slightly fissured
2 - Ground which is stratified and slightly vertically fissured
3 - Ground which is irregular, lenticular, or fissured in several directions.

- the term "degree of stressing" covers the degree of fissuring of the ground on two levels:

. the periphery of the working (superficial ground). The decompression which accompanies heading is often reflected in superficial loosening (spalling and flaking) which must be contained by the bolting system.
. the solid surrounding the working over several metres. This fissuring is unavoidable due to the great depth, and the role of rock bolting is to fight against and adapt to the deformations within the rock.

This analysis may be made by calculations on mathematical or physical models, or else entirely by measurements in situ such as those taken by flat jacks or by core-sampling, or even more simply on the basis of data drawn from observation and measurement of deformations. It enables conclusions to be drawn as to the stability or potential instability (degree of fissuring) of the ground, distinguishing between the two levels of influence exerted by the bolts, in other words the superficial level (about a metre around the working) and the deeper level.

This has led to the use of 3 categories of stressing:

1 - Stability at depth and superficial stability (workings in competent ground and at shallow depth)
2 - Stability at depth and superficial instability
3 - Instability both at depth and superficially.

4.2 Bolting parameters

The variations to be given to the different parameters have been simplified to the maximum, and only the main ones are quoted here:

- the type of anchorage, the choice lying between point anchorage whether it be mechanical or chemical and full-column anchorage whether the grouting consists of resin or cement. In the case of point anchorages, we make a distinction between strong anchorages and "light" anchorages, the latter being characterised by a small contact surface and limited capacities for lateral expansion (American type of anchorage).

- the type of rod, distinguishing between steel rods which are highly deformable before breaking (type A60) and which are particularly well suited to point anchorage, and rods which can provide a rigid support system, a prime example of which is the full-column bonded scrolled-steel type.
- the length, distinguishing between bolts which are short, average or long. Short bolts are those which measure about one metre in length. Average bolts are those whose length does not in general exceed one-third of the width (or apparent diameter) of the working, in general measuring between 1.80 and 2.40 metres. Long bolts are those whose length is between 1/3 and ½ (or more than half) of the width of the working.
- the diameter, making a distinction between small, average and large diameters. Small diameters are those below 20 mm, average diameters between 20 and 25 mm and large, above 25 mm.
- density, with 3 categories: low, average and high. Low densities are those of less than 0.5 bolts per m^2, average densities are generally in the range between 0.5 and 1 bolt per m^2 and high densities above 1 bolt per m^2.
- lagging can prove useless and when it must be installed the choice may be made between light or heavy lagging.

A light lagging is made up of several metal strips or perforated plates if discontinuous lagging is required, or a flexible wire-mesh such as is used for fencing, if a continuous lagging is required.

Heavy lagging is made up of strong wire-mesh or welded lattice-work, possibly in conjunction with shotcrete.

5. THE USES OF ROCK BOLTING IN COALMINES OVER RECENT YEARS - CONCLUSIONS

In 1982, Charbonnages de France used about 1 500 000 bolts in their underground excavations. About a quarter of these bolts were used as the sole means of support in the roadway (integral bolting). The great majority were used for reinforcing the crown or the wall, with load-bearing supports, or for anchorage purposes, and the rules for utilisation are not then covered by the dimension parameters we have put forward. More than 800 000 bolts are used each year in the support system for the mining of steep-seam rise faces with stowing, according to a pattern which has already been published.

It is interesting to note nevertheless that more than 400 000 bolts are used each year in the Charbonnages de France as the sole means of support in roadways, in spite of the difficult ground behaviour conditions generally encountered. This can be achieved with safety due to previous selection of the workings to be bolted and systematic monitoring of the bolted ground when the latter is subject to considerable stresses in the course of excavation. The need for checking certainly acts as a curb to the development of this support technique, but 15 years' experience with the latter shows that this monitoring acts as an essential safeguard.

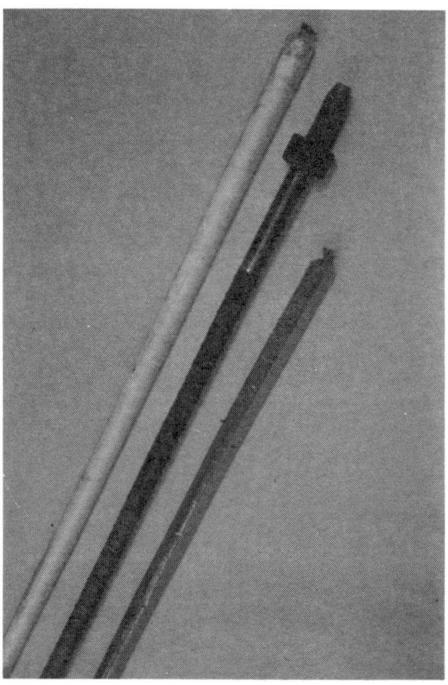

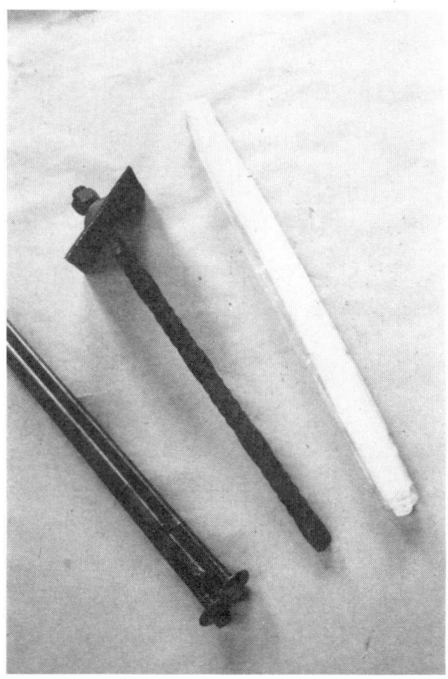

Figure 1

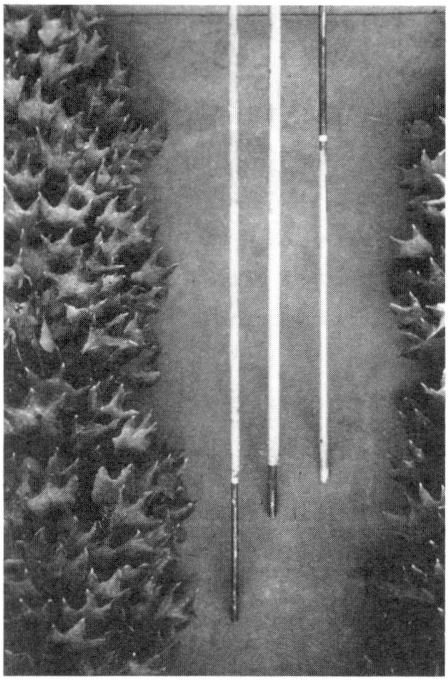

Figure 2

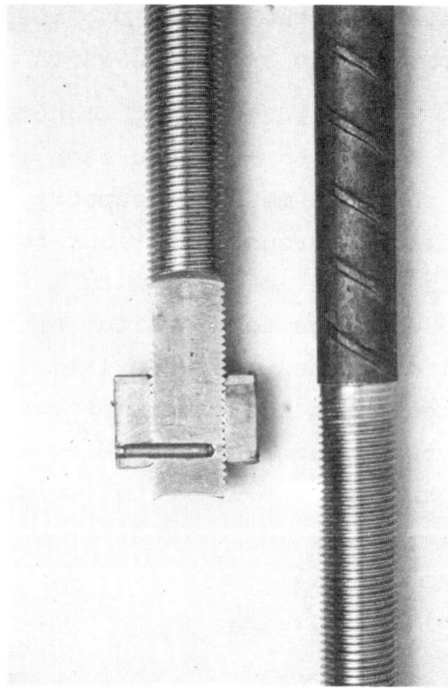

Figure 3

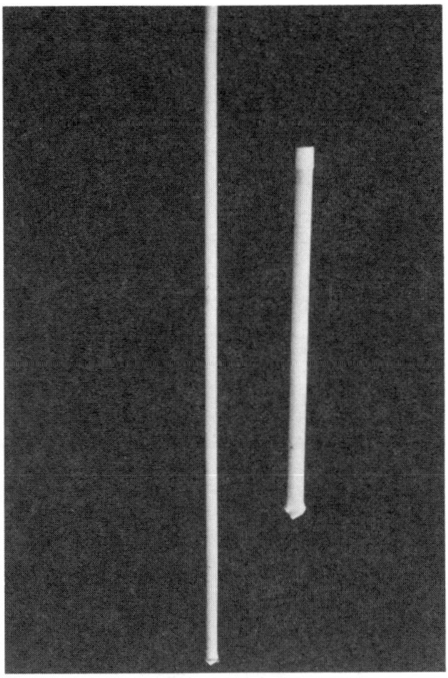

Figure 5

Figure 4

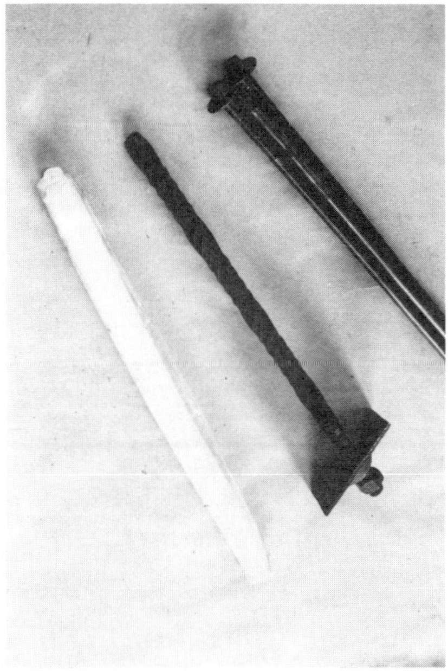

Figure 6

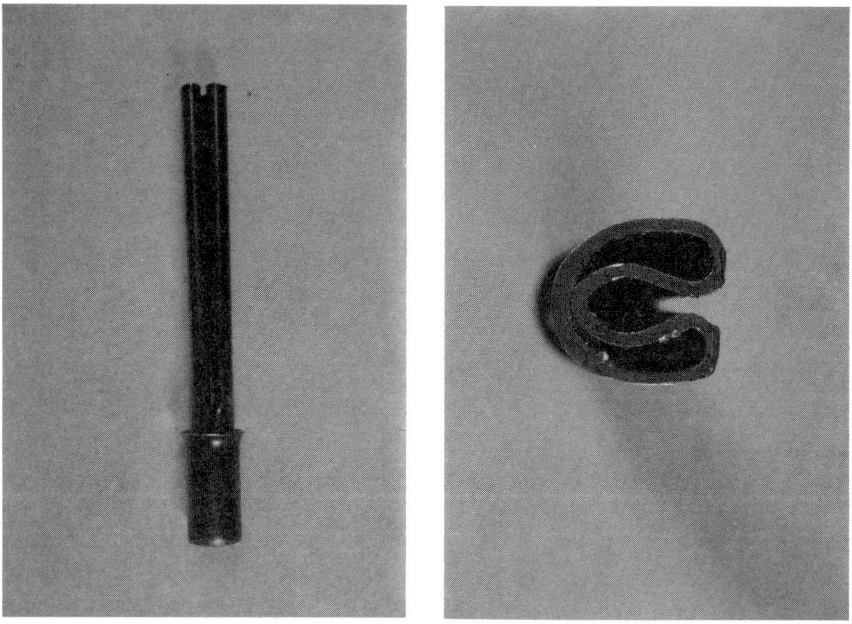

Figure 8

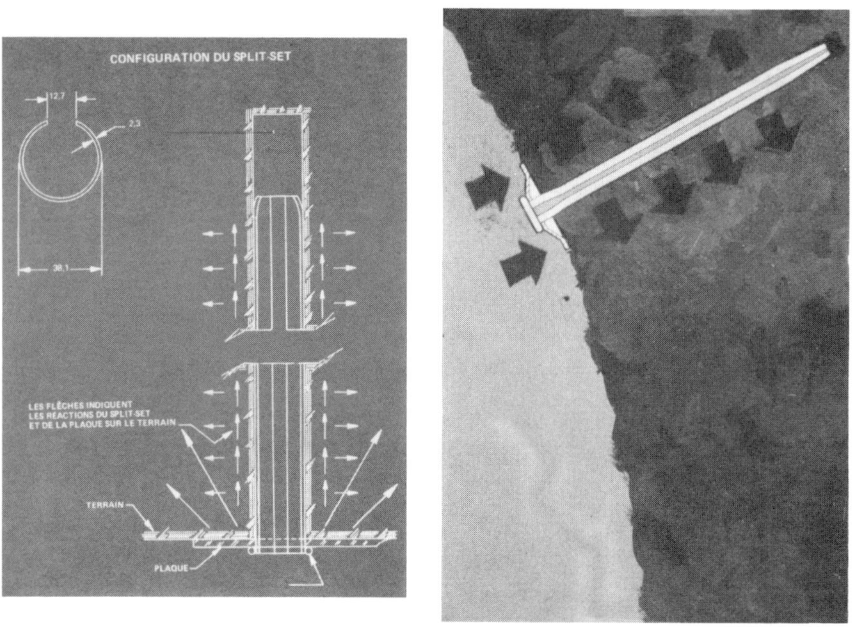

Figure 7

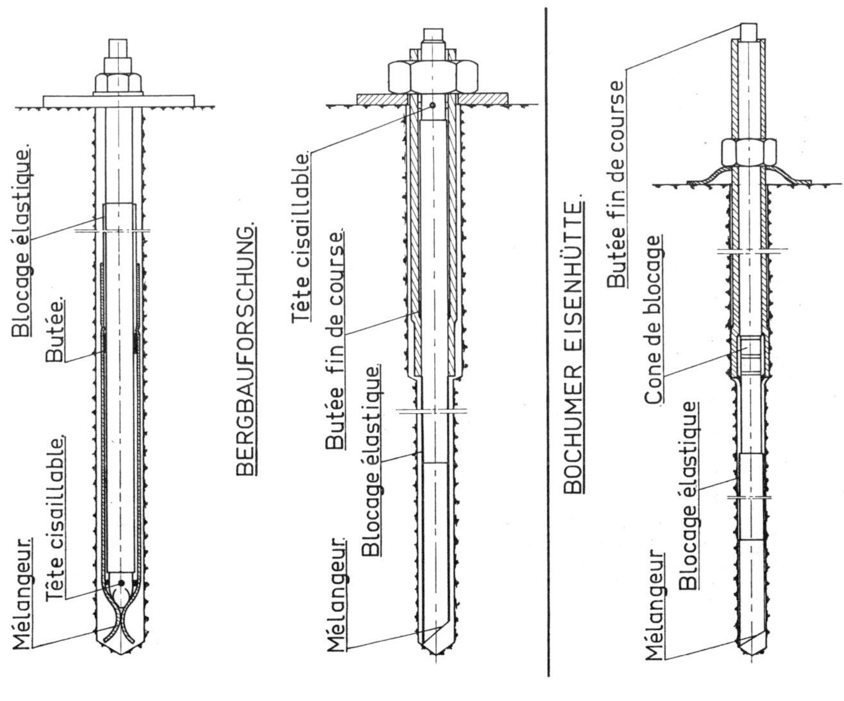

Figure 9

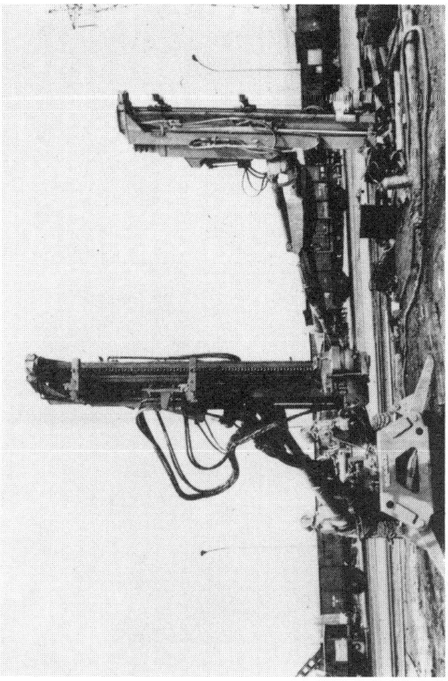

Figure 10

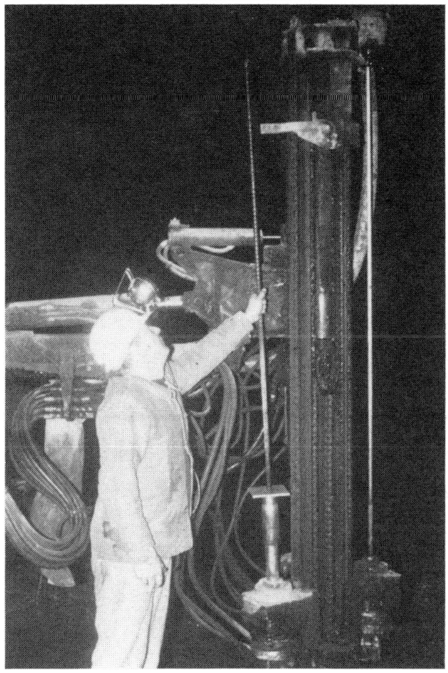

Figure 11

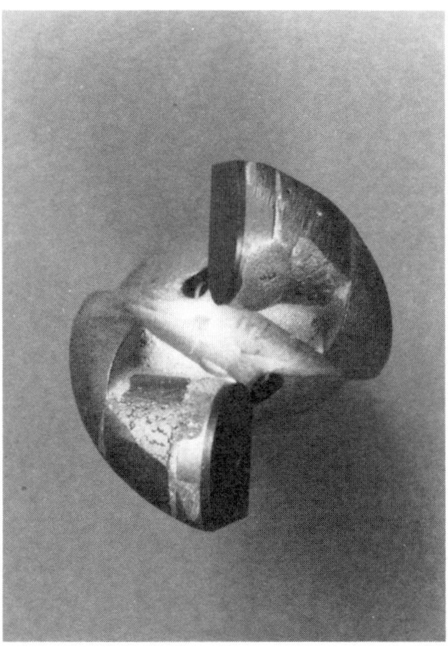

Figure 12

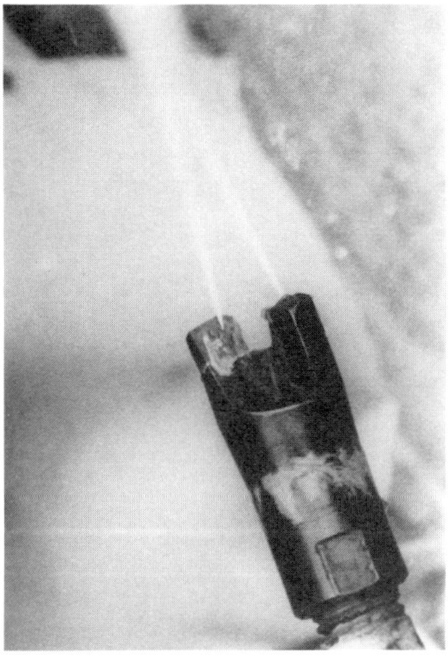

Figure 13

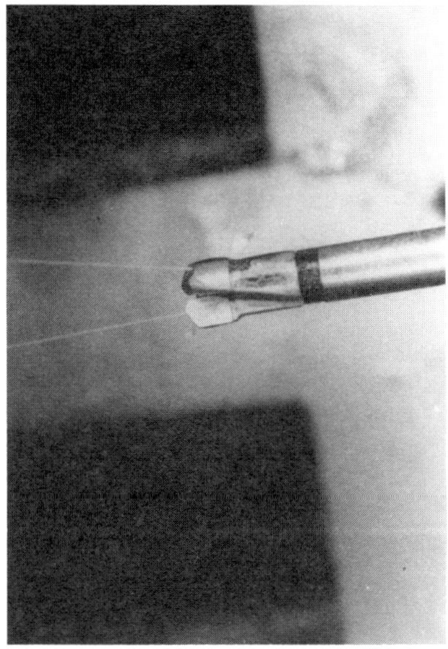

Figure 14

FIGURE 15. CHOICE OF MAIN PARAMETERS FOR BOLTING PATTERN

Conditions	Nature of strata	Bolting type	Rod type	Length	Diameter	Density	Lagging
Deep and superficial stability	all types	point (1) or full-column	immaterial	short	small	wide	light*
Deep stability, superficial instability	uniform	point or full-column	deformable	medium	small	medium	light
	stratified irregular lenticular	full-column	deformable	medium	medium	medium	light
Deep and superficial instability	uniform	point (strong)	deformable	medium	medium	close	heavy
	stratified	point	deformable	long	large	close	heavy
	lenticular irregular	point or full-column	deformable	long	large	close	heavy

* When deterioration matters, a grouting layer may suffice.

Figure 15

USE OF CEMENT PRODUCTS UNDERGROUND

J.E. WOOD, O.B.E.

Area Director

North Nottinghamshire Area,

National Coal Board

The use of cement products underground in the U.K. is relatively limited and the paper gives some measure of experience in the North Nottinghamshire Area of the National Coal Board, where cement products have been used to solve a variety of problems.

The paper briefly relates our experiences with backfilling, spraying and strata replacement for support on coal faces and roadways, with shotcrete used as a sealant for spontaneous combustion prevention, with the filling of large cavities and the reduction of resistance to airflow along roadways.

The Appendix describes briefly the equipment and materials used.

INTRODUCTION

Under this heading cement products have been satisfactorily used with associated equipment, for support, air sealing, reducing resistance to air currents and strata replacements underground. This is summarised in the body of the paper. Technical detail is appended.

I. EXPERIENCE WITH CEMENT PRODUCTS IN ROADWAYS

1.1. Support

A site for underground use of grout filled envelopes was chosen at Thoresby Colliery in the North Nottinghamshire Area of the National Coal Board as an added means of supporting an intake roadway to a working district. (95's District, Top Hard Seam). Shotcrete had already been chosen as the method for sealing the same length of roadway for spontaneous combustion prevention purposes.

It was, therefore, decided to combine the two principles to provide an effective support in a length of roadway which, on account of its position in relation to faults and the face start position, was expected to be subject to extremely heavy loading. Under ordinary support, the roadway would require re-ripping with consequent destruction of the spontaneous combustion seal at the original face start line, as well as being costly in manshifts.

It was agreed that the length of roadway should be supported by arches with grout filled envelopes and mesh between the web of the arches to increase their resistance to buckling, after which shotcrete would be applied.

The equipment used to prove the principle was at the time mainly prototype and subsequent developments have produced machinery and materials which have been proved in practice to be much more efficient and suitable. These items and materials are described in Appendix I.

The results of the experiment are shown in the following slides:

1. Roadside with envelopes in place prior to shotcrete. (Fig. 1).
2. Apply shotcrete. (First application).
3. Finished roadway. (At conveyor drive).
4. Finished roadway. (Inbye conveyor drive). (Fig. 2).
5. Untreated roadway. (Adjacent to treated roadway before panel started advancing).

The method has not been used subsequently in North Nottinghamshire but would certainly be effective if extreme loading conditions were unavoidable and, as a follow up to this work, it is planned for a 50 metre length of roadway to be supported by a combination of circular H section rings with shotcrete sprayed between the webs in a district which is subject to extreme strata pressure at Warsop Colliery, North Derbyshire Area.

1.2 Sealant

In 1978 a deep seated, active spontaneous heating occurred at Harworth Colliery in the North Nottinghamshire Area and the speed of the Carbon Monoxide build up was such that the slow method being used to seal the roadway sides by boarding between the arch flanges and pumping 'Hardstem' behind the boards was proving ineffective (25 metres/shift only being achieved due to disturbed arches).

As a last resort, it was decided to use a dry process 'Shotcrete' machine to seal the roadway and 180m were sealed over the bank side pack area of the roadway in one shift. By the end of that shift the carbon monoxide readings were reducing and the whole length of suspect packside was sealed subsequently, after which the carbon monoxide content returned to normal.

Having proved the system, again with prototype equipment, it was decided that by improving the equipment and experimenting with different materials an efficient system could be established for combating active heatings quickly or, in more severe situations, using it continuously for spontaneous combustion prevention. The equipment is as described in Appendix I.

New equipment has been developed to enable shotcreting machines to be used at sites which do not have a mains compressed air supply. New materials have also been developed to do a more efficient job.

So far, 33 suspect sites have been treated for sealing purposes, none of which have resulted in heatings, and 5 active heatings at untreated sites have been combated in a much more efficient manner than with any previous methods used. To illustrate its use in a high risk spontaneous heating situation, the system in use at Welbeck Colliery of North Nottinghamshire Area is described.

1.3 Spontaneous Combustion Precautions taken at Welbeck Colliery in 102's District Roadways in the Top Hard Seam

Face Start

The face was developed, the method of roof and side support being 4.87m x 3.66m. Hollybank arches with straight girder roof beams, lagged only on the coal rib. The face was advanced sufficiently to allow the face side arch legs to be replaced under the first 6 arches set at the entrance to each end of the face. These entrances were sealed off by pumping liquid cement by a Varimix placing machine between 2 wood wool retaining walls, 4 metres between the walls at the return end of the face, and 6 metres at the Loader Gate.

Roof coal was being left and because of air leakage through this, it was decided to try lining the road side pack of the intake and return roadways with a coating of sprayed concrete, (shotcrete), in order to effectively seal any air paths. (Fig. 3).

Steel lagging was used behind the arches except for the area 250mm above and below the seam roof level which was lined with small panels of 76.2m x 152.4mm x 5.38 diameter weldmesh.

The coating of concrete mix adhered to the steel lagging and the steel meshed area sealed quite easily with an application of 'Supaspray' or 'Bucshot'.

Spraying was carried out as near to the face as practicable. For the first 50 to 60 metres advance of the face it created problems when spraying too close to the working face due to the amount of equipment in the gate. It was therefore decided to allow the face to advance some 80 metres in the intake gate and the coating to take place immediately behind the electric switches on the pantechnicon.

In the return gate shotcreting was carried out some 40 metres behind the face advance, as there was less equipment. Spraying is carried out on the one shift only, 3 men spraying with a fully compressed air machine. One machine in each gate. The sprayed area is checked daily and a full examination weekly. Should any lining be found not to be airtight the area is resprayed. (Fig. 4).

For any extended holiday period the sprayed area is linked up to the face entry in both gates.

Observations and Recommendations

1. Compressed air to be available for spray machines that are used

permanently, and not worked in conjunction with a Blower Unit.

2. Machine team to have access to protective clothing, airstream helmet, etc.

3. Instructions must be given to the machine operators on correct machine maintenance by the machine manufacturer; this is very important, as any excessive wear or incorrect operations produces an airborne dust problem.

No problems have been encountered in these very high risk situations even though a major fault has passed through the face. The principle has been adopted at other sites.

1.4 Cavity Filling

Major cavities have been of concern for many years as their filling has been a long, inefficient and messy job. Different products have been used for this purpose over the years with varying success.

The National Coal Board Research Establishment have, however, in conjunction with a cement manufacturer and a pump manufacturer, succeeded in providing a system whereby a material called 'AQUALIGHT' (which has an expansion ratio of 14:1, a strength of 0.1 MPa within 10 minutes of pumping and the advantage of not being a free flowing material (i.e. it will not seep out from behind normal arch covering) can be pumped into cavities very effectively and efficiently.

A cavity of some 90 cubic metres was left over a loader gate of a working panel after a major fall.

After the forward ground had been re-established 4 - 50mm diameter pipes were forced through the steel sheet covering to within 1 metre of the top of the cavity. The 'AQUALIGHT' was pumped through the pipes in turn along the length of roadway (15 metres) and the material pumped into the cavity (using a modified hardstem pump). (Fig. 5).

318 x 25 kg. bags of 'AQUALIGHT' were pumped into the cavity which was then full - giving back pressure at the pump (4 x 25 kg. bags = 1 cubic metre).

No material at all came through the covering into the roadway. The equipment and material has been used extensively and is described in Appendix I.

1.5 Re-ripping in Running Ground

The Deep Soft Seam Return Drivage at Welbeck Colliery, 4.87m x 3.66m in size and 2,200 metres in length, had serious weighting problems at the 538 metre mark for a length of 100 metres. A re-ripping programme was drawn up but on withdrawing the first arches the ground was so broken the stone ran like sand giving a filling problem and a resultant cavity over the arches.

It was decided to shotcrete the full periphery of the roadway in its crushed state first having pushed 50mm diameter injection pipes through to the cavity and the loose strata. (Fig. 6).

The concrete was allowed to set, then 'POZAMENT' pumped over the top of the arches through the 50mm diameter pipes to the required depth, allowed to set then the arches withdrawn and replaced under a consolidated cement stone roof. (Fig. 7).

Equipment and material are described in Appendix I.

1.6 Reducing Air Resistance of Roadways

The slides illustrate the roadway after shotcreting and shows similar application at a junction.

A permanent 'whitening' is achieved by adding 'SNOWCEM' at a ratio of 1:10 to the shotcrete mix into the machine. Equipment and material are described in Appendix I.

1.7 Face Recovery by Strata Replacement

A major fall some 80 metres long and 15 metres high occurred on the face line of a coal producing unit at Blidworth Colliery, North Nottinghamshire Area. The recovery of the face was attempted by:-

(a) Long and short hole wood dowelling

(b) Resin injection under pressure

(c) Carefully 'bleeding' the stone and forepoling R.S.J. forward.

These methods were not successful.

Strata replacement techniques had been used at a colliery in Doncaster Area and it was decided to utilise this method to recover the face, using 'Packbind'. 'Packbind' is a cement product combining quick setting properties with adequate strength, yet capable of being cut by a power loader.

The principle used was to wooden-shutter the coal face in such a way that the cement would form a new roof using the loose stones over the chock to help form an aggregate and to build up a new 'coal/cement face' for the shearer to cut through and establish a new roof to set the advancing chocks to providing a new environment. (Fig. 8). The shearer successfully cut through the cement face, established a new roof which the chocks (on reduced pressure) were set to, and continued to advance normally. (Fig. 9).

This system is a successful means of re-establishing extensive fall areas on producing units and has been used again in a less severe area at Bevercotes Colliery. AQUALIGHT has since been used in a similar application with success and a reduced transport and pumping problem due to its expansion ratio. The equipment and technical detail is described in Appendix I.

1.8 Lining of Circular Roads with Concrete Segments

The use of circular concrete segments as a means of roadway support is new to U.K. coal mines but is likely to become more widespread in long life tunnels or those passing through heavily loaded ground. Grouts are being developed to fill the cavity between the segments and the strata which will allow some movement of the strata to take place initially without damage to the concrete lining.

1.9 Back filling with Aqualight

This has been successfully achieved at Coventry Colliery in the South Midlands Area to infill any overbreak of surrounding strata.

At Ollerton Colliery in North Nottinghamshire Area, a Markham (Chesterfield) Roadhead Shield Support has been used in a solid drivage (Fig. 10), one problem was backfilling of the arch supports and steel covering after the shield advances, leaving a space around the arch of approximately 15 cms.

Trials have been carried out with Aqualight to back fill this space, prior to introducing a similar machine to the return gate ripping of a production panel, and it appears the solution to the problem. (Fig. 11).

APPENDIX I

EXPERIENCE WITH CEMENT PRODUCTS IN NORTH NOTTINGHAMSHIRE AREA OF THE NATIONAL COAL BOARD

The items of equipment are briefly described in Part I and the branded products are listed and briefly described in Part II.

PART I

Shotcrete Machine

Two types are in use currently the Markham 'E' machine (Fig. 10) and the D.P. Engineering '343' Shotcrete Machine.

Both machines require 14 cubic metres/minute at a pressure of 4 bar to pneumatically project the dry concrete mix through the hose system and to power the material feed motor. (If the feed motor is electrically powered the air requirement is 10 cubic metres/minute at a pressure of $1\frac{1}{2}$ bar).

The 'Drum' Blower

A blower unit capable of delivering 10 cubic metres/minute at $1\frac{1}{2}$ bar for the electric driven feed motor machine has been approved for underground use - the M.D.400. (Fig. 11).

Driven by a 40 h.p. flameproof electric motor (optional diesel or hydraulic drive) the machine is oil free, the only regular maintenance being attention to 2 grease points. In addition to the full flow air relief valve two other pre-set sensing devices are used to shut down the unit should either temperature or pressure exceed the pre-determined limits.

This machine makes the 'shotcrete' package reasonably portable by replacing a bulky compressor with its associated problems by a 2,600mm long x 900 mm wide x 945mm high 'blower' unit.

Grout Pump

Two types are in use currently the Varimix Pump (British Gypsum Ltd.) and the Hunt Pump (Langley Mining Engineering).

The mix is fed into the hopper at the top of the unit from which it feeds at a constant rate into the mixing chamber by means of an auger. Water is added in the mixing chamber at a controlled rate, the mixture then being pumped by a Mono pump unit mounted on the mixing chamber outlet. This pump also delivers 'packbind' for strata replacement and is slightly modified to deliver 'AQUALIGHT' for cavity filling or strata replacement.

Grout Envelopes

Bidim FP U34 non woven polyester treated with Mobil Anti blaze 19 (Butyl Products Limited) and Bondina PP16-39A non woven 100% 2.8 denier thermouyl PVC Fibre with acrylate bonding (Bondina Ltd.).

Concrete Reinforcement - Weldmesh

Weldmesh panels to reinforce sprayed concrete by BRC Engineering Ltd. 25.4mm x 25.4mm x 1.63mm diameter weldmesh.

PART II

Hardstop/Hardstem

Quick setting 'Gypsum' products. (British Gypsum Ltd.).

Shotcrete Mixes

Two products are used, one 'fine' and one 'course'. 'Fine' is a mixture of 3 parts zone 3 sand by weight to 1 part ordinary cement with a setting accelerator. (Pozament Cement Ltd.). 'Course' is a mixture of 3 parts zone 2 sand by weight to 1 part ordinary cement with no setting accelerator. (B.F.C. Packed Products Ltd.).

Grout for Envelopes

Equal parts ordinary cement and pulverised fly ash with 5% Long Chain Polymer to aid pumpability and 5% High Alumina Cement. 28 day strength is 15H/mm^2. (TGT/BB by Pozament Cement Ltd.).

Packbind

Specially formulated to produce quick setting properties with adequate strength for support (0.5 H/mm^2) yet easily cut by a power loading machine. (Blue Circle Cement Ltd.).

Aqualight

Used for cavity filling or strata replacement.

A specially formulated cement product which is admixed with water, the resultant 'slurry' being pumped through 60m of hose to the placing site. During passage through this hose the entrained air aided by a foaming agent disperses throughout the mixture in fine bubble form. The product then emerges from the hose in the form of an aerated 'gel' which can be placed accurately and easily.

The 'gel' is capable of bridging a gap of at least 50mm between roadway lining sheets and is therefore 'clean' in use, there being no waste or hazard from splashing.

The aeration gives rise to an expansion ratio of up to 15 times the original volume of the 'slurry'.

Expansion ratio of 15:1 gives a support strength of 0.1 MPa.

Expansion ratio of 7:1 gives a support strength of 0.7 MPa.

The expansion ratio is controlled simply by the water quantities.

Figure 1 : Grout envelopes in place prior to shotcrete

Figure 2 : Finished roadway inbye conveyor drive

Figure 3 : Sealing roadside pack with shotcrete

Figure 4 : Sealing near to face line for holiday period (Note reasonable environment)

Figure 5 : 'Aqualight' filling a cavity

Figure 6 : Shotcreted roadway (injection pipes in position)

Figure 7 : Consolidated cement/stone roof over arches

Figure 8 : Shearer establishing a new roof, to set the self advancing supports to

Figure 9 : Re-established coal face after first cut

Figure 10

Figure 11

Figure 12: Markham 'E' shotcrete machine

Figur 13: M.D.400 air blower unit

LININGS FOR CIRCULAR ROADWAYS

H. van Duyse

Head of Research Section
Institut National des Industries Extractives

Summary

In roadways driven by blasting, the use of circular linings is mainly limited to the Campine Coalfield's stone drifts where the use of prefabricated concrete segments has met with a great deal of success, in terms of both strength and cost. One important problem - to which we are currently seeking the most satisfactory solution - is the best method of filling the cavity between the strata and the lining. In the Campine Coalfield the cost of driving a circular roadway, 4.60 m in diameter, in virgin rock, is identical with that of driving a roadway lined with yielding arch supports, of cross-section 17 m^2, in virgin rock. Then again, in roadways driven by tunnelling machine a circular lining is virtually the norm. In the German collieries these roadways are lined with steel arches which take 15 to 30 min. to erect. The presence of radially-anchored gripper shoes means that at the roadhead there has to be a space of at least 75 cm between arches, which in the least favourable strata conditions requires the support to be reinforced at the roadhead either by means of strata-bolting or injection, or behind the tunnelling machine by means of extra arches or by guniting. Conditions specific to the Campine Coalfield necessitate the use of circular panel linings installed close to the roadhead, which in turn raises the crucial question of accommodating the thrust exerted at the roadhead. We chose the Bouygues tunnelling machine with a limited thrust of 120 t. At Cadley Hill colliery in Great Britain circular supports made of prefabricated concrete segments are installed at the rear end of a tunnelling machine shield.

Various support techniques can be used in circular roadways:

- yielding or rigid circular steel arches;
- prefabricated sections made of reinforced or non-reinforced concrete;
- concrete poured in situ directly behind the lining;
- shotcrete, timber units; and
- strata-bolting.

Most of these support techniques provide a continuous lining which reduces pressure-drops caused by friction of the airflow against the roadway walls.

Until only a few years ago these circular roadways, like all other roadways, were driven by means of blasting. Today, a large proportion of circular roadways are driven by tunnelling machines. With this heading technique a circular roadway cross-section is almost always the norm.

1. ADVANTAGES OF CIRCULAR LININGS FROM THE STRATA CONTROL POINT OF VIEW

It is undeniable that roadways with a circular or an elliptical cross-section offer better resistance to strata pressure than those which are horse-shoe shaped or trapezoidal. Circular supports are the only type to offer some resistance to floor-heave in the roadway.

The primary advantage of circular supports lies in their great resistance to strata pressure; for example, a circular roadway, internal diameter 4.10 m, lined with blocks 50 cm wide of non-reinforced concrete can withstand prior to breakage a load of 5250 t/m of roadway, when pressure is applied to one half of the circumference and 1600 t/m of roadway when pressure is applied to one quarter of the circumference.

In the case of conventional-type supports, which are open at the base, this breaking load is well below 200 tonnes per metre.

It should, however, be stressed that if in the case of high-pressure due to winning operations or the presence of faults, substantial damage is caused to these circular supports, maintenance costs are much higher than with supports which are open at the base. Thus in the case of circular roadways lined with ferro-concrete panels, the cost of back ripping operations is 2 to 3 times higher than the cost of heading an ordinary cross-section roadway. On the other hand if in both these instances we had used supports which were open at the base the number of back ripping operations necessary to keep the roadway open would also have been prohibitive.

If we are to dispense completely with back ripping operations, the strength of the circular supports must be further improved:

- by better infilling of the cavity between lining and strata;
- by making the strata surrounding the roadway react more closely to the resistance offered by the support;
- by adapting, where possible, mining plans to avoid all prohibitive high-pressure areas in roadways.

Since strata pressure is not the theme of this symposium, we will not expound further on the major advantages gleaned by the use of circular linings; there is currently a growing interest in this type of support in European mining industries, as the depth of underground workings increases.

It should, however, be pointed out that driving circular roadways lined with prefabricated concrete sections does not necessarily work out more expensive than driving roadways of large cross-section lined with conventional arch-type supports. It is, therefore, not necessary to restrict the use of circular linings to instances where substantial strata pressure is predicted.

2. HEADING CIRCULAR ROADWAYS BY BLASTING

When driving a roadway by conventional blasting techniques, the choice of circular (or elliptical) supports produces, by its very nature, two major drawbacks:

- the need to load and remove the spoil lying below haulage level; and

- stowing of the floor section to produce a haulage surface of sufficient width, which naturally reduces the finished cross-section of the roadway.

2.1 Lining the roadways with prefabricated concrete sections as used in the Campine coalfield

On account of the substantial strata pressure which occurs in the Campine collieries (partially due to the presence of 500 to 600 metres of sandy, aquiferous overburden) the majority of stone drifts in this Coalfield are circular and lined with prefabricated sections of concrete.

From 1935 to 1970 these linings consisted of non-reinforced concrete blocks 50 to 56 cm thick, weighing 80 to 120 kg, which entailed the handling and positioning of some 138 to 236 blocks per metre of roadway. With sound organisation and semi-mechanised operations, mining engineers managed to achieve advances of 2 m/day with a three-shift system and 4 men per shift.

Figure 1 shows how the concrete blocks are positioned: those of the floor section by means of a monorail, the side blocks by means of a manoeuvrable mast and those of the crown section by means of a hoist and a steel framework. It takes 4 hours 36 mins. to set this type of support in a roadway with an internal diameter of 4.50 m for a 2 m run.

To this total must be added the time taken for temporary timbering at the crown, which is carried out whilst the spoil is being loaded out.

From 1967 onwards the concrete blocks were gradually replaced by reinforced concrete panels 20 cm thick. This limited the number of units requiring positioning (each weighing 1 t) to 7 per metre of roadway (Figure 2). Compressible wood-fibre inserts 4-8 cm thick are placed between the 5 panels which make up a complete ring.

Using this new type of lining has made it possible to achieve advance rates of 4-6 m/day in 3 shifts with 4 men/shift.

Panel-setting operations

The panel-setting machines consist of a basic chassis to which is fixed a platform on which is mounted a telescopic boom for panel-handling. The base may be either a track-mounted loader chassis or a crawler-mounted gathering arm loader chassis (Figure 3). In the latter instance a sixth panel is placed on top of the panels of the floor section in order to provide a horizontal haulage surface and prevent wear on the crawler axes.

The end of the telescopic boom is fitted with a knuckle-joint whose internal spherical concave surface is designed to secure the spherical screwhead of the metal rod for panel-handling (Figure 3). Each panel has a central hole passing right through it; this hole is lined with plastic tubing terminating in a screwnut on the outer side of the panel. This screwnut forms the basis of the panel-handling system: the units are positioned by means of a metal rod which is threaded at one end to screw home onto the screwnut and has a spherical screwhead at the other. The centre of this screwhead is strategically positioned at exactly the centre of gravity of each panel so that when the screwhead is secured in the knuckle-joint with the same internal diameter the panel can then be placed effortlessly in any position required.

The end of the telescopic boom can reach any position required for any respective handling operations on account of the following features:

- the platform can rotate through 360° on the chassis;
- the boom can be raised and lowered in the vertical plane; and
- the easy sliding movement of the boom's internal telescopic section.

Panel-setting operations take 55-90 mins. for a 1.33 m run and approx. 2 hours for a 2 m run. Temporary timbering at the crown of the roadway is again carried out whilst the spoil is being loaded out, (this takes 2 men approx. 30 mins.).

Table I gives details of the various operations involved in panel-setting with the two types of setting machine. The heading run is 1.33 m in the case of the track-mounted machine and 2 m in the case of the crawler-mounted machine.

	track-mounted machine 1.33 m run in min/m/s (4 men)		crawler-mounted machine 1.33 m run in min/m/s	2 m run (5 men)
	Pit A	Pit B		
Transport (inserts, timber, planks, sacks)	58.9	38	53.5	53
Bringing setting-machine & panels to heading area	70.4	17	71	192
Positioning & setting of panels	123.1	152	199.5	224
Assembly of floor area	24.8	15	18	40
Final dressing of panels	21.8	7		44
	299.0	229	342	553

Table I: Details of support-setting operations.

Taking into account the men employed on heading operations, actual panel-setting operations plus auxiliary tasks take 57 to 90 mins. in the case of the track-mounted machine and 68 to 111 mins. in the case of the crawler-mounted machine, which give respective panel-setting times of 43 to 68 mins. per metre of roadway in the case of machine type I and 51 to 55 mins. in the case of machine type II. Table II shows the average panel-setting times for individual panels in the 2 or 3 rings which make up the respective 1.33 and 2 m runs.

Strictly speaking panel-ring-setting takes 27 mins. per metre of roadway in the first case and 30 mins. per metre of roadway with the second type of machine.

Infilling of the cavity between support and strata

However, sound infilling of the cavity between support

and strata constitutes the decisive factor in determining whether this 20 cm-thick support can withstand high strata pressure.

The infilling may be carried out in one or more stages either at, or behind, the roadhead. The volume of the cavity can be relatively large and on average amounts to 3 to 4 m^3/m. Tests are continually under way to determine the best infilling products and to develop the most effective infilling techniques.

Operation	Type of panels	Track-mounted machine - 1.33 m run Pit B		Crawler-mounted machine 2 m run	
		in sec	in min	in sec	in min
Floor-section panels	large	245		164	
	small	235		108	
	large	250		241	
	special			89	
	small	265		135	
	special			93	
	large			176	
	small			206	
	special			98	
total		995	16.58	1310	21.87
Side panels		155		144	
				157	
		255		146	
		160		190	
		150		176	
				325	
total		720	12.00	1211	20.18
Crown panels		135		419	
		375		299	
				367	
		510	8.50	1085	18.08
General total		2225	36.08	3606	60.13

Table II. Time involved in panel-setting operations.

The following products have been tried out to date:

Gravel, lightweight aggregates, mortar, fly ash, concrete,

synthetic anhydrite, limestone/cement and limestone/synthetic anhydrite. The aim is for these products to remain in their plastic state long enough to adapt to - but not crack under - the initial deformation of the rocks surrounding the roadway.

The product must be capable of being transported in bulk in order to reduce the laborious aspects of the task, increase the efficiency of the operation and achieve as sound an infilling as possible.

It may take one or more stages to inject the various products, using a variety of different pumps such as the Putzmeister USI 139, the Netsch monopump, etc.

In the early days the heading team stowed lightweight aggregates at the start of each heading run, whilst 100 m further back an auxiliary team completed the infilling by mortar injection.

Nowadays it is deemed preferable to infill the cavities closer to the roadhead in 20 m stretches; injection operations which are the responsibility of the heading team are carried out in such a way that the flow of injected matter solidifies at, or near, the actual roadhead.

Prior to this the vertical joints between panel rings are sealed with polyurethane-based foam.

These infilling operations represent a substantial proportion of the conventional roadway drivage costs. Thus, if the panel support costs 15 870 BF/m and panel setting operations 4400 BF/m, the cost of the current method of infilling (using bags of synthetic anhydrite) is 12 835 BF/m (of which 9300 BF/m represents the products and 3535 BF/m the wages costs).

The aim of current tests is to drastically reduce the infilling costs: it is hoped to cut the cost to less than 8000 BF/m by means of a new product which can be delivered and transported in bulk.

So that this lining method can be compared with others,

Table III shows the average cost per metre of ferro-concrete panel-lined roadway (where the average advance rate is 50 m/month) with an internal diameter of 4.20 m (prices as at 1 January 1983).

By way of comparison we should point out that the cost of driving a roadway lined with the usual type of TH arches is:
66 420 BF/m in the case of a roadway of cross-section 12 m^2 driven in virgin rock and
81 000 BF/m in the case of a roadway of cross-section 17 m^2 driven in virgin rock.

	Cost price in BF -	Total
1. Wages - at the roadhead - for filling operations - supervision - auxiliary tasks	 23 580 3 535 9 960 6 510	
Sub-total		43 585
2. Support equipment - panels - inserts - timber - synthetic anhydrite - sealing of joints between rings	 15 600 270 920 9 000 300	
Sub-total		26 090
3. Equipment for drilling, shotfiring	8 876	8 876
4. Plant hire - for supports - for filling operations - other equipment (loaders, drilling machines, ventilation ducts, etc.)	 900 138 1 411	2 449
TOTAL		81 000

Table III. Cost of a panel-lined roadway.

2.2 Circular steel arch supports

At present in Belgium and Germany circular steel arch supports are no longer being used in roadways driven by means blasting; some applications are planned in Great Britain.

Compared with reinforced concrete panel linings, steel arch supports:

- give a slightly slower rate of advance;
- offer a distinctly lower breaking strength (when the load is applied to one quarter of the circumference the breaking strength of 2 x 36 kg/m steel arches is 320 t/m of roadway as opposed to 900 t/m in the case of a reinforced concrete panel lining);
- work out more expensive to buy (the cost price of driving 2 x 36 kg/m steel arches, 4.80 m in diameter, in virgin rock is 48 000 BF (not including lagging) whilst it costs 15 900 BF per metre of panel-lined roadway not including the infilling of the cavity between strata and panels);
- when using steel arches lagging is often a difficult business (especially when the roadway has been driven by blasting); and
- the steel arch supports do not offer a smooth roadway-wall surface for the airflow to travel along.

2.3 Non-circular, but completely closed steel support units

In order to eliminate all risk of floor-heave in certain roadways some collieries in the Ruhr use the conventional arch-type supports at the actual roadhead and further back (40 to 150 m behind the roadhead) support units are sunk into the floor section to provide a non-circular, but completely closed unit. The two ends of these floor units can slide along the lateral uprights.

3. CIRCULAR ROADWAY DRIVAGE BY TUNNELLING MACHINE

Whilst steel arches are normally used to line circular roadways driven by tunnelling machine, Cadley Hill colliery in Great Britain has carried out mechanised heading operations with prefabricated concrete sections and it is also proposed to conduct a second trial with this type of lining in the Campine coalfield.

One of the distinct advantages of using a tunnelling machine for roadway drivage is that it gives improved support behaviour and on account of the accurate profile it cuts into the rock:

- the support can be braced against the strata;
- continuous contact between lining and rock can be achieved; and
- no fissuring occurs in the surrounding strata.

3.1 Circular steel supports

Where the roadways are mechanically driven in the German coalfields (using Robbins, Demag and Wirth tunnelling machines) circular steel arches, either rigid or yielding, weighing between 34 and 44 kg/m are the usual type of support. Some designs of tunnelling machine enable the last arch to be set less than 1.75 m from the actual roadhead. During support-setting operations, a short shield or 'cap' is fixed to the tunnelling machine in order to protect the crown of the roadway.

It takes 15 to 20 mins. to install a rigid arch, complete with lagging, and from 20 to 30 mins. to install a yielding arch; these operations require 3 or 4 men per shift (not accounting essential personnel responsible for transportation of the arches). The rigid arches have to be assembled in a special way which allows them space to yield a little (often restricted to 20 cm).

These arches constitute an 'active' form of support as they are braced against the rock under pressure from jacks.

In slaty strata, setting the support can take longer than driving the actual roadway (which can be achieved in under 11 mins. per run). A recent setting device makes it possible to store the yielding arch already assembled; once there is adequate clearance the arch is moved forward to its final position where it is extended and tensioned.

As all tunnelling machines used in Germany are anchored radially against the roadway wall (case a) Figure 4) the distance between two arches has to be greater than the width of the anchor-plates. In adverse strata conditions an increase in support density can only be achieved by installing additional arches once the back anchor-plates have been moved forwards, which requires extra men.

In poor strata conditions the area between arches is frequantly sprayed with shotcrete; this is always carried out behind the machine; occasionally too, the strata around the tunnel is reinforced by injections of cement or magnesian binder, to a depth of 6 m, under pressures of up to 60 bar.

Another method of additional reinforcement is to insert strata bolts very near the roadhead, using drill slides attached to the tunnelling machine. At Lohberg colliery, 4 slides are used to insert 12 anchor bolts, 2.80 m long, in the upper quarter of the roadway (these bolts are installed at a distance of 6 to 20 m behind the roadhead).

We should point out that in the case of roadways lined with yielding arches, instances of convergence of up to several centimetres are often reported, a short distance behind the roadhead. At Lohberg colliery, the convergence ranges from between 10 and 20 cms and 3.50 m from the roadhead - i.e. at the point where the arches are being set.

3.2 Circular supports made up of prefabricated concrete segments

Where prefabricated concrete segments are used to form a circular support, the problem arises of accommodating the thrust exerted by the roller bits against the roadhead.

The maximum thrust exerted against the roadhead by the cutting head, equipped with roller bits, of the Demag, Robbins, Wirth, etc. tunnelling machines can be as great as 1000 t; if this thrust has to be accommodated by the radial anchorage force of the gripper shoes (Figure 4a) it must be as great as 2400 t. In the case where the shoes bear against the strata between 2 arches, no problem occurs, except where the strata is very weak and the shoes sink in to a large extent.

Where the prefabricated concrete segments are installed only a short distance away from the roadhead, this reinforced concrete support would not be sufficient to withstand major strata reactions, especially in the case of localized cavities caused for instance by movements of coal or soft shale.

For the purpose of illustrating this problem, we have placed the gripper shoe of a tunnelling machine against the inner side of a single panel and simulated cavities of increasing dimension (Figure 5) between the testing frame and the supports against which the outside of the panel rests.

Table IV shows the widths of these cavities plus the thickness of the wooden blocks inserted between the steel girders and the 20 cm thick panel; the Table also provides the loads at which initial cracking occurred.

Test c represents the rare instance in which the bearing surfaces at either side of a small cavity consist of beds of very strong sandstone. In the other cases slightly compressible wooden blocks are used in the simulation.

Test	Free space between supports of testing frame - in m	Thickness of wooden blocks in mm	load at which initial cracking occurs in t
a	0	45	40
b	500	52	70
c	500	4	270
d	920	60	25
e	1000	28	50
f	1000	28	40
g	1190	30	30

Table IV. Initial cracking load.

These tests show that in carboniferous strata substantial radial stress cannot be applied to prefabricated concrete supports which are installed less than 2 m away from the roadhead.

Figure 4 shows several possible solutions: case b) represents a frequently-selected instance where the forward thrust is accommodated by a large number of rams arranged tangentially to the roadway pushing against the forward edge of the panels. Tests carried out on Campine panels (Figure 6) show that it is possible, without causing the concrete to crack, to apply pressure around the ring's circumference from 30 x 35 t rams, each ram having a bearing surface of 40 cm.

Solution adopted in the Campine coalfield (Belgium)

Another possible solution is to select a tunnelling machine which only exerts a limited thrust against the roadhead. The Campine collieries opted for this solution in choosing the Bouygues tunnelling machine; this machine has only 4 roller bits which limits the maximum thrust at the roadhead to 120 t and the radial thrust exerted on the support to only 300 t. This radial load is accommodated by 10 gripper shoes, each with a maximum load-bearing capacity of 50 t; 6 of these shoes have a bearing surface of 50 x 50 cm^2, angled downwards at 45°, and the other 4 have a bearing surface of 50 x 80 cm^2 which is angled slightly sidewards along the axis of the tunnel.

The support is made up of panels identical with those described earlier in the paper, the only change being an increase in their thickness, from 20 to 30 cm.

Each of the 5 panels which form the ring is positioned by means of a radial ram mounted on the cutting head which can occupy any position desired in the vertical plane. In order to insert the last panel (at the crown) the cutting head is advanced at the roadhead and then retracted, leaving the panel in position.

Solution adopted at Cadley Hill colliery (Great Britain)

The solution adopted by Cadley Hill colliery most closely resembles case (g) in Figure 4.

Figure 7 shows the 2-piece shield with its two longitudinal slots which enables the upper half of the shield to be braced against the roof strata whilst the lower half is pressed against the ground by means of a set of vertical rams. This type of machine complete with shield is the first of its kind to be used at great depth underground. The shield is 6.70 m long.

The cutter unit which appears in Figure 7 is attached to the shield by means of a series of rams.

The thrust exerted at the roadhead is accommodated by:

- 10 horizontal rams pushing against the forward edge of the lower 150° of the prefabricated concrete support. The maximum thrust exerted by all rams together is 750 t; the maximum load exerted against the concrete of the panels is 120 kg/cm^2.
- 6 rams pressing the upper half of the shield against the strata; this vertical thrust can be as much as 800 t with a maximum load against the rock ranging from 0.5 to 2.7 kg/cm^2 (125 mm play).

The support, made up of 7 or 21 concrete panels, is positioned at the rear section of the shield; the panel-setting machine is mounted on a revolving cylinder which can reach any point in the ring. The panels which form the ring are each bolted to one another. In the case of a support made up of 21 panels, the latter are pre-assembled in groups of three. An additional key piece is placed first at the base of the ring. Manoeuvrable erector arms and the rams attached to the shield hold the upper panels in place whilst the ring is waiting to be completed. The panels are 25 cm thick, 1 metre wide and weigh 1.25 t. The inside diameter is 4.925 m and the gap between the strata and the roof of the ring is 110 mm.

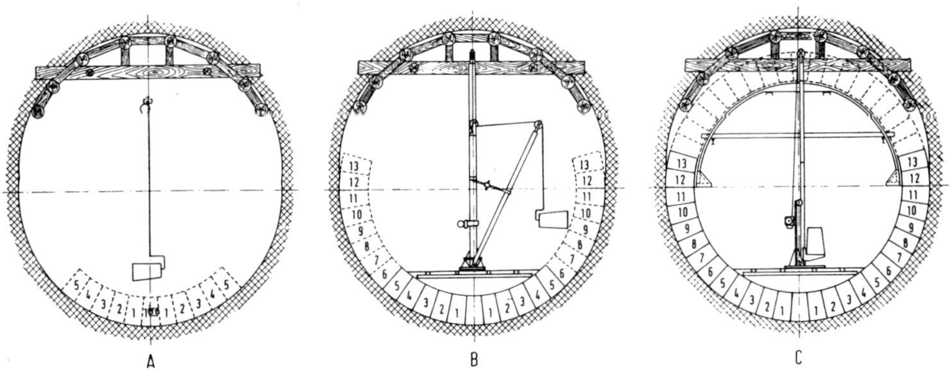

a those of the floor section, by means of a hook fixed to a monorail
b the side blocks, by means of a manoeuvrable mast
c those at the crown by means of a hoist which lifts and installs them above a steel framework
Figure 1. General view of how the concrete blocks are positioned.

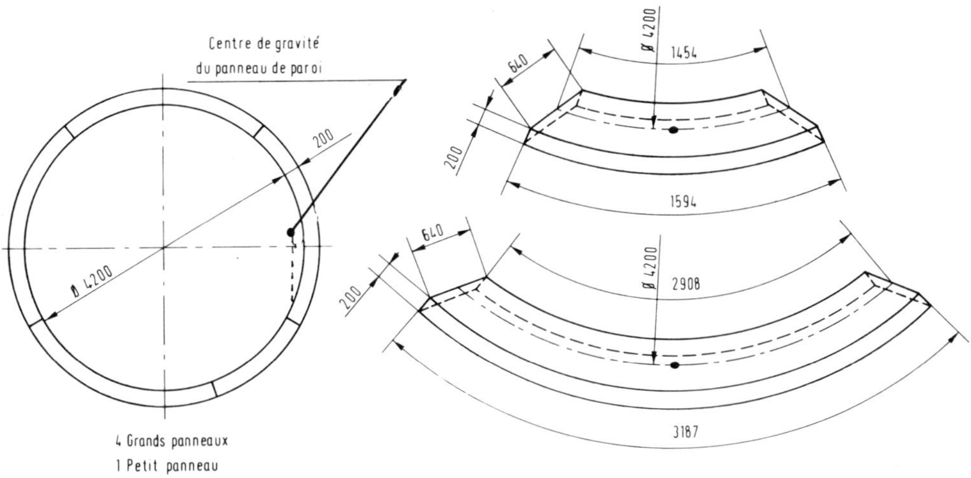

Centre de gravité du panneau de paroi	centre of gravity of side panel
4 grand panneaux	4 large panels
1 petit panneau	1 small panel

Figure 2. Circular roadway lining made up of 4 large (weight 1 t) and 1 small reinforced concrete panels. It can be seen from the diagram that the vertical projection of the centre of gravity of each of the 2 side panels does not fall on the inside of its bearing surface onto the floor-section panel, which ensures stability and eliminates the need for support girders.

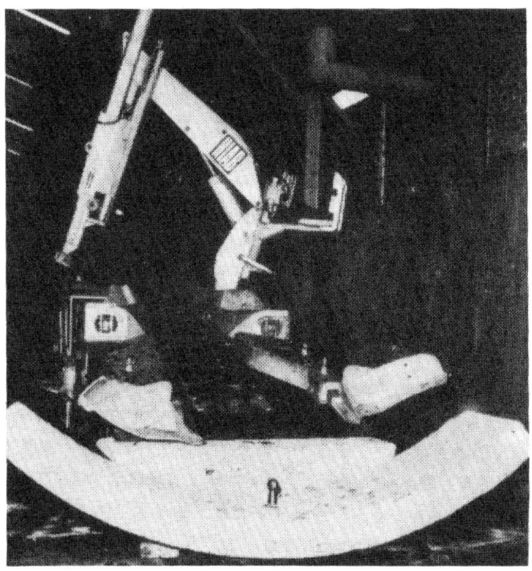

Figure 3. Equipment designed by Zolder colliery with a panel-handling boom, mounted on a 2 PNB 2 gathering arm loader chassis. In the foreground can be seen a panel into which the metal rod for panel-handling, terminating in a spherical screw-head, has been inserted.

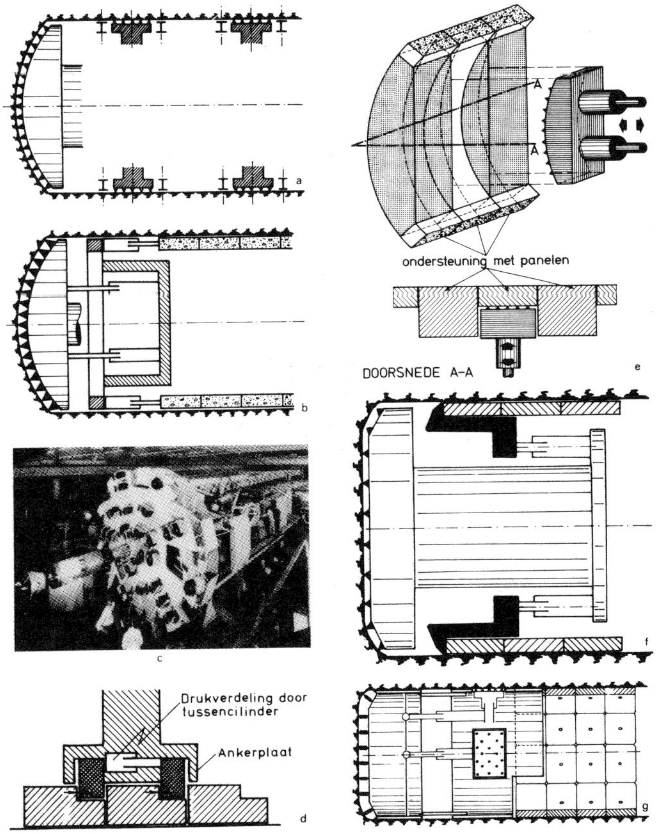

a the system used in the German collieries: the radially-arranged gripper shoes which bear against the strata between two steel arches
b tangential reactions by means of a large number of rams arranged all around the front edge of the panels
c the Lawrence solution where the machine is anchored against the sides of a hole drilled ahead of the tunnelling machine's drum
d tangential reactions accommodated by grooves in the panels
e tangential and radial reactions also accommodated by special panels (concrete sectors)
f tangential reaction by the gripper shoes pushing against the front edge of the panels
g radial reactions by the shoes passing through openings provided in tunnelling machines with shield
Figure 4. Various systems of anchoring tunnelling machines, which are essential for the purpose of accommodating the thrust exerted at the roadhead.

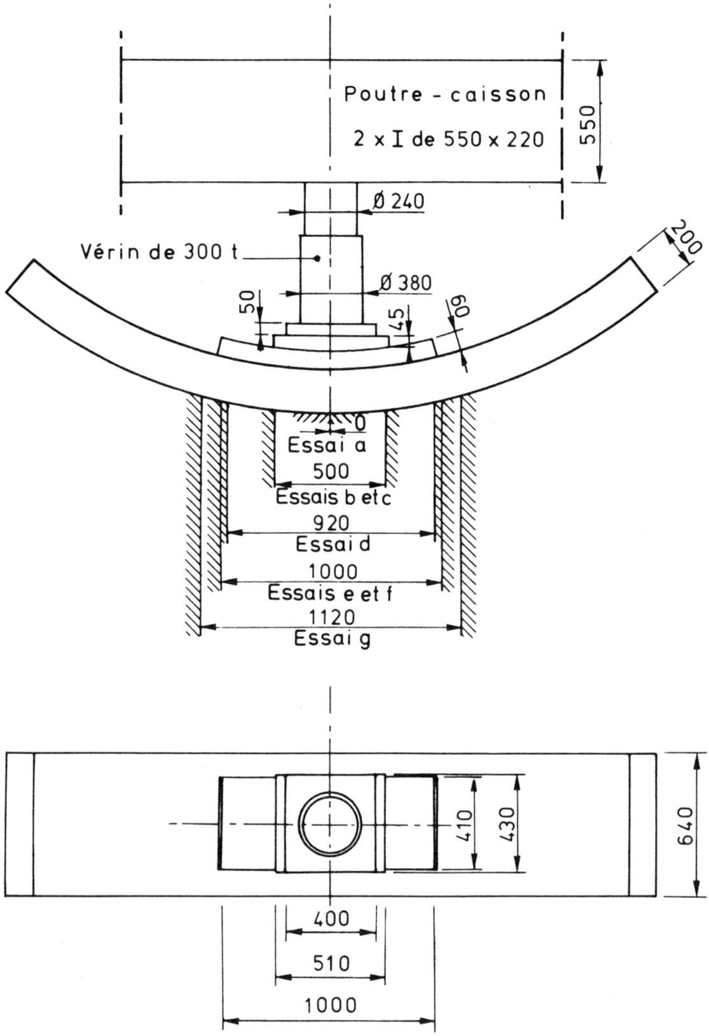

Poutre-caisson	box girder
Vérin de 300 t	300 t ram
Essai a, essai b et c, essai d	test a, tests b and c, test d etc.

Figure 5. Procedure followed for testing panels of current design (20 cm thick):
- inner side, thrust exerted by a 300 t ram via the intermediary of a 41 x 100 cm gripper shoe;

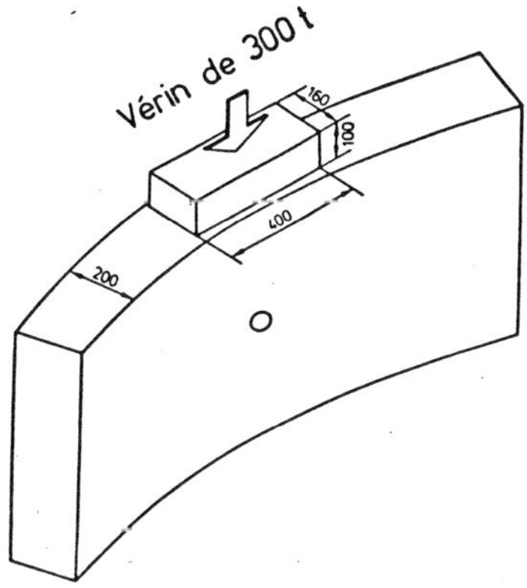

- outer side, no cavity at all (test a)
- cavities increasing in size from 500 to 1190 mm.

Figure 6. Load distribution on the side of a reinforced concrete panel.

Figure 7. Mechanised stone-drift heading machine developed by the NCB's Mining Research & Development Establishment at Bretby; It has a 2-piece shield section featuring two longitudinal slots.

MECHANIZED SETTING OF ROADWAY SUPPORTS

by W. Reiff

Head of HQ Department 'Technology Underground'
at Ruhrkohle AG (BAG Lippe)

Summary

Almost 40% of the 700 and more roadways being driven in West German coal mines feature mechanical systems for setting the arch supports. This report presents a survey of the various types of support-setting equipment used for individual heading techniques.

In shotfired drivages one still finds a predominance of power-driven working platforms which, as roadway profiles increase, represent a considerable step forward in terms of safety and the sequence of work. Multi-purpose platforms have established themselves as the pinnacle of mechanized arch setting. These platforms can transport and set a fully-assembled crown section, undertake forepoling and provide effective protection at the heading face. The mechanized systems available for roadheaders range from the simple boom-mounted arch-setter to the self-advancing roadhead support frame and roadway heading shield. In full-face drivages mechanized systems are employed to set and hydraulically tension complete ring segments, usually without the need for backfilling.

1. INTRODUCTION

In 1982 some 520 km of level and inclined roadways, including 100 km of stone drivages, were driven in West German coal mines. More than 700 heading projects are in operation at any one time.

The average winning depth, and hence the depth of the in-seam roadways, is around 870 m. The average depth of stone drivages is 1060 m. The problems associated with such depths have led to a constant increase in the excavated cross-section; in stone drivages the average section is now 24.4 m^2 and in gate-roads and in-seam roadways 20.6 m^2.

Arch-section roadways are the accepted rule. Circular roadways tend to be associated with full-face roadheading machines while rectangular-section roadways play a minor role for mechanized drivages in thick seams. Yielding trough-section arches dominate and continue to supersede rigid roadway supports. In stone roads the average interval between arches is 75 cm and in coal roads 80 cm.

The manpower expenditure on support work accounts for 40 to 55% of the working cycle, depending on the degree of mechanization of the other tasks which go to make up the heading operation. Support setting is a physically-exhausting, high-risk job which until now has benefited the least from mechanization. From the point of view of ergonomics, safety and efficiency the mechanization of support setting has now become a matter of urgency.

From the economic viewpoint, any system of mechanized support setting must be measured against the increase in the setting rate as well as against the amount of time which can be cut from the working cycle. This can best be achieved by the physical separation and simultaneous implementation of the various operations which comprise the heading project. Taking the roadheading machine as an example, Figure 1 shows the extent to which the various operations at the roadhead can be segregated using various types of mechanized support system.

The traditional sequence of support work can be broken

down as follows:
- 40% setting of arches and struts
- 40% lagging and building of stone packs
- 15% support work at the heading face
- 5% preparatory and finishing work.

Practically all of these tasks can be mechanized in one way or another. The support-setting aids which have been developed and tested can achieve this to varying degrees of success depending on the type and range of the device and the extent to which it is used (Figure 2).

According to a survey, approximately 40% of all heading operations being conducted in April 1983 made use of a support setting device. In shotfired drivages the degree of mechanization ranged from power-driven working platforms to multi-purpose platforms, while in mechanized drivages the equipment ranged from simple boom-mounted arch-setters to roadway heading shields. At the time of the survey there were no powered roadhead support frames or heading shields in operation (Figure 3).

2. MOBILE WORKING PLATFORMS

In shotfired drivages power-driven platforms represent the first level of support mechanization (Figure 4). There has been a rapid rise in the number of these devices in recent years. A wide variety of designs, some based on the modular construction principle, are available from several manufacturers. Collieries and specialist mining companies also design their own versions. Support setters with pneumatic or hydraulic drive feature friction wheels or, for working in steep drivages, rack or gripper-type advancing systems. Vertical movement is achieved by hoisting gear or in the case of the latest versions by hydraulic cylinders which give finer control, produce less noise and provide more clearance beneath the platform. Rail-mounted self-driven platforms have the following advantages over earlier systems:
- less lifting work for the support sections
- no manual transport of support material in the roadhead area

- improved safety due to reduced height of drop in the event of falls of coal and stone from the roof, side-walls or heading face
- safer working at the side-walls by using telescopic base extensions
- division of the roadway profile into 2 levels with a limited possibility for parallel drilling, loading and support work
- easier hand-held drilling and auxiliary work.

Power-driven working platforms can be used in roadway sections of 16 m^2 and over in conjunction with side-tipping loaders and low-height drilling rigs (1.5 m).

When driving the gate-road in line with the face a special platform can be used which is connected to the AFC drive as its power source. Such platforms operate in conjunction with impact rippers and drum shearer loaders.

3. ARCH TRANSPORTING AND SETTING DEVICES

Transporting and setting devices can be used for bringing up and setting individual arches or groups of crown sections. Different versions are available to provide various degrees of mechanization.

3.1 Devices for individual arch sections

In mechanized drivages the supports are usually installed arch by arch behind the heading machine. Lack of space alongside and lack of standing room on top of the roadheader makes support work both difficult and dangerous.

3.1.1 Machine-mounted devices

The most basic type of mechanical arch setter is the machine-mounted device which is swung forward manually when required. This device has now been superseded in most drivages by hydraulically-powered arch setters (Figure 5).

The arches are pre-assembled behind the roadhead area and are brought forward on a rail-mounted transporter to be transferred to the arch setter. Working in conjunction with the machine boom the arch setter positions the support section in the roadway crown below pre-installed wire-mesh panels. This is then followed by the manual work of bringing up and installing the struts and arch legs which is usually done without the use of horsehead girders.

These boom-mounted setting devices and arch transporter systems alleviated the work of transporting support sections and helped mechanize the setting of arch crown sections. The use of a pre-assembly table meant that part of the support work was transferred away from the roadhead area and this in turn reduced the number of machine stoppages caused by support work.

The disadvantages of this system are that the cutting work associated with the placing of each arch is interrupted and that the duration of this interruption is increased by the need to transfer the arch section from the transporter to the setting device.

3.1.2 Rail-mounted power-driven arch transporting and setting devices

Lack of space prevents the installation of a working platform on top of the heading machine in order to improve support setting. As a result, self-driven transporting and setting devices, mounted on single or twin rails, were developed from the original arch transporter systems (Figure 6).

By suspending the device from two rail tracks the arches can be pre-assembled at the side of the roadway and brought forward along the side of the drivage. By using a sluable carrying arm and pick-up device the arch can also be transported in line with the roadway axis, a major advantage especially when complete arch supports are being brought up. When the fully-assembled support has been lined up correctly the elevated arch legs can be lowered into their proper position. These arch transporter and setting devices have an active forepoling effect and as a result horsehead girders are not needed.

The use of self-driven arch transporting and setting devices enables the arch to be brought up and set in one move. If pre-assembly also includes the arch legs, these too can be advanced and installed mechanically. As a result, the degree of parallel working can be increased in this sector of the heading project. The work of strutting, bolting and lagging, on the other hand, remains a manual task and it too is fully incorporated into the working sequence. As with the setting of the crown section and side legs, the aforementioned tasks are carried out in the unsupported roadhead area.

3.2 Transporting and setting devices for pre-assembled sets of arches

In shotfired drivages it is not feasible, given the sequence of work, to bring up pre-assembled individual arches. Such a proposal was dropped as a result of underground trials with a working platform equipped with an arch setting device. Only in in-line headings is it possible to use platforms which are mounted on the coal-face conveyor and can be fitted with a manipulator for lifting arch crowns.

In shotfired drivages performance has been increased by means of a multi-arch method whereby a group of arch sections, pre-assembled to match the length of the pull, is brought up and installed. Three different types of setting device have been developed for this technique:

The self-driven working platform with setting table for multi-arch setting (Figure 7) is lowered to ground level behind the roadhead area. While heading debris is being loaded-out the struts are bolted to the arch structure and the wire mesh lagging put in position. The platform is then raised and advanced, thus bringing the ready-assembled arch structure and other support material up to the roadhead. Here the structure is raised to the roadway crown, secured with horsehead girders and connected to the existing arch support, the arch legs being set after loading-out has been completed.

A special version of this type of platform and setting table is that which provides a positive forepoling action, thus eliminating the need for horsehead girders. Such a system can be designed as a multi-purpose platform without drill carriage and will be discussed later in the paper.

The self-driven platform with support manipulator provides a greater degree of freedom than the first-mentioned device with the result that the arch structure can be easily and rapidly connected to the last arch in the row (Figure 8).

Both types of platform are a great improvement on the basic working platform. The multi-arch structure can be preassembled behind the roadhead area and this can be carried out in parallel with other types of work. Safety is improved in that men no longer have to work below an unsupported roof.

Only the work of bolting, strutting and lagging of the side legs remains a manual task. As with the advancing of the horsehead girders, these duties can be fully incorporated in the working cycle.

The fully-independent transporting and setting device for multi-arch structures (Figure 9) is basically a higher-performance version of the transporter-setter device for single arches. Such a system is capable of accommodating multi-arch structures and can undertake forepoling tasks depending on its design and the length of the pull. This device can also be equipped with a heading-face protection arrangement or with an underslung platform.

4. MULTI-PURPOSE SETTING PLATFORMS

To date the maximum degree of mechanization for shotfired drivages with the greatest possible number of tasks being undertaken in parallel has been achieved by using the multi-purpose platform system. Originally intended for shotfired in-seam drivages conducted independently of the winning operation, this type of platform has now proved effective in crosscut drivages.

The multi-purpose setting platform is composed of a traverse structure which supports a multi-tool system and runs on a special carrier rail. The tool system includes a lifting platform with active forepoling for multi-arch sets, a raise/lower and side-extension working platform, 2 hydraulic drill carriages, a hydraulic heading-face support system and a hydraulic arch-tensioning device (Figure 10).

With the exception of loading-out, the tool system of the multi-purpose platform can carry out all the tasks needed for shotfired drivages while the installation itself only takes up the space of one single machine. There is no need for various machines to be manoeuvred on the floor of the heading with the result that floor space is available for a second loading system.

The organisation of the working sequence, as depicted in Figure 11, illustrates how the main working tasks overlap and how the ancillary work can be undertaken in parallel with the main sequence.

This example shows that in a cross-cut of some 20 m^2 finished cross-section a heading team of 7 men can operate without any scheduled stoppages. This creates a high-performance working pattern with a 3 m pull being completed per shift. Over a period of several months an average heading advance of 9 m/d was achieved in this particular cross-cut drivage.

With this system the heading team is able to work under the protection of a wire mesh cover, so that no-one is ever in an unsupported area even when the arches are being connected, while at the same time the team is protected against falls of stone and coal from the heading face.

This high degree of overlapping and simultaneous working in the heading sequence demands a well-trained and highly-organised team, a good team leader and a supervising official who is present at all times.

5. SELF-ADVANCING ROADHEAD SUPPORT SYSTEM

With self-advancing roadhead support systems, which are hydraulically-powered, it is possible to separate the cutting process from the support-setting work in mechanized drivages. The roadheading machine cuts under the protection of the roadhead support unit which is either constantly advanced or 'leap-frogged' forward segment-fashion. The permanent arch supports are set behind the roadheader under cover of the final section of the self-advancing support system. As soon as the self-advancing unit is moved forward the permanent arch support must be brought into immediate contact with the rock using a backfilling technique.

The self-advancing roadhead support unit is now the state-of-the-art for mechanized support setting in roadheader drivages. Cutting and support setting remain physically separate and fully simultaneous tasks.

Over the past ten years a number of self-advancing roadhead support systems have been designed, built and introduced. Some proved to be too complex and their development was not pursued.

At Camphausen and Sophia-Jacoba collieries a 'roving roadhead support system' has been operating in conjunction with roadheading machines. The system is composed of a number of independent support frames. A 'leapfrog' method of advance is employed whereby the rear frame, which is next to the last permanent arch, is released and brought up to the heading face by a transport device mounted on the roadheader; it is then re-set by means of hydraulic jacks (Figure 12). It is planned to introduce more of these systems.

At Heinrich-Robert colliery 2 roadways, each some 1100 m in length, have been driven with the blade-shield system (Figure 13).

This roadheading system is composed of a heading shield equipped with forward canopies and trailing blades; the shield is braced against the walls of the drivage and incorporates a roadheading machine. The roadheader is coupled to a pantechnicon which comprises a scissor-action arch setter, a support manipulator and a backfilling system. The blade-shield roadheader has already shown itself capable of a heading advance of 2.20 m per day with 0.8 m between arches.

6. SUPPORT METHODS USED IN FULL-FACE DRIVAGES

The circular roadway profile created in full-face heading operations means that conventional support techniques cannot be employed. Instead, ring supports are used, these tending to be of yielding trough-section construction when greater roadway depths are involved. Because of the dimensions of the roadheader's stabilizer pads the support setting interval cannot be reduced at will, with the result that the sectional weight of the rings must be increased. This, together with the high heading performance, means that mechanized ring-setting is essential if the support work is to keep pace with the full-face tunneller.

The support material is basically supplied in packages for one complete ring. Special transport devices on the tunneller, either belt conveyors or trolleys, bring the support elements to the rear of the cutting head in the correct sequence. Full-face tunnellers are now equipped with support setting systems which can rotate around the full circumference of the roadway profile. There are 3 different types of system in use:

In the case of rigid supports the ring is connected together above the machine body under cover of the rear roof shield behind the tunnelling head. With the exception of the final segment, all the ring segments feature special socket connections and are laid individually on the rotating support setter. A loose connection is first made between the first and last segments by means of an adjuster plate. After the struts have been set and the roof shield advanced the support ring is put under hydraulic tension and the adjuster plates bolted up to full load. In this way a rigid support ring can be set in 15 to 20 minutes.

When the first type of setting system is used to place yielding supports the ring segments are raised above the machine body and placed in the setting device. For each ring segment the placing procedure is immediately followed by the making of the overlap joint and the bolting-up of the connector clamps. This process requires upwards-facing handholds and the use of an impact wrench also in an upwards direction. Before the placing of the final segment the ring is removed from the setting device. Once installed in its final position the ring is then completed, strutted, tensioned and bolted up. With this method a yielding support ring can be installed in 23 to 28 minutes.

Yielding supports can be installed more quickly with the second type of setting system. By means of a different system below the machine body the ring segments can now be inserted in the rotating setter to the exact degree of overlap. The connector clamps can then be mounted using downward handholds and downward-facing impact wrenches. The complete ring is assembled inside the setting device, after which it is withdrawn, strutted, tensioned and bolted up. With this modified system the work of setting yielding support rings is less complex and involves less physical effort, with the result that setting times are comparable with those achieved with rigid rings.

In full-face tunnelling the support work is physically separated from the cutting operation, while remaining a parallel procedure in which the men are fully protected against falls of coal and stone. Strutting and lagging still remain manual tasks. By using state-of-the-art technology support work need no longer be the bottle-neck area in drivages of this type.

7. OUTLOOK

Tried-and-tested equipment, which for the most part has attained a considerable level of technical proficiency, is now available for mechanized arch setting. Further improvements are needed, however, if we are to make full use of the performance capability of roadheading systems.

The greatest progress in the field of mechanized support setting has been achieved in full-face drivages with whole ring segments being mechanically transported and installed without backfilling. Here no further basic improvements can be made to the support setting procedure. As far as machine design is concerned, however, more attention should be paid to the support setting aspects.

As regards roadheading machines, further progress could be made with the self-advancing roadhead support system by the introduction of less complex and more reliable designs. The roadheading shield has now been developed to its full technical potential and its application usually depends on the economic aspects of the heading operation.

In shotfired drivages there will be a trend in favour of platform systems which can adapt to the working sequence and advance and set a pre-assembled multi-arch canopy corresponding to one complete pull. Such platforms should be designed for forepoling so that the use of horsehead girders can be eliminated. Shotfiring is mainly used in short drivages where high advance rates cannot be achieved. For such operations the support setting system should be as simple as possible so that its low assembly and machine-rental costs can justify its introduction on an economic basis.

Further rationalisation possibilities are currently being examined by way of preliminary trials aimed at including the arch legs in the crown and multi-crown assembly stage. Standard arch shapes do not lend themselves to such a move and work would be needed to develop and test suitable arch profiles and connectors.

In spite of the increasing use of rock bolting methods, arch supports will remain the predominant roadway support system. Further mechanization in this sector remains a high priority for reasons of safety, ergonomics and efficiency.

For key to illustrations see end of the report.

Figure 1

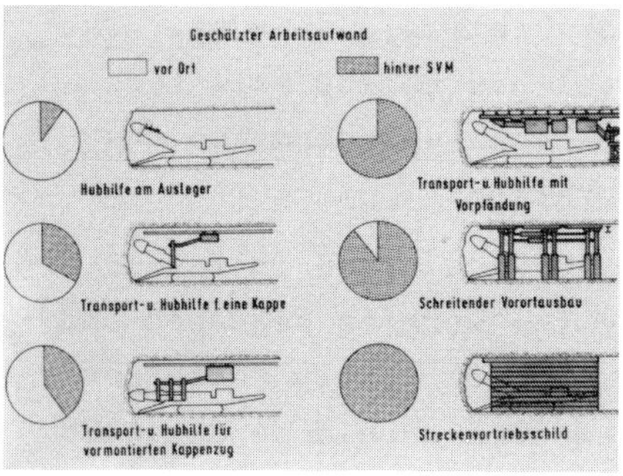

Figure 2

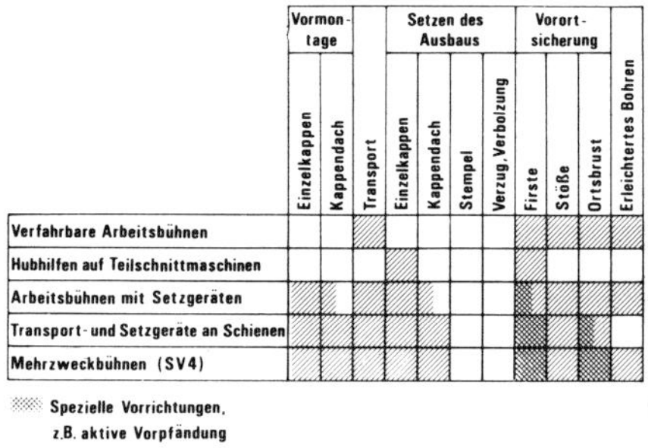

Figure 3

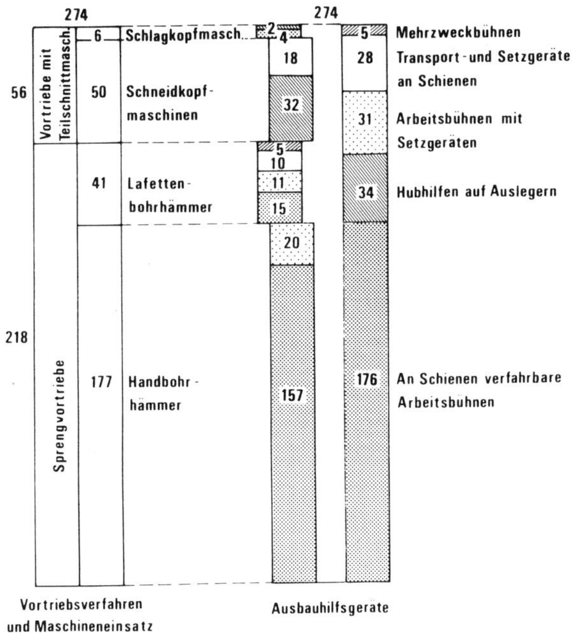

Figure 4

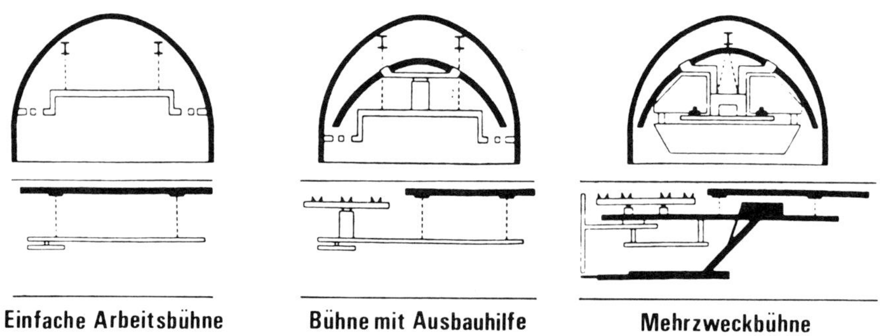

Figure 5

Figure 6

Figure 7

Figure 8

Figure 9

Figure 10

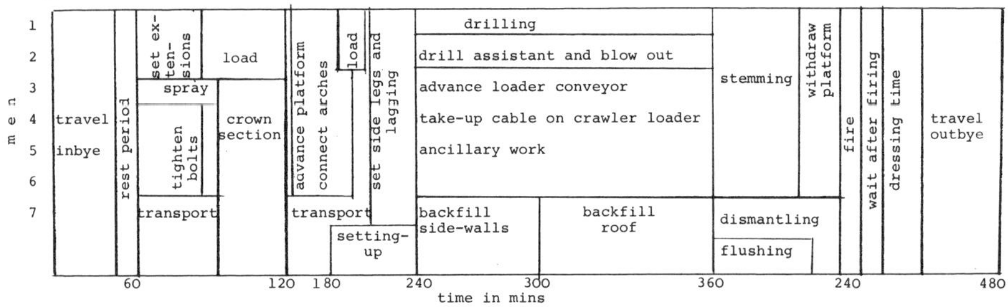

Figure 11 : Plan of organization of drivage with multi-purpose platform.

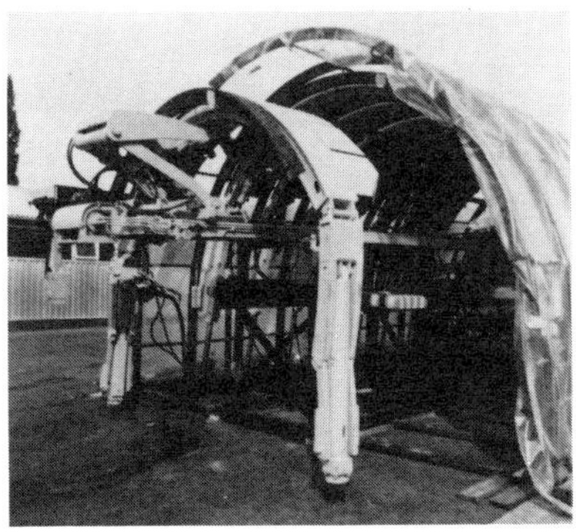

Figure 12

Figure 13

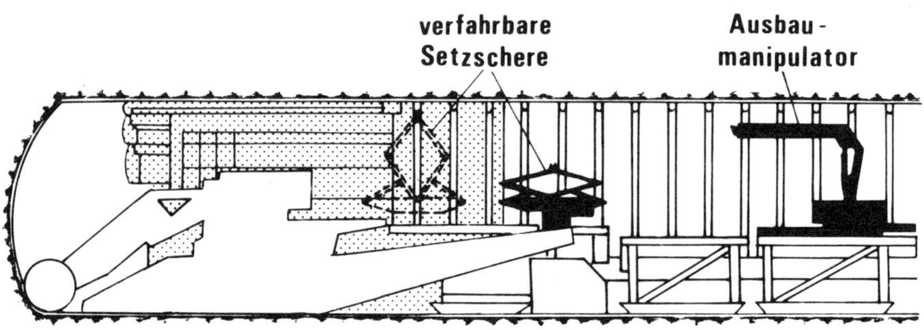

estimated worktime expenditure

at roadhead	behind roadheader
boom-mounted arch setter	transporting and setting system with forepoling
transporting and setting device for single arches	self-advancing roadhead support
transporting and setting device for pre-assembled multi-arch sets	roadheading shield

Figure 1. Mechanized support setting in roadheader drivages.

	pre-assembly			support setting				road-head support			
	single arches	multi-arches	transport	single arches	multi-arches	side legs	lagging, strutting	roof	side-walls	heading face	assisted drilling
mobile working platforms											
arch setter on roadheader											
platform with arch setter											
rail-mounter transporter/ setter											
multi-purpose platforms (SV4)											
special equipment, e.g. active forepoling											

Figure 2. Mechanization of roadway support stages.

mechanized drivages

impact rippers

roadheaders

multi-purpose platforms

rail-mounted transporter-setters

working platforms and arch setters

hammerdrill carriages

boom-mounted arch setters

shotfired drivages

hand-held hammerdrills

mobile rail-mounted setting platforms

roadheading method and machine system

support setting aids

Figure 3. Arch setting systems in use as at April 1983.

Simple working platform

Platform with arch setter

Multi-purpose platform

Figure 4. Degrees of mechanized support setting with platforms in shotfired drivages.

Figure 5. Hydraulic arch setter mounted on roadheader boom.

Figure 6. Independently-powered transporting and setting device for arch crowns.

Figure 7. Working platform with setting table for multi-arch sets.

Figure 8. Working platform with support manipulator for multi-arch sets.

Figure 9. Transporting and setting system for multi-arch sets.

Figure 10. The SV 4 multi-purpose platform.

Figure 12. Roving roadhead support system.

verfahrbare Setzschere — scissor-action arch setter

Ausbaumanipulator — support manipulator

Figure 13. The blade-shield roadheading system.

SESSION IV - SHOTFIRING AND LOADING, SHAFTS

- Recent developments in drilling techniques
- Use of trackless vehicles in roadway drivage in the Aquitaine Coalfield
- The loading out of stone in annular panels
- Latest developments in shaft construction technology
- Comparision - Shafts versus drifts
- Closing address

RECENT DEVELOPMENTS IN DRILLING TECHNIQUES

J. Delcruzel

Charbonnages de France

Summary

Conventional rotary drilling is well enough suited to relatively soft and non-abrasive rock.

But for rocks having a high degree of hardness and abrasivity conventional rotary drilling is inadequate and one has to resort to rotary-percussive drilling, a mode which has many drawbacks, especially relating to working conditions.

In order to extend the range of application of rotary drilling, two directions have been marked out:

- Drilling with diamond bits,
- Drilling backed with high-pressure water jets.

Drilling tests with diamond bits have shown relatively high drilling speeds to be feasible in abrasive rock, but the high cost of the bits and their fragility in disparate rock do not hold out the prospect of extensive development in this type of equipment.

Drilling assisted by high-pressure water jets has given remarkable results in CERCHAR laboratories, in both hard and abrasive rock. The equipment already developed has a high reliability which makes for cost-effective employment underground. One such installation is being brought into service at the Nord et Pas-de-Calais coalfield.

RECENT DEVELOPMENTS IN DRILLING TECHNIQUES

Rotary drilling has seen a great deal of development in recent years. This type of drilling is well suited to relatively soft and not appreciably abrasive rock. With rock possessing a high degree of hardness or abrasivity, the mechanical equipment and the cutting tools are subjected to great stress and wear, which limits the use of rotary drilling. Recourse is then had to rotary-percussive drilling with all its familiar drawbacks, including noise, vibration and hard physical effort.

Extending the range of application of rotary drilling has been sought in two directions:

- utilising materials with better performance than tungsten carbide for the cutting tools,
- assistance for the cutting tools by high-pressure water jets.

1. DRILLING WITH DIAMOND BITS

Of all the mechanical characteristics which limit the range of application of rotary drilling abrasity is the major one. In practice, conventional tungsten carbide tools will not ensure a constant calibre in the holes drilled in abrasive rock and the bits have to go for resharpening after only a few metres of drilling, which makes rotary drilling not an operational proposition.

In view of the commonly used rotary speeds (from 400 to 1500 rpm) and the high thrusts needed to obtain satisfactory advance rates, a temperature rise occurs through contact between rock and tool. If the temperature of the tool exceeds a given value, the mechanical properties of tungsten carbide are diminished down to 80% of the original properties.

In order to enable tools to work properly at high temperatures without a drop in cost-effectiveness, there is at present only the diamond which could replace tungsten carbide.

The use of diamond-encrusted drill bits has the invaluable advantage of being usable with conventional rotary drilling equipment.

Tests have been carried out with diamond bits at the CERCHAR laboratory and also underground in Lorraine collieries.

Several types of bits have been tested:

- A 28 mm diameter bit for bolting,
- A 36 mm diameter bit for shotfiring (Figure 1).

These tests have shown that it is possible to obtain encouraging results in rock in which conventional rotary drilling was either not possible at all or gave advance rates which were below those regarded as economically acceptable, namely 1 m/min.

In abrasive and relatively hard rock it was established that it was necessary to work at a rotation speed over 1000 rpm and to lower the penetration effort from 70 to 80%. This made it possible to have a small depth of penetration of the tool and consequently very low wear on the bit. When the bit retains a virtually constant diameter, the calibre of the drilled borehole remains good.

It was also important for the water throughput to exceed 15 l/min to provide adequate tool cooling.

The rates of advance obtained in these conditions are excellent. It proved possible to drill at speeds in excess of 1.50 m/min in hard rock where conventional tools would fail to reach operational drilling rates. On the other hand, it was possible to carry out rotary drilling at speeds over 1 m/min in abrasive rock, in which until now only rotary-percussive drilling was possible, in noise level terms something of an advantage.

The effective life of diamond bits, where the material to be drilled is homogenous, is greatly superior to that of conventional bits of tungsten carbide. In the conditions under which the underground tests were carried out it was possible to drill more than 150 m with one diamond bit. In the same conditions a tungsten carbide bit only managed 10 m and even that with three resharpenings.

These results seem particularly encouraging for the future but diamond bits do bring some serious drawbacks.

For one thing, the diamond bit is a fragile tool. It stands up badly to discontinuity in the rock. In particular, care has to be taken when the borehole is collared. The diamond bit is sensitive to heterogenous rock which leads to flaking of the diamond-encrusted part (Figure 2). To avoid the intermittent shocks during drilling it is essential to use a pressure throughput regulator.

Secondly, the current price is very steep, working out at twice that of a conventional bit per metre drilled.

Although having excellent intrinsic performance the diamond bits are not yet economically acceptable tools.

A reduction in their fragility and their price remains to be achieved before their use could be justified in economic terms.

2. ASSISTED ROTARY DRILLING

The purpose in assisting rotary drilling by means of high-pressure water jets is to open up three areas for rotary drilling:

- in soft rock, to attain drilling speeds twice as high or even more,
- in abrasive rock, to reduce wear in tools,
- in hard rock, to provide an economically acceptable drilling rate wherever conventional tools are powerless.

The first trials have shown that, at pressures below 2000 bar, very notable improvements could be made to the functioning of cutting tools.

Theoretical studies and first test of cutting with water under pressure have also shown that the water jet lost a lot of its effectiveness the further the point of attack on the rock became from the jet outlet duct. This phenomenon may be explained by the drop in jet stream coherence reducing the energy transmitted to the rock. Consequently, the attack on, or the preliminary cutting into the rock by water jets under pressure needs to occur at the very outlet of the duct, or as close as possible to it.

This injection of water under pressure is directed to the bottom of the hole from accurately dimensioned orifices (0.3 to 0.4 mm in diameter) fashioned in the tungsten carbide inserts, or just behind these. These water streams break the rock at exactly the spot where the lips of the bit pass. The combination of tool/water is thus used in an optimal way, provided the jet attacks the rock at the outlet of the duct fashioned in the bit, to provide good energy transmission. The tool then works only in a rock whose mechanical properties have been weakened.

These new conditions in which the drilling tool works lead to the following changes:

- Increase of the advance rate,
- Appreciable reduction in wear on tools,
- Less vibration in the mechanical equipment,
- Better evacuation of drillings, with jamming in the bottom of the hole ruled out,
- Suppression of dust and any possible sparks,
- Continuous cooling of the tool during drilling,
- Accuracy and alignment of borehole, making bolting or shot-firing easier,
- Adapting rotary drilling to relatively hard and abrasive rock.

2.1 LABORATORY TESTS

2.1.1 Drilling equipment

- The drill (Figure 3)

The equipment needed for tests was manufactured by CMM (Construction de Matériels de Mines) within a cooperation project with CERCHAR.

One rotary drill and one slide providing the forward travel of the tool were supplied from this project.

The drill had two water circuits available for drilling:

- one low-pressure circuit for flushing out the drillings,
- one high-pressure circuit to back up the drill bit.

The drill is provided with a swivel joint of good reliability to give sealing at the point of water injection at high pressure. The sealing has been proved over a period of time at pressures going up to 2000 bar.

The working life of this joint is excellent, up to 1000 hours of operation without a single replacement.

Rotation of drill and forward travel of the tool are hydraulic.

- Drill rod

A drill rod was also supplied within the collaboration project with CMM.

It was adapted to the requirement of this study to enable high water pressures to be used in the tests.

- Drill bits

The manufacture of experimental drill bits was carried out in collaboration with METAFRAM, a specialist firm developing industrial products from metal powder.

This firm produces, in particular, drilling tools with welded-on tungsten carbide inserts.

The bodies of experimental drill bits, especially the steel mountings, were machined at CERCHAR workshops.

The high-pressure water jet outlets were positioned:

- either in the carbide insert, when these are manufactured,
- or in the steel ducts, set behind the carbide inserts in the body of the bit.

Several types of bits have been tested to correspond with the fields of application envisaged:

37 mm diameter bits

- single-lip with 0.3 mm ducts
- twin-lip with 0.3 mm ducts
- twin-lip with 0.3 mm bore in the carbide inserts

• crown with 0.3 mm and 0.4 mm ducts (Figure 4).

22 mm diameter bits

• twin-lip with 0.3 mm, 0.35 mm, 0.4 mm ducts
• single-lip with 0.3 mm and 0.4 mm ducts.

For comparison, tests were also carried out with drill bits but without high-pressure water jets (conventional rotary drilling).

2.1.2 Measurement

A mini-computer linked to the test rig was used to record the readings and evaluate the results.

2.1.3 Main rock types tested

Assisted rotary drilling was tested on a number of typical rocks:

Rock	Compressive strength (bars)	Abrasivity (CERCHAR index)
Reconstituted concrete	600	1.6
Mazaugues limestone	1900	0.5
Granite	1700	5
Uranium-bearing sand-stone	350	2
Silver-bearing sand-stone	600	4.3
Lorraine sandstone	1600	3

2.1.4 Results

Laboratory work made it possible to gain numerous positive results. The first of these was that the drill as manufactured by CMM proved effective and reliable: its use underground should, therefore, not create any particular problems. The drill rods, perfected during the tests performed equally well throughout the research programme.

With regard to the drill bits, which were more difficult to get right, there remain a number of points to clarify but some positive findings can already now be singled out, both in relation to shotholes and bolting holes:

- Back-up by high-pressure water jets effectively helps the

work of bits to break up the solid: at a constant rate of advance the thrust to be exerted is less great and the energy needed by the drill is smaller. This leads to less rapid wear on equipment and especially on bits.

- Jets play an important part in cooling the tool, a factor greatly favouring wear reduction as well.
- At constant thrust it is possible to attain drilling speeds which are distinctly higher, provided of course that the geometry of the bit allows drillings to be evacuated: especially the outlet ports need to be adapted.
- Improvement in drilling performance rises in step with water pressure and throughput. It will be limited by the level of additional power one is prepared to install on the drilling jumbo to supply water under pressure. It will be noted that, if the specific energy required for cutting can be appreciably reduced by the introduction of water jets, the total specific energy (the energy consumed by the drill and water jets to extract one unit of the solid) is higher than that needed if no assistance is used: our view is that this discrepancy will more often than not be acceptable, because of the advantages enjoyed in terms of higher drilling speeds, longer service life of equipment and, of course, being able to use rotary drilling in strata until now worked by rotary-percussive drilling.

The permutations of duct features (position, inclination, diameter, number, type, water pressure) are numerous. Tests have allowed certain rules to be formulated:

- The inclination of the jets is slight but they do have to operate in different grooves so as to shape a central shoulder which can then be readily broken by the tool.
- Assisted drill bits have to be so designed that the jets attack the rock close to the duct outlets, and one jet is aimed at the periphery while the other is directed at the centre of the hole.

It is not clear that the value of additives, proven and indexed in the case of jets cutting at very high pressure, is by any means as certain in the case of assistance. Even so, this is a point that will have to be examined because it would obviously be very valuable to be able to reduce water pressure levels while maintaining performance.

Although the effective life of ducts is adequate compared with the life of a drill bit, it will be necessary to compare the effectiveness of assistance using ducts fashioned by borings in the steel, in artificial stones or in tungsten carbide inserts.

When all parameters are optimised excellent drilling speeds are attained. Such optimisation is clearly a function of the rock to be drilled.

The following results merit a mention:

22 mm drilling (Figure 5)

In abrasive rock of slight hardness and in average abrasive rock of medium hardness, in which rotary drilling will yield useful rates of advance up to 2 m/min, high pressure jet back-up has made it possible to multiply these rates fourfold with twin-lip bits provided with two 0.3 or 0.4 mm diameter ducts. The single-lip bit is not suited to this type of rock, with or without high-pressure jets.

In hard and abrasive rock an unassisted twin-lip bit will give rates of 1.4 m/min, with a very short working life of 3 to 6 m; but with assistance the rates have been stepped up to 2.6 m/min and a working life to 30 m.

The best result has been achieved in these rocks with a single-lip bit, with advance rates of 3.5 m/min and working lives over 30 m.

37 mm diameter drilling

In rock of 1600 bar hardness and high abrasivity, such as Lorraine sandstone, unassisted rotary drilling is impossible, but a twin-lip bit with jet assistance has enabled drilling at rates of up to 1.20 m/min.

But the best results have been attained in such rock with a crown bit specially designed with ducts bored through the inserts. Rates of 2.5 m/min were achieved (Figure 6).

2.2 UNDERGROUND TESTS

The examination of rotary drilling with jet assistance will be continued with a test programme, to be carried out very soon underground at the Nord et Pas-de-Calais coalfield.

A drilling jumbo has been prepared for these tests to accommodate the CMM drill. The hydraulic pack with Hydraulique-Paul pressure booster will give a throughput of 12 l/min at 2500 bar.

In order to carry out comparative tests, several types of bits have been manufactured in 22 mm for bolting and in 37 mm for drilling shotfiring holes. They will make it possible to carry out endurance tests on site and to observe the effect of their geometry, the type of ducts and diameter of ducts.

Jet-assisted drilling will be tried in shales and sandstone shales, then in hard and abrasive sandstones, whose compressive strength is 1200 bar and abrasivity 3.8 points on the CERCHAR scale.

3. CONCLUSION

While diamond bits are still tools which are vulnerable in use and of a price that is hardly economic, the first tests with jet-assisted rotary drilling seem to be very promising.

The underground tests which are due to begin should confirm that it is possible to extend the range of application of rotary drilling to the majority of rock types we encounter in our mines.

Apart from the increased productivity to be expected from this technique, it will bring an improvement of working conditions, an important advance in itself.

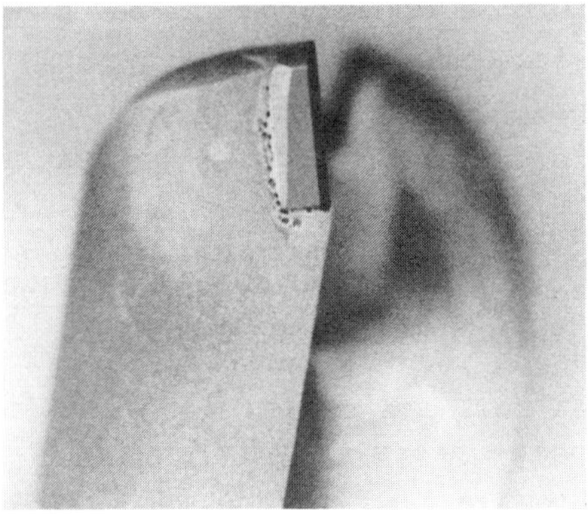

Figure 1.

Diamond bit of 36 mm diameter.

Figure 2. Flaking on a diamond bit.

Figure 3. CMM drill on the CERCHAR test rig.

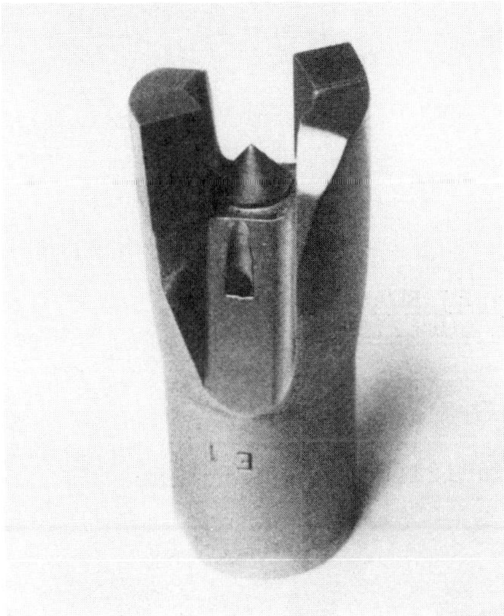

Figure 4.

Crown bit of 37 mm diameter.

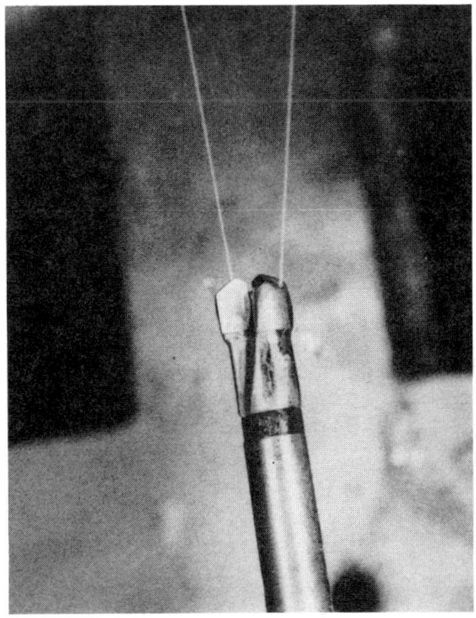

Figure 5. Twin-lip bit of 22 mm diameter with two 0.3 mm diameter ducts.

Figure 6. Crown bit of 37 mm diameter with two 0.4 mm ducts.

USE OF TRACKLESS VEHICLES IN ROADWAY DRIVAGE IN THE AQUITAINE COALFIELD

M. Blanc

Chief Production Engineer, Houillères d'Aquitaine

Summary

Twelve years after the introduction of trackless vehicles in the Aquitaine Coalfield, this paper provides a survey of their use in the field of roadway drivage. Various parameters relating to the operation of LHD's are discussed (production curves, maximum permissible elongation, etc.). The mechanised materials transport chain, which is based on the use of specialised Diesel vehicles throughout, is also described.

1. INTRODUCTION

The techniques used for driving roads during preparatory work have been tailored to the very special characteristics of the coal measures at Carmaux.

1.1 Characteristics of the measures

The measures are in the form of a trough made up to 20 seams (of which 14 have been worked), with 300 metres of Carboniferous overlaid by some 100 metres of Tertiary. The seams already worked range between 1.8 and 10 metres in thickness. A very dense pattern of faults breaks the measures up into small panels each giving an average extraction of 37 000 t. Planned output in 1983 is 540 000 t saleable; men on books (underground) total 893 and OMS is 3209 kg. The measures are classified as dusty but not gassy.

1.2 General comments on preparatory work

The Table below shows the order of magnitude of the lengths of roadway driven in 1983 in the rock and in coal, in the case of both district preparatory work and preparatory work in the panels themselves, according to their use.

	Type	Rock		Coal		Total	
		Metres	m/1000 t	Metres	m/1000 t	Metres	m/1000 t
District Preparatory Work	Service levels	690		420		1110	
	Return airways	340		400		740	
	General access Road	100		200		300	
	Total	1130	2.09	1020	1.89	2150	3.98
Preparatory work in panels	Access to one panel	300		1700		2000	
	Exploratory work	0		480		480	
	Panel entry roads	70		3500		3570	
	Total	370	0.69	5680	10.52	6050	11.20
Total preparatory work		1500	2.78	6700	12.41	8200	15.19

2. MACHINES USED AND PERFORMANCE ACHIEVED IN 1982

2.1 Tunnelling machines:

Manufacturer	Type	Number in pool	Number used in prep.	Weight (t)	Height (m)	Width (m)	Length (m)	Rated capacity (kW)
Alpine	AM 50	6	6	24	1.65	2.50	7.50	155
Alpine	F 6 A	13	3	9.5	1.52	2.00	6.82	75

	m^3 extracted by machine	
	per day's operation	per shift operation
Alpine AM 50	49	22
Alpine F 6 A	56	20

It should be noted that three of the Alpine AM 50's are fitted with on-board, wet-type dust extractors. The Alpine AM 50's are reserved for difficult areas of work, notably preparatory work, which explains why their performance is worse than that of the Alpine F 6 A's.

2.2 LHD's

Manufacturer	Type	Number used in prep. work	Number in pool	Max. payload (t)	Height (m)	Width (m)	Length (m)	Diesel motor Capacity (kW)
Equipement minier	CT 3500	2	5	3.5	1.54	1.80	7.45	102
Equipement minier	CT 2500	2	5	2.5	1.45	1.60	6.70	63
Wagner	ST2B	3	18	2.5				
Wagner	HST1	4	9	1.3	1.83	1.22	5.28	38

	Running time per machine (hours)	
	per day's operation	per shift operation
CT 3500	13.6	4.8
CT 2500	10.7	3.8
HST1	6.2	2.5

All the LED's are fitted with foam-type fire extinguishers. All the CT 3500's have a coal bucket without a scraper blade, capacity 3000 L (water). Most of the ST2B's have a coal bucket with a scraper blade, capacity 2300 L (water).

2.3 Trackless service vehicles

Manufacturer	Type	Number in pool	Payload (t)	Height (m)	Width (m)	Length (m)	Diesel motor Capacity (kW)
Equipement Minier	TM 40	5	7	2.05	1.95	7.00	63
Brimont	Panda 350	1	0.4 1.8	2.23	1.73	3.16	38
Brimont	Panda 380	2	0.6 2.8	1.80	1.75	3.40	38

The TM 40 materials transporters are fitted with a hydraulic skip loading and unloading device (manufactured by GUIMA) which makes them totally autonomous. The Brimont all-terrain vehicles are equipped with a crane with a grapnel and stabilising jack legs.

3. APPROACH TO ROADWAY DRIVAGE OPERATIONS

Given that 75% of district preparatory work consists of roads driven in rock and that 85% of preparatory work in panels consists of roads developed in coal, this paragraph deals only with the approach to drivage work in these two situations.

3.1 In-seam drivage operations

Once the cross-cuts leading from the spine roads have been driven into the panel, thus forming an initial ventilation loop, the areas which are particularly important for the trouble-free operation of a trackless district are fitted out. These consist in particular of the following:

- the district vehicle shop, which is equipped with a pit for vehicle repairs and (occasionally) maintenance;
- the fuelling station, which is located in a fire-proof zone; fuelling takes place here at the beginning of each shift;
- the district equipment stores (cross section 23 m^2) where the LED's are loaded with supplies;
- the explosion proof chamber.

Starting from this basic situation, the layout of the panel depends on the method of extraction (conventional longwall or sub-level caving). However, in all cases there are at least two gate roads, one for product clearance and one a supply road. These two roads are generally developed in parallel so as to minimise service needs. They are linked by one or more cross-entries.

The section of the drivages varies from 14 to 16 m^2. The roadways are supported by TH arches, with 0.6 to 1 m between arches, and are lagged with wood coverboards (Photo 1).

3.11 Clearance of products behind the winning machine

The base road of the panel is driven with continuous clearance. During the drivage of the tail gate and the cross-entries, the product is cleared by LHD's which unload onto an armoured conveyor connected into the continuous clearance system in the base road. In order partially to dissociate the operation of the winning machine from that of the LHD, an intermediate stock with a capacity of some 10 m^3 is provided by installing a suspended armoured conveyor 8 m long behind the winning machine (Figure 1).

By analysing the result of remote monitoring of the LHD's for over 100 shifts, we have been able to establish certain norms for their use in drivage operations:

3.11.1 Working time to be allowed for the various operations carried out during one shift by LHD's behind a winning machine:

Coal clearance	3 h 30 min	60%
Materials transport	1 h	20%
Fuelling	30 min	10%
Electro-mechanical repairs/maintenance	30 min	10%
Total	5 h 30 min	100%

3.11.2 Production curves of CT 3500's and CT 2500's as a function of the gradient:

- Travelling speed of the LHD's:

Uphill gradient 20%	3 km/h
Uphill gradient 10%	5 km/h
Other gradients	7 km/h

- Time included in the total transport cycle:

Loading of the bucket	30 s
Unloading of the bucket	20 s
Manoeuvring	40 s

- Fill coefficient of the bucket: 0.85.

3.11.3 Maximum extension up to which the operations of the winning machine are not slowed down by product clearance; the time taken for cutting the coal and for setting a TH arch, with 1 m between arches and lagging with wood coverboards, is shown in the following Table:

Winning machine	Alpine F6A		Alpine AM 50	
Section of drivage (m^2)	13.3	16	13.3	16
Coal cutting time (min)	25	40	20	25
Support setting time (min)	40	55	40	55
Length of cycle (min)	65	95	60	80

Given that 75% of the useful running time of LHD's is used for product clearance, the maximum extensions for various gradients are as follows:

Winning machine	Alpine F6A		Alpine AM 50	
Section of drivage (m^2)	13.3	16	13.3	16
Average gradient 5%	300 / 200	350 / 250	260 / 160	300 / 200
5% Average gradient 15%	225 / 140	275 / 190	180 / 120	225 / 140
15% Average gradient 20%	130 / 90	160 / 110	110 / 75	130 / 90

CT 3500 / CT 2500

Maximum extension for a single journey

It will be seen that gradient is important in influencing the performance of LHD's. The condition of the travelling surface is also very important, but this is more difficult to quantify. Concreting over travelling surfaces remains the most effective solution to the problem of water and ruts on the surface.

Once the maximum extension has been reached, it is possible either to deploy to LHD's operating in parallel or in series to provide clearance from the face, or else to adapt a materials transport vehicle. The second solution is that generally adopted when this vehicle can serve several drivages in the same district.

3.12 Transport of equipment for drivage operations

The equipment is loaded at the pit top into skips with moveable stanchions having a payload of five tonnes (3.5 m x 1.65 m x 0.95 m). The skips are brought down to the 260 m level via a surface drift, on pitcars travelling on a metric track. Using a travelling gantry crane they are then transferred to the vehicles deployed for materials transport (TM 40's - see Photo 2). The TM 40's transport the skips to the various district equipment stores. There they are unloaded by a Panda Diesel tyre-mounted vehicle fitted with a crane and a grapnel (Photo 3). This compact and highly manoeuvrable vehicle, which can turn through 180° on the spot, is also used to load the buckets of the LHD's which come to the store in order to collect supplies. In the case of roads developed with continuous clearance, these vehicles are usually HST1's, otherwise they are the LHD's providing clearance from the face or specially adapted vehicles which may or may not be fitted with a fork instead of the bucket.

The transport arrangements do not allow skips to be left underground, as there is insufficient room for them at the stores and there are too many skips. This means that all skips must be unloaded as soon as they reach the store.

The materials transport chain is, therefore, completely mechanised from the surface to the winning machine in the majority of districts. However, transfer from the bucket of the LHD to the face is still done by hand.

3.2 Drivage operations in the rock

Product clearance is by means of 2500 type LHD's, in some cases fitted with side-tipping buckets, as for in-seam drivage work; however, a different approach is adopted, given that when tunnels are blasted in the rock, the product loading phase is at a critical point in the cycle, making it essential to minimise the clearance time. For this reason the armoured conveyor fitted with a crusher onto which the LHD unloads is kept at least 50 m from the roadhead. The installation is, therefore, extended every 40 s. The readjustment of the clearance arrangements takes three shifts and 26 manshifts. The armoured conveyor, the crusher and the sledge-mounted winch are moved forward by two LHD's.

Under these conditions, the period taken to clear the debris after a blast 2 m deep in a 16 m^2 section ranges from 70 to 110 minutes, depending on the distance from the unloading point and the gradient of the roadway. This enables the cycle to be completed within a shift.

Materials supply is carried out by the LHD working at the face in the same way as for in-seam drivage operations.

4. RESULTS

Under the particular conditions prevailing at Carmaux, using the machines described and allowing for the face that the organisation of the workings was not yet entirely as described in section 3 of this paper, especially as regards materials transport, the average results recorded in 1981 and 1982 were as follows:

Result of preparatory work in the solid	Ratios	1981	1982
in seam	cm/hp at face	60	64
	m^3/hp at face	8.8	9.3
	cm/hp total	18	21
	advance per face and per day's operation	4.3	4.2
in the rock	cm/hp at face	21	23
	m^3/hp at face	3.1	3.8
	cm/hp total	14	13
	advance per face and per day's operation	1.8	2.3

Photo 1. Alpine AM 50 with clearance by a CT 3500.

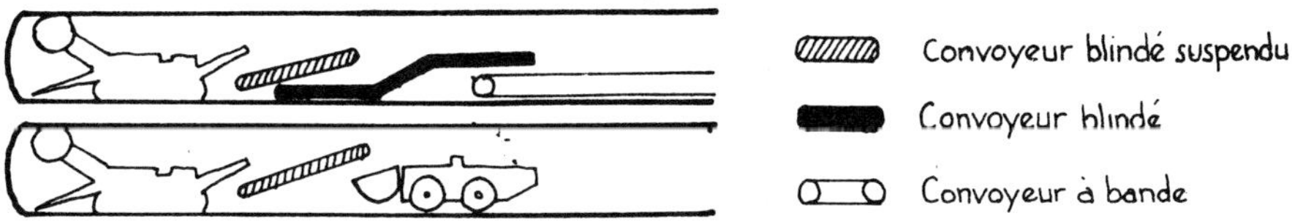

Convoyeur blindé suspendu — suspended armoured conveyor
Convoyeur blindé — armoured conveyor
Convoyeur à bande — belt conveyor

Figure 1. Clearance arrangements.

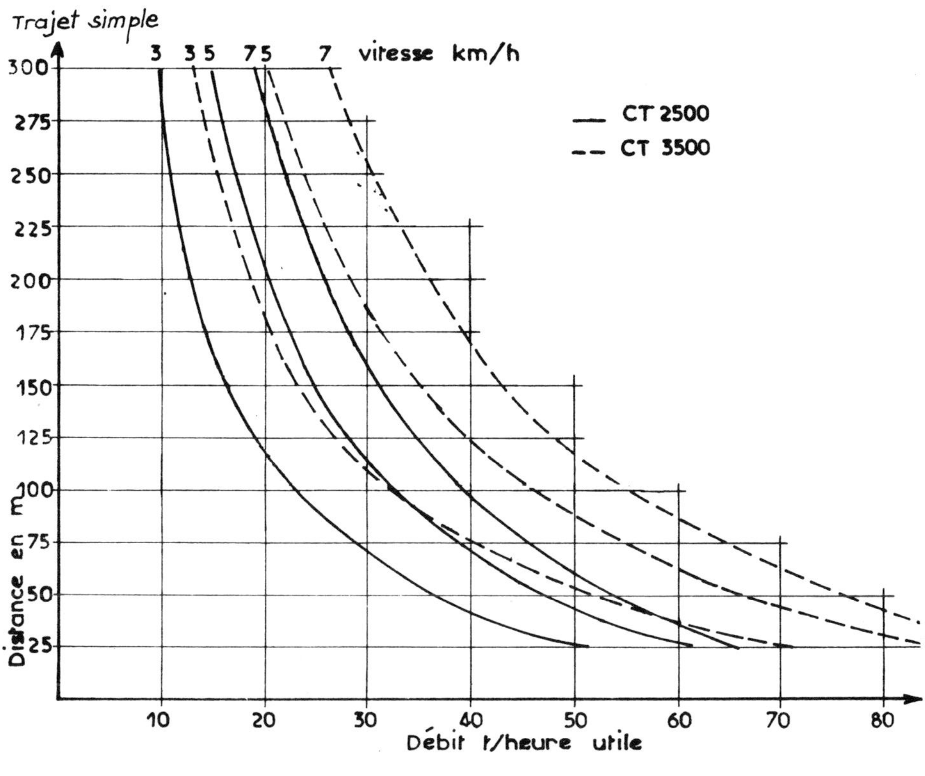

Figure 2. Production curves of the CT 3500's and CT 2500's.

Photo 2. Panda 380 vehicle.

Photo 3. TM 40 materials transport vehicle.

THE LOADING OUT OF STONE IN ANNULAR PANELS

C. CLAES

Head of the Preparatory Work Department,
Winterslag Colliery, N.V. Kempense Steenkolenmijnen

Summary

The horizontal main stonedrifts in the Campine coal basin are lined during driving operations with standardized annular concrete panels. The drifts contain a double set of rail tracks for mine tubs.

In view of the requirments thus imposed, the loading of debris was until 1980 carried out exclusively using track-mounted shovel loaders.

Although results in terms of production costs and output were satisfactory, greater daily advances were desirable. However, the constantly increasing distances to shafts and shorter working hours impaired output and shortened the drivage cycle.

From 1980, therefore, studies and tests were carried out in order, among other things, to shorten the important phase of rock loading in the drivage cycle. Heavier track-mounted shovel loaders were used, with caterpillar track grab loaders being operated at certain faces.

This report gives a general outline of the installation of the equipment and of the organization of the drivage cycle. The potential, improvement, advantages and drawbacks of the loading systems are also discusses.

1. INTRODUCTION

Coal winning in the Campine basin has always been hampered by high strata pressures. The main reasons for this are the considerable thickness of the overburden, the soft surrounding rock an the predominance of deep deposits.

Fairly early on, the horse-shoe shaped yielding arches used to line the stone drifts were replaced by annular supports consisting of concrete blocks. Depending on the useful diameter of the drift, these supports consisted of 40 - 50 non-reinforced concrete blocks. Between the blocks, which could be positioned using rudimentary equipment, were compressible wood-fibre boards. These supports were satisfactory from the technical point of view. However, the constant changes in the mechanization and organization of the drivages did not remedy the alow rate of advance. This was 2 m, with output up to 0.16 m or 2.6 m^3 per manshift.

From 1968 the type of lining used was substantially changed: the concrete blocks were replaced by reinforced concrete panels. A closed ring is made from 5 panels, which are positioned completely mechanically with specially designed cranes.

In the Campine basin more than 80 km of stonedrifts have been driven using panels. A a result of the new technique, the daily rate of advance doubled to 4 m, with an output of 0.32 m or 4.5 m^3 per manshift. Progress was thereafter limited by the loading and removal of the debris an obstacle which prompted further tests.

2. THE DRIVAGE : EQUIPMENT AND WORKING CYCLE

In the Campine basin all annular stonedrift supports have a standardized internal diameter of 4.2 m. Each support consists of 5 prefabricated reinforced concrete panels, the dimensions of which were determined in the light of functional requirments.

The sections excavated measure 17.5 m^2. The useful section of the stone-drift, after the laying of the ballast for two sets of tracks, is 13 m^2. Since mine tubs have become widely used for transport, virtually all drifts are driven horizontally. This means that in most drivages various stone strata are out successively.

The depth or pull in a drivage depends on the type of stone and the equipment available at the face. As far as possible, each pull is planned so as to be completed within one shift. With conventional drilling equipment in normal drifts a pull of two rings, or 1.30 m, is possible per shift. At faces equipped with a drilling jumbo, the target is three annular supports or 2 m. Futhermore, where simple equipment is used, a very large percentage of all the workers are engaged in actual driving operations.

The following are the main phases of the driving cycle:

PHASE 1 : The drilling of shot-holes and shot-firing

At half the faces, compressed air hammer drills on extendable air-legs are used. At other faces it is impossible to use a twin-boon drilling jumbo fitted with compressed air hammer drills. The installation may be mounted on a self-propelled undercarriage or suspended from a specially constructed monorail.

PHASE 2 : The loading out of the stone

The loading out of the stone accounts for a quarter to one third of the drivage cycle and is the longest phase.The use of shovel loaders makes the process non-continuous and the machines have rather low loading capacities. However, the preparation is very simple and the debris is removed very safely. With continuous loaders, which have considerable loading capacities, a great deal of preparation is required in handling the rather cumbersome transport equipment. There is this a loss in simplicity and safety.

PHASE 3 : The positioning of the panels

The concrete panels are positioned by means of a crane constructed by Kempense Steenkolemijnen. This is a HIAB-typ hydraulic crane mounted either on a self-propelled undercarriage or on a grab loader. The way in which the concrete panels are transported to the heading face depends on the type of loading machine used.

Some of the operations are not strictly tied to the cycle, e.g. the lenghtening of tracks, the extension and shortening back of continuous conveyors, the injection of building materials to fill the gaps between the panels and the drift, and the lengthening of pipes. The method of loading the stone affects these operations and determines the extent to which they lug behind the drivage proper.

3. THE USE OF LOADERS IN THE CAMPINE

Those in normal use are the overhead and side discharge loaders which are mounted on tracks and are powered and controlled by compressed air.

The traction motor provides the thrust required by the bucker to penetrate the pile of stones, for bringing the empty tubs into position at the face and removing them when filled to the area served by the general transport system.

The bucher motor develops maximum power during lifting and discharging and maximum speed during lowering of the empty bucher.

Following the suivelling movement the central position is automatically resumed and the machine is locked in that position. It can be regarded as simple to operate; the particular technical features are given in table I.

3.1. Organisation of loading installation of equipment

Fig. I liste the necessary equipment - the drill carriage, shovel loader and crane. The tracks at the heading are layed out in such a way as to make it easy to handle and replace equipment and tubs (used to remove rock and supply panels).

The volume of rock removed is 30 m^3 per m stonedrift. This means that for a depth of pull of 2 or 3 rings 15 to 30 tubs will be loaded, depending on their volume.

The stone is loaded in two distinct phases:

- that lying above the level of the tracks is loaded by means of the shovel loader (95 %);
- that lying below the level of the tracks is loaded by means of the grab attached to the crane(5 %).

3.2. Analysis of loading performance

3.2.1. Volume of rock loaded by the two processes

Since shovel loaders load faster than grabs, as much rock as possible must be loaded using the shovel loader. This is achieved by lowering the ends of the track leading to the panelling. The normal track level is approached gradually and is reached at the points. Ninety-five per cent of the rock is thus excavated by means of shovel loaders. A shovel loader can easily push a full tub up the sloping tracks. On the horizontal section the shovel loader is powerful enough to move a train of loaded tubs to the general transport system.

3.2.2. The decisive influence of compressed air pressure

In pits with central compressed air supply the large distances to the shafts have meant that the air pressure at the headings is reduced from 6 bar to 4.5-4 bar. This is obviously detrimental to all phases of the driving cycle, in particular in rock loading. Consequently, tubs were only 60 % full and uneven loading occured regularly.

A neat and profitable solution to this problem was found by using an Atlas Copco SG 3 screw compressor at the workface. Compressed air pressure rose to around 7 bar, and the tubs were filled to over 75 %. The penetration force of the grabs increased and changeover of tubs became more rapid. The drilling phase also became much shorter.

Figure II shows this development in quantitative terme, the overall reduction amounting to 19 %. It is clearly important to allow a certain number of drivage machines to operate under optimum conditions as regards compressed air supply.

3.2.3. Use of high-capacity loaders

Where compressed air pressure is satisfactory, shovel loaders of a greater capacity, the Atlas Copco LM 70, have been installed. On account of its size, this machine is optimum in moving and placing other drivage equipment.

With this shovel loader the rock loading phase is shortened by 40 or 50 % depending on whether a shovel of 0.26 m^3 or 0.4 m^3 is used. This is possible since tub loading improves further to over 95 %.

This process is illustrated in Fig. II.

3.3. Practical experience

Day-to-day practical experience of stonedrift driving using shovel loaders indicates the following:

1. the simple and robust design of the loader ensures trouble-free operation and low maintenace costs;
2. apart from the laying of extension tracks, there are no temporary installations and arrangements; final fitting-out of the stonedrift is carried out immediately after driving;
3. crew numbers can remain the same for all phases of the driving cycle, with no significant delays slowing down part of the crew; crews normally consist of four workers and one supervisor;
4. side loading and interruptions result in the theoretical loading capacity being greatly under-utilised;
5. an adequate compressed air supply is very important.

4. USE OF THE GRAB LOADER IN THE CAMPINE BASIN

The grab loader is an electrically powered, hydraulically operated loader mounted on caterpillar tracks. Two moving loading arms push the stones onto a sectional movable gate-end loader, which discharge them onto the subsequent means of transport - tubs, belt conveyors or armoured conveyors. It is essential for the loading operations to be carried out continuously.

One of the first times such a loader was used at the Zolder colliery for the drivage of a belt incline of -4°, where tubs obviously could not be used.

The grab loader selected for the job was the Russian machine 2 PNB 2, on which was mounted a HIAB crane. The belt and monorail installations needed for the drivage were installed permanently.

This method was a precursor of the method now being used by the firm FORAKY at the Eisden colliery, though in this case normal stonedrifts are being driven.

For the characteristics of loader 2 PNB 2, see Table II.

4.1. Installation and adaptation of drivage equipment

A general outline of equipment is given in Fig. III. All the transport equipment used in this case is temporary and will be replaced by the permanent infrastructure at approximately 250 m from the face. The loader selected is the 2 PNB 2. The following additional arrangements had to be made:

4.1.1. Laying of an artificial running surface for the grab loader

In order to protect the outer edge of the caterpillar track during movement in the circular stonedrift lining, an additional temporary concrete panel is placed on the ground in each ring. These panels are the same shape as a segment of the circle and are as long as the track is wide. The stones can thus be loaded without the machine having to move sideways.

The amount of ballast subsequently used for fill can be reduced. However, more rock must be loaded by the grab.

4.1.2. Adjustment of the grab loader arm

To prevent some of the stones on the curved walls of the lining from remaining there, the grab loader arms must be adapted to the curvature of the lining. This is accomplished by attaching a curved metal extension piece to the arms.

4.1.3. Adaptation of the HIAB crane

Since no tracks are laid near the working face and since the grab loader is a wide machine which is difficult to manoeuvre:

- the HIAB crane is mounted on the 2 PNB 2 loader;

- the twinboom drilling jumbo is suspended by a special frame from two parallel monorails.

4.1.4. Removal of the stones

The stones are finally removed by means of two overlapping gate-end loaders. These are attached to the lining and rest on metal supports. The drive of the second gate-end loader is raised so that the tubs can be brought into position beneath the discharge point. At the face side the first armoured conveyor is lengthened by three or four pans depending on progress.

4.1.5. Transport of concrete segments

The concrete segments are transported in two phases - in track-mounted cradles up to about 30 m from the working face, after which they are removed, one segment at a time, by a compressed air traction unit suspended from a monorail to within the range of the HIAB crane.

4.2. Description of the lading operations

After the loader has been brought out from the position which it occupies during shotfiring, the stones are continuously loaded ou up to the level of the machine's running area. Net loading times are short, barely more than half of those achieved using the shovel loader. However, delays due to transport hitches are more frequent than with the other procedure. Loading times thus amount to 2/3 of those obtained with the LM 56 and are the same as those obtained in drivages equipped with the LM 70.

The first armoured conveyor is lengthened after each advance of approximately 5 m. After approximately 100 m drivage, the 2nd conveyor is dismantled and relocated as the 1st conveyor, with a start of about 25 m. The points are changed and the loading area is re-established. A normal crew working 6 shifts are needed for the changeover.

4.3. Practical experience

The use of grab loaders for the driving of stonedrifts offers undeniable advantages:

1. loading is continuous and rapid;
2. the loader and crane from a single piece of equipment, thus simplifying handling;
3. they can be used in inclined stonedrifts.

The following are the disadvantages:

1. stone transport is complex: the armoured conveyors must be lengthened daily and shortened periodically;
2. track-laying and the relocation of points are time-consuming since the HIAB crane cannot be used for this tasks;
3. the monorail system (3 tracks) increase the complexity of operations; it is moreover a temporary arrangement since it does not form part of the permanent fixtures;
4. the making and laying of the floor segments involve extra work and cost;
5. crews consist of 5-6 workers and one supervisor.

It is still too early to make a final comparative assessment of this system and the shovel loader system: further efforts have to be made to perfect various features.

5. CONCLUSION

The latest generation of shovel loaders is satisfactory if allowance is made for the constraints imposed by roadway cross-sections, limited manoeuvreability on each side and length of throw.

Owing to their increased output and loading capacity, the latest shovel loaders not only reduce loading times considerably but also improve the filling percentage of the tubs.

Crew numbers can be kept to a minimum, i.e. not more than two workers engaged in cutting, two engaged in hauling operations and one supervisor/ shotfirer per shift. All other activities, such as the periodic lengthening of the tracks and the raising of the track level, are carried out quickly by the same crew.

In view of the robustness and simplicity of the equipment and of its trouble-free operation, investment and maintenance costs are minimal.

To ensure optimum performance of the shovel loader, an air pressure of at least 6 bar is essential at the working face.

In special cases, such as inclined stonedrifts, where normal tub transport is not possible, and in horizontal drifts where the rate of advance must take priority over ouput, the use of a grag loader should be considered. The net loading time is considerably shorter, and air pressure is less important. However, this system requires greater capital outlay, more personnel and very strict planning. Work can easily be disrupted by all sorts of incidents owing to the complexity of the equipment.

Other alternatives such as the Hägglund loader with shuttle-cars and the Salzgitter bunker conveyor have been examined, but because of high capital outlay no tests have been carried out on such equipment.

As far as large-scale projects are concerned, work in the future will be geared towards the complete mechanization of stonedrift operations involving the use of tunneling machines.

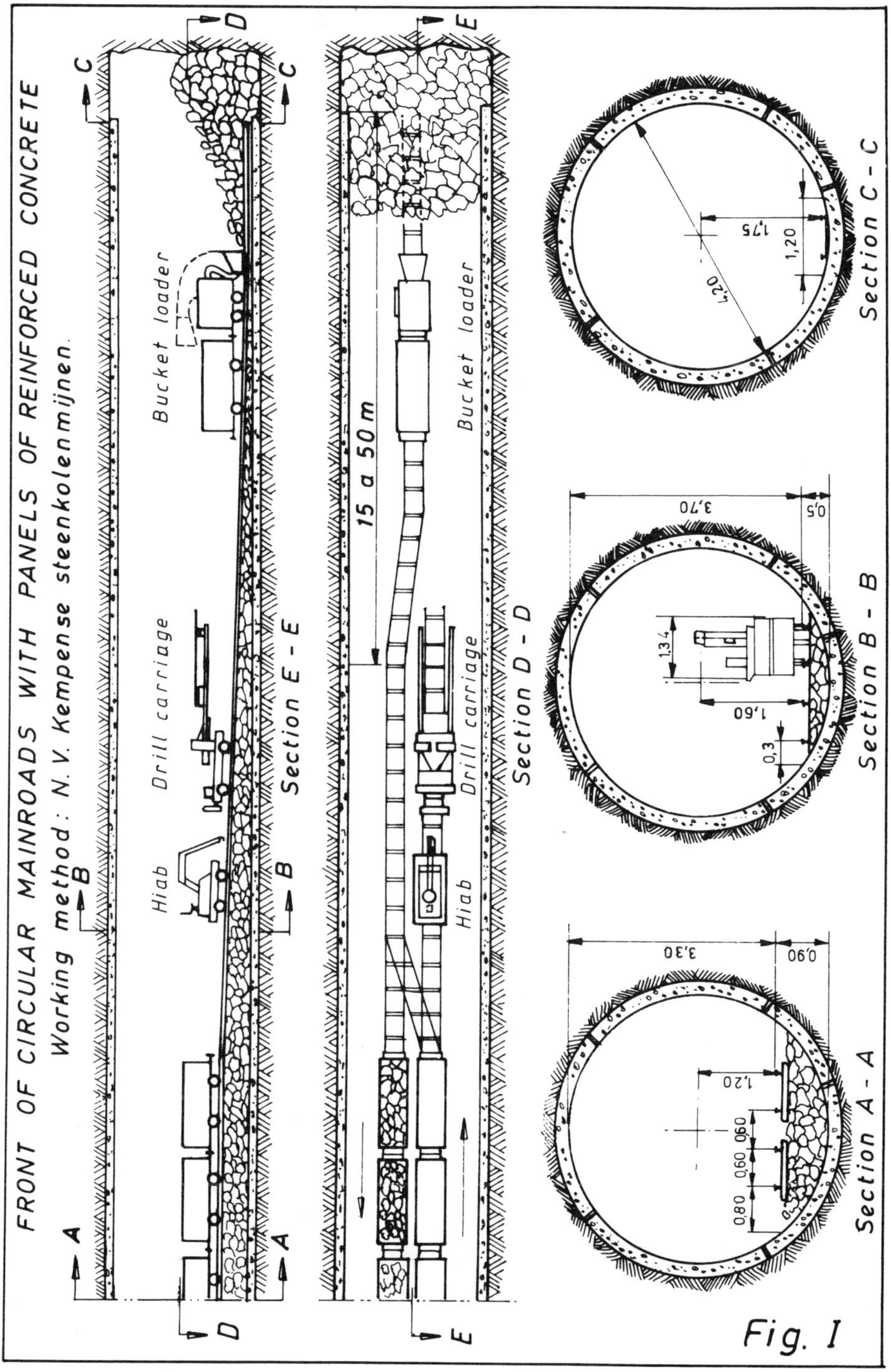
FRONT OF CIRCULAR MAINROADS WITH PANELS OF REINFORCED CONCRETE
Working method: N.V. Kempense steenkolenmijnen.
Bucket loader
Drill carriage
Hiab
Section E - E
15 a 50 m
Section D - D
Section C - C
Section B - B
Section A - A
4,20
1,75
1,20
3,70
0,5
1,34
1,60
0,3
3,30
0,90
0,60
0,80

Fig. I

SPECIFICATIONS of LOADERS

OVERHEAD LOADERS

		Atlas Copco		Salzgitter	
		LM56H	LM70	HL 221	HL 400
Dimensions:					
width	mm	1040	1340	1200	1475
height	mm	1450	1750	1450	1630
lenght	mm	2340	2720	2600	3050
Weight	kg	3150	4600	3600	6000
Clean-up range	mm	2850	3300	2660	2800
Bucket					
capacity	l.	260	400	250	400
width	mm	1000	1100	1000	1200
Motor rating at 6 bar					
tractionmot.	kW	10	10	9,5	10,3
bucketmot.	kW	17,5	2 × 11	9,5	17,6
Loading capacity	m³/h	60	80	60	108

Tab. I

GATHERING ARM L.

		2PNB2
Dimensions:		
width	mm	1800
height	mm	2350
lenght	mm	10990
Weight	kg	14260
Clean-up range	mm	1800
Motor rating (Elek)		
gather. arms	kW	2 × 15
conveyer	kW	1 × 20
undercarriage	kW	1 × 20
Hydraul oil		5 l/min.
fl. capac - press.		10 MP
Loading capacity	m³/h	140

Tab. II

DECREASING of the CYCLE TIME

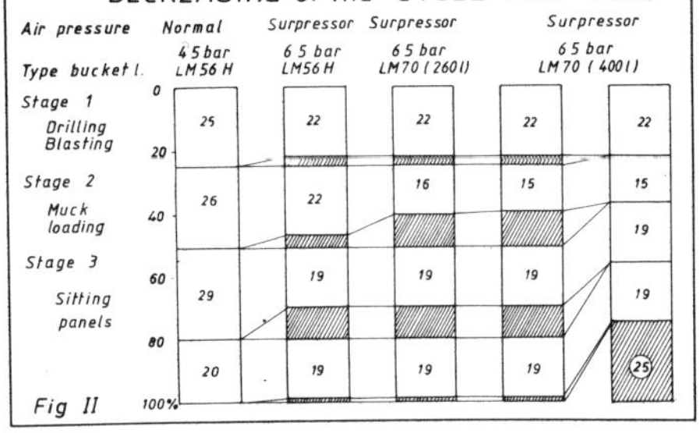

Fig II

FRONT OF CIRCULAR MAINROADS WITH PANELS OF REINFORCED CONCRETE

GENERAL EQUIPMENT: GATHERING ARM LOADER AND DRILL CARRIAGE

WORKING METHOD: N. V. FORAKY - COLLIERY EISDEN

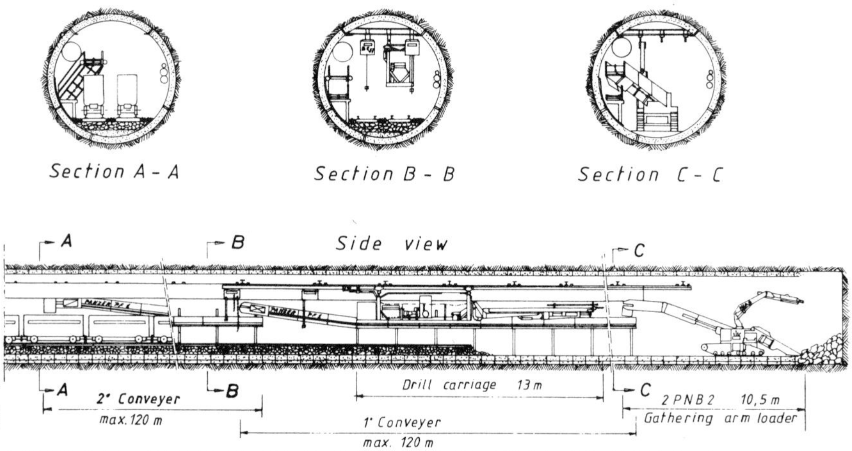

Fig.III

LATEST DEVELOPMENTS IN SHAFT CONSTRUCTION TECHNOLOGY

H.-G. Riess

Colliery Director
Gewerkschaft Sophia-Jacoba, Hückelhoven
Bituminous coal mine

Summary

The vertical, surface-connected shaft will, in future, remain the artery of a colliery. Its necessity as regards layout underground, manwinding, transport and, of course, ventilation remains unchallenged. Progress in the construction of surface-connected shafts has been one of the reasons behind the upturn in the meters of shaft sunk per year in the West German coalmining industry. Heavier-duty sinking equipment, new ways of mechanically breaking the strata and the perfecting of shaft lining techniques have all played their part. Aspects of shaft construction techniques of the future and their requirements are being studied in project-related R & D programmes.

1. SURVEY

Shaft construction has received a new lease of life in recent years with the increase in the number of shaft sinking projects. The last decade has seen the emergence of a new generation of surface-connected shafts in the deep coal mines of the Federal Republic of Germany (Figure 1). Shaft construction techniques have had to change considerably in the light of the greater depths involved, the wider diameters and the more stringent demands made on sinking operations. Remarkable performances have been put in using the latest sinking equipment and the provision of tailored sinking techniques and various styles of shaft lining has become the order of the day for sinking contractors. Vertical shafts are long-life structures and the underground network to which they connect gives a particular technical and economic importance to shaft sinking operations. Shaft construction techniques will no doubt continue to improve in the futre. In 1983 alone, work was proceeding on 6 surface-connected shafts; (Figure 2) shows the position as at April 1983. Major shaft projects are being planned or are in hand in other countries, particularly in France and England.

2. SINKING METHODS

The major factor affecting the choice of sinking method for underground excavations is the geology. Prior to actual sinking operations, a foreshaft is made, a kind of start pipe for subsequent sinking equipment. Depths now involved are such that conventional sheet piling can rarely be used. Deep wall slotting or shell-less concrete piling are more commonly used to secure the shaft top. Civil engineering plant is usually employed with the open pit work being carried out with two skins with or without a slip joint as conditions dictate. Plastic membranes have been incorporated as water-seals down to a depth of 100 m. The usual technique for stable rock is still drill and blast. Multi-boom drills have improved drilling work, dirt grabs capable of carrying up to 800 l have speeded up mucking out, while 7 m^3 kibbles - used for the first time at Haus Aden 6 (Figure 3) - have increased sinking performances (1).

The preferred winder for sinking operations is still the electric twin flat-rope drum winder with automatic cotter pinning for rope compensation. Motor ratings are as high as 1000 to 2000 kW. A 2 x 1160 kW twin drum winder was used for the first time on the Lauterbach shaft; these are conventional for shafts in English-speaking countries. The flat-rope drum centres, which are normally determined by the distance between the kibbles in the shaft, was modified at Polsum 2 shaft by angling the drum axis. Depending on the sinking sequence (sinking and lining alternating or in parallel), sinking scaffolds are selected or additional sinking plant brought in.

Where the freeze technique is used to consolidate non-competent overburden, blasting operations must be far more controlled so as to avoid damage to the ice wall and to the freeze pipes. At the Voerde shaft, for example, the explosive used was Donarit. Compressed air hammers also have to be used to trim the shaft walls within the frozen section. Discounting the preparatory freeze hole drilling and the measurement techniques required for this, the sinking equipment corresponds to that for hand-dug shafts. If the freeze shaft is to be lined from the bottom upwards by slip forming, it is the lining technique which determines how sinking shall be carried out in the frozen section of the shaft.

Two surface-connected shafts have been sunk in the Federal Republic since 1974 using full diameter shaft drills; these were the Asse shaft in salt and ventilation shaft 7 for Sophia-Jacoba mine (2). Where large diameters are involved, large diameter drilling still appears to be less economic than the freeze technique. It would not only be a step forward for freeze shaft construction if shaft sinking, which even today involves heavy physical work, could be carried out with the assistance of fully mechanized drivage systems. Trials are at present under way with a new shaft drill with hydraulic mucking out for use in stable rock (staple pit, Heinrich Robert

Colliery).

3. NEW METHODS OF BREAKING UP THE ROCK IN FREEZE SHAFTS

A vital factor in sinking performance, in addition to time-consuming loading out, is the selection of the correct breaking technique for the rock. Percussive work using compressed air hammers is out of the question for both performance and health reasons. On the other hand, a hydraulic hammer has been used successfully in salt deposits where disturbance from shotfiring is particularly significant (3). Experience has shown that this method is far superior to conventional jackhammer operations.

Another way of introducing mechanization, as has been done on the Voerde shaft, is to use a dinter loader with a powered shovel. The shovel is equipped with several, hydraulically powered pick hammers which bite horizontally into the rock. The broken (frozen) material was loaded by the shovel directly into the kibble, cutting out the need for a cactus grab.

Heading machines have been used experimentally for sinking through the frozen section of Polsum 2 shaft. Once initial experiments had proved the suitability of a roadheader, a few improvements were made to it before it was used on shaft construction work (5). The machine, whose direction of cutter head rotation was reversed (Figure 4) cut helically and backwards with a 40 cm undercut, producing an even, gently inclined floor. The average daily rate of advance was 1.4 m (machine used down to 90 m). The significance of the trial were the labour saving (3 instead of 6 men in the shaft sump) and the absence of all shotfiring. Valuable insight was gained into the best pick array for the cutter head. Despite some shortcomings, it would seem worthwhile pursuing this avenue and developing the system to technical and economic maturity.

4. LINING TECHNIQUES

The choice and size of lining is particularly important

when constructing a shaft. Rockbolts and wiremesh are used to pin together rock strata which might be subject to rock falls during sinking operations and injure the men. Non-reinforced concrete is almost always used to support the walls in stable strata. From a design point-of-view, the most reliable lining system is with concrete rings up to 4.20 m high, 40 cm thick with a horizontal joint of usually 30 cm (open or lined with light-weight blocks). This concrete cylinder formed by individual rings allows a certain amount of movement about the shaft axis should mining occur near the shaft and has its advantages for out-of-true situations.

Matters are far more complex when deciding on a lining system for freeze shafts, since freeze depths in future are going to extend down to 700 m (6). There will be numerous problems with constructing freeze shafts at such depths. Drilling precision for the freeze holes, high support resistance for shafts with sliding shuttering and the more massive ice walls required with the concomitant stress on the freeze pipes are the problems on which success depends. On the Voerde project, the concrete block inner lining (multi-skin in places) for the first time has been used to help the ice wall support the water and strata pressure. Low specific gravity bitumen can be used which forms a reinforced concrete/steel composite lining in the Voerde shaft. Good results were obtained with a new application of pre-cast concrete segments for sectionalized cribbing (Figure 5) which made redundant the old, time-consuming construction of bricked cribbing. In addition to load-bearing and water-proofing, the main function of the shaft lining in non-competent overburden is to provide stability, particularly where coal is to be extracted in the shaft vicinity. Present sinking rates in freeze shafts, including those sections requiring special lining, are approximately one meter a day, which is only 30% of the rates achieved in hand-got sections of shaft.

5. NOTABLE SHAFT CONSTRUCTION PROJECTS (TABLE 1)

The surface-connected shafts in the East of the Ruhr

coalfield do not require any special sinking techniques because the overburden is usually competent. One of the most recent shafts is Haus Aden 6 whose finished diameter is 7.50 m. For the first time in the Ruhr, the more than 40 m deep foreshaft was sunk using a small excavator on the shaft floor and a truck crane to hoist the direct kibbles. Since the foreshaft was so deep, the rest of the shaft could be sunk immediately with the concreting scaffold and underdeck mucking unit. At the beginning of 1980, Gesteins- und Tiefbau GmbH, the sinking contractors, set up a new European shaft sinking record, when they completed 131 m through the Emscher marls within 25 working days. Two heavy-duty winders carrying the largest kibbles at that time were used to complete the 1388 m deep shaft, the deepest at any Ruhr coal mine.

Sinking of the 8.0 m finished diameter Polsum 2 shaft had to start within an ice wall because the overburden was not competent. Thyssen-Schachtbau GmbH installed a modified Voest-Alpine AM 50 roadheader on the 100 m deep floor of the frozen section. A further mechanical and electrical innovation was the use of a specially developed twin flat-rope drum winder (2 x 920 kW). Sliding shuttering was used on the frozen shaft, followed by concrete rings in the hand-got shaft which reached the final depth of 774 m at the end of 1981.

Part of the extension programme for the "Houilleres de Provence" combined mine at Gardanne is the construction of the new shaft "Y". The contract was awarded to Deilmann-Haniel and the French company Entreprises Quillery. Shaft "Y" is downcast and will be used for manwinding and materials. Its finished diameter is 10 m and it will be 1500 m deep, making it the deepest shaft in the European mining industry at the present time (Figure 6).

The strata are competent and mainly comprise limestone and sandy marls. B 25 cast-in-situ concrete is used for the 50 cm thick shaft lining. The very large shaft section - more than 50% more than for an 8 m shaft - paid a particular role in the selection of the shaft sinking equipment. The sinking frame is designed to cope with large loads at great depths and is such that a 7 m^3 kibble can be used right down to 1500 m. Almost 80% of the shaft is now complete and the equipment has proved fully satisfactory.

Work is still continuing on the freeze shaft with the deepest lining section installed by sliding shuttering, viz. the 581 m freeze section of the Voerde shaft. The final lining installed by Gewerkschaft Walter in the 6 m diameter shaft is a technical innovation for shaft construction. Some 12 million kJ/h refrigeration capacity were required, making this the largest refrigeration plant used at any freeze site to date. The final depth of 1060 m should be reached during 1985, by which time further new techniques will have been applied.

6. R & D WORK

If the best shaft construction technique is to be used and the ice wall economically dimensioned, then it is necessary to know the in-situ behaviour of the rock which is to be artificially frozen and its soil mechanics characteristics. A research project "Deep, frozen shafts", funded by the Federal Ministry for Economic Affairs, has helped extend the frontiers of knowledge with numerous measurement programmes and rock mechanics investigations on the Voerde shaft project (Bergbau AG, Niederrhein, Ruhrkohle AG), the deepest freeze shaft at that time (7). In addition to novel inclinometer measurements in special research boreholes, temperature and convergence measurements on the ice wall provided much insight into the stressing/deformation pattern for the macro-system.

The calculation of cylindrical ice bodies and the prediction of potential wall shaft for the ice wall are one aspect of the recent Bergbau-Forschung GmbH project "New shaft construction techniques", which receives funding from the Federal Ministry for Research and Technology. Work is in conjunction with various specialist mining contractors and covers as yet unresolved themes and problems associated with deep freeze shafts, together with the technical refinement of shaft drilling. The new freeze shaft Lohberg 4 is also included in the project whereby strain gauge measurements directly on the freeze pipes should provide an indicator for the first time of the actual stressing of these pipes and help clarify the mining consequences when constructing surface-connected shafts of this type.

Though the output of bituminous coal may vary in the course of time, we still need to access replacement reserves through surface-connected shafts.

Shaft sinking performance at Selby

Sinking through coal measure strata - the best results were obtained at Stillingfleet and at Whitemoor.

At Stillingfleet, progress of 128 metres in a month was achieved at a depth of 400 metres.

At Whitemoor, this was exceeded by the sinking of 132 metres in a month at a depth of 600 metres. In both cases the lining was concrete 305 metres thick.

In frozen ground, sinking through Bunter Sandstone at North Selby, an average advance of 12.75 metres per week was maintained over a 16 week period, with a lining thickness partly of 600 millimetres, and partly of 800 millimetres.

All the shafts are 7.315 metres diameter.

The contractor at Stillingfleet and Whitemoor was Thyssen (GB) Ltd. The contractor at North Selby was Cementation Mining Ltd.

Bibliography

1 Micke, H.: "Records in shaft sinking in European mining". Glückauf 117 (1981) No. 14, p. 891.

2 Vassen, P. and Knickmeyer, W.: "New shaft sinking techniques". Glückauf 115 (1979) No. 9, p. 383.

3 Dobert, H.: "Interim Report - Sinking of Riedel airshaft". Unser Betrieb, Deilmann-Haniel GmbH, No. 26 (1980) p. 26.

4 Bittner, F.: "Voerde: 400 m in iced strata". Für unsere Mitarbeiter, Gewerkschaft Walter, June (1982) p. 17.

5 Wiesenfeldt, L.: "Mechanised sinking of Polsum 2 freezer shaft". Mitarbeiterbrief, Thyssen-Schachtbau GmbH, No. 2 (1982), p. 10.

6 Weber, H.: "Future mining problems of the German coalmining industry and their solution". Glückauf 117 (1981) No. 16, p. 1006.

7 Klein, J.: "Deep freezer shafts". Final Report, Bergbau-Forschung GmbH, Project No. 591 214 RAG 412 (1983), p. 43.

Kenndaten der erwähnten Tagesschächte

		Haus Aden 6	Polsum 2	"Y" Gardanne	Voerde
Endteufe	m	1388	774	1150	1060
Durchmesser licht	m	7,5	8	10	6
Ausbau wasserdicht bis	m	93	106	-	617
Fördermaschine	kW	2000	1840	1600	1600
Ausbaustärke wasserdicht	cm	Stahlbeton B 25 45	Stahlbeton bis B 45 55	-	Stahlbeton bis B 55/ Stahlverbund St 52 70
Karbon	cm	Stahlbeton B 25 45	Stahlbeton B 25 40	Stahlbeton B 25 50	Stahlbeton B 25 40
Ausführende Firma		Gesteins + Tiefbau GmbH	Thyssen Schachtbau GmbH	Deilmann-Haniel GmbH + Entreprises Quillery	Gewerkschaft Walter

Table 1. Data on surface-connected shafts reviewed.

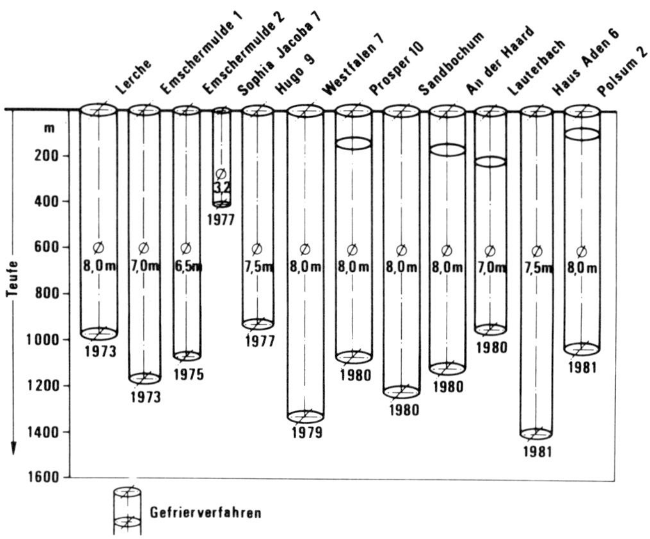

Figure 1. Complete surface-connected shafts in the coal mining industry.

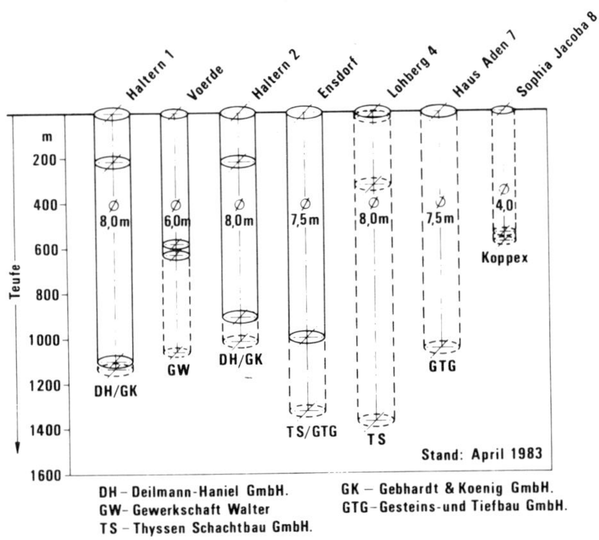

Figure 2. Surface-connected shafts under construction in the coal mining industry.

Figure 3. Sinking headframe above Haus Aden 6 shaft.

Figure 4. Modified roadheader in Polsum 2 shaft.

Figure 5. Crib of pre-cast concrete sections for the outer lining in the Voerde shaft.

Figure 6. Gardanne shaft.

COMPARISON - SHAFTS VERSUS DRIFTS

D.J. STEPHENSON

Chief Planning Engineer, National Coal Board

Summary

This paper gives a brief history of shafts and drifts as means of accessing coal measures, and describes the merits of each from the point of view of safety, geology and transport. The effects of modern technology and environmental considerations on the choice between the two methods are examined, the equipment available for each is described, and the costs of the two systems are compared.

1. INTRODUCTION

The British Mining Industry has experienced two major periods of expansion within recent memory, these coming in the 1950's and after the oil crisis of the early 1970's. During these periods it became necessary to conduct major reconstructions of existing mines and to embark upon the development of entirely new mines. In each period the fundamental problem of comparing access to reserves by drift or shaft arose and became the subject of much debate among mining engineers.

Prior to the 19th Century coal mining in Britain was generally limited to the working of the shallowest reserves in the exposed coal-fields by drift or shallow shaft mines. The absence of high capacity haulage systems limited the adoption of drift mining. As mines became deeper in the 19th Century and the early 20th Century developments were almost exclusively by shaft and general mining practice dictated that mine workings more than 300 m in depth could not be economically accessed by inclined drifts.

In the more recent periods of development in Britain the mining engineer has been faced with three distinct types of mining problem, namely:-

(a) to find methods of accessing new reserves at existing collieries;
(b) to access shallow seams from new mines; and
(c) to access deep seams from new mines.

One must recognise that in the U.K. there are some significant differences in mining practices as compared with the mainland of Europe. British mines are in general designed to work within the seam, the seams for the most part being reasonably flat. Horizon mining is, therefore, unusual in the United Kingdom. When new seams are accessed from existing mines, it is common to drive inclined drifts from the current pit bottom level or workings, upwards or downwards, to the new seam. In this way developments can take place with the least effect on the current operational efficiency of that mine. Further development is then carried out by main roadways driven in-seam.

The factors that are generally taken into account when considering methods of access at new mines are as follows:-

(a) Safety
(b) Geography and geology
(c) Method of formation
(d) Ventilation
(e) Transport
(f) Cost

This paper discusses each of these major factors and indicates how they have influenced recent British practice and how they continue to influence future plans.

2. SAFETY IN SHAFTS AND DRIFTS

With regard to manriding the potential for a major accident in a shaft mine is considerable as there is no real preventative measure yet available to arrest a falling cage. In a drift, arrestors can be strategically placed and manriding trains provided with braking systems.

In a shaft mine in the very remote event of a major failure of the existing facility (i.e. two shafts out of action at the same time) there is no exit. However, a drift provides a walkable outlet from the mine.

Despite these apparent advantages of drifts, the number of serious accidents occurring in shafts and main drifts has been so small that there are no meaningful statistics to indicate an advantage for either method of access. Decisions between the two means of access to reserves are therefore unlikely to be based on safety aspects but upon operational criteria.

3. GEOGRAPHY AND GEOLOGY

The sinking of a new mine in the United Kingdom is a major event attracting the attention of many interested groups outside of those normally connected with the coal mining industry itself. The new mine would be a scheme with a capacity of not less than 2m saleable tonnes per annum; new mine proposals giving output below this level are unlikely to be considered. The environmental lobby is very strong and the siting of the surface of a new mine is therefore the subject of public planning enquiries. The major issues considered at such inquiries relate to changes in the appearance of the countryside, the disposal of mine waste, the development of local infrastructure,the effects of mine subsidence and the overall impact on the life of the community. The engineer therefore has to make many compromises between the ideal surface position for access to the new reserves and that which is acceptable to the community at large. It may be necessary to access the reserves from a surface position outside of the plan area of the reserves to be worked. This in itself will influence the decision on the type of access that is finally adopted. Where the access is within the plan area of the reserves the sterilised area left to protect the stability of the drift or shaft must be taken into account. The pillar to protect a drift will be far greater in area than for a shaft. In certain circumstances this may dictate the choice of shaft access (see Figure I).

Most new access arrangements for coal reserves in the concealed part of the British coalfields are required to pass through water bearing strata of varying thicknesses. A knowledge of the thickness and strength of this strata and potential water flows is essential prior to any mining work taking place. Mining work through these beds is difficult and expensive, and the presence of these beds alone has led to decisions being taken for shaft rather than drift access, as the drift will have a longer driven length through the affected zone.

The depth of the deposits was once regarded as the single most significant factor in the choice between shaft and drift, and was once the limiting factor with regard to the use of drifts. However the development of conveying and haulage systems suitable for modern requirements (Figure II) has meant that access drifts are now being driven to reserves up to 800 m in depth.

4. METHODS OF FORMATION

Between the late 1950's and the mid-1970's shaft sinking experience was extremely limited in Great Britain. Once the new surge of activity took place in the 1970's engineers began to re-learn established methods and to examine ways of increasing shaft sinking rates. Conversely inclined drifting practice, which has so much in common with normal roadway drivage techniques underground was the subject of continuous development throughout the 50's, 60's and 70's, when all underground activities in coal mines were being mechanised and improved. Traditional drill and fire techniques are being rapidly replaced in all but the strongest strata by cutting techniques.

In both drifts and shafts it is necessary to find a suitable method of treating the surrounding strata when driving through wet or very soft ground. In 1977, 53.5% of our development drivages were being driven by roadheaders. By 1983, this proportion had increased to 68.3%. As average advance rates for roadheading work are double those for more conventional methods, this trend is likely to continue.

As roadheading techniques have developed the need has arisen for heavy duty roadheaders capable of making major drivages through strata up to 17,000 p.s.i. in Uniaxial Compressive Strength. The Board employ 26 of these machines in its expansion programmes and new mine developments for cross-measure drivage work. Experience is also being gained with a full face tunnelling machine of the Robbins design at Selby.

One of the most recent developments in the field of shaft sinking has been the blind shaft boring technique which has originated in the United States, firstly as part of the Defence programme, but latterly as a means of very quickly and safely driving shafts (Figure III). The advantages of such a technique are many. Construction takes place with no men working below the collar until formation and lining are completed. The first claims suggest sinking rates of three times those for more conventional methods. In New Mexico, a 6.1 m diameter shaft to a depth of 1067m, which would take four years to complete with conventional methods, can be drilled and lined in one year. One major advantage of this technique is that no prior treatment of weak or water bearing ground is required as the drilling fluid density can be varied to prevent wall collapse or water inrush (Figure IV). More severe water makes can be treated by the formation of cement plugs and re-drilling. Costs of blind shaft drilling are claimed to be equal to or below conventional methods, but the advantage in time does mean an earlier return on capital invested.

5. VENTILATION

Ventilation requirements for new mines mean that access drifts or shafts will have to carry considerable quantities of air. A balance has to be achieved between air velocity, cross-sectional area, resistance and power requirements.

Where surface drifts can replace shafts and long lengths of underground airways, for example in accessing undersea reserves, they become more attractive both in terms of initial cost and ventilation power costs.

However, drift dimensions must ensure acceptable air velocities in coal conveying roads. In shafts, velocities up to about 8m/s are acceptable with rope guides and about 13m/s with rigid guides, and the dust aspect does not arise. If drifts are used in place of shafts then ideally they should be of greater total cross-sectional area than the equivalent shafts in order to limit the velocity in the mineral conveying road.

The following table compares ventilation requirements between a shaft and a drift for a 600m depth.

Assumptions:-

(a) Pair of circular shafts, 7m diameter, concrete lined fitted with buntons, pipe ranges, etc.

(b) Pair of drifts, smooth lined and of various dimensions and functions as stated. Length 2400m.

(c) Required airflow = 250m^3/S

(d) Ventilation power cost equivalent to £270/KW/year and overall fall/drive efficiency of 80%.

		D R I F T S				
	Shafts	5m Circular Unobstructed	5m Circular with Max Im. Infill and Coal Conveying		5.5m x 4.25m Arched with Coal Conveying	
			Conv. Drift	Other	Conv. Drift	Other
Maximum Area(M^2)	38.5	19.6	16.8	16.8	19.4	19.4
Available Area(M^2)	33.0	19.6	15.3	16.8	17.9	19.4
Velocity (M/S)	7.6	12.75	16.34	14.88	13.97	12.89
Pressure Loss per 100m (kPa)	0.023	0.039	0.085	0.067	0.065	0.054
Pressure Loss for each Shaft/Drift(kPa)	0.138	0.938	2.032	1.608	1.569	1.283
Total Pressure Required (kPa)	0.276	1.876	3.640		2.852	
Air Power (kW)	69	469	910		712	
Elect.Power(kW)	86	586	1138		890	
Annual Cost	£23,220	£158,220	£307,260		£240,300	

NOTE - With the drift dimensions used the resulting velocities become important, particularly in conveyor roads. For example, in the circular tunnel the air velocity of over 16m/s (3200 ft/min) would be excessive. Conveying in the return reduces the relative velocity by the belt speed but does not affect the "belt stood" situation and introduces additional problems of conveying through a surface air-lock.

6. TRANSPORT

Surface drifts have clear advantages for the transport of mineral as a belt conveyor allows high capacity and continuity of flow. The flow can be regulated by conveniently sited bunkerage both at drift bottom and between drift mouth and coal preparation plant.

Some measure of the high capacities involved are illustrated by the proposal for the Selby coalfield. The National Coal Board's plans for Selby illustrate an interesting compromise between shaft and drift access, allowing minimisation of travelling times to different parts of the coalfield while retaining the advantages of a concentrated and continuous coal clearance system (Figure V).

The coalfield will be worked as one large integrated mine. Over the take, five twin shaft sites will provide ventilation, manriding and materials transport to specific areas of the coalfield. Two drifts, both 14,800m long, known as the "Spine Roads" are being driven from a surface position to the South-West of the coalfield, dissecting the field and taking all the output. The vertical lift over the length of the roads is 1000m and the conveyors in both drifts are being designed to carry in excess of 2,000 tonnes/hour. The South Drift conveyor, to a design by Anderson Strathclyde Limited, will be a 1,300mm wide steel cord belt. The power unit, at 8,000 kw will provide various speeds up to 7.08m/sec. The North Drift conveyor, to a design by Cable Belt Limited, will be 1,050m wide. The power unit will input 7,260kw to give a maximum belt speed of 6.35m/sec.

While the Selby complex with its large surface area will employ drift access, the normal form of access at these depths is still by vertical shaft. One example is the new shaft being driven for access to the lower seams at Maltby Colliery in South Yorkshire. The size of the take and the central surface position of the existing installation precluded the use of an inclined drift to uprate colliery output by giving access to the lower levels at 1000m depth (Figure VI). This 8m diameter shaft will be required to wind a R.O.M. output of 3,667,000 tonnes per annum with a peak capacity of some 20,000 tonnes/day. The shaft will be equipped with 4 x 22 tonne capacity skips powered by two separate 3000 kw winder drivers (Figure VII).

With respect to materials transport, drifts have advantages, especially for long or large loads, as slinging and double handling are avoided. Materials can also be properly secured in the vehicles which transport them from stockyard to destination. In a long drift where the journey cycle time is extensive the interference between manriding and materials activities may cause problems unless there is adequate train load capacity. A shaft with its short cycle time does not normally have this problem.

7. COST

A recent analysis of drivage costs and the advance rates expected for both shafts and drifts is given in Table 1 below.

MARCH 1983 PRICES

	Drifts 4.90m dia. or 17 x 13 Arched	Shafts 7.3m diameter
A. Dry Strata		
i Cost of lined shaft	£1,300/m (best) £2,200/m (Average)	£5,000/m (best) £7,000/m (average)
ii Rate of Advance		131.23 m in 31 days (British record)
	30 m/wk (best) 16 m/wk (average)	30 m/wk (best) 15 m/wk (average)
B. Water Bearing Strata		
i Cost of Drivage or sinking	£9,600/m (best) £13,000/m (average) (costs include lining materials)	£8,500/m (best) £10,000/m (average) (costs include concrete lining)
ii Ground treatment	£10,000/m (freezing) £3,000-£10,000/m (grouting)	£8,000/m (freezing) £3,000-£16,000/m (grouting)
iii Rates of Advance	28 m/wk (best) 12 m/wk (average)	15 m/wk (best) 8 m/wk (average)

8. CONCLUSIONS

There has been an increased adoption of drift access to greater depth in the United Kingdom for the following reasons:-

(i) Faster drivage rates are available from mechanised mining systems.

(ii) Lining usually consists of colliery arches.

(iii) Surface equipment is environmentally more acceptable.

(iv) Drifts have a higher capacity and scope for continuous minerals transport.

(v) A walkable outlet has safety advantages.

(vi) Improved conveying and haulage technology are now available.

However, drifts have the following attendant disadvantages:-

(i) A longer length of drivage is required.

(ii) Larger sterilised pillars of coal are created.

(iii) Ventilation constraints exist.

(iv) New drivage techniques for shafts may make them far more economic.

(v) Ground treatment requirements are far greater than for shafts.

There may be conflict between cost of formation and cost of operating which influences final decisions.

It would appear that to design a Mining complex combining the use of shafts and drifts gives the best operational efficiency and lowest running costs. Also to overcome the increasing environmental pressures being imposed on our industry we must look at underground connections with facilities for men and materials to be accessed at several points from the surface and the coal to be brought out at one point only. The development of the Selby coalfield provides an ideal example of a mining complex combining the best aspects of shaft and drift access.

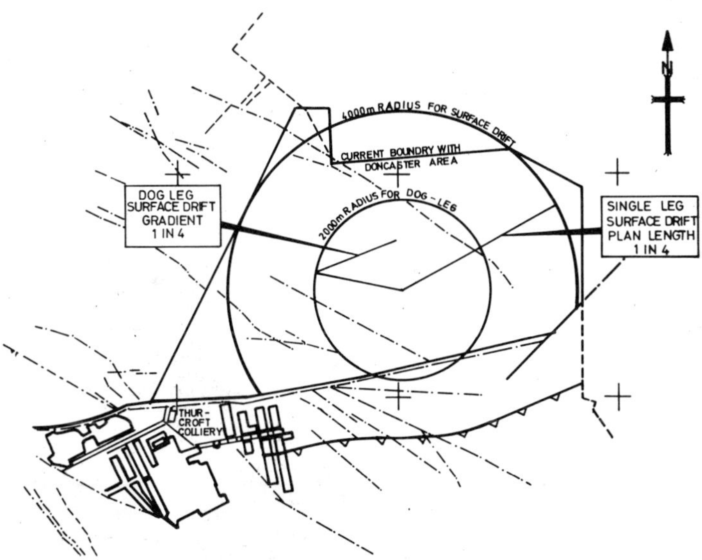

Figure 1

Figure 2

Figure 3

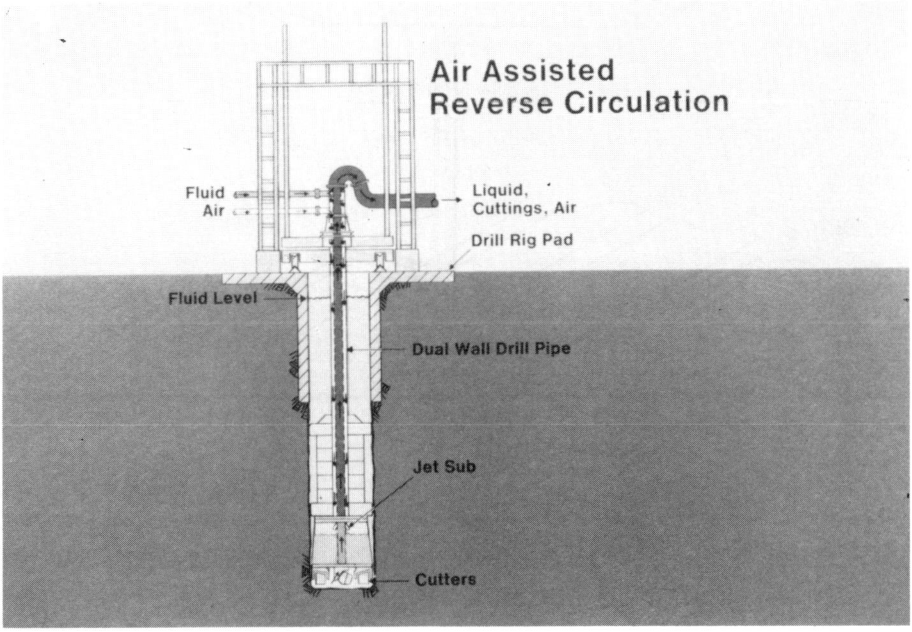

Figure 4

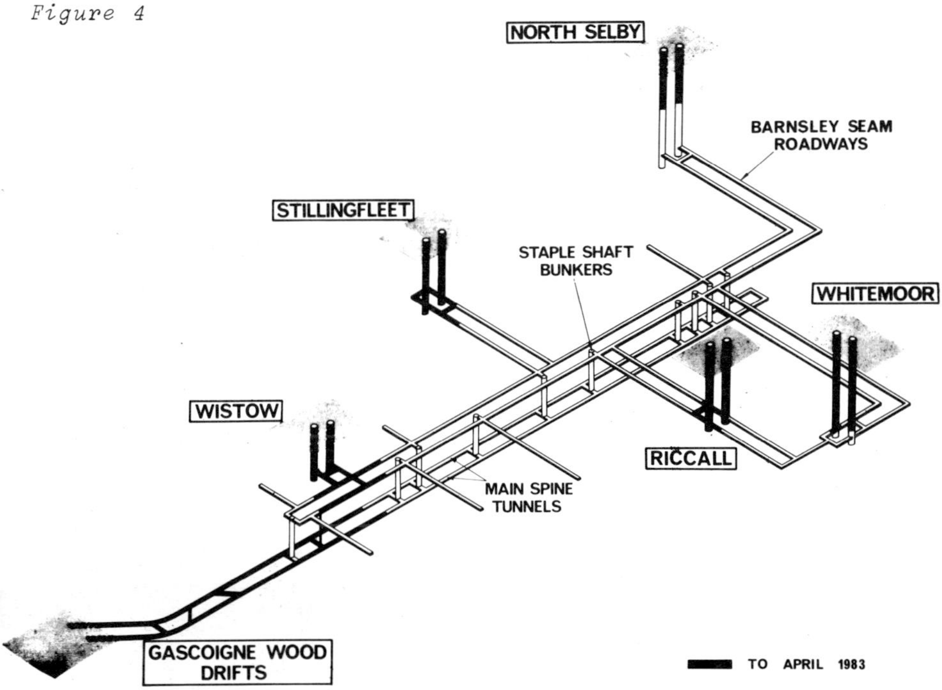

Figure 5

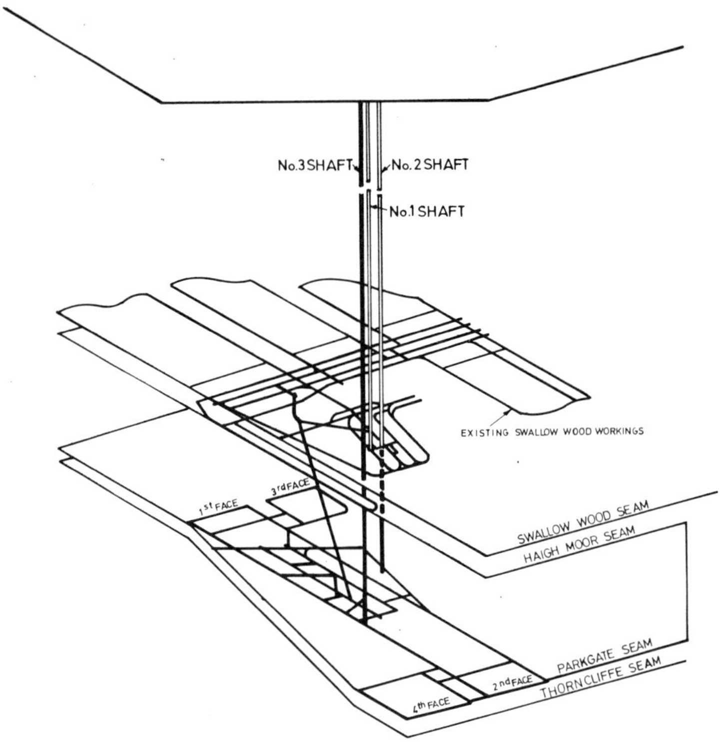

Figure 6

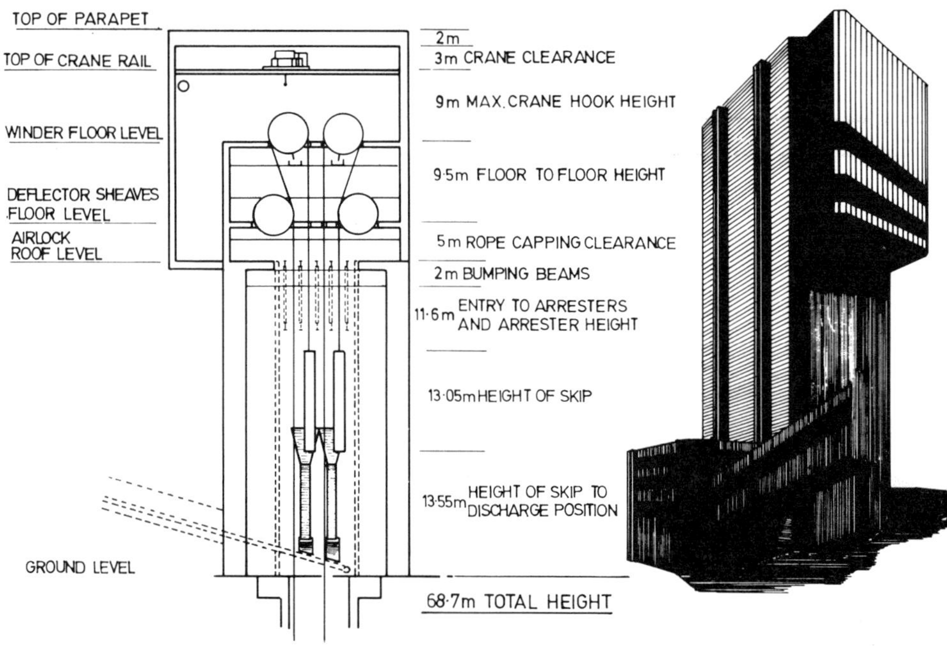

SHAFT TOWER

Figure 7

CLOSING ADDRESS

André DE GREEF

Head of Technology Division
Directorate-General "Energy"
Commission of the European Communities

Ladies and gentlemen,

Please allow me to address a few final words to you as this information symposium draws to a close.

First of all, I should like to thank, on behalf of the Commission of the European Communities and in my own name, all those who have contributed from near and from far to the success of this conference, whether it be in their participation as delegates behind the scenes or in the interpreting booths. Without them and their cooperation it would have been impossible to stage such a symposium and to make pass on to the mining world the latest information or roadway drivage techniques and technology in the coal mines of the European Community.

During these two days, which are still not quite over, we have been able to exchange our expertise in an area of vital importance in coal mining, namely roadway drivage and other preparatory work, an area where extensive progress must still be made as it is the prime element of coal-getting. This afternoon you will be able to continue your exchange of experience from a perhaps more specialized angle at the round table discussion organized for this purpose.

These meetings have enabled each one of you to assess the state of the art in an atmosphere which has been both pleasart and marked with mutual cooperation, in a town which knows how to offer good hospitality. During the course of this conference you will certainly have renewed old acquaintances hips and had the opportunity to make new ones.

It is my deep conviction that all of the research work carried out in a real spirit of Community cooperation in this area of roadway drivage techniques has helped to boost coal production in the Community. But apart from the technical progress, I believe that these two days, our hard won research results of, our common effort, contribute something in their own way to the sometimes difficult construction of a Europe in which our children will grow up tomorrow. The success of this symposium is concrete and tangible proof of this, and each one of us here will have the feeling that he has done something to build the Europe of today and the future.

I should like to express my thanks, once again, to all the participants and in particular to those who have contributed to its success, notably Mr. K. Reichert, for his presentation of the coal policy of the Community, Mr. John Mills, Deputy-Chairman to the National Coal Board, for his brillant introductory talk, the Chairman of the various sessions, those who presented papers and those who led the discussions, the Scientific and Technical Preparatory Committee, the Directorates-General "Information Market and Innovati and "Administration", the Conference Services and interpreters and lastly, Miss Goebels, Mr. Linster and Mr. Fred Kindermann who have each played their own part with dedication.

I now declare this information symposium closed and wish you all a safe journey home to your respective countries and families.

ROUND TABLE DISCUSSIONS

- ROUND TABLE A - FULL-FACE TUNNELING MACHINES
- ROUND TABLE B - BOOM HEADERS
- ROUND TABLE C - MECHANIZATION OF SUPPORT
- ROUND TABLE D - FUTURE DEVELOPMENTS

ROUND TABLE DISCUSSION A - FULL-FACE TUNNELING MACHINES

Chairman : P. PHILIPPRON, Ingénieur principal,
siège Zolder, N.V. Kempense Steenkolenmijnen,
Zolder

1. INTRODUCTION

Mr. PHILIPPRON

The purpose of the discussion was to consider a number of specific points in connection with the use of full-face tunnelling machines for stone drivages.

The subject to be discussed was important and covered a broad field. Its importance reflected the significance of the objective to be achieved and the means employed while its broadness resulted from the wide range of technical options currently available, with the very rapid developments currently taking place in drivage technology.
The objective to be attained was to drive roadways which met certain criteria such as :

1. ability to withstand strata pressure without excessive deterioration ;
2. ability to satisfy ever-greater requirements in respect of ventilation and increasingly rapid transport ;
3. acceptable cost so that they met the requirements of reasonable economic management ;
4. complete safety during the drivage operation ;
5. rapid drivage to conform with the deadlines imposed by the mining programmes (rapid drivage could in fact affect the viability of working a deposit).

At the previous symposium on drivage techniques in 1973, only the German engineers had reported satisfactory results with full-face tunnelling machines. At that time, only one drivage of this type was at the planning stage in the United Kingdom and none was planned in the French or Belgian mines.

Ten years later the situation was as follows :

- Federal Republic of Germany : 7 tunnelling machines in service ;
- United Kingdom : 2 tunnelling machines in service ;
- France : 1 drivage planned for December 1983 ;
- Belgium : 1 drivage planned for April 1984.

2. SUMMING UP OF POINTS RAISED BY THE AUTHORS OF THE PAPERS

Mr. PHILIPPRON

2.1. Mr. HOEVELHAUS had given a very full review of trends and results to date in the Federal Republic of Germany. The following points were to be noted :

1) an increase in cut diameter from 4.80 m to 6.50 m ;

2) a tendency to use a cutterhead with a flat face (on the inbye side) ;

3) a trend towards single-disc cutters ;

4) mechanized support setting as close as possible to the heading face ;

5) increasing use of lasers and microprocessors for steering in the vertical and horizontal planes ;

6) a preference for smaller machine components for ease of transport and assembly ;

7) a thorough investigation of the problems of ventilation, safety and fire precautions ;

8) a very concise review of the problems involved in negotiating faults and other geological disturbances.

2.2. Mr. MASSEY had described the Robbins tunnel boring machine used for a long inclined drivage at the Selby complex and had drawn attention to the following problems :

- cutter replacement ;
- filling of the invert (infill dirt or ballast) ;
- use of a laser ;
- supports ;

- ventilation problems ;
- negotiation of a fault zone.

2.3. Mr. BLANC had set out the reasons for using a full-face machine in France and had then described the Bouygues tunnel boring machine, which was of novel design.

3. DESCRIPTION OF THE IMPROVEMENTS MADE IN VARIOUS TYPES OF FULL-FACE TUNNELLING MACHINE

The various representatives attending gave details of the latest innovations introduced by their firms.

3.1. Demag tunnelling machine

Mr. MERTENS (Steinkohlenbergauverein) mentioned that Demag was currently developint a new type of cutterhead which would allow all the cutters to be replaced from the rear with no need for men to work between the heading face and the cutterhead.

Demag was also developing a device for drilling strata bolting holes.

3.2. Wirth tunnelling machine

Mr. VOERCKEL (Wirth) gave a brief description illustrated with slides of the Wirth tunnelling machine (6.50 m in diameter) recently commissioned at the Lohberg Mine, where it had driven 2100 m of roadway to date with a mean rate of advance of 15 m/day. This machine was fitted with a circular support erector for mechanized setting of the circular TH steel arches (using a section weighing 44 kg/m) without any need to interrupt cutting and loading. The supports were preassembled, working up from the foot section by section, the bolts of all the joints except one were tightened and they were slid into their final position before being set against the ground. The tunnelling machine was also fitted with a roof bolting drill (it was currently not possible to set the bolts radially). It took about 30 minutes to cut the ground and set supports for and advance of 75 cm. No risk was incurred when changing cutters since the cut chamber (6.50 m in height and 0.80 m in depth) was protected by a safety shield which was advanced by rams and pressed against the heading face and by a forepoling system which protected the exposed roof in front of the cutterhead.(This second device, which could also be advanced was not yet entirely reliable and installation was fairly time-consuming). On two tunnelling machines currently under construction it would be possible to change the cutters in complete safety from behind the cutterhead.

Hydraulic devices had been developped to ensure that the diameter cut after changing the worn peripheral cutters would not be greater than that obtained before fitting the new cutters.

The cutters used were of the single-disc type and could exert a static thrust of 20 t. The cutter fastening arrangements had been modified.

3.3. Jarva tunnelling machine

Mr. KARNELO (Atlas Copco-Jarva) stated that several of this company's tunnelling machines operating in mines in the USA had been equipped with microprocessors to regulate thrust and rotation speeds in the light of the experience acquired by the company during drivage by Raise Borer. Tests were still being carried out in order to improve this technology and to optimize the operation of the tunnelling machines by reducing the influence of the human factor, i.e. the development workers. Trials were being carried out with a small Mini Fullfacer in order to obtain further information.

3.4. Bouygues tunnelling machine

Mr. MONTACIE (Entreprises Bouygues) mentioned that two machines built by this company were to be commissioned in coal mines in 1984, one in the Valenciennes Group (with steel supports weighing 29 kg/m) and the other in Campine (with concrete segments).

He gave a general description of these tunnelling machines, which were fitted with four arms, each carrying one disc. The thrust applied could be as much as 30 t per disc, i.e. a total of 120 t. The peripheral speed of the discs was fairly high (about 6 m/s) ; the single -disc cutters were 400 mm in diameter.

A major advantage of this type of machine was that it could be assembled without any need for an erection chamber and that all the components could be transported by the usual transport equipment of the mine.

For the machine to be used in the Nord/Pas de Calais Area, steel supports were to be set 1.20 m from the heading face without interruption of the cutting process. It was hoped that an advance rate of 2 m/h would be achieved. The supports were to be at 1.20 m centres. Tt was anticipated that 30 minutes would be required to set each support, this being the time required for the corresponding amount of cutting and loading. The components of the circular arches were set by means of a chain system. Changing of the cutters was carried out in a zone protected by the supports already set, the tunnelling machine having been moved back from the face. In the machine to be used in Campine, the forward thrust was to be transmitted to 10 radial grippers, each applying a maximum load of 50 t to the RC segment supports. The machine weighed 110 t and was fitted with 4 motors, each of 200 KW.

Each of the five segments, which were 30 cm thick and 64 cm wide, was set by means of a radial ram mounted on one of the cutting arms. In order to set the last (roof) segment, the revolving head was advanced towards the heading/face and then drawn back to put the segment in place. This support erector was rotated at low speed.

It was estimated that actual cutting for an advance of 64 cm would take between 20 and 25 minutes and that the support could also be set in 20 to 25 minutes, the anticipated rate of advance thus being 70 cm/h.

Mr. MERTENS asked how work was distributed between the different cutters, whether there was interference between the cutters and whether the arms were all of the same length.

Mr. MONTACIE explained that the arms were all of the same length. Each described a torus, revolving within a clearly defined annular zone so that the different arms did not interfere with each other. As soon as the cutters had reached the limits of their respective ranging zones, rotation of the arms was halted and the revolving head was advanced by between 15 and 25 mm and depending on the type of ground. Rotation of the arms was then resumed for the next cut. It was thus clear that thecutter arm was performing more cutting work than any of the others. The linear speed of the cutters increased progressively from the inner to the cutter arm. The arm cutting the central ring broke only 1/16 of the total surface area. The service life of the central cutters was much greater than that of the outer cutters. The unit of advance was 15 to 25 mm.

Mr. REULENS (International Mine Service) mentioned that his company was involved in a drivage with a Demag tunnelling machine in a coal mine at Ostrava, where the firedamp concentration was fairly high (giving rise to a safety problem)..

Mr. PHILIPPRON asked the participants not to depart from the subject currently under discussion, i.e. description of the latest innovations with regard to tunnelling machines. Safety problems would be discussed in due course as item 8. He asked Mr. REULENS to make his contribution at that point.
Mr. PHILIPPRON mentioned the various reasons for choosing a Bouygues tunnelling machine in Campine :

a) the components could be readily transported in the various shafts and main roadways of the Kempense Steenkolenmijnen ;

b) the machine could be viable for drivages of as little as 1500-2000 m, since no erection chamber was required ;

c) the supports could consist of segments set less than 2 m from the heading face ;

d) the tunnelling machine could be moved back inside the supports ;

e) access to the heading face when disturbed zones had to be negotiated, was very easy.

4. FORM OF THE CUTTERHEAD

Mr. PHILIPPRON

The papers read had emphasized the quality of the equipment currently available. This still left room for continuous development, which provided food for thought.

The authors of the papers had spoken of the form of the cutterhead. The trend seemed to be towards a flat head. What were the basic reasons : to ensure better ground control or greater safety during cutter replacement ?

Mr. DUERRFELD (Gesteins- und Tiefbau), who was involved in the drivage operations of 4 or 5 of the tunnelling machines in service in the Ruhr area, said that the problem of whether to adopt a flat cutterhead had not been finally settled. The question had been under discussion for many years and it had not yet been possible to arrive at a definite decision as to the best form.

In the coal measures, the type of ground changed very rapidly and the problems involved became increasingly complex as workings progressed beyond a depth of 1000 m.

At a depth of 1200 m, for example, there was a tendency to overcut in the upper half of the roadway in front of the cutterhead.

A flat head made it easier to replace the cutters, while a curved head facilitated vertical and horizontal steering of the machine. In short, there was an ideal cutterhead geometry for every given type of ground. As yet there was no clearly defined trend.

5. CUTTERS

Mr. PHILIPPRON asked where the current trend was leading. Was there a tendency to increase or reduce the number of cutters? What was the most common diameter? What problems arose with regard to wear?

Mr. MERTENS mentioned that cutter costs varied. Each manufacturer had his own design and his own manufacturing technique. Wear undoubtedly increased with the hardness of the rock.

Fastenings and hard-metal discs had no longer been a problem for a number of years (at least on the tunnelling machines currently used in the Ruhr area).

6. TYPE OF ANCHORAGE

Mr. PHILIPPRON asked how the problems of radial grippers had been solved with close-set arches.

Mr. MASSEY (National Coal Board) said that in difficult cases he had increased the number of struts between the arches and had then applied a thick coat of shotcrete between and on the arches. In such cases it should be possible to press the gripper pads against the supports without difficulty, even if the arches were very close-set.

He also pointed out that the protective roof shield should be as short as possible and should possess a degree of flexibility.

Mr. DUERRFELD mentioned that the use of radial gripper pads was constraint on the improvement of supports.

7. SUPPORTS

Mr. DUERRFELD (Gesteins- und Tiefbau) said that it was not easy to set arches more closely ; arches were generally set (in the Ruhr area) at 75, 80 or 90 cm centres. He also pointed out that as a result of the overlap between the sections of yielding arches and the presence of fastenings and lagging boards it could be estimated that only 50% of a circular steel arch bore directly on the ground, even if pressed hydraulically against the roadway side. The contacts were thus more or less point contacts. The problem was less serious when rigid arches were used. It had been observed that fine dust produced by the heading operation accumulated in the gaps (about 50 mm wide) between the ground and the supports, especially at the roadway sides, where the rock was cut in an upwards direction.

In most drivages there was fairly substantial closure in the first metres driven by mechanized means with consequent bed separation and some destressing of the rock.

Because of the presence of the tunnelling machine it was not easy to carry out shotcreting in the immediate vicinity of the heading face. Further research into shotcreting was required in order to find an acceptable solution.

Mr. MERTENS pointed out that while the support-setting devices which had been fitted to the various tunnelling machines made work at the head end much easier there remained a number of difficult and strenuous tasks to be carried out manually under arduous conditions such as transport and handling of support components, placing and tightening of bolts for the clamps of various joints lagging operations etc.

Further mechanization of the transport and setting of supports was therefore required. It was also important to ensure better support/rock contact, especially in future drivages at greater depth, by better guniting or setting (The American Foster-Miller method, for example, should be investigated). Experiments should also be carried out with additional strata bolting and suitable drilling machines should be developed allowing the bolts to be placed as rapidly as possible in a suitable pattern.

In short, while the current performance of tunnelling machines was satisfactory, further consideration should be given to the problem of supports (further mechanization of arch transport and setting, study of shotcrete, suitable bolting arrangements etc).

Mr. DUERRFELD said that strata bolting could also be improved if a fully satisfactory drilling machine were available. What has required was the capability to place long bolts rapidly.

Further research should be carried out on supports.

Mr. PHILIPPRON listed the reasons which had led the Campine mines to use 30 cm reinforced concrete segments to support full-face machine-cut roadways,

a) since 1968 the Belgian Campine Area had specialized in the use of reinforced concrete segments and because of its high strength this type of support was suitable for the difficult conditions obtaining in these mines.

b) The Kempense Steenkolenmijnen workforce had acquired considerable experience in installing this type of support (80 km of roadway to data).

c) The constraints on the diameter and thickness of the components, which were the most negative aspect of the use of segments and arose from the present method of segment handling, no longer applied.

d) The design of the tunnelling machine allowed easier support setting than the current support.

e) Placement of segments seemed to be faster than arch setting since no lagging was required.

f) This type of support with its smooth inner face reduced friction losses of of air along the walls.

g) The cost of such supports was very reasonable(about 22 000 Bfrs/m for 30 cm segments).

8. SAFETY

Mr. PHILIPPRON said that safety problems had been repeatedly mentioned in the many papers read on tunnelling machines. These problems related mainly to two aspects :

1. work between the cutting head and the heading face - mention had been made of the possibility of placing a shield against the heading face and roof and of the use of air cushions ; in addition, cutters were generally replaced from the rear ;

2. fire precautions comprizing automatic powder extinguishers, jets of high pressure water, nitrogen, mobile explosion barriers and moveable refuge chambers.

Mr. DUERRFELD said he was familiar with these problems, which he had encountered on various occasions at the Westfalen and other mines. He thought the fire hazard arose when repairs had to be carried out on the front of the tunnelling machine with a flame torch when a coal seam was present in the heading face. In such situations certain precautions had to be taken, e.g. provision of a reserve of dry powder and a water supply so that action could be taken at once if necessary.

A major hazard was seizing of cutter bearings or blockage of a cutter by a slab of rock.

A further ignition source could result from the simultaneous presence of coal and a bed of very hard sandstone. The violent impact of the cutters against this sandstone bed produced large number of sparks. Under such circumstances the normal water spraying arrangements were inadequate and the water flow rate had to be considerably increased (the resulting mud caused problems with debris clearance and the quantity of water penetrating the rock could result in floor heave).
One solution was optimum control of the ventilation. Suitably positioned methanometers could be connected to microprocessors which were intended to detect any potential hazard in good time and cut off electric power immediately if the methane concentration reached 1%. Such microprocessors had to permit early detection of any increase in the methane concentration. The CO concentration also had to be measured.

As soon as the microprocessors detected a methane concentration in excess of the limit, an alarm was triggered and a powder extinguisher and a substantial water supply had to be available for immediate application. Further improvements were necessary in the efficiency of these devices (or at least that of powder extinguishers).

Mr. MERTENS mentioned that after a fire involving a tunnelling machine at the Westfalen Mine the colliery management, research workers, mines inspectorate and central rescue stations had drawn up a list of measures to be taken innediately in the event of an incident. They had also drawn up an inventory of equipment and measures to be considered when designing new tunnelling machines.

Mr. MERTENS described some of these measures with the aid of a set of slides: installation of pipes to deliver water to the head end, fitting of an extinguisher capable of discharging powder to the cutters and designed to operate automatically as soon as ignition occurred.
A barrier was also erected behind the cutterhead to protect the machine in the event of a potential fire hazard and to allow the rescue brigade to reach to the heading face.
Much remained to be done in order to enable preventive measures, firefighting and other rapid countermeasures to be taken when such accidents occurred. These new safety measures should be incorporated in the design of future tunnelling machines.

Mr. PHILIPPRON asked about the current methane limits in the Federal Republic of Germany.

Messrs DUERRFELD and MERTENS said that an initial alarm was triggered at a methane concentration of 0.8% and that the second alarm operated at 1% methane, immediately cutting off all electric power.

Mr. ASTON (Coal Research, Canmet) mentioned that a Lovat tunnelling machine of 7.6 m diameter had recently been brought into service in the Canadian coal mines of Nova Scotia, where it was to drive an 8 km in-seam roadway. Many ventilation and safety studies had been carried out before the machine was commissioned. This work was to be published at a congress to be held in June 1984.

Mr. KARNELO (Jarva-Atlas Copco) said that the fire hazard increased with the peripheral tool speed and asked whether this had been taken into consideration when the Bouygues tunnelling machine was selected.

Mr. PHILIPPRON replied that this problem had certainly been considered, indeed it would have been irresponsible not to do so. While the peripheral speed of the outer cutter of the Bouygues machine was 6 m/s, the vigorous mixing of air by the cutting and deflector arms would prevent any local firedamp accumulation at the head end. Measurements would also be taken to provide a continuous record of the firedamp concentrations at various critical points in the drivage, the results being transmitted to the surface control room. Due account would be taken of the experience of other countries with regard to automatic tripping as soon as abnormal air conditions developed.

Mr. MONTACIE pointed out that in addition to the anti-methane measures mentioned by Mr. PHILIPPRON, the machine had an advantage with regard to the danger of cutter bearing seizure. This risk was proportional to the interval between inspections and regular examinations of the cutters. Whereas a given cutter on the Robbins, Demag or Wirth machines was replaced after a long service life, the two outer cutters of the Bouygues machine had to be replaced after cutting for only a short time, with the result that they could be checked more frequently. He mentioned that when a Bouygues machine was used to drive a tunnel at Vianden a cutter had been blocked because a seal was in poor condition. This fault had been detected immediately.
He also mentioned that in the usual tunnelling machines the total thrust could be concentrated on a single cutter, resulting in a loading of e. g. 800 t, whereas the maximum overload possible in a Bouygues machine was 120 t.

Mr. MONTACIE thought that in future the maximum linear speed could be increased slightly to over 6 m/s.

9. CONCLUSIONS

Mr. PHILIPPRON thanked the participants for their contribution to the success of the discussion. He hoped that they could not have to wait another ten years before they had the opportunity to compare notes on the results of the experiments and trials carried out in the various centres.

He hoped that it would soon be possible to bring into service an ideal European tunnelling machine which would meet the following requirements :

- easy transport of the machine components ;
- easy assembly without the need to excavate a launching chamber ;
- ability to drive large-diameter tunnels and to cut hard rock ;
- support setting close to the heading face ;
- use of a laser beam and a microprocesssor for vertical and horizontal steering ;
- absolute safety ;
- acceptable cost.

So that further research could be done towards this aim, all those involved hoped for further and increased support from the European Communities.

ROUND TABLE DISCUSSIONS - B - BOOM HEADERS

CHAIRMAN: **Mr. P. M. Davies, B. Eng. (Hons.), M. Phil., F.I.Min.E., C. Eng.**

SECRETARY: **Mr. K. G. Presley, A.C.I.S.**

The Chairman welcomed participants to the round table discussion on boom headers and referred to papers presented to the Information Symposium which were relevant to the subject. Firstly, Messrs. Tregelles and Morris had referred to the development and improvement of roadheaders and Herr Sander had presented a paper on the first practical experiences with the Paurat E.200 roadheader in stone drivages. Mr. Harris presented a summary of all the relevant types of machine and experience in the South Midlands Area in Great Britain and Monsieur Geraud, talked of the experience with the S.S.C.M. machine in the Lorraine coalfield. Herr Reiff addressed the symposium on the subject of mechanised support setting and in part of his paper referred to its relevance with regard to boom headers. Monsieur Blanc from Aquitaine referred to boom headers where they were used in conjunction with trackless vehicles.

There being no immediate response from the meeting, the Chairman stated that in his opinion in developing roadheaders an approach had been followed typical of engineers over many years of solving problems by producing larger and more powerful machines on the assumption that they would be more efficient. As any make of motor car was only as good as the tyres on which it ran, so a roadheader was only as good as the tungsten carbide picks with which it cut the stone. The machines had become larger and more powerful and, significantly, more expensive in an attempt to increase the range of utilisation of the equipment. In practice the increase in machine cost severely limited these attempts to increase the utilisation range. The Chairman stated that some 12 months ago he had become interested in a development whereby high pressure water was being utilised to cut material. The first attempt to develop the system in the U.K. was on a standard Dosco roadheader probably the least expensive road heading machine in use in the country. By converting the Dosco with the use of high pressure water it became capable of cutting material that would normally have required a roadheader five times as costly. The Chairman believed that the limit of the tungsten carbide tipped pick had been virtually reached unless some completely new approach were introduced.

Mr. A. H. Morris, Chief Project Manager, Mining Research and Development Establishment, National Coal Board stated that over the past four or five years two quite separate lines had been followed in an attempt to obtain efficient cutting with boom type machines. One of the approaches had been to attempt to obtain more rigid anchorage of the machine in the roadway, much stiffer booms and a much slower linear speed at the cutting pick together with much more force at the cutting pick at the lower speeds. The principal was similar to that followed in machining metal and although the method appeared to be effective, the results were only obtained by paying for them. If there was adequate anchorage in the roadway, if the booms were bigger, stiffer and more powerful and if the special gear boxes were installed to obtain the low linear speed and the high torques, the expenditure was obviously high. The system did work however, and was capable of cutting rock at more than 2,000 bar.

The second approach in a parallel line of development related to high pressure water jets assisting cutting. Initially trials had taken place using very high pressure water jets by themselves to cut rock but the operation was too expensive. The installed horse power required to raise pressure sufficiently high to cut the rock solely with water was extremely high and was not a practical proposition. When applied, high pressure water was used to assist mechanical cutting and it was discovered that pressures as low as 700 bars could be effective. This level was acceptable both in engineering and cost terms. There had been a long development period both at the test site and underground on a production machine and it could now be stated that the use of high pressure water jets did enable some of the lighter and medium weight machines to cut quite hard rock. Although the reasons were not fully understood, the use of high pressure water jets reduced vibration which was a critical factor and protected the picks as well as reducing the power requirement. Engineers would be in a better position to make a judgment in 6 months time when more field experience with the high pressure water jet system would have been obtained.

With reference to the heavier machines and the new technology associated with them, Mr. R. E. Collins, Deputy Managing Director of Eimco (G.B.) Ltd., stated that his company, together with the National Coal Board had developed a heading machine to this latest technology. It was essential that the machines were anchored to enable cutting to take place without vibration and Mr. Collins screened a number of slides illustrating the anchoring system. (See Plate 1). The first slide illustrated the front view of the machine where the staking arrangements could be clearly seen. There were 6 booms on the machine which anchored it into the side of the roadway and the booms were graduated so that they could grip according to the hardness of the rock. Obviously if the rock was soft and total pressure was applied it would force the booms further into the rock and so a graduated system was employed whereby they gripped the roadway and anchored the machine positively. In addition to the side anchors the machine had a total weight of 110 tonnes. The boom itself worked at 7 r.p.m. but had a very high torque enabling it to cut harder rocks. Rigidity in the boom was needed and as there was a fixed leg arrangement the whole boom moved as a total carriage and would sump up to 1 metre. It was not an extending boom which would introduce engineering weaknesses.

The principle of cutting was to cut in a semi-circle about the centre so that a clean profile was obtained enabling supports to be easily set and minimising point loading of the supports. In order to cut the floor an electronic device was installed which gave sufficient control to cut a level floor. The same machine could be used to cut a circular tunnel if it were fitted inside a shield. The slides illustrated the conveyor and the gathering arm arrangement and the cutter boom powered by its own 200 h.p. motor. The remainder of the machine was also powered by a 200 h.p. motor. The plan of the machine showed the stakers, the gathering apron which was capable of swinging to either side of the roadway and the feed through onto the conveyor. The machine must cut a half cusp in the side of the roadway which the stakers were designed to fit into thereby anchoring the machine. Adequate anchorage was absolutely essential with roadheading machines of the future.

Herr Swoboda of Saarbergwerke A.G. stated that all the boom headers used by the N.C.B. appeared to have only axial cutting booms. This type was also used in Germany where it had been found that in heavy rock machines had to be disconnected because it was extremely difficult for them to resist the various forces. They had come to the conclusion that when there was extremely heavy rock, transverse head cutting was better especially with a view to anchoring the machine. Herr Swoboda enquired whether any tests had been carried out in the U.K.

In reply to Herr Swoboda's question, Mr. Morris stated that in the British coal mining industry transverse heads had never been used only axially mounted heads. There was however a test rig at Swadlincote where the effects of a transverse head against an axially mounted head were being measured. He was not sure however if the results would be any different from those obtained in France and Germany where it was understood that the transverse heads gave much greater stability if working was taking place in hard rock. One of the reasons for vibration and movement of the machine in Great Britain was possibly because of the type of head being used. Colleagues in Germany and France had assured Mr. Morris that much greater stability would be achieved with the transverse head.

Mr. Morris was asked a question in relation to the use of a water jet of approximately 700 bar supporting mechanical cutting in relatively hard rock. The questioner asked what were the properties of the rock in relation to hardness and resistance to cutting.

In reply Mr. Morris stated that work being carried out in the laboratory should be disregarded because although valuable, it was usually done at slow linear speeds which did not simulate actual machine conditions. The only experience which could be quoted using high pressure water jets on actual machines were at the test mine where the rock varied in compressive strength between 700 and 1200 bar. The underground experience also went up to approximately 1200 bar but it could be stated categorically that with conventional light duty machines, the rock could not have been cut satisfactorily without the water jets. It was believed that if very hard rock were to be cut, higher water pressures than presently being used would be needed. Pressures of up to 700 bar were presently being used as this was the most practical engineering compromise. Higher pressures would be preferred if they could be handled however. In Great Britain only about 15% of all the strata had a compressive strength greater than 1300 bar so even if further developments did not take place, there was still a high exploitation potential. With regard to the hardness of the rock in the test mine, the material was limestone and was very homogeneous. In the one underground site the hard material occurred in bands and was very tough.

Mr. D. Johnson, Sales Director of Dosco Overseas Engineering Ltd., referred to the merits of transverse against axial heads. Whilst there were no machines with transverse heads operating in the U.K., his company many years ago carried out trials with both transverse and axial heads. The company's experience had been that in hard cutting, an axial head was far more efficient but there were advantages of using a transverse head in softer materials. Although manufacturers naturally tended to support the type of head which they produced in the experience of Dosco the axial was more efficient in the harder cutting.

Herr Kleinert from Bergbau-Forschung GmbH, engaged in mining research, gave details of the experience that had been gathered relating to the use of transverse and axial heads. In hard cutting a transverse head had been preferred as it could cope better with the reaction forces in the rock and provided better stability of the machine. With regard to axial heads those used in Germany were bigger than those currently used in the U.K. and they also had a greater number of picks. To work effectively with a cutting head greater power was required with more force at the pick. In the German situation attempts were being made to drill the whole length of the cutting head into the stone and this was quite difficult for an axial head when it was working on very hard rock.

With regard to the use of high pressure water Herr Kleinert considered that it was a technique which could be used but results similar to those obtained in the U.K. were not yet available. In the view of Herr Kleinhert high pressure water jets should be used in places where cutting was not economically viable with present-day techniques. In other words high pressure water jets should be used where the wear of picks and equipment was high and the advance was slow. In applying the high water pressure technique it would be necessary to calculate the amount of water required dependent upon the rock to be cut based upon samples. When particularly hard stone, for example sandstone was to be cut, water pressures higher than 700 bars would be required. On the test site in Essen a trial had been undertaken where it was possible to have water pressure up to 2500 bars and although it was not necessary to go to this level, effective cutting into the sandstone was a possibility at 1500 bars.

On the question of high pressure water jets, Mr. Morris stated that he was aware of the research work being undertaken in Germany. At the Research Establishment at Bretby it was hoped that in the second stage of development, pressures somewhat higher than 700 bars would be used. Restrictions on water pressure in the past had been due to engineering practicability. Two of the manufacturers had already incorporated the pressure raising equipment onto the machine itself to give pressures up to 700 bars. To obtain the increased pressure higher horse power motors would be required and it was an important factor that the equipment should be accommodated on the machine itself. Whilst Mr. Morris was confident that higher water pressure could be incorporated into the system, he reiterated that the results being obtained with the present trials made the technique valuable and worth exploiting.

Herr Kleinert raised the question of the equipment in use for obtaining the high pressure water jets. In Federal Germany a number of manufacturers produced equipment capable of pumping water safely up to a pressure of 2000 bars. The pumps could be assembled directly onto the machine or be accommodated behind it. Herr Kleinert was inclined to the view that the pumps should not be installed directly onto the roadheader. With regard to adjustments to the water jets it was reported that a great many problems had been encountered in obtaining components that functioned correctly. This was the main reason for lack of progress in the studies. It was not clear what the situation would be when the technique was applied operationally underground. It was hoped that some headway would be made with existing trials and that more detailed information would be available later.

Monsieur R. Resseler with N.V. Kempsense Steenkolenmijnen (Belgium) replied to a written question on behalf of Monsieur J. J. Stassen Professeur à L'Université de Liège. The question related to the paper presented by Monsieur J. Legrande on mechanised drivage of narrow headings. The questioner asked the advantages of the ESA machine against the boom type roadheaders. Monsieur Resseler outlined the main benefits of the ESA machine with the use of slides. The first slide demonstrated that the machine was short and this was considered to be a major advantage as it could easily move to the heading. As the machine was well anchored, stability was assured and the drum and cutter could be moved easily. The machine operated at 60 kw and the pressure that the boom could exercise was 10 tonnes. When the boom was lifted there was still a pressure of 8 tonnes on the drum. The ESA machine itself weighed approximately 6 tonnes and could be easily dismantled being moved from one point to another in just 2 shifts. Monsieur Resseler summarised the advantages of the ESA machine which were high power at 60 kw, it was small in size, it could be easily dismantled and the price was not exhorbitant.

Replying to a question from Mr. D. Johnson, Sales Director of Dosco Overseas Engineering Ltd., Monsieur Resseler stated that the ESA machine had been used in coal at a thickness of 3.20 metres. The ESA machine had not been used for cutting stone. In reply to a question from Herr Kindermann from the European Commission Monsieur Resseler stated that approximately 20 machines were available of which 17 were actually deployed. In the coal heading however, only one machine had been tested and three trials had been carried out, one of 100 metres, one of 300 metres and one of 600 metres.

The Chairman extended the discussion from boom headers to methods of clearance of mineral. Over many years virtually nothing had been used to clear mineral other than continuous conveying systems. Monsieur Blanc had referred to a recent experience in the Acuitaine coalfield where this type of roadheader had been used with a clearance system using rubber tyred vehicles where the advantage of a floor stock pile behind the machine gave some bunkerage. The system meant that the machine did not necessarily stop cutting when the main mineral transport system was halted. The roadheader was allowed to be more continuous and was not dependent upon many other items of plant between the heading face and the mine surface. Another advantage was that where any admix of dirt created a marketing problem, it was possible to use a clearance system involving an intermittent discharge by load haul dump or shuttle car. It was therefore possible to introduce to the conveyor system dirty material in such a manner as it could be later segregated without the need for complicated coal preparation processes.

It was pointed out from the floor that transportation of the material depended to a large extend on its condition. The material was obviously much heavier if it were wet and the major consideration must continue to be the cutting process at the face.

An application for some form of vehicle behind the roadheader was confirmed by Mr. D. Johnson, Sales Director of Dosco Overseas Engineering Ltd. A further problem area was when a machine was breaking away on a junction and some form of free steered vehicle could offer many advantages over conveyors in that situation.

Mr. R. E. Collins, Deputy Managing Director of Eimco (G.B.) Ltd., stated that there had been a number of cases in the U.K. where load haul dumps had been used behind boom heading machines. Mr. Collins screened a number of slides illustrating the schemes. The first related to Bevercotes which was one of the larger collieries sunk by the National Coal Board after Nationalisation and it was necessary to develop into new seams. A new pit bottom was required with a view to obtaining the most economic method of material transport and handling and manriding. In the Bevercotes case, boom heading machines and also drill blast and load were used and the load haul dump was utilised behind both the roadheading machines and from the blasted headings. A cable reel machine of 3 cu. yards capacity was used and it was capable of developing the layout as illustrated. (See Plate 2). The machine had to run approximately 400 metres and there was no necessitity to install conveyors because the control box feeding the machine could be strategically placed so that it could feed onto existing conveyors within the network. In the Bevercotes case the machine was operated by contractors rather than the National Coal Board's own employees.

The second illustration related to a pit bottom at Harworth Colliery also in the North Nottinghamshire Area. Again a new pit bottom had to be developed with a smaller machine being utilised with a cable reel machine of approximately 100 metres of cable, enabling 200 metres of run to be made. The load haul dump had a 1 cu. yd bucket and the whole of the pit bottom was driven by this machine working behind a roadheader. (See Plate 3). In the Harworth example no conveyors were installed and no dirt was sent out of the mine as it was all stowed in old workings clearly illustrating that a flexible system could be established with the boom header and the load haul dump. If the conveyor system to which the dirt was eventually being fed stopped, the mineral could be put to a strategic stock pile and loaded out at a time when it would not contaminate the coal. The machine became truly a continuous boom header. A further scheme illustrated by Mr. Collins related also to the North Nottinghamshire Area of the National Coal Board where a diesel machine had been used. Again there was a series of curves involved and some longer distances than would be possible with the cable reel machines. The diesel machine was used to give greater flexibility. (See Plate 4). In some places drill and blast methods were used but one machine was capable of handling the spoil from the headings be they blasted or be they boom headers.

Finally Mr. Collins referred to possible developments in the future. In other parts of the world trucks were now being used to haul the dirt over much longer distances and he screened a slide showing such a machine. He asked for comments from colleagues within the European mining scene on the application of this type of machine where conveyors would be less and less used on the development of roadways.

The Chairman stated that he had been Manager for some four years of a mine where rubber tyred vehicles had been used extensively. The advantages had been local bunkerage in terms of development drivages, segregation of dirt and coal and the equipment could be used for materials transport to the site. The Chairman also considered that as the system involved no conveyors, no rails and locomotives, the heading machine was free to flit and go to work elsewhere in a matter of hours. This was an important feature in improving the utilisation of boom headers which may cost of the order of £500,000. He believed that to some extent the fixed ideas of mining engineers was stifling this type of development.

Herr Lehmann of BAG Niederrhein stated that at his colliery load haul dumps had been used for special drivages of large areas, shafts, haulage dumps, etc. A large scale project was planned for a new shaft and also for the drivage of a large filling station. It was the intention to use the load haul dump techniques in combination with conventional drivage methods of drilling and shotfiring. For the kind of application referred to, tests had been carried out and it had been concluded that to a large extent the use of mobile vehicles particularly for longer roadway lengths was out of the question. It had emerged quite clearly that the use of the free steered vehicles should be limited to special cases.

Herr A. Zitz, a design engineer with Osterreich Vobst/Alpine raised another point against the use of the load haul dumps. He stated that in those countries where load haul dumps had been used successfully, particularly the U.S.A., Australia, and South Africa, the seams were thick and coal could be mined under relatively favourable conditions. In other less favourable conditions, bottle-necks could arise which could hamper the continuity of cutting. With a smaller or medium sized machine it may be possible to achieve virtual continuity but this was doubtful with the larger machines. It may be prudent to search for a more continuous conveying system which could follow the machine.

Mr. Collins of Eimco (G.B.) Ltd., agreed that to develop continuous conveyors which would operate behind machines would be an ideal arrangement. Such a system had been tried particularly with continuous miners where large quantities of mineral were involved but they had not yet been successful. He did not consider that there were any boom heading machines with the exception of full faced tunnelling machines whereby the load haul dump could not comfortably handle the amount of material which was cut. Mr. Collins added however, that if a roadway were being worked and it was necessary to install a conveyor and that conveyor was not subsequently needed in the roadway, a great deal of expense was involved both in its installation and its removal together with the actual cost of the conveying system. The principle of free steered vehicles did therefore, have a lot to offer.

Mr. Collins was asked a question by a member of the German delegation regarding problems with movement of roadway floor. There were big movements of the floor in deeper mines leading to restrictions in the possible use of load haul dumps. In reply, Mr. Collins stated that his company had gained a great deal of experience with load haul dumps in the circumstances described. There was a good deal of prejudice against rubber tyred vehicles in coal mines throughout the whole of the western world. Such vehicles were now widely used in Australia and South Africa and in some cases their conditions were no better than those typical in Europe. When the machines were first introduced into Western Europe they were adapted to suit certain circumstances and were not specifically designed. The machines now being operated had been designed for use in coal mines and the results were clearly coming to light in their operation. Initially, small wheels with standard car tyres were used and the manufacturers were not interested in designing tyres specifically for these mines. However, it was realised that there was a big market world wide and tyres have since been specifically designed for this purpose. In the conditions which have been referred to where there was a lot of water, the hard rock lug type of tyre would work adequately. There had been cases in the U.K. where track mounted dinting machines were not capable of working in the conditions but free steered vehicles fitted with the correct tyres were quite satisfactory. Coal mining engineers tended to feel that load haul dumps were designed specifically for hard rock mines and the terminology was now coming to the fore of calling them free steered vehicles. Load haul dumps were relative to hard rock mines, free steered vehicles had been designed specifically for coal mines and offered great advantages.

Mr. Johnson of Dosco stated that he believed that a continuous conveying system would be the ideal solution but this was difficult to achieve and work being carried out in the United States, South Africa and Australia was not too successful. In South Africa where the output per machine shift was higher than anywhere else in the world, material in excess of 3,000 tonnes was being conveyed from the back of the machine in one shift by a free steered vehicle. Whilst the use of free steered vehicles behind a boom heading machine where volumes of material were relatively low, may not be completely suitable, there were other applications which were.

Further experience in the use of free steered vehicles was given by Mr. J. H. White, a Production Manager with the North Derbyshire Area of the National Coal Board. His recent experience was in relatively difficult strata with soft floor conditions some 900 metres deep where a load haul dump machine was utilised to take out 1.5 metres of floor which had virtually closed the roadway. He was convinced that if the load haul dump had been installed earlier, the problems associated with 1.5 metres of floor lift could have been solved by taking it off incrementally during the working of the panel.

The Chairman stated there were two schools of thought in relation to the operation of boom headers. On the one hand there was the machine which was advanced and never retracted, cutting from one position. On the other hand there was the roadheader that was able to cut from more than one position and had the advantage of being more mobile and could be withdrawn easily. The Chairman had had experience of operating with both systems and he sought the view of the round table on the relative advantages and disadvantages.

One of the views expressed from the floor related to utilisation of the machines. The first machines introduced particularly in Europe had great manoeuvrability and it was possible to cut very large profiles in the cutting path. The present generation of machines were larger and heavier and could do more work with the resultant advantages. It was considered easier to keep the machines in one position rather than to manoeuvre them; there was also an advantage in setting supports. The small machines were utilised for between 40%/50% of the total time available giving a high degree of utilisation. The more powerful and larger machines were used only between 25%/35% as they could complete the work in a much shorter time and were delayed by the other activities away from the site. Machine utilisation could be increased if the system as a whole were improved.

Herr Swoboda from Saarbergwerke considered that a clear definition of utilisation should be agreed. It could be interpreted as the degree of utilisation when the machines were actually used for heading or the time when the machine was actually available down the mine. He considered that the concept should be agreed before any solutions could be discussed. Mr. J. H. White referred to his experience in driving 1,000 metres with the Paurat E200 known as the Titan in the U.K. The machine had been available for work 100% of the time working downhill in a cross measure drift of 25% grade and it had never failed. By comparison the lighter machines which were flexible and more manoeuvrable in operation were available for only approximately 70% of the time due to component failure. The reasons for the down time were associated with the flexibility and manoeuvrability of the machines which resulted in damage to tracking equipment, tracks, engines, etc. The heavier duty machine remaining in a straight line operating from one position had given much greater reliability and thereby utilisation.

During the discussion on utilisation of boom headers, it was pointed out that actual cutting time was probably as low as 30%. If the work were fairly well organised, 50% of the time would be required for setting arches, 20% in preparing jobs and allowing for breakdowns leaving a balance of 30% as cutting time. This figure was unsatisfactory and steps should be taken to improve the utilisation.

Herr Swoboda raised the question of the degree of usability. He stated that over a 24 hour period a machine may be available for 14.4 hours bearing in mind repairs etc., giving a 60% usability which was Utopia. What should also be borne in mind was the time taken to move a machine from point to point. Ought not availability to be calculated from the moment the machine was delivered to a given point. Did 60% usability mean time spent in cutting or the time that the machine was available. A figure of 60% for availability was not sufficiently high bearing in mind that actual time spent in driving was only 30%.

The Chairman summarised the wide ranging round table discussion on boom headers. The discussion had ranged from the advantages of transverse versus axial heads and although there was little experience in the U.K. to judge one from the other, there was a measure of support from Western

Germany and France in favour of transverse heads as they gave greater stability. Although one British manufacturer had been in favour of axial heads, final conclusions could not be drawn.

The round table had discussed developments with the use of high pressure water and there were numerous possibilities ahead. The use of larger and more powerful cutting machines had been discussed together with the advantages of anchoring them, rather than simply relying on the bulk of the machine to give it stability at the cutting point. Transport alternatives had been discussed including the argument for free steered vehicles compared with conveyors. The discussion had then moved to the relative advantages of cutting from a fixed rather than a moveable position. It had been appropriate to move the machines when they were light but as they became heavier and given the reliability of the track system, there were considerable advantages in operating from a fixed position. Finally, the round table had discussed machine utilisation. The Chairman concluded by thanking members for their attendance and contributions.

PLATE Nº 1.

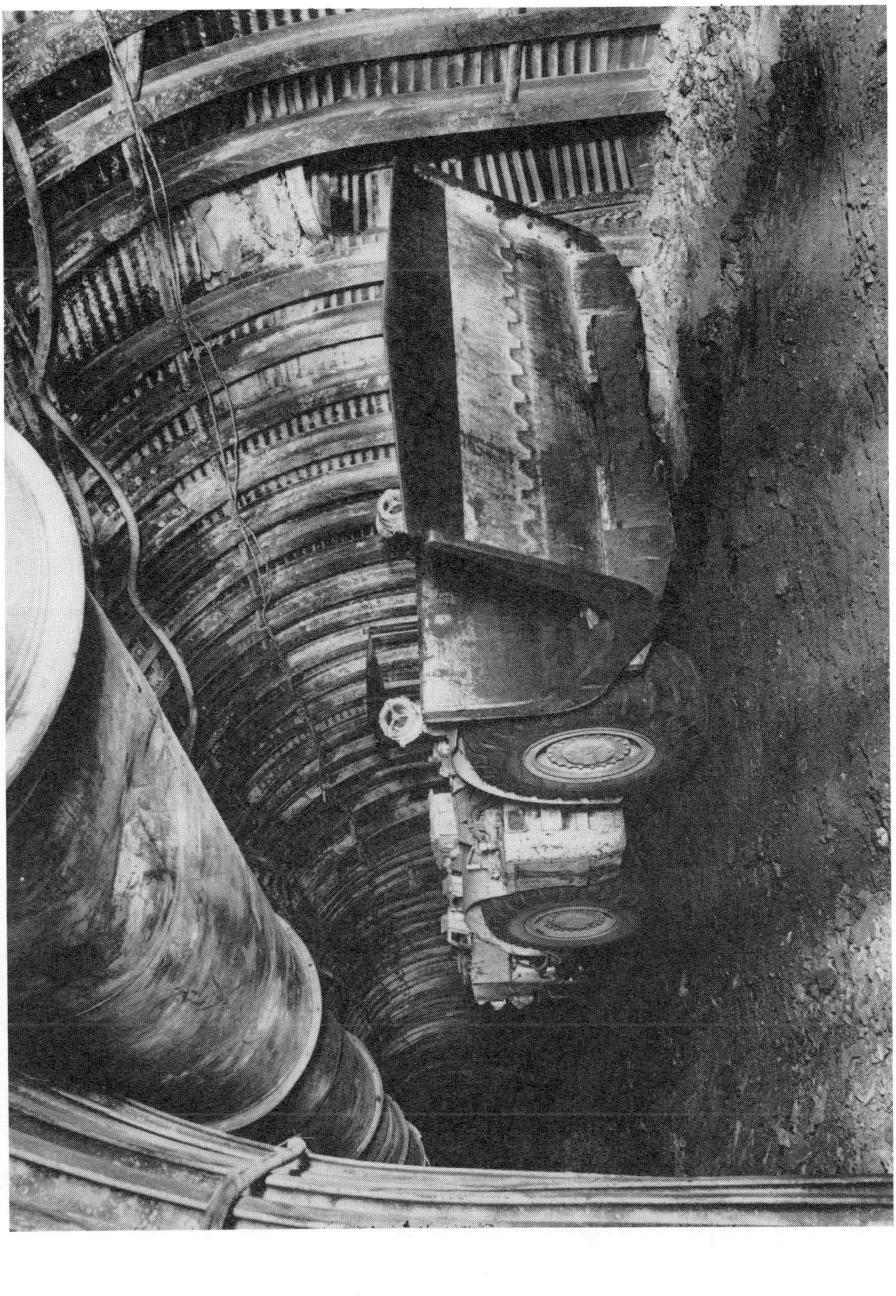

PLATE Nº2.

PLATE Nº 3.

PLATE Nº 4.

ROUND TABLE DISCUSSION C - MECHANIZATION OF SUPPORT

Chairman : M. Naquet (HBL, France)

Sekretariat : M. Tinchon (HBL, France)

1. - THE CHAIRMAN OPENED THE SESSION AND STATED THE POINTS FOR DISCUSSION

- Rock bolting
- Shotcrete - filling
- Concrete panels
- Mechanical aids

These subjects will be handled in the above order.

2. - ROCK BOLTING

Mr. SPIESS. During the morning session we were shown a cable bolt which provides a very special type of anchoring. I have two questions on this :

- what is the breaking strendgth of this bolt and
- now is the section emerging from the hole assembled ?

Mr. TINCHON. The cable bolt has been used in France for 15 years. In its initial form, i.e. with a diameter of 20 mm, it has a tensile strength of approximately 20 tonnes. One of the problems with this type of bolt is rotating it through the resin capsules, which has been solved by reversing the direction of lay.

To ensure effective bolting the bolt hole must be cleaned before the resin is inserted. To fulfil specific requirements for selective heading machines in steep seams, the diameter of the cable has been increased to 22 and 25 mm to raise the tensile strenght to 30 and 40 tonnes respectively. The advantage of the cable bolt design is that it is independent of the direction of load and the direction of the bolt hole.

In answer to the second question the part of the bolt emerging from the hole comprises a crimped 3/4 -inch high-pressure (60 bar) tube, 80 mm long, fitted with a nut as a retainer. The unit is designed as a self-locking assembly which is crimped with a wedge inside the cable and has not presented any problems.

Mr. NAQUET. I would like to add something. The cable bolts used in the Lorraine area have a special application ; they are not used for support but, for instance, as anchoring points for machines, in particular in steeply inclined workings for securing a plate to which the machine haulage chain is attached.

Mr. TINCHON. The cable bolt can be used asa means of support, in particular in thin seams. Part of the bolt penetrates the capsule directly, with mixing taking place when the second part is inserted. Information on the characteristics of this type of bolt can be forwarded to potential users.

Mr. X. Have you experimented wiht cement cartridges where the water is incorporated in the cartridge in the form of droplets encased in wax ? I think that these cartridges are manufactured in the United Kingdom but I don't know whether thay have been tested or not.

Mr. TINCHON. Cement cartridges with water have been marketed in France by CELTITE. We have tested these cartridges and have not encountered any problems.

Parallel to these tests we examined wet cemented capsules and discovered that the space around the bolt played an important role. To obtain results comparable to those with resin, the diameter of the holes was increased from 28 to 32 mm for a 20 mm diameter bolt.

We did not continue the tests as the method was not cheaper than using resin. Recently we restarted after discovering that, in highly fissured zones, the resin flows away into the fissures whilst cement does not and thus ensures efficient anchoring. This could be extremely useful in steeply inclined workings and highly stratified surrounding rock.

Mr. X. A question on this subject. What anchorage lengths have you achieved and what do you think the limits are for artificial cements or resins ?

Mr. TINCHON. Manual and mechanical setting have to be considered separately here.
With the current characteristics of our resins the maximum length of a correctly set bolt is 3 m. Bolts of 4 to 5 m can be set if the catalyser is modified to increase the curing time.
With slow curing cement, percussive insertion of the shank allows bolts of 4 to 5 m to be set.
For still longer shanks injection equipment (for resin or cement) has to be used.

Mr. FEISTKORN. What interest me is the way in which anchoring properties are calculated so as to obtain the strength, length of the shanks, etc., in short, the way in which a bolting pattern is produced. How is the placing of various bolts determined ?

Mr. TINCHON. Dr. GOETZE has made an extremely detailed study on this subject which is considered extremely valuable in the mining profession. Highly scientific methods have been used in processing the various parameters affecting strata life in an elastic environment or, still nearer to reality, an elastic/plastic environment, by means of a mathematical model. This method is a little too theoretical for miners and has the disadvantage of only presenting the situation which has been created by data input. Apart from the selection matrix presented this morning Dr. GOETZE's studies provide the answers to your question.

Mr. REIFF. At the morning session the pratical application of this type of support was not underlined sufficiently. In the Federal Republic of Germany the Niederberg miners have considerable experience in bolting and it would be helpful if Dr. LUDWIG would tell us something about its application.

Dr. LUDWIG. It would take all day to cover all the experiments made at Niederberg. Since 1978 30 km of roadway have been tested. It took a long time before initial results were obtained. Niederberg is well placed for these tests as work is only carried out at depths of between 450 and 700 m, which is not excessive for testing.

In the meantime coal is being won with 5 km of roadways which provides us with a certain amount of empirical data, and we can now say that the roadways have proved themselves as far as single-unit winning is concerned : face entries were excellent and a large number of man-shifts were saved. These roadways were driven by selective heading machines with integral drilling machines. Up to recently only a few bolts were set in the front of the machine, the remainder being set behind it. Experience showed that drilling boom could be installed behind the machine on a monorail. We have not gained a great deal of experience in this area so far.
The crews took some time to master the cycle with simultaneous bolting and at present 10 to 11 m per day are being driven. The savings are enormous - some 500 DM per metre advanced. We have very little experience in shotfiring drivage where twin-boom drilling jumbos are being used to drill bolting and shotfiring holes alternately.

Mr. X. In the Federal Republic of Germany bolting is based on Dr. GOETZE's theory whilst in France it has been used for a long time. On what basis do you apply bolting, and is it experience on which your technique is based ? This supplements the question which has just been raised. Calculations are preferred as a basis for the technique, but do you not think that experience can also be used as a selection criterion ?

Mr. TINCHON. First of all I would like to thank the Niederberg miners who have contributed a great deal to bolting over the last two years and, in particular, have broken the taboo against ripping into the seam roof for bolting as sole support. In answer to your question we have been using bolting regularly for 20 years and I would like to ask Mr. PIGUET to tell you about the experience derived from an enormous number of measurements which have given rise to simple rules for application. The measurements were made in situ.

Mr. PIGUET. As an engineer at Cerchar, I am responsible for teaching rock mechanics at the College of Mines at Nancy. As far as the questions on

calculation methods and bolting pattern are concerned, I would like to answer simply by referring to Mr. TINCHON's paper this morning in which a selection matrix for the various bolting parameters was described. This is a method which has been presented on various occasions and is based on an initial analysis of the type of stability expected around a bolted area. The first three lines reflect the type of deep or surface stability which are interconnected.
These three classes are determined by using the method of finite limits of elasticity or, to remain closer to rock behaviour, elasticity/plasticity. They are also based either on an assessment or on measurements of the initial stress conditions before drivage. Given the initial stress values before drivage, the mechanical characteristics of the rock and a calculation method for determining stability or instability, it is possible to find the right place on the matrix for bolting parameters. I think our method for determining these parameters is best understood by concentrating on this matrix.

Mr. KRONENBERGER. As regards the theory of bolting, have you any preference for either rigid or yielding bolts ?

Mr. PIGUET. Speaking of this matrix in abstract terms, if one of the columns covers delayed or non-delayed rock behaviour, the development or non-development of deformations, it is on these terms that the choice between fixed anchoreage to ensure rigid support or yielding anchoreage to allow deformation is based.

Mr. REIFF. In German coal mines where Dr. GOETZE's calculation method is used, 30% of the cases in the Ruhr area have to be heated with rigid bolts. The use of yielding bolts involves economic problems.

Mr. FINK. My question is to Cerchar. Apart from the calculation programmes, do you use in-situ measurements to determine the strata tension in fissured or non-fissured zones ?

Mr. PIGUET. The Cerchar analysis is based on theoretical studies and on experiments carried out in situ in all coal mines to provide analyses and checks of rock behaviour. They chiefly consist of measurements of expansion and convergence, the measurements of expansion giving information on internal strata deformation. There are also more complicated measurements of tension.

Mr. LEYHE. To what extent do you use anchorage as a safety measure in roadways ?

Mr. TINCHON. Rock bolting is either used as sole support, where it is necessary to check its effectiveness by measurements of expansion or convergence, or as an additional source of reinforcement to make the face/gate junction safer.

Mr. NAQUET. May I add that we use bolting for safety in some coal winning districts for working through faults, repairing the roof and providing additional support in steeply inclined workings.

Mr. FEISTKORN. Mr. LEYHE was most concerned with the face/gate junction where part of the support must be removed to allow work to be carried out. In this case bolting can constitute a replacement of the support. This system has proved itself in our coal mines ; have you used it ?

Mr. TINCHON. To be really effective, bolting has to be applied when the roadway is driven. Supplementary bolting is mainly used to maintain support when one of the elements is removed but we have not attempted to dispense with all reinforcement in a bolted area at the face/gate junction in the Lorraine area. Results at Niederberg have been excellent in thin seams but great care is required in extending application to medium and thick seams. Anyway, supplementary support extends roadway life.

Mr. LUDWIG. In a retreating working at Niederberg the top road was bolted and the original support was removed, which did cause some problems. As a solution to this problem we planned to lay the arches mechanically but this has not yet been tried out. As regards the face/gate junction, the partial withdrawal time is relatively short, only several hours, during which bolt anchorage has to provide support. Bolting should be regarded as an alternative to or replacement for normal support methods. In the case of Niederberg where heading rates can reach 10 m a day, the period is short and the advantages are commensurately large (there is much more room and freedom of movement in the roadway).

Mr. TINCHON. This method can certainly not be applied everywhere as rock behaviour in the roadway must be appropriate before the face passes into the zone subjected to stress from winning.

Mr. LEYHE. Two comments ; this method is not only used at Niederberg, as supplementary bolting is widely used, and the above method is only applied after the situation has been appraised and tests made.

Mr. NAQUET. This is an interesting theme as the number of questions show ; let us now go on to the second theme dealt with this morning, which was the problem of lagging supports.

Mr. REIFF. Mr. WOOD's paper described the method well - what interests us is advance rates in roadways. A lot of efforts are being made in this field which are costly and entail a great deal of work. In my paper I pointed out that it was not just a question of setting mechanical supports but that various methods of lagging behind the arches should be discussed. Mr. LAYHE has a great deal of experience in this field as has Mr. BOHNENKAMP with Bullflex.
Mr. SCHUMANN has developed a new process used at the Victoria mine and I fell that all these people should give their comments. Perhaps there are other specialists present.

Mr. WOOD. Dr. REIFF, I have seen Bullflex being set in the Ruhr area over approximately 3500 m of roadway. This is a cement-based material. We drove 300 m of roadway with a tunnelling shield, where support operations were dangerous, involving 78 accidents during setting operations behind the shield. The space between the support and the strata can both enable firedamp to accumulate and the crown section to deteriorate. Research was carried out to find a material which could be injected in the annular spaces and 'Aqualite' appeared to be the right solution.
This is a special cement which can be pumped over distances of up to 60 m when mixed with water. The expansion of the product depends on the volume of water used which is calculated in accordance with the strength required. The aim of the operation is to fill the space between the arches and the roof to distribute the stress on the supporting members evenly. This material, 'Aqualite', was developed at Bretby.

Mr. REIFF. After your paper, I asked you a question about this product and I have just found out that a pilot experiment was planned at the Lobberg mine. We could thus make a comparison between the materials used traditionally in our country.

Mr. GREEN. A 15 to 1 mix of water to Aqualite cement provides a strength of 0.1. MP. This product is cheap, very clean as it does not run and looks rather like toothpaste.

Mr. REIFF. Perhaps you could give Mr. SCHUMANN the floor, as he has also worked with this 'toothpaste'.

Mr. SCHUMANN. I am not familiar with Aqualite but find this extremely interesting, particularly if parallel tests could be carried out in the United Kingdom and the Federal Republic of Germany.
We have developed a cement-type material which, when mixed with water, forms a paste which can be foamed by an air jet to be used to fill cavities. The performance of Aqualite does not mean anything to me at the moment as I have no parameters for comparison.
Could you tell us the compress on strength of the set product in newtons per cm^2?
A strength of between 5 and 7 newtons/cm^2 appears sufficient to avoid any problems necessitating adjustment of the support equipment.

Mr. STUBBS. Our tests started approximately six months ago and we have now found that the mix used prevents falls.
As regards the strength of the product I cannot convert it into your system of measurement.

Mr. SPIES. Mr. WOOD, to what extent have you succeded in combining filling with drivage ; do you need to stop the machine ? Can you combine it with drilling ?

Mr. WOOD. We do not stop work, but set the arches behind the machine and do not need to expose men to apply the product in the area left free after the shield has advanced. The numerous accidents which had occurred in this area have been eliminated entirely.

Mr. REIFF. This mechanical filling technique calls for material which is technically suited, has early curing and early strength properties and, for drivage purposes, can be injected in large quantities over a short period of time. I would like to know now many cubic metres per hour you have managed to inject.

Mr. WOOD. In my report I stated that 1 m^3 of product can be produced with 25 kg of Aqualite. Assuming the void to be 15 cm one can line 2 m in 18 minutes. In addition, transport problems are reduced. We have also carried out a test on the fully enclosed product to the satisfaction fo the user. This test will be continued. The tests on the product at the face to fill gaps left by roof falls was not satisfactory and new experiments are planned.

Mr. LEYHE. Mr. WOOD, a major problem is the viscosity of the product - it should not run. On a sketch you showed that you work at two levels - could you give us some examples, in particular for the measures taken against heatings. What is the distance between the two levels ?

Mr. WOOD. The upper section where the shield is located has mesh lagging to maintain and increase protection against spontaneous combustion and grouting is continuedto the roof to block the passage of air completely.

Mr. OTTO. Mr. WOOD, if I have understood correctly, the purpose of this lining is to preclude accumulations of firedamp and roof falls. We would also like to reduce convergence which exerts stress on the face.

Mr. WOOD. In the example that I saw in the Federal Republic of Germany the lining was complete and fulfilled its purpose - to reduce deformation - extremely effectively. We expect that our material will have the same effects in distributing pressure more evenly to the support members.

Mr. X. I would like to make one comment as I think that these are very different materials. The strengths stated for Aqualite correspond to our Isoschaum.

Mr. WOOD. That is true, this is relatively light material but by varying the quantity of water we can achieve far higher strengths With a mix of

7 to 1 the strength is 7 bar. The problem is to find the mix appropriate to the application without raising transport problems.
The proportions to be used will be based on experience.

Mr. NAQUET. Two more themes must be dealt with, first of all circular support elements.

Mr. THOMAS. The slides this morning showed panels for lining shafts. What is the procedure for traversing aquiferous zones ?

Mr. CLAES. The procedure is the same as usual, the panels being set and material injected behind them. Water makes do not halt work.

Mr. NAQUET. Let us go on to the last subject : mechanical setting of supports.

Mr. REIFF. This theme is of great importance. As regards drivage, support work represents 40 to 60% of the time involved whatever the method of drivage, be it mechanical or shotfiring.
As time was too short, I could not finish my paper this morning and in particular I did not deal with setting supports using various machines.
We have one continuous machine which has reached a high level of development. Safety problems have been solved and the machine is not halted to set the supports.
In the little time that I had I tried to indicate some of the methods we use and to show now mechanization can allow supports to be set in safety.
We use mobile platforms of which Mr. LEYHE can tell you more. In shotfiring workings this platform is extremely important and permits work on two levels.
In the Federal Republic of Germany powered supports did not prove particularly effective. The pumping action they generated gave rise to roof falls and they constitute an extremely expensive method of support. The tunnelling shield which Mr. WOOD spoke of has proved itself but requires large capital investment which is why it is no longer used.
Mr. LEYHE will show you a series of slides explaining the use of mobile platforms.

Mr. LEYHE. Annexes 1 and 2 show a shotfiring drivage in various phases using a working platform and shield.

<u>Mr. NAQUET</u>. Can this equipment be used in inclined roadways and if so to what gradient ?

<u>Mr. LEYHE.</u> This platform has been used in rising workings at up to 12 gon and dipping workings to 14 gon.
In the Saar it is being used in greater inclines but the hydraulic traction unit has been modified and an additional monorail installed. The arches are not backfilled directly at the roadhead - this is one of the faults of this method. Tests have been carried out using an ash cement mix, however the slides which I will show you now show the installation on the surface of a continuous type of transport system for backfilling products. On the surface there are silos of 30 m^3 with a mixer and a nitrogen pump at the base. The product can thus be transported down the mine up to 4.5 km (Annex 3).
The transported material comprises 40% water, is rendered semi-solid by the mix of nitrogen and air and is placed behind the support (Annex 4).

<u>Mr. NAQUET</u>. Mr. LEYHE, may I thank you in the name of all the participants for an extremely comprehensive paper. The number of questions raised this afternoon show the importance that you attach to these problems of support mechanization. We should bear in mind that support represents 40 to 45% of the time involved and in recent years attention has been devoted to winning problems, largely ignoring support systems.
Your work, and in particular the work touched on this morning by the four speakers, show that the problems of drivage are now being tackled seriously.
I would now like to leave the technical aspects and thank you for your attention over the last two days and in particular at this round table.

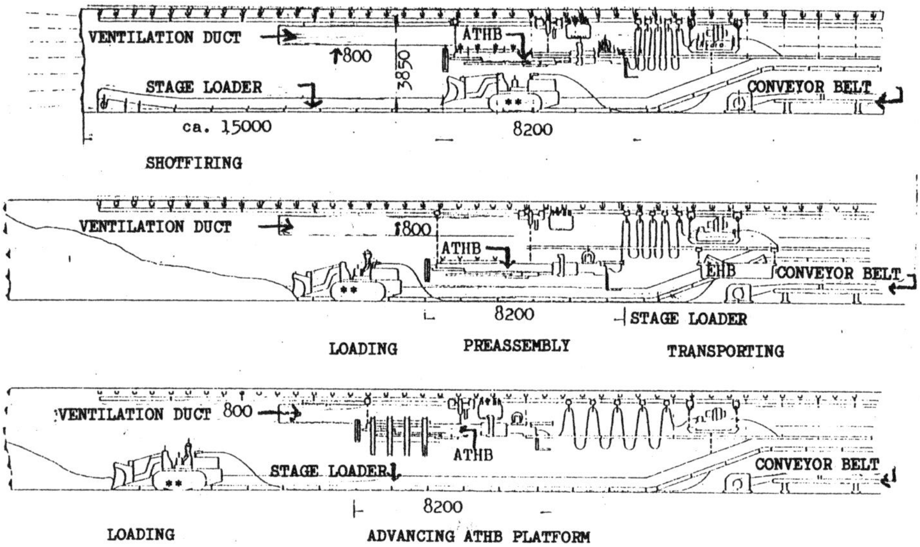

** SIDE-TIPPING LOADER

ANNEXE 1

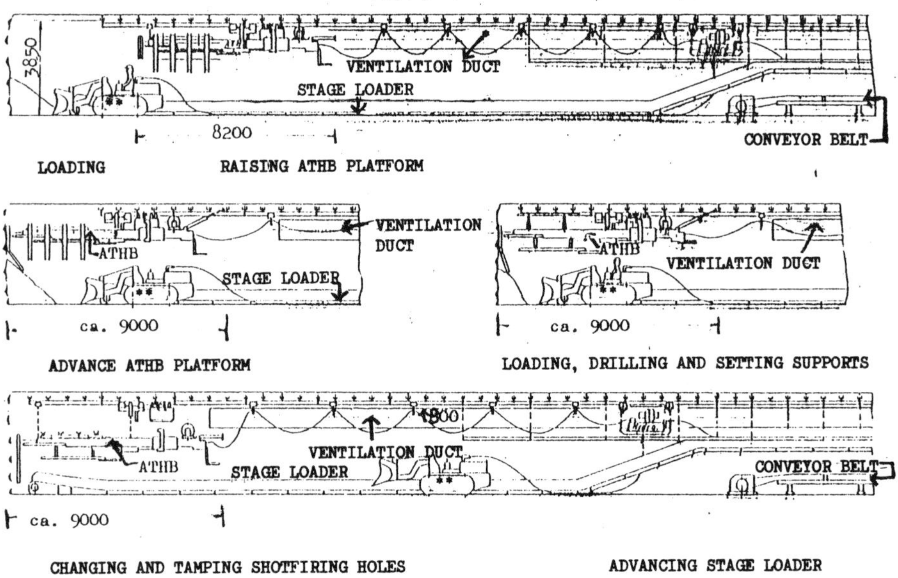

** SIDE-TIPPING LOADER

ANNEXE 2

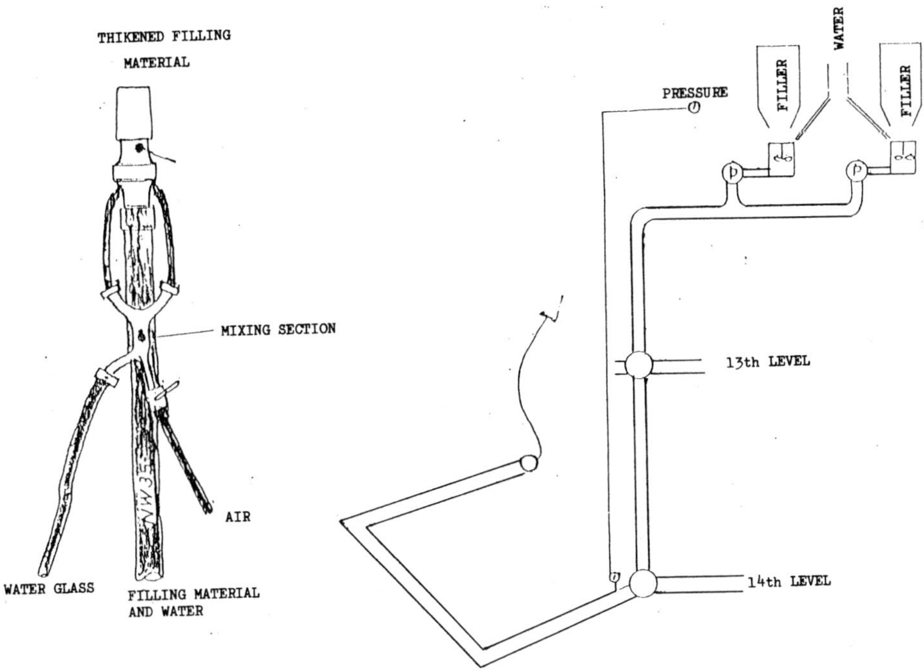

ANNEXE 3

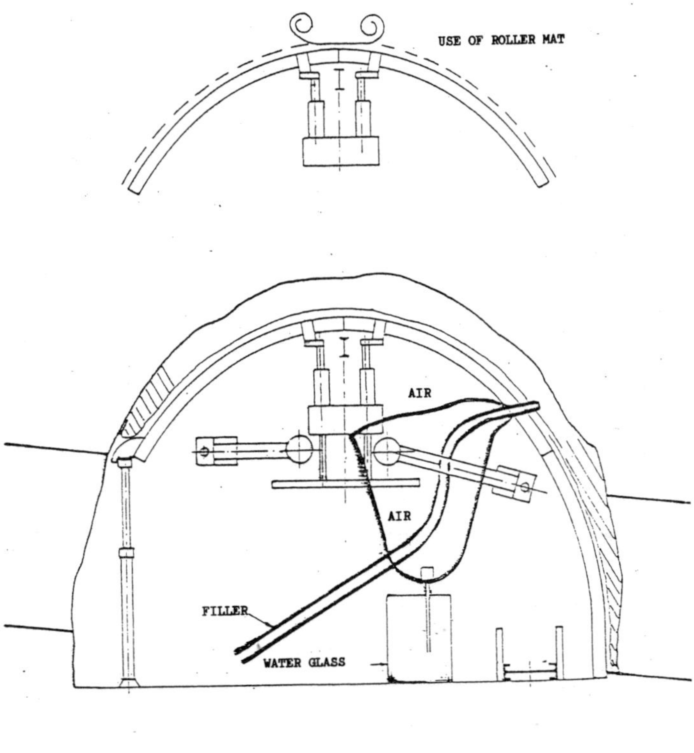

ANNEXE 4

Round Table Discussion D - 'Future Developments'

Chairman : W. KNICKMEYER, Bergbau-Forschung Gmbh, Essen

Secretary : L. BAUMANN, Bergbau-Forschung Gmbh, Essen

Mr. KNICKMEYER (Germany) opened the discussion and welcomed the participants. He referred to the papers read on the previous days. These had shown that while the heading machines used in the coalmining industry were technically advanced, the degree of drivage mechanization was still inadequate. The aim of future work should therefore be to extend the applications of heading machines and to enable the various work processes in conventional drivage, to be mechanized to a large extent.

Three important steps were improvement of machine utilization, optimization of heading systems and development of new techniques and processes for roadway drivage.

The discussion was to cover the following topics :

1. technical and operational monitoring ;
2. computerized control of machines and operations ;
3. new techniques for breaking ground.

Item 1 of the agenda

Mr. PAURAT (Germany) mentioned the different ways in which measurements signals were processed in Germany and Britain. In Germany the signals were processed on the spot whereas in Britain they were transmitted to the surface for processing. A major factor was the rapid transmission time.
Mr. KNICKMEYER pointed out that a large amount of data had to be sent to the surface for the purposes of straightforward monitoring of operations. Roadheaders, for example, were connected to the mine control rooms to enable machine running time and down time to be recorded.
Mr. PAURAT thought that the ultimate technical constraint would lie in data transmission as rapidly changing data required a high carrier frequency with current intensities which lay outside the intrinsically safe range.

Item 2 of the agenda

Mr. ZOLLMAN (Britain) showed a film of an automatic profile control system for roadheaders.

Mr. KNICKMEYER thanked Mr. ZOLLMAN and drew attention to the problem of overcut and accurate profiling when headings were driven by selective cut machines. Every new idea could be a step towards the solution of these difficulties.

Mr. HEISING (Germany) asked Mr ZOLLMAN how necessary it was for the cutting operation in the technique illustrated in the film to be fully automated.

Mr. ZOLLMAN pointed out that the automatic system led to a smooth profile and made life much easier for the machine operator, who still had to react to a large number of measurement signals.

Mr. RAFFOUX (France) asked whether the program took account of the position of the laser beam.

Mr. SCHEELE (Germany) asked whether fully automated control was expected to yield advantages as compared with manual control.

Mr. ZOLLMAN answered that no spectacular results could be reported on the basis of the limited experience acquired. The drivage rate, however, had not fallen and there were some advantages with regard to safety.

Mr. PAURAT expressed the view that automatic cutting had to be an objective for the future. He reported that on a machine in Ensdorf, the way in which the operator moved the cutting head was displayed on a plotter. By refining the system and directly analyzing the process on the monitor a machine operator could reduce the 175 m of cutting head travel currently required for a cross-section of 22 m^2 to about 80 m.

Mr. BEER (Germany) asked whether a time/study had been carried out during the research in order to compare automated and manual operation.

Mr. SOLLMAN replied that this had not been done as the tests had been carried out under artificial conditions only.

Mr. BEER then described a lightweight tunnel boring machine with automatic steering (Appendix 1).

Mr. PAURAT argued that roadheader manufacturers should build robust machines which were as compact as possible. The electronics should be an integral part of such machines.

Item 3 of the agenda

Mr. KNICKMEYER observed that the last two contributions had indicated one approach to reducing the weight of heavy heading machines with a consequent enlargement of their field of application. A further possibility was to improve the cutting efficiency of the tools and to increase the specific cutting performance. A number of projects were being carried out in the Community with this objective, aspects of particular interest were the use of high-pressure water, which was being intensively studied in the United Kingdom, France and the Federal Republic of Germany, and, as a long-term aim, the use of activated cutting tools.

Mr. BAUMANN (Germany) reported on new techniques for breaking ground - roadway profiling (Appendix 2) - and showed a film.
Referring to the film shown by Mr. BAUMANN, Mr. PAURAT Junior asked what the specific energy requirement was, how laminar flow conditions were maintained, why the high pressure zone was necessary and how the cuttability of rock by jet-assisted methods was assessed.

Mr. BAUMANN said that specific energy consumption was fairly high when cutting by high-pressure water jets alone was practised. When water jets were used in conjunction with mechanical cutting tools, the energy balance improved.
With regard to water pressure, Mr. BAUMANN stated that in order to fracture the rock the water had to be at a pressure which was generally above the rock's compressive strength. In the profiling technique shown, the optimum water pressure proved to be 1500 bars, the results being less satisfactory at lower pressures
Mr. BAUMANN argued that water jets alone were not viable as a means of cutting stone. No detailed information was yet available on the action of the water jet and tool.

Mr. KNICKMEYER also thought that the water pressures for jet cutting would have to be reduced. Possible approaches were the use of additives and optimization of nozzle design. The Bergbau-Forschung Gmbh was primarily interested in situations in which conventional hard metal could not be used and in types of rock which could only be cut at the cost of high pick wear.

Mr. PAURAT enquired about the effect of water on clay seat earths and on floors susceptible to heave.

Mr. KNICKMEYER replied that the quantities of water required for high-pressure jet cutting were less than those used for cooling of the picks or dust suppression on both roadheaders and full-face machines.

Mr. HAMM (Germany) asked about coordination of operations below ground as even the best machines were of little use unless there were suitable arrangements for stone clearance and setting of supports.

Mr. KNICKMEYER pointed out that this question belonged to the field of support technology and that the present discussion was primarily concerned with optimization of drivage systems.

Mr. HEISING asked about the possibility of combining high-pressure water jets and drilling bits with polycrystalline diamond inserts.

Mr. KNICKMEYER replied that diamond inserts of the requisite size cost DM 1000 apiece. The same performance as was achieved with these expensive bits was also possible with normal hard metal bits whose wear was reduced by the use of high pressure water.

Mr. BEERMANN (Germany) mentioned that the de Beer concern had been engaged for about 8 years in the development of drilling bits which were less susceptible to wear. The first tests would be carried out later that year using silicon nitrides.

Mr. WELLER (Germany) drew attention to the development of new alloys which would be ready for service in about two years.

Mr. RAFFOUX supplemented Mr. BAUMANN's remarks with a brief description of a high-pressure water jet technique employed in France. Further information would be provided on request.

Mr. KNICKMEYER closed the meeting and expressed the hope that when the next symposium was held in a few years' time it would be possible to report initial successes.

Appendix 1

Lightweight tunnel-boring machine

Reporteur : H. BEER

Figure 1 shows a machine specially designed for underground coal mines in the Federal Republic of Germany with a cut diameter of 5,5 m, which is the optimum size for the conditions obtaining in that country. A simple fixed anchorage is provided at the end of the machine with a slide immediately behind the cutterhead. The machine is extremely compact, its length being only about 5 m. It weighs about 150 t and can be assembled in about 14 days in a normal roadway, i.e. without any special launching chamber. The size of the individual machine components (weight 8 t and maximum dimensions 1 x 1.5 x 3.5 m) is such that they can readily be transported through narrow underground mine workings, even to remote working areas, e.g. by means of overhead monorails.

It is possible to gain access to the cutterhead of this machine from the inside for maintenance of the roller cutters. There is thus no need for men to work in the cut chamber where there is a risk of falls of ground.

A special application is the location of a coal seam in the heading face during roadway drivage in deep coal mines. The movement of the instrumented roller cutters reveals marked changes in the cutting forces at the coal/stone interfaces as a result of the differences in hardness between the coal and the adjacent strata. These force pulses are used to display the position of the coal seam to the machine operator on a television monitor. By suitable preprogramming of the microprocessor it is thus possible to ensure automatic compliance with specific drivage requirements such as the position of the seam within the roadway cross-section. The tunnel boring machine can thus follow closely the often undulating course of the coal seam. (Fig. 2).

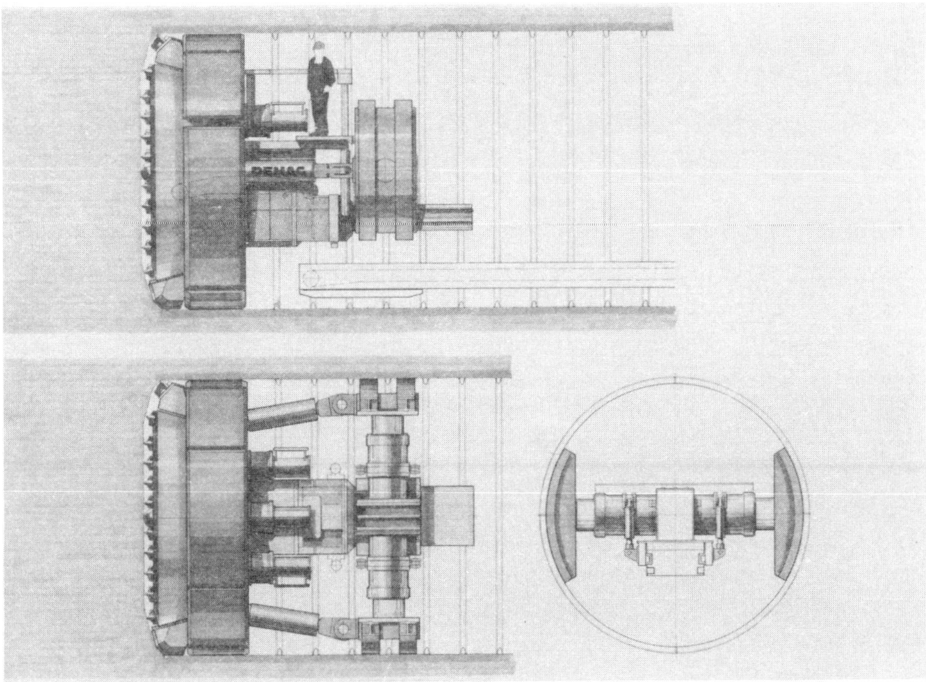

Figure 1

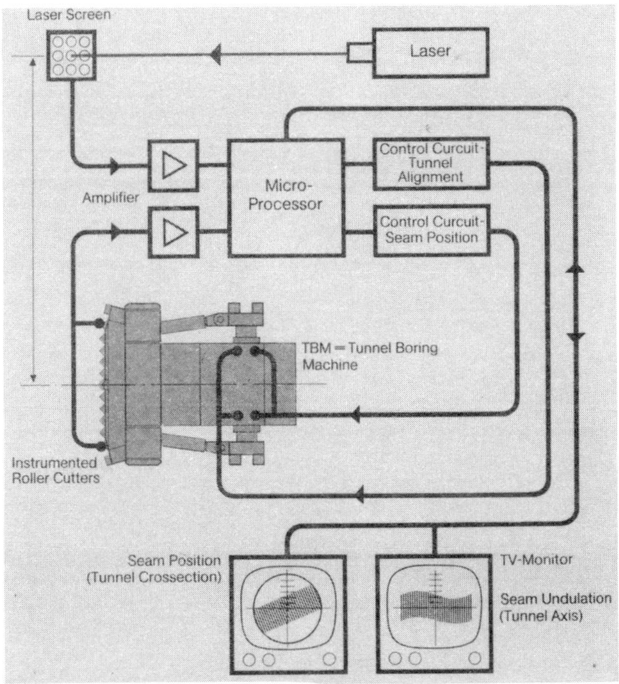

Figure 2

Appendix 2

New techniques for breaking ground - roadway profiling

Reporteur : L. BAUMANN

The degree of mechanization in roadway drivage and especially in gateroad drivage is still fairly low. It is hoped to improve the situation by developing a lighter roadway profiling system which will increase roadway stability by cutting true to profile and minimizing strata degradation and reduce the delay before setting of supports, thus permitting economic drivage of gateroads.

A roadway profiling and heading system is being developed and tested in a project subsidized by th Land North Rhein Westphalia. The technique minimizes damage to the strata and gives a very accurate profile (Figure 1).

Tests were successfully conducted in hard sandstone, in softer bedded stone and in the transition zone between the coal face and the stone (Figure 2). They showed that the forces acting on hard-metal picks can be reduced by an average of about 50% by means of grouped high-pressure water jets. The tests also demonstrated that the pick service life can be considerably increased when high-pressure water jets are used, thus allowing new applications of hard-metal tools in the cutting of hard rock.

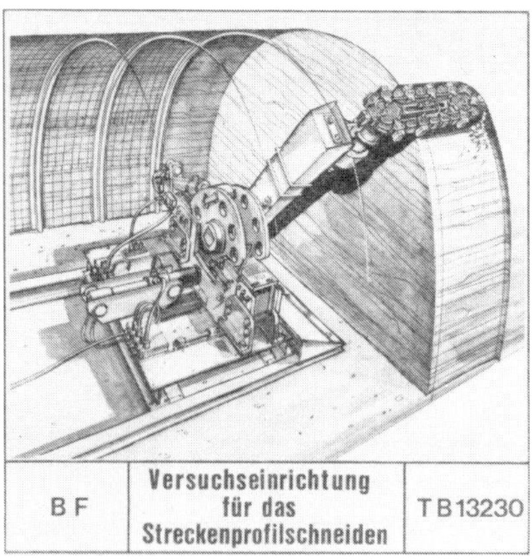

Fig. 1. BF/Test rig for roadway profiling/TB 13230

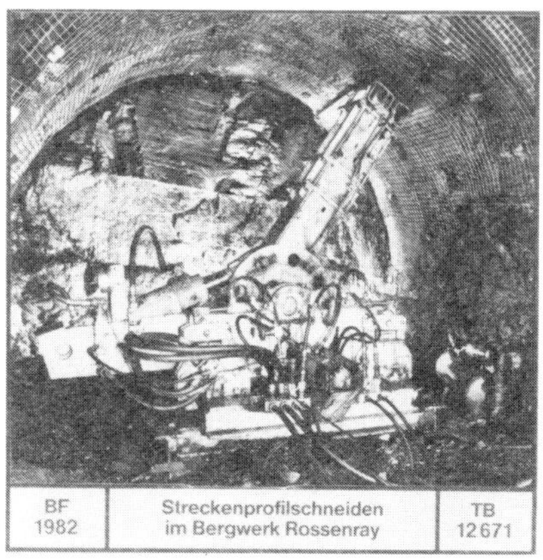

Fig. 2. BF 1982/Roadway profiling in the Rossenray Mine/TB 12671

LIST OF PARTICIPANTS

AHCAN, R.
Professor
Faculty of Technologie
Askerceva ul. 20
YU - 61000 LJUBLJANA

ALLMANG, R.
Ingénieur
H.B. Lorraine
20 A, rue de Forbach
F - KERBACH

AMBERG, R.
AMBERG Ing. Büro für Tunnelbau
CH - 7320 SARGANS

AMLER, F.
Rudolf Hausherr und
Söhne & Co.KG
Postfach 1240
D - 4322 SPROCKHOEVEL 1

AMTHOR, W.
Redakteur
Verlag Glückauf GmbH
Postfach 10 39 45
D - 4300 ESSEN

ASTON, T.
Research Scientist
Coal Research - CANMET
210 George St
Canada - SYDNEY, Nova Scotia, B1P 1J3

AUER, A.
Ingénieur
Lenoir et Mernier
14, rue Pasteur
L - 4276 ESCH-SUR-ALZETTE

BACHMANN, K.D.
Verkaufs-Ingenieur
Deutsche Montabert GmbH
Wasserbank 9
D - 5810 WITTEN 3

BAERTLING, F.
Assessor des Bergfachs
Preuss AG
D - 4530 IBBENBUEREN 1

BARCHANSKI, B.
Bergingenieur
Berg-und Hüttenakademie AGH
Al. Mickiewicza 30
POLEN - 30-059 KRAKOW

BARRIQUAND, E.
Ingénieur
SOFREMINES
59, rue de la République
F - 93104 MONTREUIL CEDEX

BATTAREL, J.
Ingénieur
SOFREMINES
59, rue de la République
F - 93104 MONTREUIL CEDEX

BAUER, H.D.
Techn. Direktor
Bergbau-BG
Hunscheidtstrasse 12
D - 4630 BOCHUM 1

BAUMANN, J.
Vorstandsmitglied
Gewerkschaft Walter
Stauderstr. 213
D - 4300 ESSEN 1

BAUMANN, L.
Dipl.-Ing.
Bergbau-Forschung
Franz-Fischer-Weg 61
D - 4300 ESSEN 13

BECHEM, K.G.
Ingénieur
Bechem GmbH
D - 5800 HAGEN-BERCHUM

BEER, K.
Projekt-Ingenieur
Mannesmann Demag
Bergwerktechnik
W.-Reuter-Platz, 41
D - 41 DUISBURG 1

BELL, H.
Wiss. Angestellter
Institut für Bergbaukunde II
Lochnerstrasse 4/20
D - 5100 AACHEN

BLACK, J.C.
Managing Director
Cementation Mining Limited
Bentley House, P O Box 22
GB - DONCASTER, S. Yorksh. DN5 OBT

BLANC, A.
Ingénieur en Chef
Houillères du Bassin du Nord et du Pas-de Calais - Charbonnage de France
64, rue des Minimes
F - 59505 DOUAI CEDEX

BLANC, M.
Ingénieur
Houillères d'Aquitaine
2, rue du Gaz
F - 81400 CARMAUX

BOAST, G.E.
Group Head Tunnelling
Mining, Research & Development
National Coal Board
Ashby Rd, Stanhope
GB - BURTON-on-TRENT, Bretby Staffs DE15 OQD

BOCOLINIS, S.
Commission des Communautés européennes
Direction générale "Personnel et Administration"
B.P. 1907
L - 2920 LUXEMBOURG

BOHNENKAMP, G.
Dipl.-Berging.
Gesteins- und Tiefbau GmbH
Karlstr. 37-39
D - 4350 RECKLINGHAUSEN

BOLDT, H.
Bergwerksdirektor RAC
Rheurdter Str. 64
D - 4174 ISSUM

BONIS, A.
Ingénieur des Mines
M.F.I.
Rue du Champ de Mars
F - 57202 SARREGUEMINES

BORGMEIER, E.
Wiss. Angestellter
Institut für Bergbaukunde II
Lochnerstrasse 4/20
D - 5100 AACHEN

BORNEMANN, E.
Managing Director
Thyssen (GB) Ltd
Bynea
GB - LLANELLI, Dyfed, SA14 9SU

BOROWSKI. K.H.
Prokurist
Leiter der Technik und Forschung
Gebrüder Eickhoff
Hunscheidtstr. 176
D - 4630 BOCHUM 1

BOULICAULT, M.
Ingénieur en Chef des Mines
Ministère de l'Industrie et de la Recherche
97, rue de Grenelle
F - 75700 PARIS

BOUNAKOFF, S.
Ingénieur
Ets. MONTABERT
23, ave Marceau
F - 75116 PARIS

BRABBINS, M.
Sales Director
Gullick Dobson Limited
P.O. Box 12
Ince
GB - WIGAN, Lancashire WN1 3DD

BRISTOW, N.
Technical Director
Perard Torque Tension
Brittain Drive,
Codnor Gate Industrial Estate
GB - RIPLEY DE5 3QB

BRUEMMER, K.H.
Geschäftsführer
Deilmann-Haniel GmbH
Postfach 130220
D - 4600 DORTMUND 13

BRUNE, F.
Assessor des Bergfachs
Thyssen Bergbautechnik
Ehinger Str. 80
D - 4100 DUISBURG 28

BUHLES, A.
Dipl. Ing.
H. KRUMMENAUER
Wellesweilerstr. 95
D - 6680 NEUNKIRCHEN

BUNNISS, A.
UK Sales Manager
Dosco Overseas Engineering Limited
British Fields
Ollerton Road, Tuxford
GB - NEWARK, Nottinghamshire, NG22

BUSCHMANN, H.E.
Assessor des Bergfachs
Bergbau AG
Niederrhein
Postf. 170154
D - 4100 DUISBURG 17

BUTTIENS, E.
Engineer
B.E.P.
Hof van Delftlaan 15
B - 2070 EKEREN

CAILLET, A.
Technical Representation
Dosco Overseas Eng.
Av. de Roovere 16 - B 42
B - 1080 BRUSSELS

CALDOW, G.
Colliery Manager
National Coal Board
Scottish Area, Green Park
Green End
GB - EDINBURGH EH17 7PZ

CARIVEN, D.
Directeur des Serv. Techniques
Charbonnages de France
B.P. 396-08
F - 75360 PARIS CEDEX 08

CARTA, M.
Prof. Univ.
Università Cagliari
Dipart. Inc. Miner Università
I - 09100 CAGLIARI

CASSINI, P.
Ingénieur
CERCHAR
B.P. 2
F - 60550 VERNEUIL

CAUNT, R.A.
Chief Mining Engineer
National Coal Board
South Nottinghamshire Area
GB - BESTWOOD, Nottingham NG6 8UE

CECIL, W.G.
Colliery Manager
National Coal Board
"Hillcrest" Highlang Crescent
GB - PONTLLANFRAITH, Gwent

CHAUDHURI, D.B.R.
Editor,"Urja" Indian
Journal of Energy
Post Box 3008, G-82
Sujan Singh Park
India - NEW DELHI 110003

CLAES, C.
Engineer
N.V. Kempense Steenkolenmijnen
Oostlaan 59
B - 3600 GENK

COLLINS, R.E.
Deputy Managing Director
EIMCO (Great Britain) Ltd
Earlsway, Team Valley
GB - GATESHEAD, Tyne and Wear NE11 OSB

COLLOMB, F.R.
Ingénieur
M.F.I.
37, Av. P. Doumer
F - 75116 PARIS

COX, W.
Vice President, P. Eng.
Beaver Underground Structures
1300 Hymus Blvd.
CANADA, QUEBEC - MONTREAL

CREPLET, J.P.
Directeur
DEHEZ S.A.
65, ave de la Corniche
B - 1310 LA HULPE

DANIELS, W.H.
Verbandsgeschäftsführer
Vereinigung der Bergbau-
Spezialgesellschaften e.V.
Rüttenscheider Str. 14
D - 4300 ESSEN 1

DAREAU, M.
Responsable Dept. Mines
Le Burin Commercial SA
27, rue Péclet
F - 75015 PARIS

DAVIES, P.M.
Senior Mining Engineer
National Coal Board
North Derbys Area, Bolsover
GB - CHESTERFIELD

DAVIES, P.M.
Senior Mining Engineer
National Coal Board
North Derrys Area, Bolsover
GB - CHESTERFIELD

DE GREEF, A.
Chef de Division "Technologies"
Direction générale "Energie"
Commission des Communautés
européennes
200, rue de la Loi
B - 1049 BRUXELLES

DE SAINT MELEVC, N.
Ingénieur
Charbonnages de France
9, ave Percier
F - 75360 PARIS CEDEX 08

DEEMING, J.K.
Mine Manager
N.C.B.
BIRCHLEA, 1 Coppice Grove, Hatfield
GB - DONCASTER DN76 NU

DEGUELDRE, G.
Professeur
Faculté Polytechnique
de Mons
9, rue de Houdaing
B - 7000 MONS

DELCRUZEL, J.
Ingénieur
Charbonnages de France
9, ave Percier
F - 75008 PARIS

DELFORGE, A.
Ingénieur
Sté des Explosifs TITANITE
F - 21270 PONTAILLER SUR SAONE

DELIAC, E.
Directeur Adjoint du CMI
Ecole des Mines de Paris
35, rue Saint Honoré
F - 77305 FONTAINEBLEAU

DELIGNE, A.
Directeur de Société
SA Andre Deligne
48, rue de Cartier
B - 6030 MARCHIENNE-AU-PONT

DEQUET, G.
Ingénieur
Sté Celtite
Vielverge
F - 21270 PONTAILLER SUR SAONE

DICHTL, R.
Marketing Service Manager
Boart HWF GmbH & Co. KG
Hartmetallwerkzeugfabrik
Städeweg 18
D - 6419 BURGHAUN 1

DINGERKUS, O.
Rudolf Hausherr und
Söhne GmbH & Co.KG
Postfach 1240
D - 4322 SPROCKHOEVEL 1

DOERELL, P.
Editor
International Mining
123 A, Ch. de Charleroi
B - 1060 BRUXELLES

DOM, P.
Ingénieur - Dir. Général Adjoint
FORAKY S.A.
13, place des Barricades
B - 1000 BRUXELLES

DUBUGNON, O.
Physicien
Institut CERAC SA
Ch. des Larges Pièces
CH - 1024 ECUBLENS

DUCLOS, L.
Ingénieur
Montabert
23, ave Marceau
F - 75116 PARIS

DUERRFELD, W.
Ingenieur
Gesteins- und Tiefbau GmbH
Karlstr. 37-39
D - 4350 RECKLINGHAUSEN

ECKER, R.
Dipl.-Ing.
KFA/PLE Jülich
Postfach 1913
D - 5170 JUELICH

EDELHOFF, P.
Dipl.-Ing., Wiss. Angestellter
Institut Für Bergbaukunde II
Lochnerstrasse 4/20
D - 5100 AACHEN

ENGELBOS, J.M.
Ingénieur Civil
Corps des Mines
Prins Albert laan 22
B - 3800 ST TRUIDEN

EULGEN, K.
Dipl.- Bergingénieur
R A G Essen
Am Habeweik 79
D - WALTROP-OBERWIESE

EVANS, J.W.
Deputy Director (Mining)
National Coal Board
Staffordshire House, Berryhill Road
GB - FENTON, Staffs

EVANS, G.
Mining Engineer
Perard Torque Tewsion Ltd
Brittain Drive
Codnor Gate Industrial Estate
GB - RIPLEY DE5 3QB

FEISTKORN, E.
Ingenieur
Bergbau-Forschung GmbH
Franz-Fischer-Weg 61
D - 4300 ESSEN 13

FIELD, B.R.
B. Sc.
PAURAT GmbH
Vortriebsmaschinen
Postfach 1220
D - 4223 VOERDE 2 (Friedrichsfeld)

FILLAUX, J.
Ingénieur
GOGEMA
2, rue Paul Dautier Velizy
F - 78140 VILLACOUBLAY

FINK, F.
Ing. (Bergbau)
SAARBERGWERKE AG, Abt. E-FF
Postfach 1050
D - 6600 SAARBRUECKEN

FISCHBACH, E.
Export-Direktor
Gebr. Eickhoff
Hunscheidtstrasse 176
D - 4630 BOCHUM 1

FOURMAINTRAUX, D.
Ingénieur Géotechnicien
Electricité de France
EDF-TEGG
3, rue de Messine
F - 75384 PARIS CEDEX 08

FRANCOIS, G.
Président Directeur général
LE BURIN S.A.
171, rue de Paris
F - 93012 BOBIGNY CEDEX

FRENAY, C.
Directeur Divisionnaire des Mines
9, rue de la Chapelle
B - 4623 MAGNEE

FRENYO, P.
Ingenieur
PAURAT GmbH Vortriebsmaschinen
Postfach 1220
D - 4223 VOERDE 2 (Friedrichsfeld)

FREYTAG, L.
Assessor d.B. Westfalia Lünen
Postfach
D - 4670 LUENEN

FROEHLICH, K.
Dipl.-Ing.
Dollendorfer Str. 8
D - 5300 BONN 2

FU, G.
Stagiaire
Université Libre de Bruxelles
B.P. 8, rue St. Georges 102
B - 1050 BRUXELLES

GAILER, G.
Dipl.-Ing
Deilmann-Haniel GmbH
Haustenbecke 1
D - 4600 DORTMUND 13

GARLAND, J.
Senior Sales Executive
Gullick Dobson Limited
P.O. Box 12, Ince
GB - WIGAN, Lancashire WN1 3DD

GERAUD, C.
Ingénieur
Houillères du Bassin de Lorraine
Charbonnages de France
2, rue de Metz
F - 57802 FREYMING-MERLEBACH

GETHING, R.
Director
Rota Engineering Ltd.
Wellington Street
GB - BURY BL8 2BD

GOFFIN, J.
Engineer
N.V. Kempense Steenkolenmijnen
Onderwijslaan 25
B - 3600 GENK

GOLDSBY, M.
Senior Mining Engineer
National Coal Board
South Nottinghamshire Area
GB - BESTWOOD, Nottingham NG6 8UE

GOTTSCHALK, G.
Dipl.-Ing.
Bochumer Eisenhütte
Heintzmann GmbH & Co. KG
Bessemerstr. 80
D - 4630 BOCHUM 1

GOULDING, C.
Colliery General Manager
National Coal Board
Staffordshire House, Berryhill Rd.
GB - FENTON, Staffs

GRANT, I.C.
Minging Engineer
Andonson Strathclyde PLC
47, Broad Street
GB - GLASGOW

GRAUMANN, K.
Wissenschaftl. Mitarbeiter
WBK Bochum
Herner Str. 45
D - 4630 BOCHUM

GUERIN, G.
Dir. Marketing
EIMCO-SECOMA
BP 46
F - 69881 MEYZIEU - CEDEX

GULDIMANN, H.J.
Bau. Ingenieur
Perslucht Centr. NED.B.V.
P.B. 2069
NL - 2301 CB LEIDEN

GURREY, P.
Ingénieur des Mines
Compagnie MECO
15, place de la Madeleine
F - 75008 PARIS

HAARMANN, K.R.
Bergwerksdirektor
Ruhrkohle AG
D - 4712 WERNE

HAMM, E.
Assessor des Bergfachs
Ruhrkohle AG
Rellinghauser Str. 1
D - 4300 ESSEN

HARRISON, W.
Chief Engineer
Anderson Strathclyde PLC
Board Street 47
GB - GLASGOW G40 2QW

HARRIS, L.
Deputy Director (Mining)
National Coal Board
Coleorton Hall
GB - COLEORTON, Leics. LE6 4FA

HAUSHERR, H.R.
Dipl.-Berging.
Rudolf Hausherr & Söhne GmbH
& Co.KG
Postfach 1240
D - 4322 SPROCKHOEVEL 1

HEGEMANN, J.
Dipl.-Ing.
Vorstandssprecher
Gewerkschaft Walter
Stauderstr. 213
D - 4300 ESSEN 1

HEISING, F.
Steinkohlenbergbauverein
Franz-Fischer-Weg 61
D - 4300 ESSEN 13

HERRMANN, A.
Senior Adviser
Breloer Weg 11A
D - 4322 SPROCKHOEVEL 1

HERZBRUCH, P.
Dipl.-Berging.
PAURAT GmbH
Nordstr. 73
D - 4223 VOERDE 2

HESSBRUEGGE, F.W.
Wissenschaftl. Mitarbeiter
WBK BOCHUM
Herner Str. 45
D - 4630 BOCHUM

HEYMANN, M.
PAURAT GmbH
Vortriebsmaschinen
Postfach 1220
D - 4223 VOERDE 2 (Friedrichsfeld)

HILL, J.
Sales Manager
EIMCO (Great Britain)Ltd.
Earlsway, Team Valley
GB - GATESHEAD, Tyne & Wear NE11 0SB

HIRENGEN, P.
Product Design Engineer
Atlas Copco MCT AB
S - 104 84 STOCKHOLM

HOEVELHAUS, H.W.
Betriebsdirektor, Bergassessor a.D.
Bergbau AG Lippe
Bergwerke Zollverein und Nordstern
Postfach 1145
D - 4690 HERNE 1

HUCK, D.
Ingénieur chercheur
Ecole des Mines de Paris
35, rue Saint Honoré
F - 77305 FONTAINEBLEAU

HUREL, F.
Ingénieur
CERCHAR
BP 2
F - 60550 VERNEUIL EN HALATTE

HURTEVENT, D.
Ingénieur
HBWPC
59, rue des Foulon
F - DOUAI

HUYGENS, H.
Juriste
40, rue de Fingig
L - 4966 CLEMENCY

IMMHOFF, B.
Bergass. a.D., Prokurist
Montanbüro
Sevinghauser Weg 100
D - 4630 BOCHUM 6

IRELAND, J.T.
Sales Director
Webster Machine Co.
Wortley Road
GB - ROTHERHAM S61 1LZ, S. Yorks.

JACOBS, F.
S.A. Andre Deligne
Rijksweg 560
B - 3630 MAASMECHELEN

JENKINS, C.B.
Production Manager
National Coal Board
Tondu, Aberkenfig
GB - BRIDGEND CF32 9HB, S. Wales

JOHNSON, J.
Mine Manager
National Coal Board
"Beacon View" Common Lane, Upton
GB - NR PONTEFRACT, Yorks. WF91DE

JOHNSON, D.
Sales Director
Dosco Overseas Engineering Limited
British Fields
Ollerton Road, Tuxford
GB - NEWARK, Nottinghamshire NG22 OPQ

JOHNSON, S.T.
University of New Castle Upon Tyne
Dept. of Geotechnical Engineering
GB - NEW CASTLE, Upon Tyne

JONES, C.T.
PAURAT GmbH
Vortriebsmaschinen
Postfach 1220
D - 4223 VOERDE 2 (Friedrichsfeld)

KABONGO, N.
Ing. Civile des Mines
Gécamines
15, rue de la Loi
B - 1040 BRUXELLES

KALLENBERG, H.H.
Dipl.-Ingenieur
Dr. Boing KG
Tilgelstr. 6
D - 5800 HAGEN

KARNELO, E.
Physicien
Atlas Copco Jarva AG
Postfach 20
CH - 3601 THUN

KETTELER, H.
Bergwerksdirektor
Bergbau Prosper-Haniel
Fernewaldstr.
D - 4250 BOTTROP

KINDERMANN, F.
Kommission der Europaäischen
Gemeinschaften
Generaldirektion "Energie"
200 rue de la Loi
B - 1040 BRUXELLES

KLEIN, J.
Ingenieur
Bergbau-Forschung GmbH
Postfach 13 01 40
D - 4300 ESSEN 13

KLEIN, E.
Masch.-F.Stgr.
SAKRET GmbH
Im Grünenkraut 3
D - 6624 GROSSROSSELN

KLEINERT, H.-W.
Dipl.-Berging.
Bergbau-Forschung GmbH
Abt. Vortriebstechnik
Franz-Fischer-Weg 61
D - 4300 ESSEN

KNICKMEYER, W.
Dezernent, Dipl.-Ing.
Steinkohlenbergbauverein
Franz-Fischer-Weg 61
D - 4300 ESSEN 13

KNIGHT, R.
Sales Director
Holly Bank Eng. Co. LTD
Ollerton Rd, Tuxford
GB - Nr NEWARK, Notts. NG22 0PQ

KOCH, K.
Dipl.-Ing.
KFA/PLE Jülich
Postfach 1913
D - 5170 JUELICH

KONDER P.
Bergdirektor
Oberbergamt Saarbrücken
D - 6600 SAARBRUECKEN

KOPKE, M.
Gewerkschaftssekretär
IG Bergbau und Energie
Alte Hattinger Str. 19
D - 4630 BOCHUM 1

KRAEMER, T.
Physicien
Institut CERAC SA
Ch. des Larges Pièces
CH - 1024 ECUBLENS

KRAUSE, J.
Dipl.-Ingenieur
Frölich & Klüpfel
Postfach 100 853
D - 4650 GELSENKIRCHEN

KRONENBERGER J.
Ingénieur
A.T.C.
29, route du Vin
L - 5450 STADTBREDIMUS

KUEPPERS, H.J.
Stellvertr. Betriebsführer
Gewerkschaft
Sophia-Jacoba
Postfach 1320
D - 5142 HUECKELHOVEN

KULLAS, H.
Director
Amalgamated Construction CO.
P.O. Box 1, Whaley Rd
GB - BARNSLEY, S. Yorks

KULNIG, H.
Betriebsdirektor
Barbara Rohstoffbetriebe GmbH
Zur Fliethe 23-25
D - 5603 WUELFRATH-WILHELMSHOEHE

LACHMAN, K.
Technical Director
P.R.G. - KATOWICE
ul. Damrota 16
Poland - 40021 KATOWICE

LAMONT, G.
Mining Engineer
Preuss AG Kohle
D - 5430 IBBENBUEREN

LAMOURIC, P.
SEMAFOR
B.P. 12
F - 78730 ST. ARNOULT

LECLERCQ, G.
Engineer
N.V. Kempense Steenkolenmijnen
Dennenstraat 14
B - 3600 GENK

LEGRAND, J.
Engineer
N.V. Kempense Steenkolenmijnen
Minderbroedersstraat 30
B - 3550 HEUSDEN-ZOLDER

LEHMANN, H.
Dipl.-Ing.
BAG Niederrhein
Roosenstr. 5
D - 4133 NEUKIRCHEN-VLUYN

LIESNER, M.
Assesor des Bergfachs
Ruhrkohle AG
Postfach 10 32 62
D - 4300 ESSEN 1

LEISTENSCHNEIDER, J.
Directeur des Ventes
Forges & Boulonneries
ARS/MOSELLE
68, rue G. Clémenceau
F - 57130 ARS SUR MOSELLE

LEMMES, F.
Dipl. Geologe
H. Krummenauer
Wellesweilerstr. 95
D - 6680 NEUNKIRCHEN

LEONET, O.
Ingénieur
SAMIFER
B.P. 29
F - 54152 BRIEY CEDEX

LEROY, J.
Directeur général
S.A. Armand Colinet
Faubourg de Mignault 17
B - 7078 LE ROEULX

LETHUAIRE, R.
Directeur de Société
Lenoir & Mernier
14, rue Pasteur
L - ESCH/ALZETTE

LEYHE, K.
Berging.
BAG LIPPE
Postfach 1145
D - 4690 HERNE 1

LOECHTE, K.
Dipl.-Ing
Atlas Copco Deutschland GmbH
Ernestinenstr. 155
D - 4300 ESSEN 1

LONGELIN, R.
Président Directeur général
Perforex
95/99, Bvd d'Alsace Lorraine
F - 93115 ROSNY-SOUS-BOIS

LOUBELE, J.
Engineer
N.V. Kempense Steenkolenmijnen
Swennenblook 14
B - 3600 GENK

LUDWIG, G.
Techn. Sachbearbeiter
BAG Niederrhein
Oldenburger Str. 14
D - 4200 OBERHAUSEN

LUECKER, A.
Dipl.-Berging.
Gesteins- und Tiefbau GmbH
Karlstr. 37-39
D - 4350 RECKLINGHAUSEN

LUTHE, W.
Deputy Managing Director
Thyssen (GB) Ltd
Bynea
GB - LLANELLI, Dyfed. SA14 9SU

LUX, J.
Adjoint au directeur
Services Techniques
Charbonnages de France
9, ave Percier
F - 75008 PARIS

MACKOWIAK, J.
Chef de Service
S.A. GOLDENBERG
B.P. 90
F - 67702 SAVERNE CEDEX

MAINJOT, M.
Ingénieur principal
divisonnaire des Mines
Administration des Mines
30, rue J.A. De Mot
B - 1040 BRUXELLES

MAIRHOFER, E.
Marketing Voest-Alpine/Austr.
AUTRICHE

MALBRECHT, D.
Studienrat z. A.I.E.
BERGFACHSCHULE
Herdinger-Str. 74
D - 4130 MOERS

MANCA, P.P.
Ingénieur
Université de Cagliari
Piazza d'Armi
I - 09100 CAGLIARI

MARX, J.P.
Ingénieur
Mines de Potasse d'Alsace S.A.
(M.D.P.A.)
11, ave d'Altkirch
F - 68055 MULHOUSE CEDEX

MASSEY, C.T.
Deputy Director (Mining)
National Coal Board
P.O. Box 13, Allerton Bywater
GB - CASTLEFPRD, YORKSHIRE

MATHIS, R.
Verk. Ing.
ALIVA AG
Postfach 69
CH - 8967 WIDEN

MENDE, H.
Bergwerksdirektor
Bergbau AG Lippe
WD Schlägel & Eisen
Postfach 1420
D - 4352 HERTEN/WESTF.

MENZEL, W.
PAURAT GmbH
Vortriebsmaschinen
Postfach 1220
D - 4223 VOERDE 2 (Friedrichsfeld)

MERMINOD, A.
Ingénieur
Institut CERAC SA
Ch. des Larges Pièces
CH - 1024 ECUBLENS

MERTENS, V.
Dipl.-Berging.
Steinkohlenbergbauverein
Franz-Fischer-Weg 61
D - 4300 ESSEN 13

MILLER, W.
Colliery Manager
National Coal Board
Green Parc, Scottish Area
Green End
GB - EDINBURGH EH17 7PZ

MILLS, L.J.
Deputy Chairman
National Coal Board
Hobart House, Grosvenor Place
GB - LONDON SW1X 7AE

MINNE, J.
Ingénieur délégué
S.A. C B M
1A, rue A. Degrâce
B - 7230 FRAMERIES

MITCHELL, J.
Marketing Manager
Anderson Strandcyne
47 Beoan St.
GB - GLASGOW

MOEBUS, D.
Ing. Grad.
Bochumer Eisenhütte
Heintzmann GmbH & Co. KG
Bessemerstr. 80
D - 4630 BOCHUM 1

MONIQUET, J.P.
Ingénieur Civil des Mines
Foraly S.A.
Place des Barricades 13
B - 1000 BRUXELLES

MONTACIE, M.
Ingénieur
Bouygues
97, avenue General Leclerc
F - 75014 PARIS

MOORE, K.
Mining Engineer
National Coal Board
MRDE, Bretby
GB - BURTON-ON-TRENT, Staffs.

MOORE, D.
Mining Engineer
A.M.C.O.
P.O. Box 1, Whaley Road, Barugh
GB - BARNSLEY, S. Yorks

MORRIS, A.H.
Chief Project Manager
Tunelling and Transport Branch
National Coal Board
Ashby Road, Stanhope Bretby
GB - BURTON-ON-TRENT, Staffs DE15 0QD

MUELLER, G.
Sales Manager
WIRTH GmbH
Kölner-Str. 71-78
D - 5140 ERKELENZ 1

MUELLER, H.
Dipl. Ing.
Krupp Widia
Abt. Hvb.
Münchener Str. 90
D - 4300 ESSEN 2

McDERMOTT, J.
Regional Sales Manager - Export
Dosco Overseas Engineering Limited
British Fields
Ollerton Road, Tuxford
GB - NEWARK, Nottinghamshire NG22 0PQ

McDONALD, H.
Sales Manager
WINSTER Group
Manners Av.
GB - ILKESTON, Derbys DE7 8EF

McKAY, A.M.
Managing Director
EIMCO (Great Britain) Ltd
Earlsway, Team Valley
GB - GATESHEAD, Tyne & Wear NE11 0Sb

NAQUET, J.
Ingénieur Principal
Houillères du Bassin de Lorraine
Charbonnages de France
2, rue de Metz
F - 57802 FREYMING-MERLEBACH

NAUD, C.
Ingénieur
H.B.L. Siège Wendel
Houillères du Bassin de Lorraine
F - 57802 FREYMING-MERLEBACH

NICHOLAS, H.
Dep. Chief Insp.
Mines & Quarries
Health & Safety Executive
Regina House, Old Marylebone Road
GB - LONDON NW1

NOACK, K.
Wissenschaftl. Mitarbeiter
WBK, Bochum
Herner Str. 45
D - 4630 BOCHUM

OBST, W.
Bergassessor a. D.
Hauptverwaltungsrat
Kommission der Europäischen
Gemeinschaften - DG V
Beschäftigung, soziale
Angelegenheiten und Bildung
L - 1907 LUXEMBOURG

OERINGE, B.
Product Manager
Atlas Copco MCT AB
S - 10484 STOCKHOLM

OLIVAR, J.
Dir. Général des Ventes
EIMCO-SECOMA
2, rue de Clichy
F - 75009 PARIS

OTTO, K.
Assessor des Bergfachs, Dipl.-Ing.
Bergbau AG Lippe
Shamrockring 1
D - 4690 HERNE 1

OUILL, H.-J.
Dipl.-Ing.
Klöckner-Becorit GmbH
Werkstättenstr. 7 - 13
D - 4350 RECKLINGHAUSEN,
Nordrhein Westf.

PANICH, D.
Junior Research Associate
University of Newcastle/Tyne
44, Briardene
Cres. Kenton
GB - NEWCASTLE/TYNE NE3 4RY

PAREE, J.C.
Ingénieur Civil des Mines
Administration des Mines
Demerstraat 81, 4de verdieping
B - 3500 HASSELT

PARET, A.
Ingénieur
Houillère Provence
B.P. 1
F - 13520 MEYREUIL

PASTOR, G.
Représentant
Etablissements PASTOR
B - 4121 NEUVILLE-EN-CONDORZ

PAURAT, F.W.
PAURAT GmbH
Vortriebsmaschinen
Postfach 1220
D - 4223 VOERDE 2 (Friedrichsfeld)

PAURAT, R.
Dipl.Ing.
PAURAT GmbH
Vortriebsmaschinen
Postfach 1220
D - 4223 VOERDE 2 (Friedrichsfeld)

PELKA, W.
Dipl.Ing., Verkaufsingenieur
Gebr. Eickhoff
Hunscheidtstr, 176
D - 4630 BOCHUM 1

PEREZ, T.
Product Specialist
ATLAS Copco Belgium
Brusselse Stwg. 346
B - 1900 OVERYSE

PHILIPPRON, P.
Engineer
N.V. Kempense Steenkolenmijnen
Brelaarstraat 48
B - 3530 HOUTHALEN-HELCHTEREN

PIGUET, J.P.
Ingénieur
CERCHAR
Laboratoire de Mécanique
des Terrains, ENSMIN
F - 54042 NANCY

PILGRIM, K.O.
Assessor d. Bergfachs
Landesoberbergamt NW
Goebenstr. 25
D - 4600 DORTMUND 1

POLMAN, H.A.
Adj. Directeur
Persl. Centr. NED.B.V.
P.B. 2069
NL - 2301 CB LEIDEN

POOLE, G.M.
Regional Sales Manager
I.C.I. Nobel's Explosives
Station Road, Dunscroft
GB - DONCASTER, S. Yorks DN7 4DZ

PRINZ, J.
Dipl.-Ing.
Leiter des Sprengtechn. Dienstes
Karl-Marx-Str. 32
D - 4600 DORTMUND 1

RAFFOUX, J.F.
Chef du département
"Techniques Minières"
CERCHAR
B.P. 2
F - 60550 VERNEUIL EN HALATTE

RAUBER, M.
Bergwerksdirektor
Saarberg
Hauptstr. 193
D - 6633 ENGDORF

REGELMANN, H.
Dipl. Ing.
Krupp Widia
Abt. HVB.
Münchener Strasse 90
D - 4300 ESSEN 1

REICHERT, K.
Directeur "Charbon"
Direction générale "Energie"
Commission des Communautés européennes
200, rue de la Loi
B - 1049 BRUXELLES

REIFF, W.
Hauptabteilungsleiter
BAG - Lippe
Hernerstr. 234
D - 4690 BOCHUM

REKITTKE, H.
Dipl.-Ing.
Tamrock (Deutschland) GmbH
Industriestr. 36
D - 4250 BOTTROP

RENARD, A.
Ingénieur Civil des Mines
8, Station Laan
B - 3700 TONGEREN

RENARD, D.
Marketing Relations
85, Ch. de Bruxelles
B - 4342 HOGNOUL-AWANS

RESSELER, R.
Engineer
N.V. Kempsense Steenkolenmijnen
P. Amideuslaan 6
B - 3550 HEUSDEN-ZOLDER

REULENS, J.S.
Directeur de Ventes
Intl. Mine Service Co.
416, ave Louise,
B - 1050 BRUXELLES

REYNARD, P.
Ingénieur
CERCHAR
BP 2
F - 60550 VERNEUIL EN HALATTE

RICHTER, P.
Verkaufsleiter
Boart HWF GmbH & Co. KG
Hartmetallwerkzeugfabrik
Städeweg 18
D - 6419 BURGHAUN 1

RIESS, H.G.
Bergwerksdirektor
Gewerkschaft
Sophia-Jacoba
Postfach 1320
D - 5142 HUECKELHOVEN

RIPANUCCI, G.
Ing. (Minerario)
INAIL
Via Nomentana 74
I - 00161 ROMA

RIX, P.
Akademischer Oberrat
Institut für Bergbaukunde II
Lochnerstrasse 4/20
D - 5100 AACHEN

ROBERTSON, J.T.
Director
Cementation Mining Limited
Bentley House P.O. Box 22
GB - DONCASTER, S. Yorks. DN5 0BT

ROESEN, W.
Verkaufs-Ing.
Tamrock (Deutschland) GmbH
Industriestr. 36
D - 4250 BOTTROP

ROLSHOVEN, H.J.
Werksleiter
Dyckerhoff Zementwerke AG
Werksgruppe Geseke
D - 4787 GESEKE

ROTONDO', P.P.
Direction générale "Marché de
l'information et innovation"
Commission des Communautés
européennes
B.P. 1907
L - 2920 LUXEMBOURG

RUETHER, W.
Ass. d. Bergfachs
Tamrock (Deutschland) GmbH
Industriestr. 36
D - 4250 BOTTROP

RZONZEF, L.
Ingénieur en Chef
Directeur des Mines
Administration des Mines
30, rue J.A. De Mot
B - 1040 BRUXELLES

SAHUC, L.
Ingénieur divisionnaire des TPE
Ministère de l'Industrie
et de la Recherche
97, rue de Grenelle
F - 75700 PARIS

SANDERSON, G.B.
Sales Manager
EIMCO (Great Britain) LTD
Earlsway, Team Valley
GB - GATESHEAD NE11 OSB

SANDER, R.
Ass. d. Bergfachs
Betriebsdirektor - RAG
Bergwerkstr.
D - 4130 MOERS 1

SCHARNWEBER, U.
Betriebsleiter
Barbara Rohstoffbetriebe GmbH
An der Erzgrube 9
D - 4952 PORTA-WESTFALICA

SCHEELE, K.
Ingenieur
RWTUV
Postfach 103261
D - 4300 ESSEN 1

SCHELTER, H.
Präsident
LOISA NW Bergbehörde
Apl. Schulstr. 11
D - 4600 DORTMUND

SCHICK, U.
Wiss. Angestellter
Institut für Bergbaukunde II
Lochnerstrasse 4/20
D - 5100 AACHEN

SCHICKHOFF, F.
Sales Manager
Eickhoff
31, Parsonswalk
GB - WIGAN WN1 1BX

SCHLIESING, G.
Bergwerksdirektor
Saarbergwerke AG
Postfach 10 30
D - 6600 SAARBRUECKEN

SCHMIDT, E.G.
Abt. Leiter Vortrieb
B.A.G. Westfalen
Beylingstr. 43
D - 4600 DORTMUND 14

SCHOENFELD, H.
Assessor des Bergfachs
Bergbau AG Lippe
Postfach 11 45
D - 4690 HERNE 1

SCHOKNECHT, H.
Dipl.-Ing
Deilmann-Haniel GmbH
Haustenbecke 1
D - 4600 DORTMUND 13

SCHORRADT, G.
Verkaufsleiter
Hermann Hemscheidt
Machinenfabrik Gmbh u. Co
Bornberg 97-103
D - 5600 WUPPERTAL 1

SCHRANCK, N.
Ingénieur
Lenoir et Mernier
14, rue Pasteur
L - 4276 ESCH/ALZETTE

SCHROEDER, H.
Dipl. Ingenieur RAG
Lülfstr. 41
D - 4350 RECKLINGHAUSER

SCHULPEN, C.
Techn. Commerc. Dir.
NV A. DELIGNE
Rijksweg 560
B - 3630 MAASMECHELEN

SCHULTE, S.
Dipl.-Berging.
Salzgitter Maschinen
und Anlagen AG
Hafenstrasse 215
D - 4300 ESSEN 11

SCHULZ, E.
Betriebsdirektor
Gewerkschaft Walter
Stauderstrasse 213
D - 4300 ESSEN 1

SCHUMANN, K.
Diplomingenieur
Thyssen Schachtbau GmbH
Ruhrstr. 1
D - 4330 MUELHEIM-RUHR

SCHUSTER, A.
Directeur
Inspection du Travail
et des Mines
BP. 27
L - 2010 LUXEMBOURG

SCHWEINBERGER, P.
Grubeninspektor
Eschweiler Bergwerksverein
D - AHLEN/WESTFALEN zu HAMM

SEPPI, M.
Verkaufsleiter
Boart HWF GmbH & Co. Kg
Hartmetallwerkzeugfabrik
Städeweg 18
D - 6419 BURGHAUN 1

SHAW, L.
Public Servant
Dept. Supply + Service - Canada
Place du Portage
CANADA - OTTAWA K1A 0S5

SHEPHERD, C.
Chief Mining Engineer
National Coal Board
"The Knoll" Tankersley
GB - NR BARNSLEY, S. Yorks

SIEBERT, H.
Kfm. Werksleiter
Dyckerhoff Zementwerke AG
Werk Neubeckum
D - 4720 BECKUM

SIMON, W.
Geschäftsführer
Eimco GmbH
Postfach 153
NL - 6300 AD VALKENBURG a/d GEUL

SINCLAIR, J.
Senior Inspector of Mines
Department of Labour
Mespil Road
Ireland - DUBLIN 4

SIROU, J.
Directeur général
Le Burin Commercial SA
27, rue Péclet
F - 75015 PARIS

SISSON, K.T.
Director
Hollybank Engineering Co. LTD
Ollerton Road, Tuxford
GB - NEWARK, Notts. NG22 OPQ

SPENCE, A.M.
Managing Director
Winster Group
Manners Avenue
GB - ILKESTON, Derbys DE7 8EF

SPIES, K.
Institutsdirektor
Institut für Bergbaukunde II
Lochnerstrasse 4/20
D - 5100 AACHEN

STASSEN, J.J.
Professeur à l'Université de Liège
Inspecteur général des Mines
28/091, bvd Piercot
B - 4000 LIEGE

STEPHENSON, D.J.
Chief Planning Eng.
National Coal Board
Beeches' Waterlane, Tickhill
GB - DONCASTER

STIMPFLING, G.
Technicien-démonstrateur
Concentall-Ateliers de Carspach
44, rue de la Sinne
F - 68100 MULHOUSE

SURKE, H.
Représentant
Firma Eickhoff
214, Chaussée
B - 7030 ST SYMPHORIEN

SWOBODA, C.H.
Dipl.-Ing.
Saarbergwerke AG
Postfach 1030
D - 6600 SAARBRUECKEN

TAIT, A.
Head of Shaft Sinking Branch
National Coal Board
The Lodge, South Parade
GB - DONCASTER, S. Yorkshire DN1 2DX

TAMO, P.
Ingénieur Civil des Mines
ACEC Charleroi
Avenue E. Rousseau
B - 6001 MARCINELLE

TARKO, H.
Representative
DHA MACIEJ CZARNECKI iS-Ka
ul. A. Mickiewicza 26/3
Poland - 40092 KATOWICE

THEUWISSEN, J.
Salesman
ATLAS Copco Belgium
Brusselse Stwg. 346
B - 1900 OVERYSE

THIELEN, G.
Ltd. Bergdirektor
Oberbergamt für das Saarland
und das Land Rheinland-Pfalz
Am Staden 17
D - 6600 SAARBRUECKEN

THOMAS, N.S.
Manager
Fairclough Civil
Eng. Limited
Cold Meece, Swynnerton
GB - NR STONE, Staffordshire ST15 0QX

TINCHON, L.
Ingénieur Divisionnaire
Houillères du Bassin de Lorraine
Charbonnages de France
2, rue de Metz
F - 57802 FREYMING-MERLEBACH

TREGELLES, P.G.
Director
National Coal Board
Mining Research and Development Esta
Ashby Rd., Stanhope Bretby
GB - BURTON-ON-TRENT, STaffs. DE15 0

TUCKER, R.
Head of Mining Development &
Tunnelling
National Coal Board
The Lodge, South Parade
GB - DONCASTER DN1 2DX

TUFFS, G.W.
Regional Director
Thyssen (GB) Ltd
Langthwaite Grange Industrial Estate
South Kirkby
GB - PONTEFRACT, West Yorkshire WF9

TUNNICLIFFE, J.F.
Professor, Head of Mining Dept.
University of Newcastle upon Tyne
Dept. of Mining Eng.
GB - NEWCASLTE UPON TYNE, NE1 7RU

TWEEDY, D.M.
Mining Engineer
Radmark Eng.
Phoenix House
GB - SUTTON-IN-ASHFIELD, Notts.

UNGER, E.
Oberingenieur
Westfalia Lünen
D - 4670 LUENEN

VAILLEMANS, W.
Persplucht Centr. NED.B.V.
P.B. 2069
NL - 2301 CB LEIDEN

VALGAERTS, J.
Ingénieur
N.V. Kempense Steenkolenmijnen
1 Meilaan
B - 3650 DILSEN

VAN DUYSE, H.
Ingénieur
INIEX
200, rue du Chera
B - 4000 LIEGE

VAN ESPEN, R.
Ingénieur
N.V. Kempense Steenkolenmijnen
Koolmijnlaan 190
B - 3560 BERINGEN

VAN GESTEL, K.
Tech. Comm. Director
S.A. Andre Deligne
Rijksweg 560
B - 3630 MAASMECHELEN

VANDERPUTTE, J.
Directeur général
Kempense Steenkolenmijnen
Grote Baan 27
B - 3530 HOUTHALEN-HELCHTEREN

VAN LEEUWEN, P.
Ing.- Professor
Tech. Hogeschool Delft
Mynbouwstr. 120
NL - 2628 RX - DELFT

VASSART, A.
Ingénieur Civil des Mines
Service Exploitation des Mines
Université Libre de Bruxelles
50, ave Roosevelt
B - 1050 BRUXELLES

VERBESSELT, J.
Secrétaire général
C.E.P.C.E.O.
168, ave de Tervueren, bte 11
B - 1150 BRUXELLES

VERCOLLIER, J.
Ingénieur/Pdt du Direct.
M.F.I.
37, av. Paul Doumer
F - 75116 PARIS

VERHAEGEN, H.J.
Ingénieur des Mines
Intl Mine Service Co.
416, ave Louise
B - 1050 BRUXELLES

VESCHKENS, M.
Ingénieur Civil, Chargé de Recherches
INIEX
200, rue du Chéra
B - 4000 LIEGE

VOERCKEL, M.
Sales Engineer
WIRTH GmbH
Kölner Str. 71-78
D - 5140 ERKELENZ

VON PELSER-BERENSBERG, B.
Ass.d.B. Prokurist
Klöckner-Becorit
Werkstättenstr. 7 - 13
D - 4350 RECKLINGHAUSEN,
Nordrhein Westf.

WAGENKNECHT, A.
Bergdirektor, Dipl.-Ing
Oberbergamt für das Saarland
und das Land Rheinland-Pfalz
Am Staden 17
D - 6600 SAARBRUECKEN

WALKER, L.F.G.
Director
Anderson Strathclyde P.L.C.
Broad Street 47
GB - GLASGOW

WALKER, P.
Directorate General "Employment,
Social Affairs and Education"
Commission of the
European Communities
P.O.B. 1907
L - 2920 LUXEMBOURG

WEBSTER, S.
Webster Machine Co.
Wortley Road
GB - ROTHERHAM S61 1LZ, S. orkshire

WEGRZYK, Z.
Senior Economic Officer
UN Economic Commission for Europe
UNECE Secretariat
Palace of Nations
CH - 1211 GENEVE

WELLER, L.
Marketing Leiter
KENNAMETAL GmbH
Max-Planck-Str 13
D - 6382 FRIEDRICHSDORF

WELZEL, J.
Geschaeftsführer
Hermann Hemscheidt
Machinenfabrik Gmbh u. Co
Bornberg 97-103
D - 5600 WUPPERTAL 1

WESSOLOWSKI, U.
Dipl. Bergingenieur
Deilmann-Haniel GmbH
Haustenbecke 1
D - 4600 DORTMUND 1

WETZCHEWALD, K.H.
Produktmanager und Prokurist
Boart HWF GmBH & Co.KG
Hartmetallwerkzeugfabrik
Städeweg 18
D - 6419 BURGHAUN 1

WHEELER, A.
Director of the Scottish Area
National Coal Board
Green Park, Green End
GB - EDINBURGH EH17 7PZ

WHITE, J.H.
Production Manager
National Coal Board
North Derrys Area, Bolsover
GB - CHESTERFIELD

WIEDEMANN, G.
Dipl.-Ing
Atlas Copco Deutschlang Gmbh
Ernestinenstr. 155
D - 4300 ESSEN 1

WILLIAMS, I.
Colliery General Manager
National Coal Board
Riccall Mine, Selby Road
GB - RICCALL, York YO4 6QR

WIRTH, H.
Dipl.-Ing.
WBK - BVS
Beylingstr. 65
D - 4600 DORTMUND 14

WOLLHAUPT, D.
Ingenieur
SAKRET Productions
Münsterland, Postfach 4480
D - 4400 MUENSTER, Westfalen

WONNERTH, G.
Direktor E.A.
Kommission der Europäischen
Gemeinschaften
Borndorferstr. 8
D - 8121 PAEHL

WOOD, J.E.
Area Director
National Coal Boad
North Notts. Area, Edwinstowe
GB - MANSFIELD, Notts. NG21 9 PR

WORT, K.G.
Sales Director
Markham & Co. LTD.
Broad Oaks Works, S41 ODS
GB - CHESTERFIELD, Derbyshire

WUELFING, F.
Bergass. a. D.
Bochumer Eisenhütte
Heintzmann GmbH & Co. KG
Bessemerstr. 80
D - 4630 BOCHUM 1

ZADROZNY, M.
Technical Director
ZJEDNOCZENI BUDOWNICTWA
GORNICZEGO
ul. Damrota 16
Poland - 40021 KATOWICE

ZIMMERMANN, H.
Prokurist
Montanbüro GmBH
Sevinghauser Weg 100
D - 4630 BOCHUM 6

ZITZ, A.
Konstrukteur
Voest/Alpine
A - 8740 ZELTWEG

ZOLLMAN, P.M.
ZED Instruments Ltd.
Molesey Ave
GB - WEST MOLESEY, Surrey KY8 ORY

Self-Management of Asthma

LUNG BIOLOGY IN HEALTH AND DISEASE

Executive Editor

Claude Lenfant
Director, National Heart, Lung and Blood Institute
National Institutes of Health
Bethesda, Maryland

1. Immunologic and Infectious Reactions in the Lung, *edited by Charles H. Kirkpatrick and Herbert Y. Reynolds*
2. The Biochemical Basis of Pulmonary Function, *edited by Ronald G. Crystal*
3. Bioengineering Aspects of the Lung, *edited by John B. West*
4. Metabolic Functions of the Lung, *edited by Y. S. Bakhle and John R. Vane*
5. Respiratory Defense Mechanisms (in two parts), *edited by Joseph D. Brain, Donald F. Proctor, and Lynne M. Reid*
6. Development of the Lung, *edited by W. Alan Hodson*
7. Lung Water and Solute Exchange, *edited by Norman C. Staub*
8. Extrapulmonary Manifestations of Respiratory Disease, *edited by Eugene Debs Robin*
9. Chronic Obstructive Pulmonary Disease, *edited by Thomas L. Petty*
10. Pathogenesis and Therapy of Lung Cancer, *edited by Curtis C. Harris*
11. Genetic Determinants of Pulmonary Disease, *edited by Stephen D. Litwin*
12. The Lung in the Transition Between Health and Disease, *edited by Peter T. Macklem and Solbert Permutt*
13. Evolution of Respiratory Processes: A Comparative Approach, *edited by Stephen C. Wood and Claude Lenfant*
14. Pulmonary Vascular Diseases, *edited by Kenneth M. Moser*
15. Physiology and Pharmacology of the Airways, *edited by Jay A. Nadel*
16. Diagnostic Techniques in Pulmonary Disease (in two parts), *edited by Marvin A. Sackner*
17. Regulation of Breathing (in two parts), *edited by Thomas F. Hornbein*
18. Occupational Lung Diseases: Research Approaches and Methods, *edited by Hans Weill and Margaret Turner-Warwick*
19. Immunopharmacology of the Lung, *edited by Harold H. Newball*
20. Sarcoidosis and Other Granulomatous Diseases of the Lung, *edited by Barry L. Fanburg*

21. Sleep and Breathing, *edited by Nicholas A. Saunders and Colin E. Sullivan*
22. *Pneumocystis carinii* Pneumonia: Pathogenesis, Diagnosis, and Treatment, *edited by Lowell S. Young*
23. Pulmonary Nuclear Medicine: Techniques in Diagnosis of Lung Disease, *edited by Harold L. Atkins*
24. Acute Respiratory Failure, *edited by Warren M. Zapol and Konrad J. Falke*
25. Gas Mixing and Distribution in the Lung, *edited by Ludwig A. Engel and Manuel Paiva*
26. High-Frequency Ventilation in Intensive Care and During Surgery, *edited by Graziano Carlon and William S. Howland*
27. Pulmonary Development: Transition from Intrauterine to Extrauterine Life, *edited by George H. Nelson*
28. Chronic Obstructive Pulmonary Disease: Second Edition, Revised and Expanded, *edited by Thomas L. Petty*
29. The Thorax (in two parts), *edited by Charis Roussos and Peter T. Macklem*
30. The Pleura in Health and Disease, *edited by Jacques Chrétien, Jean Bignon, and Albert Hirsch*
31. Drug Therapy for Asthma: Research and Clinical Practice, *edited by John W. Jenne and Shirley Murphy*
32. Pulmonary Endothelium in Health and Disease, *edited by Una S. Ryan*
33. The Airways: Neural Control in Health and Disease, *edited by Michael A. Kaliner and Peter J. Barnes*
34. Pathophysiology and Treatment of Inhalation Injuries, *edited by Jacob Loke*
35. Respiratory Function of the Upper Airway, *edited by Oommen P. Mathew and Giuseppe Sant'Ambrogio*
36. Chronic Obstructive Pulmonary Disease: A Behavioral Perspective, *edited by A. John McSweeny and Igor Grant*
37. Biology of Lung Cancer: Diagnosis and Treatment, *edited by Steven T. Rosen, James L. Mulshine, Frank Cuttitta, and Paul G. Abrams*
38. Pulmonary Vascular Physiology and Pathophysiology, *edited by E. Kenneth Weir and John T. Reeves*
39. Comparative Pulmonary Physiology: Current Concepts, *edited by Stephen C. Wood*
40. Respiratory Physiology: An Analytical Approach, *edited by H. K. Chang and Manuel Paiva*
41. Lung Cell Biology, *edited by Donald Massaro*
42. Heart–Lung Interactions in Health and Disease, *edited by Steven M. Scharf and Sharon S. Cassidy*
43. Clinical Epidemiology of Chronic Obstructive Pulmonary Disease, *edited by Michael J. Hensley and Nicholas A. Saunders*
44. Surgical Pathology of Lung Neoplasms, *edited by Alberto M. Marchevsky*
45. The Lung in Rheumatic Diseases, *edited by Grant W. Cannon and Guy A. Zimmerman*

46. Diagnostic Imaging of the Lung, *edited by Charles E. Putman*
47. Models of Lung Disease: Microscopy and Structural Methods, *edited by Joan Gil*
48. Electron Microscopy of the Lung, *edited by Dean E. Schraufnagel*
49. Asthma: Its Pathology and Treatment, *edited by Michael A. Kaliner, Peter J. Barnes, and Carl G. A. Persson*
50. Acute Respiratory Failure: Second Edition, *edited by Warren M. Zapol and Francois Lemaire*
51. Lung Disease in the Tropics, *edited by Om P. Sharma*
52. Exercise: Pulmonary Physiology and Pathophysiology, *edited by Brian J. Whipp and Karlman Wasserman*
53. Developmental Neurobiology of Breathing, *edited by Gabriel G. Haddad and Jay P. Farber*
54. Mediators of Pulmonary Inflammation, *edited by Michael A. Bray and Wayne H. Anderson*
55. The Airway Epithelium, *edited by Stephen G. Farmer and Douglas Hay*
56. Physiological Adaptations in Vertebrates: Respiration, Circulation, and Metabolism, *edited by Stephen C. Wood, Roy E. Weber, Alan R. Hargens, and Ronald W. Millard*
57. The Bronchial Circulation, *edited by John Butler*
58. Lung Cancer Differentiation: Implications for Diagnosis and Treatment, *edited by Samuel D. Bernal and Paul J. Hesketh*
59. Pulmonary Complications of Systemic Disease, *edited by John F. Murray*
60. Lung Vascular Injury: Molecular and Cellular Response, *edited by Arnold Johnson and Thomas J. Ferro*
61. Cytokines of the Lung, *edited by Jason Kelley*
62. The Mast Cell in Health and Disease, *edited by Michael A. Kaliner and Dean D. Metcalfe*
63. Pulmonary Disease in the Elderly Patient, *edited by Donald A. Mahler*
64. Cystic Fibrosis, *edited by Pamela B. Davis*
65. Signal Transduction in Lung Cells, *edited by Jerome S. Brody, David M. Center, and Vsevolod A. Tkachuk*
66. Tuberculosis: A Comprehensive International Approach, *edited by Lee B. Reichman and Earl S. Hershfield*
67. Pharmacology of the Respiratory Tract: Experimental and Clinical Research, *edited by K. Fan Chung and Peter J. Barnes*
68. Prevention of Respiratory Diseases, *edited by Albert Hirsch, Marcel Goldberg, Jean-Pierre Martin, and Roland Masse*
69. *Pneumocystis carinii* Pneumonia: Second Edition, Revised and Expanded, *edited by Peter D. Walzer*
70. Fluid and Solute Transport in the Airspaces of the Lungs, *edited by Richard M. Effros and H. K. Chang*
71. Sleep and Breathing: Second Edition, Revised and Expanded, *edited by Nicholas A. Saunders and Colin E. Sullivan*
72. Airway Secretion: Physiological Bases for the Control of Mucous Hypersecretion, *edited by Tamotsu Takishima and Sanae Shimura*

73. Sarcoidosis and Other Granulomatous Disorders, *edited by D. Geraint James*
74. Epidemiology of Lung Cancer, *edited by Jonathan M. Samet*
75. Pulmonary Embolism, *edited by Mario Morpurgo*
76. Sports and Exercise Medicine, *edited by Stephen C. Wood and Robert C. Roach*
77. Endotoxin and the Lungs, *edited by Kenneth L. Brigham*
78. The Mesothelial Cell and Mesothelioma, *edited by Marie-Claude Jaurand and Jean Bignon*
79. Regulation of Breathing: Second Edition, Revised and Expanded, *edited by Jerome A. Dempsey and Allan I. Pack*
80. Pulmonary Fibrosis, *edited by Sem Hin Phan and Roger S. Thrall*
81. Long-Term Oxygen Therapy: Scientific Basis and Clinical Application, *edited by Walter J. O'Donohue, Jr.*
82. Ventral Brainstem Mechanisms and Control of Respiration and Blood Pressure, *edited by C. Ovid Trouth, Richard M. Millis, Heidrun F. Kiwull-Schöne, and Marianne E. Schläfke*
83. A History of Breathing Physiology, *edited by Donald F. Proctor*
84. Surfactant Therapy for Lung Disease, *edited by Bengt Robertson and H. William Taeusch*
85. The Thorax: Second Edition, Revised and Expanded (in three parts), *edited by Charis Roussos*
86. Severe Asthma: Pathogenesis and Clinical Management, *edited by Stanley J. Szefler and Donald Y. M. Leung*
87. *Mycobacterium avium*–Complex Infection: Progress in Research and Treatment, *edited by Joyce A. Korvick and Constance A. Benson*
88. Alpha 1–Antitrypsin Deficiency: Biology • Pathogenesis • Clinical Manifestations • Therapy, *edited by Ronald G. Crystal*
89. Adhesion Molecules and the Lung, *edited by Peter A. Ward and Joseph C. Fantone*
90. Respiratory Sensation, *edited by Lewis Adams and Abraham Guz*
91. Pulmonary Rehabilitation, *edited by Alfred P. Fishman*
92. Acute Respiratory Failure in Chronic Obstructive Pulmonary Disease, *edited by Jean-Philippe Derenne, William A. Whitelaw, and Thomas Similowski*
93. Environmental Impact on the Airways: From Injury to Repair, *edited by Jacques Chrétien and Daniel Dusser*
94. Inhalation Aerosols: Physical and Biological Basis for Therapy, *edited by Anthony J. Hickey*
95. Tissue Oxygen Deprivation: From Molecular to Integrated Function, *edited by Gabriel G. Haddad and George Lister*
96. The Genetics of Asthma, *edited by Stephen B. Liggett and Deborah A. Meyers*
97. Inhaled Glucocorticoids in Asthma: Mechanisms and Clinical Actions, *edited by Robert P. Schleimer, William W. Busse, and Paul M. O'Byrne*
98. Nitric Oxide and the Lung, *edited by Warren M. Zapol and Kenneth D. Bloch*

99. Primary Pulmonary Hypertension, *edited by Lewis J. Rubin and Stuart Rich*
100. Lung Growth and Development, *edited by John A. McDonald*
101. Parasitic Lung Diseases, *edited by Adel A. F. Mahmoud*
102. Lung Macrophages and Dendritic Cells in Health and Disease, *edited by Mary F. Lipscomb and Stephen W. Russell*
103. Pulmonary and Cardiac Imaging, *edited by Caroline Chiles and Charles E. Putman*
104. Gene Therapy for Diseases of the Lung, *edited by Kenneth L. Brigham*
105. Oxygen, Gene Expression, and Cellular Function, *edited by Linda Biadasz Clerch and Donald J. Massaro*
106. $Beta_2$-Agonists in Asthma Treatment, *edited by Romain Pauwels and Paul M. O'Byrne*
107. Inhalation Delivery of Therapeutic Peptides and Proteins, *edited by Akwete Lex Adjei and Pramod K. Gupta*
108. Asthma in the Elderly, *edited by Robert A. Barbee and John W. Bloom*
109. Treatment of the Hospitalized Cystic Fibrosis Patient, *edited by David M. Orenstein and Robert C. Stern*
110. Asthma and Immunological Diseases in Pregnancy and Early Infancy, *edited by Michael Schatz, Robert S. Zeiger, and Henry N. Claman*
111. Dyspnea, *edited by Donald A. Mahler*
112. Proinflammatory and Antiinflammatory Peptides, *edited by Sami I. Said*
113. Self-Management of Asthma, *edited by Harry Kotses and Andrew Harver*
114. Eicosanoids, Aspirin, and Asthma, *edited by Andrew Szczeklik, Ryszard J. Gryglewski, and John R. Vane*
115. Fatal Asthma, *edited by Albert F. Sheffer*

ADDITIONAL VOLUMES IN PREPARATION

Inflammatory Mechanisms in Asthma, *edited by Stephen T. Holgate and William W. Busse*

Biology of Lung Cancer, *edited by Madeleine A. Kane and Paul A. Bunn, Jr.*

Pulmonary Edema, *edited by Michael A. Matthay and David H. Ingbar*

The Lung at Depth, *edited by Claes Lundgren and John N. Miller*

Lung Tumors, *edited by Christian Brambilla and Elizabeth Brambilla*

Five-Lypoxygenase Products in Asthma, *edited by Jeffrey M. Drazen, Sven-Erik Dahlén, and Tak H. Lee*

Diagnostic Pulmonary Pathology, *edited by Philip Cagle*

Immunotherapy of Asthma, *edited by Jean Bousquet and Hans Yssel*

Rhinitis, *edited by Robert M. Maclerio, Stephen R. Durham, and Niels Mygind*

Interleukin-5: From Molecule to Drug Target for Asthma, *edited by Colin Sanderson*

Viral Infections of the Lung, *edited by Raphael Dolin and Peter F. Wright*

Pediatric Asthma, *edited by Shirley J. Murphy, H. William Kelly, and Bernie C. McWilliams*

LAM and Other Diseases Characterized by Smooth Muscle Proliferation, *edited by Joel Moss*

Control of Breathing in Health and Disease, *edited by Murray D. Altose and Yoshikazu Kawakani*

Physiological Basis of Ventilatory Support, *edited by John J. Marini and Arthur S. Slutsky*

Complexity in Structure and Function of the Lung, *edited by Michael P. Hlastala and H. Thomas Robertson*

Human Immunodeficiency Virus and the Lung, *edited by Mark J. Rosen and James M. Beck*

The opinions expressed in these volumes do not necessarily represent the views of the National Institutes of Health.

Self-Management of Asthma

Edited by

Harry Kotses
Ohio University
Athens, Ohio

Andrew Harver
University of North Carolina at Charlotte
Charlotte, North Carolina

MARCEL DEKKER, INC. NEW YORK • BASEL • HONG KONG

Library of Congress Cataloging-in-Publication Data

Self-management of asthma / edited by Harry Kotses, Andrew Harver.
p. cm. -- (Lung biology in health and disease ; vol. 113)
Includes bibliographical references and index.
ISBN 0-8247-9858-9 (alk. paper)
1. Asthma--Popular works. 2. Self-care, Health. I. Kotses, Harry. II. Harver, Andrew.
[DNLM: 1. Asthma. 2. Self Care. W1 LU62 v. 113 1998 / WF 553 S465 1998]
RC591.S44 1997
616.2'38--dc21
DNLM/DLC
for Library of Congress 97-49681
CIP

This book is printed on acid-free paper.

Headquarters
Marcel Dekker, Inc.
270 Madison Avenue, New York, NY 10016
tel: 212-696-9000; fax: 212-685-4540

Eastern Hemisphere Distribution
Marcel Dekker AG
Hutgasse 4, Postfach 812, CH-4001 Basel, Switzerland
tel: 44-61-261-8482; fax: 44-61-261-8896

World Wide Web
http://www.dekker.com

The publisher offers discounts on this book when ordered in bulk quantities. For more information, write to Special Sales/Professional Marketing at the headquarters address above.

Current printing (last digit):
10 9 8 7 6 5 4 3 2 1

PRINTED IN THE UNITED STATES OF AMERICA

INTRODUCTION

> The object you propose, the recovery of your health, is one of vast moment to you and is deserving of your highest efforts. Let me urge you, in the first place, to have confidence in your remedies, and in the second, to commence their use with the highest caution.
>
> *Samuel Sheldon Fitch (1853)* (1)

The *Lung Biology in Health and Disease* series has included several volumes on asthma and its treatment. More are in preparation and will soon be published. Why so many volumes? Simple! Asthma prevalence is increasing worldwide and, although the disease can be quite severe, it need not be fatal if appropriately treated. Today's patients have a great advantage over their predecessors: our knowledge of the pathogenesis of this condition and our options for treating it are remarkable, largely because of the interest and productivity of the scientific community.

Although asthma and psychological factors have been associated for almost as long as asthma has been recognized, it was not until a decade or so ago that a derivative of psychological factors—self-management—was introduced as an important component in the armamentarium of asthma care.

Surely, Samuel Fitch was thinking about such approaches when he admonished his patients to give their highest efforts to the recovery of their health and to have confidence in their remedies.

Until the publication of this volume, *Self-Management of Asthma*, this series of monographs had not considered this approach to asthma management. Today, we know that it works and that the patient will be the first beneficiary of his or her highest efforts and self-confidence in applying appropriate remedies. Furthermore, society and the health care system will benefit as well. This volume provides general practitioners, asthma specialists, and patients with the theory and practice of asthma self-management.

Drs. Kotses and Harver are experts in this area. They have assembled a roster of contributors who have studied and published on various aspects of self-management; their expertise is widely recognized. This volume fills a significant gap in the series, and it will be a guide to all health professionals treating asthma patients.

I am grateful to the editors and authors for this major contribution to this series.

Claude Lenfant, M.D.
Bethesda, Maryland

Reference

1. Fitch, S. S. Six Discourses on the Functions of the Lungs. S. S. Fitch and Co., New York, 1858.

PREFACE

In Western medicine, the link between asthma and psychology has been debated for over two thousand years. Initially, the debate was focused on the psychological consequences of an asthma attack. Physicians in antiquity knew that the disease led to a decline in the quality of subjective experience, a conclusion that enjoys support today. But the mirror image of this conclusion, that asthma is affected by psychology, was reached only in recent years after extensive discussion. These two lines of inquiry demonstrated beyond a doubt that asthma not only affects—but is affected by—psychology.

The relationship between psychology and asthma has been promoted by important developments in their respective fields. In psychology, a shift in the emphasis from subjective to behavioral events set the stage for objective measurement of psychological quantities. This ushered in the scientifc era in psychology and made possible the application of psychological principles to other disciplines. In medicine, the increasing acceptance of behavioral considerations in the development and control of disease permitted inclusion of psychological factors in treatment. Asthma self-management is the beneficiary of these trends. Its aim is to bring under control those behaviors that reduce the likelihood of frequent, or severe, asthmatic episodes.

This volume is a compilation of behavioral elements that play a role in the control of asthma. It deals with a number of the traditional asthma self-management topics that have immediate implications for management of asthma. These include avoidance of asthma triggers, recognition of asthma symptoms, and behavioral strategies for the reduction of asthma. Among other things, the latter covers procedures for improving adherence to medication recommendations. In addition, the volume contains discussions of scientific interest to asthma self-management: cognitive/behavioral theory, measurement considerations, and the role of both personality and emotion in asthma self-management. Finally, there is a presentation of topics that are at the forefront of research on asthma self-management: psychophysics, the scaling of respiratory sensations, and efforts to train patients in perception of respiratory events. The latter topics are directed toward early detection of asthma exacerbation, a potentially valuable addition to the armamentarium of asthma self-management. The volume, including as it does a wide range of contributions from both physicians and behavioral scientists, will be of interest to clinician and researcher alike.

But the benefit of this volume will accrue neither to the clinician nor to the researcher so much as it will to the asthma patient. Anyone who has followed research in asthma self-management over the past 25 years knows that although asthma self-management reduces symptoms, it is not widely applied. It is unfortunate that relatively few patients receive self-management training, a tragedy that will be compounded if the number of asthma patients continues to increase as it has in recent years. It is our opinion that widespread application of self-management training can provide enormous benefits in managing an ever-increasing number of patients. Therefore, our hope for this volume is that it will encourage physicians who treat asthma to consider use of self-management training. If this happens, even on a modest scale, the preparation of the volume will have been worth the effort.

The editors gratefully acknowledge the contributions of three institutions that provided facilities for our use during the preparation of this volume. H. K. was supported by both Ohio University and the University of South Florida and A. H. was supported by the University of North Carolina–Charlotte. Without their generous support, work on this volume could not have been undertaken.

Harry Kotses
Andrew Harver

CONTRIBUTORS

Lauren Allen, M.S. Department of Psychology, Ohio University, Athens, Ohio

William C. Bailey, M.D. Professor of Medicine and Director, Lung Health Center, School of Medicine, University of Alabama at Birmingham, and the Veterans Affairs Medical Center, Birmingham, Alabama

John C. Baird, Ph.D. Professor of Psychology, Dartmouth College, Hanover, New Hampshire

Bruce G. Bender, Ph.D. Psychologist and Senior Faculty Member, Department of Pediatrics, National Jewish Medical and Research Center, Denver, Colorado

Noreen M. Clark, Ph.D. Marshall H. Becker Professor of Public Health, School of Public Health, University of Michigan, Ann Arbor, Michigan

Thomas L. Creer, Ph.D. Professor, Department of Psychology, Ohio University, Athens, Ohio

Susan L. Davies, M.Ed. Clinical Coordinator, Department of Nutrition Sciences, University of Alabama at Birmingham, Birmingham, Alabama

Gregory K. Fritz, M.D. Professor and Director of Child and Adolescent Psychiatry, Department of Psychiatry and Human Behavior, Brown University School of Medicine and Rhode Island Hospital, Providence, Rhode Island

Andrew Harver, Ph.D. Associate Professor, Department of Psychology, University of North Carolina at Charlotte, Charlotte, North Carolina

Elizabeth F. Juniper, M.C.S.P., M.Sc. Associate Professor, Department of Clinical Epidemiology and Biostatistics, McMaster University Medical Centre, Hamilton, Ontario, Canada

Edward S. Katkin, Ph.D. Professor, Department of Psychology, State University of New York at Stony Brook, Stony Brook, New York

Ad A. Kaptein, Ph.D. Psychologist, Department of Medical Psychology, Leiden University, Leiden, The Netherlands

Mary D. Klinnert, Ph.D. Psychologist and Associate Faculty Member, Department of Pediatrics, National Jewish Medical and Research Center, Denver, Colorado

Connie L. Kohler, Dr.PH. Assistant Professor, Department of Health Behavior, School of Public Health, University of Alabama at Birmingham, Birmingham, Alabama

Harry Kotses, Ph.D. Professor, Department of Psychology, Ohio University, Athens, Ohio

Deirdre A. Levstek, M.S. Clinical Psychology Doctoral Candidate, Department of Psychology, Ohio University, Athens, Ohio

Donald A. Mahler, M.D. Professor, Department of Medicine, Dartmouth Medical School, and Director, Pulmonary Function and Cardiopulmonary Exercise Laboratories, Dartmouth Hitchcock Medical Center, Lebanon, New Hampshire

Russ V. C. Reynolds, Ph.D. Optimum Behavioral Health, Inc., Memphis, Tennessee

Michael S. Taitel, Ph.D. Adjunct Assistant Professor, Department of Psychology, Ohio University, Athens, Ohio

Marianne Z. Wamboldt, M.D. Director of Pediatric Behavioral Health, Department of Pediatrics, National Jewish Medical and Research Center, Denver, Colorado

Andrew G. Weinstein, M.D. Clinical Associate Professor of Pediatrics, Department of Allergy/Immunology, Jefferson Medical College, duPont Hospital for Children, Wilmington, Delaware

John A. Winder, M.D. Allergy and Asthma Research Center, Toledo, Ohio

CONTENTS

Part One

THE PSYCHOLOGY OF ASTHMA: CONCEPTUAL DEVELOPMENTS

1

The Impact of Asthma on the Patient, Family, and Society

MICHAEL S. TAITEL, LAUREN ALLEN, and THOMAS L. CREER

Ohio University
Athens, Ohio

I. Definition of Asthma

Asthma has a long history, as evidenced by its inclusion in some of the earliest medical writings. In the West, the writings of Hippocrates (460–370 B.C.) contained references to asthma. During the second century A.D., Aretaeus the Cappadocian provided a detailed description of the symptoms, physical changes, and psychological states associated with asthma attacks. Ancient Chinese writings, dated as early as 2697 B.C., offered a diagnosis and prescription for asthma (1): "When the lungs suffer from the obstruction of the upper respiratory tract, one should quickly eat bitter (food) which will disperse the obstruction and restore the flow."

Asthma is characterized by episodes of widespread narrowing of the airways caused by the following components: constriction of the bronchial smooth muscles, swelling of the bronchial tissue, increased mucus secretion, infiltration of inflammatory cells, or a combination of these factors (2). These physiological responses restrict the flow of air into and out of the lungs. During an asthma attack the patient may experience a number of symptoms, such as scratchy throat, chest tightness, shortness of breath, wheezing, difficulty in breathing, or congested cough. These primary symptoms may also lead to fear, panic, depression, and fatigue. During a severe attack, asthma can cause total airway obstruction and death. The symptoms of asthma occur within the context of three reliable characteristics of the disease: intermittence, variability, and reversibility.

Characteristics of Asthma

Intermittency

The number of asthma attacks experienced varies from patient to patient and, for a given patient, from time to time. A patient may experience a series of attacks in a span of a few days and then go several months or even years between episodes. Another patient may have perennial asthma and experience episodes most days throughout the year. The frequency of attacks experienced by patients is a function of a number of variables. These include: (1) physiological variables, including the number and variety of stimuli that trigger episodes, and the degree of hyperreactivity of the airways, (2) health care variables, such as access to asthma specialists and the degree of medical control established over their disorder, and (3) patient variables, such as medication compliance and expectations acquired about asthma (3).

Variability

The term *variability* refers both to the severity of a patient's asthma and to the intensity of discrete attacks. Use of the term has long been confusing in that it has not always been clear whether severity was being used to define the general condition of patients or to the specific attacks they experience. This was certainly the case two decades ago (4). Both the National Asthma Education Project (5) and the International Consensus Report (6) resolved some of the ambiguity concerning the variable nature of asthma by offering both a general categorization of asthma severity and guidelines for classifying mild, moderate, or severe attacks.

The guidelines used by practitioners are often viewed as value-free. However, patient expectations associated with the perceived severity of their asthma are usually subjective. This may present a problem. Renne and Creer (7) pointed out that patients acquire different expectations as a function of the severity of their attacks. If their episodes are mild, patients tend to expect that all future attacks will be mild and are unprepared to cope with more severe attacks if and when they occur. A single severe attack can not only lead to enduring psychological and behavioral reactions that can exacerbate a given attack, but the episode can alter, sometimes unrealistically, the patient's expectations regarding future attacks (3). On the basis of a systematic analysis of the cause of the problem, Creer (8) found that most children who panic during attacks acquired the behavioral pattern during a single incident when they or their parents became frightened. On the other hand, expectations that they will experience only mild attacks of asthma can lead patients to delay seeking medical attention for severe episodes when they occur (3,7). Failure to promptly seek aggressive medical care for attacks that rapidly intensify in severity is an oft-cited factor contributing to deaths from asthma (9).

Reversibility

In 1980, McFadden (10) referred to reversibility as the sine qua non of asthma; it distinguished the condition from other respiratory disorders, particularly emphysema, which is irreversible. The National Institute of Allergy and Infectious Diseases (NIAID) Task Force Report (4) was more cautious on this point in noting that airway obstruction was often found when patients were asymptomatic. These comments suggest that, as with other characteristics of asthma, the issue of reversibility is somewhat ambiguous. First, reversibility may be relative. While most patients show complete reversibility of their asthma with appropriate treatment, there is considerable evidence that many patients exhibit airway obstruction when they are supposedly asymptomatic (11). Second, even intensive therapy fails to reverse the asthma in some patients (12). This point was emphasized in both the reports from the National Asthma Education Project (5) and the International Consensus Report (6). Finally, the spontaneous remission of some attacks makes it impossible to prove a cause-effect relationship between changes in a patient's asthma and treatment for the disorder (13). This point has ramifications not only for outcome data regarding a particular treatment, but for how such changes are perceived and interpreted by patients.

II. The Impact of Asthma on Society

Asthma is a chronic disorder that afflicts millions of Americans. For many, asthma is an easily treated inconvenience; for others, the disease can be debilitating and even fatal. Over the past decade asthma morbidity and mortality have increased. Factors that may have contributed to increases in asthma morbidity and mortality include a general increase in the prevalence and severity of the disease and certain combinations of asthma drugs. Behavioral factors such as a failure by patients to perceive or appreciate the severity of their symptoms contribute to the overall problem (14).

A. Prevalence

There is evidence that because of either the increased incidence of asthma or improved diagnoses and reporting, the disorder is becoming more common.

The growing prevalence of asthma has demanded that the combined armature of medical and behavioral sciences be synthesized to produce effective self-management approaches to control the disorder. A National Institute of Allergy and Infectious Diseases (NIAID) Task Force Report, published in 1979, reported that 35 million Americans had asthma and other allergic diseases; of this total, 8.9 million or 4% of Americans were estimated to have asthma (4). These data gave impetus to the Division of Lung Diseases of the National Heart, Lung, and Blood Institute to issue a request for applications (RFA) to develop self-management programs for childhood asthma. While there had been earlier investigations of asthma self-management, it was both the response to the RFA and the research it generated that stimulated the systematic study of asthma self-management with both pediatric and adult asthma (15).

Prevalence data regarding asthma is far more striking today than it was two decades ago. Approximately 80 million Americans—one third of the U.S. population—suffer from a respiratory disease, a category that includes asthma. This translates to a conservative estimate of 10–15 million Americans with asthma, although some experts place the figure at closer to 25 million Americans (16). On the basis of National Health Interview Survey data from 1980 through 1990 (17), the age-adjusted prevalence rate of self-reported asthma increased by 38%; this included a 50% increase for females and a 27% increase for males. From 1981 through 1988, the annual prevalence rate for black females increased by 45%; the annual prevalence rate for white females increased by 63% (17). An update from the Centers for Disease Control (CDC) recently reported that in people aged 5–34 years (considered the most stable population to study), the rate of asthma increased 52% from 34.6 to 52.6 per 1000 from 1982 to 1992. The rate increased by 29% for males and 82% for females (18).

On the basis of the National Health Interview Survey, Taylor and Newacheck (19) reported a statistically significant increase from 3.2% in 1981 to 4.3% in 1988 in the prevalence of asthma in children under 18 years of age. It is known that racial or ethnic minorities who are poor and live in urban environments are at high risk for asthma (20). A number of factors are believed to be associated with increased asthma prevalence and morbidity of asthma in children from minority groups. Identified risk factors include poverty, environmental, psychological, family, and physical variables (21,22). Poverty has an impact on asthma by limiting access to quality health care, particularly preventive health care. Environmental risk factors that appear to increase asthma risk factors in minority groups include exposure to cigarette smoke, air pollution, dust mites, and cockroach parts; all these stimuli can trigger asthma. Psychological risk factors for asthma include those associated with poverty, crime, single-parent families, and other asthma precipitants pervasive in these settings. Gang-related activities, high rates of drug and alcohol abuse, and a lack of community support make adherence to medication regimens, regular medical care, or other asthma-prevention efforts a low priority in the day-to-day existence of patients who live in an inner city. Familial factors that increase asthma risk include increased family size, as well as limited parenting skills. The latter is particularly a problem when the age of the mother was less than 20 years at the time of the child's birth (21). Physical factors associated with increased asthma risk include lower birth weight of children and heredity risk factors. Socioeconomic level, correlated with the variables described above, is a primary factor that explains differences in asthma morbidity and prevalence rates between minority and nonminority populations. The collective impact of the poverty-related environment appears to account for the growing risk of asthma in minority populations (21,22). Weiss and colleagues (20) note that we know little about those who live in rural areas, mainly because these populations are rarely studied. However, a report prepared for Congress (24) found that chronic illness and disability affect a greater proportion of the rural than the urban population (14% vs. 12%, respectively). These findings highlight the fact that we have only epidemiological data about pockets of patients with asthma and, therefore, likely underestimate both the prevalence and the impact of the disorder.

B. Morbidity

Morbidity data reflect the qualitative and quantitative states or conditions affected by asthma. The NIAID Task Force Report (4) noted that asthma ranked among the leading causes of physician visits; indeed, the report continued, asthma accounted for one-third as many physician visits as essential hypertension and half as many as for diabetes. Annually, there were approximately 27 million patient visits for the treatment of asthma. In 1975, there were 1.15 million first visits to private physicians because of asthma. These data also spurred on initial attempts to develop and evaluate asthma self-management.

Today, there is a broad array of data available on the morbidity of asthma. The National Ambulatory Medical Care Survey indicated that physician visits for asthma as a first-listed diagnosis increased from 6.5 million in 1985 to 7.1 million in 1990 (17). Gergen and Weiss (25) found that between 1979 and 1987, hospitalizations for asthma among children aged 0–17 years increased 4.5% per year, with the largest increase among children 0–4 years of age. Among the latter group, blacks had approximately 1.89 times the increase of whites. Data from the CDC (18) indicated that the age-adjusted discharge rate for asthma as the primary diagnosis for patients between the ages of 5 and 34 was constant in both 1982 and 1992. Rates of asthma for females were consistently higher than for males; rates for blacks were consistently higher than for whites.

Costs of asthma have increasingly become a significant factor in assessing morbidity from the disorder. The NIAID Task Force Report (4) noted that the total annual cost of allergic diseases in the United States, including asthma, was more than $1 billion. In 1968, for example, asthma and hay fever accounted for 134,000 hospital discharges, with an average hospital stay of 8.3 days. The total cost for these services was $62 million. In 1972, direct costs for physician services, hospital care, and drugs for treatment of allergic diseases, including asthma, were $850 million. According to 1975 industry estimates of medication costs, $224.2 million were spent on bronchodilators; $24.7 million for corticosteroids; $43 million for over-the-counter asthma remedies; and $248.5 million for antihistamines and combination drugs containing antihistamines.

The above data can be compared with recent information. Weiss et al. (26) estimated that costs of asthma amounted to $6.2 billion in 1990. Direct costs, which included expenditures for hospital care, physician services, and medications, approached $3,637.6 million. Inpatient hospitalization, representing the largest single single medical expenditure for asthma, was projected to be $1,559.6 million in 1990. The cost of outpatient hospital care was estimated to be $190.3 million; expenditure for asthma care provided in hospital emergency rooms was approximately $295 million. While the National Asthma Education Program (5) considered asthma as a condition that could be treated on an outpatient basis, the combined costs for outpatient services were approximately $537.3 million, while the combined costs for inpatient services were estimated at $2.000.6 million (26). Medication costs added approximately $1,009.7 million to the total bill. Indirect costs totaled $2,568.4 million. The largest component of indirect costs ($899.7 million) was attributed to the loss of more than 10 million school days among children 5–17 years of age. The latter data reinforce previous findings, going back to Schiffer and Hunt (27), that pinpointed asthma as the leading cause of school absenteeism because of chronic illness. Other indirect costs included loss of work ($849.3 million) and mortality ($819.3 million) (26).

Information concerning the economic costs of asthma under managed health care are even more bleak, especially for childhood asthma. Data collected from 25,614 patients with asthma enrolled in a national HMO indicated that members 18 years and under consumed 47% of direct medical expenses for asthma-related illness (28). If these expenses were extrapolated to the total U.S. population, Stempel and coworkers calculated, direct medical costs for asthma would exceed $5 billion annually. This is 40% higher, corrected for inflation, than was projected in 1990.

The impact of asthma costs becomes more relevant when presented within the context of family income. One of the most recent studies in this area, reported in 1985, found that the combined direct and indirect costs of asthma care consumed 6.4% of a family's income in families who participated in a pediatric asthma self-management program (29). The percentage of family income consumed by asthma care has likely increased as health care costs have soared over the past decade.

C. Mortality

In the NIAID Task Force Report (4), it was estimated that 2000–4000 Americans die annually because of asthma. However, the report continued, the death rate from asthma had been declining and was only one-third the rate reported for 1960. Deaths from asthma were reported to rise slightly with increasing age; in addition, there were striking race-related differences in death rates from asthma. Below the age of 65, mortality among whites was consistently below that of nonwhites for both sexes; after the age of 65, however, the death rate of whites was higher than that of nonwhites. Race-related differences in deaths among children were striking, noted the NIAID report, although nothing further was added.

While the death rate due to asthma is low when compared to deaths from other disorders or factors, e.g., cancer, heart disease, accidents, an increasing rate of deaths due to asthma has been of concern to both clinicians and patients. From 1982 through 1991, the annual age-adjusted death rate for asthma increased 40% from 13.4 per million (3154) to 18.8 per million (5106). During that period, the rate increased 59% for females and 34% for males. Data showed that for people aged 5–34 years, the rate increased 42% (18). The annual death rate was consistently higher for blacks than for whites. During the period, the rate increased 41% for females and 43% for males. Deaths among patients aged 5–34 were examined by Weiss and Wagener (30). They reported that during the 1970s, mortality attributable to the condition declined by 7.8% a year in the United States. During the 1980s, however, asthma mortality rose by 6.2% a year, increasing faster among those aged 5–14 years than among those aged 15–34 years. These findings are conservative in that asthma mortality rates, determined through death certificate information, underestimate actual asthma-related mortality (31).

D. Demographics

Data concerning demographics related to asthma only began to emerge two decades ago. Demograpics are are the characteristics of the asthma population and population segments, particularly those that require particular attention. Few data were noted in the NIAID report (4) concerning demographics. It was noted that genetic characteristics play a role in asthma, but, as is still the case today, little was known about the exact relationship of these risk factors to asthma. It was also noted that an increased incidence of asthma among minority children was beginning to be reported. As described, considerable demographic information has been gathered on this topic during the past two decades.

Other data emerged two decades ago concerning the increased number of children with asthma whose parents were divorced. In a 10-year study that traced this trend at a residential treatment facility for asthma, it was noted that the percentage of patients with divorced parents increased from 15% in 1969 to 61% in 1979 (32). From 1969 through 1974, an average of 16.3% of the residents had experienced the divorce of their parents; at that time, this rate was near the average for all children in the United States between 6 and 16 years of age. However, in the 5-year period between 1975 and 1979, there was an average of 45.4% of children with divorced parents. The findings were higher than those for all children in the United States at that time and reinforced the impression by health care providers that many children with asthma lived in single-parent homes, often with a mother who worked outside the home. The finding suggested many asthmatic children would be required to treat any attacks they experienced; these results helped stimulate the movement towards the development of self-management programs for pediatric asthma (8). There is still a lack of information regarding demographic factors and their role in asthma. It is known that 38% of all American children will have been involved in divorce by the time they reach 18 years (33); this not only places children at risk for developing behavioral problems, but puts children with asthma at an additional risk for controlling their disorder. Such risk factors reinforce the necessity of self-management programs for children with asthma.

III. The Impact of Asthma on the Individual

> As a child, I was not allowed to have friendships with anyone who had a cat or dog because the visits could not be reciprocated. I wasn't allowed to go on school field trips, overnight camp, or sleepovers. I was not allowed to participate in any team sports or neighborhood games because of my "delicate" condition so I sat and watched and the other kids made fun of me and I felt ostracized. To this day I don't participate in any team sports because they terrify me. So in an indirect way, asthma has probably affected my social skills. I don't think it had to be this way, I think my parents were a little over-protective. In their defense, I was an amazingly sick kid—I missed 40+ school days every year due to illness (usually respiratory infections, flu, and pneumonia) and did have lots of asthma attacks. On the plus side, I read a lot and always excelled at schoolwork.
>
> An asthma attack is about the scariest thing I can think of. With me, it usually starts slowly. I get really hot, and slightly out of breath, and my chest

> tightens. Breathing becomes labored and tight. This is when I need to stop whatever I'm doing and get medicine and get away from whatever's causing it. If that doesn't happen, this stage of asthma can last for a long while. You can't lay down, you can't move. Every step or extra motion robs you of breath. You burn up, you need to use your hands to help support yourself for every breath. You lean forward trying to help pull air in. Air isn't moving in or out of your chest. The wheezing may go away, which is a bad sign, because it means your airways are even further restricted. At this point you're in real trouble. Saying "I'm having an asthma attack, please get me help" is too much. Talking and moving is nearly impossible. If panic sets in, the situation is life threatening. (Internet discussant, Anne G., age 28)

This section is concerned with how asthma interferes with the lives of patients. Clearly, the most salient aspects of asthma are the result of physiological mechanisms: shortness of breath, chest tightness, wheezing, and coughing. Dyspnea, or difficult breathing, is a pervasive symptom and is reported to be the most common reason for emergency room visits (34–36). But the impact of asthma on patients is not limited to its physical symptoms. Concern about dyspnea, which can occur unexpectedly, may have a profound effect on the patient. In some cases, concern affects psychological factors such as depression, anxiety, or panic; in others, concern may impose activity limitations. In addition, the common asthma medications, effective in controlling the disease, may also introduce a number of unwanted reactions.

A. Effect of Asthma on Activity

As a result of activity limitations, many individuals describe a lower quality of life (8,37–39). The inability to breathe comfortably can prevent participation in a wide variety of physical activities. Among the most serious activity limitations for adults with asthma are those related to work. Asthma is the third leading cause of work loss, behind only intervertebral disk and back disease (40). The relationship between degree of asthma severity and work loss is orderly: the two quantities are positively related. The findings of surveys concerned with asthma and work loss are corroborated by longitudinal studies. A large, longitudinal study, conducted in Great Britain (41), showed that asthma had an adverse effect on employment of young asthmatics of both sexes. By age 23, men and women with current or past asthma were more likely to have been unemployed, had spent less time employed, had more jobs, and had spent less time in their most recent full-time jobs than healthy individuals. In addition, adults with asthma seemed to have a significantly lower social class grouping. The author suggested that a reason for this may be that it is difficult to find a job that doesn't aggravate asthma. Job opportunity may be limited further by actions of employers who may not be inclined to hire an individual with asthma for fear that he or she may be absent more often than other employees. All in all, asthma may have devastating effects on the patient's employment.

Analogous to the problem of work loss in adult asthmatics is the problem of school absenteeism in children with asthma. As noted, asthma is the leading cause of absenteeism among school children (42). The absenteeism associated with asthma may be due to frequent doctor visits, symptoms that restrict activity, or psychosocial problems brought on by the disease. The latter may include a parent's belief that the child is too vulnerable to participate in certain activities or the stress that is often associated with chronic illness. The result of school absenteeism is not only poor academic performance but also poor social development (43), including poor peer relationships and increased parent-child conflict.

One problem that many asthmatic children face is in dealing with teachers who do not know enough about asthma and its treatment to reduce some of its consequences. As a result, many asthmatic children face unnecessary restrictions or inadequate resting periods at school. Behavioral problems at school often may be controlled either by parent-teacher communication or by some knowledge of asthma on the part of the teacher. Education of teachers in these matters would go a long way toward control of asthma during school periods.

B. The Psychological Impact of Asthma

In the 1930s, at the Chicago School of Psychoanalysis, Alexander and French studied asthma and defined it as a psychosomatic illness. They reported that a relationship existed between personality type and the acquisition of the disease. In particular, the researchers cited that a failed mother-child relationship was usually the basis for developing asthma. Although additional studies failed to support this notion, until recently a belief in the dynamics of the mother-child relationship continued to play a role in the illness (7). Recently, the belief that the etiology of asthma is psychological has lost popularity; however, few would deny that, as with other chronic diseases, psychological factors play a part (38,44). Depression, panic, and anxiety are the most widely studied psychological factors associated with asthma.

Depression

At the outset of this discussion, it should be noted that depression is no more prevalent among asthma patients than it is among healthy individuals (8,45). The occurrence of symptoms, however, may have a depressing effect on asthma patients. (34,35,45,46). In a study of 715 individuals, researchers reported a relationship between depression and asthma symptoms such as breathlessness after physical activity and waking with attacks of breathlessness (45). Depression in asthma increases with perceived danger and severity of the asthma (35), individuals who believed their asthma to be severe and life-threatening suffered more with depression than those who did not.

Asthma may cause depression in a number of ways. Nocturnal wheezing and attacks at night, common symptoms of asthma, disrupt normal sleep patterns and lead to or aggravate depression in patients (34). Perceived dependence upon asthma medications may also contribute to depression. Asthma also negatively affects quality of life, sense of control, self-esteem, and sense of worth. A reduction of any of these factors may be associated with an increase in depression (34,35,38). The multiple sources of depression in asthma ensure that each patient experiences some degree of depression as a result of the disease. If there is an optimistic note in any of this, it is simply that depression largely resolves with amelioration of symptoms.

Panic and Anxiety

Chest pains, breathing difficulty, sweating, faintness, and shaking are all symptoms of panic (47) that are also related to asthma attacks. The panic can be a response to the medical environment, including equipment, procedures, and personnel (8), or the result of a lack of the knowledge and confidence needed to control the attack (36,46). Unfortunately, asthma knowledge is reported as being low even among patients and their families, which indicates that many panic attacks might be preventable (49).

Not only are panic symptoms a source of discomfort for the person experiencing them, panic and phobic responses can also complicate the asthma attack itself. By interfering with the ability to make decisions and respond effectively to symptoms, panic can prompt an attack and worsen an existing episode (2,48). If the panic increases and the attack severity increases, hospitalization or even death can occur (50).

Anxiety, another psychological factor associated with asthma, can develop before, concurrently with, or following an asthma attack (8). Aside from panic, anxiety is the most commonly reported emotion accompanying the dyspnea associated with an asthma attack (36,46). It is speculated that the anxiety results from the unpredictability of the occurrence of an attack (38) and the fear of death (8,38). However, such speculation is unnecessary if one only considers the following: What is more frightening than not being able to breathe? As with depression, a lack of self-management skills or medical treatment to minimize severity, can increase or even cause anxiety during the attack (34).

Other Psychological Reactions to Asthma

One of the more debated issues associated with asthma is aggressiveness (8,46). A study of adults with asthma did find an increased rate of hostility during high dyspnea (46). In addition to hostility, it is suggested that aggressive behaviors may stem from feelings of frustration and fear, particularly among children (8). When aggression is exhibited, it is often seen as either verbal or physical abuse, often directed at members of the medical team (8).

A more frequently reported aspect of asthma is denial (8, 38). The denial can be seen as a way of life where the patient continually refuses to comply with medication and physician recommendations, or the denial can be more sporadic and appear only in certain situations where the individual ignores symptoms in order to pursue a pleasing activity (8). Although it will ultimately interfere with an individual's state of health, it is speculated that denial is a way of coping and adapting to the disease (8) or a way of preserving self-esteem and independence (38).

No one would deny that a persistent problem can lead to frustration. Asthma is no exception (8,39). The constant effort required to control coupled with the inevitable and repeated exacerbation of the disease can try the patience of even the most stable individual. Dealing with asthma can be a constant battle with frustration. But the source of frustration may not be limited to efforts to control the disease. One report states that the frustration associated with the restriction of physical activities and thus the decrease in the quality of life is greater than the frustration associated with the discomfort experienced from the symptoms of the disease (39).

Some people with asthma report feelings of shame and embarrassment (35). In our society, much emphasis is placed upon remaining in control and not depending on others. According to some people with asthma, a visit to the emergency room is a major concession to the illness demonstrating a loss of control; they must learn to live with the resulting shame and embarrassment.

C. Treatment Side Effects

Asthma medications are powerful drugs (5,6). Considered in the aggregate, they consist of anti-inflammatory agents, antiallergic drugs, and smooth muscle relaxants. The side effects of these preparations may range from serious to mildly annoying. It is probably safe to assume that, in some measure, every asthma patient must deal with unwanted reactions to these drugs. By far the most serious side effects are associated with the anti-inflammatory agents. These include, among others, weight gain, change in body fat distribution, change in adrenal function, cardiac arrhythmia, and change in sensory function. Fortunately, the use of inhaled rather than oral steroids reduces side effects enormously. Other categories of drugs are associated with less serious side effects, but even these can be a continuing source both of concern and of annoyance for the patient. The list of what might be called "less serious" side effects is long. It includes nausea, anorexia, vomiting, nervousness, irritability, dizziness, confusion, sleep disturbance, sweating, headache, blood pressure elevation, and respiratory irritation. The addition of one or more of these conditions to the respiratory symptoms of asthma can contribute substantially to the impact of the disease on the individual patient.

IV. Reducing Asthma Costs

Clearly, asthma extracts a heavy toll both in its costs to society and its affect on the individual. And despite progress in the treatment of the disease, asthma costs have risen in recent years. One way of reversing the trend may be widespread use of asthma self-management techniques. These procedures not only reduce asthma symptoms in patients but also reduce the financial costs of asthma. More will be said of this, especially of the reduction of symptoms, in subsequent chapters. But efforts have also been made to document the effect of asthma self-management on the costs of treating the disorder. One way to show this is by cost-benefit and cost-effectiveness analysis (CBCE) (51). Cost-benefit (CB) compares the monetary costs of conducting a program with the monetary benefits of the program. Benefits are often accrued as a result of decreased use of health services. Cost-effectiveness (CE) compares the cost of the program with the change in the desired behavior or health status.

Since CBCE analyses in health care are relatively new, there have only been a few attempts at evaluating asthma-treatment programs. Deter (52), working in Germany, evaluated the economic impact of psychosomatic group therapy for asthma. The one-year therapy consisted of asthma education, breathing and relaxation techniques, group support, and psychoanalytic interpretation of group dynamics. The treatment group improved significantly more than a control group on the cost of hospitalization and the number of work days lost to asthma. Deter reports that the benefit from this improvement compared with the cost of the therapy yielded approximately a 1:5 cost-benefit ratio.

In a cost-benefit analysis of a self-management program for adult asthma, direct and indirect cost data from 47 subjects who participated in the self-management program were analyzed (53). In particular, costs incurred by the subjects one year prior to participation were compared to costs incurred one year after participation. The cost-benefit analysis indicated that the program was beneficial by reducing the cost of asthma by $475.29 per patient. The benefit came primarily from reduced hospital admissions—from $18,488 to $1,538—and less income lost due to asthma—from $11,593 to $4,589. With a program cost of $208.33 per patient, the asthma self-managementm program yielded a 1:2.28 cost-benefit ratio.

In view of the medical and financial benefits associated with the use of asthma self-management, incorporating self-management procedures into a plan for the control of asthma may be one of our most powerful tools in reducing the overall costs of asthma.

V. Conclusions

This chapter paints a dark picture concerning the impact asthma has upon patients, their families, and society. Even more foreboding is the fact that there is no evidence of any immediate amelioration of the disorder. Not only is the prevalence of the asthma increasing year by year, but the costs, both in terms of morbidity of mortality, are soaring. There is no "magic bullet" on the horizon for controlling asthma. Asthma, along with other respiratory disorders, promises to become even a major health concern in the twenty-first century. Murray et al. (54), for example, recently estimated that there are 1.5 billion people in the world infected with the tubercle bacillus, and, of these, there are 20 million cases of active disease; moreover each year there are 8 million new cases of tuberculosis and 3 million deaths from the disease. Peto (55) analyzed trends in deaths from smoking. He noted that 3 million people worldwide now die each year of smoking-related disorders; it is estimated that there will be 10 million deaths annually attributable to smoking by the year 2025.

The tragic irony of the rise of respiratory disorders is that we have the technology to control them. The progress made in the past two decades to develop medical strategies for controlling asthma, for example, has been remarkable (5,6). As asthma is a reversible disorder in most patients, it should not be a major health concern. It continues to be a concern, however, because its control depends not only on medical progress but also on behavioral changes. The latter, involving teaching and motivating patients to achieve the behavioral skills that only they can control, is a significant component of the management and control of any respiratory disorder.

References

1. Veith I. The Yellow Emperor's Classic of Internal Medicine. Berkeley, CA: University of California Press, 1970.
2. Creer TL, Kotses H, Reynolds RVC. A Handbook for Asthma Self-Management: A Leader's Guide to Living with Asthma. Athens, OH: Ohio University Press, 1991.
3. Creer TL, Bender BG. Asthma. In: Gatchel RJ, Blanchard EB, (eds.) Psychophysiological Disorders: Research and Clinical Applications. Washington, DC: American Psychological Association, 1993:151–208.

4. NIAID Task Force Report. Asthma and the other allergic diseases. Bethesda, MD: National Institutes of Health, 79-387, 1979.
5. National Asthma Education Program. Executive Summary: Guidelines for the Diagnosis and Management of Asthma. Washington, DC: National Institutes of Health, 3042, 1991.
6. International Consensus Report on Diagnosis and Treatment of Asthma. Bethesda, MD: National Institutes of Health, 1992.
7. Renne CM, Creer TL. Asthmatic children and their families. In: Walraich ML, Routh DK, (eds.) Advances in Developmental and Behavioral Pediatrics. Greenwich, CT: Jai Press, 1985:41–81.
8. Creer TL. Asthma Therapy: A Behavioral Health Care System for Respiratory Disorders. New York: Springer Publishing, 1979.
9. Sheffer AL, Buist AS. Proceedings of the asthma mortality task force. J Allergy Clin Immunol 1987; 80:361–514.
10. McFadden ER Jr. Asthma: airway reactivity and pathogenesis. Sem Respir Med 1980; 1:287–296.
11. Creer TL, Levstek D, Winder JA. Home monitoring of lung function measures. In: Kotses H, Harver A, (eds.) Behavioral Contributions to the Self-Management of Asthma. New York: Marcel Dekker, 1997.
12. Loren ML, Leung PK, Cooley RL, Chai H, Bell TD, Buck VM. Irreversibility of obstructive changes in severe asthma in children. Chest 1978; 60:639–643.
13. Creer TL. Asthma. J Consult Clin Psychol 1982; 50:912–921.
14. Parker SR, Mellins RB, Sogn DD. Asthma education: a national strategy. Am Rev Respir Dis 1989; 140:848–853.
15. Creer TL, Winder JA. Asthma. In: Holroyd KA, Creer TL, (eds.) Self-Management of Chronic Disease. Handbook of Clinical Interventions and Research. Orlando, FL: Academic Press, 1986:269–303.
16. Marwich C. Inner-city asthma control campaign under way. JAMA 1995; 274: 1004.
17. Asthma—United States, 1980–1990. MMWR 1992; 41:733–735.
18. Asthma—United States, 1982–1992. MMWR 1995; 43:952–955.
19. Taylor WR, Newacheck PW. Impact of childhood asthma on health. Pediatrics 1992; 90:657–662.
20. Weiss KB, Gergen PJ, Crain EF. Inner-city asthma: the epidemiology of an emerging U.S. public health concern. Chest 1992; 101:362S-367S.
21. Evans R III. Asthma among minority children: a growing problem. Chest 1992; 101:368S-371S.
22. Weitzman M, Gortmaker S, Sobol A. Racial, social and environmental risks for childhood asthma. Am J Dis Child 1990; 144:1189–1194.
23. Weiss KB, Gergen PJ, Hodgson TA. An economic evaluation of asthma in the United States. N Engl J Med 1992; 326:862–866.
24. Health Care in Rural America. U.S. Congress, Office of Technical Assistance, OTA-H-434, Washington, DC, U.S. Government Printing Office, 1990.
25. Gergen PJ, Weiss KB. Changing patterns of asthma hospitalization among children: 1979 to 1987. JAMA 1990; 264:1688–1692.

26. Weiss, KB, Gergen PJ, Hodgson TA. An economic evaluation of asthma in the United States. N Engl J Med 1992; 326:862–866.
27. Schiffer CG, Hunt EP. Illness Among Children. Washington, DC: Children's Bureau Publication No. 405, 1964.
28. Stempel DA, Sturm LL, Hedblom EC, Durcanin-Robbins JF. Total cost of asthma care. J Allergy Clin Immunol 1995; 95:217.
29. Marion RJ, Creer TL, Reynolds RVC. Direct and indirect costs associated with the management of childhood asthma. Ann Allergy 1985; 54:31–34.
30. Weiss KB, Wagener DK. Changing patterns of asthma mortality: identifying target populations at high risk. JAMA 1990; 264:1683–1687.
31. Hunt LW Jr, Silverstein MD, Reed CE, O'Connell EJ, O'Fallon WM, Yunginger JW. Accuracy of the death certificate in a population-based study of asthmatic patients. JAMA 1993; 269:1947–1952.
32. Creer TL, Ipacs J, Creer PP. Changing behavioral and social variables at a residential treatment facility for childhood asthma. J Asthma 1983; 20:11–15.
33. Fishman KD. Behind the One-Way Mirror: Psychotherapy and Children. New York, Bantom Books, 1995.
34. Janson-Bjerklie S, Ferketich S, Benner P. Predicting the outcomes of living with asthma. Res Nurs Health 1993; 16:241–250.
35. Janson-Bjerklie S, Ferketich S, Benner P, Becker G. Clinical markers of asthma severity and risk: importance of subjective as well as objective factors. Heart Lung 1992; 21:265–272.
36. Janson-Bjerklie S, Carrieri VK, Hudes M. The sensation of pulmonary dyspnea. Nurs Res 1986; 35:154–159.
37. Agle DP, Baum GL. Psychological aspects of chronic obstructive pulmonary disease. Med Clin North Am 1977; 61:749–758.
38. Alt H. Psychiatric aspects of asthma. Chest 1992; 101:415S-416S.
39. Quirk FH, Jones PW. Patients' perception of distress due to symptoms and effects of asthma on daily living and an investigation of possible influential factors. Clin Sci 1990; 79:17–21.
40. Blanc, Paul D., M. Jones, C. Besson, P. Katz, and E. Yelin. Work disability among adults with asthma. Chest 1993; 104:1371–1377.
41. Sibbauld B, Anderson HR, McGuigan S. Asthma employment in young adults. Thorax 1992; 47:19–24.
42. Freudenberg N, Feldman C, Clark N, Millman, E, Valle I, Wasilewski Y. The impact of bronchial asthma on school attendance and performance. J School Health 1980; 50:522–526.
43. Weitzman M. School absence rates as outcome measures in studies of children with chronic illness. J Chronic Dis 1986; 39:799–808.
44. Creer TL. Is your asthmatic patient his own worst enemy? J Respir Dis 1986; 7:27–32.
45. Janson C, Bjornsson E, Hetta J, Boman G. Anxiety and depression in relation to respiratory symptoms and asthma. Am J Respir Crit Care Med 1994; 149:930–940.
46. Gift A. Psychologic and physiologic aspects of acute dyspnea in asthmatics. Nursing Res 1991; 40:196–199.

47. Argyle N. The nature of cognitions in panic disorder. Behav Res Ther 1988; 26:261–264.
48. Brooks CM, Richards JM Jr, Bailey WC, Martin B, Windsor RA, Soong S. Subjective symptomatology of asthma in an outpatient population. Psychosom Med 1989; 51:102–108.
49. Bucknall CE, Robertson C, Moran F, Stevenson RD. Management of asthma in the hospital: a prospective audit. BMJ 1988; 296:1637–1639.
50. Friedman MS. Psychological factors associated with pediatric asthma death: A review. J Asthma 1984; 21:97–117.
51. Warner KE, Luce BR. Cost-Benefit and Cost-Effectiveness Analysis in Health Care. Ann Arbor, MI: Health Administration Press, 1982.
52. Deter HC. Cost-benefit analysis of psychosomatic therapy in asthma. J Psychosom Res 1986; 30:173–182.
53. Taitel MS, Kotses H, Bernstein IL, Bernstein DI, Creer TL. A self-management program for adult asthma. Part II: Cost-benefit analysis. J Allergy Clin Immunol 1995; 95:672–676.
54. Murray JF, Erarson DA, ATS Committee on World Lung Health. World lung health: a concept that should become a reality. Am Rev Respir Rev 1992; 146:818–822.
55. Peto R. Smoking and death: the past 40 years and the next 40. BMJ 1994; 309:937–939.

2

Personality Correlates of Asthma

AD A. KAPTEIN

Leiden University
Leiden, The Netherlands

I. Introduction

Creer noted in his editorial about emotions and asthma that "any discussion of emotions and asthma is apt to become very emotional" (1). In a similar vein, one can argue that any discussion of personality and asthma is apt to become very personal. Over the past 50 years, heated debates have taken place about specific personality traits as causative of asthma, as consequences of asthma, or as modifying the responses of asthma patients to their disorder. The asthma patient has been made to shoulder the burden of developing a psychogenic disease, of exhibiting distinct personality patterns linked to asthma, and of contributing to a poor medical outcome due to failure to cope appropriately with the disease.

Recent updates show that asthma patients themselves attribute the disorder to a large extent to personality factors (2,3). Physicians to this day continue to identify personality characteristics as contributors to undesirable patient behavior such as noncompliance or deficient self-management behavior (4,5). Psychiatrists persevere in focusing on psychopathology in patients with asthma (6,7). Even the public at large holds unfavorable views concerning the personality of asthma patients (8,9). In view of this background, a skeptical approach is warranted in the study of interactions between personality and asthma.

In this chapter, I review research relating to personality correlates of asthma. First, I discuss how personality factors contribute both to health and illness. Second, I provide a description and an evaluation of psychosomatic views and theories regarding asthma. Third, I review the work on "psycho-maintenance of asthma" begun by the Denver research group. Finally, I conclude the chapter with an attempt to integrate the empirical data on personality correlates of asthma with theoretical models of personality factors and illness.

II. Personality in Relation to Health and Illness

There is renewed interest in whether personality traits or factors contribute to the development and course of physical disorders. Suls and Rittenhouse outlined three models concerning this issue (10). In their first model, "the personality-induced hyperreactivity model," personality traits are said to influence the appraisal of stressors, possibly leading to heightened physiological reactions that may increase health risks. The emphasis on stress makes this model consistent with both type A behavior and vital exhaustion formulations (11,12). In their second model, the "constitutional predisposition model," both personality traits and physical illness are considered consequences of an inborn physical weakness or organ system(s) abnormality. The physical weakness serves to increase disease susceptibility. Their third model, described as "personality as precipitator of dangerous behaviors," focuses on relationships between personality and behaviors that threaten health (hazardous behaviors) and expected health outcome.

Suls and Rittenhouse (13) maintained that "even if a personality dimension is important only because of its association with certain health-related behaviors or genetic predisposition, the dimension itself may be significant because, in some cases, the dispositional variable may be more identifiable than the behavior (or future behavior). Hence, the focus on dispositional variables may enable biobehavioral researchers and medical scientists to target the most "at risk" groups for prevention or intervention.". Holroyd and Coyne, however, cautioned against optimism about the explanatory power of personality factors either in the etiology of illness or as determinants of medical outcome of physical disorders and illnesses (14). Similar concerns were raised by Krantz and Hedges (15) and by Creer (16), who concluded after reviewing the literature on personality and asthma that the focus on personality research in asthma may not have been helpful.

Despite the pessimism of many authorities concerning the value of personality research, a great deal of study has been devoted to the personality correlates of asthma. It is helpful to organize the work along theoretical lines. Three approaches have been used in an effort to understand how personality and asthma are related: psychodynamics, psychomaintenance, and health psychology. In the following sections, I review research related to each of these formulations.

III. Psychodynamic Theory

The influence of personality on precipitation of asthma attacks and on the course of asthma was noted in the oldest references to asthma. As early as 400 B.C., Hippocrates recognized that anger and hostility influenced the occurrence of an "asthmatic" attack (17,18). With the advent of psychoanalysis, Freud alluded to the concept of "asthma nervosum" (19); asthma was viewed as an example of a somatic dysfunction, which was the expression of an underlying unconscious conflict. As such, asthma was merely a symptom that symbolically expressed an anxiety neurosis. However, it should be noted that psychoanalytic thinking and therapy did not include a unifying theory of asthma.

The first model proposed by Suls and Rittenhouse for personality factors in health and illness is a psychodynamic model. It is typified by the work of Alexander and Dunbar. These authors were influenced profoundly by medical advances based on the discovery of both the causes and the treatment of bacterial disease. The framework afforded by an understanding of how bacteria led to disease provided the model used to explain how psychological factors led to disease. As Knapp remarked (20), investigators "often follow an implicit model of the germ theory and try to find, in one or another facet of the syndrome, a 'bacillus asthmaticus psychosomaticus.'" Ouellette Kobasa, in a more recent publication on personality and health (21), remarked, not without irony, that in modern theories on personality in health and illness "personality emerges simply as a new kind of germ."

Dunbar (22) reviewed the psychoanalytical and psychosomatic literature on somatic disorders, and in so doing, noted similarities in personality, behavior, family background, and early experience between asthma patients. She concluded that personality profiles of individuals with a certain physical disease showed considerable correspondence and were causally related to diseases of specific organ systems: one personality profile leads to asthma, while another leads to a different disease. About asthma, Dunbar (23) wrote that "the element of separation from the mother was important. The demand for love coexisting with a fear of love and revolt against it seemed to keep . . . patients in a constant state of vibration between activity and passivity, between love and hate. They feared both love and loss of love; all spoke of fear of being held too tight—'smothered with kisses'—as well as of fear of sexual activity and of separation and rejection."

Alexander, a contemporary of Dunbar and a chief proponent of the psychodynamic view of asthma, conducted research and developed a theoretical view of asthma as a psychosomatic disease. In a series of publications, he and his colleagues at the Chicago Institute for Psychoanalysis described their observations on the effects of emotional factors on the disease process and formulated a theory of psychosomatic disease (24–26). According to Alexander and colleagues, stress arising out of everyday contacts with the environment caused, in the individual, a chronic disturbance of organ systems. Based on Cannon's work on the autonomic system, Alexander considered disturbance of vegetative functions to result either when fight-or-flight behavior is blocked or when nurturing behavior is denied. Blocking fight-or-flight behavior leads to sustained excitation of the sympathetic nervous system, disturbance in neuroendocrine systems, and, ultimately, to diseases such as hypertension and migraine headaches, diseases that are related to sympathetic arousal. On the other hand, blocking the "being taken care of" wish, by either internal denial or external circumstances, leads to an increase in neuroendocrine-mediated, parasympathetic activity and to diseases such as peptic ulcer or asthma, diseases which are related to parasympathetic arousal. Regarding the problem of specificity of emotional factors and somatic disturbances, Alexander (24) stated "it is not the presence of any one or more of these psychological factors (e.g., anxiety, repressed hostile and erotic impulses, frustration or dependent craving) that is specific, but the presence of the dynamic configuration in which they appear." With regard to asthma, Alexander (24) wrote "the nuclear psychodynamic factor is a conflict centering in an excessive unresolved dependence upon the mother . . . it is the wish to be protected—to be encompassed by the mother or the maternal image. . . . Everything which threatens to separate the patient from the protective mother or her substitute is apt to precipitate an asthmatic attack."

Alexander subsequently provided a somewhat revised set of three requisite conditions for development of organic disease: (1) a constitutional or acquired organ vulnerability, (2) a characteristic emotional-conflict pattern, and (3) a precipitating life event that leads to a breakdown of the patient's psychological defenses. Additional work was conducted relating to the first two of these conditions. Concerning the first condition, organ vulnerability, the model gave Alexander the opportunity to incorporate the ever-increasing knowledge of allergy into his psychosomatic theories. Disruption of physiological mechanisms relating to allergy mechanisms constituted organ vulnerability (26).

Alexander's second condition for the development of an emotional-conflict pattern was the subject of research. The most extensive study of conflict theory regarding asthma was conducted by Alexander et al. (25). They studied 70 patients who suffered from one of seven psychosomatic disorders, and on the basis only of taped recordings of recollected material tried to determine each patient's disorder. Alexander et al. maintained that "it is possible to make the correct diagnosis of any one of seven specific organ disorders, without the help of any organic-medical clues, merely on the basis of the formulation of psychiatric . . . interview protocols." Overall, psychosomatic diagnoses were ascertained by listening to recorded recollections of the patients, but with considerable error. Asthma was correctly identified in only 50% of the cases.

Evidence for the specific conflict hypothesis is meager, and the idea may be criticized on the basis of other considerations. First, the patients studied in conflict research may have been grossly unrepresentative of the general population of asthma patients; relatively few asthma patients come to the attention of psychiatrists. Second, specific conflict theory is not helpful in understanding cause-and-effect relationships. For example, it is silent about the mechanisms by which psychological and physiological factors are linked. Third, the dependency conflicts, those thought to be causal in the development of asthma, may very well arise as a consequence of having asthma. Creer (16) provided a critical analysis of research done by investigators who adopt a psychosomatic orientation in the study of asthma: "there is the sad conclusion that a psychosomatic conceptualization has added nothing to the treatment and management of patients, especially those afflicted by asthma or other chronic obstructive respiratory disorders. In fact, a very strong case could be made that such a model has contributed more to the distress and anguish of patients, especially by nurturing myths about their diseases, than to providing them with any enduring benefit." His arguments are persuasive. Myths, however, may have long lives. Unfortunately, the legacy of these myths continues to plague not only the asthma patient, but also the physician who treats the disease as well as the social scientist who studies it.

IV. Psychomaintenance of Asthma

A 1973 publication (27) triggered an impressive series of articles on what later came to be called psychomaintenance of asthma. Subsequently, approximately 60 papers were published on studies performed at the National Jewish Hospital and Research Center in Denver. The research dealt with how psychological factors characteristic of asthma patients affect outcome of the medical treatment. Kinsman, the initiator of the research, had been impressed with differences between U.S. Air Force personnel in their reactions to the stress of weightlessness (R. A. Kinsmans, personal communication, 1979). He took these observation as the starting point for his psychological research in asthma.

As a first step, Kinsman and colleagues constructed an instrument, called the Asthma Symptom Checklist (ASC), to quantify and classify the subjective symptomatology accompanying asthma attacks (28). They discovered that one category of symptoms on the ASC, labeled panic-fear symptomatology (ASC-pf), was related to aspects of treatment outcome. Patients who scored low on the panic-fear scale were prescribed less potent medication at discharge than patients with high scores. The relationship was shown to be independent of longitudinal pulmonary function measurements during hospitalization, age at onset of asthma, sex, age at admission, and duration of asthma (27). Scores on the panic-fear symptomatology scale were shown to be unrelated to two other aspects of medical outcome: length of hospitalization and rate of rehospitalization (J. F. Dirks, unpublished, 1978).

To further explore the relationships between psychological factors and medical outcome in asthma, Dirks et al. (29) reported the development of an empirically derived subscale of the Minnesota Multiphasic Personality Inventory (MMPI), labeled panic-fear personality (MMPI-PF). In contrast to panic-fear symptomatology, which indicated anxiety associated with a specific situation—i.e., an asthma attack—panic-fear personality reflected a traitlike personality characteristic. Patients scoring high on this scale were described as fearful, withdrawn in a hostile manner, emotionally labile, and giving up easily in the face of difficulty. Patients who scored low on the scale were described as being unusually free of anxiety, extremely calm, and unusually self-controlled (J. F. Dirks, unpublished, 1978).

The panic-fear personality scale appeared to measure neither a somatic component of asthma nor a psychological reaction to asthma; scores on this scale were not related to age at asthma onset, age at admission, duration of asthma, and spirometric measures. However, scores on the panic-fear personality scale were related to various aspects of treatment outcome. They were related positively to length of hospitalization (31) but not in all cases (32). They were also related to rehospitalization but in a curvilinear fashion: low, moderate, and high panic-fear personality patients exhibited rates of hospital readmission in a 2:1:2 ratio (J. F. Dirks, unpublished, 1978). Differences in readmission rates could not be attributed to the severity of asthma, as indexed either by pulmonary function measurements or by physician ratings during long-term treatment. In addition, panic-fear personality scores were related to potency of medication prescribed at discharge: high panic-fear personality scores were associated with more potent medication (i.e., corticosteroids); low panic-fear personality scores were associated with less potent medication (J. F. Dirks, unpublished, 1978).

In addition to illness-specific symptomatology and the patient's general personality characteristics, patient attitude toward illness represents another potential index of treatment outcome. Kinsman et al. (32) constructed a scale for measuring attitudes toward respiratory illness and hospitalization. This scale, the Respiratory Illness Opinion Survey (RIOS), measures six attitude categories: optimism, negative staff regard, specific internal awareness, external control, psychological stigma, and authoritarian attitudes. Staudenmayer et al. (33) studied the relationship between the six categories and personality, symptomatology, and response to treatment. Results indicated that patients who scored low on the optimism dimension were significantly more likely to be rehospitalized during a 6–month follow-up period than the total sample. In addition, combinations of attitudes were also predictive of various aspects of medical outcome. For example, patients with a combination of low optimism and high psychological stigma showed a rehospitalization rate of more than twice that of the overall sample.

In reviewing their research, Kinsman et al. concluded that situational responses to asthma, attitudes, and personality characteristics predicted whether the disruptive nature of asthma is reduced or maintained by the patient (34). In their words, psychomaintenance referred to "how psychologic and behavioral factors maintain and increase both perceived severity and medical intractability of the illness once it has already developed."

Psychomaintenance was the basis for a subsequent research on psychological factors in asthma. After Kinsman and colleagues compiled the Battery of Asthma Illness Behavior (BAIB), a questionnaire that combined the ASC, the MMPI-PF and the RIOS, they documented a link between BAIB scores and medication usage patterns (35), and found that personal style could potentially predict the worst possible asthma treatment outcome, death (36). In 1983, the Kinsman group extended their work on subjective symptomatology to new categories of respiratory patients, those with either chronic bronchitis or emphysema (37,38). Sadly enough however, the record of the Denver group terminated that same year.

Others have followed Kinsman's lead. Baron and colleagues adapted the BAIB to children with asthma and found strong associations between personality traits (trait-anxiety and panic-fear personality) and excessive use of asthma medication (39,40). Belloch and his associates found in adult asthmatics that length of hospitalization, medication at discharge, and rate of rehospitalization were associated with panic-fear personality score and vigilant attitudes toward asthma symptoms (41). Brooks et al. (42) verified the ASC factor structure and demonstrated a statistically significant correlation between panic-fear symptomatology and health care utilization in a large sample of adult asthmatics. Janson-Bjerklie et al. (43) applied the panic-fear symptomatology subscale of the ASC in a study on patient perceptions of asthma severity and danger from asthma: panic-fear symptomatology correlated significantly with perceived severity and danger. In the Netherlands, Kaptein replicated some of the findings of Kinsman and colleagues in a sample of hospitalized asthma patients (44,45). In a study by Mawhinney and colleagues, BAIB scores were associated with adherence to medication: panic-fear personality was associated with arbitrary medication use (46).

The lack of a relationship between personality and health also has been reported. In particular, Northup and Weiner found no relationship between panic-fear personality scores and hospitalization rates (47). But work relating to psychomaintenance continues. Perrudet-Badoux et al. produced a French translation of the ASC and reported reductions in scores on panic-fear symptomatology after combined relaxation and respiratory training (48). Clearly, Kinsman inspired many asthma researchers in the asthma area, and his extension of personality research to chronic bronchitis and emphysema (37,38) continues to inspire researchers up to the present time (49).

V. Personality Factors and Asthma in Health Psychology

In health psychology, the influence of personality factors on physical disorders has been considered within the context both of coping theories and of cognitive-behavioral theories. Surprisingly, coping with asthma has not been the focus of many empirical studies. An early exception was research by de Araujo et al. (50) who were concerned with coping as it related to medication usage. But coping with chronic illness was not to be elevated to the status of a major area in health psychology until the appearance, some years later, of research by Lazarus and colleagues (51). In this and other work, underlying theory included personality as one determinant of coping. This is exemplified by the diagram in Figure 1. Consistent with this approach is the work of Scheier and Bridges (52), who presented a model on person variables (i.e., personality dispositions and transient psychological states) related to multiple health outcomes in various illnesses. They showed that three categories of person variables were related to health outcomes in coronary heart disease, cancer, and AIDS: hostility/anger, suppression of emotion, and a disengagement cluster comprising depression, pessimism/fatalism, fighting spirit/active coping, and vital exhaustion. Their findings, consistent in some ways with research on psychomaintenance, supported a multidimensional view of personality factors related to coping with chronic disease. This point of view is in contrast with earlier, unidimensional approaches to coping.

Recently, Turk and Salovey (53) emphasized the change in how responses by patients to chronic physical illness have been conceptualized over the past few decades. Instead of viewing responses of patients to physical illness on a continuum of illness behavior, with an anchor point labeled as abnormal, these authors maintained that (53):

> subjective interpretations and appraisal of symptoms as well as social reinforcement contingencies affect physical and psychological functioning based on an individual's unique illness schema. Thus, for all patients with a chronic condition,

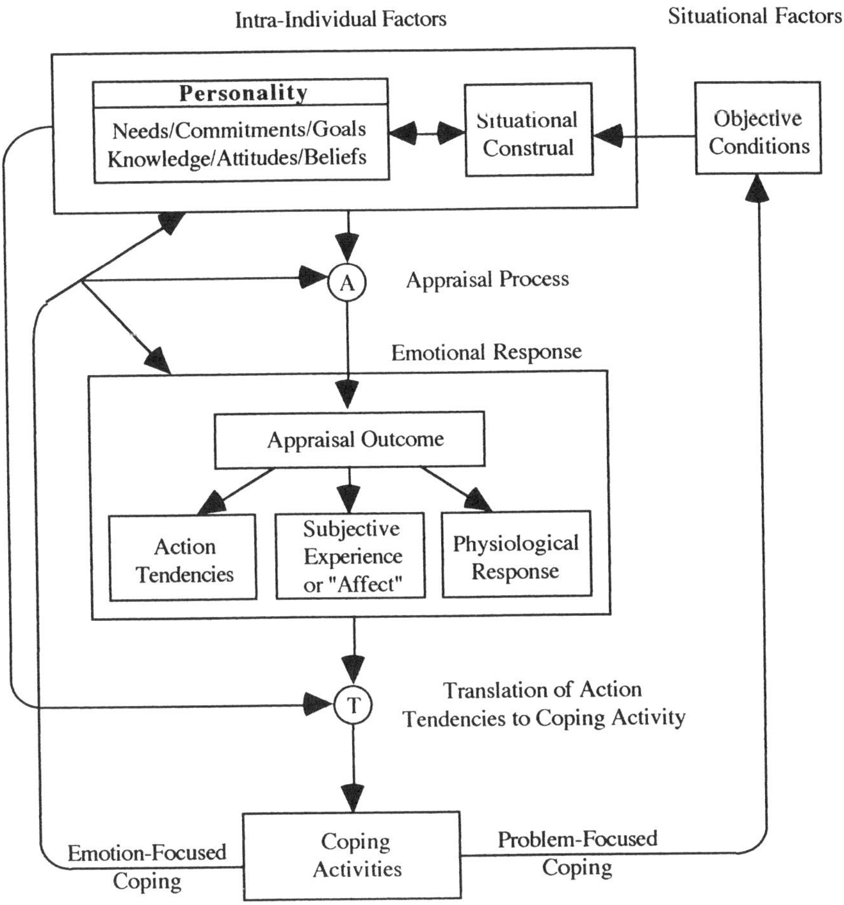

Figure 1 A model of the cognitive-motivational-emotive system. (From Ref. 51.)

> one must attend to individuals' own conceptualizations of their symptoms, disease, and the responses from significant people in their environment because all of this may affect physical and psychological responses. Such responses in turn influence physical and psychological functioning as well as response to treatment and rehabilitation. In contrast to some, we do not refer to "abnormal" illness behavior, which, by definition, assumes that we know what normal and abnormal responses to illness should be. We focus our discussion primarily on psychologically healthy people confronted with physical impairments and chronic diseases."

Concepts from health psychology are being studied increasingly in behavioral research on asthma. Carr et al. combined panic-fear, a concept from psychomaintenance research, with cognitive-behavioral models and provided empirical evidence for associations between panic-fear personality, illness perceptions, and coping with asthma (54). Concepts such as self-efficacy (55,56), negative affectivity (57), and symptom perception (58–61) are promising avenues to understanding psychological factors in asthma (62). At a more molecular level, as exemplified by work in psychoneuroimmunology, personality factors may be shown to affect asthma through their control of physiological and immunological function (63,64).

VI. The Future of Personality Research

In this chapter, I discussed issues concerning personality and asthma from a historical perspective. But what is the future of personality research in asthma? The answer is no more clear today than it was in 1979 when Creer (16) wrote, "finally, the question that must be asked is: 'Of what value to a health care system are studies of personality patterns in asthmatic patients?' Up until now the answer is that such investigations have made a meager contribution: they have mainly served to divert attention away from the behavioral aspects of asthma." But it is not farfetched to suggest that knowledge of the personality correlates of asthma may be valuable insofar as it concerns asthma self-management (65). In particular, identification of personality variables that modulate effectiveness of asthma self-management would be invaluable. Research along these lines would promote not only an understanding of how personality contributes to health care but also how best to achieve control over asthma.

Acknowledgment

The author thanks Barbara d'Arnaud for expert technical assistance and Manon Waterreus for linguistic help. This work was supported in part by grants from the Dutch Asthma Foundation (project numbers 90.37 and 92.31).

References

1. Creer TL. Emotions and asthma. Editorial. J Asthma 1993; 30:1–3.
2. Carter B. An asthmatic in 'Wonderland': a patient's perspective of accident and emergency. Acc Emerg Nurs 1995; 3:139–141.
3. Williams SJ. Chronic Respiratory Illness. London: Routledge, 1993.
4. Bosley CM, Fosbury JA, Cochrane GM. The psychological factors associated with poor compliance with treatment in asthma. Eur Respir J 1995; 8:899–904.

5. Strunk RC, Mrazek DA, Wolfson Fuhrmann GS, LaBrecque JF. Physiologic and psychological characteristics associated with deaths due to asthma in childhood. JAMA 1985; 254:1193–1198.
6. Campbell DA, Yellowlees PM, McLennan G, Coates JR, Frith PA, Gluyas PA, Latimer KM, Luke CG, Martin AJ, Ruffin RE. Psychiatric and medical features of near fatal asthma. Thorax 1995; 50:254–259.
7. Yellowlees PM, Haynes S, Potts N, Ruffin RE. Psychiatric morbidity in patients with life-threatening asthma: initial report of a controlled study. Med J Aust 1988; 149:246–249.
8. Brook U, Kishon Y. Knowledge and attitude of healthy high school students toward bronchial asthma and asthmatic pupils. Chest 1993; 103:455–457.
9. Coombs RH, Chopra S, Schenk DR, Yutan E. Medical slang and its functions. Soc Sci Med 1993; 36:987–998.
10. Suls J, Rittenhouse JD. Models of linkages between personality and disease. In: Friedman HS, ed. Personality and Disease. New York: Wiley, 1990:38–64.
11. Meesters C, Appels A. An interview to measure vital exhaustion. II. Reliability and validity of the interview and correlations of vital exhaustion with other personality characteristics. Psychol Health 1996; 11: 573–581.
12. Smith TW, Anderson NB. Models of personality and disease: an interactional approach to type A behavior and cardiovascular risk. J Pers Soc Psychol 1986; 50:1166–1173.
13. Suls J, Rittenhouse JD. Personality and physical health: an introduction. J Pers 1987; 55:155–167.
14. Holroyd KA, Coyne J. Personality and health in the 1980s: psychosomatic medicine revisited? J Pers 1987; 55:359–375.
15. Krantz DS, Hedges SM. Some cautions for research on personality and health. J Pers 1987; 55:351–357.
16. Creer TL. Asthma Therapy: A Behavioral Health Care System for Respiratory Disorders. New York: Springer, 1979.
17. Abramson HA. Psychosomatic aspects of hay fever and asthma prior to 1900. Ann Allergy 1948; 6:110–121.
18. Ziment I. Five thousand years of attacking asthma: an overview. Respir Care 1986; 31:117–136.
19. Freud S. Gesammelte Werke. London: Imago, 1952:I,II.
20. Knapp PH. The asthmatic child and the psychosomatic problem of asthma: toward a general theory. In: Schneer HI, ed. The Asthmatic Child. New York: Harper & Row, 1963:234–255.
21. Ouellette Kobasa SC. Personality and health: specifying and strengthening the conceptual links. In: Rev Pers Soc Psychol 1985; 6:291–311.
22. Dunbar F. Emotions and Bodily Changes. New York: Columbia University Press, 1935.
23. Dunbar F. Psychoanalytic notes relating to syndromes of asthma and hay fever. Psychoanal Q 1938; 7:25–68.
24. Alexander F. Psychosomatic Medicine: Its Principles and Applications. London: Allen & Unwin, 1952.

25. Alexander F, French TM, Pollock GH. Psychosomatic Specificity. Chicago: University of Chicago Press, 1968.
26. Alexander F, Selesnick ST. The History of Psychiatry. New York: Harper & Row, 1966.
27. Kinsman RA, O'Banion K, Resnikoff P, Luparello TJ, Spector SL. Subjective symptoms of acute asthma within a heterogeneous sample of asthmatics. J All Clin Immunol 1973; 52:284–329.
28. Kinsman RA, Luparello T, O'Banion K, Spector S. Multidimensional analysis of the subjective symptomatology of asthma. Psychosom Med 1973; 35:250–267.
29. Dirks JF, Jones NF, Kinsman RA. Panic-fear: a personality dimension related to intractability in asthma. Psychosom Med 1977; 39:120–126.
30. Dirks JF, Kinsman RA, Jones NF, Spector SL, Davidson PT, Evans NW. Panic-fear: a personality dimension related to length of hospitalization in respiratory illness. J Asthma Res 1977; 14:61–71.
31. Dirks JF, Horton DJ, Kinsman RA, Fross KH, Jones NF. Patient and physician characteristics influencing medical decisions in asthma. J Asthma Res 1978; 15:171–177.
32. Kinsman RA, Jones NF, Matus I, Schum RA. Patient variables supporting chronic illness. J Nerv Ment Dis 1976; 163:159–165.
33. Staudenmayer H, Kinsman RA, Jones NF. Attitudes toward respiratory illness and hospitalization in asthma. J Nerv Ment Dis 1978; 166:624–634.
34. Kinsman RA, Dirks JF, Jones NF. Psychomaintenance of chronic physical illness: clinical assessment of personal styles affecting medical management. In: Millon T, Green C, Meagher R, eds. Handbook of Clinical Health Psychology. New York: Plenum Press, 1982:435–466.
35. Dirks JF, Kinsman RA. Bayesian prediction of noncompliance: as-needed (PRN) medication usage patterns and the battery of asthma illness behavior. J Asthma 1982; 19:25–31.
36. Dirks JF, Kinsman RA. Death in asthma: a psychosomatic autopsy. J Asthma 1982; 19:177–187.
37. Kinsman RA, Fernandez E, Schocket M, Dirks JF, Covino NA. Multidimensional analysis of the symptoms of chronic bronchitis and emphysema. J Behav Med 1983; 6:339–357.
38. Kinsman RA, Yaroush RA, Fernandez E, Dirks JF, Schocket M, Fukuhara J. Symptoms and experiences in chronic bronchitis and emphysema. Chest 1983; 83:755–761.
39. Baron C, Lamarre A, Veilleux P, Ducharme G, Spier S, Lapierre J-G. Psychomaintenance of childhood asthma: a study of 34 children. J Asthma 1986; 23:69–79.
40. Baron C, Veilleux P, Lamarre A. The family of the asthmatic child. Can J Psychiatry 1992; 37:12–16.
41. Belloch A, Perpiñá M, Paredes T, Giménez A, Compte L, Banos R. Bronchial asthma and personality dimensions: a multifaceted association. J Asthma 1994; 31:161–170.
42. Brooks CM, Richards JM, Bailey WC, Martin B, Windsor RA, Soong S-J. Subjective symptomatology of asthma in an outpatient population. Psychosom Med 1989; 51:102–108.

43. Janson-Bjerklie S, Ferketich S, Benner P, Becker G. Clinical markers of asthma severity and risk: importance of subjective as well as objective factors. Heart Lung 1992; 21:265–272.
44. Kaptein AA. Illness Behavior of Patients with Asthma. Meppel, The Netherlands: Krips Repro, 1982.
45. Kaptein AA. Psychological correlates of length of hospitalization and rehospitalization in patients with acute, severe asthma. Soc Sc Med 1982; 16:725–729.
46. Mawhinney H, Spector SL, Heitjan D, Kinsman RA, Dirks JF, Pines I. As-needed medication use in asthma usage patterns and patient characteristics. J Asthma 1993; 30:61–71.
47. Northup L, Weiner MF. Hospitalization, life change and ability to cope with asthma. J Psychosom Res 1984; 28:177–183.
48. Perrudet-Badoux A, Chauvelle MT, Guinnepain MT. Evaluation de la symptomatologie subjective chez les asthmatiques par "l'Asthma Symptom Checklist": son application aux effets de la kinésithérapie respiratoire. Rev Allergol 1987; 27:133–136.
49. Graydon JE, Ross E. Influence of symptoms, lung function, mood, and social support on level of functioning of patients with COPD. Res Nurs Health 1995; 18:525–533.
50. Araujo G de, Arsdel P van, Holmes TH, Dudley DL. Life change, coping ability and chronic intrinsic asthma. J Psychosom Res 1973; 17:359–363.
51. Smith CA, Lazarus RS. Emotion and adaptation. In: Pervin LA, ed. Handbook of Personality—Theory and Research. New York: The Guilford Press, 1990:609–637.
52. Scheier MF, Bridges MW. Person variables and health: personality predispositions and acute psychological states as shared determinants for disease. Psychosom Med 1995; 57:255–268.
53. Turk DC, Salovey P. Cognitive-behavioral treatment of illness behavior. In: Nicassio PM, Smith TW, eds. Managing Chronic Illness—A Biopsychosocial Perspective. Washington, DC: American Psychological Association, 1995:245–284.
54. Carr RE, Lehrer PM, Hochron SM. Predictors of panic-fear in asthma. Health Psychol 1995; 14:421–426.
55. Clark NM, Starr-Schneidkraut J. Management of asthma by patients and families. Am J Respir Crit Care Med 1994; 149:554–566.
56. Wigal JK, Stout C, Brandon M, Winder JA, McConnaughy K, Creer TL, Kotses H. The knowledge, attitude, and self-efficacy asthma questionnaire. Chest 1993; 104:1144–1148.
57. Priel B, Heimer D, Rabinowitz B, Hendler N. Perceptions of asthma severity: the role of negative affectivity. J Asthma 1994; 31:479–484.
58. Boner AL, De Stefano G, Piacentini GL, Bonizzato C, Sette L, Banfi F, Hindi-Alexander MC. Perception of bronchoconstriction in chronic asthma. J Asthma 1992; 29:323–330.
59. Fritz GK, Yeung A, Taitel MS. Symptom perception and self-management in childhood asthma. Curr Opinion Pediatr 1994; 6:423–427.

60. Harver A. Effects of feedback on the ability of asthmatic subjects to detect increases in the flow-resistive component to breathing. Health Psychol 1994; 13:52–62.
61. Rietveld S, Kolk AMM, Prins PJM, Van Beest I. The influence of external stimulation on airflow detection by children with and without asthma. Psychol Health (in press).
62. Weinman J, Petrie KJ, Moss-Morris R. The illness perception questionnaire: a new method for assessing the cognitive representation of illness. Psychol Health 1996; 11:431–445.
63. Busse WW, Kiecolt-Glaser JK, Coe C, Martin RJ, Weiss ST, Parker SR. Stress and asthma. Am J Respir Crit Care Med 1995; 151:249–252.
64. Mrazek DA, Klinnert M. Asthma: psychoneuroimmunologic considerations. In: Ader R, Felten DL, Cohen N, eds. Psychoneuroimmunology. San Diego: Academic Press, 1991:1013–1035.
65. Creer TL, Bender BG. Asthma. In: Gatchel RJ, Blanchard EB, eds. Psychophysiological Disorders: Research and Clinical Applications. Washington, DC: American Psychological Association, 1993:151–203.

3

Emotional Precipitants of Asthma

HARRY KOTSES

Ohio University
Athens, Ohio

In 1886, Mackenzie (1) reported eliciting a condition called "rose cold" in a female patient by bringing her into contact with an artificial rose. The patient had sought relief from a variety of allergic and asthmalike symptoms that she experienced annually throughout the period between late spring and early fall. Brought under control with a combination of treatments, the symptoms reappeared abruptly when the patient was shown an artificial rose. At the appearance of the rose, Mackenzie (1) noted that his patient immediately exhibited symptoms of rhinitis which were followed in a few minutes by a "feeling of oppression in the chest . . . with slight embarrassment of respiration." To this he added that by assuring his patient that her symptoms could be controlled, he helped her overcome her reaction to roses.

So began the modern era of inquiry into the relationship between asthma and emotion. In his brief case study, Mackenzie anticipated not only the view that emotion may precipitate asthma, but also the idea that psychological intervention may reduce the symptoms of asthma. In addition, he suggested that, in some cases, asthma symptoms may be an outgrowth of an association of ideas, a conjecture that, in subsequent years, was discussed within the context of learning as it applied to the control of physiological responses.

More than a century later, the influence of Mackenzie's observations is still apparent in studies of the relationship between asthma and emotion. In the intervening years since his report, dozens of investigators have documented a link between emotion and the occurrence of asthma symptoms. Dozens more have demonstrated that the inhibition of emotion by one technique or another reduces symptoms in asthma patients. Considered as a group, these studies reveal an orderly relationship between emotional arousal and asthma and document the ability of behavioral and psychological strategies to reduce symptoms of asthma.

In this chapter, I examine the research relating emotion to asthma. I am concerned chiefly with emotion as a precipitant of asthma, not as a consequence of asthma. Following a brief discussion of the nature of asthma, I review research on the effects of both potentiation and attenuation of emotion on asthma. From a consideration of this work, I describe how variation in emotion can affect mechanisms of asthma and lead to variation in the disease.

I. The Nature of Asthma

A description of asthma is well beyond the scope of this chapter. A discussion of the fundamental aspects of asthma, however, is helpful in formulating hypotheses relating emotion to asthma. In general terms, the airways of asthma patients are more irritable than the airways of healthy individuals. Augmented airways irritability, called hyperreactivity or hyperresponsiveness, is exhibited by asthma patients in response to many stimuli, including chemical agents, viral infections, exercise, cold air, hyperventilation, allergens, and emotion. The response of patients to these stimuli is reversible. Patients often appear to be healthy between attacks of asthma, although evidence of disease usually can be found at the cellular level. Healthy individuals also exhibit airways responses to the stimuli that provoke asthma in patients, but the responses of healthy individuals are limited and usually are not a cause for concern.

An aspect of hyperresponsiveness that complicates the description of asthma is variability. Asthma patients exhibit several forms of variability in their reactions to airways irritants. First, asthma patients vary in the number of stimuli to which they respond: some patients respond to a large number of stimuli, and others to only a few. Second, asthma patients vary in the degree to which they respond to a stimulus. Variation in amplitude runs the gamut from mild responses, in which difficult breathing is minimally disruptive of other activities, to severe responses, in which the use of powerful drugs and, in the most serious cases, hospitalization is required. Variation in both range and amplitude of response occurs not only between individuals, but also within individuals. As a consequence of its variability, asthma defies clear-cut description, and a definition of asthma has been difficult to formulate (2,3).

The result of hyperresponsiveness is bronchoconstriction, a narrowing of the airways. Bronchoconstriction is due primarily to two events: inflammation and airway smooth muscle contraction. The significance of inflammation and smooth muscle contraction in asthma has been recognized for over a century (4), but their relative importance has been a matter of controversy. Until recently, the term bronchospasm was used almost synonymously with asthma (3). Now, greater emphasis is placed on the inflammation component (5). It is relevant to our discussion that the major interest in the question of how emotion affects asthma coincided with the period during which bronchospasm rather than inflammation was considered to be the primary contributor to bronchoconstriction. One additional feature of asthma that may help explain how it is affected by emotion is airways structural change associated with the disease. Structural change in asthma involves thickening of the walls of the airways. Although thickening does not augment hyperresponsiveness, it contributes to bronchoconstriction and, in so doing, interacts with the effects of inflammation and smooth muscle contraction to retard air flow. The consequences of these three elements of asthma—inflammation, airways smooth muscle contraction, and structural change—delineate the fundamental dimensions of the disease. They are described briefly in the following paragraphs.

Inflammation is a very complicated response, and it affects air flow in asthma patients in a number of ways. First, due to the congestion of blood and substances associated with the inflammatory response, the walls of the airways enlarge and expand into the airways, and thereby interfere with air flow. Second, mucus, the production of which is increased during inflammation, is secreted into the airways and causes a further reduction of space through which air may pass. Third, air flow may be completely eliminated in some airways by the formation of mucus plugs, a phenomenon brought about by a reduction in mucus hydration. Fourth, inflammation causes airways smooth muscle contraction and consequent airways narrowing via the action of several of its mediators including, among others, histamine and some leukotrienes (6). Thus, an interaction exists between inflammation and bronchospasm. A full understanding of the effects of inflammation on bronchoconstriction may reveal not only a clear picture of asthma, but also clues as to the causes of the disease.

Airways smooth muscle contraction, the second factor affecting bronchoconstriction, has been studied extensively. From normal physiology, it is known that airways caliber is dependent upon both cholinergic and adrenergic mechanisms (7). Bronchomotor tone is largely determined by cholinergic influences arising from parasympathetic vagal activity (8). Electrical stimulation of vagal efferents to the lungs results in the release of acetylcholine and immediate bronchoconstriction. The rapid onset of bronchoconstriction implicates smooth muscle contraction as its cause, rather than slower-acting humoral agents (9). There is ample evidence also that the vagus participates in a variety of neural reflexes, which result in bronchoconstriction. The stimuli initiating these reflexes include mechanical and chemical irritants applied to the upper respiratory tract (8). In addition, the bronchoconstrictive effects of exercise and cold air (10), and even emotion (11), may, to some extent, be due to vagal activity. The importance of cholinergic mechanisms in the control of airways smooth muscle tone is well established.

Adrenergic control of airways smooth muscle is no less important than cholinergic control. Both direct and indirect adrenergic mechanisms have been identified. Direct adrenergic control is due not so much to neurotransmitter release from sympathetic axons, but to the action of circulating catecholamines, the chief of which is epinephrine (12). The effects of epinephrine are mediated by receptors, which are located on the surface of effector cells in the airways. Several types of receptors have been implicated in the control of airways smooth muscle in humans: β_2, α_1, and α_2. Activation of alpha and beta receptors by epinephrine produces opposing effects of the airways: activation of beta receptors results in airways smooth muscle relaxation, but activation of alpha receptors results is muscle contraction (13). Overall, however, epinephrine dilates the airways, owing possibly to the greater density of beta as compared with alpha receptors in the airways (12). Indirect adrenergic control of airways caliber is the result both of modulation of acetylcholine release by norepinephrine and of stabilization of airways mast cells (12). The latter represents an effect of nervous system activity on the immune response.

There is also evidence of a third autonomic system that may participate in the regulation of airways smooth muscle: the nonadrenergic noncholinergic system (12). Activation of this system relaxes airways smooth muscle, but not via an adrenergic mechanism. The neurotransmitter involved has not been identified. It is possible that this system contributes in an important way to the regulation of airways smooth muscle, but at the present time little is known about it, especially insofar as human physiology is concerned.

Theories of asthma have been proposed that emphasize abnormalities of autonomic control of airways smooth muscle. The most well-established theories are concerned with augmentation of either cholinergic (14) or alpha-adrenergic (15) function or reduction of beta-adrenergic function (16). All these mechanisms shift the balance of airways smooth muscle control toward bronchospasm resulting in airways narrowing. Despite their appeal, recent years have witnessed a decline of interest in theories that focus on these mechanisms. A principal reason for this is the recent change in emphasis in the way asthma is viewed, from a disorder of airways smooth muscle control to an inflammatory disease. But there are more specific reasons for the de-emphasis of autonomic malfunction in asthma (17). First, abnormalities in autonomic control of airways smooth muscle are not limited to asthma: they are present in patients with other lung diseases including cystic fibrosis, emphysema, bronchitis, and some allergic rhinitis and atopic dermatitis patients. Therefore, asthma is not distinguished by autonomic abnormalities. Second, autonomic abnormality in asthma may be a result of the drugs used to control asthma. Third, the effect of inflammatory responses on airways smooth muscle tone make it difficult to demonstrate unequivocally that the airways in asthma are abnormal. At the very least, it may be said that the question of abnormality of autonomic control of airways smooth muscle in asthma remains unanswered (9,13).

In contrast to the uncertainty about the importance of autonomic abnormality in asthma, the significance of structural change is not debatable. As noted earlier, the structural change that affects asthma is thickening of airways walls. The thickening is due both to deposits of material associated with the inflammation response on the walls of the airways and to hyperplasia of airways smooth muscle. The increase in the amount of smooth muscle tissue is substantial, representing a doubling or tripling of the number of cells (18). Thickened airways walls reduce the caliber of the airways and are obviously related to air flow. But changes in the caliber of the airways are not linearly related to alterations in air flow: resistance to air flow in the airways is related inversely to the fourth power of the radius (3). Thus, the same degree of bronchoconstriction will result in vastly different reduction in air flow depending on the initial caliber of the airways. A stimulus that minimally affects resistance to air flow when the airways are open affects air flow profoundly when the airways are obstructed. In this way, thickening of airways walls in asthma potentiates increases to air flow resistance produced both by inflammation and by smooth muscle contraction.

II. Potential Mechanisms of Emotional Involvement in Asthma

The routes by which emotion may influence asthma are suggested by the abnormalities of the disease. Emotion may affect asthma via nervous, endocrine, respiratory, or immune system mechanisms. But the question of how emotion affects asthma is too general to be useful. A better question, one that is more specific, is how emotion affects airways caliber. One way to answer this question is to examine the overlap between the physiological correlates of emotion and the control of airways caliber.

Nowhere is the overlap between emotion and airways control mechanisms greater than it is in the autonomic and endocrine systems. Autonomic and endocrine involvement both in emotion and in control of airways smooth muscle is unquestioned. An abnormality of bronchial smooth muscle control may be exploited by emotion-related responses. Any such effect of emotion may be amplified further by the thickening of airways wall that takes place in asthma; as noted earlier, airways thickening increases the effect of smooth muscle contraction on air flow. It should come as no surprise then that interest in the effects of emotion on asthma peaked some time ago, when the role of airways smooth muscle contraction in asthma was dominant.

Emotion may also affect airways caliber via some aspect of the immune response. Control of airways caliber in this way involves both the nervous and immune systems. The evidence for this is less impressive than that concerning direct emotional control of airways smooth muscle. But several promising avenues of neural control of immune function have been suggested by animal studies. One of these concerns increased mast cell release of histamine by vagal stimulation (19). Presumably, histamine-mediated control of airways could be effected by vagal activity related to emotion. Other nervous system effects on immune function may be mediated by substance P and vasoactive intestinal peptide (20). The latter is a potential neurotransmitter candidate of the nonadrenergic noncholinergic system. The significance of these observations for human physiology and their applicability to the relationship of emotion to asthma is still ambiguous. Until a clear picture emerges, the importance of immune function control by emotion in asthma remains uncertain.

At present, the most obvious route by which emotion may affect asthma is via control of airways smooth muscle. This assumption underlies research on control of air flow by emotion, an area of investigation in which changes in smooth muscle tension are usually inferred from fairly rapid changes in air flow. In my review, I consider research relating both to asthma patients and to healthy individuals.

III. The Influence of Emotion on Asthma

In this section, I describe research showing that emotion affects asthma. The research I examined consisted of experiments in which the independent variable was an aspect of emotion and the dependent variable was a measure of asthma. Consistent with my emphasis on air flow, I limited my review to those studies that included air flow as a measure of asthma severity. This had both a positive and a negative consequence. On the positive side, air flow measures provided objective evidence of airways variation and thereby minimized both experimental error and interpretation. On the negative side, measures of air flow did not distinguish between airways changes and asthma. Although asthma always involves changes in air flow, not all air flow change is related to asthma. This point has not always been appreciated in research dealing with the relationship between emotion and asthma. As a consequence, some of the findings of research in this area have been misinterpreted.

For convenience, the studies in this area may be divided into those in which level of emotional arousal is increased, and those in which arousal is reduced. For lack of a better term, I used the term "stress" to designate the independent variable in studies in which emotion was aroused. In addition, I grouped the studies by manner in which stress was brought about. It should be noted that for the majority of studies examined in this section, the stress applied was of short duration. I also included a discussion of the effects of long duration stress on air flow. Little is known about this area, but it may prove to be important in the control of asthma. I used the term "relaxation" to describe experimental operations that involve a reduction in affect. As I did with the stress studies, I grouped the relaxation studies by method.

A. The Effects of Stress on Airflow

In the first half of the twentieth century, the idea that emotional disturbance caused asthma, at least in some patients, was held almost universally. The supporting evidence, however, was weak. In a 1953 review of the early literature, Leigh (21) concluded that the notion of psychological causation was interesting but speculative and lacked scientific foundation. Since that time, experimental evidence has accumulated that emotion can cause air flow changes, but there is still no evidence that emotion causes asthma.

Evidence that air flow change may be caused by emotion derives from studies of stress, a concept that may be confusing in the description of research. There are two potential sources of confusion: first, the meaning of stress varies between investigators, and second, the term itself fails to convey information about experimental methods. For these reasons, this review focuses not so much on stress, but on the ways in which stress was produced. In the studies reviewed, stress was aroused by aversive stimulation applied in one of three ways: anticipation of aversive stimulation, imagination of aversive stimulation, and application of aversive stimulation.

Anticipation of Aversive Stimulation

A series of inhalation studies was begun in 1968 on the so-called "suggestion" effect. The term suggestion was used to describe the studies because asthma patients who participated in the research were led to believe they would experience transient bronchial change as a result of their research participation. The expectation of respiratory difficulty was established by telling patients they would inhale aerosolized solutions containing allergens or respiratory irritants. In some cases, patients were asked to inhale a neutral substance but were told it was an allergen or an irritant. In other cases, bronchodilators or bronchoconstrictors were administered along with suggestions that were either consistent or inconsistent with the pharmacological effects of the substance administered. The results of these experiments established that verbal stimuli, called suggestions, had an effect on the airways. The effect was consistent with what patients were led to expect: suggestion of improved respiration was followed by bronchodilation, and suggestion of respiratory difficulty was followed by bronchoconstriction.

In the earliest study of suggestion, aerosolized saline was administered to individuals who were told it was either an allergen or a respiratory irritant (22). Forty asthma patients and a group of 40 subjects who were either bronchitis patients, restrictive airways disease patients, or healthy individuals participated in the research. Because the focus of the study was on asthma and not on suggestion, the nonasthma subjects were considered to be controls. The experimental procedure itself was not controlled; saline inhalation without suggestion was not conducted. Measurements of resistance to air flow were taken immediately prior to inhalation and at intervals of 1 and 4 minutes after inhalation. As compared to the baseline measurement, resistance to air flow increased in asthma patients after the inhalation task. Of the 40 asthma patient studied, 19 exhibited a response to suggestion plus saline. Air flow resistance following inhalation of saline also increased in the other groups, but the increases of individuals in these groups were not great and were not statistically reliable. Comparisons between the groups, however, were difficult to make because the groups differed from one another in several ways. One further aspect of this study was that the bronchoconstriction that developed in some patients was successfully reversed by inhalation of aerosolized saline that was labeled a bronchodilator.

Both the reliability and a potential mediator of the suggestion effect were examined in a second study of asthma patients (23). In this case, resistance to air flow was measured in 29 asthma patients both before and after a saline-inhalation task. As in earlier work, no provision was made to control the experimental procedure. The patients were told that the saline was either an allergen or an irritant. Fifteen of the 29 patients evidenced an increase in resistance to air flow after saline inhalation. The individuals who responded to the saline repeated the procedure during a second laboratory session held about a week later. Of the 15 patients who returned for a second testing, 13 exhibited higher air flow resistance after saline inhalation as compared to their baseline values, thereby documenting both the reliability of the suggestion effect and a degree of habituation of the response to repeated stimulation. In 5 patients who underwent extended testing, the suggestion effect was abolished by administration of atropine sulfate, a cholinergic blocker. This was evidence that the response to suggestion was mediated by a cholinergic mechanism.

Procedural improvements, concerning design issues, were made in subsequent studies of neutral substance inhalation. These were important in establishing the validity of the suggestion effect. Although there was no doubt that air flow changed in earlier suggestion studies, it was not clear that the change was due to suggestion. This was corrected in a study of 25 asthma patients who were given a suggestion procedure on one day and a similar control procedure on a second day (24). The use of the two procedures controlled for task administration but not for task order. Bronchoconstriction, measured as change in one-second forced expiratory volume (FEV_1), occurred in response to saline plus suggestion but not to saline alone. Other findings in this study included a reversal of the drop in FEV_1 by bronchodilation suggestion and elimination of the suggestion response by administration of ipratropium bromide. The latter provided further evidence of a cholinergic mediator of suggestion. Both task administration and task order were controlled in an inhalation study in which aerosolized saline was presented as either a neutral substance or a substance that could cause bronchoconstriction (25). Bronchoconstriction followed when the saline was given with the bronchoconstriction suggestion but not when saline was given with a neutral suggestion. Other research showed that suggestion could be used to abolish the effects of verbal instructions (26): suggestion-induced bronchoconstriction in 12 adult asthma patients was abolished by prior inhalation of a neutral aerosol described as a drug, but not by prior inhalation of a neutral aerosol described as a nonactive substance. Thus, fairly good evidence of the suggestion phenomenon emerged from studies involving inhalation of saline.

More elaborate studies of the suggestion effect have been concerned with responses to pharmacologically active substances. In general, these studies have shown that the effects of pharmacological agents may be modified by suggestion. Of particular interest are changes in the response to bronchoactive substances given in conjunction with the suggestion that breathing will become more difficult. The bronchoconstrictor, carbachol, was more effective in producing bronchoconstriction when described as a bronchoconstrictor than when described as a bronchodilator (27). And the bronchodilator, isoproterenol, was less effective in producing bronchodilation when described as a bronchoconstrictor than when described as a bronchodilator (27,28). The suggestion effect was even demonstrated in the absence of verbal stimuli. A flavored neutral substance previously paired with methacholine produced bronchoconstriction when administered alone (29). On the other hand, altered response to methacholine did not follow suggestion in one study that involved multiple inhalation maneuvers (30). It is possible that the repeated testing in the experiment caused habituation of the bronchial response (31).

Suggestion effects on air flow response have also been noted in healthy individuals, thereby raising the possibility that control of respiration by suggestion is not limited to asthma patients. In a study designed to examine air flow response in healthy females, bronchoconstrictive suggestion resulted in reduced air flow but neutral suggestion had no effect (32). Air flow, in this case, was measured by total respiratory resistance. As compared to earlier research, the inhalation procedure used in this study was simplified: bronchoconstrictive suggestion was not accompanied by aerosolized saline; each subject was told that a substance would be introduced into the apparatus that measured air flow at the time of her test. Air flow during suggestion was compared to air flow measured just prior to suggestion. The findings were confirmed by the results of two subsequent studies of healthy females (33,34). In both of these studies, all subjects exposed to bronchoconstrictive suggestion exhibited reduced air flow, a fairly strong demonstration of the reliability of the phenomenon. The air flow reductions of healthy individuals, however, were smaller than those reported for asthma patients in earlier research.

For the sake of completeness, it is worth noting that evidence of the suggestion effect was provided in several experiments designed primarily to investigate other phenomena (35,36). Such studies were not considered in detail owing to incomplete description of their suggestion procedures. In addition, the suggestion phenomenon has not been observed universally. Failure to observe the suggestion effect is often linked to the use of unusual methodologies. These include (1) the use of an atypical subject population, children with asthma who had been exposed to bronchial challenges repeatedly in earlier research (37), and (2) the use of atypical procedures such as paced respiration (38), failure to discontinue asthma medications (38), and inhalation trials without suggestion immediately prior to suggestion trials (39).

Imagination of Aversive Stimulation

The imagination of aversive events by itself may reduce air flow in asthma patients. Unlike the studies described in the previous section, the studies in this category do not involve an inhalation task—they rely only on an imagination task. Typically, investigators have taken an air flow measure both before and after the imagination task.

The most ambitious research concerning the effects of imagination on air flow was conducted on children with asthma (40). A total of 60 children, whose ages ranged between 10 and 15 years, participated in three fear- and three anger-imagination sessions. A measure of FEV_1, taken before and after imagination, revealed that air flow was reduced by the task. Imagination of travelog scenes did not result in an analogous FEV_1 reduction. Adult asthma patients exhibited a similar phenomenon. Three adult patients under hypnosis were asked to imagine cough, fear, anger, and asthma, either individually or in combination (41). Both the imagination of asthma and the simultaneous imagination of asthma, fear, anger, and cough resulted in air flow reduction as measured by FEV_1. Fear and anger alone produced FEV_1 decreases that were not statistically significant. The outcome of studies in which FEV_1 was used as a measure of air flow were confirmed by a study in which air flow resistance was the dependent variable: in two female asthma patients, airway resistance increased during suggested asthma but returned to normal during suggested relaxation (42).

There is also evidence that both healthy individuals and respiratory patients who do not have asthma exhibit air flow change in response to an imagination task. Imagination of asthma resulted in reduction of both FEV_1 and FEV_3 not only in asthma patients but also in healthy individuals (43). The reductions occurred primarily in those individuals with high hypnotic susceptibility. Imagination of aversive events caused bronchoconstriction in a patient with an esophageal spasm (44). As determined by bronchoscopy, the patient exhibited immediate narrowing of the bronchial lumen when asked to imagine an unpleasant situation in his life. These observations provide evidence that air flow response to imagination of aversive events may be a characteristic of the population and not limited to asthma patients.

Application of Aversive Stimulation

If either the anticipation or the imagination of aversive stimuli leads to air flow changes, it should come as no surprise that the application of aversive stimulation has a similar effect. Anecdotal evidence of this was provided some time ago by Dekker and Groen (45). They reported that asthma was unexpectedly elicited in patients by stimuli not usually associated with asthma. One female patient reported experiencing asthma when she read newspaper stories of cruelty or disaster, a second when she entered an elevator, a third when she heard loud noises, and a fourth when looking at a picture of an asthmatic child. What brought about breathing difficulty in these individuals is not known. The breathing changes were idiosyncratic, and they may have been related to earlier associations between asthma and the precipitating stimuli. Whatever the developmental history of the responses, it seems likely that the eliciting stimuli affected breathing by an emotional route. Because all the stimuli were, in one way or another, unpleasant, they may be considered examples of aversive stimuli, and their effect on breathing may be taken as early evidence that application of aversive stimulation affects air flow.

More systematic work concerning the effects of aversive stimuli on air flow is consistent with anecdotal findings. Research of the phenomenon is limited to two paradigms: exposure to visual material of a stressful nature and performance of mental arithmetic. Children with asthma exhibited decreases in air flow when they watched a motion picture depicting pediatric asthma patients during a severe asthma attack (46). Neither a shock avoidance task nor several auditory tasks elicited as much change in air flow as did the film of children experiencing asthma.

Adult asthma patients also responded to stressful films. Like their pediatric counterparts, adult patients evidenced a drop in air flow when they watched stressful films. Films depicting pediatric asthma attacks, industrial accidents, and maternal separation resulted in decreased air flow (47). In addition, a film showing burn accidents, burn treatment, and autopsy procedures resulted in decreased air flow in asymptomatic adult asthma patients (48). Even a simpler form of visual stimulation, still scenes depicting stressful interpersonal relationships, caused reduced respiratory function in adult asthma patients (49). The scenes shown were tailored to the interpersonal histories of individual patients.

Another form of aversive stimulation, performing mental arithmetic, reduces air flow in asthma patients and probably in healthy adults. Asthma patients were studied during mental arithmetic combined with several other stress-producing procedures (48). The mental arithmetic task consisted of paced subtraction of serial sevens with criticism of errors. This was a complex stimulus involving not only the stress of performing the arithmetic calculation, but also that of performance evaluation and reaction to criticism. In addition, the vocalization of the results of computations may have affected respiratory measures. A simpler evaluation of the effects of mental arithmetic on air flow was conducted exclusively with healthy individuals (50), who performed either an easy or a difficult mental arithmetic task. The easy task was subtraction of serial fives starting from a number ending in 5, and the difficult task was subtraction of serial seventeens starting from the same number. To avoid introducing the complicating effects of verbalization on the respiratory measure, the subjects were asked to report the result of their computation only at the end of the arithmetic period. The individuals exhibited decreases in air flow, as measured by total respiratory resistance, to difficult but not to easy mental arithmetic. One inconsistent finding, that air flow increased following mental arithmetic performance in healthy individuals, also has been reported (48). The outcome of the latter research, conducted within a hospital setting, may have been affected by the unfamiliarity of healthy individuals with hospital routines.

The Effect of Stress Over a Longer Period

Information about stress applied over a period of time was obtained from recent studies of the relationship between mood and air flow. Mood is of interest, in this context, because it shares elements with stress. The terms used to describe mood (e. g., tension, depression, anxiety) are similar to descriptors of stress. Under the assumption that mood affects air flow, Steptoe and Holmes (51) examined fluctuations in both quantities over an extended period. They correlated peak expiratory flow (PEF) with measures of mood state in both asthma patients and in healthy individuals daily for 24 days. They found that PEF correlated with mood state more frequently in asthma patients than in healthy individuals. But the relationship was not completely limited to asthma patients. And the specific nature of the relationship was idiosyncratic: although high PEF was associated most frequently with positive moods such as calmness and relaxation, it was associated in one asthma patient with anger and depression. Subsequent research (52) questioned the direction of causality of the PEF-mood relationship: it is just as plausible that air flow is affected by mood as it is that mood is affected by air flow. Some evidence that air flow affects mood was obtained from a study of one patient who in response to drug treatment experienced air flow improvement followed by an improvement in mood (53), but the data were not strong (54).

The work relating air flow to stress applied over the long term is limited. Not much work has been done up to the present time, and all of it is correlational; experimental studies have not been conducted. An additional limitation is that the period during which measurements are made—one day—is not long by some standards. Nor is it clear that mood is affected by air flow; causality may proceed in the opposite direction. Despite these obstacles, progress has begun in an area that may ultimately prove significant.

B. The Effects of Relaxation on Air Flow

Interest in psychological treatments for reduction of asthma symptoms has paralleled the course of research on the effects of stress on air flow. Psychological treatments have emphasized relaxation. Following clinical reports of several asthma patients successfully treated by relaxation procedures (55,56), relaxation research has grown to include systematic study of a wide variety of relaxation procedures. The research has demonstrated that relaxation improves air flow, but only modestly. Because of its modest therapeutic effect, the role of relaxation procedures in the management of asthma has been questioned (57). Some have gone so far as to call the findings unimportant because the air flow changes effected by relaxation and related procedures are not clinically significant (58,59). The concept of clinical significance, however, has little to do with an understanding of the relationships between behavioral and physiological quantities. Studies of relaxation are important for what they reveal about stress and asthma and about the physiological components of emotion.

Relaxation procedures vary substantially. This is because relaxation is not generally considered to be a procedure, but a subjective variable which may be affected by variety of techniques. If there is a predominant relaxation procedure used with asthma patients, it is progressive relaxation training (PRT). In this section, I separate studies that include PRT from those that do not.

Progressive Relaxation Training

Initially proposed by Jacobson (60) in 1938, PRT is one of the most widely used psychotherapeutic procedures. It has been applied extensively for the reduction of anxiety and stress. An abbreviated version of PRT is commonly used with asthma patients (61,62). The procedure involves directing the patient to alternately tense and relax each of the major muscle groups in a prescribed order. The patient is also instructed to attend to differences in subjective states associated with muscular tension and relaxation. Presumably, recognition of the subjective states serves to inform the patient when relaxation is achieved. Ultimately, the instruction to tense is withdrawn, and the patient is encouraged to relax. After a few sessions, typically three to six, the therapist-initiated prompts to relaxation are withdrawn. After this, patients relax on their own.

The effects of relaxation procedures on air flow are difficult to interpret. As an experimental procedure, relaxation may affect variables throughout a wide range. Relaxation affects not only reports of subjective experience, but also a variety of physiological activity. In PRT, there is the added complication of widespread muscle tension variation. If air flow change follows a relaxation procedure, the change may be mediated either by central nervous system factors relating to experience or by peripheral factors that may include both muscular and autonomic activity. Despite these complexities, the first step in the study of relaxation is to show that it does have an affect on air flow.

The effects of PRT on air flow were initially investigated in children with asthma (63). Air flow, expressed as peak expiratory flow rate (PEFR), was measured in 20 children before and after relaxation sessions and in 16 children before and after a control procedure consisting of sitting quietly. Six 20–minute sessions were administered. The children in the relaxation group exhibited PEFR improvement, but the children in the quiet sitting group did not. These findings were confirmed in a second study of relaxation in which children with asthma served as their own controls (64). In this case, the quiet sitting condition preceded the relaxation condition, and the number of sessions of each was reduced to five. The results, however, were consistent with the earlier findings: children evidenced PEFR improvement following relaxation training but not sitting quietly. Interestingly, the changes in PEFR were not related to variations in anxiety. Uncontrolled investigations of the effect of relaxation on air flow in adult asthma patients have yielded mixed results: in one study, PRT was followed by both an increase in PEFR and a decrease in anxiety (65); in a second experiment, PRT resulted in no change in air flow expressed as FEV_1 (66).

Subsequent study of PRT effects on air flow involved combining it with other procedures. Most of these studies documented a modest improvement in air flow following relaxation training. A combination of PRT and autogenic training, a procedure which involves imagination both of tranquil scenes and sensations of warmth, led to air flow improvement in one case (67) and arrested a decrease in air flow that was initiated during control procedures in another case (68). Combinations of PRT with either guided imagery (69) biofeedback (70), reciprocal inhibition (67), or systematic desensitization (71) led to air flow improvement in asthma patients. In some cases (67,70), the effect of combination treatment air flow was greater than the effect of PRT alone. A combination of PRT with biofeedback and systematic desensitization yielded no FEV_1 improvements but resulted in increased methacholine tolerance for treated subjects (72).

Other Forms of Relaxation

Unlike PRT, other forms of relaxation do not involve the routine of actively tensing and relaxing muscles. Nevertheless, relaxation of a more passive nature also affects air flow. Most of the relaxation work not involving PRT may be organized around two techniques: hypnosis and reciprocal inhibition.

The most extensive study of the effect of hypnosis on air flow succeeded in showing only that FEV_1 improvement was correlated with hypnosis (73). The study, which involved 252 asthma patients and lasted over an entire year, included a hypnosis control condition, which consisted of relaxation training. Patients in both groups improved slightly in FEV_1 during the course of the experiment. Better information about the effects of hypnosis was obtained in a study of children with asthma who demonstrated improvement in FEV_1 following quasi-hypnotic instructions to relax (40). A hypnosis effect also has been shown with resistance to air flow as the dependent variable instead of FEV_1 (42), and a combination of hypnosis and relaxation suggestion attenuated the FEV_1 drop that is usually associated with exercise in asthma patients (74).

Reciprocal inhibition has also been linked with increased air flow. The related observations are few, but they are consistent with one another. When applied by itself, reciprocal inhibition increased both FEV_1 and methacholine tolerance (75). When applied in conjunction with an imagination task, it increased FEV_1 (76) and PEFR (77). The imagination task itself was ineffective in changing air flow (76,77).

IV. Facial Muscle Tension: A Possible Mediator of Emotion-Induced Changes in Air Flow

Air flow change in response to emotion is well established. As outlined in the foregoing discussion, variation in air flow induced by stress and relaxation has been observed frequently under a variety of experimental conditions. The findings have consistently shown that air flow and emotional arousal are related. But the research has been relatively silent as to the mediators of the relationship. One reason for reluctance to address the mediation issue is that the effects of experimental conditions leading to change in stress and relaxation affect a large number of variables. Not only do stress and relaxation procedures lead to a change in emotional arousal, they also lead to changes in cognitive variables and changes in an array of physiological quantities. Any one of these quantities may effect change in air flow. As a consequence, the manner in which emotion controls air flow remains obscure. Resolution of the mediation issue may follow, however, from experimental techniques whose effects are more limited.

One experimental procedure whose effects may be limited is electromyographic (EMG) biofeedback. The technique involves the display of a signal to the subject that varies in accordance with changes in muscle tension. The subject is told how to interpret the signal and instructed as to the goals of the procedure. Interest in the use of EMG biofeedback for control of air flow developed as a result of successful application of the procedure in conjunction with PRT (78). In one of the earliest studies of the use of biofeedback to control symptoms of asthma (70), PRT plus EMG biofeedback of facial muscle tension produced a greater increase in air flow than did PRT alone. These findings generated interest in the role of facial muscle tension in the control of air flow.

A. Facial Muscle Tension and Air Flow

We investigated the effect of changes in facial muscle tension on air flow both in pediatric asthma patients and in healthy adults. In general, we found that facial muscle relaxation led to air flow improvement and that an increase in facial muscle tension resulted in decreased air flow. Although these findings were consistent with those of studies in which emotion was shown to control air flow, it was unlikely that our procedures affected emotion.

In studies of children with asthma, we limited training to facial relaxation. An initial study in our program (79) was designed to control for several sources of error. The procedure included giving children three series of laboratory sessions, each series consisting of two sessions of biofeedback training followed by a generalization session. Feedback during the generalization sessions was omitted to avoid the possibility that the feedback itself might elicit a systematic response. Muscle tension measurements, made during the generalization sessions, revealed a steady progression of muscular relaxation in children in the experimental group. A group of children in a control condition, treated like children in the experimental group but given noncontingent feedback instead of accurate feedback, exhibited no muscle tension change during the generalization sessions. Daily PEFR scores, taken daily prior to each meal by individuals who were blind as to group assignments, increased in experimental group children but not in control group children. The PEFR scores of children in the control condition were similar to those of children in a third group, a group that did not participate in biofeedback.

We obtained support for these findings in a study in which we took PEFR measures immediately before and after each of nine facial muscle training sessions (80). We took the pulmonary measurement before and after facial training sessions to minimize the likelihood that they might be influenced by factors not related to the experiment. We observed facial muscle relaxation in experimental group children but not in control group children given noncontingent feedback. Facial relaxation was related to air flow improvement: for children in the experimental group, PEFR scores taken immediately after training were higher than those taken just before training began; for children in the control group, PEFR scores were not affected by biofeedback.

The use of noncontingent control procedures in EMG biofeedback has been criticized. It is conceivable that subjects given false feedback are unsuitable biofeedback controls because they either lose their motivation to perform the biofeedback task or are frustrated by it (57). Evaluation of these possibilities, however, has shown that subjects who receive noncontingent feedback, even when prompted, do not verbalize awareness of their noncontingent status (81,82). But to be on the safe side, we designed a procedure to examine facial relaxation control of air flow in children with asthma that did not involve a noncontingent feedback (83). We used a control procedure for muscle relaxation training that involved rewarding a stable level of tension. Children received feedback signaling successful performance of the task so long as their tension level remained close to their initial level. The signal for successful accomplishment of the task was the same as that signaling success for children in the muscle relaxation condition. After a series of eight training sessions, we found evidence of clear differences in facial muscle tension between experimental and control group children (84). The children in the two groups differed also in pulmonary function scores: the children in the facial relaxation group exhibited higher ratios of FEV_1 to forced vital capacity (FEV_1/FVC) than did children in the control group. Over a 5-month follow-up period, the children in the facial relaxation group also exhibited better attitudes toward asthma than children in the control group.

Studies of facial muscle control of air flow in children with asthma are consistent with studies of healthy individuals. Like children with asthma, healthy adults trained to relax the facial musculature evidenced PEFR improvement (85). The amplitude of air flow change in healthy adults was stable between studies and comparable to that of children with asthma (80,85). But studies of air flow control by facial tension in healthy individuals are not limited to relaxation procedures. There is evidence as to the effects of increased facial tension on air flow in healthy adults. Air flow, measured as either PEFR (85, 86) or total respiratory resistance (87), decreased in response to increase in facial tension, suggesting that facial tension is related to air flow throughout a fairly broad range of values. The same cannot be said for tension in nonfacial muscles. In this case, no relationship between variation in nonfacial muscles and air flow was found: increases (85,86) and decreases (85) in forearm tension had no effect on air flow. These findings implied that air flow control by skeletal muscle is limited to the facial musculature.

Although EMG biofeedback studies showed that facial tension training affects air flow, they did not rule out other potential sources of air flow control. It could be argued that air flow, in biofeedback studies, was controlled not by facial tension but by another quantity that varied systematically as a function of biofeedback training. This is less likely with biofeedback training than with other forms of relaxation training, but it is still a possibility.

To address this argument, we evaluated a range of variables that could be affected by facial EMG biofeedback training and found little evidence that they change systematically. The variables we examined included tension in nonfacial muscles, respiration rate, heart rate, and cognitive quantities. We found that changes in facial tension brought about by biofeedback training did not generalize to other muscles of the body (88). To the contrary, we found that facial training reduced covariation between the facial muscles and other muscles. Our findings were consistent with predictions from conditioning theory: as training proceeds, the conditioned response becomes more specific (89). In addition, we found no evidence that facial muscle training affects cardiovascular response or respiration rate: biofeedback-induced increases in facial tension were not related to heart rate or respiration rate change (86). Finally, we found no evidence that facial muscle training affects subjective quantities such as relaxation, anxiety, frustration, or suspiciousness (81,82). From these observations, we concluded that the air flow changes we observed following facial muscle tension changes were not the result of changes in either the tension of nonfacial muscles, cardiovascular events, respiration rate, or cognitive events, but were due to variation in facial tension. In an important way, facial tension is related also to emotion.

B. Facial Muscle Tension and Emotion

Interest in the connection between facial tension and emotion is rooted in the relationship between emotion and facial expressions. Charles Darwin (90) was among the first to study facial expressions of emotion in a systematic manner. He argued that emotions were linked in a consistent way with specific facial expressions throughout the entire human population. Recent work has confirmed Darwin's early observations. In a cross-cultural study, expressions related to emotions such as happiness and fear were identified similarly by members of different cultures (91). Electromyographic recordings of facial muscle tension added an additional dimension to studies of facial expressions. In these studies (92–94), heightened emotional arousal was accompanied by increased facial tension. This line of evidence showed that facial expression or, more specifically, patterns of facial muscle tension are good indicators of emotion. But recently, the role of facial tension in emotion was reevaluated. Findings suggested that patterns of facial muscle tension may be an important factor in the generation of emotional responses. Specifically, muscle-by-muscle coaching to produce patterns of facial configurations similar to those occurring naturally during emotion elicited both the subjective experience and the autonomic activity typical of the associated emotion (95). This research supported the argument that specific changes in the facial musculature are antecedent to both the subjective and the autonomic components of emotion and led to the proposal that sensory feedback from the facial musculature is instrumental in the development of emotional responses (96,97). It should be noted, however, that facial tension variation may be a sufficient but not a necessary condition for the elicitation of emotion. It is likely that emotion arises not only as a result of variation in facial tension but also via the activation of other mechanisms.

C. Emotion and Air Flow

Evidence that facial tension controls both air flow and emotion suggests that it mediates the relationship between them. A model of emotion and air flow that includes facial tension would predict that emotion associated with an increase in facial tension reduces air flow and that emotion associated with facial relaxation increases air flow. It is important to note the model assumes only a consistent relationship between facial tension and bronchomotor tone. It does not assume a consistent relationship between emotion and bronchomotor tone. Not all emotional responses involve an overall change in facial tension; some may involve only a change in the pattern of tension. But it is likely that the majority of emotional responses are associated with either an increase or a decrease in the overall level of facial tension.

The relationship between facial tension and air flow may be mediated by a cholinergic mechanism. The evidence for this is indirect. It includes both the demonstration that bronchoconstriction in response to suggestion was blocked by administration of anticholinergic drugs (23,24) and the observation that the effects of stress on air flow developed rapidly (50). The latter finding seems to rule out involvement of slower systems, such as adrenergic and immune response mechanisms, in psychological control of air flow. Both adrenergic and immune control of airways rely on circulating substances whose effect on the airways is slower than that of neural factors associated with cholinergic control.

Additional elements controlling air flow may include trigeminal afferent and vagal efferent activity. Afferent activity from the facial muscles is transmitted centrally via the trigeminal (98) and efferent activity modulating bronchomotor tone is transmitted peripherally via the vagus (8). These two pathways are involved also in a number of reflexive mechanisms linking facial stimulation and bronchomotor tone (99). Considered in their entirety, these observations support a relationship between emotion and bronchomotor tone, which is mediated by facial muscle tension and associated cholinergic activity.

The mechanisms underlying the effect of emotion on bronchomotor tone apply equally to healthy individuals and asthma patients. Air flow change in response to emotion, however, has been reported to be more pronounced in asthma patients than it is in healthy individuals (22,47). This difference may be due to either a cholinergic abnormality, possibly a hypersensitivity, characteristic of atopic disease (17), or a structural change associated with asthma. Of the two explanations, the cholinergic abnormality is the more controversial (9). Structural changes, on the other hand, including airways thickening with its associated reduction of airways caliber has been well documented in asthma (18), and has been implicated in the control of air flow. As noted earlier, the geometry of the airways affects air flow (3): the same degree of bronchoconstriction will have different effects on resistance to air flow depending upon initial airways caliber. It is entirely possible that heightened air flow response to emotion in asthma patients as compared to healthy individuals is the result of reduced airways caliber caused by asthma-related thickening of airways walls.

V. Conclusion

There is no evidence that emotion is a precipitant of asthma. Nor is there convincing evidence that emotional response is abnormal in asthma, beyond the contribution of structural factors that might interact with emotionally induced air flow changes. However, it is clear that emotion may exacerbate symptoms of asthma and that reduction of emotional arousal reduces asthma symptoms. In view of these considerations, efforts to control emotion may pay dividends in the management of asthma. For this purpose, procedures that reduce stress and promote relaxation are recommended. Special consideration might be given to those procedures that bring facial tension under control. The latter are not limited to biofeedback training; other relaxation procedures, including PRT, are effective in achieving facial relaxation and have the advantage of not requiring the use of costly electronic feedback devices.

The air flow changes brought about by emotion are modest. But this should not minimize the importance of relaxation training as a way of keeping stress under control. Even a small improvement in air flow may interrupt the series of events leading to a full-blown asthma attack. In addition, relaxation procedures may be practiced at any time to effect an improvement in air flow. They may be especially important in situations in which other measures used to combat asthma are unavailable. One common situation of this sort is occurrence of asthma when a patient's medications are not readily available.

Finally, it should be noted that emotion has a role in asthma apart from precipitant: emotion may be a consequence of asthma. In particular, the emotional conditions of anxiety (100,101) and depression (102) have been associated with asthma. But in this, asthma may not be different from other disorders. It is widely recognized that physical disorders are associated to a great degree with psychological abnormality. In chronic disease, psychological factors may wax and wane in response to variations in physical symptoms (103). Comparisons of asthma patients and patients with other chronic diseases usually have failed to reveal distinct group differences in psychological quantities (104). Nevertheless, the fact that emotional factors might exacerbate any air flow limitation makes control of emotional consequences of asthma an important research question.

References

1. MacKenzie JN. The production of the so-called "rose cold" by means of an artificial rose. Am J Med Sci 1886; 91:45–57.
2. Creer TL, Kotses H. Asthma. In: Ollendick TH, Hersen M, eds. Handbook of Child Psychopathology. New York: Plenum, 1989:341–357.
3. Busse WW, Reed CE. Asthma: definition and pathogenesis. In: Middleton E, Reed CE, Ellis EF, Atkinson NF, Yunginger JW, eds. Allergy: Principles and Practice, 3rd ed. St. Louis: Mosby, 1988:969–998.
4. Osler W. The Principles and Practice of Medicine. New York: Appleton-Century-Crofts, 1892.
5. Reed CE. New therapeutic approaches in asthma. J Allergy Clin Immunol 1986; 77:537–543.

6. Barnes PJ, Rodger IW, Thomson NC. Pathogenesis of asthma. In: Barnes PJ, Rodger IW, Thomson NC, eds. Asthma: Basic Mechanisms and Clinical Management. London: Academic Press, 1988:415–444.
7. Szentivanyi A, Krzanowski JJ, Polson JB. The autonomic nervous system. In: Middleton E, Reed CE, Ellis EF, Atkinson NF, Yunginger JW, eds. Allergy: Principles and Practice, 3rd ed. St. Louis: Mosby, 1988:461–493.
8. Nadel JA. Autonomic regulation of airway smooth muscle. In: Nadel JA, ed. Physiology and Pharmacology of the Airways. New York: Marcel Dekker, 1980: 217–257.
9. Gross NJ. Cholinergic control. In: Barnes PJ, Rodger IW, Thomson NC, eds. Asthma: Basic Mechanisms and Clinical Management. London: Academic Press, 1988:381–393.
10. Boushey HA, Holtzman MJ, Sheller JR, Nadel JA. State of the art: bronchial hyperreactivity. Am Rev Respir Dis 1980; 121:384–413.
11. Kotses H, Glaus KD. Applications of biofeedback in the treatment of asthma: A critical review. Biofeedback Self-Regul 1981; 6:573–593.
12. Nadel JA, Barnes PJ. Autonomic regulation of the airways. Ann Rev Med 1984; 35:451–467.
13. Ind PW, Barnes PJ. Adrenergic control of the airways in asthma. In: Barnes PJ, Rodger IW, Thomson NC, eds. Asthma: Basic Mechanisms and Clinical Management. London: Academic Press, 1988:357–380.
14. Nadel JA, Barnes PJ, Holtzman MJ. Autonomic factors in hyperreactivity of airway smooth muscle. In: Macklem PT, Mead J, eds. Handbook of Physiology. Vol 3, part 2. Bethesda, MD: American Physiological Society, 1986:693–702.
15. Richardson JB, Bolande J. Nonadrenergic inhibitory nervous system in human airway. J Appl Physiol 1976; 41:764–771.
16. Szentivanyi A. The beta-adrenergic theory of atopic abnormality in bronchial asthma. J Allergy 1968; 42:203–232.
17. Lemanske RF, Kaliner MA. Autonomic nervous system abnormalities and asthma. Am Rev Respir Dis 1990; 141:S157–161.
18. Hogg JC. Normal and abnormal airway cell structure. In: Barnes PJ, Rodger IW, Thomson NC, eds. Asthma: Basic Mechanisms and Clinical Management. London: Academic Press, 1988:1–9.
19. Leff AR, Stimler NP, Munoz NM, Shioya T, Tallet J, Dame C. Augmentation of respiratory mast cell secretion of histamine caused by vagus nerve stimulation during antigen challenge. J Immunol 1986; 136:1066–1073.
20. Barnes PJ. Airway neuropeptides. In: Barnes PJ, Rodger IW, Thomson NC, eds. Asthma: Basic Mechanisms and Clinical Management. London: Academic Press, 1988:395–413.
21. Leigh D. Asthma and the psychiatrist. A critical review. Int Arch Allergy 1953; 4:227–246.
22. Luparello T, Lyons HA, Bleecker ER, McFadden ER. Influences of suggestion on airway reactivity in asthmatic subjects. Psychosom Med 1968; 30:819–825.
23. McFadden ER, Luparello T, Lyons HA, Bleecker E. The mechanism of action of suggestion in the induction of acute asthma attacks. Psychosom Med 1969; 31:134–143.

24. Neild JE, Caneron IR. Bronchoconstriction in response to suggestion: its prevention by an inhaled anticholinergic agent. Br Med J 1985; 290:674.
25. Horton DJ, Suda WL, Kinsman RA, Souhrada J, Spector SL. Bronchoconstrictive suggestion in asthma: a role for airways hyperreactivity and emotions. Am Rev Respir Dis 1978; 117:1029–1038.
26. Butler C, Steptoe A. Placebo responses: an experimental study of psychophysiological responses in asthmatic volunteers. Br J Clin Psychol 1986; 25:173–186.
27. Luparello T, Leist N, Lourie CH, Sweet P. The interaction of psychologic and pharmacologic agents on airway reactivity in asthmatic subjects. Psychosom Med 1970; 32: 509–513.
28. Strupp HH, Levenson RW, Manuck SB, Snell JD, Hinrichsen JJ, Boyd S. Effects of suggestion on total respiratory resistance in mild asthmatics. J Psychosom Res 1974; 18:337–346.
29. Spector S, Luparello T, Kopetzky MT, Souhrada J, Kinsman RA. Response of asthmatics to methacholine and suggestion. Am Rev Respir Dis 1976; 113: 43–50.
30. Pastorello EA, Codecasa LR, Gerosa A, Buonocore E. Sillano M, Zanussi C. The role of suggestion in asthma. II. Effects of a bronchoconstrictor drug on bronchial reactivity under bronchoconstrictor or bronchodilator suggestion. Ann Allergy 1987; 59:339–340.
31. Miller DJ, Kotses H. Habituation of total respiratory resistance responses. Psychol Rep 1990; 67:1139–1145.
32. Kotses, H, Rawson JC, Wigal JK, Creer TL. Respiratory airway changes in response to suggestion in normal individuals. Psychosom Med 1987; 49:536–541.
33. Wigal JK, Kotses H, Rawson JC, Creer TL. The effects of suggestion on the total respiratory resistance of nonasthmatic female subjects. J Psychosom Res 1988; 32:409–416.
34. Wigal JK, Kotses H, Creer TL, Stout C. The effects of suggestion on total respiratory resistance of high- and low-suggestible females. Psychophysiology 1991; S61.
35. Kahn A. The effects of biofeedback counterconditioning in the control of bronchial asthma. J Psychosom Res 1977; 21:97–104.
36. Khan, A, Staerk M, Bonk C. Hypnotic suggestibility compared with other methods of isolating emotionally prone asthmatic children. Am J Clin Hypn 1974; 17:50–53.
37. Weiss JH, Martin C, Riley J. Effects of suggestion on respiration in asthmatic children. Psychosom Med 1976; 32:409–415.
38. Lewis RA, Lewis MN, Tattersfield AE. Asthma induced by suggestion: Is it due to airways cooling? Am Rev Respir Dis 1984; 129:691–695.
39. Pastorello EA, Codecasa LR, Pravettoni V, Zara C, Incorvaia C, Froldi M, Zanussi C. The role of suggestion in asthma. I. Effects of inactive solution on bronchial reactivity under bronchoconstrictor or bronchodilator suggestion. Ann Allergy 1987; 59:336–338.
40. Tal A, Miklich DP. Emotionally-induced decreases in pulmonary flow rates in asthmatic children. Psychosom Med 1976; 38:190–200.
41. Clarke PS. Effect of emotion and cough on airway obstruction in asthma. Med J Aust 1970; 535–537.

42. Smith MM, Colebatch HJ, Clarke PS. Increases and decreases in pulmonary resistance with hypnotic suggestion in asthma. Am Rev Respir Dis 1970; 102: 236–242.
43. Thorn DE, Fisher AG. Hypnotically suggested asthma. Int J Clin Exp Hypnosis 1978; 26:92–103.
44. Faulkner WB. Influence of suggestion on size of bronchial lumen: bronchoscopic study and report of one case. Northwest Med 1941; 40;367–368.
45. Dekker E, Groen J. Reproducible psychogenic attacks of asthma. J Psychosom Res 1956; 1:58–67.
46. Weiss JH, Lyness J, Molk L, Riley J. Induced respiratory changes in asthmatic children. J Psychosom Res 1976; 20:115–123.
47. Levenson, RW. Effects of thematically relevant and general stressors on specificity of responding in asthmatic and non-asthmatic subjects. Psychosom Med 1979; 41:28–39.
48. Mathé AA, Knapp PH. Emotional and adrenal reactions to stress in bronchial asthma. Psychosom Med 1971; 33:323–340.
49. Singer EM. The effect of personalized emotional stimuli on asthmatic reactions. Dissert Abstr Int 1975; 35:4664–B.
50. Kotses H, Westlund R, Creer TL. Performing mental arithmetic increases total respiratory resistance in individuals with normal respiration. Psychophysiol 1987; 24:678–682.
51. Steptoe A, Holmes R. Mood and pulmonary function in adult asthmatics: a pilot self-monitoring study. Br J Med Psychol 1985; 58:87–94.
52. Hyland ME. The mood-peak flow relationship in adult asthmatics: a pilot study of individual differences and direction of causality. Br J Med Psychol 1990; 63:379–384.
53. Browne S, Hyland ME. Mood and peak flow in asthma. Lancet 1992; 339:118–119.
54. Geddes DM. Mood and peak flow in asthma. Lancet 1992; 339: 302.
55. Cooper AJ. A case of bronchial asthma treated by behavioral therapy. Behav Res Therapy 1964; 1:351–356.
56. Rathus SA. Motoric, autonomic, and cognitive reciprocal inhibition in a case of hysterical bronchial asthma. Adolescence 1973; 8:29–32.
57. Alexander AB. Behavioral approaches in the treatment of bronchial asthma. In: Prokop CK, Bradley LA, eds. Medical Psychology: Contributions to Behavioral Medicine. New York: Academic Press 1981:373–394.
58. King NJ. The behavioral management of asthma and asthma-related problems in children: a critical review of the literature. J Behav Med 1980; 3:169–189.
59. Knapp TJ, Wells LA. Behavior therapy for asthma: a review. Behav Res Therapy 1978; 16:103–115.
60. Jacobson E. Progressive Relaxation. Chicago: University of Chicago Press, 1938.
61. Paul GL. Insight versus desensitization in psychotherapy: an experiment in anxiety reduction. Stanford, CA: Stanford University Press, 1966.
62. Alexander AB. Chronic asthma. In: Williams RB, Gentry WD, eds. Behavioral Approaches to Medical Treatment. Cambridge: Ballanger, 1977

63. Alexander AB, Miklich DR, Hershkoff H. The immediate effects of systematic relaxation training on peak expiratory flow rates in asthmatic children. Psychosom Med 1972; 34:388–394.
64. Alexander AB. Systematic relaxation and flow rates in asthmatic children: relationship to emotional precipitants and anxiety. J Psychosom Res 1972; 16:405–410.
65. Freedberg PD, Hoffman LA, Light WC, Kreps MK. Effects of progressive muscle relaxation on the objective symptoms and subjective responses associated with asthma. Heart Lung 1987; 16:24–30.
66. Erskine J, Schonell M. Relaxation therapy in bronchial asthma. J Psychosom Res 1979; 23:131–139.
67. Moore N. Behaviour therapy in bronchial asthma: a controlled study. J Psychosom Res 1965; 9:257–267.
68. Alexander AB, Cropp GJA, Chai H. Effects of relaxation training on pulmonary mechanics in children with asthma. J Appl Behav Anal 1979; 12:27–35.
69. Hock RA, Rodgers CH, Reddi C, Kennard DW. Asthmatic children: an evaluation of physiological change. Psychosom Med 1978; 40:210–215.
70. Davis MH, Saunders DR, Creer TL, Chai H. Relaxation training facilitated by biofeedback apparatus as a supplemental treatment in bronchial asthma. Psychosom Res 1973; 17:121–128.
71. Miklich DR, Renne CM, Creer TL, Alexander AB, Chai H, Davis MH, Hoffman A, Danker-Brown P. The clinical utility of behavior therapy as an adjunctive treatment for asthma. J Allergy Clin Immunol 1977; 60:285–294.
72. Lehrer PM, Hochron SM, McCann B, Swartzman L, Reba P. Relaxation decreases large-airway but not small-airway asthma. J Psychosom Res 1986; 30:13–25.
73. Meyer-Loughnan GP, Kinsley BJ. Hypnosis for asthma—a controlled trail. Br Med J 1968; 4:71–76.
74. Ben-Zvi Z, Spohn WA, Young SH, Kattan, M. Hypnosis for exercise-induced asthma. Am Rev Respir Dis 1982; 125:392–395.
75. Philipp RL, Wilde GJS, Day JH. Suggestion and relaxation in asthmatics. J Psychosom Res 1972; 16:193–204.
76. Yorkston NJ, McHugh RB, Brady R, Serber M, Sergeant HGS. Verbal desensitization in bronchial asthma. J Psychosom Res 1974; 18: 371–376.
77. Yorkston NJ, Eckert E, McHugh RB, Philander DA, Blumenthal MN. Bronchial asthma: improved lung function after behavior modification. Psychosomatics 1979; 20:325–331.
78. Scherr MS, Crawford PL, Sergent CB, Scherr CA. Effects of bio-feedback techniques on chronic asthma in a summer camp environment. Ann Allegy 1975; 35:289–295.
79. Kotses H, Glaus KD, Crawford PL, Edwards JE, Scherr MS. Operant reduction of frontalis EMG activitiy in the treatment of asthma in children. J Psychosom Res 1976; 20:453–459.
80. Kotses H, Glaus KD, Bricel SK, Edwards JE, Crawford PL. Operant muscular relaxation and peak expiratory flow rate in asthmatic children. J Psychosom Res 1978; 22:17–23.

81. Kotses H, Segreto-Bures J. Subject expectancy effects in frontal EMG conditioning. Biol Psychol 1983; 17:97–104.
82. Segreto-Bures J, Kotses H. Experimenter expectancy effects in frontal EMG conditioning. Psychophysiol 1982; 19:467–471.
83. Harver A, Segreto J, Kotses H. EMG stability as a biofeedback control. Biofeedback Self-Regul 1992; 17159–164.
84. Kotses H, Harver A, Segreto J, Glaus KD, Creer TL, Young GA. Long-term effects of biofeedback-induced facial relaxation on measures of sthma severity in children. Biofeedback and Self-Regulation 1991; 16:1–21.
85. Glaus KD, Kotses H. Facial muscle tension influences lung airway resistance; limb muscles do not. Biol Psychol 1983; 17:105–120.
86. Harver A, Kotses H. Pulmonary changes induced by frontal EMG training. Biol Psychol 1984; 18:3–10.
87. Kotses H, Miller DJ. The effects of changes in facial muscle tension on respiratory resistance. Biol Psychol 1987; 25:211–219.
88. Glaus KD, Kotses H. Generalization of conditioned muscle tension: a closer look. Psychophysiol 1979; 16: 513–519.
89. Kotses, H Glaus KD. Generalization of conditioned muscle tension: sharpening the focus. Psychophysiol 1982; 19:498–500.
90. Darwin C. The Expression of the Emotions in Man and Animals. New York: Appleton, 1873.
91. Ekman P, Sorenson ER, Friesen WV. Pan-cultural elements in displays of emotion. Science1969; 164: 86–88.
92. Schwartz GE, Ahern L, Brown SL. Lateralized facial muscle response to positive and negative emotional stimuli. Psychophysiol 1979; 16:561–571.
93. Schwartz GE, Brown SL, Ahern L. Facial muscle patterning and subjective experience during affective imagery: sex differences. Psychophysiology 1980; 17:75–82.
94. Schwartz GE, Fair PL, Salt P, Mandel MR, Klerman GL. Facial muscle patterning to affective imagery in depressed and nondepressed subjects. Science 1976; 192;489–491.
95. Levenson RW, Ekman P, Friesen WV. Voluntary facial action generates emotion-specific nervous system activity. Psychophysiology 1990; 37:363–384.
96. Izard CE. Differential emotions theory and the facial feedback hypothesis of emotion activation: Comments on Tourangeau and Ellsworth's "The role of facial responses in the experience of emotions." J Person Soc Psychol 1981; 40:350–354.
97. Tompkins SS. Affect theory. In: Ekman P, ed. Emotion in the Human Face. New York: Cambridge University Press, 1982:353–359.
98. Carpenter MB, Suten J. Human Neuroanatomy. 8th ed. Baltimore: Williams and Wilkins, 1983.
99. Tomori Z, Widdicombe JG. Muscular, bronchomotor, and cardiovascular reflexes elicited by mechanical stimulation of the respiratory tract. J Physiol 1969; 20:25–49.
100. Butz AM, Alexander C. Anxiety in children withasthma. J Asthma 1993; 30:199–209.

101. Janson C, Björnsson E, Hetta J, Bowman G. Anxiety and depression in relation to respiratory symptoms and asthma. Am J Respir Crit Care Med 1994; 149:930–934.
102. Janson-Bjerklie S, Ferketich S, Benner P, Becker G. Clinical markers of asthma severity and risk: importance of subjective as well as objective factors. Heart Lung1992; 21:265–272.
103. Gift A. Psychologic and physiologic aspects of acute dyspnea in asthmatics. Nurs Res 1991; 40:196–199.
104. Siegel WM, Golden NH, Gough JW, Lashley MS, Sacker IM. Depression, self-esteem, and life events in adolescents with chronic diseases. J Adolesc Health Care 1990; 11:501–504.

4

Psychological Correlates of Asthma Severity and Treatment Outcome in Children

BRUCE G. BENDER and MARY D. KLINNERT

National Jewish Medical and Research Center
Denver, Colorado

I. Overview: The Psychological Impact of Asthma

An estimated 12 million Americans, including 4 million children, have asthma. The impact of this chronic disease is large. Annually, asthma accounts for 479,000 hospitalizations, 1.2 million emergency room visits, 15 million physician visits, and, for children, 10 million missed school days (1). Asthma-related deaths have risen since 1982, surpassing 5000 per year, with rates significantly higher in African Americans (2). The dollar costs of asthma in 1990 exceeded $6.2 billion, including $3.6 billion direct costs (1).

The psychological, social, and educational consequences of this disease are likewise significant (3,4). As with other chronic conditions, asthma presents a financial burden to the family, restricts the child's physical activities and development, impairs the child's development of social connections and adaptive resources, and contributes to general disruption in the family (5). Still, the psychological ramifications of chronic childhood asthma are neither obvious nor universal. A simple, linear relationship between disease severity and psychological disorder has not been established. Many asthmatic children have no psychological difficulties. Further, a single causal direction does not exist in the relationship between disease and psychopathology; children with psychological disorder, or from dysfunctional families, may have more severe, poorly controlled asthma largely because of poor management of the illness and, in many cases, blatant nonadherence. Poverty and other life stressors further undermine the medical and psychological stability of children with high-risk asthma.

The relationship between asthma and psychological disorder in pediatric patients is examined in this chapter. The frequency and severity of psychopathology are reviewed. The relationship between severity of medical and psychological disorders is explored. The psychosocial factors that contribute to poorly controlled disease, including individual and family dysfunction, are assessed. We propose that children with severe, poorly controlled asthma make up a subgroup of patients with this chronic illness for whom psychopathology and family dysfunction are significantly increased, compliance is poor, risk of medication side effects is high, use of the health care system is inappropriate and costly, and risk of complication including death is high. Intervention with this group should therefore be individualized, intensive, and multidisciplinary and must include an innovative and comprehensive psychosocial component.

II. Psychological Functioning in the Population of Asthmatic Children

Children with asthma are generally believed to have more psychological problems than healthy children. The overall picture of the psychological status of asthmatic children, however, is complex. The determination of whether asthmatic children have more psychological problems than other children depends on the characteristics of the asthmatic children studied, the informant reporting about the children's psychological functioning, the measures of psychological problems used, and the nature of the group with which asthmatic children's psychological functioning is compared. When these complexities are sorted out, it is concluded that only children with severe, poorly managed asthma demonstrate a significant increase in psychological disorders.

The prevalence of psychopathology among asthmatic children was investigated a quarter of a century ago in two classic epidemiological studies. In a study on the Isle of Wright, Graham et al. (6) concluded that the amount of psychiatric disorder found in 9- to 11-year-old asthmatic children was a "trivial and statistically insignificant" increase over that found in the rest of the children studied. In another study in Australia, McNicol et al. (7) concluded that 14-year-old asthmatics as a group did not show more behavioral or emotional disturbances than nonasthmatic children. However, both investigators documented increased behavioral disturbances among the children with severe, continuing asthma.

In contrast to the earlier epidemiological studies, a number of recent studies have found increased levels of behavior problems among asthmatic children. Some investigators have compared children with asthma to nonasthmatic children (5,8), and others have compared them to normative samples (9,10). Many of those studies have employed the Child Behavior Checklist (CBCL) (11), a standardized questionnaire regarding behavior problems, which is filled out by the parent, usually the mother. Studies using the CBCL have frequently found that asthmatic children were elevated on the broad-band internalizing scale, but not on the externalizing scale (5,9), although in one study elevations on both scales were found (8). High internalizing scores are indicative of anxious, depressed, and overcontrolled behavior, whereas high externalizing scores reflect acting-out, aggressive, undercontrolled behavior. Findings have been mixed regarding parent reports of social competence among asthmatic children, with some studies finding that children with asthma have relatively poor social competence (9), and others finding no difference between asthmatic and control children on social competence or on self-concept (8).

Based on maternal reports, then, children with asthma are more likely to have problems with anxiety, depression, inhibition, and perhaps social competence. However, mothers' reports provide only one perspective on their children's adjustment. Mothers' CBCL responses have only moderate agreement with responses made by the children's fathers, teachers, or clinicians (12). Several studies using reports about child functioning from informants other than the parents, either from teachers or directly from the child, are less consistent regarding increased levels of psychopathology in asthmatic children. For example, Graham et al. (6) obtained teacher reports, in addition to parent reports, about the children's behavior problems. Although the parents of the asthmatic children had reported higher levels of behavior problems than did the parents of healthy children, the teachers of the asthmatic children did not observe more problems among the asthmatic as compared to the nonasthmatic children. In a more recent investigation, Kashani et al. (8) included a structured child psychiatric interview to directly assess 7- to 16-year-old asthmatic children. Again, although the parents' questionnaire responses indicated more problems among the children with asthma, the child interviews did not show increased problems. The child measure revealed that number of symptoms was not significantly increased among the asthmatic children when compared to a group of matched controls, although there was a nonsignificant trend toward more psychiatric diagnoses among the asthmatic children. The parent respondent version of this interview indicated significantly more symptoms, mostly of the overanxious type, reported for the children with asthma than for the controls. However, with this instrument even the parent report data did not reveal differences between the asthmatic children and the controls in number or type of diagnoses.

In a study of 6- to 7-year-old asthmatic children, Klinnert et al. (13) found that parents' reports of child problems contrasted with child interview and behavioral observation reports. This study included asthmatic children with a range of illness severity as well as a group of healthy controls. The CBCL was completed by the children's mothers, and the children were interviewed using the Semi-Structured Clinical Interview for Children (SCIC) (14). In addition, the children's behavior was rated, both alone and during interactions with their mothers. Using the CBCL, the mothers of the asthmatic children reported higher levels of total problem behavior as well as higher scores on the internalizing scale than the mothers of the healthy controls reported for their children. In contrast, the child interviews did not reveal differences between the asthmatic and nonasthmatic children in observable psychopathology. The behavioral observation ratings also showed no differences in child negativity or in the children's ability to regulate affect, nor were there differences in the quality of the parent-child interaction. This pattern of findings is similar to those of a number of other investigators, who have found that while mothers of asthmatics report increased problems in their children, when other methods of assessment are used, no differences are found between asthmatic and healthy children.

There are a number of possible reasons for the contradictions among these findings. First, some of the results may vary with the methodological approach. Some of the studies that have found elevated problem scores among asthmatic children compared the scores to normative samples (9,10). These investigations assessed asthmatic children only, and compared their scores to those of a large group of children previously assessed during the development of the measure. It is possible that the samples were not appropriately matched on relevant variables such as socioeconomic status, race, family configuration, or geographic location. However, studies that included healthy controls have also found differences in maternal reports of problems, particularly internalizing ones (5,8,13).

A second reason for the differences between maternal reports of increased psychological problems, particularly internalizing ones, and other methods of assessing child adjustment may have to do with the level of disorder being reported by the mothers. In most studies that have found more problems among asthmatic than nonasthmatic children, the mean differences reported are small, even though asthmatic means are significantly different from control group or normative means. Although the asthmatic groups invariably are at least slightly higher than comparison groups, the difference is usually within 5 T-score points of the comparison group mean (standardized T-scores have a mean of 50, and a standard deviation of 10). Clinical problematic cut-offs are often judged to be 1.5 or 2 standard deviations above the mean (scores of 65 or 70). Thus, the group mean increases, while consistent, may not be clinically significant. The minimal elevation among asthmatics in most of the studies led Hamlett et al. (5) to remark that the "levels and patterns of behavior problems were not in the clinical range. . . . They may be interpreted as reflecting problems of adaptation or adjustment."

It is also possible that the elevated CBCL scores primarily reflect the maternal perception that their asthmatic children have more problems, particularly anxiety and depression. Mothers' reports of increased problems may be related to their expectations that their asthmatic children will have problems or to their own emotional status. Mothers of asthmatics have been found to have lower perceived support than mothers of healthy children (15), and Hamlett et al. (5) found that the adequacy of the mothers' social support was a significant predictor for the Internalizing scores of asthmatic children. It is also possible that the higher scores reflect difficulties in the parent-child relationship that are not captured in the observations of the interactions or the mothers' more intimate knowledge of real physical and emotional discomfort or social problems that are experienced by their children.

With all of these methodological issues taken into account, it is concluded that asthmatic children as a group are not more psychologically disturbed than other children. While there is some evidence that children with asthma have slight increases in adaptational problems, evidence for increased psychopathology is weak, resting on small mean increases on dimensional scales filled out by parents. Information provided by individuals other than parents is not consistent with the parent-report information, suggesting instead lower levels of problems than indicated by parent reports. The small increases in mean scores for the entire group of children with asthma may be due to greater psychological problems among a subgroup of asthmatics with severe, poorly controlled disease. As discussed below, psychological problems are increased among those children with severe disease. It follows that the average behavior problem score for any group of children with asthma would be related to the proportion of that group with severe asthma. In evaluating the group mean scores for behavior problems, it would be important to know how many of the children were classified as severely asthmatic, and on what basis. Psychological functioning in asthmatic children must necessarily be considered in light of the severity of the disease.

III. The Relationship Between Severity of Asthma and Psychological Adjustment

Many studies examining psychological functioning in asthmatic children have attempted to determine whether such problems are increased among children with more severe asthma. In their early study, Graham et al. (6) looked at the psychological adjustment of the children in relation to the severity of their asthma. Asthma severity was measured by obtaining parent and teacher reports of school days missed and restricted activity. These investigators concluded that there were more problems among the severely affected children, even though there was not an increase in adjustment problems among asthmatic children in general. McNicol et al. (7) also found that behavioral problems were significantly higher only in the small group of children with severe and continuing asthma. To clarify the relationship between asthma severity and psychological problems, Norrish et al. (16) assessed asthma severity using two components: the level of medications required for controlling asthma symptoms and the extent to which the asthma symptoms were controlled. They found that psychological problems were unrelated to required medication levels, but they were related to poor control of the asthma.

More recent investigations of the relationship between children's asthma severity and psychological functioning have produced mixed results, with some studies finding severity level related to psychological impairment and some finding little or no relationship. Perrin et al. (17) studied psychological functioning in a group of 5- to 16-year-old asthmatic children using an "objective composite" measure of severity based on medication level, frequency of attacks, and days missed from school. In addition, maternal ratings of the children's asthma severity (mild, moderate, severe) were obtained. Mothers used the Health Resources Inventory to rate their children's social competence and resilience in meeting new tasks or challenges. The results indicated a nonlinear relationship between the mothers' ratings of their children's asthma severity and their adaptation level. The children in the mild and severe asthma groups received lower adaptation scores, while the children with moderate asthma received higher scores, suggesting a U-shaped relationship between asthma severity and psychological functioning. In contrast to the severity ratings made by the mothers, the objective composite measure did not reveal differences in adaptation among the three severity levels. However, when one component of the objective measure, medication level, was utilized to grade severity, the severely affected group had poorer social competence and resilience than the mild and moderate groups.

MacLean and colleagues (9) assessed illness severity in a group of 6- to 14-year-old asthmatic children using the Perrin et al. (17) composite scale (medication level, attack frequency, and missed school days). The mothers completed the CBCL, and the children reported on stressful life events they had experienced. The severely asthmatic children received significantly higher overall problem behavior scores compared to those with moderate asthma, and they received lower social competence scores than the remainder of the children. Importantly, when negative life events and illness severity were entered into a regression equation, asthma severity did not add to predictability of behavior problems over and above negative life events. In another study of the relationship between disease severity and psychological disturbance, the mean Behavior Problem Index scores were significantly higher among children with severe asthma (7.43) when contrasted with mildly asthmatic (5.40) or unaffected (5.43) children (18). Severe asthma was again defined in terms of functional control, in this case as the presence of activity limitations.

In the Klinnert et al. (4) study, the severity of the children's asthma was graded on a continuous scale, which combined medication level and symptom control. Within the group of asthmatic children, both maternal report of behavior problems and clinician ratings of psychological symptoms were significantly related to asthma severity. Maternal reports showed a linear increase in problems on both the internalizing and externalizing scales of the CBCL, with number of reported problems increasing with asthma severity. The scores from the clinician-rated SCIC interviews showed the same linear increase with increasing severity. However, when the continuous scale was divided into three levels reflecting mild, moderate, and severe asthma, posthoc comparisons showed that the significant differences were between the children with severe asthma and the remainder of the group.

The Kashani et al. (8) group also examined the relationship between asthma severity and psychological functioning. Following Norrish et al. (16), these investigators measured severity in terms of medication level only. They found no relationship between asthma severity measured in this way and the children's psychological functioning, whether measured by parent report or by the structured child interviews.

At first glance, the contradictions in the findings from these studies are puzzling. Some studies find that psychopathology is increased only among the most severe asthmatics, some find a linear relationship between asthma severity and psychological disturbance, some find a U-shaped relationship, and some find no relationship at all. However, a close examination of the methodology of the studies reveals that the method by which asthma severity was measured had an important influence on the distribution of psychopathology found. Asthma is complex, and the severity of the condition for an individual is determined by a number of components. At one time or another, each of these components has been used as a measure of asthma severity. Frequently measured components of severity are medication levels, poor control of symptoms, and measures of daily functioning such as school absence. An examination of each of these components reveals a probable relationship with psychological disturbance.

The use of a functional status estimate as part of the measurement of asthma severity can be confounded when the focus of investigation is adaptive functioning. The inclusion of functional variables such as school absence in asthma severity ratings may be circular and lead to inflated estimates of relationships between severity and psychopathology. School absence, for example, may be as much an indicator of how the asthma is managed by the family as of the severity of the asthma itself. It will be recalled that when Graham et al. (6) found a tendency to more psychopathology among severely asthmatic children, severity was defined in terms of missed school and restricted activities. Klinnert et al. (13) found that family functioning early in a child's life was not predictive of asthma severity by age six when measured by medication level and symptom control, but was predictive of functional status as measured by days of restricted activity and school absence. Similarly, work by Stein and Jessop (19) showed that psychological adjustment among chronically ill children, half of whom had respiratory conditions, was related to school absence, although it was not related to morbidity measures such as days in bed or hospitalizations.

The use of medication level as an index of asthma severity has a long history (7). The medication level required by asthmatic children has the appearance of being an "objective" measure of asthma severity. The use of this measure is based on the assumption that children with asthma take their medication, and thus the level of medication required for control of symptoms indicates the severity of the disease. This assumption is flawed, however, because of the pervasiveness of medication nonadherence (20–22). The fact that many prescribed medications are not taken, leading to poorly controlled asthma, has an influence on the level of medication that is subsequently prescribed. For example, the regular use of oral corticosteroids is usually considered a marker of severe asthma. However, it is now known that the use of oral corticosteroids may result from poor adherence with other preventive medications (23). To the extent that nonadherence is associated with psychological disturbance, the use of oral corticosteroids may be associated with poor adaptation and adjustment problems. To further complicate the issue, the development of newer medications means investigators must adjust the medication level measure of asthma severity. Such adjustments may change the constituency of the subgroup labeled severe. Kashani et al. (8), who found no relationship between psychological disturbance and asthma severity, defined their severe group as children taking corticosteroid medications, most of which were inhaled. With this questionable definition, the severe group became the largest in the study (mild = 17, moderate = 15, severe = 24), and the resulting proportion of severe to moderate and mild asthmatics contrasted sharply with the usual estimate of 5% severe asthmatics among the asthmatic population. Additionally, a group of children on inhaled corticosteroids likely has a very different set of characteristics when compared to a group of children who utilize regular oral steroids. Considering these factors, then, it is not surprising that this study found no relationship between asthma severity, as the investigators defined it, and psychological dysfunction.

The use of symptom control as a measure of asthma severity has many of the difficulties described for the use of medication level as a measure. While frequency and severity of wheezing attacks provides some indication of the severity of the disease, there are numerous other determinants of these indices. Recent guidelines for medical management are clearly aimed at symptom prevention and attack management (24). However, many health care providers continue to treat acute symptoms only and do not provide their patients with treatment regimens aimed at prevention or with information regarding environmental control. These patients will appear to have more severe disease if severity is measured in terms of attack frequency or severity. Similarly, patients who are provided optimal management plans may be nonadherent with medications, as well as other recommendations, and continue to have more symptoms. Thus, while some patients with severe asthma have inherently difficult to control disease even with optimal medical care and adherence, it is more common that a measure reflecting frequency of asthma attacks indicates poor management of the disease.

It is concluded that there is increased psychological dysfunction among severely asthmatic children. The contradictions among studies as to the nature of the relationship between asthma severity and psychological functioning can be understood in light of the various ways in which asthma severity has been measured. Whether asthma severity is measured by medication level, symptom control, or functional impairment, a single conclusion can be drawn. With each measure, asthmatic children with significant asthma who are nonadherent with their medications will find their way into the group labeled severe. It well may be, then, that the often-seen correlation between asthma severity and psychological dysfunction can be accounted for by the inclusion in that severe group of a sizable subset of nonadherent, poorly controlled asthmatics.

IV. Other Factors Affecting the Relationship Between Severe, Poorly Controlled Asthma and Psychological Disturbance

The relationship between asthma and psychological disturbance is bidirectional. Asthma, particularly severe asthma, can be a chronic stressor, which taxes individual and family psychological resources and induces or exacerbates preexisting psychological problems. Conversely, individual and family dysfunction frequently undermine good asthma care. Families with notable disorganization and conflict, in particular, often have great difficulty performing the behaviors necessary to good asthma management, while sick children in such a system may be more easily drawn to denial of symptoms or medication nonadherence as an expression of their anger (25).

The psychological and disease risk factors, therefore, are interactive. Other factors may further increase the danger inherent in this group of asthmatic children at risk for dangerous, out-of-control illness. Michael Rutter and colleagues, who have written extensively about the "protective factors" that insulate children against the risk of psychopathology (26), have concluded that the child's risk is potentiated by the combining of risk factors. The co-occurrence of risk factors is particularly detrimental because their effects can be multiplicative rather than simply additive (27). These risk factors may include poverty, parental psychopathology, lower intelligence, chronic illness, and the occurrence of other life stressors (28).

The same models of risk and resiliency apply to children with chronic illness, including asthma. The factors identified which increase risk for poorly controlled asthma in children include family dysfunction, poverty, presence of other illness, negative life events, temperament and parental criticism, and medication side effects.

Family dysfunction appears to be a particular hallmark of severe, poorly controlled asthma and is likely the source of much nonadherence. Not all children with severe, difficult-to-control asthma come from dysfunctional families or are seriously nonadherent. However, the frequency of these problems among severe, poorly controlled asthmatic patients is markedly increased. Underuse of aerosolized corticosteroids was associated with decreased communication and organization in the families of 24 asthmatic children (29). In other studies, high levels of family conflict were correlated with theophylline nonadherence (30,31). When asthma severity, poor medical adherence, and serious family disturbance reach an extreme, asthma-related death becomes a particular risk. In a classic study of the circumstances surrounding asthma death in children, Strunk and colleagues reviewed the cases of 21 patients who later died from their asthma, using a stepwise discriminant analysis to evaluate the predictive role of 57 physiological and psychological variables (32). The asthma death risk profile which emerged included families with histories of conflict between parents and their child and with medical staff, depressive symptoms in the child, family dysfunction (including parental psychopathology, intense marital conflict, or alcoholism), disregard of asthma symptoms, and deficient self-care and compliance. Other investigators, generally reporting on fewer cases and a less standardized format for categorizing histories, have similarly concluded that deaths from pediatric asthma are associated with denial of symptoms, poor asthma management including medication nonadherence, and increased psychological dysfunction (33–36).

Poverty and membership in urban minority groups are associated with more frequent and more severe asthma. Asthma mortality and hospitalization rates are greater among nonwhites (37) and low-income patients (38), possibly for the reasons that such individuals may have decreased access to health care, incomplete use of asthma medications, failure to recognize illness severity, poor treatment adherence, and greater exposure to respiratory infections and environmental conditions that tend to exacerbate asthma, such as air pollution, dust mites, indoor molds, and cigarette smoke (39). Low socioeconomic status has been found to be associated with increased psychological disorder (40) and decreased academic performance (41) in asthmatic children.

The co-occurrence of other illness in addition to asthma may further stress the psychological adaptation of the child. Results from the 1988 U.S. National Health Interview Survey on Child Health, a national cross-sectional survey of 17,710 children, found that the most common chronic illnesses were respiratory disorders, including asthma and related allergies (42). About 20% of the sampled children had one chronic condition. For those 5% of children who had two or more chronic health conditions, a threefold increase for risk of learning difficulties and school absenteeism was reported. Children with more than one physical illness (43) or physical disability (44) are more likely to develop psychopathology than children with only one such condition. In one study of 551 asthmatic children, those children with asthma and another chronic medical condition demonstrated more behavioral/psychological problems than children with asthma alone (18).

Negative life events, such as frequent moves, parental job loss, or loss of a family member, increase the psychological duress for a child with a chronic illness. In a study of behavior problems and social competence in 81 asthmatic children ages 6–14, such negative life events predicted increased behavioral problems on the internalizing score from the Child Behavior Checklist, reflecting increased withdrawal and depression, while illness severity did not (9). "Undesirable life events" were also predictive of psychopathology in a study of 36 asthmatic children and played a greater role in predicting psychopathology than other factors such as illness severity (40).

Garmezy (45) identified the child's disposition, or temperament, as a primary predictor of risk. Children with difficult or irritable temperaments are more likely than other children to be the targets of parental hostility or criticism (26). Asthmatic children with irritable temperaments, an outcome not infrequently resulting from the effects of the illness itself, may similarly induce a greater degree of parental irritability than a healthy child. Kim et al. (46) compared temperament profiles of asthmatic children with children with other chronic illnesses and with normals, and found that the asthmatics were characterized by significantly lower rhythmicity, lower adaptability, lower intensity of reaction, more negative mood, and lower persistence. Parents of asthmatic children have been found to be more critical of their children than parents of healthy children (47,48). Illness improvement is greater when asthmatic children (49) or adolescents (50) are removed from critical homes for treatment of their disease.

Medication side effects may serve as yet another perturbation to the psychological adaptation of asthmatic children. Because children with severe, poorly controlled asthma are likely to receive more medications at larger doses, the risk factors once again increase to the greatest degree in this segment of the asthmatic population. While the psychological effects of theophylline are not as severe as once believed (51), the caffeinelike effects of theophylline have been documented and included a slight trend towards more anxiety and hand tremor in treated versus untreated children, accompanied by slightly increased attention and/or verbal memory (52–54). Classic antihistamines, such as diphenhydramine, have been found to cause drowsiness in some children (55). Individual case studies have reported an association between sympathomimetics and a variety of unusual behaviors in children, including aggressiveness and visual hallucinations (56). Asthmatic children stabilized on high doses of oral corticosteroids (40–80 mg/day) can experience significant depression, anxiety, and impairment of long-term memory compared with children on low maintenance doses (2–20 mg/day) (57). However, steroid-induced psychosis, which has been reported in adults, has not been documented in children. Of particular interest is the finding that children with a preexisting history of emotional difficulties were more likely to become anxious and depressed while receiving high doses of prednisone than children without such histories (52), further supporting the conclusion that the combining of risk factors heightens the potential for maladaptation. Finally, low intelligence, a factor that undermines psychological adaptiveness in the general population of children (58), may increase the risk of psychological side effects occurring with theophylline use (54).

V. Nonadherence, Poor Symptom Control, and Treatment Failure

The finding that inadequate symptom control is associated with psychological dysfunction in the child or the family indicates that health care behavior is a primary determinant of disease severity and health care cost. Clearly, appropriate management of asthma requires considerable effort on behalf of patients and their families, and nonadherence to a prescribed self-management regimen undermines disease control. That asthmatic patients do not take their medications is a well-established finding in children (21,22,59) and adults (20,60), in prn (61) and fixed-dose medications (62), and in the use of tablets (30), as well as aerosolized medications (20,56). Frequently, the degree of nonadherence is startling. In one survey of asthmatic children treated at a children's hospital, subtherapeutic theophylline levels were found in 70% of outpatients (63).

It is appealing to assume, based on logic, that patients would become more adherent as the severity of their asthma worsens, in which case anxiety about daily symptoms should increase health behavior motivation. However, for many patients this has not been found to be the case. One study followed the daily beclomethasone use of 10 moderately severe asthmatic adults using a metered dose inhaler chronolog equipped with an electronic recording system, which recorded date and time of each use. Changes in asthma severity over the 9-week protocol were evaluated with measurements of peak expiratory flow rates, symptoms, and albuterol use, none of which were significantly associated with inhaled steroid adherence. Adherence decreased as time progressed, but even after controlling for the effect of time, inhaled steroid adherence did not increase with increasing asthma severity (64). Others have similarly reported no increased adherence in patients with greater airways obstruction (65). Intriguingly, while physicians who witness symptom exacerbation and follow expert guidelines (24) typically increase dosing frequency of aerosolized anti-inflammatory agents, patients respond by taking a decreasing proportion of their medications (21).

Evidence is increasing that many severely asthmatic patients are dramatically nonadherent. Spilker (66) cites evidence across various chronic diseases that adherence improves as disease severity increases from mild to moderate illness, but appears to reverse with severe illness. Hospitalizations and emergency room visits, it might be assumed, would dramatically boost adherence motivation because they are frequently unpleasant and costly experiences, which signal to patient and caregiver that a serious health decline has occurred. Strikingly, this assumption is frequently incorrect. Nonadherence is often high among patients who appear in the emergency room or hospital, and there is no evidence that such contacts with emergency care serve to increase and sustain subsequent adherence. Theophylline-treated asthmatics presenting at an emergency room have been found to have subtherapeutic theophylline levels (67). In a 90-day longitudinal study, 8 children requiring urgent care visits resulting in oral steroid bursts (2 of whom required subsequent hospitalization) had been dramatically less adherent with a daily regimen of inhaled steroids than a group of 16 patients with stable symptom control (23).

Perhaps the most revealing evidence regarding the consequences of nonadherence in severely asthmatic patients emerges from studies attempting to alter these patterns. A number of studies provided psychoeducational intervention without altering availability or cost of medical care and medications. The degree to which the course of severe asthma was altered in these programs reflects the extent to which inadequate health care behavior contributes to poorly controlled asthma. Adult patients participating in a 7-week asthma self-management class were found to have significantly reduced asthma symptoms at 1-year follow-up (68). In another study (69), 104 adults with a history of asthma-related hospitalizations participated in a program emphasizing "teaching patients aggressive self-management strategies in case of marked asthma exacerbation" resulting in a two-thirds reduction in readmissions. Similarly, 53 children receiving standardized care in the Harvard Community Health Plan, but with a history of high use of emergency rooms and hospitalization, were included in a program providing them and their parents individualized instruction in "asthma management, medications, triggers, and use of inhalers and peak flow meters" (70) along with regular telephone contact from an outreach nurse. Emergency ward admissions were reduced from 72 to 15, and hospitalizations from 35 to 5 in the follow-up year (70). If the behavioral/educational interventions in these studies can be presumed to be only partially successful in improving health care behavior, then it is likely that between 50 and 100% of hospital and emergency room use by asthmatic patients is attributable to insufficient medical adherence.

VI. Treatment Adherence and Health Care Costs

Asthma is a costly disease. The direct and indirect cost of asthma in 1990 was estimated to be $6.2 billion. The $3.6 billion direct medical costs included physician visits and medication, but the largest portion (almost $3 billion) resulted from emergency room visits and hospitalizations (1). Since only a small proportion of asthmatic patients require hospitalization, much of the cost of asthma is created by a relatively small group of patients. Approximately 5% of asthmatics account for more than 70% of the total cost of asthma (71). Many of the asthmatic patients who present at the emergency room or hospital have been there before (1). A large portion of the total cost of asthma care, then, is unnecessary and traceable to poor health care behavior.

As reviewed above, interventions aimed at changing health care behavior have demonstrated some success in improving symptom control and reducing use of the health care system. When emergency room visits and hospitalizations decrease, costs decrease significantly. Reductions in emergency care use in the 53 patients from the Harvard Community Health Plan resulted in a total savings of $87,000, or an average of $1,642 per patient, in the year following intervention (70). The potential for savings, given more refined interventions tailored to specific patient groups, is even greater.

Clark and colleagues (72) have demonstrated that those patients with severe, out-of-control asthma have the most to gain from improved health care behavior. Two hundred and ninety urban asthmatic children were randomized to a control group or to an experimental group participating in an asthma education program designed to improve self-management. No change in health care use or cost was found in the follow-up year when contrasted to the baseline year. However, when outcome data were examined separately for those children with histories of hospitalization and emergency room use in the baseline period, significant reductions were documented in both types of emergency health care contacts. Similarly, a negative cost-effectiveness was documented for the entire group (i.e., for every $1.00 spent on the program, only $.62 was saved in reduced health care cost); for the high-risk patients, however, a significant savings was realized ($11.22 was saved for each $1.00 spent on the program).

In summary, among children with mild to moderate asthma there is little evidence of an increase in psychological disorder. Treatment nonadherence is a problem for the entire asthma population, but in this milder patient group adherence appears to increase with increasing symptom exacerbation. The characteristics of children with histories of severe, poorly controlled asthma, however, appear to be different. In this group, adherence to good medical management is often extremely poor and contributes to their high-risk profile. In fact, the evidence reviewed above indicates that "severe" asthma might in fact be "moderate" if some of these patients were better managed. These patients are excessive and inappropriate users of the health care system and are responsible for a significant portion of the huge costs associated with asthma. Increased individual and family psychological dysfunction likely contribute to an atmosphere of chronic stress and disorganization that conflict with an appropriate asthma management system. In the worst cases, the combination of family dysfunction and inadequate asthma care can result in death.

VII. Psychosocial Intervention with Severe, Poorly Controlled Asthma

It is argued here that severe, high-risk asthma is frequently associated with child or family psychological dysfunction and treatment nonadherence, creating an interactive spiral often resulting in treatment failure. If a child's asthma is to be better controlled, the factors underlying their poorly controlled asthma must be addressed. While ample evidence shows that changing patients' health care behavior can dramatically alter the course of chronic disease, such change is often difficult to effect. Further, interventions designed to change health care behavior based upon inaccurate assumptions seldom succeed. In the case of high-risk asthma, such assumptions have included that poor health care behavior is largely the result of inadequate information and that a single, brief intervention is sufficient.

Asthma education has been widely endorsed as the preferred intervention in health care behavior. Numerous proponents have noted the virtues of asthma education programs, viewed by many as the answer to an epidemic of increasing asthma-related morbidity and mortality (73). Sound education about the disease and its treatment can be helpful to asthma patients. The erroneous assumption inherent in many such programs, however, is that poor self-management and medication adherence is largely the result of inadequate knowledge about the disease. However helpful they may be, asthma education efforts have frequently failed to change patient behavior or have demonstrated only temporary changes.

Approaches to asthma education have been quite varied, ranging from a brief series of group education classes (74) and "mail-in" education booklets (75) to more comprehensive programs involving several months of classes (76), telephone follow-up (60,70), home visits, and even a 5-day inpatient program (77). Comparisons of effectiveness across these education intervention studies is particularly difficult given great variation in patient sampling, interventions, outcome measures, and study design. Nonetheless, one group of investigators has reported the results of a carefully conducted meta-analysis of 12 such studies, selected out of a group of 23 randomized clinical trials, which met the following criteria: each included asthmatic children and a clearly documented "teaching intervention," a trained instructor (as opposed to the exclusive use of written materials), and outcome variables related to asthma morbidity, including frequency of asthma attacks, emergency room visits, hospitalizations, and school absenteeism. Despite the conclusions by some individual investigators that these educational interventions achieved the desired change, particularly when measured by paper-and-pencil tests which assessed the acquisition of new knowledge, the meta-analysis demonstrated that, overall, these "teaching programs have surprisingly little influence on morbidity outcomes" (78). In a recent study of patient behavior, which evaluated both peak flow meters and inhaled medication use, adherence to the treatment regimen was only 40% following a formal asthma education program, which included videos, written materials, and face-to-face instruction and demonstrated mastery of treatment knowledge and skills (79).

Increased asthma knowledge among patients with limited initial understanding may result in some improvement in self-management. In a study of the relationship between knowledge and behavior in 91 asthmatic children, increasing asthma knowledge from "little" to "average" levels was correlated with behavior change, while further increases in asthma knowledge had little impact (80). Even where new information can improve outcome, such changes are often short-lived. Medical care use and activity limitations due to asthma were reduced at 4 but not 12 months following three one-hour asthma education classes in a study of high-risk patients recruited from an emergency room (74). A group of 31 asthmatic children demonstrated increased knowledge but no improvement in asthma control following a three-session education class

(81). Even among those children from disorganized families with a documented history of poor management, no symptom improvement was achieved. The Bolton et al. (74) and LeBaron et al. (81) studies show that for both high-risk adults and children, educational classes alone are often unable to change self-management of asthma.

Why do some education programs demonstrate successful changes in health outcomes while others do not? While study design and patient sample differences undoubtedly account for some of the outcome variance, a key difference appears to lie not in the content of the education program but in the relationship between "educator" and patient. While often evaluated under the general heading of "education intervention," many education programs move beyond simply imparting information to patients. Such "psychoeducational" programs, whether or not they are so described, involve psychological interventions that usually take the form of the introduction of a special relationship not available to patients outside the intervention program. For example, the "education program" employed by Greineder and colleagues (70) included face-to-face and telephone counseling from a nurse. Given the well-established importance of the patient-caregiver relationship in assuring treatment adherence (66), the inclusion of a new and sympathetic caregiver may very well affect health care behavior more than the imparting of factual information regarding asthma.

For patients with severe, poorly controlled asthma, introduction of a "one size fits all" intervention is inadequate to change the sometimes powerful underlying factors which contribute to poor treatment adherence. Asthma education is best viewed as one component in a multidimensional intervention program. Additional psychological services are frequently necessary to address child and/or family psychopathology. As noted above, some asthma education programs may have affected behavior change by the introduction of a sympathetic caregiver who provided the patient not only educational information but also a "therapeutic" relationship. When that relationship involves a trained mental health provider able to assess and target psychological issues that interfere with appropriate health care behavior, such intervention can be expected to be even more effective. In fact, the use of family therapy in the treatment of childhood asthma has been demonstrated to improve symptom control (82–85).

While these studies demonstrate some value in applying traditional psychological interventions with chronically ill patients, they do not provide a clear differentiation as to the specific components of psychotherapy which are effective in altering the course of asthma. Traditional forms of individual or family psychotherapy may not be equally effective across the range of asthma severity. Further, if the psychotherapist is not knowledgeable about the disease and is not working in close collaboration with the medical caregiver, the possibility of directly affecting control of the disease is greatly diminished. Psychological interventions with the severely asthmatic patient are most likely to be effective when:

1. They are provided by a mental health practitioner knowledgeable about the disease: Understanding the physiology involved during rapidly developing asthmatic episodes, as well as the components of good asthma management, greatly enhances the therapist's ability to decrease psychological defenses and increase problem-solving skills.
2. They are provided in a health care setting in order to allow a direct collaboration between the mental and medical health care providers: Close communication between these two providers allows ongoing refinement of the medical and psychological treatment plans and increases the capacity of both to confront nonadherence.
3. They are directly focused upon health care behavior: While a variety of intrapsychic and interpersonal conflicts are necessarily targeted by traditional psychotherapy, assessment and intervention must involve the patient's willingness and capacity to change health care behavior if they are to directly impact disease outcome.
4. They are tailored to the individual patient or, in the case of a child, the patient's family: While numerous asthma education/self-management programs have been developed for broad-based distribution to groups of asthmatics, patients with severe, poorly controlled asthma, many of whom have co-morbid psychological disorders, will require a more intensive, individualized, multidisciplinary approach.

Assessment of the psychological characteristics of the patient and the family, of their degree of understanding of asthma, of their capacity to manage the illness, and of the barriers that prevent appropriate management must precede the introduction of a planned intervention. When the assessment and intervention are conducted by a multidisciplinary team specialized in the care of asthma, employing a well-focused and structural approach, dramatic changes in asthma morbidity may be accomplished cost effectively and in relatively short time (76,86). Following is a more detailed discussion of such an approach for asthmatic children and their families.

VIII. A Family Asthma-Management Approach

For children, asthma management occurs within the context of a family. By their nature, children are unable to completely "self-manage" their asthma. Whatever the composition of the family, children and adolescents require the involvement of one or more parents or caretakers. The behaviors of asthma management occur within the parent-child interaction. The parents of young children take complete responsibility for monitoring symptoms and administering medications. The "balance of responsibility" shifts with development, and older adolescents should be able to take complete responsibility for managing their own asthma (87). The shift occurs over time in the context of innumerable parent-child interactions related to the various tasks involved in asthma management. The manner in which transactions occur around asthma is representative of the manner in which parenting occurs in general. The quality of parenting in a family thus has a major impact on children's adherence with medication regimens and on their attitudes about managing their illness. Besides socializing their children, parents are responsible for providing for the family and interacting with the broader social systems. Inadequacies or dysfunction in fulfilling these responsibilities have a major impact on families' abilities to effectively manage asthma. Thus, children with asthma develop within a dynamic family system, and parental functioning and parent-child interaction strongly impact the effectiveness with which the asthma is managed. An effective intervention designed to have a lasting effect on controlling asthma symptoms then takes into consideration the nature of the relationships within the family.

An integrated and comprehensive approach to psychological treatment for children with severe, poorly controlled asthma must focus on the entire family system and on the entire range of behaviors required for effective asthma management. Adherence in asthma involves not only taking medications on time and using good technique, but also working in a partnership with a physician, having good environmental controls of asthma triggers, and recognizing and treating asthma symptoms when they occur. The specific tasks to be accomplished for effective asthma management have been catalogued and are extensive (88). Because the required behaviors are complex and because families are so varied, individualized assessments of the barriers which impede effective asthma management must precede the design of a treatment tailored to overcome those barriers.

Central to a family asthma management approach is a behaviorally focused assessment conducted with the parents together with the asthmatic child. This assessment includes an analysis of the family system and, in particular, the problem areas that serve as barriers to good asthma management. The assessment focuses on all aspects of the child's asthma, such as symptom patterns, triggers, monitoring of symptoms, medications prescribed, decision-making processes regarding administering medications, the relationship with the physician, and the distribution within the family of asthma-related tasks. The family's knowledge of asthma and management behaviors, their capacity to follow through with illness-management procedures, and deficiencies undermining asthma control are determined. An exploration of the family's existing asthma management system uncovers impediments to more effective management. The family's current behavior deficits may be related to a lack of knowledge about asthma, counterproductive health beliefs or attitudes, organizational problems that prevent adherence, or interpersonal or relational difficulties. It is not uncommon to discover more systemic problems, such as parent-child interaction problems, spousal conflict, limited intellectual abilities, or financial limitations.

Once problem areas have been identified, the specific problems within an individual family can be addressed. For example, a family may reveal their confusion about the circumstances under which they should treat attacks at home as opposed to calling their physician for guidance. Such a family would benefit from help in communicating with their physician and obtaining a clear asthma-management plan. Another family may be unwilling to administer inhaled steroids. Exploration may reveal a lack of understanding of the inflammatory basis of chronic asthma or misconceptions about side effects of steroids. Individualized education may be sufficient to address the problem. However, fears about side effects of medication are often based not only on ignorance but also on previous emotionally complex experiences. An examination of the source of the fears is necessary in order to dispel them and clear the way for a more enlightened view. For yet another family, the child's wheezing symptoms may meet with high anxiety, overestimation of symptom severity, and inappropriate attack management. Exploration may reveal that the anxiety underlying the inappropriate responses is related to a relative's death from asthma. In this case, it would be necessary to address the loss of the relative, while also educating regarding symptom assessment and rehearsing appropriate responses to symptoms. Finally, for many families with children with severe, poorly controlled asthma, overwhelming problems in living take precedence over attending to the specific needs of their asthmatic child. These families are often "multiproblem" families with a variety of financial, social, and psychological needs. Intervention requires the involvement of mental health providers who can help to address the organizational and programmatic problems that face these families. With such families, behavioral contracting provides a focus on specific, manageable problems and an arena within which to address the family's motivation to change.

It should be clear from these examples that all interventions are conducted within the family context, because this is where the interactions comprising asthma management occur. This is asthma-focused family therapy and should be distinguished from traditional family therapy in its targeting of asthma management behaviors. Asthma-focused family therapy must be provided by a psychosocial clinician who is highly knowledgeable about asthma and who has a working relationship with the patient's physician. At times, it may become apparent that a more traditional form of therapy is indicated. This typically becomes the case when inadequate progress is seen in the asthma-focused family therapy and in the course of which other areas of serious conflict become apparent. For example, a particular family may reveal marital conflict focused on the appropriateness of utilizing preventive medication. Further exploration may indicate that the marital conflict is severe and enduring, and the disagreement about medication is only a symptom of a spousal problem. In another family, a mother may be unable to summon the energy to monitor her child's symptoms. As a result of the evaluation, it may be clear that she suffers from depression. Such problems are clearly barriers to effective asthma management, and traditional psychological or psychiatric therapies are indicated in order to make effective asthma-management possible.

IX. Summary and Conclusions

Asthmatic children bear a burden that can impede physical, psychological, and social development. Still, the majority of asthmatic children do not demonstrate psychological dysfunction. Among severe, poorly controlled asthmatic children, however, psychological disorders are increased. The relationship between psychological and physical disorders is bidirectional. Just as asthma increases psychological stress, the presence of a psychologically dysfunctional child or family contributes to severe, poorly managed asthma. Treatment nonadherence, which is pronounced in families whose disorganization and conflict undermine good asthma management, leads to excessive and costly use of the health care system. Other factors that increase the risk for dangerously out-of-control asthma include poverty, negative life events, difficult temperament, medication side effects, and other illness. For patients with severe, poorly controlled asthma, intervention must be multidisciplinary and should include a psychosocial component. While asthma education is an important part of such a treatment program, as a sole intervention it cannot be expected to result in significant change in health care behavior in this group. Effective psychological interventions must be (1) provided by a mental health practitioner knowledgeable about the disease, (2) in a health care setting where mental and medical health care providers can collaborate, (3) focused upon health care behavior, and (4) tailored to the individual patient and family. For the child with asthma, assessment should be directed at the family's understanding of the asthma and its management and on areas of dysfunction that interfere with appropriate management. This assessment can then be followed by asthma-focused family therapy aimed at eliminating management barriers and increasing problem-solving skills related to asthma care.

References

1. Weiss K, Gergen P, Hodgson T. An economic evaluation of asthma in the United States. N Engl J Med 1992; 326:862–868.
2. Centers for Disease Control. Asthma—United States, 1982–1992. MMWR 1995; 43:952.
3. Perrin JM, MacLean WE, Perrin EC. Parental perceptions of health status and psychologic adjustment of children with asthma. Pediatrics 1989; 83:26–30.
4. Klinnert MD. The psychology of asthma in the school-aged child. In: Kember P, Bemporand J, eds. Handbook of Child and Adolescent Psychiatry: The Grade School Child. New York: John Wiley & Sons (in press).
5. Hamlett KW, Pelligrini DS, Katz KS. Childhood chronic illness as a family stressor. J Pediatr Psychol 1992; 17:33–47.
6. Graham PJ, Rutter M, Yule W, Pless IB. Childhood asthma: a psychosomatic disorder? Br J Prev Soc Med 1967; 21:78–85.
7. McNicol KN, Williams HE, Allan J, McAndrew I. Spectrum of asthmas in children—III, Psychological and social components. Br Med J 1973; 4:16–20.
8. Kashani JH, Konig P, Shepperd JA, Wilfley D, Morris DA. Psychopathology and self-concept in asthmatic children. J Pediatr Psychol 1988; 13:509–520.
9. MacLean WE. Perrin JM, Gortmaker S, Pierre CB. Psychological adjustment of children with asthma: effects of illness severity and recent stressful life events. J Pediatr Psychol 1992; 17:159–171.
10. Carson DK, Schauer RW. Mothers of children with asthma: perceptions of parenting stress and the mother-child relationship. Psychol Rep 1992; 71:1139–1148.
11. Achenbach TM, Edelbrock CS. Manual for the Child Behavior Checklist and Revised Child Behavior Profile. Burlington, VT: University of Vermont, 1983.
12. McQuaid EL, Klinnert MD, Mrazek D. Behavior problems in young children: Is a clinical interview useful? (Submitted)
13. Klinnert MD, McQuaid EL, Bryant NE, Walker DA. Psychopathology in asthmatic children. Presented at the Eighth Biennial Retreat of the Developmental Psychobiology Research Group (DPRG), Estes Park, CO, May 1994.
14. Achenbach TM, McConaughy SH. Semi-structured Clinical Interview for Children Aged 6 to 11. Burlington, VT: University of Vermont, 1989.
15. Kagan J, Snidman N, Julia-Sellers M, Johnson MO. Temperament and allergic symptoms. Psychosom Med 1991; 53:332–340.
16. Norrish M, Tooley M, Godfrey S. Clinical, physiological, and psychological study of asthmatic children attending a hospital clinic. Arch Dis Child 1977; 52:912–917.
17. Perrin JM, MacLean WE, Perrin EC. Parental perceptions of health status and psychologic adjustment of children with asthma. Pediatrics 1989; 83:26–30.
18. Bussing R, Halfon N, Benjamin B, Wells KB. Prevalence of behavior problems in U.S. children with asthma. Arch Pediatr Adolesc Med 1995; 149:565–572.
19. Stein REK, Jessop DL Relationship between health status and psychological adjustment among children with chronic conditions. Pediatrics 1984; 73:169–174.
20. Spector SL, Kinsman R, Mawhinney H, Siegel SC, Rachelefsky GS, Katz RM, Rohr AS. Compliance of patients with asthma with an experimental aerosolized

medication: implications for controlled clinical trials. J Allergy Clin Immunol 1986; 77:65–70.

21. Coutts JAP, Gibson NA, Paton JY. Measuring compliance with inhaled medication in asthma. Arch Dis Child 1992; 67:332–333.
22. Milgrom H, Bender B, Sarlin N, Leung DYM. Difficult to control asthma: the challenge posed by non-compliance. Am J Asthma Allergy Pediatricians 1994; 7:141–146.
23. Milgrom H, Bender B, Ackerson L, Bowry P, Smith B, Rand C. Children's compliance with inhaled asthma medications. J Allergy Clin Immunol 1995; 95:217.
24. National Asthma Education Program. Executive Summary: Guidelines for the Diagnosis and Management of Asthma. U.S. Department of Health and Human Services, 1991.
25. Miller BD. Depression and asthma: a potentially lethal mixture. J Allergy Clin Immunol 1987; 80:481–486.
26. Rutter M. Psychosocial resilience and protective mechanisms. Am J Orthopsychiat 1987; 57:316–331.
27. Pellegrini DS. Psychosocial risk and protective factors in childhood. Dev Behav Pediatr 1990; 11:201–209.
28. Luthar SS, Zigler E. Vulnerability and competence: a review of research on resilience in childhood. Am J Orthopsychiat 1991; 61:6–22.
29. Bender B, Milgrom H, Bowry P, et al. Asthmatic children's adherence with aerosolized medications. Am J Respir Crit Care Med 1995; 151:A352.
30. Christiannse ME, Lavigne JV, Lerner CV. Psychosocial aspects of compliance in children and adolescents with asthma. Dev Behav Pediatr 1989; 10:75–80.
31. Wamboldt FS, Wamboldt MZ, Gavin LA, Roesler TA, Brugman SM. Parental criticism and treatment outcome in adolescents hospitalized for severe chronic asthma. J Psychosom Res (in press).
32. Strunk RC, Mrazek DA, Wofson Fuhrmann GS, LaBrecque JF. Physiological and psychological characteristics associated with deaths from asthma in childhood: a case-controlled study. J Am Med Assoc 1985; 254:1193–1198.
33. Fritz GK, Rubenstein S, Lewiston NJ. Psychological factors in fatal childhood asthma. Am J Orthopsychiatry 1987; 57:253–257.
34. Kravis LP. An analysis of fifteen childhood asthma fatalities. J Allergy Clin Immunol 1987; 80:467–471.
35. Lanier B. Who is dying of asthma and why? J Pediatr 1989; 115:838–840.
37. Sly RM. Mortality from asthma, 1979–1984. J Allergy Clin Immunol 1988; 82:705–717.
38. Wissow LS, Gittelsohn AM, Sazklo M, Starfield B, Mussman M. Poverty, race, and hospitalization for childhood asthma. Am J Public Health 1988; 78:777–782.
39. Marder D, Targonski P, Orris P, Persky V, Addington W. Effect of racial and socioeconomic factors on asthma mortality in Chicago. Chest 1992; 101:426S-429S.
40. Steinhausen H, Schindler H, Stephan H. Correlates of psychopathology in sick children: an empirical model. J Am Acad Child Psychiatry 1983; 22:559–564.

41. Gutstadt LB, Gillette JW, Mrazek DA, Fukuhara JT, LaBrecque JF, Strunk RC. Determinants of school performance in children with chronic asthma. Am J Dis Child 1989; 143:471–475.
42. Newacheck PW, Stoddard JJ. Prevalence and impact of multiple childhood chronic illnesses. J Pediatr 1994; 124:40–48.
43. Kopp CB, Krakow JB. The developmentalist and the study of biological risk, a view of the past with an eye toward the future. Child Dev 1983; 54:1086–1108.
44. Cadman D, Boyle M, Szatmari P, Offord DR. Chronic illness, disability, and mental and social well-being: findings of the Ontario Child Health Study. Pediatrics 1987; 79:805–813.
45. Garmezy N. Stress-resistant children: the search for protective factors. In: Stevenson JE, ed. Recent Research in Developmental Psychopathology. Oxford: Pergamon, 1985: 213–233.
46. Kim SP, Ferrara A, Chess S. Temperament of asthmatic children. J Pediatr 1980; 97:483–486.
47. Hermanns J, Florin I, Dietrich M, Rieger C, Hahlweg K. Maternal criticism, mother-child interaction, and bronchial asthma. J Psychosom Res 1989; 33:469–476.
48. Schobinger R, Florin I, Zimmer C, Lindemann H, Winter H. Childhood asthma: paternal critical attitude and father-child interaction. J Psychosom Res 1992; 36:743–750.
49. Purcell K, Brady K, Chai H, Muser J, Moir L, Gordon N, Means J. The effect on asthma in children of experimental separation from the family. Psychosom Med 1969; 31:144–164.
50. Wamboldt FS, Wamboldt MZ, Gavin LA, Sarlin NS, White JJ. Effects of expressed emotion on medical outcome in severe adolescent asthmatics. Presented at conference on Physical Illness, Psychologic Development, and How the Twain Shall Meet, Keystone, CO, June 1993.
51. Rachelefsky G, Wo J, Adelson J, Mickey MR, Spector SL, Katz RM, Siegel SC, Rohr AS. Behavior abnormalities and poor school performance due to oral theophylline use. Pediatrics 1986; 78:1133–1138.
52. Bender BG, Lerner JA, Polland JE. Association between corticosteroids and psychologic change in hospitalized asthmatic children. Ann Allergy 1991; 66:414–419.
53. Bender B, Milgrom H. Theophylline-induced behavior change in children: an objective evaluation of parents' perceptions. J Am Med Assoc 1992; 267:2621–2624.
54. Schlieper A, Adcock D, Beaudry P, Feldman W, Leikin L. Effect of therapeutic plasma concentrations of theophylline on behavior, cognitive processing, and affect in children with asthma. J Pediatr 1991; 118:449–455.
55. Vuurman EF, Veggel VL, Uiterwijk MM, Leutner D, O'Hanlon, JF. Seasonal allergic rhinitis and antihistamine effects on children's learning. Ann Allergy 1993; 71:121–126.
56. Bender BG, Milgrom H. Neuropsychiatric effects of medications for allergic diseases. J Allergy Clin Immunol 1995; 95:523–528.

57. Bender B, Lerner J, Kollasch E. Mood and memory changes in asthmatic children receiving corticosteroids. J Am Acad Child Adolesc Psychiatry 1988; 6:720–725.
58. Werner EE, Smith RS. Vulnerable but Invincible. A Longitudinal Study of Resilient Children and Youth. New York. McGraw-Hill, 1982.
59. Baum D, Creer TL. Medication compliance in children with asthma. J Asthma 1986; 23:49–59.
60. Bailey WC, Richards JM, Brooks CM, Soong S, Windsor RA, Manzella BA. A randomized trial to improve self-management practices in adults with asthma. Arch Intern Med 1990; 150:1664–1668.
61. Mawhinney H, Spector SL, Kinsman RA, Siegel SC, Rachelefsky GS, Katz RM, Rohr AS. Compliance in clinical trials of two nonbronchodilator, antiasthma medications. Ann Allergy 1991; 66:294–299.
62. Rand CS, Wise RA, Nides M, Simmons MS, Bleecker ER, Kusek JW, Li VR, Tashkin DP. Am Rev Respir Dis 1992; 146:1559–1564.
63. Cox S, Webster M, Ilett K, Walson PD. Audit of theophylline plasma level monitoring in a pediatric hospital. Ther Drug Monit 1993; 15(4):289–293.
64. Mann MC, Eliasson O, Patel K, ZuWallack RL. An evasion of severity-modulated compliance with q.i.d. dosing of inhaled beclomethasone. Chest 1992; 102:1342–1346.
65. Horn CR, Clark TJH, Cochrane GM. Compliance with inhaled therapy and morbidity from asthma. Respir Med 1990; 84:67–70.
66. Spilker B. Methods of assessing and improving patient compliance in clinical trials. In: Cramer JA, Spilker B, eds. Patient Compliance in Medical Practice and Clinical Trials. New York: Raven Press, 1991.
67. Birkhead G, Olfaway NJ, Strunk RC, Townsend MC, Teutsch S. Investigation of a cluster of deaths of adolescents with asthma: evidence implicating inadequate treatment and poor patient adherence with medications. J Allergy Clin Immunol 1989; 84:484–491.
68. Kotses H, Bernstein IL, Bernstein DI, Reynolds RVC, Korbee L, Wigal JK, Ganson E, Stout C, Creer TL. A self-management program for adult asthma. Part I: Development and evaluation. J Allergy Clin Immunol 1995; 95:529–540.
69. Mayo PH, Richman J, Harris HW. Result of a program to reduce admissions for adult asthma. Ann Intern Med 1990; 112:864–871.
70. Greineder DK, Loane KC, Parks P. Reduction in resource utilization by an asthma outreach program. Arch Pediatr Adolesc Med 1995; 149:415–420.
71. Todd WE. New mindsets in asthma: interventions and disease management. J Care Management 1995; 1:2–8.
72. Clark NM, Feldman CH, Evans D, Levison MJ, Wasilewski Y, Mellins RB. The impact of health education on frequency and cost of health care use by low income children with asthma. J Allergy Clin Immunol 1986; 78:108–115.
73. Boulet LP, Chapman KR, Green LW, FitzGerald JM. Asthma education. Chest 1994; 106:184S-196S.
74. Bolton MB, Tilley BC, Kuder J, Reeves T, Schultz LR. The cost and effectiveness of an education program for adults who have asthma. J Gen Intern Med 1991; 6:401–407.

75. Osman LM, Abdalla MI, Beattie JAG, Ross SJ, Russell IT, Friend JA, Legge JS, Douglas JG. Reducing hospital admission through computer supported education for asthma patients. Br Med J 1994; 308:568–571.
76. Creer TL, Backial M, Burns KL, Leung P, Marion RJ, Miklich DR, Morrill C, Taplin PS, Ullman S. Living with asthma. I. Genesis and development of a self-management program for childhood asthma. J Asthma 1988; 25:335–362.
77. Trautner C, Richter B, Berger M. Cost-effectiveness of a structured treatment and teaching programme on asthma. Eur Respir J 1993; 6:1485–1491.
78. Bernard-Bonnin A, Stachenko S, Bonin D, Charette C, Rousseau E. Self-management teaching programs and morbidity of pediatric asthma: a meta-analysis. J Allergy Clin Immunol 1995; 95:34–41.
79. Chmelik F, Doughty A. Objective measurements of compliance in asthma treatment. Ann Allergy 1994; 73:527–532.
80. Rubin DH, Bauman LJ, Lauby JL. The relationship between knowledge and reported behavior in childhood asthma. Dev Behav Pediatr 1989; 10:307–312.
81. LeBaron S, Zeltzer LK, Ratner P, Kniker WT. A controlled study of education for improving compliance with cromolyn sodium (Intal®): the importance of physician-patient communication. Ann Allergy 1985; 55:811–818.
82. Gustafsson PA, Kjellman NM, Cederblad M. Family therapy in the treatment of severe childhood asthma. J Psychosom Res 1986; 30:369–374.
83. Gustafsson PA, Kjellman NM, Ludvigsson J, Cederblad M. Asthma and family interaction. Arch Dis Child 1987; 62:258–263.
84. Liebman R, Minuchin S, Baker L. The use of structural family therapy in the treatment of intractable asthma. Am J Psychiatry 1974; 131:535–540.
85. Lask B, Matthew D. Childhood asthma. A controlled trial of family psychotherapy. Arch Dis Child 1979; 54:116–119.
86. Taitel MS, Kotses, H, Bernstein IL, Bernstein DI, Creer TL. A self-management program for adult asthma. Part II: Cost-benefit analysis. J Allergy Clin Immunol 1995; 95:672–676.
87. Klinnert MD, Tedesco J. Developmentally-based self-care in chronically ill children. Workshop presented at the meeting of the Association for the Care of Children's Health, Boston, 1985.
88. Wilson SR. Patient and physician behavior models related to asthma care. Med Care 1993; 31:MS49–MS60.

Part Two

ASTHMA SYMPTOMS: ASSESSING SUBJECTIVE EXPERIENCE

5

Quality-of-Life Measures

ELIZABETH F. JUNIPER

McMaster University Medical Centre
Hamilton, Ontario, Canada

I. Introduction

The effectiveness of asthma interventions—pharmaceutical, rehabilitative, and educational—has traditionally been assessed by measuring clinical indices such as airway caliber, symptoms, and hospitalizations. Although these primarily evaluate the status of the target organ, the airways, there has tended to be the assumption among clinicians that they also provide a good estimate of how patients function physically, emotionally, socially, and occupationally in their day-to-day lives as a result of their asthma. It has also been assumed that if there were changes in the conventional clinical outcomes, the patient's day-to-day life experiences would reflect these changes. However, recent evidence from a number of high-quality clinical studies from different investigators in asthma quality-of-life research (1–4) has shown that the associations (correlations) between health-related quality of life (HRQL) and the clinical outcomes of asthma severity are only weak to moderate.

On reflection, this is not very surprising. For instance, a patient may experience clinically important changes in airway caliber or airway hyperresponsiveness but may not report any changes in HRQL. This may occur because the patient is not perceptive to changes in airway caliber or leads a lifestyle in which there is little provocation of bronchoconstriction. Let us take two patients with the same improvement in airway caliber and airway hyperresponsiveness as a result of an intervention. One patient works in a high-pressure job, she has to attend meetings in smoky rooms, is also an athlete, once experienced a life-threatening asthma attack, and is generally an uptight person. The second patient works at home and can easily regulate her life to avoid situations that might exacerbate her asthma, she leads a very sedentary lifestyle, has never had a serious asthma attack, and is generally a very laid-back person. Improvement in HRQL as a result of intervention is likely to be much greater in the former patient.

In contrast, some patients may experience improvements in their ability to carry out daily activities and in how they feel as a result of an intervention, but these may not be captured by the instruments used to measure airway caliber, symptoms, etc. This failure to record clinical changes may occur because the actual changes experienced by the patient are not expressed in any of the outcomes being measured or because the instruments are insufficiently sensitive to detect the changes.

HRQL research in asthma has highlighted the functional impairments that patients experience (5–8). Adults with asthma are not only troubled by the symptoms themselves, such as shortness of breath, wheeze, and cough, they also are troubled by limitation of daily activities (occupational, social, and physical), sleep impairment, emotional problems, such as anxiety and frustration, and problems associated with exposure to atmospheric stimuli such as cigarette smoke. The burden of illness for children with asthma is similar. In addition, children experience important quality-of-life impairments because of their inability to integrate fully with their peers (9–12).

If an investigator is going to include an evaluation of HRQL in a clinical study, it is important to select the instrument most appropriate for the task. There is no best instrument since they all have strengths and weaknesses, and therefore the investigator should first determine which properties and measurement characteristics are going to be essential to address the particular study question. In this chapter, I discuss the strengths and weaknesses of generic and disease-specific instruments, comment on the measurement characteristics required for different types of clinical studies, and review the instruments in which measurement properties have been evaluated in asthma patients.

II. Types of Health-Related Quality-of-Life Instruments

A. Generic: Health Profiles

Generic health profiles (Table 1) are designed to be applicable to patients in all health states. The most commonly used and the best validated are the MOS

Table 1 Types of Health-Related Quality-of-Life Instruments

Instrument type	Strengths	Weaknesses
Generic		
Health profiles	Comparison across conditions possible	Do not focus adequately on areas of interest May not be responsive
Utilities	Single number representing quality of life Cost-utility analysis possible	Do not allow examination of different aspects of quality of life May not be responsive
Disease-specific	Clinically sensible More responsive	Comparison across conditions not possible

Short Form 36 (SF-36) (13), the Sickness Impact Profile (SIP) (14), the Nottingham Health Profile (15), and the McMaster Health Index Questionnaire (16). The main advantage of health profiles is that impairments can be compared across different medical conditions. Although each health profile attempts to measure all important aspects of HRQL, they achieve this in different ways. For instance, the SF-36 contains three major health attributes (functional status, well-being, and overall evaluation of health), which are not combined into one overall score. In contrast, the SIP has two domains (physical and psychosocial), which can be combined into a total.

Since they need to be applicable in all diseases, a limitation of these instruments in any one particular condition is that they may not measure areas of impairment that are characteristic of that condition and are important to patients. For instance, the SIP tends to focus on quite severe impairments (feeding, dressing, etc.) that may not be applicable to most patients with asthma (6,24). In addition, health profiles may not be responsive to small but important changes when used in clinical trials.

B. Generic: Utilities

These instruments evaluate preferences for various health states on a scale of 1 to 0, where 1 represents perfect health and 0 represents death. The strength of these generic HRQL instruments is that they meet the criteria of utility theory and the data may be used in cost/utility analyses. The best known of the instruments for evaluating patient preferences are the Standard Gamble (17), the Time Trade Off (17), and the Feeling Thermometer (17). The Multi-attribute Health Status Classification Systems determine preferences from the perspective of society (18). Data from both types of instruments may be used to generate Quality Adjusted Life Years (QALYs) (17) for health economic research and policy making.

C. Disease-Specific

Recognition that the generic instruments are often insufficiently responsive to changes or differences in HRQL that are important to patients has led to the development of disease-specific instruments for asthma. The strength of the disease-specific instruments is that they focus on the areas of function that are most important to patients and are intuitively sensible to clinicians. However, the weakness of disease-specific instruments is that the degree of impairment cannot be compared across different conditions.

III. Instrument Measurement Properties

A. Content Validity

When selecting an instrument for a particular study, the investigator should first ensure that it has content validity, that is, the items in the questionnaire have been selected using recognized procedures that ensure that they capture all areas of function considered important by patients. Questionnaires in which items have been selected by clinicians rarely meet this criterion. There are two schools of thought as to how items should be selected for quality-of-life questionnaires. The first approach, known as the "impact" or "importance" method, asks patients to identify items of dysfunction that they have experienced as a result of their condition and to score each item according to its importance. Items identified most frequently and scoring the highest are selected for the final questionnaire. The second approach, often known as the "factor analysis" method, uses psychometric methods to evaluate patient responses to items in order to determine the relationship between them. Items are selected for the final questionnaire on the basis of mathematical modeling. Initial studies suggest that the two approaches may produce fairly similar final questionnaires (19). There is certainly some heterogeneity in the impairments patients experience as a result of a specific condition; both methods select items likely to be appropriate to the majority of patients.

B. Types of Instruments

Having determined that the selected instrument has content validity, the investigator should assess the measurement properties. Instruments which are to be used in cross-sectional studies (surveys) require different measurement properties from those used in longitudinal studies (clinical trials) (20).

C. Discriminative Instruments

These instruments are used to distinguish between individuals or groups of patients, for example, between individuals who do or do not have asthma or, within asthma patients, between those who have mild, moderate, or severe impairment. They are most commonly used for screening and in cross-sectional surveys. The measurement properties required of a discriminative instrument are reliability and cross-sectional construct validity (20).

Reliability

A simple method of deciding what measurement properties are going to be needed in a particular situation is to consider the signal to noise ratio (Table 2). The "signal" is the true difference or change that is to be measured, and the "noise" is variance, unrelated to the true signal, that is going to interfere with the detection of the signal. Discriminative instruments need to be able to detect differences between individuals or patients. Therefore, the signal is the between-subject difference at one point in time. The noise that is going to tend to mask this signal is the within-subject variance. The test statistic usually used to express the relationship between the signal and the noise for discriminative instruments is the intraclass correlation coefficient (ICC), which relates the between-subject variance to the total variance. (Cronbach's alpha, which measures the internal consistency, does not give an indication of this property.)

Cross-Sectional Construct Validity

When there is no gold standard against which to demonstrate that the instrument is actually measuring what it purports to measure, the developer puts

Table 2 Measurement Properties Necessary for Evaluative and Discriminative Instruments

	Discrimination	Evaluation
Signal	Between-subject differences	Within-subject differences related to true within-subject change
Noise	Within-subject differences	Within-subject differences unrelated to true within-subject change
Signal-to-noise ratio: descriptive term	Reliability	Responsiveness
Construct validity	Cross-sectional	Longitudinal

forward hypotheses or constructs, which, if they are met, provide evidence that the instrument is valid. The approach frequently used is to demonstrate that the various domains of the new HRQL instrument correlate in a predicted manner with other indices of clinical asthma severity and with other HRQL instruments.

D. Evaluative Instruments

These instruments are required to measure longitudinal change within an individual or group of patients and are the type used in clinical trials. The measurement properties required of an evaluative instrument are responsiveness and longitudinal construct validity (20).

Responsiveness

Evaluative instruments must be responsive to small but clinically important changes that occur either spontaneously or as the result of an intervention. The signal in an evaluative instrument is the true within-subject change over time, and the noise is the within-subject variance unrelated to the true within-subject change. The relationship between the two demonstrates the responsiveness of the instrument (21).

Longitudinal Construct Validity

The second measurement property necessary for an evaluative instrument is longitudinal validity. This addresses the question: When there is a change in score, does this reflect a true change in the patient's HRQL? Longitudinal validity is usually demonstrated by showing that changes in the various domains of the new HRQL instrument correlate in a predicted manner with changes in other appropriate outcome measures, such as clinical asthma severity and generic HRQL.

E. Evaluative versus Discriminative Instruments

It is common to find that developers report good reliability and cross-sectional construct validity for a new instrument and then conclude that it can be used in clinical trials. This is often erroneous. Instruments with good reliability may not be responsive, and good cross-sectional validity is no guarantee of longitudinal validity. This was demonstrated clearly in a study by Rutten (2). She compared the measurement properties of several HRQL instruments in asthma. She showed that the SIP (14) has good reliability but was insufficiently responsive to detect clinically important changes in HRQL detected by other instruments. Similarly, she found that cross-sectional validity for the SIP was sound but longitudinal validity was poor. When selecting a questionnaire for a clinical trial, the clinician should ensure that responsiveness and longitudinal validity are secure.

F. Interpretability

Repeated experience with a wide variety of physiological measures allows clinicians to make meaningful interpretation of results. For instance, the experienced clinician will have little difficulty in interpreting a 0.5-liter increase in forced expiratory volume in one second (FEV_1). In contrast, the meaning of a change in score of 1.0 on a HRQL instrument is less intuitively obvious, not only because there are no units but also because health professionals seldom use HRQL measures in clinical practice. Two fundamental approaches have been suggested for the interpretation of HRQL data (22). The first, "distribution-based," is based entirely on the statistical distribution of the results of a particular study, the most commonly used being the effect size, which is derived from the magnitude of change and the variability in stable subjects. The problem with this approach is that there is still no indication as to whether the effect is important to the patient. The second approach, which is the one we use, is referred to as "anchor-based," where the changes in quality-of-life measures are compared, or anchored, to other clinically meaningful outcomes. We have defined a minimal important difference as "the smallest difference in score in the domain of interest which patients perceive as beneficial and would mandate, in the absence of troublesome side-effects and excessive cost, a change in the patient's management" (23). In brief, using global rating of change questionnaires, patients are asked whether they have experienced changes in their asthma quality of life that they consider important. To obtain the minimal important difference, the change in asthma quality-of-life score that corresponds to change that patients consider small but important is calculated (23).

IV. Generic Quality-of-Life Questionnaires in Asthma

A. Health Profiles

Preliminary work suggests that both the SIP and the SF-36 may have satisfactory properties as discriminative instruments in asthma. The SIP has shown good reliability (ICC = 0.81) (2) and acceptable cross-sectional correlations with conventional clinical indices of asthma and disease-specific quality of life (2,4,24). The SF-36 has shown moderate cross-sectional correlations ($r = 0.30$–0.50) with clinical asthma severity scores in 5 of the 9 domains and with FEV_1% predicted in 3 of the 9 domains (25). A weakness of generic instruments is that they may not be responsive in clinical trials to small but important changes that occur as a result of the intervention. Two asthma clinical trials failed to show differences between the treatment groups using the SIP (2,26). Similarly, both we (24) and Marks (3), in the validation studies of our asthma-specific instruments, showed that the SIP is unlikely to have good responsiveness characteristics in asthma. As yet there are no published reports of asthma clinical trials in which the SF-36 has been used.

B. Utilities

So far there is only one published study in which the measurement properties of utility instruments in asthma are reported. Rutten (2) showed that both the standard gamble and the feeling thermometer (rating scale) have moderate reliability (ICC = 0.56 and 0.65, respectively), but only the feeling thermometer had acceptable cross-sectional validity. This suggests that only the feeling thermometer has adequate discriminative properties in asthma. For evaluative properties, the feeling thermometer showed moderate responsiveness but poor longitudinal construct validity. This suggests that it is sensitive to change but what it is actually measuring is unclear. The standard gamble was totally lacking in both evaluative properties.

V. Disease-Specific Instruments in Asthma

A number of instruments claim to measure HRQL in asthma but focus exclusively on symptom and physical activity limitations that clinicians consider important and do not incorporate the functional impairments that are important to patients. I have restricted this review to those instruments that measure aspects of the condition that are identified as being important by the patients themselves and for which there are at least initial reports of development methods and measurement properties. Currently, there are 10 (7 adult and 3 pediatric) disease-specific HRQL instruments that meet these criteria and are applicable to patients with asthma (Tables 3 and 4). I have summarized the instruments and the measurement characteristics that investigators may need to take into consideration when selecting an instrument.

A. Adults

The Asthma Quality of Life Questionnaire

This is a 32-item questionnaire (5,24) in which the items were selected entirely according to their importance to patients. Patients were asked which of a pool of 150 potentially important items they had experienced in the last year, and they scored the importance of each positively identified item on a 5-point scale. Items identified most frequently and scoring the highest were those included in the Asthma Quality of Life Questionnaire. The items are in

Table 3 Disease-Specific Quality-of-Life Questionnaires for Adults with Asthma: Published Measurement Properties

Instrument	Population	Number of items	Reliability	Cross-sectional validity	Responsiveness	Longitudinal validity	Clinical interpretation
Asthma Quality of Life Questionnaire (24)	Asthma	32	Yes	Yes	Yes	Yes	Yes
Asthma Quality of Life Questionnaire (6)	Asthma	20	Yes	Yes	Yes	Yes	?
Living with Asthma Questionnaire (27)	Asthma	68	Yes	Yes	?Yes	?	?
St. George's Respiratory Questionnaire (31)	COPD and asthma	76	Yes	Yes	?Yes	?	?
Respiratory Illness Quality of Life Questionnaire (33)	Asthma, COPD, and emphysema	55	?	Yes	?	?	?
Life Activities Questionnaire for Adult Asthma (8)	Asthma	71	?	?	?	?	?
Asthma Bother Profile (34)	Asthma	22	?	?	?	?	?

Table 4 Disease-Specific Quality-of-Life Questionnaires for Children with Asthma: Published Measurement Properties

Instrument	Population/age	Number of items	Reliability	Cross-sectional validity	Responsive-ness	Longitudinal validity	Clinical interpretation
Paediatric Asthma Quality of Life Questionnaire (36)	Asthma, 7–17 yr	23	Yes	Yes	Yes	Yes	Yes
Life Activities Questionnaire for Childhood Asthma (37)	Asthma, 5–17 yr	71	?	?	?	?	?
Childhood Asthma Questionnaires (38) CAQA, CAQB, CAQC	Asthma, 4–7, 8–11, and 12–16 yr	Variable	Yes	?	?	?	?

four domains (symptoms, emotions, exposure to environmental stimuli, and activity limitation). Patients respond to each item on a 7-point scale. A unique feature of the Asthma Quality of Life Questionnaire is that 5 of the 11 items in the activity limitation domain are "individualized." At the start of a study, each patient identifies 5 activities which are done regularly and which are important in the patient's day-to-day life. At each visit, the patient is asked about limitations experienced in these activities. This allows the questionnaire to capture specific activity limitations that are important for the individual patient. The instrument is in both interviewer and self-administered format and takes approximately 10 minutes to complete at the first visit and 5 minutes at follow-up.

There have been three independent studies in which the measurement properties of the Asthma Quality of Life Questionnaire have been evaluated. The first was a formal validation by the developers (24), the second was a clinical trial in an emergency department setting (4) and the third was a comparison of the measurement properties of several HRQL instruments during a salbutamol versus salmeterol clinical trial (2).

The formal validation study was an 8-week single cohort study of 39 adults with asthma. Asthma quality of life, symptoms, asthma medication use, airway hyperresponsiveness, and generic HRQL (SIP and a shortened form of the Rand General Health Survey) were evaluated at baseline, 4, and 8 weeks. Patients with inadequately controlled asthma were offered additional inhaled steroids. Using global rating of change questionnaires, patient health status was classified as having remained stable or having changed for each time period (baseline to 4 weeks and 4–8 weeks). In patients who remained stable, the instrument demonstrated good reliability both for overall quality of life and for individual domains with intraclass correlation coefficients (ICC) ranging from 0.89 to 0.94. Responsiveness was demonstrated by the instrument's ability to detect change in overall quality of life and all domains in those patients who changed ($p < 0.001$) and in differentiating successfully between those who changed and those who remained stable ($p < 0.001$). Longitudinal and cross-sectional correlations between asthma quality of life and both conventional clinical asthma measures and generic HRQL were close to predictions made before the analysis. In the longitudinal correlations, changes in asthma quality of life were highly correlated ($r > 0.50$) with changes in clinical asthma control, and generic quality of life as measured by the Rand; moderately correlated ($r = 0.35$–0.50) with peak flow rates and medication use; and weakly correlated ($r = 0.20$–0.35) with FEV_1%-predicted airway responsiveness and the SIP. In the cross-sectional correlations, asthma quality of life was highly correlated with clinical asthma control and the Rand, moderately correlated with the SIP, but not correlated with FEV_1% predicted peak flow rates or airway responsiveness.

In the emergency department clinical trial (4), 43 adults were treated for asthma exacerbations and reevaluated after 7–10 days. Asthma quality of life, spirometry, and generic HRQL (SIP) were assessed at each visit. As in the first study, global rating of change questionnaires were used to determine whether patients changed their health status. In the stable patients, the intra-class correlation coefficient for overall quality of life and for each of the domains ranged from 0.90 to 0.96. Responsiveness was clearly demonstrated by significant changes in all domains of asthma quality of life. In cross-sectional correlations, asthma quality of life correlated well with the SIP ($r = 0.49$) and asthma symptoms ($r = 0.74$) but not so well with the FEV_1% predicted ($r = 0.27$). In the longitudinal correlations, change in asthma quality of life correlated well with global rating of change ($r = 0.78$) and moderately with change in FEV_1 ($r = 0.43$). Changes in the SIP were not reported.

Using the global rating of change questionnaires, both studies (4,24) showed that for the Asthma Quality of Life Questionnaire a change in score of 0.5, both for overall quality of life and for each of the individual domains, represents a minimal important difference. They also demonstrated that a change in quality-of-life score of approximately 1.0 represents a moderate change and a change in score greater than 2.0 represents a large change in HRQL. Using the responsiveness index, which is derived from the minimal important difference and the pooled within-subject standard deviation, clinical study sample sizes have been calculated for various alpha and beta error rates (24). For example, a parallel group design clinical trial with the alpha error rate (two-tailed) set at 0.05 and the beta error rate at 0.1 requires a sample size of 35 patients per group.

The results of the instrument comparison study are presented in detail later in the chapter. In brief, the results showed that the Asthma Quality of Life Questionnaire is very reliable (ICC = 0.93)—the most responsive of all the instruments tested—and has good construct validity.

In a multicenter, double-blind randomized cross-over trial of salmeterol, salbutamol, and placebo in mild to moderate asthma, the Asthma Quality of Life Questionnaire was able to detect, with high statistical significance ($p < 0.001$), differences between salmeterol and salbutamol for all domains and for overall quality of life (1). It was also able to demonstrate that only the differences in the symptom and emotional function domains could actually be considered clinically important.

The Asthma Quality of Life Questionnaire

The Asthma Quality of Life Questionnaire (AQLQ) (3,6) is a self-administered questionnaire containing 20 items in 4 domains (breathlessness and physical restrictions, mood disturbance, social disruption, and concerns for health).

Patients rate each item on a 5-point scale. The investigators selected items using both importance as rated by patients and psychometric techniques. The questionnaire takes approximately 5 minutes to complete.

The instrument has been thoroughly tested for both discriminative and evaluative properties. Reliability was assessed in 58 patients with stable asthma and resulted in an intraclass correlation coefficient of 0.80. Cross-sectional construct validity was evaluated in a survey of 87 patients. Asthma quality of life showed moderate correlation with the amount of medication used ($r = 0.38$) but no correlations with either FEV_1% predicted ($r = -0.20$) or airway responsiveness ($r = -0.16$). Evaluative properties were examined in a cohort of 44 adults with asthma followed for 3–4 months. In the 19 subjects who improved according to clinical criteria, the instrument was able to measure changes in quality of life and show that these changes were greater than those observed in stable patients ($p = 0.007$). In the longitudinal correlations, asthma quality of life was moderately correlated with asthma symptoms ($r = 0.37$) and airway hyperresponsiveness ($r = 0.38$) but not with peak flow variability ($r = 0.12$) or the SIP ($r = 0.18$).

The Living with Asthma Questionnaire

The Living with Asthma Questionnaire (7,27) is a 68-item instrument with 11 domains (social/leisure, sport, holidays, sleep, work, colds, morbidity, effects on others, medication use, sex, and dysphoric states and attitudes). Items, identified from patient focus group discussions, were selected for the questionnaire using psychometric techniques and factor analysis. There were four independent surveys during the item reduction process. After each survey, items were discarded if they were poor discriminators (more than 70% of patients endorsed a response), showed low factor loading, or wording was a problem. Unlike the Juniper (24) and Marks (6) instruments, impairments experienced as a direct result of asthma symptoms are not included. Responses are given using a 3-point scale, suggesting that it may have acceptable discriminative properties but be less responsive to within-patient changes.

The test-retest of the instrument was evaluated in two postal surveys separated by 2 months; 95 responses were obtained and showed a correlation coefficient (r) = 0.95. For cross-sectional construct validity, the Living with Asthma Questionnaire was highly correlated with the SIP ($r = 0.66$, $n = 76$), moderately with steroid prescribing ($r = 0.35$, $n = 40$), and moderately with peak flow rates ($r = -0.44$, $n = 40$). An independent study of the measurement properties (2) provides further evidence of good discriminative properties with good reliability (ICC = 0.85) and very acceptable correlations with asthma symptoms ($r = 0.42$) and generic HRQL (SIP: $r = 0.59$). The evaluative properties are not so strong. In a 425-patient double-blind placebo-controlled clinical trial of salmeterol (28), there were highly significant differences in peak flow rates between salmeterol and placebo ($p = 0.0001$), but only 63% of patients on salmeterol showed improvement in quality of life compared to 46% on placebo. Although this difference reached significance ($p < 0.05$), these results suggest that the 3-point response options may limit the responsiveness of the instrument. Similarly, in the instrument comparison study (2), the Living with Asthma Questionnaire was unable to discern differences between salmeterol and salbutamol that were detected by other HRQL instruments.

More recent factor analysis has suggested that the items in the questionnaire represent two constructs—problems (49 items) and evaluations (19 items)—both with high internal consistency (Cronbach's alpha > 0.9) (29). In a 6-week, double-blind, randomized comparison of salmeterol (50 μg, administered twice daily) and placebo in 79 patients, improvements in the problems subscale were significantly greater on salmeterol than placebo, but there were no differences between the two treatments for evaluations (30). When analyzed in the original 11 domains, only three reached significance (social/leisure, sport, and sleep). Once again, these results suggest that the instrument may only have limited responsiveness.

The St. George's Respiratory Questionnaire

This questionnaire (31) is self-administered and is applicable to patients with both reversible and fixed airway obstruction. It contains 76 items in three domains [symptoms, activity, and impacts (on daily life)]. The methods used for the development have not yet been published, and so the criteria for item selection are unclear.

The questionnaire, when presented to 40 patients with asthma on two occasions 2 weeks apart, showed good reliability (ICC = 0.91). In a cross-sectional study of 141 patients with long-standing airflow limitation, the activity domain correlated with the 6-minute walk test ($r = 0.35$), the MRC dyspnea grade ($r = 0.50$), and the SIP ($r = 0.39$), but not with the FEV_1 ($r = 0.01$). Similar correlations were found for the impacts domain, but the symptom domain did not correlate with any of the clinical or generic HRQL measures.

In a 719-patient randomized placebo-controlled trial of nedocromil, all patients showed a significant improvement in all domains of the quality-of-life instrument after 24 weeks. However, only the impacts domain of the questionnaire showed a difference between nedocromil and placebo ($p < 0.05$) (26). This suggests that the instrument may have only moderate responsiveness. Patient opinion of treatment efficacy was used to determine what change in score of the St. George's Respiratory Questionnaire can be considered clinically important. Patients who considered that the medication was slightly efficacious had a change in score of approximately 4 units and those who considered it very efficacious had a change in score of approximately 12 units (32).

Respiratory Illness Quality of Life Questionnaire

Like the St. George's Respiratory Questionnaire, the items in this instrument (33) were selected to be applicable to both COPD and asthma patients. Initially, a pool of 221 potentially important items was generated from a literature review, existing measures, and consultation with health professionals. Two hundred and four patients identified the amount of bother experienced with each item during the previous year. Items were included if they were applicable to at least 70% of respondents and loaded highly on the first two or three factors in the factor analysis. There are 55 items in seven domains (breathing problems, physical problems, emotions, situations triggering or enhancing breathing problems, daily and domestic activities, social activities relationships and sexuality, and general activities). Cross-sectional correlations between the domains and clinical indices of disease severity (dyspnea, frequency of attacks, severity, GP visits, and absence from work: $r = 0.14–0.59$) suggest that the instrument may have good discriminative properties. However, reliability, in terms of the ICC, has not yet been reported and neither have the measurement properties necessary for use in clinical trials.

A Life Activities Questionnaire for Adult Asthma

This instrument (8) has 70 items in seven domains (physical activities, work activities, outdoor activities, emotions and emotional behavior, home care, eating and drinking activities, and miscellaneous). Items were selected on the basis of frequency of experience by patients. Responses are given on a 5-point scale, and the questionnaire is self-administered. The instrument shows high internal consistency with a Cronbach's alpha of 0.96–0.97. Other measurement properties necessary for confident use as either a discriminative or evaluative instrument have not been published. The questionnaire reports only activity limitations and does not include other aspects of HRQL that other investigators have shown are important to patients with asthma (e.g., emotional function, concerns about health, symptoms).

Asthma Bother Profile

The 22 items in this self-administered questionnaire (34) were selected from five earlier asthma and COPD quality-of-life questionnaires (3,24,27,31,35) to measure the psychological impact of asthma. Two domains measure the distress that asthma occasions in a variety of contexts (items = 15) and asthma management, which itself has three subdomains: asthma knowledge, quality of care, and confidence in self-management (items – 7). Focus group discussions with 32 asthma patients helped to refine the items and suggested that the content was relevant. However, cross-sectional construct validity has not yet been established. The management domain but not the distress domain was able to detect differences between patients who attended a self-management asthma clinic and those who did not. The reliability for the full instrument and the measurement properties necessary for clinical trials have not yet been reported.

Summary of Adult Asthma-Specific Quality-of-Life Instruments

Data available suggest that the first five adult instruments (6,24,27,31,33) have measurement properties that enable them to be used satisfactorily for discriminative purposes, but only the first two by Juniper (24) and Marks (6) have been shown to have good responsiveness and longitudinal construct validity, properties that are essential for use in clinical trials. Both the Living with Asthma Questionnaire (27) and the St. George's Respiratory Questionnaire (31) are moderately responsive. An additional advantage of both the Juniper questionnaire (24) and the St. George's Respiratory Questionnaire is that interpretation of data has been addressed. Five of the instruments are specific for asthma (6,8,24,27,34) whereas the St. George's Respiratory Questionnaire (31) and the Respiratory Illness Questionnaire (33) are for both asthma and COPD patients.

Comparison of Health-Related Quality-of-Life Instruments in Adults with Asthma

So far, there has only been one direct comparison of the measurement properties of HRQL instruments in asthma. In a 6-week double-blind randomized trial of salmeterol (50 μg, administered twice daily) and salbutamol (400 μg, administered twice daily) in 120 adults with asthma, Rutten (2) compared the measurement properties of the Asthma Quality of Life Questionnaire (24), the Living with Asthma Questionnaire (27), the SIP (14), and two utilities: the feeling thermometer (17) and the standard gamble (17). The results for responsiveness and construct validity are shown in Tables 5–7. Reliability for these instruments was as follows: Asthma Quality of Life Questionnaire (ICC = 0.93), Living with Asthma Questionnaire (ICC = 0.85), Sickness Impact Profile (ICC = 0.81), Feeling Thermometer (ICC = 0.65), and Standard Gamble (ICC = 0.56). This comparison suggests that both the Asthma Quality of Life Questionnaire and the Living with Asthma Questionnaire are reliable disease-specific instruments, with the Asthma Quality of Life Questionnaire being more responsive to change and having greater validity. Of the generic instru-

Table 5 Responsiveness to Changes Within Subjects and Differences in Changes Between Salmeterol (N = 53) and Salbutamol (N = 54)

	Change in quality of life				
	Salmeterol	Salbutamol	Difference (sm – sb)	95% CI of difference	*p*-values (sm vs. sb)
Asthma Quality of Life Questionnaire (24)	+0.49[a]	+0.27[a]	0.22	0.03–0.42	0.022
Activities	+0.57[a]	+0.22[b]	0.35	0.14–0.56	0.002
Symptoms	+0.58[a]	+0.33[c]	0.25	–0.001–0.49	0.051
Environment	+0.44[a]	+0.22[b]	0.22	–0.04–0.48	0.100
Emotions	+0.18[b]	+0.17[b]	0.01	–0.19–0.20	0.960
Living with Asthma Questionnaire (27)	–0.12[a]	–0.09[b]	0.03	–0.04–0.09	0.425
Health knowledge	–0.13[a]	–0.08[c]	0.05	–0.02–0.12	0.146
Health appraisal	–0.06[c]	–0.10	0.04	–0.12–0.03	0.290
Sickness Impact Profile (14)	–1.59	–0.93	0.66	–1.11–2.43	0.463
Physical	–0.83	–0.20	0.62	–0.73–2.00	0.363
Psychosocial	–2.82[c]	–1.91[b]	0.91	–1.35–3.16	0.429
Utilities					
Feeling thermometer (17)	+7.35[a]	+4.08[c]	3.27	0.08–6.46	0.045
Standard gamble (17)	+0.008	–0.034	0.042	–0.005–0.09	0.077

Improvements are indicated by + for the AQLQ and the utilities and by – for the LWAQ and the SIP.

[a] $p \leq 0.001$.

[b] $p \leq 0.05$.

[c] $p \leq 0.01$.

Source: Ref. 2.

Table 6 Cross-Sectional Validity: Spearman Rank Correlation Coefficients

	Quality of life				
	Asthma Quality of Life (24)	Living with Asthma (27)	Sickness Impact Profile	Feeling Thermometer	Standard Gamble Utilities
Lung function					
PEFR am	0.02	–0.13	–0.16b	0.13	0.05
PEFR pm	0.03[a]	–0.09	–0.17b	0.11	0.06
FEV_1 % predicted	0.12	–0.15	–0.17b	0.22[c]	–0.04[a]
FVC	0.13	–0.26[c]	–0.29[c]	0.10	0.06
Reversibility (%)	–0.28[c]	0.17	0.02	–0.09	0.04[a]
Symptoms					
Daytime score	–0.59[c]	0.42[c]	0.39[c]	–0.26[b]	–0.17
Nighttime score	0.70[c]	0.42[c]	0.38[c]	0.33[c]	–0.21
Quality of life					
Asthma Quality of Life		–0.73[c]	–0.48[c]	0.47[c]	0.19[b]
Living with Asthma			0.56[c]	–0.43[c]	–0.13
Sickness Impact Profile				0.59[c]	–0.15
Feeling Thermometer					0.17[b]

[a] Paradoxical direction.
[b] $p \le 0.01$.
[c] $p \le 0.001$.
Source: Ref. 2.

Table 7 Longitudinal Construct Validity: Spearman Rank Correlation Coefficients Between Change Values

Change in:	Change in quality of life				
	Asthma Quality of Life (24)	Living with Asthma (27)	Sickness Impact Profile	Feeling Thermometer	Standard Gamble Utilities
Lung Function					
PEFR am	0.36[b]	–0.04	–0.19	0.19	0.23[c]
PEFR pm	0.35[b]	–0.20	–0.28[c]	0.08	0.16
FEV_1 % pred.	0.32[b]	–0.05	–0.11	0.19	0.09
FVC	0.28[c]	–0.05	–0.12	0.21	0.24[c]
Reversibility (%)	–0.21	–0.02[d]	–0.01	–0.12	–0.03
Symptoms					
Daytime score	0.51[b]	–0.24[c]	–0.20	0.24[c]	0.08
Nighttime score	0.54[b]	–0.06	–0.04	0.23[c]	0.14
Overall assessment of efficacy[a]	0.64[b]	–0.20	–0.29	0.24	0.20
Quality of life					
Asthma Quality of Life		–0.29[c]	–0.31[b]	0.36[b]	0.29[b]
Living with Asthma			–0.31[b]	–0.11	–0.09
Sickness Impact Profile				–0.17	–0.18
Feeling Thermometer					0.16

[a] Patient's overall assessment of efficacy is not a change score.
[c] $p \leq 0.01$.
[b] $p \leq 0.001$.
[d] Paradoxical direction.
Source: Ref. 2.

ments, the SIP has discriminative properties. The feeling thermometer is responsive but has only poor longitudinal construct validity, making it questionable as to what it is actually measuring. The standard gamble has poor measurement characteristics both for discrimination and evaluation.

B. Children

Several questionnaires assess impairments in children with asthma, but only three (36–38) meet the two criteria specified for inclusion in the adult instrument review: the method of development ensured that items important to children with asthma are included, and there is evidence of the development methods and measurement properties. Usherwood (12) developed an instrument for assessing symptoms and disability in children, but the items were selected by general practitioners and mothers of children with asthma and it does not cover emotional or social function. In addition, the questionnaire is completed by the mother and there is evidence to suggest that parents may have a poor perception of their child's HRQL (39).

The Paediatric Asthma Quality of Life Questionnaire

This questionnaire (36) has been designed for children with asthma age 7–17 years. It has 23 items in three domains (symptoms, activity limitation, and emotional function); three of the activity items are individualized and identified by the patient at the beginning of the study. Items were selected on the basis of their importance to the children themselves. It is available in both interviewer and self-administered versions, has 7-point response options, and takes approximately 10 minutes to complete. Patients experience no difficulty understanding the questions or the response options. In a 52-patient validation study, the questionnaire has shown good measurement properties as both an evaluative and discriminative instrument. In patients who are stable, it has very acceptable reliability with an intraclass correlation coefficient of 0.84. It has also shown good responsiveness in being able to detect changes in patients whose health state changed ($p < 0.001$) and being able to differentiate between stable and unstable patients ($p < 0.0001$). Correlations between the instrument and both conventional clinical asthma measures and generic HRQL were close to predicted for both longitudinal and cross-sectional validation. For the longitudinal correlations, asthma quality of life was highly correlated ($r > 0.50$) with generic HRQL, moderately correlated ($r = 0.35$–0.50) with peak flow rates and beta-agonist use, weakly correlated ($r = 0.20$–0.35) with clinical asthma control, and not correlated with FEV_1% predicted. For the cross-sectional correlations, asthma quality of life was highly correlated with clinical asthma control and beta-agonist use, moderately correlated with generic HRQL, weakly correlated with peak flow rates, and not correlated with

FEV_1% predicted. Both discriminative and evaluative measurement properties were consistent across all ages. The clinical interpretation of changes in score is very similar to that found for the Asthma Quality of Life Questionnaire (24), namely, a change in score of approximately 0.5 represents a minimal important difference.

A Life Activities Questionnaire for Childhood Asthma

In an open-ended questionnaire (37), 92 children (5–17 years) with asthma were asked to identify activities in which they were limited by their asthma. Of the 246 activities identified, 71 were reported by two or more children and included in the questionnaire. Activities are in seven domains [physical, work, outdoor, emotions and emotional behavior (e.g., laughing), home care, eating and drinking, and miscellaneous]. Children are asked to recall limitations during the past week and to score these on a 5-point scale (no restriction–total restriction). Test-retest yielded a Pearson correlation of 0.76, and there was internal consistency with Cronbach's alpha equal to 0.97. Other measurement properties have not yet been reported.

Childhood Asthma Questionnaires

Three separate instruments have been developed for children aged 4–7 years (CAQA), 8–11 years (CAQB), and 12–16 years (CAQC). The questionnaires (10,38) were developed from discussions with children, but the actual method of item selection is unclear. The CAQA is in two domains: quality of living (applicable to asthmatic and nonasthmatic children, 14 items) and distress (asthma only, items = 4). Response options are four "smiley" faces, and the questionnaire is completed with the help of the parent. The CAQB has four domains: active quality of living and passive quality of living (22 items), distress (6 items), and severity (6 items). Response options are five faces. The CAQC has five domains: active and teenage quality of living, distress, severity, and reactivity. Response options are five faces. The instruments were tested in 535 children, 281 (4–16 years) with a history of asthma and 256 (4–11 years) with no history of asthma. Reliability was assessed for each instrument (CAQA: ICC = 0.59–0.63; CAQB: ICC = 0.72–0.75; CAQC: ICC = 0.68–0.84). No differences were observed between asthmatics and nonasthmatics for the quality of living domains of the CAQA and CAQB; this property was not tested in other domains or the CAQC. Lack of ability to differentiate between asthma and non-asthma subjects suggests that the instrument may have limited discriminative properties. No cross-sectional correlations between the instruments and indices of clinical asthma severity are provided. Similarly, data are not yet available for responsiveness or longitudinal construct validity.

Summary of Pediatric Asthma-Specific Quality of Life Instruments

The Paediatric Asthma Quality of Life Questionnaire (36) has demonstrated good properties both as an evaluative and discriminative instrument. The question of interpretation of data has been addressed, and it can therefore be used with confidence in both clinical trials and surveys. The Life Activities Questionnaire for Childhood Asthma (37) measures activity limitations but not other quality-of-life impairments important to children with asthma. Neither discriminative nor evaluative properties have yet been reported. Although the Childhood Asthma Questionnaires (38) show moderate to good reliability, discriminative ability may be poor, and no evidence is yet available on whether the instruments are sufficiently responsive for use in clinical trials.

C. Quality of Life of Caregivers

Chronic illness not only has an impact on the patients themselves, but also impairs the day-to-day lives of the family and in particular the primary caregiver. Many caregivers find themselves limited in their activities and emotionally distressed by the additional burden. Although there are few adults with asthma who require a caregiver or who have a significant effect on family function, children with asthma often require families, and particularly parents, to modify or limit activities and cause a great deal of concern and worry.

The Paediatric Asthma Caregiver's Quality of Life Questionnaire is a self-administered 13-item instrument with two domains (activity limitations and emotional function) (40). Responses are given on a 7-point scale, and caregivers (usually the parent) are asked to recall experiences during the past week. The instrument has been fully validated and shows both good reliability (ICC = 0.85) and responsiveness in that it was able to detect changes in caregivers whose quality of life changed ($p < 0.0003$) and able to differentiate these caregivers from those whose quality of life stayed stable ($p < 0.0001$). Both cross-sectional and longitudinal correlations were close to predicted and therefore provided evidence of the validity of the instrument.

VI. Cultural Adaptation of Quality-of-Life Questionnaires

All but one of the instruments included in this review were developed in English-speaking countries. The original language of the Respiratory Illness Quality of Life Questionnaire is Dutch. Adapting quality-of-life questionnaires to other languages and cultures is considerably more complicated than just doing a simple translation (41). Quite a number of the generic and asthma-specific quality-of-life questionnaires have been adapted into European, American, and Asian languages using well-established procedures. Information on the availability of the translations can be obtained from the developers. Cultural adaptations should not be undertaken without the permission of the developer.

VII. Conclusions

Many clinicians today feel that they would like to incorporate an assessment of HRQL into their asthma evaluations but are overwhelmed by the number of instruments available and don't know which one to choose. The first step is to formulate the question carefully and decide exactly what is to be measured. Is it differences between patients in a survey? Is it the effects of an intervention in a clinical trial? Is it impairments in asthma compared with impairments in another condition? Having decided exactly what is to be measured, the clinician should decide whether good evaluative or discriminative properties are going to be needed and whether the question is best addressed using a disease-specific or generic instrument. Having gone through these steps, I hope that armed with the information contained in this review, the investigator will then find it really quite simple to select the instrument most suited to the task.

References

1. Juniper EF, Johnston PR, Borkhoff CM, Guyatt GH, Boulet LP, Haukioja A. Quality of life in asthma clinical trials: comparison of salmeterol and salbutamol. Am J Respir Crit Care Med 1995; 151:66–70.
2. Rutten-van Molken MPMH et al. Comparison of performance of four instruments in evaluating the effects of salmeterol on asthma quality of life. Eur Respir J 1995; 8:888–898.
3. Marks GB, Dunn SM, Woolcock AJ. An evaluation of an asthma quality of life questionnaire as a measure of change in adults with asthma. J Clin Epidemiol 1993; 46:1103–1111.
4. Rowe BH, Oxman AD. Performance of an asthma quality of life questionnaire in an outpatient setting. Am Rev Respir Dis 1993; 148:675–681.
5. Juniper EF, Guyatt GH, Epstein RS, Ferrie PJ, Jaeschke R, Hiller TK. Evaluation of impairment of health-related quality of life in asthma: development of a questionnaire for use in clinical trials. Thorax 1992; 47:76–83.
6. Marks GB, Dunn SM, Woolcock AJ. A scale for the measurement of quality of life in adults with asthma. J Clin Epidemiol 1992; 45:461–472.
7. Hyland ME, Finnis S, Irvine SH. A scale for assessing quality of life in adult asthma sufferers. J Psychomatic Res 1991; 35:99–110.

8. Creer TL, Wigal JK, Kotses H, McConnaughy K, Winder JA. A life activities questionnaire for adult asthma. J Asthma 1992; 29:393–399.
9. Townsend M, Feeny DH, Guyatt GH, Seip AK, Dolovich J. Evaluation of the burden of illness for pediatric asthmatic patients and their parents. Ann Allergy 1991; 67:403–408.
10. Christie MJ, French D, Sowden A, West A. Development of child-centred disease specific questionnaires for living with asthma. Psychosom Med 1993; 55:541–548.
11. Nocon A. Social and emotional impact of childhood asthma. Arch Dis Child 1991; 66:458–460.
12. Usherwood TP, Scrimgeour A, Barber JH. Questionnaire to measure perceived symptoms and disability in asthma. Arch Dis Child 1990; 65:779–781.
13. Stewart AL, Hays R, Ware JF. The MOS short-form general health survey. Reliability and validity in a patient population. Med Care 1988; 26:724–732.
14. Bergner M, Bobbitt RA, Carter WB, Gilson BS. The sickness impact profile; development and final revision of a health status measure. Med Care 1981; 19:787–805.
15. Hunt SM, McKenna SP, McEwen J, Backett EM, Williams J, Papp E. A quantitative approach to perceived health status; a validation study. J Epidemiol Comm Health 1980; 34:281–286.
16. Sackett DL, Chambers LW, MacPherson AS, Goldsmith CH, McAuley RG. The development and application of indices of health; general methods and summary of results. AJPH 1977; 67:423–428.
17. Torrance GW. Measurement of health state utilities for economic appraisal. J Health Econ 1986; 5:1–30.
18. Feeny D, Furlong W, Boyle M, Torrance GW. Multi-attribute health status classification systems. Pharmacoeconomics 1995; 7:490–502.
19. Juniper EF, Guyatt GH, King DR. Comparison of methods for selecting items for a disease-specific quality of life questionnaire: importance versus factor analysis. Quality of Life Res 1994; 3:52–53.
20. Guyatt GH, Kirshner B, Jaeschke R. Measuring health status: What are the necessary measurement properties? J Clin Epidemiol 1992; 45:1341–1345.
21. Guyatt GH, Walter S, Norman G. Measuring change over time: assessing the usefulness of evaluative instruments. J Chron Dis 1987; 40:171–178.
22. Lydick E, Epstein RS. Interpretation of quality of life changes. Quality of Life Res 1993; 2:221–226.
23. Juniper EF, Guyatt GH, Willan A, Griffith LE. Determining a minimal important change in a disease-specific quality of life questionnaire. J Clin Epidemiol 1994; 47:81–87.
24. Juniper EF, Guyatt GH, Ferrie PJ, Griffith LE. Measuring quality of life in asthma. Am Rev Respir Dis 1993; 147:832–838.
25. Bousquet J, Knani J, Dhivert H, Richard A, Chicoye A, Ware JE, Michel FB. Quality of lite in asthma. 1. Internal consistency and validity of the SF-36 questionnaire. Am J Respir Crit Care Med 1994; 149:371–375.
26. Jones PW. Nedocromil sodium quality of life study group. Quality of life, symptoms and pulmonary function in asthma: long-term treatment with nedocromil sodium examined in a controlled multicentre trial. Eur Respir J 1994; 7:55–62.

27. Hyland ME. The living with asthma questionnaire. Respir Med 1991; 85:13–16.
28. Palmer JBD, Hyland ME. Salmeterol in clinical practice: comparator and safety studies, quality of life studies. Eur Respir Rev 1991; 1:301–303.
29. Hyland M. Antiasthma drugs: quality of life rating scales and sensitivity to longitudinal change. Pharmacoeconomics 1994; 6:325–329.
30. Hyland ME, Kenyon CAP, Jacobs PA. Sensitivity of quality of life domains and constructs to longitudinal change in as clinical trial comparing salmeterol with placebo in asthmatics. Quality of Life Res 1994; 3:121–126.
31. Jones PW, Quirk FH, Baveystock CM, Littlejohns P. A self-complete measure of health status for chronic airflow limitation; the St. George's Respiratory Questionnaire. Am Rev Respir Dis 1992; 145:1321–1327.
32. Jones PW, Lasserson D. Relationship between change in the St. George's Respiratory Questionnaire score and patient perception of treatment efficacy after one year of therapy with nedocromil. Am J Respir Crit Care Med 1994; 149:A211.
33. Maille AR, Kaptein AA, Koning CJM, Zwinderman AH. Developing a quality of life questionnaire for patients with respiratory illness. Monaldi Areh Chest Dis 1994; 49:76–78.
34. Hyland ME, Ley A, Fisher DW, Woodward V. Measurement of psyehologieal distress in asthma and asthina management programs. Br J Clin Psychol 1995; 54:601–611.
35. Guyatt GH, Bennan LB, Townsend M, Pugsley SO, Chambers LW. A measure of quality of life for elinieal trials in ehronie lung disease. Thorax 1987; 42:773–778.
36. Juniper EF, Guyatt GH, Feeny DH, Ferrie PJ, Griffith LE, Townsend M. Measuring quality of life in children with asthma. Quality of Life Res 1996; 5:35–46.
37. Creer TL, Wigal JK, Kotses H, Hatala JC, McConnaughty K, Winder JA. A life activities questionnaire for childhood asthma. J Asthma 1993; 30:467–473.
38. French DJ, Christie MJ, Sowden AJ. The reproducibility of the childhood asthma questionnaires: measures of quality of life for children with asthma aged 4–16 years. Quality of Life Res 1994; 3:215–224.
39. Juniper EF, Guyatt GH, Griffith LE, Feenly DH, Ferrie PJ. Should the parent or the child be asked about changes in the child's clinical asthma status? Am J Respir Crit Care Med 1995; 151:A589.
40. Juniper EF, Guyatt GH, Feeny DH, Ferrie PJ, Griffith LE, Townsend M. Measuring quality of life in the parents of children with asthma. Quality of Life Res 1996; 5:27–34.
41. Guillemin F, Bombardier C, Beaton D. Cross-cultural adaptation of health-related quality of life measures: literature review and proposed guidelines. J Clin Epidemiol 1993; 46:1417–1432.

6

Home Monitoring of Lung Function Measures

THOMAS L. CREER and DEIRDRE A. LEVSTEK

Ohio University
Athens, Ohio

JOHN A. WINDER

Allergy and Asthma Research Center
Toledo, Ohio

I. Introduction

At one time, almost all respiratory measurements took place in a hospital or laboratory. Portable devices that could be used in the home and other natural settings were simply unavailable. In 1959, however, Wright and McKerrow (1) described the development of the peak flow meter. The original meter was a simple device consisting of a mouthpiece attached to a meter. A person blew air through the mouthpiece against a vane, causing a spindle and pointer to rotate against spring-produced pressure until the maximum flow was attained. The spindle and pointer were kept in place by a ratchet until released by depressing a reset button. The maximum flow of air sustained for 0.01 second, referred to as a peak expiratory flow rate (PEFR), was indicated in terms of liters per minute. It is the peak or maximal rate of airflow assessed during a maximal exhalation. Values obtained with a peak flow meter are a function of instructions (2), effort (3), and the strength of expiratory muscles (4). Cross and Nelson (5) point out that the physiological determinants of the PEFR are the force of contraction of expiratory muscles, recoil forces of the lungs and chest wall, and large airway resistance. PEFR measures obstruction in the large airways and is relatively insensitive to small airway obstruction.

This chapter describes four aspects of the use of peak flow monitoring in patients with asthma. These include: (1) a history of use of peak flow meters, particularly in residential settings and in homes, (2) clinical uses of the meter, (3) research using peak flow devices, and (4) strengths and limitations of the peak flow meter.

A. History of Residential Use of the Peak Flow Meter

When the peak flow meter was introduced in England, a number of studies were conducted to establish norms and to determine potential uses of the device. An example was research by Nairn et al. (6), who used the meter to determine respiratory lung function in normal schoolchildren. Initial uses of the meter in the United States occurred primarily through the efforts of Hyman Chai, a physician at the Children's Asthma Research Institute and Hospital (CARIH), later a division of the National Asthma Center, in Denver. Beginning in the early 1960s, Dr. Chai visited England annually. During each trip, he acquired a number of meters and returned with them to CARIH in the United States. The meters were used both for clinical and research purposes. A review of uses of the peak flow meter at CARIH is relevant because these efforts set the stage for their subsequent utilization in other settings.

Peak Flow Meters and Asthma Management in Residential Environments

In the early 1960s, peak flow meters were employed in several ways at CARIH to assess and to treat children with asthma. Children in residence at the facility were asked to blow into the meter three times both morning and evening. The highest PEFR was recorded during these twice-daily maneuvers. The rumor rapidly spread among the children that peak flow rates were the "gold standard" with respect to being discharged from the facility. The higher the peak flow rate you blew over a period, the rumor continued, the more apt you were to be discharged and returned to your home. While the PEFR was but one factor considered in making discharge plans for a child, the rumor maintained a high level of motivation among residents in blowing peak flow rates. PEFR values were faithfully collected until CARIH closed in 1981.

PEFR Information to Physicians

The use of peak flow meters provided considerable information and guidance to physicians treating children at CARIH. Peak flow values were useful in predicting asthma exacerbations, and often let physicians know that a child was apt to be hospitalized with an asthma episode 2 days ahead of time. Their real value, however, came in the form of trends of changes in PEFR that could be analyzed over time. Two flow indexes are particularly noteworthy in determining the effectiveness of a patient's treatment: (1) peak to trough variability between morning and evening peak flow values, and (2) changes in the daily peak flow values over time. Arguments can be made regarding the value of either index. Decreases in peak-to-trough variability occur as bronchial lability decreases. In addition, there is a slow increase in mean PEFRs over time that can be witnessed in longitudinal observations. Both approaches to assessment have proven to be useful to physicians, as peak to trough variability is closely correlated with the degree of bronchial hyperreactivity in patients. At CARIH, physicians came in looking at whatever value was chosen, over time, and any correlations with changes or events observed by patients, such as exposure to certain stimuli or symptoms, or changes in the management of a patient's asthma. It was often found that the bronchial hyperreactivity posed such a problem for patients, as when significantly present it permitted even trivial stimuli to trigger severe reactions in the airways and lungs.

The peak flow meter was also useful in determining the presence or absence of exercise-induced asthma (EIA) in response to physical activities. Peak flow values provided meaningful data, particularly if there were difficulties in symptom discrimination or the presence of EIA needed to be documented. However, this approach to detecting EIA gradually gave way to the use of a preexercise bronchodilator or cromolyn sodium before exercise. Later, the performance of an activity without pretreatment permitted the clinician to determine both the degree of the EIA and the influence of pretreatment in controlling the exercise trigger of asthma.

Chai and Newcomb (7) noted that PEFR information was useful only if meters were used on a frequent basis. Otherwise, they continued, an incorrect notion of the state of a patient's asthma might be obtained. Their advice is as relevant today as it was over two decades ago. PEFR data gathered twice daily were also used to support other pulmonary function data gathered less frequently, such as with the spirometer or plethysmograph.

First Uses of PEFR at Home

While there was never a plentiful supply of the standard Wright Peak Flow Meters—the meters cost approximately $300 each—enough of these were obtained to permit some children to take them home during what was referred to as a "home challenge," in which the children were charged with obtaining recorded peak flow values in their homes for varying periods of time; it was one of the first times peak flow meters were used in this manner. Peak flow values obtained during a home challenge were especially useful in that they provided information about a youngster's readiness for discharge from the facility *and* allowed for contrasts of rates of peak flow values gathered in the home versus those collected at CARIH to be analyzed. Striking decreases in

PEFR taken at home were suggested to indicate the presence of stimuli in the environment that served as triggers for a child's asthma. Upon their return to CARIH, the children could be tested with both skin and bronchial challenges to determine what stimuli produced differences in PEFRs gathered in the two settings. Changes in a child's treatment, particularly with a focus on environmental changes in the home, were often recommended prior to the child's discharge from CARIH. The method of detecting different stimuli in two environments is essentially the same technique currently used in determining the role of different stimuli that may precipitate asthma in the natural environment, including the home and workplace (e.g., Ref. 8). It is important to point out that while different peak flow values may be obtained in two settings, peak flow rates by themselves do not reveal why different sets of values are obtained. Testing for differences in stimuli in the two settings is the only way to determine the reasons for the different reactions. The controlled residential environment of CARIH served as a baseline against which to compare changes in PEFR that occurred elsewhere.

Trends in Patient Response to Peak Flow Measurement

Use of the peak flow meters by the children at CARIH revealed several interesting trends. Some children became adept at predicting what their PEFR would be prior to blowing into a meter. They often came within a few liters of predicting their actual peak flow rate. In these cases, the children used the meter as an objective index that was correlated with their subjective feelings. Other children were less certain of what their PEFR would be; although they had twice-daily experience with the device, they were often inaccurate in their predictions. Many of this second type of patient appeared to have difficulty determining and detecting either objective or subjective changes in airway obstruction. In addition, they often misinterpreted other symptoms, such as feelings of anxiety, as indicative of asthma (9). There was a third type of patient, who claimed that despite the objectivity provided by PEFR data, they could better predict and detect respiratory changes by the subjective symptoms they experienced of asthma. Their reports have been echoed in other studies comparing objective signs of asthma versus subjective symptoms reported by some asthma patients (e.g., Refs. 10,11). In the study by Garrett and colleagues, for example, 75% of 352 patients with asthma appeared to prefer making management decisions based on symptoms they experienced rather than peak flow rates.

Transition to Home Monitoring

When meters were available or when parents could afford the cost of the instruments, children were asked to use them on a daily basis once they were discharged from CARIH. The peak flow meter proved a useful tool as children and their families attempted to effectively perform asthma self-management skills at home. The instruments were particularly useful to patients who had demonstrated problems at CARIH in detecting early asthma symptoms, thus delaying the onset of treatment of an episode. In addition, patients who lacked confidence in dealing effectively with asthma developed confidence or self-efficacy at asthma self-management by using a peak flow meter on a longitudinal basis. The majority of patients discharged from the facility found that if they and their families used the peak flow meter in an appropriate manner, they were able to make many clinically significant decisions regarding a child's asthma.

Monthly diaries, including PEFR data, were obtained from the youngsters after their discharge. Questions were sometimes raised about the accuracy of data returned by some children, particularly when they were asked to gather information over a prolonged period of time. It was problematic that no conclusions could be made about the accuracy of generalizations made about naturalistic data gathered from a group of patients over time (12). Some patients appeared to be vigilant at complying to directions and reporting the asthma data requested by CARIH physicians. PEFR values were likely the most reliable information obtained by these children. However, the data, including peak flow values, gathered from other patients often appeared inaccurate. It appeared as if the child or members of his or her family set down and completed an entire month's diary at one time. Individual differences among asthma patients is one of the major problems that hamper conclusions to be made about compliance, including adherence to peak flow instructions (13).

Peak Flow Meters and Asthma Research at CARIH

Peak flow meters were used in four pioneering studies at CARIH: the experimental separation investigation, a naturalistic study of emotions and asthma, the self-management of pediatric asthma, and the prediction of asthma attacks.

B. Historical Studies of Peak Flow Measurement

Experimental Separation Study

In the early 1960s, there was still considerable debate about the role of parental or familial factors and their possible influence on pediatric asthma. Much of the debate was sparked by the use of parentectomy, a procedure championed by Peshkin (14,15), which indicated that some children with a diagnosis of asthma showed a remission of their symptoms when removed from their homes. As the youngsters usually moved to an entirely different climate—particularly the higher elevation of Denver—the question was raised as to whether the changes could have been produced by differences in climatic and/or other environmental variables. To investigate this question, what was referred to as the family-separation study was conducted at CARIH by Purcell and coworkers (16).

The basis of the study was the opposite of parentectomy: rather than remove the child from the home, youngsters were separated from their families and remained in the home with surrogate parents for 2 weeks. Other family members were moved to a motel or hotel. Based upon a psychological evaluation and a profile compiled from the data, particularly information concerning the degree of involvement parents had with their asthmatic child, it was predicted that 13 children would show improvement in their asthma—the predicted positive-response group—with separation from their families; the parents had a high degree of involvement with the asthma of their children. It was predicted that 12 other children—the predicted negative-response group—would show no improvement in their asthma with separation from their families; the parents of these children were less involved with the asthma of their children. Findings of the study indicated a statistically significant improvement occurred in the asthma of the predicted positive group when they were separated from their families; improvement in only one of four dependent variables (history of daily asthma) occurred in the predicted negative group with separation.

The five dependent measures used by Purcell and colleagues (16) were clinical examinations, history, medication requirements, school activity levels, and use of the peak flow meter. Chai et al. (17) employed these same five measures in a second study, with the addition of a report of the value and weaknesses of each of the five measures as they were used in both investigations. Two important factors concerning PEFR emerged from these studies. First, Chai and colleagues reported that frequency of use of the peak flow meter was the key to objective measurement of respiratory function. They reported the quantity and quality of data obtained if the peak flow meter was used four times a day, twice a day, once daily, and once every 4 days. Chai and coworkers demonstrated that twice-a-day collection of peak flow data yielded reliable and valid PEFR information. The finding has become the "gold standard" for collection of PEFR values. In addition, the authors emphasized that "it is far better to use a simple measurement to assess asthma in children, such as the PEFR, which can be performed very frequently with very little technical knowledge, than to rely entirely on sophisticated measurements which are done at long intervals" (17). This finding essentially established the basis for frequent use of the peak flow meter, at least with childhood asthma. Chai et al.'s second result was equally significant. They found many of the outcome variables, including history and physical examination, were poorly intercorrelated with each other and with PEFR. Verbal report had the lowest correlation to other measures in providing an accurate index of asthma, prompting Chai and coworkers to conclude that changes in subjective reporting, particularly suggested improvement in breathing, is not always correlated with positive changes in PEFR. There are two ramifications to this finding: first, it reaffirmed previous work, extending back to James (18) and Hartshorne and May (19), indicating that a lack of correlation between various human response systems is common. Second, it supported the notion that there is often a lack of agreement between the symptoms of a disorder experienced by patients, including those with asthma, compared to the signs of the disorder as observed by others. With asthma patients, the poor congruence between verbal self-report and a more objective measure, such as that provided by PEFR data, illustrate the problem. As will be noted, considerable research has been directed at exploring the relationship between subjective symptoms and objective signs of asthma.

Natural Observation of Emotions and Asthma

The purpose of the study reported by Miklich et al. (20), was to provide objective data that emotional reactions may trigger asthma. The investigation was conducted over a several-year period at CARIH, a long period created by difficulties encountered in developing appropriate telemetry equipment for assessing emotional reactions in children with asthma in a natural setting. FM radio transmitters were worn by 18 boys with asthma for 4 hours daily. On a variable time schedule of 20 minutes, a peak flow measurement was obtained from each boy. The results indicated that, in most instances, emotional reactions did not lead to lowered peak flow values in the group. In some circumstances, however, anger and excitement preceded both the most frequent and largest PEFR drops in certain patients. Yelling, and especially laughing, showed significant negative correlations with peak flow values in some subjects.

Contribution to Early Asthma Self-Management

In 1978, Wright (21) reported on the development of the Mini-Wright Peak Flow Meter. The meter was tubular in shape with a spring-loaded piston through which air escaped. Initial meters were made of heavy cardboard; some gradually disintegrated when used by patients who expelled considerable mucus when blowing into the device. The dynamic characteristics of the instrument, noted Wright, were designed to make it respond only to peak flow and not to rate of rise. Later models of the instrument were shown to be almost perfectly correlated ($r = 0.99$) with the regular Wright Peak Flow meter.

The initial batch of Mini-Wright Peak Flow Meters were introduced in the United States in 1977 and used in an asthma self-management study, *Living with Asthma,* conducted by Creer and colleagues at CARIH (22). The highest of three peak flow values were recorded at home twice daily by 123 asthmatic children between the ages of 5 and 17 years. Information was collected from each child for a period of 21 months. PEFR data from all children were pooled, indicating a significant difference between the mean recorded and predicted peak flow values during baseline; (the difference was statistically significant, with a probability below the 0.025 level). A statistical analysis of data collected at the 18-month follow-up, however, revealed non-significant differences among the recorded and predicted peak flow values. In other words, the mean morning and evening peak flow rates were not statistically different from the mean predicted flow rate for the children at the end of 21 months. The conclusion was made that improved peak flow scores were a useful index to guage the performance of self-managment skills by participants (22).

While group changes in respiratory measures provide an overall indication of improvement in patients with asthma, the real value of peak flow measurement is the information it provides to individual patients and their families. In the asthma self-management program developed and evaluated at CARIH, children and their families used peak flow instruments in three ways (22).

First, children and their families used peak flow values to determine whether children were beginning to experience changes in their breathing. Many youngsters blew into a meter when they suspected they were beginning the onset of an attack. By reviewing their values over a period of time, the youngsters established a baseline against which to compare further data. They then gradually learned to compare any changes they detected in their breathing against the respiratory levels to which they had adapted. This procedure, similar to the shaping procedure described by Renne and coworkers (23), proved invaluable to many children and their parents. As noted by MacBryde and Blacklow (24), "Strictly speaking, symptoms are subjective, apparent only to the affected person. Signs are detectable by another person and sometimes by the patient himself." Through learning to correlate symptoms and signs, patients reported they detected the onset of attacks earlier and were quicker to initiate treatment (22).

Second, children and their parents used peak flow data as a set point against which to compare physical and behavioral changes that often preceded or accompanied the onset of attacks. Peak flow data provided an objective index of pulmonary function, particularly in children who were not always certain of the meaning of subjective signs of asthma.

Finally, some children and their parents compared peak flow rates against subjective signs of asthma they experienced. To these families, peak flow values validated their observations of behavioral and physical changes in a youngster. These changes, they declared, were more reliable in predicting or detecting the onset of attacks than were other indicators, including peak flow values.

Risk Prediction Using PEFR

Another way peak flow values were used at CARIH was in predicting the risk of attacks. Based upon the data gathered from children in the self-management program, *Living with Asthma*, Taplin and Creer (25) demonstrated that peak flow rates could be used to predict asthma attacks. Peak flow values gathered from two children were entered into a conditional probability equation to predict the occurrence of asthma. The base rate or prior probability of the occurrence of an attack and a critical PEFR that most increased the predictability of asthma were determined for each child. Two conditional probabilities were calculated: (1) the probability of asthma occurring in a 12–hour period following a flow rate less than or equal to the critical value, and (2) the probability of asthma occurring in a 12–hour period following a PEFR greater than the critical value. Taplin and Creer reported an approximate 300% increase over the base rate in predicting asthma using this procedure. Later, Harm et al. (26) extended the basic procedure with 25 children with asthma, including 13 involved with *Living with Asthma*, and found an average improvement of 500% in predicting the probability of asthma. Other investigations (e.g., Refs. 27,28) have also described the use of the peak flow meter in improving the predictability of asthma attacks.

II. Current Uses for Peak Flow Meters

A. Clinical Peak Flow Measurement

The major use of peak flow meters has been in the treatment of patients with asthma. There are a number of ways the meters can be used.

PEFR in the Diagnosis of Asthma

Most clinicians who diagnose and treat asthma believe that peak expiratory flow rates are not enough to confirm the diagnosis of asthma. They concur with the National Asthma Education Program report (29) that anyone suspected of having asthma should have an office spirometry performed during an evaluation for suspected asthma. Pulmonary function measures, along with medical history, physical examination, clinical impressions, and, if required, skin or provocation testing permit a diagnosis of asthma to be determined. However, peak flow monitoring has been recommended as a diagnostic tool for occupational asthma (e.g., Refs. 30–32). Different peak flow patterns in patients with occupational asthma have been described (8). Cote et al. (33), for example, found that peak flow monitoring was both sensitive and specific in diagnosing red cedar asthma. It has also been useful in evaluating and predicting airway dysfunction in swine confinement workers (34). Ryan et al.

(35) reported that variability of peak flow correlates well with the bronchial histamine provocation that reflects the hyperresponsiveness characteristic of asthma. Holgate and Church (36) issued a notable warning in the use of peak flow measurement to diagnose asthma—peak flow variability only permits a diagnosis of asthma in patients with fully reversible lung function.

PEFR in the Determination of Asthma Triggers

Peak flow meters are useful in determining the stimuli that could trigger asthma attacks in the natural environment. Patients may readily obtain a peak flow value when around stimuli suspected of precipitating the attacks they experience in the home, school, or at work. A decreased PEFR may be correlated with exposure to potential precipitants. Peak flow meters are particularly useful in identifying provocative factors for asthma, especially those caused by occupational stimuli (e.g., Refs. 5,8). Chan-Yeung (8) pointed out that peak flow monitoring is particularly useful in confirming occupational asthma in the following situations: (1) with workers suspected of having occupational asthma, (2) when a new agent is suspected of leading to occupational asthma, (3) when a challenge test in the laboratory is not possible or when results are negative but the patient history is suggestive of occupational asthma, (4) to document work exacerbation of asthma, and (5) to determine if the work environment is safe for workers with asthma.

It should be reiterated that obtaining low peak flow values in a particular setting or context is really the easiest part of identifying potential triggers of asthma. As was noted more than three decades ago at CARIH, actually correlating different stimuli to low PEFR values can be a daunting task. Recent research has helped in that decreases in peak flow have been indentified with various stimuli, including formaldehyde (37), ambient air pollution, including particle-strong acidity (38,39), and maternal smoking (40). As more and more environmental stimuli are linked to decreases in PEFR, increased information about what factors influence peak flow rates will likely result. At the same time, however, increased knowledge will also target tthe concurrent environmental variables that should be considered and, wherever possible, controlled when attempting to isolate asthma-inducing stimuli with the peak flow meter.

PEFR in the Prediction of Attacks

Earlier it was noted that peak flow values have been used by patients to predict the likelihood that they would experience asthma attacks within a prescribed period of time (25,26). The basis of this experimental work was the probability of analysis of peak flow records of individual patients. Refinement of this procedure was reported by Kotses and colleagues (41). Stimuli or responses correlated with asthma attacks, particularly decreases in peak flow rates, were determined in eight patients. The data were then entered into a mathematical formula to determine the probability that exposure to a specific stimulus or a lowered peak flow value would result in an attack within a specific period of time. Subjects were randomly assigned to an experimental or control group; patients in the experimental group received a brief individualized self-management program aimed at teaching them risk prediction to prevent attacks, and patients in the control group received their usual medical care. Results indicated a 22% decrease in the number of attacks experienced by patients in the experimental group. In a larger study using risk prediction, Kotses et al. (42) randomly assigned a group of 36 patients to one of three conditions: (1) individualized self-management, (2) group asthma self-management, and (3) control. Results indicated that peak flow rates improved in both intervention groups. However, there was a 38% decrease in the number of attacks experienced by patients assigned to the individualized self-management group. This indicated that the patients were able to predict many attacks and prevent them from occurring.

Cross and Nelson (5) concluded that the attack prediction procedure was too sophisticated for real-world practice. They based their assertion on the complex nature of the statistical analysis used to develop the procedure. In reality, however, prediction of the probability of asthma attacks is relatively simple: if patients keep records of peak flow values over time, they can easily determine the likelihood that a particular value may be a precursor to an attack. For example, one patient reviewed his peak flow record and found that he had obtained a PEFR value of 150 L/min on 20 different occasions. After 10 readings of 150 L/min or less, the patient invariably experienced an attack during the next 12 hour period. Thus, the individual concluded that when his PEFR was 150 L/min or lower, he had a 50:50 chance of having an attack in the upcoming 12–hour period. Knowing the likelihood of experiencing an asthma episode was of considerable benefit to the patient. He could not only take steps to avoid the possible occurrence of an attack, but he was better prepared to manage an attack when one occurred. More research needs to be conducted on the promising possibility of using the peak flow meter to predict asthma attacks. The procedure may help patients to avoid some attacks and to promptly treat incipient asthma episodes when they occur.

PEFR in the Discrimination of Asthma Attacks

The discrimination of attack onset is a function of the interaction of several types of variables (43). These include: (1) lack of knowledge, (2) misinterpretation or misperception of environmental/physical stimuli as asthma triggers,

(3) failure to detect symptoms, (4) misidentification of physiological signs, and (5) failure in interpretation. For patients who have difficulty detecting the onset of asthmatic episodes, peak flow monitoring is crucial. The low correlation of PEFR values with symptoms indicates that patients might not be able to detect any problems by monitoring symptoms alone; peak flow monitoring can identify patients who may not perceive airway obstruction appropriately. Peak flow meters then become invaluable as monitoring devices to teach patients to discriminate the onset of asthmatic episodes. Identification of patients who report few symptoms despite severe airway obstruction also permits the development of individualized treatment plans that allows patients to more rapidly control their asthmatic episodes (4,43).

PEFR Measurement During Asthma Exacerbations

Several clinicians (e.g., Refs. 5,7) have emphasized that peak flow monitoring is particularly useful both in detecting deterioration of asthma and in predicting acute exacerbations of attacks. As Cross and Nelson (5) remarked, "Availability of peak flow measurements not only allows formulations of a management plan with criteria for both intensification of therapy and recourse to medical assistance but also facilitates telephonic communication with health care providers. Regular measurement of peak flows allows objective determination of the effect of therapy." Incorporating PEFR values into a treatment protocol in this manner is a strong argument for regular collection of peak flow rates.

There are a number of other clinical uses of peak flow rates, including those discussed earlier. Cross and Nelson (5) described the value of using peak flow meters to determine morning dipping in patients with asthma. Lower morning PEFRs, like nocturnal asthma, are common in patients with relatively stable asthma. Cross and Nelson suggested that morning dipping of PEFR values may have the same significance as nocturnal asthma. Data from England (44,45) suggested that morning dipping may be implicated in the deaths of patients between midnight and early morning hours.

PEFR in Prediction of Exacerbations of Asthma

Monitoring of peak flow rates may be useful in measuring acute exacerbations of asthma in the emergency room. Cross and Nelson (5) observed that the initial value obtained from a patient may be used to determine whether that patient requires arterial blood analysis. In addition, they asserted that peak flow monitoring of the early response to therapy may provide an indication that hospitalization may be required. While a spirometer has often been used in the emergency room, it is often easier in a busy emergency room to obtain periodic PEFR values from the patient than to repeat spirometry tests.

B. Self-Management Using Peak Flow Measurement

As noted earlier, the initial use of mini peak flow meters in the United States occurred in a self management study initiated at CARIII in 1977 (22). Since that time, a number of self-management programs have adopted the peak flow meter as an assessment device. Mendoza (46) and Plaut (47) have been strong advocates of using peak flow meters for home monitoring of asthma. They suggest a system of zones be established and followed. The zones can be developed as a function of either a patient's personal best or predicted PEFR value, whichever is highest. The emphasis of such programs, noted the expert panel of the National Asthma Education Program (29), is on the variability patients exhibit from their personal best or from one reading to another. The expert panel recommended that peak measurements be made both in the morning and in the evening and both before and after inhaled medications are taken; the intermittent use of the peak flow meter, they continued, should be reserved only for cases of highly stable asthma.

Traffic Light Zone System

As has occurred with other health-related problems, such as dieting for weight control (e.g., Refs. 48,49), the zone system has been adapted to fit a traffic light system. The system described in the report by the National Asthma Education Project (29) and the International Asthma Program (50) is representative of such a system.

Green Zone

A PEFR value in the green zone represents 80–100% of the personal best peak flow value blown by a patient. Green indicates that no asthma symptoms are present; therefore, the routine treatment plan for maintaining control can be followed by the patient. According to the report, consistent readings in the green zone for patients on maintenance medications may indicate that medications can be reduced.

Yellow Zone

A PEFR in the yellow zone indicates that the peak flow rate blown by the patient is in a range of 50–80% of the personal best peak flow value blown by the individual. It suggests that caution should be taken; a point where treatment should be initiated (action point) is usually set by the physician. An acute exacerbation of asthma may be present, and a temporary increase in medications may be required. The alternate suggestion is that the patient's asthma may not be under sufficient control, and maintenance therapy may be initiated or increased.

Red Zone

A value in the red zone represents a patient's PEFR that is below 50% of the patient's personal best peak flow. According to the guidelines set out in the report, a bronchodilator should be immediately taken by the patient. In addition, his or her physician should be notified if peak flow values do not rebound immediately and remain in the yellow or green zones.

The aim of such a system is to permit patients and medical personnel to approach asthma management in a preventative manner by keeping a patient's peak flow level in the green zone (46). Others, including Beasley et al. (51), suggest that peak flow values below 70% of personal best require increased use of medications, values below 50% require the use of oral corticosteroids, and readings below 150 L/min require urgent medical attention.

Problems with the Zone System

There are five major flaws in the zone system outlined by the National Asthma Education Project (29) and the International Consensus Report (50). First, there is a paucity of empirical evidence to support use of the system. Lloyd and Ali (52) found that peak flow values were useful in teaching children and their parents to recognize severe asthma attacks (e.g., when PEFR were 60% of that predicted). However, only 20% of the families reported that their meters were useful in detecting impaired respiratory function at lower percentages of personal best. Brandon (53) compared a group of asthma children taught to use the zone system versus a control group who were taught to use meters but not the zone system. Both groups reported significantly improved peak flow readings over the course of the study, but no differences were found in asthma management between the groups. Finally, Jones et al. (54) were unable to replicate the findings of Beasley and coworkers (51) in a systematic replication of the latter study. This was a major setback for both the National Asthma Education Project (29) and the International Consensus Report (50) and their endorsement of the traffic light zone system.

Second, the sensitivity of the asthma zone system was questioned in a study by Gibson et al. (55), who compared the asthma zone system against a quality-control analysis of PEFR based on standard deviations from a patient's mean PEFR. Results showed that the zone plan had poor operating characteristics. Gibson and coworkers noted that the success rate was 35% when the action point was a PEFR value less than 60% of the patient's best peak flow. The success rate improved to 88% when the action point was a PEFR less than 80% of the patient's personal best. However, continued Gibson and colleagues (55):

> Peak flow decreased to below the published action points during a stable period of asthma in 7% to 51% of patients studied. Action points defined using quality-

> control analysis did significantly better. A peak flow value less than 3 standard deviations below the patient's mean peak flow detected 84% of exacerbations and had a low failure rate (19%). Other quality-control tests had sensitivities of 91% and 71%. Quality-control action points could detect exacerbations up to 4.5 days earlier than conventional methods.

The results of this study are too impressive to be ignored. It describes a method of using the peak flow meter that is far more sensitive in the detection of airway obstruction than anything yet developed for use by individual patients. In addition, the system presented by Gibson and colleagues (55) is based upon the use of central tendency and variability, statistical procedures that are staples of research and evaluation. The approach merits attention and further research.

Third, the asthma zone system produces many false-positive findings. False-positive PEFR values occur as a function of how a personal best peak flow is determined. For example, one patient noted that her personal best peak flow was 400 L/min. This occurred on a clear day when the woman was outdoors. However, she continued, when she was at work, her personal best never exceeded 300 L/min. Was she experiencing asthma at work? The woman said not when she blew 300 L/min. She only experienced symptoms when her readings at work fell to 250 L/min or less. This points out a problem repeatedly found by those who use peak flow meters in the treatment of asthma: personal best peak flow rates vary according to the situation in which they are obtained. The situation-specific nature of the personal best peak flow is a function of the setting and, in all likelihood, the context within which the value is obtained. It means that if patients adhered to the zone system of the National Asthma Education Program (29) and the International Asthma Program (50), many would overmedicate themselves.

Fourth, there is conflicting data about the interpretation of peak flow data, particularly in the management of childhood asthma. Although research at CARIH, described earlier, found the peak flow meter to be of value in the treatment of pediatric asthma, the accuracy of different meters, particularly the smaller or minimeters, has been questioned. This point is illustrated in a study by Sly et al. (57), who reported that not only the inaccuracy of absolute values of peak expiratory flow rates obtained with mini flow meters, but that the clinical message provided by such instruments could be inaccurate. Sly and colleagues noted that they did not wish to imply that home monitoring of PEFR had no role in the management of asthma, but that the values obtained with the meters should be interpreted carefully.

The final point is perhaps the most important. Rather than use a standard zone system applicable to all patients, as proposed both by the National Asthma Education Program (29) and the International Asthma Program (50), there is a need to tailor zone programs that incorporate the peak flow meter to individual patients. This was the conclusion reached by a number of studies described (e.g., Refs. 54,55), and by the work of Kotses and colleagues (41,42) discussed earlier. Ignacio-Garcia and Gonzalez-Santos (58) also reported research using an intervention program tailored for individual patients that incorporated peak flow monitoring and educational training. Compared to a control group, those in the intervention group lost fewer days from work, experienced fewer acute asthma attacks, consulted less with physicians, and had fewer emergency room visits. This study, along with the others reviewed, indicate that the peak flow meter can play a significant role in the management of asthma. The National Asthma Education Program (29) and the International Asthma Program (50) both advocated tailoring treatment programs for individual patients. In order to achieve this aim, adequate criteria for assessing PEFR and relating these values to the treatment received by individual patients must be available. Recent research, particularly the study by Gibson and coworkers (55), suggest that the goal of establishing empirically validated standards for use of the peak flow in treating asthma is attainable.

C. Research with Peak Flow Measurement

Several areas of research have been spawned by the widespread adoption of the peak flow meter in assessing asthma. Three areas of particular interest include the relationship of pulmonary values and subjective symptoms of asthma, teaching patients to estimate their peak flow values, and comparisons between various peak flow meters.

Comparison of Peak Flow Values to Subjective Signs of Asthma

Observations of symptoms by patients and signs by physicians have often been the standard for monitoring asthma. However, as noted by Chai and colleagues (17), there is a often a lack of agreement between the changes in response systems. A number of studies have elaborated on the problem. McFadden and coworkers (59) found that patients recovering from asthma perceived they were asymptomatic when their pulmonary function was only 40–50% of their normal peak flow values. Differences between wheezing and PEFR in detecting asthma were noted by Shim and Williams (60). Another investigation reported that over a 16-week period, airway obstruction was found in number of children considered asymptomatic (61). Kendrick and colleagues (62) asked 255 asthma patients to estimate the severity of their asthma on a visual analog scale; PEFR data were concurrently gathered. No significant correlation was found between the patients' perceived severity of their asthma and PEFR values in 60% of the patients. A few studies have provoked airway obstruction in asthma patients with methacholine or histamine. Patient symptoms and pulmonary signs of asthma are then correlated. The classic study using this approach was reported by Rubenfeld and Pain (63) who found that 15% of their subjects were unable to perceive severe airway obstruction (FEV_1 was less than 50% of predicted values). A similar study, conducted by Burdon and coworkers (64), suggested that asthma patients with the greatest degree of bronchial hyperresponsiveness were less likely to perceive airway obstruction.

Based upon a review of the literature, Li (4) concluded that the correlation between subjective symptoms and objective measures of airway obstruction indicated that one-third to one-half of asthma patients were unable to estimate airway obstruction accurately by the asthma signals they detect. A number of reasons have been put forward as to why there is often low or no correlation between respiratory symptoms experienced by patients and measures of pulmonary function (e.g., Refs. 4,17). First, patients may have difficulties detecting changes in the level of breathing to which they have adapted. This was demonstrated by Burki and coworkers (65), who found a wide variety in responses when asthma patients were deliberately subjected to different resistive loads. Second, hyperinflation of the lung may be more important to detecting signs of breathlessness than airway obstruction (66). Third, the rate of decrease of pulmonary function may affect the perception of breathlessness (67). If a patient has adapted to a particular level of breathing, a sudden change would more likely be noted than a gradual change in level of breathing adaptation. Finally, Hudgel et al. (68) found that psychological factors play an important role in the detection of breathing changes. For example, the past history of the patient often dictates whether patients are apt to quickly discriminate breathing changes and attempt to abort the attack (9,43). Cognitive, physiological, and contextual variables can influence whether patients readily detect these changes. Children with asthma who are involved in an enjoyable activity, for example, may be impervious to the audible wheezing noted by others.

Li (4) summarized the problem presented by the lack of correlation between symptoms and signs of asthma when he cautioned that: "because poor perception of airway obstruction may be an important risk factor for life-threatening asthma, physicians should identify patients with such poor perception and aggressively use other measures to assess the severity of asthma, such as spirometry or PEFR determination." The problem is exacerbated by the knowledge that many of the questions raised by James (18) concerning the relationship of various response systems to one another remain unanswered (69).

Patient Ability to Estimate Peak Flow Values

A number of investigators have reported that they were able to teach patients to predict their peak flow values. Renne and colleagues (23) at CARIH described how they taught children to match subjective impressions of their asthma with their peak flow data. The training involved the use of classical psychophysical techniques in that the children were taught to match the magnitude of one response system (their subjective symptoms) to the magnitude of a second response system (their PEFR). However, cautioned Renne and coworkers, their subjects had ample experience in using the peak flow meter prior to the study. Others have reported stronger findings in teaching patients to perceive their PEFR values. Silverman and colleagues (70) reported that 18 asthma patients, aged 7–43, were easily trained to accurately predict PEFR over a wide range of flow rates. In addition, the subjects reportedly retained this ability over a period when home recordings were made of estimated and measured PEFR rates. Similar findings in training patients to predict their pulmonary functions were later reported by Silverman et al. (71).

On the other hand, some investigators have reported a lack of success in teaching asthma patients to predict their pulmonary function. Sly et al. (72) described an attempt to use the peak flow meter to teach children with asthma to predict their PEFR. The children were unable to accurately predict their degree of airway obstruction even following completion of a learning procedure. The perceptions of the children's parents were also inaccurate; in addition, recording of PEFR values at home did not improve the childrens' perception of their asthma. More recently, Sly and colleagues (57) compared measurement of airway obstruction with a spirometer and different brands of peak flow meters in 12 boys with asthma. The relation between changes in lung function measured with the spirometer and those with the peak flow meters was generally poor.

Three conclusions can be made based upon efforts to teach patients to estimate pulmonary function values, particularly PEFR. First, it appears that some patients can be easily taught to make such predictions. Why the skill appears relatively easy for them to acquire and how large a subgroup of asthma patients can be taught this skill remains to be determined. Second, it also appears as if some patients are unable to learn to make predictions concerning their peak flow rates. Why they encounter problems in acquiring this competency, as well as how large a subgroup of patients is involved, remains a mystery. Because of methodological differences among the articles where patients have been taught to estimate their peak flow rates, there are no proven guidelines that can be used to teach others to acquire these skills. Finally, considering the bipolarity of the findings, the teaching of patients to estimate peak flow values cries out for more research. The detection of airway changes is imperative if asthma is to be managed. How this perception can be improved in all asthma patients is worthy of investigation.

Comparative Studies of Lung Function Measures

Since their introduction, peak flow meters have been compared to other measures of pulmonary function or to other peak flow meters. Chai and colleagues (17), for example, regularly compared PEFR data against spirometric data. Indeed, most experts (e.g., Refs. 4,5,7) routinely suggest that such comparisons be made. In addition, however, comparisons have been made between the regular and mini peak flow meter (21,73), mini peak flow meters and pocket flow meters (74), and among the various peak flow meters (e.g., Refs. 57,75–79). As might be expected, differences emerge. In a study by Sly and coworkers (57), for example, 26 asthmatic episodes of clinically significant deterioration were detected by the spirometer. Of the four mini peak flow meters used in the study, one detected 6 of 19 episodes, one detected 6 of 15, one detected 6 of 18, and one detected 3 of 21.

There are several reasons to compare peak flow meters to other pulmonary instruments or to other meters. First, each model of a peak flow meter has a characteristic profile that reliably overestimates or underestimates peak flow in low-flow, mid-flow, or high-flow ranges. As Li (4) pointed out, agreement among various peak flow meters is moderate, approaching discrepancies of 50 L/min. Second, as there is no gold standard for peak flow meters, comparison of data collected from various meters yields useful information. There is always the hope that a gold standard will emerge from among available meters. Third, as more and more meters enter the market, data on how they compare to existing meters are of value both to health care workers and to consumers. Fourth, only testing of meters will reveal whether peak flow meters meet standards suggested by the National Heart, Lung, and Blood Institute (80). These criteria recommend that peak flow meters be accurate over their full range within a plus or minus of 10% the readings and that they have good reproducibility within 10 L/min or plus or minus 5% of the reading. Finally, Li (4) noted that the reliability of PEFR readings from individual devices or among similar devices made by a company is good. Generally, he continued, the variation from a single peak flow meter is only about 1–4%. However, when the regular peak flow meter was initially introduced 35 years ago, a standard practice was to periodically calibrate the instrument against a more sophisticated pulmonary function device, usually the spirometer. The practice of calibrating peak flow meters seems to have disappeared, although there are reports both of increased malfunction (7,76) and of deteriorating accuracy of a meter over time (79). The lack of periodic calibration of peak flow meters could present a serious problem. While a meter is apt to be used by a single individual over a period of time, any changes that occur could be the result of changes in the meter and not the medical condition of the patient. There is a need not only to reinstitute the regular calibration of peak flow meters, but to determine how best to calibrate different brands of meters. If it is impossible to do so—and this could well be the case with some inexpensive peak flow meters—then medical personnel and patients should be provided guidelines to know when to replace a defective meter.

III. Advantages and Limitations of Peak Flow Meters

Peak flow meters, particularly the smaller or mini versions, have increasingly found a niche in both treatment and research of asthma. The occupy a prominent place in the expert panel report on asthma issued by the National Asthma Education Program (29) and the International Asthma Program (50). There are both advantages and limitations to the use of peak flow meters with asthma.

A. Advantages of Peak Flow Meters

A number of reviews (e.g., Refs. 4,5,46,81,82) and the National Asthma Education Program (29) enumerated the advantages to using the peak flow meter.

1. As noted by the expert panel of the National Asthma Education Program (29), PEFR provides a simple, quantifiable, and replicable measure of airway obstruction that can be obtained using either regular peak flow meters or smaller and more portable models. Basically, PEFR is an objective measure analogous to the assessment of blood pressure with a sphygmomanometer.
2. When done correctly, PEFR measurements generally correlate well with FEV_1 and other spirometry measures that are more sensitive to variations in airway obstruction (5). In a study by Vaughan et al. (83), the correlation between FEV_1 and PEFR was $r = 0.74$. In a study of 630 subjects, Connolly and Chan (84) reported a correlation of $r = 0.85$ between the PEFR and the FEV_1. Considering the results, the latter authors suggested that spirometry added little over the peak flow meter in patient evaluation.
3. The simplicity of these devices permits them to be readily adapted to the needs of those who require assessment of respiration but who lack extensive training in pulmonary physiology. This makes the meters ideal for assessment of airway obstruction by a broad array of users, including medical and behavioral scientists as well as patients with asthma.
4. PEFR offers an objective index of pulmonary obstruction that is particularly important in cases where a patient's subjective assessment of his or her condition is unreliable. Peak flow rates offer a standard against which patients can judge whether or not they are experiencing asthma.

5. Peak flow monitoring sometimes permits patients to determine factors that may trigger their asthma outside the laboratory. This makes the instrument invaluable in the home, school, or worksite. Moscato and colleagues (82) also noted that PEFR testing is feasible in nonspecialized centers where specific challenges cannot be performed. The portability of the meters, particularly what are referred to as minimeters, permit use in a variety of settings, including homes. Newer models can be easily carried in a purse or backpack.
6. It is possible that, as Cross and Nelson (5) suggest, higher peak flow values may improve medication compliance, although there is a limited amount of data suggesting this to be the case.
7. Peak flow values can be used by patients to predict the likelihood that they will experience asthma within a prescribed period of time (e.g., Refs. 25,26). When used in this manner, patients can quickly calculate the probability that based upon past PEFR data, they are apt to experience asthma in a specific period. This knowledge can be used both to prevent some attacks and to keep other episodes from increasing in severity.
8. Peak flow values are useful in differentiating between nonoccupational and work-related asthma (82). When performed, interpreted, and carefully documented, peak flow measures may stand by themselves in providing necessary information to determine if a person's job is a significant factor in his or her asthma. Preshift, midshift, and postshift measurements may reveal patterns, such as a sharp trend downward in PEFR over the day, that suggest that the worksite contains stimuli that trigger or exacerbate a patient's asthma. Comparing workday patterns of PEFR with patterns of PEFR values obtained when the individual has a day off may also show peak flow trends that steadily deteriorate as the work week progresses only to rebound when the person is away from work.
9. Moscato and coworkers (82) suggested that if specific bronchial challenges are negative because a person is away from work too long, sequential peak flow values provide a way to establish the relationship between asthma and the workplace.
10. As noted by the expert panel of the National Asthma Education Project (29):

> When patients learn how to take PEFR measurements at home, the clinician's ability to provide effective treatment is improved. Daily monitoring of PEFR helps, for example, in detecting early stages of airway obstruction; assessing circadian (day-night) variations in lung function (which reflect degree of airway hyperreponsiveness); providing objective criteria in planning, initiating, or terminating treatment; facilitating communication between patient and clinician; and investi-

gating specific allergens or school or working exposure that may exacerbate symptoms."

B. Limitations of Peak Flow Meters

While the peak flow meter is invaluable in the overall management of asthma, there are a number of limitations to the instruments:

1. While the meter is adequate for home use, it is does not provide as much information as other equipment, particularly the spirometer. PEFR data is adequate only within a narrow spectrum of pulmonary physiology; it lacks the sophistication of the spirometer, for example, in that the FEV_1 reflects airway caliber in both the large- and medium-sized airways (4).
2. Like spirometric information, PEFR information relies upon the motivation of patients. Adherence to a suggested schedule for use of the meter may be low when patients are asked to repeatedly gather PEFR data twice a day over a period of time (17). Other patient variables, including mem- ory, expectations, and self-confidence, can influence PEFR gathering (22).
3. Some patients may have difficulty plotting the data they gather. However, a number of user-friendly graphs has largely overcome this barrier to collecting information with the peak flow meter.
4. When used in homes or other natural settings, peak flow meters are rarely calibrated to the extent respiratory instruments are tested in the laboratory. Meters distributed to patients may be dropped and broken. They may not be cleaned as often as required. Consequently, the meters may yield inaccurate and misleading information. The problem is further exacerbated in that little attention has been paid to determining the extent of the problem of peak flow meter calibration. In many cases, medical personnel last see meters when they are distributed to patients. The result may be changes in peak flow values that are a function of changes in the meter, not the patient.
5. Moscato et al. (82) note that the poor sensitivity of peak flow self-monitoring leads to a high percentage of false-negative results. A number of factors could contribute to this finding, including patient motivation.
6. Difficulties may occur in interpreting PEFR data. Moscato et al. (82) pointed out that reading of peak flow data is not totally objective and that there are both unique patterns and borderline results.
7. Moscato and coworkers (82) enumerated a number of problems encountered in using peak flows in assessing occupational asthma.

These include: (1) reexposing patients to stimuli in a workplace, while necessary, is potentially dangerous; (2) it takes a relatively long time to obtain data; and (3) there may be organizational difficulties involving gathering PEFR data in the workplace.

8. While PEFR monitoring may yield a correlation between asthma symptoms and occupational exposure, it is impossible to relate the symptoms to a specific agent unless the patient is exposed to only one stimulus. As Moscato and colleagues (82) pointed out, peak flow values cannot distinguish between occupational asthma (i.e., asthma caused by a specific agent) and work-related asthma (i.e., bronchoconstriction in patients related to nonspecific exposure to precipitants in the workplace).
9. Peak flow meters may be useless in the case of late reaction asthma. PEFR values may show us changes to stimuli in a setting; however, exposure to stimuli in that setting may trigger late-reaction asthma, which is not detected in PEFR.
10. The value of a peak flow meter rests entirely upon its use in an overall assessment and treatment plan. The meter is nothing more than a device calculated to assess a physical attribute. In this sense, it is identical to a sphygmomanometer that is used to measure blood pressure or bathroom scales used to assess weight. All a peak flow meter provides is a value that can be used in the management of asthma.

IV. Summary and Conclusions

This chapter traced the history of the peak flow meter with asthma patients in residential treatment centers and in their homes, clinical uses of the meter, research with PEFR values, and strengths and limitations of the peak flow meter. The basic conclusion of the review is that the peak flow meter is an indispensable instrument for the assessment of airway obstruction by asthma patients, particularly in the environment that constitutes their daily lives. Peak flow devices permit access by patients and physicians to information that can be incorporated into tailored treatment plans designed to alleviate and control the patients' asthma.

Three characteristics regarding the peak flow meter merit discussion. First, there are many interpretations of the exact function of the devices. To some, the meter is a treatment tool; to others, is may be a monitoring or training device (85). In actuality, the peak flow meter is the same as a sphgmomanometer or set of bathroom scales: it is an assessment device that provides a numerical value. How that value is obtained and interpreted is the key for using peak flow meters with asthma patients.

Second, perhaps more than other instruments used in respiration, peak flow meters are heavily dependent upon the effort of individual asthma patients. If they wish, patients can adhere to the instructions provided by medical or research personnel and gather accurate PEFR data on themselves. This is the ideal. At the other end of the spectrum, however, there are patients who generate inaccurate information. Difficulties arise both in identifying patients who are apt to collect accurate PEFR information and in specifying exactly along the spectrum of adherence where a patients falls at any given moment. As patients with asthma are apt to comply to peak flow meter instructions to the same degree as they adhere to medication directions, their PEFR values likely vary from time to time, perhaps from reading to reading, along the spectrum of compliance (86). This leads to the caveat that, like a computer, the meaning of peak flow data is a function of what went into observing and gathering the information. Good data can generate accurate interpretations; bad data can generate inaccurate findings. The difficulty comes in determining the reliability and validity of whatever information is obtained from individual patients.

Finally, the value of PEFR data depends upon the context (whether monitoring or treating asthma) within which peak flow meters are used. The context is not static, but a dynamic matrix characterized by a continual reciprocal interaction among environmental stimuli, physiological changes, cognitive variables, and behavior. In addition, the actions of investigators who attempt to determine both ways to effectively use peak flow instruments and how to interpret PEFR data appropriately add another set of variables to the context. These factors, in turn, interact with the personal and environmental variables experienced by patients. The conclusion is that while much has been learned about peak flow meters in their 35-year history, much more research needs to be conducted. Lord Kelvin was quoted (87) in his *Popular Lectures and Addresses* as saying: "When you can measure what you are speaking about and express it in numbers, you know something about it; but when you cannot measure it, when you cannot express it in numbers, your knowledge is of a meager and unsatisfactory kind; it may be the beginning of knowledge, but you have scarcely, in your thoughts, advanced to the stage of *science*." While we have accumulated considerable numbers via the use of peak flow meters, particularly in the home, we have yet to determine exactly what the data mean and how they can be best used for the welfare of the patient. We have not yet achieved a scientific understanding of PEFR.

References

1. Wright BM, McKerrow CB. Maximum forced expiratory flow rate as a measure of ventilatory capacity with a description of a new portable instrument for measuring it. Br Med J 1959; 2:1041–1047.

2. Harm DL, Marion RJ, Creer TL, Kotses H. Effects of instructions on performing forced expiratory maneuvers on pulmonary function measures. J Asthma 1985; 22:289–294.
3. Harm DL, Marion RJ, Kotses H, Creer TL. Effect of subject effort on pulmonary function measures. J Asthma 1984; 21:295–298.
4. Li JTC. Home peak expiratory flow rate monitoring in patients with asthma. Mayo Clin Proc 1995; 70:649–656.
5. Cross D, Nelson HS. The role of the peak flow meter in the diagnosis and management of asthma. J Allergy Clin Immunol 1991; 87:120–128.
6. Nairn JR, Bennet AJ, Andrew JD, MacArthur P. A study of respiratory function in normal school children: The peak flow rate. Arch Dis Child 1961; 36:253–263.
7. Chai H, Newcomb RW. Pharmacologic management of childhood asthma. Am J Dis Child 1973; 125:757–765.
8. Chan-Yeung M. Assessment of asthma in the workplace. Chest 1995; 108:1084–1117.
9. Creer TL. Asthma Therapy: A Behavioral Health Care System for Respiratory Disorders. New York: Springer Publishing Company, 1979.
10. Fritz GK, Klein RB, Overholser JC. Accuracy of symptoms perception in childhood asthma. Dev Behav Peds 1990; 11:69–72.
11. Garrett J, Mitchell E, Fenwich JM, Rea H, Taylor G. Peak expiratory flow meters (PEFMs)—Who uses them and how does education affect the pattern of utilization? Aust NZ Med 1994; 123:488–492.
12. Renne CM, Creer TL. Asthmatic children and their families. In: Wolraich ML, Routh DK, eds. Advances in Developmental and Behavioral Pediatrics, Vol. 6. Greenwich CT: Jai Press, 1985:41–81.
13. Creer TL, Levstek DA. Adherence to asthma regimens. In: Gochman DS, ed. Handbook of Health Behavior Research, New York: Plenum (in press).
14. Peshkin MM. Asthma in children. IX. Role of environment in the treatment of a selected group of cases. A plea for a "home" as a restorative measure. Am J Dis Child 1930; 39:774–781.
15. Peshkin MM. Intractable asthma of childhood rehabilitation at the institutional level with a follow-up of 150 cases. Int Arch Allergy 1959; 15:91–112.
16. Purcell K, Brady K, Chai H, Muser J, Molk L, Gordon N, Means J. The effect of asthma in children of experimental separation from the family. Psychosom Med 1969; 31:144–164.
17. Chai H, Purcell K, Brady K, Falliers CJ. Therapeutic and investigational evaluation of asthmatic children. J Allergy 1968; 41:23–46.
18. James W. Principles of Psychology. Vols. 1 and 2. New York: Holt, 1890.
19. Hartshorne H, May MA. Studies in the Nature of Character: Studies in Deceit. New York: MacMillan, 1928.
20. Miklich DR, Chai H, Purcell K, Brady K. Naturalistic observation of emotions preceding low pulmonary flow rates. J Allergy Clin Immunol 1974; 53:102.
21. Wright BM. A miniature Wright peak-flow meter. Br Med J 1978; 2:1627;1628.
22. Creer TL, Backial M, Burns KL, Leung P, Marion RJ, Miklich DR, Morrill C, Taplin PS, Ullman S. Living with asthma: Part I. Genesis and development of a self-management program for childhood asthma. J Asthma 1988; 25:335–362.

23. Renne CM, Nau E, Dietiker K, Lyon R. Latency in seeking asthma treatment as a function of achieving successively higher peak flow rate criteria. Paper presented at the 10th Annual Convention of the Association for the Advancement of Behavior Therapy, New York, 1976.
24. MacBryde CM, Blacklow RS. Signs and Symptoms: Applied Pathologic Physiology and Clinical Interpretation, 5th ed. Philadelphia: Lippincott, 1970.
25. Taplin PS, Creer TL. A procedure for using peak expriatory flow rate data to increase the predictability of asthma episodes. J Asthma Res 1978; 16:15–19.
26. Harm DL, Kotses H, Creer TL. Improving the ability of peak expiratory flow rates to predict asthma. J Allergy Clin Immunol 1985; 76:688–694.
27. Stout C, Kotses H, Carlson BW, Creer TL. Predicting asthma in individual patients. J Asthma 1991; 28:41–47.
28. Pinzone HA, Carlson BW, Kotses H, Creer TL. Prediction of asthma episodes in children using peak flow rates, medication compliance, and exercise data. Ann Allergy 1991; 67:481–486.
29. Executive Summary: Guidelines for the Diagnosis and Management of Asthma. Washington, DC: National Institutes of Health (NIH Publication No. 91–3042), 1991.
30. Burge PS, O'Brien IM, Harris MG. Peak flow rate records in the diagnosis of occupational asthma due to isocyantes. Thorax 1979; 34:317–323.
31. Burge PS. Use of serial measurements of peak flow in the diagnosis of occupational asthma. Occup Med 1993; 8:279–294.
32. Eisen EA, Wegman DH, Kriebel D. Application of peak expiratory flow in epidemiologic studies of occupation. Occup Med 1993; 8:265–177.
33. Cote J, Kennedy S, Chan-Yeung M. Sensitivity and specificity of PC20 and peak flow rate in cedar allergy. J Allergy Clin Immunol 1990; 85:592–598.
34. Quinn TJ, Donham KJ, Merchant JA, Schwartz DA. Peak flow as a measure of airway dysfunction in swine confinement operators. Chest 1995; 107:1303–1308.
35. Ryan G, Latimer KM, Dolovich J, Hargreave FE. Bronchial responsiveness to histamine: relationship to diurnal variation of peak flow rate, improvement after bronchodilator, and airway caliber. Thorax 1982; 37:423–429.
36. Holgate ST, Church MK. Allergy. London: Gower Medical Publishing, 1993.
37. Kriebel D, Sama SR, Concanour B. Reversible pulmonary responses to formaldehyde. A study of clinical anatomy students. Am Rev Respir Dis 1993; 148:1509–1515.
38. Higgins BG, Francis HC, Yates CJ, Warburton CJ, Fletcher AM, Reid JA, Pickering CA, Woodcock AA. Effects of air pollution on symptoms and peak expiratory low measurements in subjects with obstructive airways disease. Thorax 1995; 50:149–155.
39. Neas, LM, Dockery DW, Koutrakis P, Tollerud DJ, Speizer FE. The association of ambient air pollution with twice daily peak expiratory flow rate measurements in children. Am J Epidemiol 1995; 141:111–122.
40. Frischer T, Kuhr J, Meinert R, Karmaus W, Urbanek R. Influence of maternal smoking on variability of peak expiratory flow rate in school children. Chest 1993; 104:1133–1137.

41. Kotses H, Stout C, Wigal JK, Carlson B, Creer TL, Lewis P. Individualized asthma self-management: a beginning. J Asthma 1991; 28:287–289.
42. Kotses H, Stout C, McConnaughy K, Winder JA, Creer TL. Evaluation of individual asthma self-management programs. J Asthma 1996; 33:113–118.
43. Creer TL. Self-management psychology and treatment of childhood asthma. J Allergy Clin Immunol 1983; 72:607–610.
44. Cochraine GM, Clark TJH. A survey of asthma mortality in patients between ages 35 and 64 in the Greater London Hospitals in 1971. Thorax 1975; 30:300–303.
45. Hetzel MR, Clark TJH, Branthwaite MA. Asthma: analysis of sudden deaths and ventilatory arrests in hospital. Br Med J 1977; 1:808–811.
46. Mendoza GR. Peak performance: A Strategy for Asthma Self-Assessment. Fairfax, VA: Mothers of Asthmatics Inc, 1988.
47. Plaut TF. Children with Asthma: A Manual for Parents, 2nd ed. Amherst, MA: Pedipress, 1988.
48. Epstein LH, Wing RR, Koeske R, Andrasek F, Ossip DS. Children and parent weight loss in family-based behavior modification program. J Clin Consult Psychol 1981; 49:674–685.
49. Epstein LH, Squires S. The Stoplight Diet for Children. Boston: Little, Brown, 1988.
50. International Consensus Report on Diagnosis and Treatment of Asthma. Bethesda, MD: National Institutes of Health (Publication No. 92–3091), 1992.
51. Beasley R, Cushely R, Holgate ST. A self-management plan in the treatment of adult asthma. Thorax 1989; 44:200–204.
52. Lloyd BW, Ali MH. How useful do parents find home peak flow monitoring for children with asthma? Br Med J 1992; 305:1128–1129.
53. Brandon M. The stepwise treatment of asthma in children and adolescents. Dissert Abst Int 1992; 53–10B, 5439.
54. Jones KP, Mullee MA, Middleton M, Chapman E, Holgate S, and the British Thoracic Society Research Committee. Peak flow based asthma self-management: a randomized controlled study in general practice. Thorax 1995; 50:851–857.
55. Gibson PG, Wlodarczyk J, Hensley MJ, Murree-Allen K, Olson LG, Saltos N. Using quality-control analysis of peak expiratory flow recordings to guide therapy for asthma. Ann Intern Med 1995; 123:488–492.
56. Isaac S, Michael WB. Handbook in Research and Evaluation. 3rd ed. San Diego: EdITS, 1995.
57. Sly PD, Cahill P, Willet K, Burton P. Accuracy of mini peak flow meters in indicating changes in lung function in children with asthma. Br Med J 1994; 308:572–574.
58. Ignacio-Garcia JM, Gonzalez-Santos P. Asthma self-management education program by home monitoring of peak expiratory flow. Am J Respir Crit Care Med 1995; 151:353–359.
59. McFadden ER Jr, Kiser R, DeGroot WJ. Acute bronchial asthma: Relationships between clinical and physiologic manifestations. N Engl J Med 1973; 288:221–225.
60. Shim CS, Williams MH Jr. Relationship of wheezing to severity of obstruction in asthma. Arch Intern Med 1983; 143:890–892.

61. Ferguson AC. Persisting airway obstruction in asymptomatic children with asthma with normal peak expiratory flow rates. J Allergy Clin Immunol 1988; 82:19–22.
62. Kendrick AH, Higgs CM, Whitefield MJ, Laszio G. Accuracy of perception of severity of asthma: patients treated in general practice. Br Med J 1993; 307:422–424.
63. Rubinfeld AR, Pain MCF. Perception of asthma. Lancet 1976; 93:882–884.
64. Burden JG, Juniper EF, Killian KJ, Hargreave FE, Campbell EJ. The perception of breathlessness in asthma. Am Rev Respir Dis 1982; 126:825–828.
65. Burki NK, Mitchell K, Chaudhary BA, Zechman FW. The ability of asthmatics to detect added resistive loads. Am Rev Respir Dis 1978; 117:71–75.
66. Lougheed MD, Lam M, Forkert L, Webb KA, O'Donnell DE. Breathlessness during acute bronchoconstriction in asthma: pathophysiologic findings. Am Rev Respir Dis 1993; 148:1452–1459.
67. Turcotte H, Boulet LP. Perception of breathlessness during early and late asthmatic responses. Am Rev Respir Dis 1993; 148:514–518.
68. Hudgel DW, Cooperson DM, Kinsman RA. Recognition of added resistive loads in asthma: the importance of behavioral styles. Am Rev Respir Dis 1982; 126: 121–125.
69. Cacioppo JT, Tassinary LG. Inferring psychological significance from physiological signals. Am Psychol 1990; 45:16–28.
70. Silverman BA, Mayer D, Sabinsky R, Williams-Akita A, Feldman J, Schneider AT, Feldman J, Schneider AT, Chiaramonte LT. Training perception of air flow obstruction in asthmatics. Ann Allergy 1987; 59:350–354.
71. Silverman B, James C, Shibita M, Schneider AT, Chiaramonte LT. Training perception of acute airflow obstruction. Ann Allergy 1990; 64:373–375.
72. Sly PD, Landau LI, Weymouth R. Home recording of peak expiratory flow rates and perception of asthma. Am J Dis Child 1985; 139:479–482.
73. Harver A, Kotses H, Creer TL. On comparing the mini- and standard Wright meters. Am Rev Respir Dis 1983; 127:135.
74. Frankland AW, Atkinson B, Morris RW, Davison JR. Evaluation of Wright peak flow meter with mini-Wright meter in children and adults. Respir Med 1993; 87:551–553.
75. Burns KL. An evaluation of two inexpensive instruments for assessing airway flow. Ann Allergy 1979; 43:246–249.
76. Eichenhorn MS, Beauchamp RK, Harper PA, Ward JC. An assessment of three portable peak flow meters. Chest 1982; 82:306–309.
77. Harm DL, Kotses H, Creer TL. Portable peak-flow meters: Intrasubject comparisons. J Asthma 1984; 21:9–13.
78. Jackson AC. Accuracy, reproducibility and variability of portable peak flow meters. Chest 1995; 107:648–651.
79. Shapiro SM, Hendler JM, Ogirala RG, Aldrich TK, Shapiro MB. An evaluation of the accuracy of the Assess and the mini-Wright peak flow meters. Chest 1991; 99:358–362.
80. Cherniack RM, Chatburn R, Gardner RM, Katz RM, Malveaux FJ, Sander N, et al. Statement on Teaching Standards for Peak Flow Meters. Bethesda, MD: National Institutes of Health (Publication No. NIH 92–2113a), 1992.

81. Mendoza GR. Peak flow monitoring. J Asthma 1991; 28:161–177.
82. Moscato G, Godnic-Cvar J, Maestrelli P. Statement on the self-monitoring of peak expiratory flows in the investigation of occupational asthma. J Allergy Clin Immunol 1995; 96:395–301.
83. Vaughan TR, Weber RW, Tipton WR, Nelson HS. Comparison of PEFR and FEV-1 inpatients with varying degrees of airway obstruction: effect of modest altitude. Chest 1989; 95:558–562.
84. Connolly CK, Chan NS. Relationship between different measurements of respiratory function in asthma. Respir 1987; 52:22–23.
85. Reeder KP, Dolce JJ, Duke L, Raczynski JM, Bailey WC. Peak flow meters: are they monitoring tools or training devices? J Asthma 1990; 27:219–227.
86. Creer TL. Medication compliance and childhood asthma. In: Krasnegor NA, Epstein L, Johnson SB, Yaffe SJ, eds. Developmental Aspects of Health Compliance Behavior. Hillsdale NJ: Lawrence Erlbaum Associates, 1993:303–333.
87. Bartlett J, Kaplan J, eds. Bartlett's Familiar Quotations, 16th ed. Boston: Little, Brown, 1992.

7

Perception of Increased Resistance to Breathing

Andrew Harver

University of North Carolina at Charlotte
Charlotte, North Carolina

Donald A. Mahler

Dartmouth Medical School
Dartmouth Hitchcock Medical Center
Lebanon, New Hampshire

I. Introduction

Asthma is a chronic inflammatory disease of the airways that afflicts about 12 million individuals in the United States and accounts for large numbers of hospitalizations, emergency room visits, and missed school days (1,2). Patients with asthma are characterized by hypersensitivity to allergens, chemicals, air temperature, dust, and other stimuli that results in variable degrees of airflow obstruction (3,4). Acute bronchoconstriction always involves increases in the resistance to breathing that are documented with tests of lung function; therapy partially or completely reverses airflow limitation. Both chronic and acute increases in the resistance to breathing are accompanied by less reliable signs and symptoms of asthma including wheeze, cough, sputum production, and shortness of breath endorsed by patients as an increased "effort" or "work" of breathing and as chest "tightness" or "constriction" (5,6).

Alarming increases in asthma morbidity and mortality (Chapter 1) have accelerated interest in disparities between levels of increased resistance to breathing on the one hand and perceived symptoms on the other hand. For example, the recent rise in deaths from asthma is attributed not only to alterations in acid-base status that follow severe bronchoconstriction, but also to factors that handicap treatment including the failure to adequately assess the severity of asthma, poor family support systems, and problems in self-management including the appropriate perception of symptoms by patients (e.g., 7–11). Many investigations confirm that patients vary widely in their ability to perceive airway obstruction and to assess changes in airway tone accurately following administration of either active or placebo agents (e.g., 12–14). For largely unknown reasons, some asthmatics seem more aware of symptoms than others despite similar degrees of airway obstruction (15,16). Asthmatic patients unable to detect acute fluctuations in airflow, those who underestimate the severity of airflow obstruction, and those with compromised lung function could be at increased risk for more frequent and severe attacks of asthma (17–20). Few studies confirm that patients with poor perception of airflow obstruction have an increased risk of death, but delays in seeking medical care for acute asthma may be common and serve to increase such risk. A survey of 50 patients attending a hospital clinic at least four times a year indicated that 72% of patients delayed seeking help in the preceding 12 months during an attack of asthma lasting at least 48 hours (21). In 10 severe cases requiring admission to a respiratory intensive care unit, the mean duration of symptoms prior to hospital admission was 24 hours; five cases reported having symptoms for over 48 hours (22).

The regulation of asthma should improve when there is concordance between levels of increased resistance to breathing and symptoms. Behavioral scientists, in concert with clinical researchers, are well positioned to consider such an hypothesis. Virtually all motivational and behavioral theories within psychology have assumed that awareness of internal states affects behavior in some ways (23–26). Empirical work in visceral self-perception is reflected in widespread attempts since the 1970s to demonstrate a relationship between the ability of human subjects to perceive autonomic activity and their ability to control or voluntarily regulate such activity (e.g., 27–32). Current health and behavior research includes efforts to uncover processes underlying the perception of symptoms (e.g., 33–35). Symptom perception refers to the way persons monitor their bodies, define and interpret their sensations, take remedial action, and utilize the health-care system (25,33,36–38). Information processing models, for example, emphasize that individuals are active and selective processors of information and operate in the environment by organizing ongoing experience into subjectively meaningful patterns (34,39,40).

In this chapter we review physiological, clinical, and behavioral studies related to the perception of increased resistance to breathing relevant to the patient with asthma. First, we examine briefly the pathophysiology of asthma by considering the mechanics of breathing, the nature of airflow obstruction in asthma, and measures of lung function. Next, we review the ability of individuals to judge changes in airway tone (i.e., levels of intrinsic resistance) and the ability of individuals to judge changes in added loads to breathing (i.e., levels of extrinsic resistance). Finally, responses to intrinsic and added loads are compared and are related to the perception of symptoms in asthma patients.

II. Mechanics of Breathing

The respiratory system is comprised of the lungs, the conducting airways, parts of the central nervous system concerned with the control of the muscles of respiration (i.e., diaphragm, intercostals, rib cage), and the chest wall (41). The main functions of the respiratory system are (1) to obtain oxygen from the environment and to supply it to the cells and (2) to remove carbon dioxide from the body produced by cellular metabolism (42,43). Respiration involves four primary processes, including pulmonary ventilation (the movement of air in and out of the lungs), diffusion (the exchange of oxygen and carbon dioxide between the alveoli and the blood), transport (the transport of oxygen and carbon dioxide in the blood to and from cells), and regulation (the neural control of breathing). Pulmonary ventilation studies focus largely on the forces that move the lung and chest wall that result in the flow of air to and from the alveoli (i.e., on the mechanics of breathing). It is the mechanics of breathing that are largely disturbed in diseases of airflow obstruction (44–46). In asthma, for example, an increased burden is placed on the muscles of respiration, which must generate sufficient tension to overcome the increase in resistance and hyperinflation that accompanies airflow obstruction (15,45–47).

Air moves into the airways because of a pressure gradient that develops between the atmosphere and alveoli during inspiration (Fig. 1). The cycle of changes in three fundamental forces, or pressures, account for the "bulk flow" of air into the lungs with each breath: atmospheric or airway opening pressure, which is typically measured near the mouth (P_{MO}); alveolar pressure (P_{ALV}); and pleural pressure (P_{PL}). At the end of expiration, when no air is flowing, $P_{MO} = P_{ALV}$ and P_{PL} is about 5 cmH_2O lower than P_{MO}. (In respiratory physiology pressure is expressed typically in units of cmH_2O; 1 cmH_2O is equivalent to about 0.75 mmHg). When the muscles of respiration contract, the diaphragm expands the thoracic cage, lowering P_{PL} further. As the airways widen and the alveoli enlarge, the distance between gas molecules increases and P_{ALV} decreases (i.e., $P_{ALV} < P_{MO}$). As alveolar pressure drops, air flows along the pressure gradient created between atmospheric pressure and the relative negative pressure developed in the alveoli. An average pressure gradient of less than 3 cmH_2O moves about 500 mL of air into the lungs.

Three variables are typically recorded in pulmonary ventilation studies: volume (the amount of air moved, in liters), flow (the rate of movement of

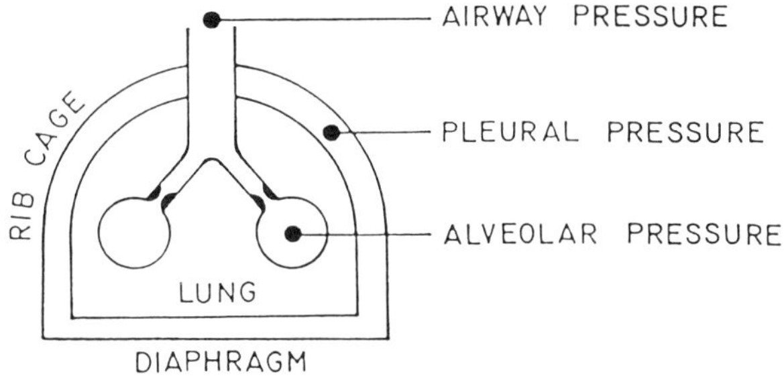

Figure 1 Pressure at different sites in the respiratory system. In healthy individuals, airway pressure is identical with mouth pressure. The mouth-alveolar pressure gradient overcomes airway resistance (indicated by the narrow tubes). (From Ref. 44.)

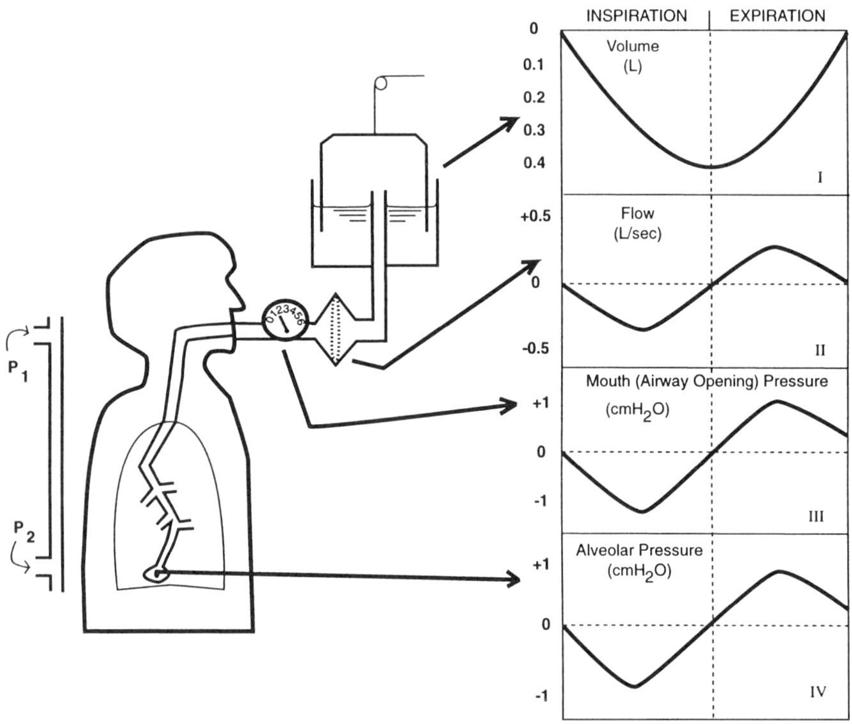

Figure 2 Respiratory volume, airflow, mouth pressure, and alveolar pressure during a single breathing cycle. See text. (Adapted from Ref. 43.)

air, in liters per second), and mouth pressure (the amount of muscle force involved in the production of chest wall movement, in cmH_2O). A schematic for the measurement of volume (with a water-sealed spirometer), flow (with a flowmeter, or pneumotachograph), and pressure (with a differential pressure transducer) is depicted in Figure 2. The pressure gradient developed during inspiration is represented also in Figure 2 simply as the difference between the airway opening (P_1) and the alveoli (P_2). The rate of airflow is dependent on the pressure gradient; an airflow rate of 1 L/sec requires a pressure drop along the airways of less than 2 cmH_2O.

The force that must be generated for flow to occur must be sufficient to overcome the impedances, or "loads," that normally oppose breathing: resistance and elastance. Pulmonary, or total, resistance is comprised of the inertia of the respiratory system (which is negligible), pulmonary tissue resistance from friction encountered as tissues slide over one another during lung expansion, and airway resistance due chiefly to the frictional resistance to airflow offered by the respiratory passageways (47). Resistive loads are due chiefly to the resistance to airflow through the tracheobronchial tree; airway resistance comprises the major component (80–90%) of pulmonary resistance. Elastic loads, due chiefly to the compliance of the lungs and chest wall to inflation, are not considered further, although both obstructive and restrictive diseases are associated with increases in the load to breathing (46).

A. Airflow Through the Tracheobronchial Tree

Air enters the mouth and nose, passes over the pharynx and larynx, and enters the trachea, which divides into two main branches, the right and left bronchi (41). The bronchi each divide again and the branching and division of the airways continues for 20–30 generations (Fig. 3). No alveoli are present in the conducting zone—the first 16 generations—but they begin to appear in the next several generations (the transitional zone). The final set of branches (the respiratory zone) are lined with alveoli, about 300 million in the two lungs (42).

The amount of respiratory pressure (in cmH_2O) required to produce a given rate of flow (in L/sec) is expressed as the ratio between the two quantities (cmH_2O/L/sec). Because the rate of airflow is determined by the mouth-alveolar pressure gradient, airway resistance (R_{aw}) is measured as the drop in pressure between the alveoli and the mouth (measured in cmH_2O), divided by flow ($\dot{V}$), or:

$$R_{aw} = \frac{P_{MO} - P_{ALV}}{\dot{V}} \tag{1}$$

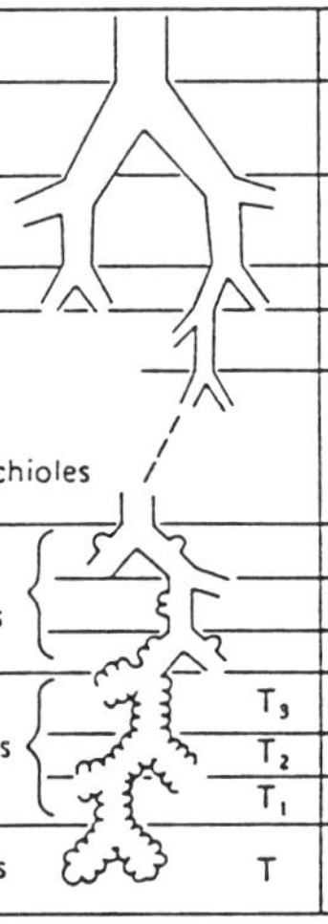

	Generation			Diameter, cm	Length, cm	Number	Total cross sectional area, cm^2
conducting zone	trachea		0	1.80	12.0	1	2.54
	bronchi		1	1.22	4.8	2	2.33
	↓		2	0.83	1.9	4	2.13
			3	0.56	0.8	8	2.00
	bronchioles		4	0.45	1.3	16	2.48
	↓		5	0.35	1.07	32	3.11
	terminal bronchioles		↓ 16	↓ 0.06	↓ 0.17	↓ 6×10^4	↓ 180.0
transitional and respiratory zones	respiratory bronchioles		17	↓	↓	↓	↓
			18				
			19	0.05	0.10	5×10^5	10^3
	alveolar ducts	T_3	20	↓	↓	↓	↓
		T_2	21				
		T_1	22				
	alveolar sacs	T	23	0.04	0.05	8×10^6	10^4

Figure 3 Airway branching in the human lung. (From Ref. 41.)

The major site of airway resistance is the medium-sized bronchi; less than 20% of total airway resistance is attributed to airways less than 2 mm in diameter. Normal airway resistance values vary between 2 and 4 $cmH_2O/L/sec$; values from 2 to 15 times as great may be observed in acute asthma.

In most physical systems, flow is assumed to be laminar (streamlined and parallel along the sides of a tube), transitional (when small eddies appear at branches in the system), or turbulent (reflected in complete disorganization of the stream lines). The French physician Poiseuille, in 1846, determined that laminar flow in a smooth, cylindrical tube was dependent not only on the driving pressure but also was directly proportional to the viscosity of the fluid (η) moving through the system as well as the length of the tube (l), and inversely proportional to the fourth power of the radius of the tube (πr^4):

$$\text{Resistance} = \frac{8\,\eta\, l}{\pi\, r^4} \tag{2}$$

Accordingly, flow through the airways is predictably dependent not only on the driving pressure but also on the viscosity of the air, the length of the airway, and the fourth power of the radius of the airway. But the irregular branching and compressible system of airways in the lungs make application of these principles complicated, especially in the analysis of airway narrowing as occurs in asthma (48,49). It is likely, for instance, that true laminar flow only occurs in the smallest airways where the velocity of airflow is low (41). In most of the tracheobronchial tree, flow is transitional (43). The addition of moisture and bronchoconstriction results in flow more turbulent.

B. Airways Obstruction

The chief problem in asthma is airway hypersensitivity to many stimuli that results in widespread narrowing of the airways and, consequently, increases in the resistance to breathing as well as marked abnormalities in the distribution of air into the lungs (3,19). The inflammatory response to allergens, infections, exercise, and other stimuli develops over minutes, hours, or days, produces variable degrees of airflow limitation both within and between individuals, and results in uneven and nonuniform degrees of obstruction throughout the lungs that can remit spontaneously or reverse with treatment (19,50,51).

Neural Regulation of the Airways

Neural control of the airways is discussed by Kotses earlier in this volume (Chapter 3). Briefly, the autonomic nerves directly regulate many aspects of airway function including airway smooth muscle tone, secretions, blood flow, microvascular permeability, and the migration and release of inflammatory cells (52). The neural regulation of the airways and blood vessels is dominated by vagal efferents from parasympathetic, cholinergic fibers; stimulation of the vagus leads to bronchoconstriction. There is virtually no direct sympathetic innervation of the airways; stimulation of sympathetic fibers results in smooth muscle relaxation but mostly through circulating catecholamines released by the adrenals (53).

Vagal afferents carry information about pathophysiological processes occurring in the lungs and airways that produce adaptive changes in the pattern of breathing as arise in coughing, swallowing, and vomiting (54). In addition, well-described mechanoreceptors are involved in the reflex regulation of the breathing pattern as well as bronchomotor tone. Slowly adapting receptors (SARs) serve as airway smooth muscle stretch receptors and are important in breath-to-breath modulation of bronchomotor tone; activation of SARs results in bronchodilation through inhibition of efferent vagal activity. Rapidly adapting receptors (RARs) are responsive to nonspecific irritants, and mechanical or chemical stimulation of RARs results in reflex bronchoconstriction. In general, however, the observed increases in the resistance to breathing in asthma are more a result of airway inflammation than of neural disruption.

Inflammatory Processes in Asthma

The early and late asthmatic response to triggers like allergens, irritants, and infections has been linked increasingly to inflammatory processes. Such processes involve the signaling and migration of inflammatory cells (e.g., mast cells) into the airways and the production of a variety of mediators (e.g., prostaglandins and leukotrienes) that induce bronchoconstriction, edema, and excessive mucus secretion (55). The "early response" in most acute asthma results from a classic antigen-antibody reaction that involves the release of specific mediators and contraction of airway smooth muscle that peaks within 10–15 minutes. Bronchoconstriction can result from direct stimulation of receptor sites on airway smooth muscle cells by mediators secreted from respiratory mast cells. Alternatively, histamine and other irritants induce bronchoconstriction through reflex stimulation of efferent parasympathetic fibers.

The "late" asthmatic response includes increased susceptibility to triggers 3–4 hours following the initial response and ensuing release and presence of mediators. For example, the inflammatory response to the presence of mediators includes membrane reactions in the submucosa that increase capillary permeability resulting in edema of the bronchial walls (19). In addition, chemical agents released by cells attract phagocytes that infiltrate the airways and release their own toxic products, which stimulates production of mucus that is typically thick and tenacious because of the presence of inflammatory cells.

Smooth muscle contraction, edema with or without inflammatory cellular infiltration, and decreases in the clearance of secretions (3) serve to increase the resistance to airflow through the tracheobronchial tree. These processes contribute to airflow limitation in varying degrees, result in uneven degrees of obstruction throughout the entire tracheobronchial system, and resolve through different means.

C. Assessment of Airflow Obstruction

Many texts, manuals, and position papers describe the assortment of tests and measures that are used to document abnormalities in lung function and gas exchange common in lung disease (e.g., 56–60). Regardless of purpose or level of technical sophistication, virtually all assessments of pulmonary ventilation depend on reliable estimates of respiratory volumes, pressures, and flows independently or in combination with one another. Perhaps the most clinically informative analysis for the presence of airflow limitation simply involves (1) a rapid, forceful, and complete expiration following maximal inhalation, and (2) a spirometer to detect the volume and rate of the "forced expiration" (60). A recording of a single forced expiratory maneuver accom-

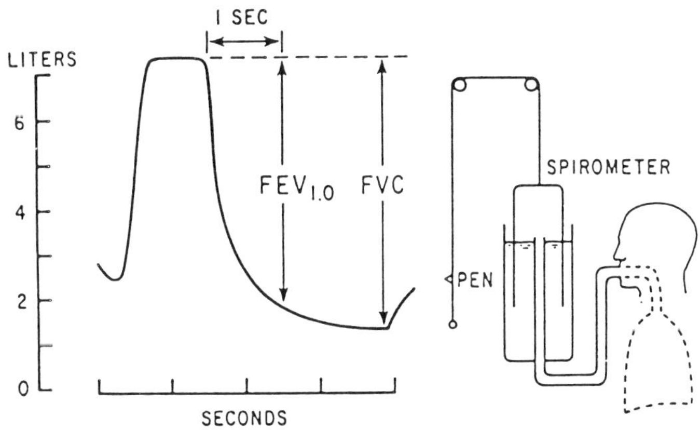

Figure 4 Measurement of forced vital capacity. (From Ref. 60.)

plished with a water-sealed spirometer is shown in Figure 4. The FVC, or forced vital capacity, represents the total volume of air expired from maximal inspiration (total lung capacity); the FEV_1, or forced expiratory volume in one second, represents the volume of air expired during the first second of the maneuver. Prolonged expiratory times and reductions in both FVC and FEV_1, and abnormalities in the FEV_1/FVC ratio, confirm the presence of acute airflow obstruction. Predicted normal values (based on gender, age, and height) for these quantities and for many other volume and flow measures derived from the spirogram are well described (e.g., 61–64).

Because airway resistance is defined as the pressure difference between the alveoli and the mouth divided by flow, direct assessment of airway resistance requires more elaborate solutions. For instance, alveolar pressure can be measured only indirectly. Such measurements are normally obtained through the technique of whole body plethysmography first described by DuBois and colleagues in 1956 (65). A subject sits in an airtight booth and pants against a flowmeter while flow, box pressure, and airway opening pressure are measured. The process is then repeated with the flowmeter occluded (i.e., when no air is flowing) so that $P_{MO} = P_{ALV}$. Finally, P_{MO} is plotted against the simultaneous changes in flow obtained during the first maneuver and against box pressure obtained during the second maneuver, and P_{ALV} is deduced through application of Boyle's law (42,57). An older method for measuring airway resistance involves rapid interruptions of airflow (10/s) during normal breathing by a shutter placed in a mouthpiece. The pressure measured at the airway opening immediately after an occlusion is assumed to provide a reasonable estimate of alveolar pressure just before occlusion

(57,66). Another method is to connect the individual to a source of forced air oscillations, such as a loudspeaker, and apply low-frequency pulses to the airway. Respiratory resistance can be determined by relating the extremes of flow and corresponding pressure variations produced by the oscillations (60,67,68).

Many authors repeatedly endorse the critical need for repeated, objective measurement of airflow obstruction in asthma. Summaries of the National Asthma Education Program have recommended that all patients "undergo objective measurements of airflow in the physician's office during each visit, and those patients with moderate-to-severe asthma, who therefore would be utilizing daily pharmacologic therapy, should have daily peak flow assessment" (51). Peak flow meters provide a simple and reliable index of airflow obstruction resistant to individual bias in symptom interpretation (69). Barnes (17) recommended that patients with poor symptom perception should be provided with written plans if peak flow levels drop by more than 30%. Despite the presumed advantage of home peak flow monitoring, only 10% of 255 patients treated in a general practice reported using a peak flow meter regularly (70). Peak flow monitoring and asthma self-management is considered fully in this volume by Creer and colleagues (Chapter 6), by Fritz and Wamboldt (Chapter 8), and by Harver and Katkin (Chapter 9).

III. Intrinsic Resistive Loads

The resistive load to breathing relates primarily to the resistance to airflow through the tracheobronchial tree. Patterns of smooth muscle contraction, edema, and mucus throughout the airways increase the resistance to breathing. The early and late asthmatic responses to triggers like allergens, irritants, and infections result in variable degrees of airflow limitation both within and between individuals. Increases in the intrinsic or normally evident levels of airway resistance give rise to symptoms.

A symptom is any morbid phenomenon or departure from the normal in function, appearance, or sensation experienced by the patient and indicative of disease (71). Symptoms are subjective, apparent only to the affected person. A sign, on the other hand, is any abnormality indicative of disease, detectable by another person and sometimes by the patient. A sign is an objective symptom of disease; a symptom is a subjective sign of disease (71). Dyspnea is a symptom; tachypnea and wheezing are signs. Some phenomena, like fever, can be considered both symptoms and signs of disease (72).

The perception of symptoms is, in part, an attribution process that is influenced by demographic variables, individual differences, and sociocultural factors (e.g., 25,38). Symptom attribution includes processes involving the ways in which persons identify and evaluate symptoms and make interpretations about their causes and consequences. People process information about symptoms according to preconceived notions they have about the relatedness among physiological, psychological, and environmental quantities (73) and form "disease prototypes"—mental models, or representations—that serve as standards against which to match and to evaluate symptoms (40). Symptoms are more likely to be recognized as increasingly troublesome, or threatening, when they relate closely to a familiar prototype. Patients may continue to emphasize symptoms that match their conception of their disease even though the underlying pathology may be altered by treatment (37). Treatment difficulties arise because people selectively attend to signals that confirm their naive diagnosis while selectively ignoring sensory inputs that are incompatible with held beliefs (74,75). Persons are especially prone to omit, substitute, or infer the importance of elemental experiences when events are ambiguous, complex, or emotionally loaded (39).

A sequence of decisions, each with a particular temporal component, can be imagined to account for processes involved in health care judgments: appraisal delay, illness delay, and utilization delay. Appraisal delay refers to the time it takes to determine that symptoms indicate disease (76). Illness delay is the time between the appraisal or self-examination processes and the decision to initiate seeking medical care. The decision to seek medical care is more likely to be delayed when symptoms are attributed to aging (76,77) or are ambiguous and appear in combination with life stressors of recent onset (78). Utilization delay refers to the time between the decision to seek medical care and entry to the health care system. Earlier we noted earlier that delays in seeking medical care for acute asthma may be common (21,22) and hypothesized that the regulation of asthma should improve when there is concordance between levels of increased resistance to breathing and perceived symptoms.

The complex pattern of symptoms experienced by patients with asthma was explored in two related investigations in which 175 patients (ranging in age from 14 to 67 years) rated the presence and frequency of 77 symptom adjectives (79,80). Five symptom clusters, or factors, emerged from analysis of patient responses to the Asthma Symptom Checklist. The most commonly reported symptoms were those of airway obstruction, or bronchoconstriction (e.g., hard to breathe, chest tightness, chest filling up, wheeze). Kinsman and colleagues reported that 91% of patients acknowledged experiencing the symptoms of bronchoconstriction frequently during asthmatic episodes (79). Four other factors were derived to account for the interrelationships among subjective experiences reported by patients with asthma: fatigue (e.g., weak, worn out, no energy), panic-fear (e.g., scared, worried, panicky), irritability (e.g., cranky, edgy, anxious), and hyperventilation (e.g., pins and needles, headache). In general, patterns of symptomatology did not vary among patients with respect to age, age at onset of asthma, or illness duration, but females tended to report more symptoms than males (80). The pattern of symptoms obtained in these investigations was confirmed in a subsequent investigation involving 132 adult asthma outpatients who rated the presence and frequency of 36 symptom descriptions or adjectives (81). Five symptom clusters were obtained also in a group of 58 adolescents between the ages of 11 and 17 years who completed the Asthma Symptom Checklist (82). Four clusters were similar to those previously observed in adults (airway obstruction, panic fear, irritability, and fatigue), but in children an unhappy-afraid cluster (e.g., unhappy, afraid of dying) was more meaningful than the one for hyperventilation.

Health-care judgments and, therefore, self-management practices in asthma relate to the pattern of symptoms experienced by patients. For example, an evaluation of 102 asthma inpatients ranging in age from 15 to 72 years indicated that nearly 20% of patients mislabeled one or more of eight nonairway obstruction symptoms (e.g., anxiety, fatigue, rapid breathing) as being an asthma attack (83). More importantly, 84.2% of those who mislabeled symptoms were rehospitalized at 6 months as compared with 39.8% of those who correctly labeled symptoms during the same period. Finally, the interaction between perceived symptoms and personality are beyond the scope of this chapter, but both panic-fear (Chapter 2) and, more recently, negative affect (i.e., undifferentiated subjective distress) are traits that serve to moderate both symptom perception and illness behavior (84–86).

A. Relations Among the Signs and Symptoms of Airflow Obstruction in Asthma

Although the relevance of symptoms in the assessment and diagnosis of asthma is well appreciated, no single measure currently assesses both the variety of experiences (wheeze, dyspnea, cough, sputum) and the different aspects of these experiences (intensity, duration, frequency) as would be required, for example, to evaluate outcomes in a clinical intervention trial (87). Moreover, the absence of symptoms does not preclude either the presence of airflow obstruction or a predisposition for airway hyperreactivity, and significant improvement in symptoms does not signal complete reversal of acute airflow limitation (e.g., 88–92).

B. Perception of Changes in Airway Tone by Patients with Asthma

Since the widely cited observations of Rubinfeld and Pain in 1976 (14), an increasing number of investigators have examined the relationship between perceived symptoms and abnormalities in lung function in patients with asthma. Three kinds of investigations have been conducted: (1) single pairs of observations are obtained from patients exhibiting a range of symptom scores and lung function values and a correlation is computed to determine the magnitude of correspondence between the two quantities; (2) multiple pairs of observations are obtained from patients and combined in a single analysis, or are used to compute separate, intraindividual correlations to index symptom awareness; and (3) the level, or intensity, of symptoms is determined for a given level of airflow obstruction on one or more occasions and, often, compared between groups of subjects with different characteristics. In these studies, values have been obtained from patients during spontaneous asthma episodes, following bronchoconstriction induced by either methacholine, antigen exposure, or exercise, or in several of these conditions. The ongoing rise in the number of such reports parallel increased apprehension that a proportion of patients are "poor perceivers" of airflow obstruction and at risk for greater morbidity and mortality. The factors that predict poor perception are of great clinical interest but remain elusive (e.g., 15,93).

Perception of Bronchoconstriction and Bronchodilation

Rubinfeld and Pain (14) compared the intensity (none, mild, moderate, severe) of symptoms (i.e., chest tightness) to percent predicted FEV_1 values in 61 patients with asthma at baseline and following methacholine challenge. Six subjects at baseline and three others at the height of the induced attack (i.e., 15% of the participants) were labeled "poor perceivers" of airflow obstruction because they reported no symptoms despite FEV_1 values less than 50% of that predicted. The correlation between the intensity of symptoms at the height of the induced attack and predicted FEV_1 values in the 61 patients, however, was significant, $r = 0.97$. Additional reports confirm that the overall relationship between symptoms and lung function in patients with asthma is robust. For example, a significant relation ($r = 0.86$) was obtained between estimated and actual peak expiratory flow rates (PEFR) recorded in 35 patients at clinic visits (94), and a significant overall relation ($r = 0.88$) was obtained in 45 patients with asthma between Borg ratings of breathlessness and reductions in FEV_1 during methacholine challenge (18). Reliable correlations were evident also between scores acquired with a 100-mm visual analogue scale (VAS) and percent predicted FEV_1 ($r = -0.532$) in 40 children ranging in age from 9 to 17 years during routine out-patient review (95) and between VAS dyspnea ratings and percent predicted PEFR ($r = -0.72$) recorded in 40 stable patients ranging in age from 16 to 60 years (96).

Correlations between pairs of observations obtained in the same individual are more variable. For example, individual correlation coefficients computed between VAS ratings and PEFR obtained twice daily during hospitalization for acute asthma in 12 patients ranged from –0.17 to –0.79 (96). In a study involving 11 general practices in and around Bristol, UK, both VAS ratings of asthma severity and PEFR were recorded up to four times daily for 14 days in 255 patients with asthma ranging in age from 17 to 76 years (70). In that investigation, 60% of participants (n = 152) were considered "poor discriminators" of asthma because, for each patient, the intraindividual correlation coefficient calculated between VAS scores and percent predicted PEFR was statistically unreliable.

Airflow limitation produced by a variety of stimuli is perceived similarly by patients. Patients with asthma experience "just noticeable symptoms" when the FEV_1 declines to around 70–75% of predicted normal values during spontaneous or methacholine induced bronchoconstriction (97). Boulet et al. (98–101) have employed a standard index to assess more directly the level, or intensity, of symptoms for a given degree of airflow obstruction (i.e., PS_{20}, or the perception score obtained with a modified Borg scale corresponding to a 20% decline in FEV_1) analogous to that used for determining levels of airway reactivity in patients with asthma (i.e., PC_{20}, or the provocation concentration resulting in a 20% decline in FEV_1). They observed that average Borg ratings of breathlessness for a 20% fall in FEV_1 induced by antigen challenge, by histamine challenge, or by exercise were equivalent in eight young adults with asthma (101). They demonstrated also that PS_{20} values to histamine provocation in 150 subjects ranging in age from 13 to 74 years (M = 3.36) are normally distributed (99), that perception scores are greater in patients with asthma (M = 2.0) than in nonpatients (M = 0.5) following methacholine challenge (98), and that perception scores are greater during the early asthmatic response (M = 2.0) compared to the late asthmatic response (M = 0.5) despite comparable decreases in FEV_1 levels (100).

Fewer investigators have assessed changes in reported symptoms following the administration of a bronchodilator. Orehek and colleagues (13) reported that asthmatic subjects without initial airflow obstruction were more likely to consider their breathing improved after salbutamol than those with initial airflow obstruction (13). Noseda and colleagues have incorporated a bipolar VAS (with the anchors "very much worse" and "very much better") to assess improvements in symptoms following administration of successive doses of terbutaline inhalations in patients with asthma and with COPD (102,103). Modest but reliable correlations were found between terbutaline-induced changes in VAS ratings and improvements in lung function in the patients with asthma. A significant relation was evident also between VAS scores and improvements in FEV_1 (r = 0.43) following nebulized salbutamol in 40 children with asthma (95).

Risk for Near-Fatal Asthma

Two recent longitudinal summaries (104,105), several editorials (e.g., 7,17, 93), recent reviews (10,15), and many discussion sections (e.g., 14,18,22,94) struggle to elucidate the factors that place a patient with asthma at increased risk for a near-fatal episode, and labor to determine the clinical significance of reduced perception of airflow obstruction in asthma self-management. Despite overall significant relationships between subjective and objective measures of airflow limitation, for example, some patients report less breathlessness than others for any given reduction in lung function (14,18,94). A recent study demonstrated that most patients with near-fatal asthma had a blunted perception of dyspnea during resistive loading and diminished chemoresponsiveness (106). However, reduced perception of breathlessness was not a significant predictor of risk for a near-fatal asthma attack in one cohort of 45 patients (105). Those who have experienced a near-fatal episode of asthma exhibit greater variability in daily peak flow scores, are less compliant with prescribed treatment, and manifest less adaptive personality traits compared to those who have never experienced such an episode (104). Recurrent risk factors in the literature include age, airway lability, delays in seeking medical attention for symptoms, denial, duration of asthma, persistent airflow obstruction, poor compliance, previous hospital admission, previous occurrence of life-threatening asthma, reduced hypoxic responses, and severe obstruction without awareness (e.g., 7,10,16–18,22,93,104–107). Many of these factors, independently and in combination, likely relate especially to the undertreatment of asthma.

It is entirely rational that reduced perception of airflow obstruction promotes undertreatment, but it must be admitted that the concept of "poor perceivers" remains ill-defined. At least three operational definitions for reduced sensitivity to increases in airway resistance are evident: a lack of symptoms for a given level of airflow obstruction, variability in the intensity of symptoms for a given level of airflow obstruction, and inconsistency in the correspondence between repeated measures of symptoms and lung function. Progress in efforts to determine the value of symptom perception in asthma self-management could hinge not only on determining why some patients are less sensitive to airflow obstruction compared to others but also on developing standard criteria for recognizing "good perceivers." Solutions to these uncertainties likely relate to understanding (1) the ways in which individuals select to organize symptoms including breathlessness into subjectively meaningful patterns of experience (34,39,40,79,80) and, perhaps most critically, (2) the behaviors individuals exhibit in response to symptoms (25,33–36,76).

IV. Added Resistive Loads

Over the course of about a decade—from 1974 to 1985—several conferences and symposia highlighted two goals that have stimulated ongoing study of resistive and elastic loads (108–110). One purpose relates to physiological interests in the regulation of ventilation (i.e., load compensation) and in the consequences of disturbances in the mechanics of breathing in lung disease on respiratory homeostasis. The other purpose relates to clinical interests in the sensation of dyspnea and in the quantification of the relationship between mechanical loads and respiratory sensations (i.e., load perception).

A convenient analogy used to characterize the effects of added loads on the "respiratory pump," or ventilatory apparatus, is to consider the effects of lifting a suitcase on the muscles of the human upper limb. Lifting a suitcase off the floor imposes a "load" on the skeletal muscles, and the size of the load depends on the mass of the suitcase. The size of the load can be expressed readily in terms of the level of tension, or force, developed by the muscles to move the suitcase a fixed distance or at a fixed rate. In addition, an individual could be asked to rate or judge the intensity of the load using verbal descriptors or numbers. External loads, in a comparable way, impose an extrinsic or "extrapulmonary" load to breathing that alters normal ventilatory parameters (e.g., the rate at which air flows to and from the lungs), and the size of the load can be expressed readily in terms of the tension, or force, developed by the respiratory muscles to move the lung and chest wall. The level of tension or force developed by the muscles of respiration is reflected acceptably in the pressure recorded at the airway opening (i.e., mouth pressure; Fig. 2). Investigators are able to determine not only the perceived magnitude of a given load by having subjects use verbal descriptors or numbers to indicate intensity, but also the ability of subjects to sense minimum or "just noticeable" increases in the added load to breathing.

Many reviews of loaded breathing experiments conducted since the 1950s in healthy subjects and patients with chronic lung disease have summarized the considerable scientific advances made in the areas of respiratory mechanics and the control of breathing (e.g., 45,46,111–113). In many instances, the limits of external loads as an analog for examining the mechanical changes of the ventilatory apparatus in disease are recognized. These limits are strikingly relevant also for discussions of the comparability of the sensory experiences in lung disease and load perception. For example, intrinsic loads are frequently associated with complex and localized stressors not produced by external loads such as inflammation, vascular changes, and hormonal and electrolyte disturbances (45,113). As discussed previously, pulmonary disease seldom effects the lungs uniformly but results in an invariant set of mechanical aberrations (45,113). Such invariance may result in unique sources of sensory information used in processing respiratory sensations that are not represented in experimental situations. In addition, mechanical loads may not fully duplicate the pattern of sensory stimulation produced by lung disease, and temporal factors (e.g., the months or years it takes for lung disease to develop) may serve to further reduce the similarity between the acute physiological and behavioral responses to extrapulmonary loads in the laboratory and those of the patient with lung disease (45,113). The predominance of evidence indicates that conscious individuals temporarily increase respiratory force when exposed to loads but not much is known about responses when exposure to loads is more prolonged. Finally, the increases in lung volume that accompany the sensation of chest tightness in asthma (i.e., hyperinflation) are not typically evident in loading studies (15,114).

A. Methodological Considerations

From a clinical standpoint, resistive loads mimic the type of flow limitation experienced by patients with obstructive lung disease including asthma by altering normally present pulmonary pressure-flow relationships. They are constructed generally of porous material (e.g., metal or plastic filters, or metal or nylon mesh screens) and connected in series, or steps, to provide a fixed range of increases in flow resistance. (Imagine breathing through a series of straws, each with a lumen more narrow than the one before.) Resistive load intensities are determined in units of $cmH_2O/L/sec$ by passing air through the resistors at various known flow rates and simultaneously measuring the drop in pressure across the resistor that results (see Eq. 1). By comparison, elastic loads are produced by requiring individuals to breathe from airtight containers of various sizes joined in parallel with tubing and normally vented to atmosphere. Breathing through an elastic circuit alters the pressure-volume relationships normally present and mimics the type of volume restriction experienced by patients with interstitial lung disease. These loads are quantified by measuring the pressure (in cmH_2O) within the container as known quantities of air (in liters) are injected or withdrawn from the circuit. The ratio between the change in pressure for a given change in volume describes the elastance of the container (cmH_2O/L). Elastic loads have been studied with the same level of intensity as resistive loads, but they are not considered fully in this chapter.

A schematic of an experimental setup to examine either resistive load compensation or load perception in an individual is shown in Figure 5. The subject breathes on a mouthpiece connected to a set of resistive elements embedded in a rigid tube while airway opening pressure and airflow are measured continuously. On most breaths, the subject's inspiration and expiration are directed to atmosphere but on selected breaths (1) a rubber stopper

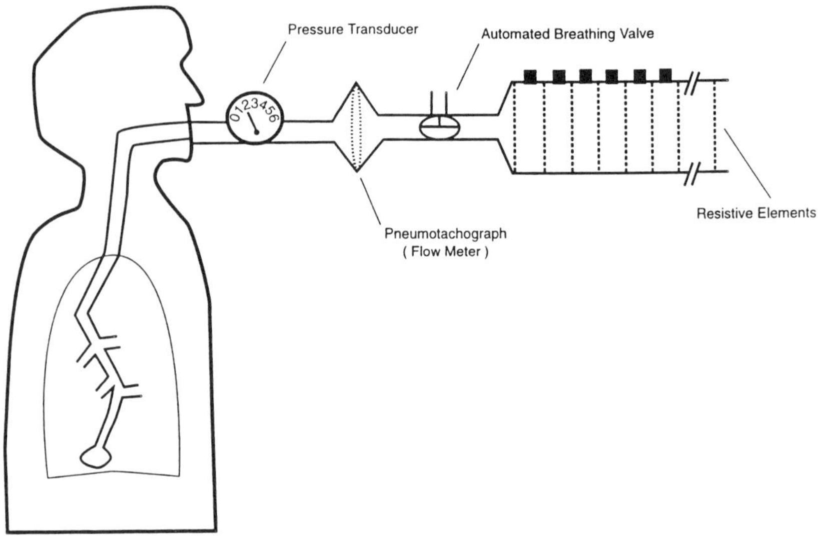

Figure 5 Schematic of an experimental setup for perception of added resistive loads to breathing.

is removed from a predetermined segment of the circuit and (2) the operation of either a mechanical or computer-operated solenoid valve redirects ventilation through the enabled resistive element(s). Loads can be presented to subjects—either with or without a warning signal—during a predetermined number of consecutive breaths, during an entire breathing cycle, or during all or part of a single inspiration or expiration (115,116). Breathing pattern responses to loading are derived from changes in the physiological measures. Immediately following the presentation of a load, subjects can be asked to decide (yes or no) whether a stimulus was present (e.g., by pressing keys on a button board as in detection tasks) or provide an estimate of the magnitude of the sensation elicited by the load (e.g., by making a mark on a VAS, pointing to a number on a modified Borg scale, or squeezing on a handgrip). In many experiments, both the level of the added load (in units of $cmH_2O/L/sec$) and peak mouth pressure (an index of the tension produced by the respiratory muscles, in cmH_2O), are used as estimates of the initiating stimulus in establishing psychophysical relations between subjective magnitude and physical intensity. Resistive loads can be applied also to the body surface (i.e., chest wall), but such testing is largely beyond the scope of this chapter.

B. Load Compensation: Breathing Pattern Responses to Added Loads

The breathing pattern is regulated by multiple reflexes acting on the brain stem in concert to determine the depth of breathing (tidal volume, V_T), the duration of the phases of the breathing cycle (inspiratory time, T_I, and expiratory time, T_E), and the lung volume at which inspiration begins (end-expiratory volume, EEV). Ventilatory responses to added loads are affected by intrinsic properties of the respiratory muscles (e.g., the velocity of shortening), by reflex mechanisms involving upper airway, chest wall, and pulmonary receptors as well as chemoreceptors sensitive to alterations in blood chemistry, and by behavioral responses following conscious perception of an added load (45,111,113,117–119). The intense physiological interest in the ventilatory responses to added loads is reflected in the nearly 200 citations surveyed in the recent chapter by Younes (113).

Both the immediate, first-breath response to an added load and steady-state responses to added loads have been well scrutinized not only in animal preparations but also in conscious and anesthetized humans (see Ref. 113). Inspiratory resistive loads induce immediate reductions in tidal volume and increases in inspiratory duration, although immediate responses in humans are extremely variable. The decreases in V_T and in breathing frequency result in reductions in ventilation. The magnitude of the decreases in V_T and increases in T_I varies with load intensity. The steady-state response to inspiratory resistive loads in conscious humans is associated with an unchanged or slightly greater V_T, lengthened T_I, reduced respiratory rate, and slightly decreased ventilation. Expiratory resistive loads induce increases in expiratory duration. Tidal volume generally increases to compensate for the decrease in frequency so that ventilation is depressed only slightly in the steady state.

A primary goal of load compensation experiments is to uncover the intrinsic muscle responses and neural mechansisms acting on brain stem that mediate breathing pattern responses to acute or chronic increases in the mechanics of breathing. Younes (113) concluded that two primary mechanisms (the intrinsic properties of the respiratory muscles and chemical feedback) and two secondary mechanisms (inceases in intercostal activity and a reflex that shortens expiration during inspiratory loads) account for the compensatory responses to added loads. The large interindividual differences in load responses observed in conscious humans, he further concluded, are due largely to "noncompensatory" behavioral (i.e., voluntary) responses following conscious perception of an added load that serve to minimize abnormal respiratory sensations. Load perception studies, therefore, which evolved from load compensation studies, serve to bridge physiological research programs with those related to the quantification of the relationship between mechanical loads and respiratory sensations and the perception of increased resistance to breathing.

C. Load Perception: The Psychophysics of Added Resistance

A large number of variables can be measured to estimate either the presence or magnitude of symptoms in asthma including numbers of physician visits, activity restriction, peak flow logs, and clinical ratings. However, none of these measures allow us to judge directly the subjective magnitude of a symptom. When we wish to predict what people will say when they try to give a quantitative description of their subjective impressions, we enter the domain of sensory psychophysics, the investigation of the correspondence between the magnitude of stimulus properties as assessed by the instruments of physics and as assessed by the perceptual systems of people (120). Excellent introductions to the field of sensory psychophysics are available (e.g., 120–124) and, in this volume, Baird (Chapter 10) reviews application of psychophysical principles to respiratory sensations.

The first investigations of the ability of human subjects to detect just noticeable increases in either the resistive or elastic component to breathing were conducted by Campbell and colleagues (125,126). Bakers and Tenney (127) first introduced direct scaling techniques to investigate the perception of respiratory pressures and volumes, and magnitude estimation was first applied to the perception of resistive and elastic loads fairly recently (128). In 1986, Zechman and Wiley (129) published a review of the field that summarized about 50 experiments (including reference to 12 abstracts) involving the quantification of respiratory sensations conducted through 1981. We have identified 73 papers published since 1962 that have examined perceptual responses to added resistive loads in humans; most (40%) were published between 1980 and1989. Several recent chapters have surveyed various numbers of these studies (130–133).

Load Detection

Experimental studies of load detection involve standard psychophysical techniques to determine the minimum level of added resistance necessary to evoke a "just noticeable" increase in the load to breathing. With few exceptions, most investigators have determined the "threshold" for added resistance with some variation on the method of constant stimuli, a versatile, accurate, and widely applicable procedure for determining sensory thresholds. Experimenters present a constant number of resistive loads to subjects (typically 5–10) for a fixed number of presentations (typically 5–10 times each). The range of stimulus intensities chosen usually span from an undetectable stimulus to

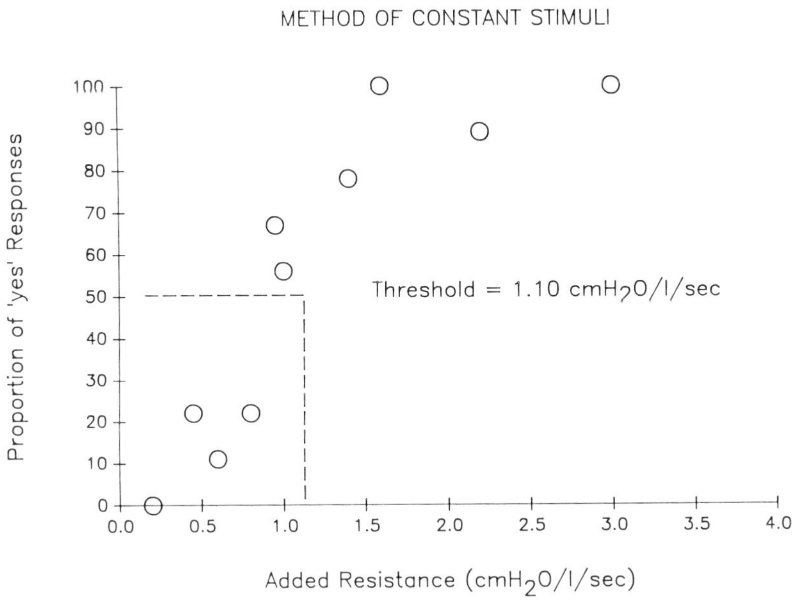

Figure 6 Determination of the threshold resistance to breathing.

one that is always detected. Loads are presented repeatedly, once every 2–5 breaths, and a psychometric function plotting the percentage of detections as a function of stimulus intensity is determined. Such a plot for determining the threshold for added resistance in a subject from our laboratory is displayed in Figure 6 and shows that the threshold value, interpolated as the stimulus intensity detected on 50% of the trials, is 1.1 $cmH_2O/L/sec$.

The just noticeable differences (JNDs) in added loads to breathing were assessed initially by Campbell and colleagues (125,126). Mean detection thresholds for resistive loads in nine subjects were 0.56 $cmH_2O/L/sec$ on one day and 0.61 $cmH_2O/L/sec$ on repeated testing. Mean threshold levels for added resistive loads in healthy adults obtained in many subsequent replications vary along a markedly narrow range, from about 0.5 to 1.5 $cmH_2O/L/sec$ (134–145) and demonstrate that detection thresholds for added loads to breathing can be assessed with a great deal of validity using techniques derived from sensory psychophysics. Despite widespread enthusiasm that load perception studies provide a means to examine variability in the perception of breathlessness, surprisingly few investigators have evaluated threshold levels for added resistive loads in adult patients with asthma (135,137,146). In these studies, detection thresholds for added resistive loads obtained in asymptom-

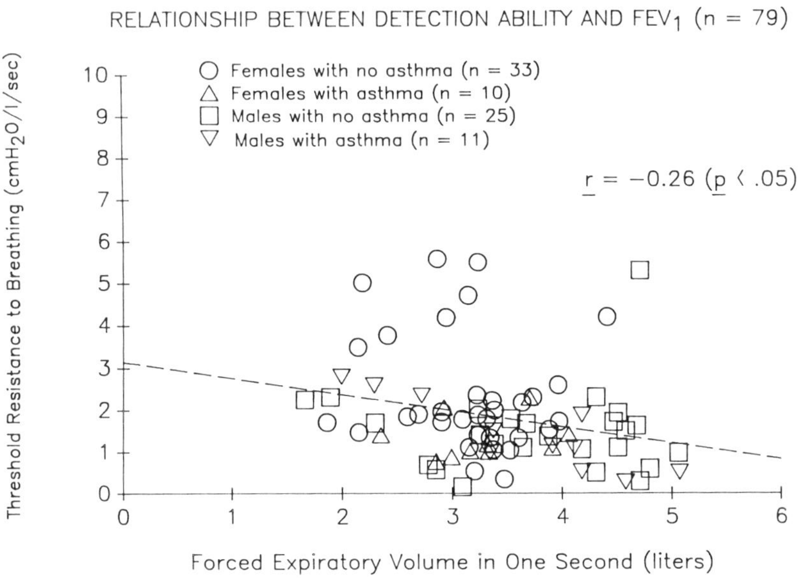

Figure 7 Scatterplot of the relationship between the threshold resistance to breathing and lung function in 79 subjects, including 21 patients with asthma.

atic patients with asthma are similar, or not significantly greater, than those obtained in healthy control subjects.

In classic psychophysics, Weber's law states that the intensity of a stimulus detected must be increased by a constant fraction of its original value in order to produce a minimum change in perceptual value. Mathematically, the relationship between the minimum change in stimulus intensity needed to produce a JND in perceptual value (ΔS) and the background sensory stimulus (S_o) is a constant (k), that is:

$$\frac{\Delta S}{S_o} = k \tag{3}$$

Threshold levels for resistive loads (ΔR) conform to such a relationship and bear a constant relation to the total background sensory stimulus, R_o, which represents the sum of an individual's intrinsic level of airway resistance, R_{aw}, and the nominal level of resistance offered by the "unloaded" breathing circuit, R_c. Bennett and colleagues (125) noted that the JND for resistive loads represented about a 25% increase in the total background resistance to breathing and subsequent experiments confirmed that the Weber fraction for resistive loads is relatively constant (range, 0.2–0.3) for a range of background inten-

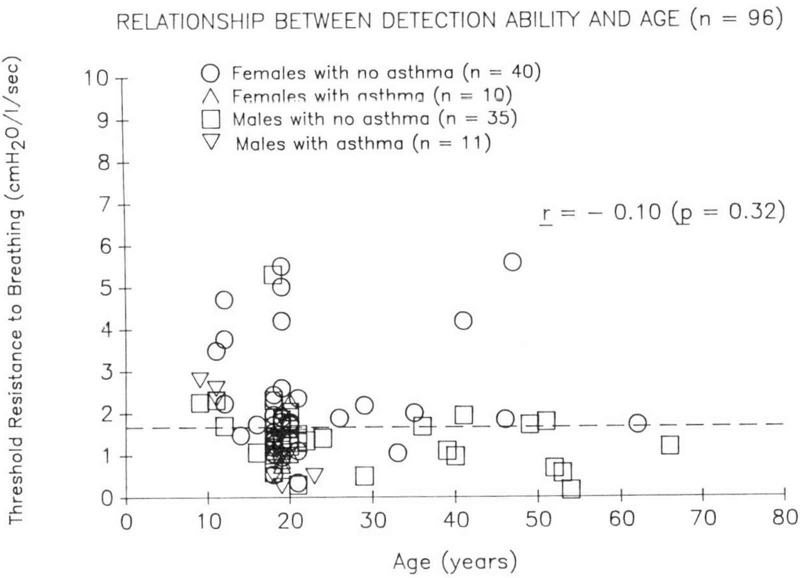

Figure 8 Scatterplot of the relationship between the threshold resistance to breathing and age in 96 subjects, including 21 patients with asthma. The mean threshold resistance to breathing (1.75 $cmH_2O/L/sec$) is represented by the dashed line.

sities (116,134,143–145). In asthma patients with increases in intrinsic levels of airways resistance, the Weber fractions for added resistive loads are similar to those for non-asthmatics (135,137,147). Both resistive load detection thresholds and Weber fractions are elevated significantly in patients with chronic airflow limitation (128,147). Consistent with the robust effects of altered mechanics on load perception, we have observed a significant inverse relation between the threshold resistance to breathing and lung function (Fig. 7).

We have also evaluated effects of age and gender on the threshold resistance to breathing by combining complete data sets from separate studies conducted previously in our laboratory involving 96 subjects, 46 males ($M = 24.72$ yr) and 50 females ($M = 21.52$ yr), including about two dozen young adults and several children with a history of asthma. As in most previous investigations, detection thresholds were obtained by regressing the added resistance of the loads against the proportion of "yes" responses and predicting the load detected 50% of the time (Fig. 6). In all our studies we elected to include "no-load" catch trials and adjust subject judgments for guessing or false alarms (148). The mean threshold resistance to breathing obtained in these subjects (1.75 ± 1.11 $cmH_2O/L/sec$) is represented by the dashed line in Figure 8, which demonstrates the largely independent relation between detection ability and age. On the other hand, male subjects exhibit significantly lower detection thresholds (M = 1.44 cmH_2O/L/sec) than female subjects (M = 2.03 cmH_2O/L/sec). In a related analysis, gender, body surface area, and lung function (FEV_1) emerged as significant and independent predictors of the threshold resistance to breathing accounting for 73% of the variance in perceptual sensitivity (149).

Although new insight into afferent nerve fibers that mediate load perception is encouraging (e.g., 133), explaining how disturbances in the mechanics of breathing give rise exactly to perceptual experiences—or, further, give rise to the uncomfortable awareness of breathing as occurs for example in asthma—remains a formidable challenge. Over 30 years ago Campbell and Howell (150) suggested that the perception of breathlessness resulted from a perceived mismatch between the ventilation demanded and the ventilation achieved or between the force or tension developed by the respiratory muscles and the resulting flow, volume, or change in volume. Evidence from many load detection studies is consistent with the general hypothesis. Load-relevant sensory information for the detection of resistive loads added to inspiration is dependent on the relation between dynamic patterns of pressure and flow generated during the early and active portion of the inspiratory cycle (130,138–140,151–153). Resistive loads added to established inspiratory flow are detected less reliably than those added at the start of inspiration, and resistive loads are detected near mid-inspiration, when flow rates are maximal, whereas elastic loads are detected near the end of inspiration, when volume is maximal (130,140,154–156). But how the central nervous system might compute "added resistance" from dynamic signals is a matter of earnest speculation.

Load Sensitivity and Response Bias

Sensory thresholds derived from classic psychophysical procedures, including the method of constant stimuli, do not take into account response bias or false alarms, the rate at which subjects indicate the presence of a stimulus when in fact none was presented. In a signal detection task the various factors that bias individual's reports are attenuated providing a relatively pure measure of discrimination (157). A subject is presented a series of trials in which the stimulus signal is either present or absent. The subject may correctly identify the presence or absence of a signal (a hit) or incorrectly identify a signal when in fact it was absent (false alarm) or miss the presence of a signal (a miss). Mathematical treatment of hits, false alarms, and misses produces two estimates of the stimulus discrimination capabilities of a subject: (1) detection ability and (2) the criterion used to judge the presence or absence of a stimulus

(120). The first of these, *d′* ("d prime"), represents the observer's capability for making a discrimination; how well the observer is able to make correct judgements and avoid incorrect ones. The second, *beta*, provides an index of the subject's criterion for acting on information arising from the discrimination. Beta is an index of response bias, or riskiness—the extent to which an observer favors one hypothesis over another independent of the evidence given.

Signal detection tasks have been used to assess the perception of increased resistance to breathing, and results from several investigations contrast those obtained in load detection studies. For example, sensitivity to added inspiratory loads was similar in subjects when loads were applied during an entire inspiration or only during the initial, accelerating phase of inspiration, was slightly but significantly improved when the background level of resistance was increased, and was decreased during chest strapping to limit chest wall movement (158–160). Males were more sensitive than females in a respiratory resistance signal detection task consistent with the gender differences observed in load detection (136). In separate studies, patients with asthma exhibited reduced, similar, or increased sensitivity to added resistance compared to healthy controls that were not attributable to differences in response bias (146,161,162). The inconsistent outcomes in these latter studies may relate, in part, to nonsensory factors that affect load perception (e.g., 163).

Load Magnitude

Direct psychophysical approaches have been derived primarily from the contributions of Stanley S. Stevens. In direct scaling, observers make judgments of sensation, which are then transformed into measures of sensory magnitude, and functions are fit to stimulus-response pairs (120). Through the use of direct scaling procedures, Stevens showed that physical intensity ratios yield corresponding sensation ratios according to a power function, where J is the sensory judgment of the stimulus, k is a scaling constant, I is the physical magnitude of the stimulus, and n is the relative sensitivity:

$$J = k\,I^{n} \qquad (4)$$

This formula (Stevens's law) states that the perceived magnitude of various sensory dimensions increases in proportion to stimulus intensity raised to a power (123,124). In the conduct of both magnitude estimation and cross-modality matching studies of added resistive loads, both the level of the added load and peak mouth pressure have been used as separate estimates of the initiating stimulus in establishing psychophysical relations between subjective magnitude and physical intensity.

We combined results provided in 13 magnitude estimation studies and determined that the average exponent for resistive loads added to inspiration in 193 subjects ranging in age from 18 to 54 years was 0.82; in four cross-modality (handgrip) matching studies involving 26 subjects, the average exponent for resistive loads added to inspiration was 0.68 (164). In six investigations, subjective magnitude was compared to the tension or force developed by the muscles of respiration (i.e., mouth pressure) and the average exponent for mouth pressure in 80 subjects was 1.2; in two cross-modality (handgrip) matching studies involving 13 subjects, the average exponent for mouth pressure was 1.06 (164). In comparison, exponents obtained for brightness are about 0.5, those for added elastic loads are about 0.95, those for line length are about 1.0, and those for force of handgrip are about 1.7.

Exponents for inspiratory resistive loads are greater than those for expiratory resistive loads, but exponents for mouth pressure in the two conditions are equivalent (165,166). These relationships, as well as others obtained between subjective magnitude and mouth pressure when either ventilation was increased or when the breathing pattern was modified, are evidence that the perceived magnitude of added loads to breathing is related to factors associated with the production of respiratory muscle force and only indirectly related to the added load (165–168). Unlike the robust finding that threshold levels for resistive loads (ΔR) bear a constant relation to the total background sensory stimulus, effects of altering background levels on exponents for added resistive loads are mixed (169,170).

A particularly relevant issue for the present discussion concerns the results of comparisons made among exponents computed for healthy young and old adults, and patients with obstructive lung disease. Gottfried and colleagues (147) reported that exponents for the perceived magnitude of resistive loads added to inspiration were significantly reduced in patients with chronic obstructive pulmonary disease (COPD) compared to healthy, young adults. Such differences, however, were not observed between patients with asthma and controls (147,171) or between asthma patients with or without a history of two more episodes of respiratory arrest (172). In a subsequent series of experiments, Tack and colleagues observed that exponents for both resistive and elastic loads were significantly reduced in older adults (>60 years) compared to young adults (<31 years) (173–175). These latter findings are consistent with the general decline in sensory and perceptual functions observed with increasing age and conclusions were made that the "blunted" perceptual performance in both older individuals and patients with COPD may be due to impairment in the central nervous system processing of separate signals related to pressure and flow (147,173,176).

Gottfried and colleagues (176) replicated their initial findings in patients with COPD and similar-aged controls, but the possibility of a confound of chronological age on psychophysical scaling in patients with lung disease was pursued in three additional studies. Ward and Stubbing (177) found that exponents for the perceived magnitude of both resistive and elastic loads computed in young adults were greater compared to those for patients, but no difference was observed between exponents for patients and age-matched controls. In more recent experiments, Mahler and colleagues (178,179) found similarities in the average size of exponents for resistive loads in patients with airflow obstruction and for elastic loads in patients with interstitial lung disease and age-matched controls. Further research is needed to resolve the way in which chronological age can mask underlying pathology in load magnitude studies.

Load Attributes

Despite the considerable literature on respiratory psychophysics, the relations between stimulus and sensation in breathing remain largely unspecified. In other words, agreement on the sensory attributes that underlie psychophysical decisions has not emerged. Physical stimulus characteristics hypothesized to mediate respiratory sensations include the "volume of breath achieved," the "volume of breath demanded," the sense of tension or force resulting from contracting respiratory muscles, chest wall movement or rate of change of movement, and the sense of effort. In most psychophysical experiments agreement on the nature of the relationship between the attribute and the stimulus has evolved. In the case of hearing, for example, it is generally accepted that the attributes of loudness and pitch are closely linked with the physical properties of sound intensity and frequency. In the case of pain, identification of attributes and stimuli has only recently been the topic of systematic investigation. In this regard, the status of research on respiration is more closely analogous to that on pain than on hearing.

A set of methods designed to help uncover the possible parameters or dimensions of sensory experience are those relying on clustering and multidimensional scaling algorithms (180). With these methods the observer represents his or her perceptual world not by judgments of perceived magnitude, but by rating the relative similarity between pairs of stimulus objects or by choosing descriptors of sensory experience from among a list that apply to specific stimulus objects. The analyses provide either clusters or maps of variables in terms of the degree of perceived similarity among the stimuli or descriptors.

Harver and colleagues (181) examined the perceived similarity among sensations elicited by nine breathing maneuvers, including inspiring and expiring against a closed airway, inspiring and expiring against two levels of resistive loads, and inspiring against two levels of elastic loads. Analyses were conducted among ratings of similarity made between all possible pairs of stimuli. The results of both cluster and mutidimensional scaling analyses were complementary and provided a context in which to consider the integration or gating of signals related to pressure or tension, flow, volume, and effort. Simon and colleagues (182) examined the similarities of sensory experience elicited by eight breathing maneuvers, including breathholding, breathing against resistive and elastic loads, and exercise in 30 healthy subjects. After each maneuver, subjects chose from among a list of 19 phrases those that described the sensations experienced during the maneuver. Nine clusters of descriptors emerged from the analysis. Some clusters were associated with different stimuli: rapid and heavy breathing were associated with exercise, and rapid and effortful breathing were associated with carbon dioxide stimulation. Breathholding was associated with air hunger. Based on the different groupings of descriptors that emerged for the different stimuli examined, the authors concluded that descriptions of breathlessness may reflect different qualities of sensation and not simply variations in the intensity or quantity of sensation.

Two subsequent papers by Schwartzstein, Mahler, and colleagues serve to extend the applicability of clustering algorithms to the study of symptoms in patients with cardiorespiratory disease (5,6). Certain statements, or descriptors, of uncomfortable awareness of breathing are more likely to be endorsed by patients with specific conditions. For example, patients with asthma select "My breathing requires effort" and "My chest feels tight" as descriptors of their breathlessness, whereas patients with COPD select "My breathing requires effort" and "My breathing requires work." The preferred descriptors of breathlessness endorsed by patients in some conditions but not in others may relate to alterations in the mechanics of breathing and to the pathophysiology of the underlying disease (6).

D. Event-Related Potentials to Added Loads

A potentially powerful method for enhancing our understanding of sensory-perceptual correlates of respiratory sensations involves analysis of central (neural) respiratory-related signals. For decades, relationships between sensory stimulation and perceptual responses have been elucidated with event-related potentials (ERPs) (183,184). ERPs are measured at the cortex and reflect specific electrical changes that occur in the brain in response to sensory and psychological variables. Exogenous, or sensory-evoked, potentials reflect the intergrity of the primary sensory pathway and vary in morphology with stimulus parameters such as intensity and frequency. Endogenous potentials, on the other hand, reflect cognitive and psychological states of the subject. They are influenced primarily by factors extrinsic to the physical stimulus such as arousal, attention, motivation, and task difficulty.

The application of noninvasive neurophysiological techniques to respiratory sensations have been concerned mostly with responses to respiratory occlusion (i.e., complete obstruction of the airway). Davenport and colleagues (185) first recorded ERPs to inspiratory occlusion and found that somato-cortical potentials (P1, N1, P2, and N2) elicited to occlusions in a passive attention paradigm were not evident on unoccluded inspirations. Revelette and Davenport (186) compared P1, N1, and P2 latencies to 64 inspiratory occlusions delivered at the onset of inspiration with ERP latencies elicited by an equal number of occlusions delivered at mid-inspiration in six adults. Mid-inspiratory occlusions resulted in potentials that were shorter in latency and larger in amplitude compared to those resulting from occlusions presented at the onset of inspiration. Most recently Davenport and colleagues (187) observed bilateral P1 and N1 responses to inspiratory occlusions in 24 children ranging in age from 10 to 13 years and speculated that such activity could arise from one of several populations of respiratory mechanoreceptors.

Harver and colleagues (188) first examined P300 (a cognitive component of the ERP) in response to respiratory occlusions in young (20–30 years) and old (50–70 years) healthy subjects, in both ignore and attend conditions. The P300 was shorter in latency and larger in amplitude in the attend condition compared to the ignore condition, in both inspiratory and expiratory tasks. Like many ERP findings in the auditory and visual modalities (e.g., 189), respiratory P300 latencies were longer, and amplitudes were smaller, in older subjects compared to younger subjects. The direction and degree of age effects, attention effects, and scalp distributions were consistent with those reported previously for a variety of oddball tasks employing exteroceptive stimuli and were consistent with the criteria recommended by Donchin and colleagues necessary to define P300 (190,191). The robust relationships between P300 latency to either inspiratory or expiratory occlusion and age recorded in our laboratory from these and other subjects, including eight patients with obstructive lung, are presented in Figure 9. Consistent with effects of altered mechanics on load perception, we have observed also a significant inverse relation between P300 latency to either inspiratory or expiratory occlusion and lung function (Fig. 10). Moreover, in separate multiple regression analyses using age and measures of lung function as predictors, both age and lung function (FVC percent predicted) emerged as significant and independent predictors of P300 latency to inspiratory occlusion (multiple $R = 0.52$), and P300 latency to expiratory occlusion (multiple $R = 0.53$).

We have also recorded endogenous ERPs to added resistive loads. In one study (192), ERPs were recorded to four levels of added resistance in young adults (0.34, 4.01, 10.4, and 57.5 $cmH_2O/L/sec$). No ERPs were elicited by the undetected, subthreshold stimulus (0.34 $cmH_2O/L/sec$) but endogenous

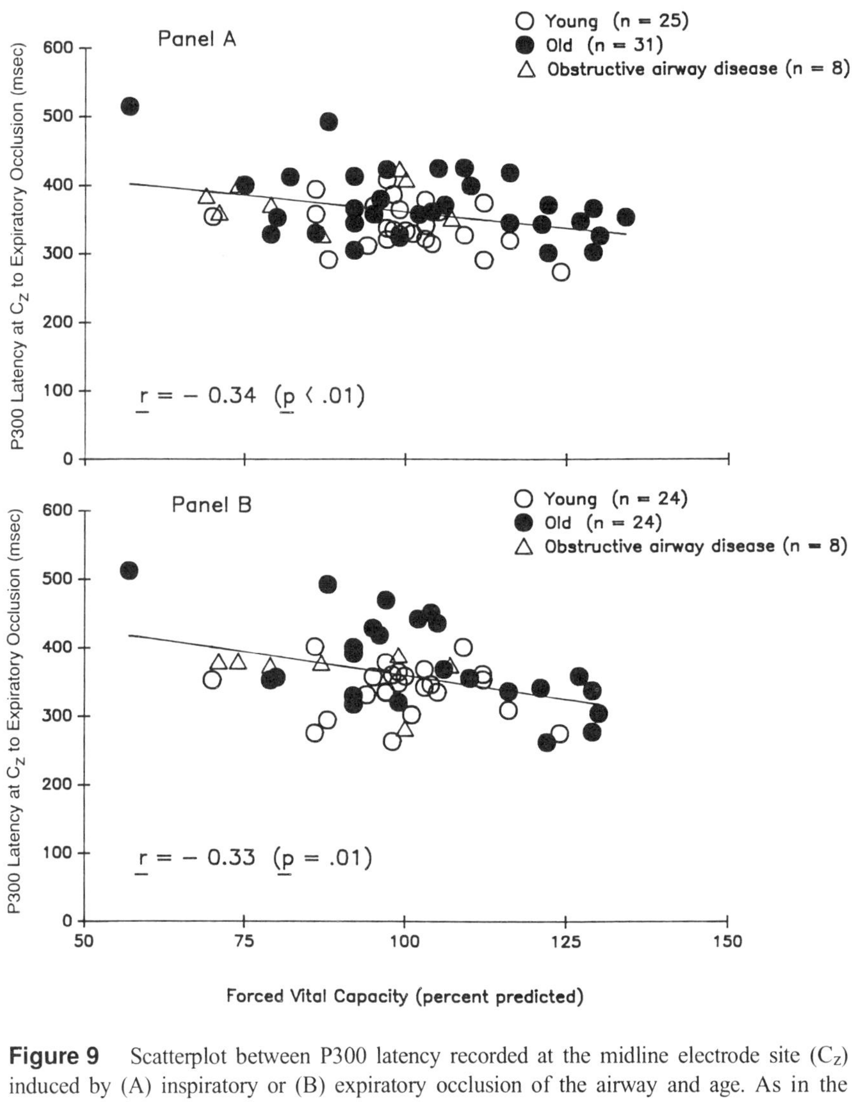

Figure 9 Scatterplot between P300 latency recorded at the midline electrode site (C_Z) induced by (A) inspiratory or (B) expiratory occlusion of the airway and age. As in the other sensory modalities, there is a systematic increase in the latency of the endogenous event-related potential to added loads with increasing age.

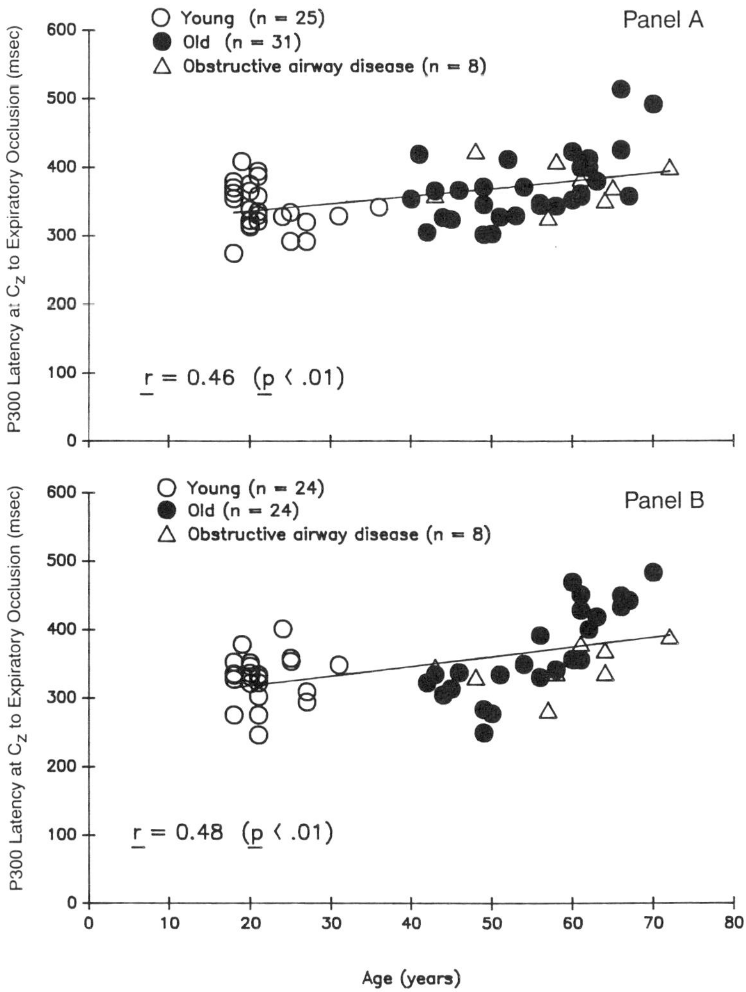

Figure 10 Scatterplot between P300 latency recorded at the midline electrode site (C_Z) induced by (A) inspiratory or (B) expiratory occlusion of the airway and lung function. There is a significant, inverse relationship between the latency of the endogenous event-related potential to added loads and lung function.

components of the ERP (P2, N2, and P300) were observed to each of the three larger stimuli and were related, in part, to behavioral judgments obtained during the stimulus periods. Both the latency and amplitude of the ERP components varied systematically as a function of stimulus magnitude and in a manner comparable to that observed in ERP paradigms using auditory and visual stimuli. In the other study, we evaluated ERPs to each of two levels of resistive and elastic loads (193). Subjects detected equivalent numbers of resistive and elastic loads equated for subjective magnitude but classified loads correctly as either "resistive" or "elastic" at chance levels. ERP latencies, however, were sensitive to load type even though resistive and elastic loads were not distinguishable subjectively: latencies for both N2 and P300 were not only shorter to larger than to smaller loads but also shorter to resistive than to elastic loads. In both experiments, loads were presented briefly (~200 msec) at the onset of inspiration.

V. Intrinsic Loads, Added Loads, and Control of Breathing

A large body of work concerned with measuring respiratory sensory experiences demonstrates substantive interactions between the perception of breathing and ventilatory control (see Refs. 113 and 133). If breathing is "a strange phenomenon of life, caught midway between the conscious and unconscious" (194), then interest in the perception of increased resistance to breathing may relate especially to hypotheses about relationships between the act of breathing and the control of breathing. We have argued previously that psychophysical outcomes and clinical ratings of dyspnea are not equivalent and therefore should not be interchanged (179). On the other hand, it is likely that sensory psychophysics will continue to provide insight to processes that account for unpleasant respiratory experience and to interactions between behavioral responses and ventilatory control.

A. Load Perception and the Control of Breathing

Many models of the respiratory system have emphasized that the pattern of breathing is set to minimize the work or energetics of breathing but the feedback mechanisms that would allow the respiratory system to be aware of the work or force of breathing are not well defined (133,195,196). Cherniack and colleagues suggested, based in part on review of the respiratory sensation literature, that ventilation may be determined to reduce respiratory sensory input and to minimize perceived respiratory discomfort (133,195–197). If respiratory sensations serve to modify the breathing pattern to minimize discomfort, as well as reduce energy costs and maintain adequate levels of gas exchange, it seems plausible that patients who adjust the pattern of breathing to minimize perceived increases in the resistance to breathing would, at the same time, attenuate those signals that give rise to symptoms. Very little research relates to the role of modification, or learning, in the regulation of breathing (133). Sears proposed that voluntary breathing can become automatic and that voluntary changes in breathing can become habitual and require little or no attention (195,198).

One explanation for the variability in levels of breathlessness for comparable degrees of airflow obstruction that accompanies acute bronchoconstriction in patients with asthma is referred to as "temporal adaptation" (e.g., 13,18). Reductions in the perceived intensity of airflow obstruction are hypothesized to relate to progressively weaker signals that result from prolonged stimulation of the structures giving rise to unpleasant sensory experiences. But neither the structures that might adapt in this context nor the source of prolonged stimulation are well defined. Heuristically, distinction may be made between "adaptation" and "habituation" (199). Adaptation refers to processes occurring in the periphery and involves changes in receptor sensitivity; habituation refers to a decrement in the magnitude of response to a novel or startling stimulus (acute bronchoconstriction?). Habituation results from a person "getting used to or ignoring" stimulation no longer judged to be of significance (199) and involves central processes. Possibly, some patients with asthma adopt breathing patterns that minimize perceived respiratory discomfort and become habitual. Such a hypothesis is speculative but the spontaneous shift from nose to mouth breathing in response to nasal congestion is not unfamiliar (133).

B. Responses to Intrinsic and Added Loads

We reviewed earlier well-recognized limits to the use of external loads as an analog for examining compensatory and perceptual response to mechanical changes of the ventilatory apparatus in disease (45,46,110–114). Direct comparisons between perceptual responses generated by increases in the intrinsic and external resistance to breathing confirm these caveats. Kelsen and colleagues reported that in both patients with asthma and healthy controls, the sense of effort was greater during methacholine challenge than with external loading at any given level of resistance (200). Taguchi and colleagues reported that in six healthy adults histamine challenge significantly increased the intensity of dyspnea over the equivalent external load (201). Both authors speculated that greater stimulation of mechanoreceptors during provocation, possibly irritant receptors, accounted for perceptual differences in the conditions.

C. Sensory Mechanisms

Unlike either for the auditory or visual systems, or for cutaneous electrical stimulation, there is little agreement on either the sensory receptors or central pathways necessary for the transduction of breathing sensations induced by either intrinsic or added resistive loads in humans. On the other hand, it has been implicit since the first experiments designed to assess the ability of humans to detect added loads to breathing that load detection and the experience of unpleasant respiratory sensations likely share common mechanisms (125,126,202,203). The "length-tension inappropriateness" theory presented over 30 years ago is perhaps the most influential model for characterizing the basis of unpleasant respiratory sensations (131,204). In its early formulations, Campbell and Howell (150) focused on the analysis of hindered, or obstructed, breathing. They suggested that the perception of breathlessness resulted from a perceived mismatch between the ventilation demanded and the ventilation achieved—between the force or tension developed by the respiratory muscles and the resulting flow, volume, or change in volume. A disturbance in the relationship between information about tension and length in the respiratory control system was envisaged in all instances that could give rise to the sensation of breathing difficulty. In the length-tension inappropriateness model, the afferent limb of sensation originates in the extravagal, gamma-efferent system. This system is hypothesized to provide information about both length (the misalignment between intrafusal and extrafusal fibers noted by the muscle spindles) and tension (sensed by tendon organs in series with contracting muscles) of respiratory fibers. Such misalignment could be transferred to higher centers and possibly to sensorimotor cortex through spinal pathways that remained largely undefined.

Respiratory sensations are likely mediated by signals arising from one or more of four systems: (1) peripheral arterial chemoreceptors, (2) the upper airways and airways of the lungs, (3) respiratory muscle afferents, and (4) central representation of a sense of effort related to a corollary discharge or motor command to the respiratory muscles (113,205). The perceptual response to immediate or acute increases in the resistance to breathing probably depends least on blood gas tension levels and humoral influences are not considered further. On the other hand, interactions between hypercapnia and hypoxia, and the sensation of an urge to breathe, or "air hunger," are widely recognized (206).

Both primary and first-order connections arising from pulmonary receptors, especially the slowly adapting pulmonary stretch receptors and rapaidly adapting receptors, and central pathways that underlie respiratory reflexes elicited by activation of vagal afferents are increasingly well defined (54). The great majority of vagal afferents terminate in the nucleus of the solitary tract, the major visceral sensory center. Although higher projections of pulmonary and airway receptors are less certain, evoked responses have been recorded in medullary reticular formation, spinal cord, cerebellum, thalamus, and cortex following stimulation of the cervical vagus nerve (54). Information from airway afferents may play a role in the sensation of chest tightness and load perception. Dynamic compression of the airways or negative intrathoracic pressure swings may stimulate pressure-sensitive mechanoreceptors sensitive either to the deformations of the airways or to the changes in pressure developed across the airways (207–209). Stimulation of irritant receptors may intensify dyspnea and alter its quality (207).

Respiratory muscles contain a rich supply of afferents (including muscle spindles, tendon organs, and slowly adapting receptors in costovertebral joints) that are sensitive to mechanical perturbations associated with loading although many fundamental questions remain concerning the influence of these muscle receptors in the control of breathing, load compensation, and load perception (e.g., 203,205,210,211). Respiratory mechanoreceptors, especially those of the chest wall, participate in postural adjustments and the compensatory response to added loads but mostly by stabilizing the chest wall or by adjusting the strength of respiratory muscle contractions through segmental, spinal reflexes (113,210). Chest wall proprioceptors seem to have little direct effect on ventilatory control under normal conditions, but neurophysiological studies show that respiratory muscle afferents project to pons, thalamus, and cortex (205). Davenport and Reep have reviewed ascending pathways for respiratory muscle afferents to account for cortical evoked potentials elicited by electrical or mechanical stimulation of phrenic and intercostal muscle afferents (212).

The fourth system used to explain respiratory sensation is based not on afferent input to the central nervous system but on central representation of the outgoing efferent motor command to the respiratory muscles—an awareness of motor output to the respiratory muscles by means of collateral discharge within the central nervous system. Neural output related to respiratory magnitude is sent from medullary centers, presumably as an efferent copy or corollary discharge, to midbrain and thalamus and probably cortex (205). In this analysis, activation of sensory cortex at the same time the respiratory muscles are signaled to contract produces the sensation of respiratory muscle effort. Eldridge and Chen (205) concluded recently that "an increase in central neural respiratory drive, with transmission to higher brain of the magnitude of medullary activity via corollary discharge, is the sine qua non for development of dyspnea."

VI. Summary and Conclusions

We have reviewed both clinical and load perception studies related to the perception of increased resistance to breathing and surveyed the relation among alterations in the mechanics of breathing, sensations, and symptoms. Our review was limited to issues relevant for the patient with asthma and not as broad based as other reviews concerned with dyspnea, an experience comprised of multiple and qualitatively distinct elements. We emphasized the abnormalities in the mechanics of breathing and the nature of airflow obstruction in asthma that give rise to symptoms and briefly reviewed measures of pulmonary ventilation, measures of lung function, and measures of sensory psychophysics that allow us to appraise directly load perception.

The ongoing rise in the number of clinical investigations concerned with the relationship between perceived symptoms and abnormalities in lung function in patients with asthma parallel increased apprehension that a proportion of patients are "poor perceivers" of airflow obstruction and at risk for greater morbidity and mortality. We argued that the concept of "poor perceivers," while convenient, remains ill-defined. Use of the term suggests an implicit range of perceptual acuity, and a judgment that good perception is more valued; but neither of these claims can be substantiated presently. We intimated that solutions to these uncertainties likely relate to greater understanding of illness behavior in patients, "the way in which persons monitor their bodies, define and interpret their symptoms, take remedial action, and utilize various sources of help as well as the more formal health-care system" (213). Our review of the literature demonstrates that the regulation of asthma should improve when there is concordance among levels of increased resistance to breathing, symptoms, and the behaviors individuals exhibit in response to symptoms.

The most prominent contribution from behavioral science to load perception to date is its methodology, including signal detection theory, magnitude scaling, and cluster analysis. Results from the many dozens of load perception studies conducted by respiratory physiologists and pulmonary clinicians using these methods serve to affirm the hypothesis that ventilation is affected by behavior and may be determined to reduce respiratory sensory input and minimize perceived respiratory discomfort. The idea that voluntary changes in breathing can become automatic or habitual suggests two important areas for future work: (1) the role of learning in the control of breathing and (2) the inexact but useful distinction between "adaptation" and "habituation" to unpleasant respiratory sensory experience. Possibly behavioral scientists are prepared to engage in the study of load perception from these more familiar attitudes.

References

1. Weiss K, Gergen P, Hodgson T. An economic evaluation of asthma in the United States. N Engl J Med 1992; 326:862–868.

2. Guidelines for the diagnosis and management of asthma. National Heart, Lung, and Blood Institute. National Asthma Education Program. Expert Panel Report. J Allergy Clin Immunol 1991; 88:425–534.
3. American Thoracic Society. Guidelines for the evaluation of impairment/disability in patients with asthma. Am Rev Respir Dis 1993; 147:1056–1061.
4. McFadden ER. Asthma: airway reactivity and pathogenesis. Sem Respir Med 1980; 1:287–296.
5. Simon PM, Schwartzstein RM, Weiss JW, Fencl V, Teghtsoonian M, Weinberger SE. Distinguishable types of dyspnea in patients with shortness of breath. Am Rev Respir Dis 1990; 142:1009–1014.
6. Mahler DA, Harver A, Lentine T, Scott JA, Beck K, Schwartzstein RM. Descriptors of breathlessness in cardiorespiratory diseases. Am J Respir Crit Care Med 1996; 154:1357–1363.
7. Barnes PJ. Poorly perceived asthma (editorial). Thorax 1992; 47:408–409.
8. Creer TL. Psychological factors and death from asthma: creation and critique of a myth. J Asthma 1986; 23:261–269.
9. Janson-Bjerklie S, Ferketich, S, Benner P, Becker, G. Clinical markers of asthma severity and risk: importance of subjective as well as objective factors. Heart Lung 1992; 21:265–272.
10. Molfino NA, Slutsky AS. Near-fatal asthma. Eur Respir J 1994; 7:981–990.
11. Parker SR, Mellins RB, Sogn DD. Asthma education: a national strategy. Am Rev Respir Dis 1989; 140:848–853.
12. Fritz GK, Klein RB, Overholser JC. Accuracy of symptom perception in childhood asthma. Dev Behav Pediatr 1990; 11:69–72.
13. Orehek J, Beaupré A, Badier A, Nicoli MM, Delpierre S. Perception of airway tone by asthmatic patients. Bull Eur Physiopathol Respir 1982; 18:601–607.
14. Rubinfeld AF, Pain MCF. Perception of asthma. Lancet 1976; 1:882–884.
15. Higgs CMB, Laszlo G. Perception of asthma. In: Adams L, Guz A, eds. Respiratory Sensation. New York: Marcel Dekker, 1996:285–309.
16. Perception of breathlessness in asthma (editorial). Lancet 1983; 1:912–913.
17. Barnes PJ. Blunted perception and death from asthma (editorial). N Engl J Med 1994; 330:1383–1384.
18. Burdon JGW, Juniper EF, Killian KJ, Hargreave FE, Campbell EJM. The perception of breathlessness in asthma. Am Rev Respir Dis, 1982; 126:825–828.
19. George RB, Owens MW. Bronchial asthma (monograph). Disease-a-Month 1991; 37:139–196.
20. Priel B, Heimer D, Rabinowitz B, Hendler N. Perceptions of asthma severity: the role of negative affectivity. J Asthma Res 1994; 31:479–484.
21. Ellis ME, Friend JAR. How well do asthma clinic patients understand their asthma? Brit J Dis Chest 1985; 79:43–48.
22. Molfino NA, Nannini LJ, Martelli AN, Slutsky AS. Respiratory arrest in near-fatal asthma. N Engl J Med 1991; 324:285–288.
23. Blascovich J, Katkin ES. Visceral perception and social behavior. In: Cacioppo J, Petty R, eds. Social Psychophysiology: A Source Book. New York: Guilford Press, 1983:493–509.

24. James W. What is an emotion? Mind 1884; 9:188–205.
25. Pennebaker JW. The Psychology of Physical Symptoms. New York: Springer-Verlag, 1982.
26. Reed SD, Harver A, Katkin ES. Interoception. In: Cacioppo JT, Tassinary LG, eds. Principles of Psychophysiology: Physical, Social, and Inferential Elements. New York: Cambridge Press, 1990:253–291.
27. Adám, G. Visceroception, awareness, and behavior. In: Schwartz GE, Shapiro D, eds. Consciousness and Self-Regulation: Advances in Research and Theory. Vol. 2. New York: Plenum Press, 1978:199–213.
28. Bergman JS, Johnson HJ. The effects of instructional set and autonomic perception on cardiac control. Psychophysiology 1971; 8:180–190.
29. Blanchard EB, Young LD, McLeod P. Awareness of heart activity and self-control of heart rate. Psychophysiology 1972; 9:63–68.
30. Brener J. A general model of voluntary control applied to the phenomena of learned cardiovascular change. In: Obrist PA, Black AH, Brener J, DiCara LV, eds. Cardiovascular Psychophysiology—Current Issues in Response Mechanisms, Biofeedback and Methodology. Chicago: Aldine, 1974:365–391.
31. Carroll D. Cardiac perception and cardiac control: a review. Biofeedback Self-Reg 1977; 2:349–369.
32. Fudge R, Adams HE. The effects of discrimination training on voluntary control of cephalic vasomotor activity. Psychophysiology 1985; 22:300–306.
33. Leventhal H. Behavioral medicine: psychology in health care. In: Mechanic D, ed. Handbook of Health, Health Care, and the Health Professions. New York: The Free Press, 1983:709–743.
34. Pennebaker JW, Brittingham G L. Environmental and sensory cues affecting the perception of physical symptoms. In: Baum A, Singer JE, eds. Advances in Environmental Psychology. Vol 4. Environment and Health. Hillsdale, NJ: Lawrence Erlbaum, 1982:115–136.
35. Becker MH, Maiman LA. Models of health-related behavior. In: Mechanic D, ed. Handbook of Health, Health Care, and the Health Professions. New York: The Free Press, 1983:539–568.
36. Mechanic D. The concept of illness behaviour: culture, situation, and personal predisposition. Psychol Med 1986; 16:1–7.
37. Pennebaker JW, Epstein D. Implicit psychophysiology: effects of common beliefs and idiosyncratic physiological responses on symptom reporting. J Personality 1983; 51:468–496.
38. Harver A, Mahler DA. Dyspnea: sensation, symptom, and illness. In: Mahler DA, ed. Dyspnea. New York: Marcel Dekker, 1998:1–34.
39. Pace TM. Schema theory: a framework for research and practice in psychotherapy. J Cognitive Psychother 1988; 2:147–163.
40. Bishop GD. Understanding the understanding of illness: lay disease representations. In: Skelton JA, Croyle RT, eds. Mental Representation in Health and Illness. New York: Springer-Verlag, 1991:32–59.
41. Levitzky MG. Pulmonary Physiology. 2nd ed. New York: McGraw-Hill, 1986.

42. Comroe JH. Physiology of Respiration. 2nd ed. Chicago: Year Book Medical Publishers, 1974.
43. West JB, ed. Best and Taylor's Physiological Basis of Medical Practice. 11th ed. Baltimore: Williams & Wilkins, 1985.
44. Haas F, Axen K. Pulmonary Therapy and Rehabilitation. 2nd ed. Baltimore: Williams & Wilkins, 1991.
45. Cherniack NS, Altose MD. Respiratory responses to ventilatory loading. In: Hornbein TF, ed. Regulation of Breathing. Vol. 17. Part II. New York: Marcel Dekker, 1981:905–964.
46. Rebuck AS, Slutsky AS. Control of breathing in diseases of the respiratory tract and lungs. In: Cherniack NS, Widdicombe JG, eds. Handbook of Physiology. Section 3. The Respiratory System. Vol II. Bethesda, MD: American Physiological Society, 1986:771–791.
47. Cherniack RM. Mechanics of breathing. Sem Respir Med 1983; 4:171–183.
48. Wiggs BR, Moreno R, Hogg JC, Hilliam C, Paré PD. A model of the mechanics of airway narrowing. J Appl Physiol 1990; 69:849–860.
49. Wiggs BR, Bosken C, Paré PD, James A, Hogg JC. A model of airway narrowing in asthma and in chronic obstructive pulmonary disease. Am Rev Respir Dis 1992; 145:1251–1258.
50. Leff A. Pathophysiology of asthmatic bronchoconstriction. Chest 1982, 82:13S–21S.
51. Sheffer AL, Taggart VS. IIIB. The National Asthma Education Program. Expert panel report guidelines for the diagnosis and management of asthma. Med Care 1993; 31:MS20–MS28.
52. Barnes PJ. Is asthma a nervous disease? Chest 1995; 3:119S–125S.
53. Leff AR. Endogenous regulation of bronchomotor tone. Am Rev Respir Dis 1988; 137:1198–1216.
54. Kubin L, Davies RO. Central pathways of pulmonary and airway vagal afferents. In: Dempsey JA, Pack AI, eds. Regulation of Breathing. 2nd ed. New York: Marcel Dekker, 1995:219–284.
55. Wenzel SE. The role of inflammation in asthma (monograph). National Asthma Education and Prevention Program. National Heart, Lung, and Blood Institute, Bethesda, MD, 1994.
56. American Thoracic Society. Standardization of spirometry—1987 update. Am Rev Respir Dis 1987; 136:1285–1298.
57. Antonisen NR. Tests of mechanical function. In: Cherniack NS, Widdicombe JG, eds. Handbook of Physiology. Section 3. The Respiratory System. Vol III. Bethesda, MD: American Physiological Society, 1986:753–784.
58. Petty TL. Pulmonary Diagnostic Techniques. Philadelphia: Lea & Febiger, 1975.
59. Ruppel G. Manual of Pulmonary Function Testing. St. Louis: C. V. Mosby, 1975.
60. West JB. Pulmonary Pathophysiology—The Essentials. 2nd ed. Baltimore: Williams & Wilkins, 1982.
61. Crapo RO, Morris AH, Gardner RM. Reference spirometric values for spirometry using techniques and equipment that meets ATS recommendations. Am Rev Respir Dis 1981; 123:659–664.

62. Knudson RJ, Lebowitz MD, Holberg CJ, Burrows B. Changes in the normal maximal expiratory flow-volume curve with growth and aging. Am Rev Respir Dis 1983; 127:725–734.
63. Kory RC, Callahan R, Boren HG, Syner JC. The Veterans Administration-Army cooperative study of pulmonary function. I. Clinical spirometry in normal man. Am J Med 1961; 30:243–258.
64. Morris JF, Koski A, Johnson LC. Spirometric standards for healthy non-smoking adults. Am Rev Respir Dis 1971; 103:57–67.
65. DuBois AB, Botelho SY, Comroe JH, Jr. A new method for measuring airway resistance in man using a body plethysmograph: values in normal subjects and in patients with respiratory disease. J Clin Invest 1956; 35:327–335.
66. Jackson AC, Milhorn HT Jr., Norman JR. A reevaluation of the interrupter technique for airway resistance measurement. J Appl Physiol 1974; 36:264–268.
67. Fisher AB, DuBois AB, Hyde RW. Evaluation of the forced oscillation technique for the determination of resistance to breathing. J Clin Invest 1968; 47:2045–2057.
68. Lándsér FJ, Nagels J, Demedts M, Billiet L, van de Woestijne KP. A new method to determine frequency characteristics of the respiratory system. J Appl Physiol 1976; 41:101–106.
69. Chai H, Purcell K, Brady K, Falliers CJ. Therapeutic and investigational evaluation of asthmatic children. J Allergy 1968; 41:23–36.
70. Kendrick AH, Higgs CMB, Whitfield MJ, Laszlo G. Accuracy of perception of severity of asthma: patients treated in general practice. Brit Med J 1993; 307:422–424.
71. Stedman's Medical Dictionary. 24th ed. Baltimore: Williams & Wilkins, 1982.
72. Blacklow RS. The study of symptoms. In: Blacklow RS, ed. MacBryde's Signs and Symptoms. 6th ed. Philadelphia: J.B. Lippincott, 1983:1–16.
73. Bishop GD, Converse SA. Illness representations: a prototype approach. Health Psychol 1986; 5:95–114.
74. Skelton JA, Pennebaker JW. The psychology of physical symptoms and sensations. In: Sanders GS, Suls J, eds. Social Psychology of Health and Illness. Hillsdale, NJ: Lawrence Erlbaum Associates, 1982:99–128.
75. Barsky AJ, Geringer E, Wool CA. A cognitive-educational treatment for hypochondriasis. Gen Hosp Psychiatry 1988; 10:322–327.
76. Safer MA, Tharps QJ, Jackson TC, Leventhal H. Determinants of three stages of delay in seeking care at a medical clinic. Med Care 1979; 17:11–29.
77. Prohaska TR, Keller ML, Leventhal EA, Leventhal H. Impact of symptoms and aging attribution on emotions and coping. Health Psychol 1987; 6:495–514.
78. Cameron L, Leventhal EA, Leventhal H. Seeking medical care in response to symptoms and life stress. Psychosom Med 1995; 57:37–47.
79. Kinsman RA, Luparello T, O'Banion K, Spector S. Multidimensional analysis of the subjective symptomatology of asthma. Psychosom Med 1973; 35:250–267.
80. Kinsman RA, O'Banion K, Resnikoff P, Luparello TJ, Spector SL. Subjective symptoms of acute asthma within a heterogeneous sample of asthmatics. J Allergy Clin Immunol 1973; 52:284–296.

81. Brooks CM, Richards JM Jr., Bailey WC, Martin B, Windsor RA, Soong S. Subjective symptomatology of asthma in an outpatient population. Psychosom Med 1989; 51:102–108.
82. Lebowitz MD, Thompson HC, Strunk RC. Subjective psychological symptoms in outpatient asthmatic adolescents. J Behav Med 1981; 4:439–449.
83. Dirks JF, Schraa JC. Patient mislabeling of symptoms and rehospitalization in asthma. J Asthma 1983; 20:43–44.
84. Cohen S, Doyle WJ, Skoner DP, Fireman P, Gwaltney JM Jr., Newsom JT. State and trait negative affect as predictors of objective and subjective symptoms of respiratory viral infections. J Pers Soc Psychol 1995; 68:159–169.
85. Dahlem NW, Kinsman RA, Horton DJ. Panic-fear in asthma: requests for as-needed medications in relation to pulmonary function measurements. J Allergy Clin Immunol 1977; 60:295–300.
86. Priel B, Heimer D, Rabinowitz B, Hendler N. Perceptions of asthma severity: the role of negative affectivity. J Asthma Res 1994; 31:479–484.
87. O'Connor GT, Weiss ST. Clinical and symptom measures. Am J Resp Crit Care Med 1994; 149:S21–S28.
88. Adelroth E, Hargreave FE, Ramsdale EH. Do physicians need objective measurements to diagnose asthma? Am Rev Respir Dis 1986; 134:704–707.
89. Baumann UA, Haerdi E, Keller R. Relations between clinical signs and lung function in bronchial asthma: how is acute bronchial obstruction reflected in dyspnoea and wheezing? Respiration 1986; 50:294–300.
90. Brand PLP, Rijcken B, Schouten JP, Koëter GH, Weiss ST, Postma DS. Perception of airway obstruction in a random population sample: relationship to airway hyperresponsiveness in the absence of respiratory symptoms. Am Rev Respir Dis 1992; 146:396–401.
91. Dales RE, Ernst P, Hanley JA, Battista RN, Becklake MR. Prediction of airway reactivity from responses to a standardized respiratory symptom questionnaire. Am Rev Respir Dis 1987; 135:817–821.
92. McFadden ER Jr., Kiser R, DeGroot WJ. Acute bronchial asthma: relations between clinical and physiologic manifestations. N Engl J Med 1973; 288:221–225.
93. Jones PW. Breathlessness perception in airways obstruction (editorial). Eur Respir J 1992; 5:1035–1036.
94. Shim CS, Williams MH Jr. Evaluation of the severity of asthma: patients versus physicians. Am J Med 1980; 68:11–13.
95. Couriel JM, Demis T, Olinsky A. The perception of asthma. Austr Paediatr J 1986; 22:45–47.
96. Dhand R, Kalra S, Malik SK. Use of analogue scales for assessment of the severity of asthma. Respiration 1988; 54:255–262.
97. Rubinfeld AR, Pain MCF. Bronchial provocation in the study of sensations associated with disordered breathing. Clin Sci Mol Med 1977; 52:423–428.
98. Boulet LP, Cournoyer I, Deschesnes F, Leblanc P, Nouwen A. Perception of airflow obstruction and associated breathlessness in normal and asthmatic subjects: correlation wiithh anxiety and bronchodilator needs. Thorax 1994; 49:965–970.

99. Boulet LP, Leblanc P, Turcotte H. Perception scoring of induced bronchoconstriction as an index of awareness of asthma symptoms. Chest 1994; 105:1430–1433.
100. Turcotte H, Boulet LP. Perception of breathlessness during early and late asthmatic responses. Am Rev Respir Dis 1993; 148:514–518.
101. Turcotte H, Corbeil F, Boulet LP. Perception of breathlessness during bronchoconstriction induced by antigen, exercise, and histamine challenges. Thorax 1990; 45:914–918.
102. Noseda A, Schmerber J, Prigogine T, Yernault JC. Perceived effect on shortness of breath of an acute inhalation of saline or terbutaline: variability and sensitivity of a visual analogue scale in patients with asthma or COPD. Eur Respir J 1992; 5:1043–1053.
103. Noseda A, Schmerber J, Prigogine T, Yernault JC. How do patients with either asthma or COPD perceive acute bronchodilation? Eur Respir J 1993; 6:636–644.
104. Boulet, LP, Deschesnes F, Turcotte H, Gignac, F. Near-fatal asthma: clinical and physiologic features, perception of bronchoconstriction, and psychologic profile. J Allergy Clin Immunol 1991; 88:838–846.
105. Ruffin RE, Latimer KM, Schembri DA. Longitudinal study of near fatal asthma. Chest 1991; 99:77–83.
106. Kikuchi Y, Okabe S, Tamura G, Hida W, Homma M, Shirato K, Takishima T. Chemosensitivity and perception of dyspnea in patients with a history of near-fatal asthma. N Engl J Med 1994; 330:1329–1334.
107. Connolly MJ, Crowley JJ, Charan NB, Nielson CP, Vestal RE. Reduced subjective awareness of bronchoconstriction provoked by methacholine in elderly asthmatic and normal subjects as measured on a simple awareness scale. Thorax 1992; 47:410–413.
108. Altose M, Cherniack N. Respiratory sensations and dyspnea. J Appl Physiol 1985; 58:1051–1054.
109. Altose MD, Zechman FW, eds. Loaded breathing: load compensation and respiratory sensation. Fed Proc 1986; 45:114–122.
110. Pengelly LD, Rebuck AS, Campbell EJM, eds. Loaded Breathing. Edinburgh: Churchill Livingstone, 1974.
111. Cherniack NS, Milic-Emili J. Mechanical aspects of loaded breathing. In: Roussos C, Macklem PT, eds. The Thorax. Part B. New York: Marcel Dekker, 1985:751–786.
112. Milic-Emili J, Zin W. Breathing responses to imposed mechanical loads. In: Cherniack NS, Widdicombe JG, eds. Handbook of Physiology. Section 3. The Respiratory System. Vol II. Bethesda, MD: American Physiological Society, 1986:751–769.
113. Younes M. Mechanisms of respiratory load compensation. In: Dempsey JA, Pack AI, eds. Regulation of Breathing. 2nd ed. New York: Marcel Dekker, 1995:867–922.
114. Martin JG, Shore SA, Engel LA. Mechanical load and inspiratory muscle action during induced asthma. Am Rev Respir Dis 1983; 128:455–460.
115. Wood MM. A technique for measurement of ability to detect an external airways resistance. J Physiol 1970; 207:46–47P.

116. Zechman FW, Burki NK. Tracking procedure to assess load detection threshold. Chest 1976; 70(suppl):165–168.
117. Daubenspeck JA. Mechanical factors in breathing pattern regulation in humans. Ann Biomed Eng 1981; 9:409–424.
118. Daubenspeck JA, Bennett FM. Immediate human breathing pattern responses to loads near the perceptual threshold. J Appl Physiol 1983; 55:1160–1166.
119. Harver A, Daubenspeck JA. Human breathing pattern responses to loading with increased background impedance. J Appl Physiol 1989; 66:680–686.
120. Baird JC, Noma E. Fundamentals of Scaling and Psychophysics. New York: J. Wiley and Sons, 1978.
121. Coren S, Porac C, Ward LM. Sensation and Perception. 2nd ed. Orlando, FL: Academic Press, 1984.
122. D'Amato MR. Experimental Psychology: Methodology, Psychophysics, and Learning. New York: McGraw-Hill, 1970.
123. Gescheider GA. Psychophysical scaling. Ann Rev Psychol 1988; 39:169–200.
124. Stevens SS. Psychophysics: Introduction to its Perceptual, Neural, and Social Prospects. New York: J. Wiley and Sons, 1975.
125. Bennett ED, Jayson MIV, Rubenstein D, Campbell EJM. The ability of man to detect non-elastic loads to breathing. Clin Sci 1962; 23:155–162.
126. Campbell EJM, Freedman S, Smith PS, Taylor ME. The ability of man to detect added elastic loads to breathing. Clin Sci 1961; 20:223–231.
127. Bakers JHCM, Tenney SM. The perception of some sensations associated with breathing. Respir Physiol 1970; 10:85–92.
128. Gottfried SB, Altose MD, Kelsen SG, Fogarty CM, Cherniack NS. The perception of changes in airflow resistance in normal subjects and patients with chronic airways obstruction. Chest 1978; 73(suppl):286–288.
129. Zechman FW Jr, Wiley RL. Afferent inputs to breathing: respiratory sensation. In: Cherniack NS, Widdicombe JG, eds. Handbook of Physiology. Section 3. The Respiratory System. Vol II. Bethesda, MD: American Physiological Society, 1986:449–474.
130. Davenport PW, Revelette WR. Perception of respiratory mechanical events. In: Adams L, Guz A, eds. Respiratory Sensation. New York: Marcel Dekker, 1996:101–123.
131. Harver A, Mahler DA. The symptom of dyspnea. In: Mahler DA, ed. Dyspnea. Mount Kisco, NY: Futura Publishing, 1990:1–53.
132. Lansing RW, Banzett RB. Psychophysical methods in the study of respiratory sensation. In: Adams L, Guz A, eds. Respiratory Sensation. New York: Marcel Dekker, 1996:69–100.
133. Shea SA, Banzett RB, Lansing RW. Respiratory sensations and their role in the control of breathing. In: Adams L, Guz A, eds. Respiratory Sensation. New York: Marcel Dekker, 1996:923–957.
134. Burki NK. Detection of added respiratory loads in patients with restrictive lung disease. Am Rev Respir Dis 1985; 132:1210–1213.
135. Burki NK, Mitchell K, Chaudhary BA, Zechman FW. The ability of asthmatics to detect added resistive loads. Am Rev Respir Dis 1978; 117:71–75.

136. Harver A, Katkin ES, Bloch E. Signal-detection outcomes on heartbeat and respiratory resistance detection tasks in male and female subjects. Psychophysiology 1993; 30:223–230.
137. Hudgel DW, Cooperson DM, Kinsman RA. Recognition of added resistive loads in asthma: the importance of behavioral styles. Am Rev Respir Dis 1982; 126:121–125.
138. Killian KJ, Campbell EJM, Howell JBL. The effect of increased ventilation on resistive load discrimination. Am Rev Respir Dis 1979; 120:1233–1238.
139. Killian KJ, Mahutte CK, Campbell EJM. Resistive load detection during passive ventilation. Clin Sci 1980; 59:493–495.
140. Killian KJ, Mahutte CK, Howell JBL, Campbell EJM. Effect of timing, flow, lung volume, and threshold pressures on resistive load detection. J Appl Physiol 1980; 49:958–963.
141. McCloskey DI. The effects of pre-existing loads upon detection of externally applied resistances to breathing in man. Clin Sci Mol Med 1973; 45:561–564.
142. Michiels TM, Light RW, Mahutte CK. Effect of ethanol and naloxone on control of ventilation and load perception. J Appl Physiol 1983; 55:929–934.
143. Stubbing DG, Killian KJ, Campbell EJM. Weber's law and resistive load detection. Am Rev Respir Dis 1983; 127:5–7.
144. Wiley RL, Zechman FW Jr. Perception of added airflow resistance in humans. Respir Physiol 1966/67; 2:73–87.
145. Zechman FW, Wiley RL. Effect of chest cage restriction on perception of added airflow resistance. Respir Physiol 1977; 31:71–79.
146. Dahme B, Richter R, Mass R. Interoception of respiratory resistance in asthmatic patients. Biol Psychol 1996; 42:215–229.
147. Gottfried SB, Altose MD, Kelsen SG, Cherniack NS. Perception of changes in airflow resistance in obstructive pulmonary disorders. Am Rev Respir Dis 1981; 124:566–570.
148. Engen T. Psychophysics. I. Discrimination and detection. In: Kling JW, Riggs LA, eds. Experimental Psychology. 3rd ed. New York: Holt, Rinehart, & Winston 1971:34.
149. Harver A, Smith DM. Gender affects the detection of resistive loads. Percep Motor Skills 1996; 82:301–302.
150. Campbell EJM, Howell JBL. The sensation of breathlessness. Brit Med Bull 1963; 19:36–40.
151. Mahutte CK, Campbell EJM, Killian KJ. Theory of resistive load detection. Respir Physiol 1983; 51:131–139.
152. Zechman FW, Muza SR, Davenport PW, Wiley RL, Shelton R. Relationship of transdiaphragmatic pressure and latencies for detecting added inspiratory loads. J Appl Physiol 1985; 58:236–243.
153. Zechman FW, Wiley RL, Davenport PW, Burki NK. Load detection latencies and temporal patterns of added load sensation. Am Rev Respir Dis 1979; 119:73–75.
154. Burdon JGW, Killian KJ, Campbell EJM. Detection latency of added loads to breathing. Clin Sci 1982; 63:11–15.
155. Burki NK. Effects of added inspiratory loads on load detection thresholds. J Appl Physiol 1981; 50:162–164.

156. Zechman FW, Davenport PW. Temporal differences in the detection of resistive and elastic loads to breathing. Respir Physiol 1978; 34:267–277.
157. Swets JA. The relative operating characteristic in psychology. Science 1973; 182:990–1000
158. Mathiot MJ, Bonnel AM, Grimaud C. Effects of airflow resistance variations on resistive load detection in normal subjects. Bull Eur Physiopathol Respir 1986; 22:99–105.
159. Narbed PG, Marcer D, Howell JBL. The contribution of the accelerating phase of inspiratory flow to resistive load detection in man. Clin Sci 1982; 62:367–372.
160. Narbed PG, Marcer D, Howell JBL, Spencer E. A signal detection theory analysis of the effect of chest cage restriction upon the detection of inspiratory resistive loads. Clin Sci 1983; 64:417–421.
161. Bonnel AM, Mathiot MJ, Grimaud C. Inspiratory and expiratory resistive load detection in normal and asthmatic subjects. Respiration 1985; 48:12–23.
162. Bonnel AM, Mathiot MJ, Jungas B, Grimaud C. Breathing discomfort in asthma: role of adaptation level. Bull Eur Physiopathol Respir 1987; 23:23–29.
163. Tiller J, Pain M, Biddle N. Anxiety disorder and the perception of inspiratory resistive loads. Chest 1987; 91:547– 551.
164. Harver A. Magnitude scaling of respiratory sensations. Unpublished manuscript.
165. Chapman KR, Rebuck AS. Inspiratory and expiratory resistive loading as a model of dyspnea in asthma. Respiration 1983; 44:425–432.
166. Muza SR, McDonald S, Zechman FW. Comparison of subjects' perception of inspiratory and expiratory resistance. J Appl Physiol 1984; 56:211–216.
167. Burdon JGW, Killian KJ, Campbell EJM. Effect of ventilatory drive on the perceived magnitude of added loads to breathing. J Appl Physiol 1982; 53:901–907.
168. Killian KJ, Bucens DD, Campbell EJM. Effect of breathing patterns on the perceived magnitude of added loads to breathing. J Appl Physiol 1982; 52:578–584.
169. Burdon JGW, Killian KJ, Stubbing DG, Campbell EJM. Effect of background loads on the perception of added loads to breathing. J Appl Physiol 1983; 54:1222–1228.
170. Revelette WR, Zechman FW Jr., Parker DE, Wiley RL. Effect of background loading on perception of inspiratory loads. J Appl Physiol 1984; 56:404–410.
171. Burki NK. Effects of bronchodilation on magnitude estimation of added resistive loads in asthmatic subjects. Am Rev Respir Dis 1984; 129:225–229.
172. Montserrat JM, Picado C, Lloberas P, Serra J, Luengo M, Agusti-Vidal A. Ability of asthmatics with and without respiratory arrest to detect added resistive loads. J Asthma 1988; 25:131–135.
173. Altose MD, Leitner J, Cherniack NS. Effects of age and respiratory efforts on the perception of resistive ventilatory loads. J Gerontol 1985; 40:147–153.
174. Tack M, Altose MD, Cherniack NS. Effect of aging on respiratory sensations produced by elastic loads. J Appl Physiol 1981; 50:844–850.
175. Tack M, Altose MD, Cherniack NS. Effect of aging on the perception of resistive ventilatory loads. Am Rev Respir Dis 1982; 126:463–467.
176. Gottfried SB, Redline S, Altose MD. Respiratory sensation in chronic obstructive pulmonary disease. Am Rev Respir Dis 1985; 132:954–959.

177. Ward ME, Stubbing DG. Effect of chronic lung disease on the perception of added inspiratory loads. Am Rev Respir Dis 1985; 132:652–656.
178. Mahler DA, Harver A, Rosiello R, Daubenspeck JA. Measurement of respiratory sensation in interstitial lung disease: evaluation of clinical dyspnea ratings and magnitude scaling. Chest 1989; 96:767–771.
179. Mahler DA, Rosiello RA, Harver A, Lentine T, McGovern JF, Daubenspeck JA. Comparison of clinical dyspnea ratings and psychophysical measurements of respiratory sensation in obstructive airway disease. Am Rev Respir Dis 1987; 135:1229–1233.
180. Dunn-Rankin P. Scaling Methods. Hillsdale, NJ: Lawrence Erlbaum, 1983.
181. Harver A, Baird JC, McGovern JF, Daubenspeck JA. Grouping and multidimensional organization of respiratory sensations. Percep Psychophys 1988; 44:285–292.
182. Simon PM, Schwartzstein RM, Weiss JW, Lahive K, Fencl V, Teghtsoonian M, Weinberger SE. Distinguishable sensations of breathlessness induced in normal volunteers. Am Rev Respir Dis 1989; 140:1021–1027.
183. Regan D. Human Brain Electrophysiology. New York: Elsevier, 1989.
184. Squires NK, Ollo C. Human evoked-potential techniques: possible applications to neuropsychology. In: Hannay HJ, ed. Experimental Techniques in Human Neuropsychology. New York: Oxford University Press, 1986:386–418.
185. Davenport PW, Friedman WA, Thompson FJ, Franzen O. Respiratory-related cortical potentials evoked by inspiratory occlusion in humans. J Appl Physiol 1986; 60:1843–1848.
186. Revelette WR, Davenport PW. Effects of timing of inspiratory occlusion on cerebral evoked potentials in humans. J Appl Physiol 1990; 68:282–288.
187. Davenport PW, Colrain IM, Hill P. Scalp topography of the short-latency components of the respiratory-related evoked potential in children. J Appl Physiol 1996; 80:1785–1791.
188. Harver A, Squires NK, Bloch-Salisbury E, Katkin ES. Event-related potentials to airway occlusion in young and old subjects. Psychophysiology 1995; 32:121–129.
189. Bashore TR, Osman A, Heffley EF. Mental slowing in elderly persons: a cognitive psychophysiological analysis. Psychol Aging 1989; 4:235–244.
190. Donchin E, Ritter W, McCallum C. Cognitive psychophysiology: the endogenous components of the ERP. In: Callaway E, Teuting P, Coslow SH, eds. Event-Related Brain Potentials in Man. New York: Academic Press, 1978:349–411.
191. Fabiani M, Gratton G, Karis D, Donchin E. The definition, identification, and reliability of measurement of the P300 component of the event-related potential. In: Ackles PK, Jennings JR, Coles MGH, eds. Advances in Psychophysiology. Volume 1. Greenwich, CT: JAI Press, 1987:1–78.
192. Bloch-Salisbury E, Harver A, Squires NK. Event-related potentials to inspiratory flow-resistive loads in young adults: stimulus magnitude effects. Biological Psychology, in press.
193. Bloch-Salisbury E, Harver A. Effects of detection and classification of resistive and elastic loads on endogenous event-related potentials. J Appl Physiol 1994; 77:1246–1255.

194. Richards DW. The nature of cardiac and pulmonary disease. Circulation 1953; 7:15–29.
195. Cherniack NS. Respiratory sensation as a respiratory controller. In: Adams L, Guz A, eds. Respiratory Sensation. New York: Marcel Dekker, 1996:213–230.
196. Oku Y, Saidel GM, Altose MD, Cherniack NS Perceptual contributions to optimization of breathing. Ann Biomed Eng 1993; 21:509–515.
197. Oku Y, Saidel GM, Chonan T, Altose MD, Cherniak NS. Sensation and control of breathing: a dynamic model. Ann Biomed Eng 1991; 19:251–272.
198. Sears TA. Breathing: A Sensori-Motor Act. The Scientific Basis of Medicine, Annual Review. London: Athlone Press, 1971:129–147.
199. Engen T. The Perception of Odors. New York: Academic Press, 1982.
200. Kelsen SG, Prestel TF, Cherniack NS, Chester EH, Deal EC Jr. Comparison of the respiratory responses to external resistive loading and bronchoconstriction. J Clin Invest 1981; 67:1761–1768.
201. Taguchi O, Kikuchi Y, Hida W, Iwase N, Satoh M, Chonan T, Takishima T. Effects of bronchoconstriction and external resistive loading on the sensation of dyspnea. J Appl Physiol 1991; 71:2183–2190.
202. Guz A. Respiratory sensations in man. Brit Med Bull 1977; 33:175–177.
203. Campbell EJM, Newsom Davis J. Respiratory sensation. In: Campbell EJM, Agostoni E, Newsom Davis J, eds. The Respiratory Muscles: Mechanics and Neural Control. 2nd ed. Philadelphia: W. B. Saunders, 1970:291–306.
204. Campbell EJM, Guz A. Breathlessness. In: Hornbein TF, ed. Regulation of Breathing. Part 2. New York: Marcel Dekker, 1981:1181–1195.
205. Eldridge FL, Chen Z. Respiratory sensation: a neurophysiological perspective. In: Adams L, Guz A, eds. Respiratory Sensation. New York: Marcel Dekker, 1996:19–67.
206. Banzett RB, Lansing RW. Respiratory sensations arising from pulmonary and chemoreceptor afferents: air hunger and lung volume. In: Adams L, Guz A, eds. Respiratory Sensation. New York: Marcel Dekker, 1996:155–180.
207. Manning HL, Schwartzstein RM. Pathophysiology of dyspnea. N Engl J Med 1995; 333:1547–1553.
208. O'Donnell DE. Breathlessness in patients with chronic airflow limitation: mechanisms and management. Chest 1994; 106:904–912.
209. Puddy A, Giesbrecht G, Sanii R, Younes M. Mechanism of detection of resistive loads in conscious humans. J Appl Physiol 1992; 72:2267–2270.
210. Shannon R. Reflexes from respiratory muscles and costovertebral joints. In: Cherniack NS, Widdicombe JG, eds. Handbook of Physiology. The Respiratory System. Section 3. Volume II. Bethesda, MD: American Physiological Society, 1986:431–447.
211. Jammes Y, Speck DF. Respiratory control by diaphragmatic and respiratory muscle afferents. In: Dempsey JA, Pack AI, eds. Regulation of Breathing. 2nd ed. New York: Marcel Dekker, 1995:543–582.
212. Davenport PW, Reep RL. Cerebral cortex and respiration. In: Dempsey JA, Pack AI, eds. Regulation of Breathing. 2nd ed. New York: Marcel Dekker, 1995:365–388.
213. Mechanic D. The concept of illness behaviour: culture, situation, and personal predisposition. Psychol Med 1986; 16:1–7.

8

Pediatric Asthma

Psychosomatic Interactions and Symptom Perception

GREGORY K. FRITZ

Brown University School of Medicine
and Rhode Island Hospital
Providence, Rhode Island

MARIANNE Z. WAMBOLDT

National Jewish Medical and Research Center
Denver, Colorado

I. Introduction

Asthma is the most prevalent pediatric chronic illness in the United States (1). Despite increased understanding of the pathogenesis of asthma and critical advances in treatment approaches, statistics reveal increasing morbidity and mortality in the past two decades. Nationally, the prevalence rate of self-reported asthma increased 42% (from 34.7 to 49.4 per 1000) from 1982 through 1992 (2). For children in particular, the estimated prevalence increased from 3.1% in 1981 to 4.3% in 1988, an increase that translates into approximately 2.7 million affected children (3). Furthermore, from 1980 to 1987, the asthma mortality rate increased 42% in children under age 20, with the highest increase in poor, urban, minority children (4,5). In addition to the risk of death, there is a high risk for morbidity from pediatric asthma. Asthma accounts for 30% of pediatric hospitalizations (1), for approximately one third of school absences (6), and for numerous other psychosocial complications including reduced participation in sports and extracurricular activities (7). The cost of illness related to asthma in 1990 was estimated to be $6.2 billion. A substantial portion (43%) of the economic impact of asthma is associated with use of emergency services, hospitalizations, and death (8). Three national health objectives for the year 2000 include decreasing hospitalization and disability due to asthma and increasing education about asthma in patients and physicians (9).

This chapter focuses on the psychological aspects of pediatric asthma, with particular emphasis on children's perception of changes in respiratory function. Fundamental aspects of pediatric asthma that distinguish it from the same illness in adults are delineated. Symptom perception is placed in an overall context of psychosomatic interactions in the course of asthma, including vulnerability to disease and influences on the quality of life with chronic asthma. A discussion of the conceptual and clinical background of pediatric symptom perception is followed by a review of measurement issues. Finally, we summarize what is known from empirical work on children's symptom perception and resistive load detection and speculate on productive directions for future research.

II. Unique Aspects of Pediatric Asthma

Although many of the basic physiological processes are similar whether asthma occurs in an adult or a child, there are sufficient differences to make the clinical picture and psychosomatic interactions unique in pediatric populations. Three aspects of pediatric asthma in particular present challenges to those attempting to understand psychophysiological interactions in children from a clinical or research standpoint: the need for a developmental perspective, the central role of the parents and family, and the large variation in physical size that occurs within the pediatric age group.

A. Developmental Issues

Although the onset of asthma may occur at any age, 80–90% of asthmatic children experience their first symptoms before 4–5 years of age (10). Those whose asthma has its onset in the first year of life may be most severely affected. Infants and toddlers obviously cannot participate significantly in their own care. However, cognitive development proceeds rapidly, and as children approach school age, they begin to assume responsibility for monitoring their symptoms; their own ability to attend to changes in respiratory status and initiate appropriate treatment becomes increasingly important (11). Children as young as 6 years old can reliably use a hand-held peak flow meter if provided with appropriate instruction and parental monitoring of the procedure (12), and school-aged children increasingly rely on internal physiological cues as signs of health difficulty (13). As children progress towards adolescence, they begin to take more responsibility not only for identifying symptoms, but for self-management. Thus, the developmental progression from the preschooler's total reliance on parental care of their asthma, to the school-aged child's active participation in symptom identification, to the adolescent's central position in asthma self-management means that the role and importance of symptom perception, as with other health behaviors, shifts with developmental stage.

B. Family Issues

The management of children's asthma is inextricably woven into the structure of family life. As with every other aspect of their lives, children are dependent on their parents for all or part of their asthma care. Although the degree of dependence decreases with development, as noted above, parents typically select the physician, schedule and transport their child to medical appointments and the emergency room, provide the required medications, support and monitor home treatment, minimize exposure to known triggers, model illness behavior, and pay the costs of medical care. The functional morbidity associated with a child's asthma may relate more to parental understanding and resources than to either the child's own coping or the particular asthma characteristics. Family disorganization, parental psychopathology, nonadherence to the medical regimen by the parent, and family poverty can directly affect the long-term course of asthma to a much greater extent with children than with adults. Clinical management approaches or research strategies that work with adults cannot, therefore, be assumed to apply readily to children.

C. Physical Growth Issues

The fact that a group of developmentally normal pediatric asthma patients 5–15 years old may vary in size from 50 to 200 lb and in height from 3 to over 6 ft, complicates both treatment and research. Lung capacity varies directly with size; intrinsic airway resistance varies inversely with size. Because every index of lung function is related to lung capacity, normal expected values vary greatly from child to child. Thus, absolute values of common pulmonary function tests (PFTs) have little meaning unless one knows the size of the child, and they must be expressed in terms of "percent-of-expected" or "percent-of-personal best" to be useful.

A particularly challenging situation exists when one attempts to study perception of added resistive loads. For most sensory modalities, the change in stimulus intensity that can just barely be perceived (the perceptual threshold) is a constant fraction of the basal stimulus. When this principle (Weber's law) is applied to a group of pediatric asthma patients whose basal resistance varies from 2–7 centimeters of water per liter per second (cm H_2O/L/sec), an added resistance of 1 cm H_2O/L/sec is a 50% increase for the largest child and a 14% increase for the smallest child. Such a discrepancy makes it impossible to meaningfully compare children with one another unless children are challenged with the same *percentage* increase over their basal resistance. To study children, internal airway basal resistance should first be measured, and the apparatus and procedure should be modified to present an individualized series of resistances that constitute a standard series of percentage increases. The research is possible but is methodologically more complex than with adults.

III. Pediatric Asthma Risks

Given the factors unique to pediatric asthma, there are several instances where psychological and family factors are particularly relevant to both the development and course of asthma symptoms. These will be reviewed briefly.

A. Environmental Risks for Developing Asthma

Current consensus views asthma as a multifactorial illness, with both genetic and environmental (i.e., infectious, allergic, mechanical, and psychosocial) factors involved in its development and course. Although asthma is fairly heritable (there is a two- to fourfold difference in the concordance rates for monozygotic and dizygotic twins), the concordance rate in monozygotic twins is only 20% (14,15). Hence, phenotypic expression of the genetic component for asthma appears dependent on substantial environmental influence.

There have been few prospective studies of the role of environmental factors in the initial onset of childhood asthma. Most of these have not found *global* psychosocial factors to be important influences, supporting instead the relevance of early infections with parainfluenza or respiratory syncytial viruses, parental smoking habits, and specific antigen exposure leading to clinical atopy as major risk factors (16,17). However, a recent study suggests that more *specific* psychosocial factors may be of importance. Klinnert and associates (18) have demonstrated an interactive effect of quality of parenting and family stress on asthma onset by 3 years of age in a group of infants at high genetic risk to develop asthma. They found that high family stress increased asthma incidence by age 3, but only when the stress occurred in the context of deficient parenting practices.

Parenting practices determine much of the environmental exposures that young children encounter, e.g., to passive tobacco smoke, to dust mites, and to allergenic animals. To prevent genetically at-risk children from developing asthma, parents need to be both educated about the potentially asthmagenic potential of these exposures and motivated to avoid them. Additionally, it is known that respiratory infections are more frequent in children from high-stress homes, and these infections may also contribute to asthma onset (19).

Thus, children at genetic risk may be more likely to develop asthma if they come from families who show deficits in keeping their children protected from the environmental stressors.

B. Emotional Precipitants

Although the role of psychological factors in the development of asthma is a topic of ongoing debate, emotional factors play a strong role in the triggering or precipitation of symptoms in a significant subset of children with asthma. Over 60 years ago, Peshkin observed that a subgroup of children with seemingly intractable asthma dramatically improved when separated from their families and placed in a hospital setting (20,21). In a most provocative study, Purcell and associates (22) used parent interviews to classify children with frequent, clinically active asthma into "family-reactive" versus "steroid-dependent" groups. "Family-reactive" children were defined as those who were frequently triggered by emotional upsets or conflict. "Steroid-dependent" children had no emotional triggers listed among their most frequent triggers, but instead had allergens or mechanical triggers, e.g., cold air. For 2 weeks all children were cared for by a parent surrogate in their own homes (to maintain a constant allergic environment); parents and siblings were moved into local hotels. For the group of 13 "family-reactive" children, all measures of asthma, including such objective indices as peak expiratory flow rate and amount of medication required, improved promptly after the separation. In contrast, only daily report of asthma symptoms improved for the 12 children predicted to be steroid dependent. However, data collected during the 4 weeks after families were reunited revealed that while "parentectomy" was an effective treatment while it lasted, it did not offer longstanding improvement in symptoms. Asthma symptomatology in the family-reactive group gradually returned to baseline levels, suggesting that patterns of reinforcement embedded in family interactions were involved in symptom exacerbation and maintenance.

This line of research has been resurrected recently, with several different groups describing children with asthma who come from families expressing high levels of criticism of the child (23,24). Preliminary investigations have indicated that families with an asthmatic child exhibit higher levels of criticism than those with a healthy child, and that the amount of criticism may be correlated with the severity of asthma. Additionally, children from highly critical families may be more likely to be noncompliant with treatment regimens than children from minimally critical families (25). In support of the idea that family conflict may play a part in triggering or maintaining asthma symptoms in certain children, several controlled studies have demonstrated the efficacy of family therapy for certain children with "intractable" asthma (26–28).

An entirely different line of research found that children with asthma could be subgrouped into those whose symptoms were more frequently, or rarely, triggered by psychological factors. In the 1970s and 1980s, efforts were made to develop behaviorally oriented approaches based on anxiety reduction through muscle relaxation, with the assumption that constriction of the airway was the primary problem in asthma. Research focused on relaxation, biofeedback, and hypnosis, with mixed results. Some studies using general relaxation with asthmatic children demonstrated that relaxation can lead to increased in peak expiratory flow rate (29) and increased lung tolerance to an inhaled irritant during a provocation test (30). There is some evidence in both of these studies that children with more emotionally determined asthma demonstrate the most benefit from relaxation.

Studies of autosuggestion and hypnosis have highlighted the power of suggestion in both symptom induction and symptom reversal in certain asthmatics. Separating extrinsic (allergic) from intrinsic (nonallergic) asthmatics, Philipp et al. (30) found that intrinsic asthmatics were more easily triggered by suggestion than extrinsic asthmatics and that their symptoms were reversed by placebo. In a review of studies using hypnosis to increase lung function, DePiano and Salzberg (31) concluded that hypnosis may have a greater effect on symptom perception and symptom control than on actual lung function.

Finally, a number of case reports and studies have suggested that psychosocial and family factors are powerful predictors of asthma deaths (32–38). In a case-controlled study, Strunk and associates (37) compared 21 children with severe asthma who were previously hospitalized on a tertiary asthma care pediatric inpatient unit, and who subsequently died of asthma, to 21 still-living control children matched for age at time of hospitalization for asthma, sex, and asthma severity. Of the 57 physiological and psychosocial variables evaluated by review of hospital records, psychosocial risk factors were most prominent in severely ill asthmatic children who subsequently died. Mediating mechanisms have been suggested to be depression or anxiety in the child with asthma, conflict within the family and between the family and health care system, noncompliance, and delayed perception, or disregard, of symptoms.

C. Adherence and Asthma Management

Since asthma is a chronic disease that waxes and wanes in response to environmental pertubations, it is essential that the patient and their family be taught how to manage the day-to-day symptoms of the illness. Patients should minimally understand the use of inhaled drugs and peak flow meters and the difference between medications to relieve symptoms and those to prevent episodes. Knowledge of individual triggers of asthma symptoms, and how to avoid these triggers, is important. A written action plan to apply when symptoms occur should be developed by the physician and given to the patient, to parents, and to school officials. Thus, parent and child education, partnerships with the physician, and the willingness and ability to modify the asthma regimen within certain guidelines are essential to good asthma management. Clearly, such psychological variables as basic intelligence, problem-solving and communication skills, motivation for behavioral change, and ability to identify exacerbation of symptoms will be associated with how well an individual patient and family cope with the illness.

There are many good educational programs available for children with asthma and their families, and these have been shown to be effective at increasing self-management knowledge and skills (39). However, as with any chronic disease, education is necessary, though not sufficient, for good self-management and prevention of morbidity. In a meta-analysis of studies of the effectiveness of pediatric asthma self-management teaching programs from 1970 to 1991, Bernard-Bonnin and colleagues (40) reviewed 23 study reports. Global assessment of the quality of the studies (Chalmers' scale) was fair; 12 flawed studies were eliminated. Based on the pooled effect size of the studies, self-management teaching did not reduce school absences, asthma attacks, acute care visits, hospitalizations, or hospital days for the overall group. This meta-analysis points out some of the limitations of education alone in decreasing asthma morbidity. More attention must be paid to attitudes and actual health behaviors, e.g., elimination of environmental triggers and timely compliance with medications, rather than just knowledge. Additionally, interventions must be tailored for specific groups, e.g., school-aged children with severe asthma. In the one study where severity of asthma was accounted for, self-management teaching had a moderate effect of reducing morbidity in those children with prior hospitalizations and emergency room visits (41).

Assessing how well asthmatic children and their families adhere to medical recommendations is critical to understanding asthma morbidity. Some studies estimate that up to 90% of pediatric asthmatics are noncompliant to some degree with their prescribed medical regimen (42,43). Nonadherence has been identified as a major factor in asthma morbidity (44) and mortality (37). Nonadherence problems include both underuse and overuse of medications. In fact, a common problem among children and adolescents is nonadherence with anti-inflammatory inhalers (e.g., steroid or cromolyn), which are prophylactic and do not yield an immediate improvement in symptoms, along with overuse of the rescue beta-adrenergics that do have an immediate effect. Some authors have cited overutilization of beta-adrenergic inhaler medication as a potential risk factor for asthma death (45).

Thus, understanding patterns of adherence with asthma regimens is critical to improving the management of asthma. One group that may be particularly salient in this context includes children who have difficulty perceiving asthma symptoms. For these children, emphasis on peak flow monitoring and self-charting may be most critical because they may not readily identify when they need to put their asthma-management skills to use. The measurement of adherence to inhaled medications has improved markedly with newer technology, e.g., the computerized MDI chronolog (46), which has two implications for our understanding of the links between symptom perception and morbidity. First, it is possible that certain nonadherent behaviors (e.g., underuse of inhaled beta-agonists) may be linked to poor symptom perception. In other words, certain patterns of medication underusage may be "passive noncompliance," a finding that has significant implications for educational intervention. Second, asthma management is likely to play an important intervening role between symptom perception and morbidity. Once an exacerbation has begun, the child's accuracy at recognizing symptoms is the first step in utilizing appropriate asthma management. Perceptual ability and asthma management likely interact such that children with poor perception and poor management skills are at the highest risk for morbidity.

D. Comorbid Psychiatric Problems

The asthma disease process and treatment can cause disruption of normal psychological and family development that place asthmatic children and their families at increased risk for secondary psychiatric problems. The often unpredictable onset of asthma symptoms, as well as the complex self-management regimen for asthma, can challenge competent families and stress marginal families beyond their limits of coping. Prior dysfunctional family dynamics can be exacerbated under this stress and can lead to increased difficulties in the child.

General Issues

There is some evidence that asthmatic families have more negative characteristics when compared to nonasthmatic families, although such studies typically fail to account for the impact of chronic illness. Block et al. (47) obtained parent descriptions of asthmatic children and observed the behavior of children alone with an examiner and with the mother. These investigators noted increased interactional difficulties between mothers and their asthmatic children. Mrazek et al. (48) studied severely asthmatic preschoolers and their interactions with their mothers and compared them with healthy children. The asthmatic children were more likely to be oppositional and the affective climate of the interactions with their mothers was more negative than those of healthy controls. In studies of slightly older children, mothers report more stress, overindulgence, and overprotectiveness in their relationships with their children (49). In observational studies, families with asthmatic children have been found to have inadequate problem-solving and conflict-resolution skills (50) as well as experiencing more chaos and difficulty making decisions than control families (51). Whether these family interaction patterns contribute to or result from the chronic asthma cannot be determined readily from these crosssectional studies.

In general, it has been suggested that children with more severe asthma have higher rates of psychopathology than either milder asthmatics or nonill children (52). However, many studies depend on parental report of asthma severity and child emotional problems, both of which are correlated with the parent's own psychological and physical status (53) but not with physician ratings of asthma severity. Thus, the emotional problems of these asthmatic children may be mediated by living with a parent who perceives the asthma to be very stressful and who is either physically ill or psychologically distressed. The perception that the asthma is severe may mean that the family is not coping adequately with the illness. Asthma that is under poor control for extended periods of time may be rated as more severe and may in actuality become more severe. We will focus on the extent to which particular forms of psychopathology may interfere with asthma symptom perception and self-management.

Anxiety and Panic Disorders

Some specific psychiatric disorders are found more frequently in children with asthma. Approximately 30% of children with asthma meet the criteria for some anxiety disorder, the most common being separation anxiety (54,55). There are a number of reasons why children with asthma may develop anxiety symptoms. Asthma often necessitates hospitalizations and medical interventions, which provoke anxiety in children. Additionally, a chronic illness where one intermittently has extreme difficulty breathing is very anxiety provoking, both as a cognitive stressor and also at a neurophysiological level. Klein (56) has postulated that central chemoreceptors monitoring arterial carbon dioxide pressure (PaCO2) can trigger anxiety and even panic, serving to warn of impeding asphyxiation. Thus, children with asthma who suffer repeated episodes of hypercarbia may be at risk of developing panic symptoms over time.

Anxiety disorders, most particularly hyperventilation and panic disorder, are intricately involved with perceptions of respiratory symptoms. The primary problem in panic disorder appears to be a hyperresponsive brainstem respiratory sensor system (57). The current leading pathophysiological model of panic disorder posits that false central "suffocation alarms" arise from the hypersensitive response of ventromedullar chemoreceptors to increasing CO_2. These "false alarms" then lead to prompt activation of the pontine noradrenergic nucleus, the locus ceruleus, thereby increasing minute ventilation two- to threefold and triggering the physiological cascade of a panic attack (56,58,59). In support of this proposition, inhalation of CO_2 or infusion of sodium lactate causes increased ventilation, and both are potent and specific precipitants of panic in susceptible individuals. Furthermore, antipanic medication has been shown to be able to block the respiratory stimulatory and panic-inducing effects of these laboratory challenges (60,61).

Available information suggests that panic attacks are common in patients with chronic respiratory illness (62,63). While panic disorder is quite rare in children, many researchers believe that other forms of anxiety, notably separation anxiety, overanxious disorder, and social phobia (64,65), may be childhood precursors to panic disorder in adults. These disorders have been linked on the basis of results from family (66) and pharmacological (67) studies as well as retrospective reports from adults diagnosed with panic disorder (68). Preliminary results from prospective studies of children with separation anxiety also reveal associations with adult panic attacks. Thus, while children with asthma may not develop full-blown panic attacks, their anxiety symptoms may be triggered by a similar neurophysiological mechanism.

Pine's work comparing children with asthma to children with congenital hypoventilation syndrome demonstrates the neurochemical mediation of anxiety (69). Children with congenital hypoventilation syndrome lack the ability to detect changes in PaCO2. Thus, even though they suffer from life-threatening episodes of hypoxia and hypercarbia, as well as endure numerous medical interventions that may lead to conditioned fears, they do not detect increased levels of CO_2. If Klein's hypothesis about detection of hypercarbia leading to anxiety and eventually panic is correct, one would hypothesize that children unable to detect PaCO2 would have less anxiety than children with other medical disorders. This is indeed what was found when carefully assessed anxiety symptoms were compared for children with congenital hypoventilation syndrome, children with asthma, children with a variety of chronic medical conditions, and healthy controls. The children with asthma had the highest rates of anxiety disorders (47%), followed by children with any medical condition (32%), community controls (24%), and children with hypoventilation syndrome (15%). Thus, children with asthma are at risk for increased anxiety disorders both because of conditioned fears concomitant with a chronic medical condition and additionally due to the nature of repeated episodes of hypercarbia triggering anxiety at a neurophysiological level.

There is some suggestion that children with asthma may have a higher genetic risk for anxiety disorders as well. Two groups have established that children with asthma have an increased rate of anxiety disorders among their first-degree genetic relatives, indicating a possible shared genetic connection between anxiety disorders and asthma (70,71). One group ascertained that all the relatives' anxiety disorders had their onset prior to the onset of the child's asthma, thus ruling out the anxiety simply as a response to having an ill child or sibling (72).

Available information suggests that panic attacks are common in patients with chronic respiratory illness (68,69). The pulmonologist treating such patients is confronted by the complex intertwining of the "somatic" anxiety disorders and asthma. Hyperventilation (even when isocapnia is maintained) can precipitate bronchial constriction. Indeed, isocapnic hyperventilation has been proposed as a highly specific measure of airways hyper responsiveness in asthma (73). Conversely, the Pa_{CO2} increase during an acute asthma attack can trigger a panic attack in patients with panic disorder (74). Although the symptoms of air hunger, shortness of breath, chest discomfort, and acute fear are similar across these conditions, the treatments are not. Hence, patients with combined panic disorder and asthma may be less adherent to their medication regimen (presumably over utilizing bronchodilators) and at risk to develop eroded confidence in their physicians who prescribe the antiasthma medications that do not fully work to alleviate their distress. Further, certain asthma treatments, such as beta-agonists, can *increase* autonomic arousal, thereby precipitating or worsening panic attacks. Thus, it is not surprising that a number of clinicians have argued for increased recognition and aggressive treatment of comorbid panic disorder in both pediatric and adult patients with asthma (70,74,75).

Depression

Children with asthma have a higher prevalence of depressive symptoms than controls. A recent meta-analysis (76) revealed that children with asthma, recurrent abdominal pain, and sickle cell anemia may be at greater risk to develop depressive disorders than children with cancer, cystic fibrosis, and diabetes mellitus. Hypotheses attempting to explain the link between depression and severe asthma have suggested that the biological derangements of anxiety and affective disorders lead to autonomic instability that exacerbates asthma. Until recently, there has been very little research to substantiate such theories. Recently, Miller and Wood (77) published the results of a study demonstrating that marked emotional and autonomic responses in asthmatic children watching an evocative movie were correlated with increased airway reactivity and decreased pulmonary function. These findings were seen as consistent with a psychophysiologically induced exacerbation of asthma mediated via cholinergic neurotransmission. The authors speculated that children whose asthma was very reactive to emotional influences could represent a subgroup at greater risk for developing increased morbidity from the negative impact of psychosocial stressors and comorbid psychiatric illness. Klinnert and associates (78), using a more precise measurement of cholinergic input, vagal tone (79), confirmed that negative affect in children both with and without asthma was associated with increased cholinergic transmission. Presumably, the asthma of children with poor affective modulation may be especially sensitive to emotional factors.

IV. The Psychophysiology of Pediatric Asthma Symptoms

As is clear from the above sections, asthma education and self-management are increasingly recognized as important components of the process to reduce asthma morbidity and mortality. Work with children and adolescents is especially important given the relative malleability of patterns in children compared to adults and the potential for long-term accrual of benefits if optimal self-management strategies are adopted early in life. Basic to the entire educational and self-management approach, but often implicit, is the patient's ability to accurately perceive the point at which it is necessary to begin self-management. *No matter how educated or skillful an individual may be, self-management skills are useless if one doesn't recognize when to apply them.* Thus, understanding the factors that play a role in symptom perception, as well as how to measure perception and possibly modify perceptions, is a critical new area of research that needs to be developed to further the optimal management of pediatric asthma.

A. Physiological Factors

Genetic Influences

To date there have been no genetic studies done on perceptual ability in children. However, in a study of 10 monzygotic and 9 dizygotic adult twin pairs, Kobayashi and colleagues (80) determined that there was no genetic component to the detection of resistive loading under normal conditions, but that there was a genetic component to the perception of resistive loading under hypoxic conditions as well as a genetic component to the hypercapnic response.

Sensitivity to Autonomic Nervous System Fluctuations

There is considerable research regarding the ability of healthy adults to evaluate various autonomic symptoms, including blood pressure, heart rate, gastric motility, skin temperature, and galvanic skin response (GSR). Adults are poor overall at perceiving discrete autonomic variations; correlations between subjective and objective measures typically fall between 0.15 and 0.25. However, most of the work on autonomic perception has been done with adults in laboratory studies using rigid and somewhat artificial paradigms. The generalizability of this approach to actual health behavior has been questioned (81). Such studies may in fact be measuring insignificant parameters with unrealistic protocols in individuals having no motivation to become perceptually skillful. When patients whose health is dependent on symptom perception (e.g., diabetics) focus on a more global physiological change such as blood glucose levels in a natural setting (the home), correlations between subjective and objective variables are found to be fairly high, averaging 0.66 (82). Hence, the autonomic perception research to date suggests that physiological changes that have adaptive utility for an individual may be perceived more accurately than rather discrete and irrelevant somatic changes. The body of research as a whole also suggests that while there is substantial variation among individuals in a group on a given perceptual ability, in adults perceptual accuracy is a stable, measurable characteristic (83). Except for our own work, symptom perception in children has been largely unstudied.

Mechanisms Underlying Respiratory Sensations

Psychophysics, the study of the relationship between physical stimulus and sensation, has provided a guiding theory for basic perceptual research in the pulmonary laboratory. Several well-established methods of psychophysics have been adapted to assess respiratory thresholds and to measure the relationship between the magnitude of a physical stimulus and magnitude of a sensation. Perception of resistive loads is of clinical interest because it may have implications for an individual's ability to detect accurately changes in pulmonary function.

There are a variety of sensory receptors that provide inputs for the perception of respiratory changes. Detection of resistance is more accurate during active respiration than passive respiration, indicating that active contraction of respiratory muscles is necessary for the detection of added resistance. If the chest is strapped or the diaphragm alone is paralyzed, perception is not impeded, indicating that both intracostal and diaphragmatic muscles play a role. The addition of elastic loads decreases adults' ability to detect changes in resistive loads, also indicating that stretch receptors on the respiratory muscles provide input to the perception of respiratory changes (84). Respiratory muscle fatigue and consequent decrease in expiratory muscle force is related to increased sensation of dyspnea as well (85). Additionally, stimulation of the oral mucosa intensifies the sensation of breathlessness (86). Finally, stimulation of afferent vagal nerve fibers found in the lungs will also increase the sensation of dyspnea, even when airway resistance and lung volumes are held constant. Thus, there are a variety of sensory inputs that appear to be integrated for perception of changes in breathing to occur. Many of these nerves can be stimulated by a variety of mechanisms; for example, the increased tidal breathing often found in anxiety attacks could stimulate stretch receptors in the respiratory musculature.

Central Nervous System Drive

Chemosensitivity refers to the brainstem-mediated response to hypoxemia or hypercapnea, produced naturally during episodes of severe bronchospasm or experimentally in the laboratory by manipulating concentrations of oxygen or CO_2 in inspired air. Chemosensitivity can be measured as the slope of the change in mouth pressure 0.1 second after airway occlusion ($P_{0.1}$, respiratory drive) per unit rise in CO_2 or fall in oxygen saturation. Most asthmatic children and adults show a steeper rise in respiratory drive in response to increased CO_2 than do normal controls (87–89). If a resistive load is added, the CO_2 response in most asthmatics is accentuated relative to controls (86). In other words, when faced with the respiratory "challenge" of increased CO_2, asthmatics demonstrate a more pronounced response to added loads. The importance of perceptual processes in the CO_2 response to external resistive loads (in the laboratory and, perhaps, clinically) is indicated by the fact that loads that are too small to be detected consciously do not elicit the chemosensitivity response (90). In addition, light anaesthesia eliminates the response both in animals (91) and in humans (92). The relationship between conscious perception of respiratory changes and the subcortical control of respiratory drive merits further investigation.

Illness Variability

Variablility of symptoms and the rapidity with which they change may also have an impact on the perception of breathlessness. In a study of 28 adults with mild asthma who were challenged with an inhaled anitigen and then tested at frequent intervals over the insuing 8 hours, subjects had poorer perception of airway changes during a late asthmatic response (LAR) than during an early asthmatic response (EAR) (93). In addition, perceptual accuracy for all subjects, during both EAR and LAR, correlated with the rate of fall in one-second forced expiration volume (FEV_1) (93). Although the authors interpreted their data to indicate it was the rate of change that affected differences in perception during EAR and LAR, it is also possible that the change in inflammatory mediators that occur during these two phases of an asthmatic response may also differentially stimulate the afferent receptors associated with perception.

B. Psychological Factors

Coping Style

How perception relates to coping or defensive styles is still controversial. Steiner et al. (94) reported an association between inability to detect respiratory sensations and a defensive style of repression, which is characterized by low anxiety despite being involved in stressful situations. Another type of coping style, defined by the Battery for Asthma Illness Behavior (95), has been shown to be predictive for various aspects of asthma outcome in adults. Persons scoring high on the panic-fear scale of this instrument are described as being "anxious dependent," and those who score low are "rigidly independent." Individuals with moderate scores are "adaptive." The high and low "panic-fear" styles were correlated with noncompliance and increased hospitalizations, as well as with hypervigilance toward or ignoring asthma symptoms. In a study of 12 asthmatic and 12 healthy adults, anxious/dependent subjects had the highest thresholds of detection for added resistive load, indicating that these anxious/dependent subjects had the worst perceptual ability of the group (96). In a replication study with 12 healthy adults, a significant inverse relationship between resting ventilatory drive and threshold detection across the three panic-fear groups was found. Again, anxious/dependent subjects had lower resting ventilatory drive and higher threshold detection than the other groups (97).

The "rigidly independent" types may be similar to repressors as a group, therefore these findings (if not due to response bias) are somewhat in contradiction to those of Steiner and colleagues (94). However, the findings in the anxious/dependent group are intriguing when considered in conjunction with the high prevalence of anxiety in asthmatics. Those subjects who were most anxiously dependent on their health care givers and others also had the least ability to acccurately perceive respiratory symptoms. Whether their anxiety interferes with the ability to perceive symptoms or results from their inability to know when their symptoms are worsening (a situation that would lead many to feel worried and uncertain) is unclear. Baron and colleagues (98) replicated the presence of high and low "panic-fear" types in children with asthma but did not study their perceptual ability. They did note that the anxious/dependent group received a significantly larger proportion of steroids than other asthmatics, however, and suggested that this was due to physicians prescribing medication based on the patient's report of dyspnea and not their objective pulmonary function tests.

Cognition

To perceive symptoms requires some degree of ability to quiet oneself enough to attend to internal stimulation, as well as the motivation to do so. These attributes may well be more difficult in young children, and clearly are more difficult in children suffering from an attention deficit disorder. The effects of attention and motivation on this process have not yet been studied.

Attention specific to physical symptoms may be another pathway through which the cognitive process of attention affects symptom perception in childhood. Barsky and associates (99) believe that persons who somatize engage in a process of "somatosensory amplification," which involves heightened attention to body sensation, overfocusing on weak bodily sensations, and reacting to these sensations with distorted cognitive processes. For children with asthma, "somatosensory clarity," i.e., accurate interpretation of symptoms, is the desired state. Thus, neither extremely high nor extremely low symptom attention and reporting is optimal, but rather a moderate, intermediate position.

In addition to the willingness to attend to perceptions, other cognitive developmental skills may be necessary for a child to accurately perceive respiratory changes. It is unclear whether young children can separate out the sensations of shortness of breath or dyspnea from emotional upsets or other dysphoric sensations. In order to more clearly delineate how children experience dyspnea, Lebowitz and colleagues (60) asked children to describe symptoms that occur during an asthma attack using the Asthma Symptom Checklist. Five significant categories of symptoms were described through factor analysis: asthma or airway obstruction, panic-fear, irritability, fatigue, and unhappy-afraid. The asthma factor, which included symptoms such as hard to breathe, shortness of breath, and wheezing, accounted for 23% of the variation. Children in this study included many emotional factors in their description of asthma attacks. Of note, there are numerous sensations of dyspnea, even in adults, and each may be triggered by different combinations of sensory inputs (100). Thus, it is not surprising that children would use a variety of constructs to describe their asthma symptoms.

Children can rate intensity of dyspnea, however, even if they cannot always distinguish dyspnea from emotional upset. Carrieri (101) studied 39 children (ages 7–13 years) with mild to moderate asthma—duration of illness 2 months to 13 years. She used three different rating scales: a visual analog scale (VAS), a shaded color scale, and a word descriptor scale ("rotten," "just okay," "good," and "great"). Children were able to use these three ratings to differentiate between good and bad breathing days, and the rating scales exhibited concurrent validity. Finally, our own studies (seen below) show that a large percentage of children are able to rate their symptoms using either a visual analog scale or estimates of objective measures of airway obstruction, e.g., peak flows.

Cultural Influences

Some cultures emphasize attributions of either physical or psychological etiology when a person is feeling distressed. For example, in Latino cultures it is more acceptable to complain of physical symptoms and problems than psychological difficulties or conflicts. Depressed persons presenting first to a primary care physician regarding physical complaints differ in some respects from those who present to a mental health clinician with psychological complaints (102,103). Some of these differences are cultural. Does the extent to which children are inculcated to describe distress in physical or psychological terms relate to their ability to accurately perceive symptoms?

V. Measurement of Pediatric Asthma Symptoms

The study of children's perception of respiratory symptoms can be approached from several different perspectives. In clinical settings, the challenge is to determine the degree to which the subjective assessment of asthma symptomatology corresponds with an objective measurement of pulmonary function. Given the high degree of asthma variability and the effort-dependent nature of pulmonary function assessment, it is desirable to determine an individual's perceptual pattern over multiple observations and at different levels of asthma symptom severity. Such research would closely relate to a child's real-life experience of asthma, but individual fluctuations of symptoms and the need to await naturally occurring asthma make clinical protocols logistically cumbersome and difficult to standardize. A laboratory procedure in which different resistive loads are repeatedly presented on a breath-to-breath basis can be rigorously controlled and allows multiple observations to be collected efficiently in a single session. At best, however, a laboratory paradigm can only approximate actual asthma in the child's natural environment. In this section, we will summarize a recent review of the options, their strengths and weakness, and our own solutions to methodological problems in studying symptom perception in pediatric asthma (11).

A. Objective Measurement of Respiratory Status

Pulmonary physiologists have developed numerous indices and ratios that use combinations of volume, pressure, and flow measurements to describe function at various points in the respiratory cycle. In pediatric practice, the number of frequently utilized indices is much smaller, but a single, validated "gold standard" does not exist. By far the most commonly used objective measure is the peak expiratory flow rate (PEFR), which is recommended as part of clinical care in the guidelines of the National Institutes of Health (104). The major reason for the ubiquity of PEFR is its ease and economy of measurement; PEFR is easily assessed with an inexpensive, hand-held peak flow meter. No other pulmonary function test approaches the technological simplicity of PEFR. Predicted normal values based on height, gender and ethnicity are available, but individually determined "personal best" values are preferable. Most children will quickly learn their typical normal range. Disadvantages of using PEFR as the objective measure of asthma include the fact that it is highly effort-dependent and that there may be considerable variability between different brands of hand-held meters (105). Additionally, our recent data suggest that because PEFR measures predominantly large airway function, a subgroup of children whose asthma involves primarily peripheral airways may not be monitored appropriately with PEFR (106).

A reasonable alternative to PEFR is FEV_1. This test is somewhat less effort-dependent than PEFR and may be less related to only large airway function. Currently, FEV_1 is obtained only through full spirometry, requiring more sophisticated and expensive equipment, careful monitoring, and physician interpretation. In 2663 pulmonary function tests done by children 8–15 years old at an asthma camp, we found 82.4% concordance between PEFR and FEV_1 (106).

The forced expiratory flow in the midportion of a forced expiration breath (FEF_{25-75}) primarily measures small airway function and is much less effort-dependent than PEFR. Like FEV_1, it can only be measured with spirometry equipment. FEF_{25-75} is an especially important index of asthma severity in the subgroup of children (approximately 18%) whose acute asthma episodes are characterized by peripheral airway compromise. Of the three indices, FEF_{25-75} was the most sensitive and specific measure of airway compromise in pediatric asthma patients over several years at a residential asthma camp (103).

B. Subjective Assessment of Respiratory Symptoms

One of the most straightforward approaches to subjective quantification of symptoms is to ask patients to predict, based on their present asthma symptoms, what their objective pulmonary function test (PFT) result would be at that moment. This is a reasonable approach to subjective assessment because PEFR monitoring is increasingly common among patients with asthma and the PEFR values are readily understood by both patients and physicians. Many asthma patients now monitor PEFR at home and are familiar with their personal range in readings. Thus, they have sufficient practice and feedback to help shape subjective assessments.

Guessing values of indices other than PEFR are problematic with children because they have no experience with them. Providing experience in a research protocol may lead to potential confusion when they return to PEFR monitoring at home. Whether children with asthma can distinguish between sensations affecting different indices has not been studied. It is likely that a PEFR guess represents a summary quantification of the perceived respiratory status at the moment. One disadvantage of the PEFR prediction approach stems from the observation that many developmentally younger children think

NAME: ____________ DATE: _____________ TIME: _____________ am/pm
How much coughing, wheezing, or shortness of breath do you feel now? The pictures below may help you rate yourself on this line. Please put a mark on the line, not on the pictures.

None at all	A tiny bit	A little	Some	Quite a bit	A lot	Very much (terrible)

Figure 1 Brown visual analog scale.

of illness in categorical terms—"healthy," "sick," "not too good," etc. To express their status in continuous numbers may be a cognitive challenge for young children. A second disadvantage is that there can be a recency effect to the prediction process, that is, some patients—especially younger, cognitively concrete children—may tend to guess the same value as they did on the last occasion. This can lead to an artifactually accurate (or inaccurate) guess that does not reflect an actual assessment of physiological state.

A second approach to subjective assessment is the use of a visual analog scale (VAS). With a VAS, a patient picks a rating from a 100 mm. line which may or may not have verbal or pictorial anchors. One such instrument, the Brown Asthma Visual Analog Scale (107) was developed for children with asthma. In this scale, both verbal anchors and cartoon depictions are used to help orient a child to the use of the scale (Fig. 1).

The advantage of a VAS is that it requires summation of symptoms and a judgment, which means that it is more global. The "gut feeling" reflected in such a global assessment may very well be the real basis for subsequent action or inaction in asthma self-management. Because it is continuous, one can look for small changes in symptomatology detected by the patient. One disadvantage of a VAS is that it requires the ability to use a combination of verbal or pictorial anchors and a proportional spatial relationship between anchors, making the approach clearly inappropriate for developmentally immature children with preoperational cognitive abilities. Further, patients develop response sets; for example, they may consistently rank extremes. Others don't use the extremes of the scales but, instead, develop a response set, which truncates the use of potential values. In these cases, the advantages of the VAS are minimal unless the patient is reeducated to the use of the scale.

A third approach to subjective assessment is categorical description. In this type of approach, a simple scale may be used. For example, five verbal descriptors might represent choices to describe asthma symptomatology, e.g., "none," "a little," "some," "pretty bad," or "awfully bad—the worst." Such categorical descriptors have an advantage because they are commonly used in everyday life. Descriptors may be more understandable than a VAS, and the wording can be changed and defined by the investigator to ensure proper understanding of the terminology by a particular population. Although categorical scales commonly employ verbal descriptors, they can also use pictorial descriptors for young children, such as the FACES scale which is used by young children to quantify pain (108). The disadvantage to the categorical approach is the limited variability; it discourages attempts at distinguishing subtle differences in symptomatology. In addition, categories may not have the same meaning across patients. There may also be intraindividual variation in rating tendencies that are more problematic when using a scale with broad categories.

C. Computational Approaches

Once the choice of appropriate objective and subjective variables is made and they are measured at a number of points to provide a pattern, the way in which the data are summarized is also critical. Very little published work on either adults or children is available, and no generally accepted convention for summarizing data currently exists. Several approaches are conceptually possible, but they do not necessarily lead to the same classification of perceptual accuracy when applied to an individual patient's data. Three analytical approaches we have used are summarized below; more detailed discussion and case examples are available elsewhere (11).

Correlational

If a sufficient number of trials are available, a child's subjective and objective ratings at each point may be statistically correlated; the correlation coefficient can be used as an indication of the magnitude of perceptual accuracy for the given child. The advantages of this correlational approach include the creation of an index of association that depends little on specific scale properties. Subject variables (e.g., VAS or "guess" of predicted PFT value) can be standardized and combined, creating a more stable subjective index to correlate with objective measurement.

Despite the appeal of such an approach, it is subject to the same limitations of most correlational methods—a sufficient number of observations are needed for a valid index, and the correlation is likely to be low if there is insufficient variability on indices measured. In the latter case, poor perceptual accuracy could be confused with statistical artifact. Further, low correlations are likely to contain no information regarding the type of perceptual error; overestimation of symptoms is indistinguishable from underestimation, even though the medical consequences are very different.

Arithmetic Differences

When the same specific measurement scales are used for objective and subjective ratings, the actual value obtained on the objective measure may be subtracted from the patient's guess that immediately preceded it according to the formula:

$$\frac{\text{Guessed value} - \text{actual value}}{\text{Predicted value}} \times 100 = \text{difference}$$

The predicted values in the formula above may be obtained from normative tables (109). Since normal PFT values depend on the size of the child, this adjustment to the raw difference score is done to approximately standardize the magnitude of the difference. These difference scores are then summed over all observations. The mean difference score summarizes the subject's average guessing error; using the absolute value of the differences avoids the artifact of errors in opposite directions canceling each other.

Unlike correlational analysis, an advantage of calculating the arithmetic differences is that the direction of inaccuracy can be meaningfully indicated (omitting the absolute value). To depict the magnitude of imprecision in either direction, the sum of overestimations (subjective PFT guess > objective PFT measure) and the sum of underestimations (subjective PFT guess < objective PFT measure) are calculated. These indices can be interpreted individually or summarized in a ratio:

$$\frac{\sum \text{overestimation}}{\sum \text{underestimation}} = \text{over/under ratio}$$

Because this method relies on mean differences, it is impossible to distinguish between a pattern of multiple small errors or a few large errors. The same mean value could result from either pattern, and a nonrepresentative outlier can have an unduly large impact on the final index. This approach is also cumbersome to apply when the subjective and objective variables are not in the same units. Finally, the arithmetic summary does not deal with the clinical relevance of the data, as differences of similar magnitude are weighted equally no matter what the child's clinical status at the time of the measurement.

Error Grid Analysis

Cox et al. (110), working with adult patients with diabetes, developed a graphic system in which a subjective estimate of blood sugar was plotted against the corresponding actual blood sugar level and then categorized in clinical groups. We have modified this creative approach for use with asthma; the resulting grid takes the arithmetic summary described above a step further because the grid is grounded on clinical specifications (11). The categories on the grid consider the actual pulmonary status (indicated by the objective measure) and the implications of decisions for self-management that would be based on the subjective estimate. Although the grid system could be applied to any pulmonary function variable, the present status of PEFR makes it the clear choice to use in the development of the error grid analytic approach.

As shown in Figure 2, the grid is divided into several zones. The diagonal is on the line of identity, reflecting perfect subject-objective agreement. The vertical line at 80% of the child's best value ("Self Best") is an arbitrary point at which PEFR can be considered to reflect clinically significant compromise. Area A ("Accurate") includes points for which the subjective estimate is within ±10% of the actual PEFR. Areas B.O. ("Benign Overestimation") and B.U. ("Benign Underestimation") include inaccuracies that will have benign consequences in self-management. Area B.O. includes clinically uncompromised points in which the subject overestimated PEFR; no intervention is needed or undertaken. Area B.U. includes points where a compromised PEFR is exaggerated in the estimate; needed intervention would be vigorously pursued but overtreatment is a theoretical risk. Area D ("Dangerous") includes overestimates of clinically compromised PEFR; it is considered dangerous because the need for self-management interventions would be overlooked. Area E ("Erroneous") reflects the potential for overtreatment since it includes points at which a normal range PEFR is perceived as compromised.

The error grid analysis provides a comprehensive display of the degree of accuracy in a child's subjective estimate, the extent and direction of inaccuracy, and the clinical risk (or lack thereof) potentially resulting from the specific inaccuracy. The approach allows the use of Area A alone to reflect overall accuracy. Alternatively, the grid can be used to generate a complete profile of accurate guesses, overestimations, and underestimations.

The main disadvantage of the error grid analysis stems from the arbitrary nature of the threshold for PFT abnormality. There is no empirical evidence for choosing 10%, 20%, or any other decrease from an individual's normal baseline level of pulmonary functioning as indicative of an asthma episode. Further, it is essential in using the error grid approach to have an individu-

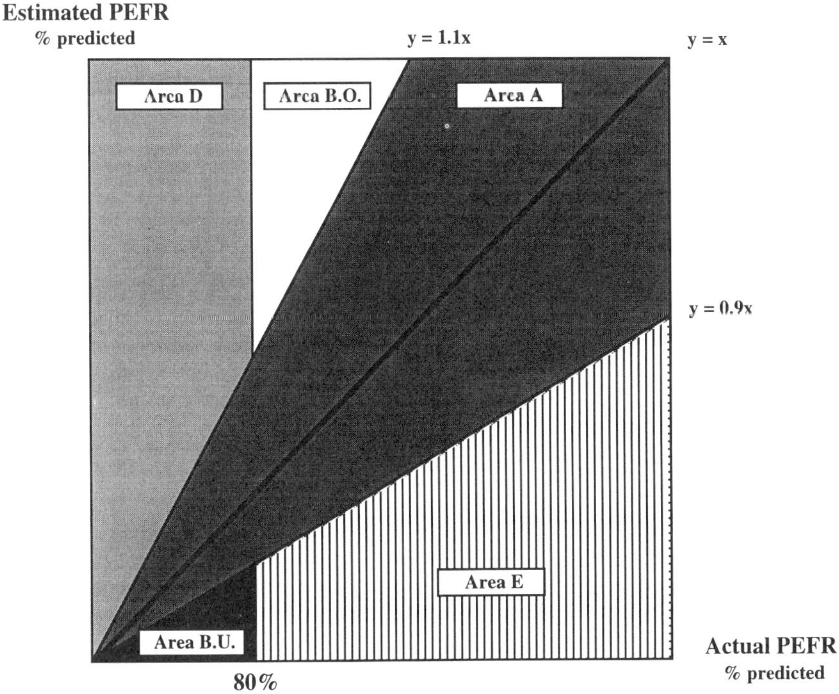

Figure 2 Error grid analysis of PEFR perception. Area A, "accurate area": subject estimateted PEFRs accurately; Area D, "dangerous area": subject overestimated PEFRs while having low PEFRs; Area E, "erroneous area": subject underestimated PEFRs while having normal PEFRs; Area B.O., "benign overestimation area": subject overestimated PEFRs while having normal PEFRs; Area B.U., "benign underestimation area": subject underestimated PEFRs while having low PEFRs.

alized baseline (the child's personal best PFT value obtained when completely asymptomatic and exerting maximal effort) to use in standardizing the objective observations. Published tables only approximate normal values for a given child due to individual anatomical and functional variation.

D. Laboratory Protocol

The use of external added resistive loads presented repeatedly to test a subject's ability to recognize obstruction to breathing has been applied with adults in a number of studies, but only recently with children. The marked differences in size of airways and level of intrinsic resistance (R_{rs}) among children necessitates an individualized series of added resistances based on the child's

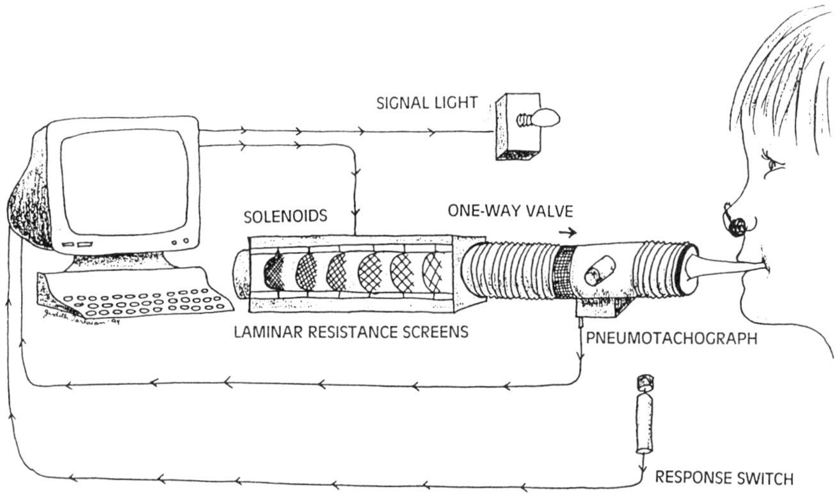

Figure 3 Resistive loading apparatus.

own intrinsic airway resistance, rather than a standard set of resistances as is commonly used with adults. Children are then compared with one another on a percentage basis ($\Delta R/R_{rs}$, the Weber fraction) in the level of resistance they are able to detect.

Although conceptually straightforward, the need to measure intrinsic resistance immediately prior to the perception protocol presents methodological challenges. R_{rs} may be measured with total body plethysysmography (the "body box"), but this procedure and equipment are cumbersome and often unpleasant for children. The method of forced oscillations is easier and noninvasive and has a history of successful use with children (111–113).

We have developed an apparatus (Fig. 3) in which six laminar flow screens of graded resistance (0.25–8.0 cm H_2O/L/sec) can be engaged in various combinations to apply predetermined inspiratory loads. A computer and a pneumotachograph determine the timing of the inspiratory loads. Following three unloaded inspirations the load is presented for two inspirations and subjects, cued by a signal light, press a button if they notice a difference. The apparatus is illustrated schematically in Figure 3.

Determination of Threshold via Tracking Protocol

The perceptual threshold can be calculated through two different approaches. A tracking procedure varies the loads based on whether the previous load was

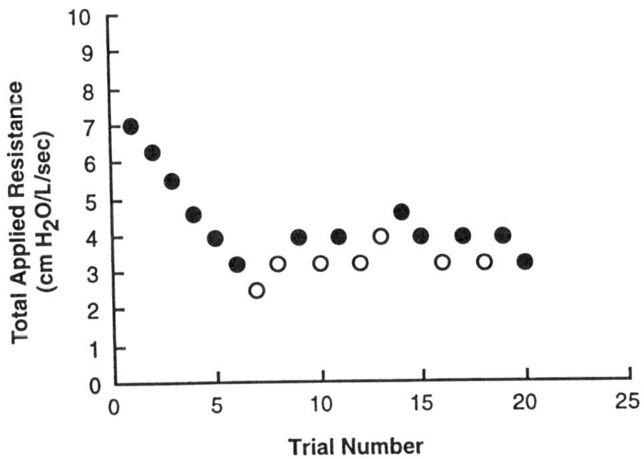

Figure 4 Tracking procedure for measuring resistive-load detection threshold: results from a 10-year-old boy. Solid dots indicate that the subject detected the applied resistance; open dots indicate that the applied resistance was not detected. The threshold is achieved by two consecutive detections and nondetections alternating around a single boundary. For this subject the threshold was measured at 3.7 cm H_2O/L/sec.

recognized: the subsequent load is reduced 10% after a load is detected or increased 10% if the subject fails to detect a load. The number of loads and the definition of a threshold are arbitrary, but depend upon repeated ascending and descending trials. We have found that 20 load presentations are not beyond the attentional capacity of most children. We consider a thresh- old to be measured if a child achieves at least two consecutive detections and two consecutive nondetections alternating around a single boundary (Fig. 4).

Determination of Threshold via Random Protocol

A random procedure presents a series of loads in a preset, nonsystematic order. In our work, we have used 20, 60, 100, and 180% of a child's R_{rs} as the loads, each presented a total of five times in the procedure. The child'a data are analyzed by a linear regression, and the threshold is defined statistically as the extrapolated resistance detected in 50% of the trials.

Our preliminary findings suggest that a laboratory protocol has promise for assessing perceptual accuracy in children. However, a significant number of children (approximately 20%) may not yield measureable thresholds with either procedure. At the present state of methodological development, it is difficult to determine whether these children are especially poor perceivers or whether their poor performance relates to problems they have with aspects of the protocol or with response bias.

VI. Symptom Perception in Children with Asthma

The preceding review and methodological discussion has emphasized the paucity of research dealing with symptom perception in pediatric asthma. Despite the relevance of studying symptom perception close to the onset of the disease, few investigators have undertaken the needed studies in a pediatric population. In this section, we will summarize what is known from empirical work on perceptual accuracy in children with asthma.

In the only published study of children's perception prior to our own work, Sly et al. (114) studied the relationship of FEV_1 to subjective ratings (visual analog scale and four–category verbal descriptor) in a group of 14 children with asthma before and after an asthma education program. This study was compromised by its reliance on only a single estimate of FEV_1 before and after the interventions. They found that the children were unable to accurately predict their degree of airway obstruction either before or after the program period.

Our clinical work with perceptual accuracy has primarily utilized an asthma camp setting to overcome the logistical problems inherent in the need to obtain multiple supervised assessments in a cohort of children with asthma. In our initial study, which employed a hand-held peak flow meter and a statistically combined PEFR guess–VAS subjective measure, 37 children were assessed an average of 34 times over the 2-week camp period. The children varied markedly in their perceptual accuracy, with correlations between subjective and objective indices ranging from 0.86 to –0.16 (115). Subsequently, our similar studies using PEFR and FEF_{25-75} (both measured with computerized spirometric equipment), larger samples, and various settings have confirmed the initial results: (1) even in children as young as 7 years, perceptual accuracy is a measurable characteristic; (2) within any group of asthmatic children, there is an extremely wide range of perceptual ability. Children at two asthma camps were less accurate on average than inpatients at the National Jewish Center in Denver. The 175 campers had a mean correlational accuracy index of 0.23 compared to 0.57 for the 97 inpatients. To date, this difference has not been satisfactorily explained.

Better perceptual accuracy in children is associated with lower functional morbidity after controlling for asthma severity. Analyses of campers' data revealed a modest but statistically significant relationship between accuracy index (using the more sensitive FEF_{25-75} as the objective measure) and both days of school missed due to asthma and emergency room visits (116).

Perceptual accuracy accounted for 17% of the variance in school absences and 14% of the variance in ER visits after severity was controlled. These findings provide empirical support for the clinical impression that perceptual ability is associated with the overall course of pediatric asthma.

The study of factors associated with better or worse symptom perception is just beginning, and scant empirical work has been focused on determinants of perceptual accuracy in pediatric asthma. We have recently reported both positive and negative findings regarding the relation of several plausible physiological and psychological variables and children's perceptual ability (116). The variables and rationale for inclusion were:

1. Asthma severity/range of pulmonary symptoms: children who have more dramatic symptom patterns may be more attuned to, and experienced at, "reading" their lungs.
2. Defensive style: unpleasant physical sensations may be managed in a manner similar to the way an inadividual manages negative affect, such that a repressor would keep asthma symptoms from awareness, resulting in reduced accuracy.
3. Intellectual capacity: since perception is a cognitive process, more intelligent children may be better at it.
4. Parental modeling: even if parents do not have asthma, a realistic approach to general physical symptoms may be more helpful to their children's accuracy than either an extremely stoic or symptom-amplifying pattern.
5. Behavior problems: children with high rates of externalizing behavioral problems may not be sufficiently attuned to internal asthma cues.
6. Anxiety level: higher anxiety impairs many psychological functions, and symptom perception may also be affected.

Extensive analyses of data from 86 children at an asthma camp in New Hampshire demonstrated that only the intelligence estimate (the sum of Vocabulary and Block Design scaled scores on the Revised Wechsler Intelligence Scale for Children (WISC-R)) was significantly and consistently related to perceptual accuracy (116). The other variables, although conceptually appealing, did not prove to be associated with symptom perception. This work needs to be replicated in different populations of asthmatic children, and other variables merit study as predictors of perceptual accuracy.

Our work with a laboratory protocol to test perception of added resistive loads in children and adolescents has documented the feasibility of such a procedure and the threshold ranges in healthy youngsters (117). We found that children's intrinsic resistance as measured by forced oscillation varied considerably (3.2–8.8 cm H_2O/L/sec) and decreased with increasing height at a rate of 0.11 cm H_2O/L/sec/cm ($R^2 = 0.50$, $p < 0.0001$) (117). These findings replicate observations obtained by the Lebeque et al. on a normal sample of children (111).

The perceptual thresholds determined by the tracking protocol and the random protocol correlated 0.56 ($p < 0.05$), providing convergent validity for this method of threshold estimation. Comparing children using the threshold obtained as a percentage of the child's intrinsic resistance (R/R_{rs}), the mean threshold was 83.05% and 90.27% for the tracking and random procedures, respectively. These first findings in children suggest that their perceptual thresholds are somewhat higher than those previously reported in adults (118). Our results did not vary systematically with age, height, or gender. Additional work assessing perceptual ability with the resistive loading procedure in pediatric asthma patients is currently underway.

VII. Further Research

Despite wide recognition of the potential clinical implications, empirical work examining symptom perception in children with asthma is just beginning. Further work developing appropriate methodologies for measuring perception that is applicable for most children across a wide age span is necessary. Replication of the above studies in disparate samples of children with varying degrees of asthma severity will help clarify the role that perception plays in asthma outcome for various types of children with asthma. There are two particularly compelling areas to pursue.

A. The Role of Symptom Perception in Adherence

As noted above, nonadherence contributes to poorly controlled asthma and has been associated with increased morbidity and risk of death due to asthma. Numerous reports evaluating the cause of childhood asthma deaths from around the world consistently list treatment nonadherence with medication and/or patient or family delay in seeking emergency care as major "avoidable" factors linked to these deaths (32,34,37,119). Despite the agreement that noncompliance among asthmatic children is widespread, there are few variables that allow clinicians to predict who will be noncompliant, nor are there adequate interventions to prevent or ameliorate the noncompliant behavior. We suspect that children with different abilities to perceive asthma symptoms may in fact exhibit different patterns in their compliance with medication usage. For example, children whose perceptions often fall in the danger zone following error grid analyses (i.e., they are not aware that their pulmonary functions are compromised) may be likely to underuse their scheduled inhalers, since they do not perceive that they are having difficulty. However, when their pulmonary functions deteriorate even more, they may then overutilize their beta-agonist inhalers in an attempt to reverse serious bronchoconstriction. Thus, they may exhibit noncompliance in two ways: underutilization of scheduled inhalers (usually anti-inflammatory inhalers, e.g., steroids or cromolyn) and periodic overutilization of as-needed inhalers (beta-agonists). As overutilization of beta-agonist inhalers has been linked to asthma death, this mechanism may additionally help to explain the link of poor perception of asthma symptoms with near fatal asthma episodes.

Conversely, children who fall into the exaggerated zone of the grid, i.e., those who perceive that their pulmonary functions are more compromised than they really are, may be likely to be very compliant with scheduled inhalers, as they may perceive the need for medications more often. These children, who also are poor perceivers, may actually have less morbidity, due to increased compliance with inhaled medications. Thus, compliance may be associated with patterns of perception, as well as explain why certain patterns of perception are linked with better or worse morbidity.

B. The Relationship Between Anxiety Disorders and Perception

As discussed earlier, the frequent co-occurence of anxiety disorders, particularly panic disorder and respiratory disorders, is intriguing. Although panic disorder in children and adolescents is not common, neither is it rare (120). Since panic disorder in particular seems to be precipitated in part by a false sensation of breathlessness or suffocation, the parallels with accurate perception of respiratory symptoms are intriguing. Further work needs to be done to clarify whether the perceptual distortions hypothesized to occur in panic disorder may be documented by poor perceptual ability as measured by either the clinical or laboratory approaches to respiratory perception. Are central nervous system mechanisms involved in both types of respiratory perception? Answers to these questions may help to clarify the higher occurrence of anxiety disorders in children with asthma and lead to treatments that might be more specific for their symptoms.

Acknowledgment

Preparation of this chapter was supported by NIH grants #RO1-HL45157 (Dr. Fritz) and #R03–MH48683 and #M01–RR0005 1 (Dr. Wamboldt).

References

1. Gortmaker SL, Sappenfield W. Chronic childhood disorders: Prevalence and impact. Pediatr Clin North Am 1984; 31:3–18.

2. Asthma—United States, 1982–1992. MMWR 1995; 43:952–954.
3. Weitzman M, Gortmaker SL, Sobel AM, Perrin JM. Recent trends in the prevalence and severity of childhood asthma. JAMA 1992; 268:2673–2677.
4. Taylor WR, Newacheck PW. Impact of childhood asthma on health. Pediatrics 1992; 90:657–662.
5. Weiss KB, Gergen PJ, Crain EF. Inner-city asthma. The epidemiology of an emerging U.S. public health concern. Chest 1992; 101:362S-367S.
6. Weitzman M, Klerman LV, Lamb G. School absence: a problem for the pediatrician. Pediatrics 1982; 69:739–746.
7. Nocon A. Social and emotional impact of childhood asthma. Arch Dis Child 1991; 66:458–460.
8. Weiss KB, Gergen PJ, Hodgson TA. An economic evaluation of asthma in the United States. N Engl J Med 1992; 326:862–866.
9. Public Health Service. Healthy People 2000: National Health Promotion and Disease Prevention Objectives. Washington, DC: U.S. Department of Health and Human Services, DHHS Pub. No. (PHS)91–50212, 1991.
10. Behrman RE, Vaughan VC, Nelson WE, eds. Nelson Textbook of Pediatrics. Philadelphia: WB Saunders, 1987:495.
11. Fritz GK, Yeung A, Wamboldt M, Spirito A, McQuaid EL, Klein R, Seifer R. Conceptual and methodologic issues in quantifying perceptual accuracy in children. J Ped Psych 1996; 21:153–173.
12. Clark NM, Evans B, Mellins RB. Patient use of peak flow monitoring. Am Rev Respir Dis 1992; 145: 722–725.
13. Hergenrather JR, Rabinowitz M. Age-related differences in the organization of children's knowledge of illness. Dev Psychol 1991; 27:952–959.
14. Edfors-Lubs ML. Allergy in 7000 twin pairs. Acta Allergol 1971; 26:249–285.
15. Hopp RJ, Bewtra AK, Watt GD, Nair NM, Townley RG. Genetic analysis of allergic disease in twins. J Allergy Clin Immunol 1984; 73:265–270.
16. Arshad SH, Hide DW. Effect of environmental factors on the development of allergic disorders in infancy. J Allergy Clin Immunol 1992; 90:235–241.
17. Horwood L, Fergusson D, Hons B, Shannon F. Social and familial factors in the development of early childhood asthma. Pediatrics 1985; 75:859–868.
18. Klinnert MD, Mrazek PJ, Mrazek DA. Early asthma onset: The interaction between family stressors and adaptive parenting. Psychiatry 1994; 57:51–61.
19. Beautrais AL, Fergusson DM, Shannon FT. Life events and childhood morbidity: a prospective study. Pediatrics 1982; 70:935–940.
20. Peshkin MM. Asthma in children, IX. Role of environment in the treatment of a selected group of cases: a plea for a "home" as a restorative measure. Am J Dis Child 1930; 39:774.
21. Peshkin MM. (1959). Intractible asthma of childhood: rehabilitation at the institutional level with a follow-up of 150 cases. Int Arch Allergy 1959; 15:91–112.
22. Purcell K, Brady K, Chai H, Muser J, Molk L, Gordon N, Means J. The effect of asthma in children of experimental separation from the family. Psychosom Med 1969; 31:144–164.
23. Hermanns J, Florin I, Dietrich M, Rieger C, Hahlweg K. Maternal criticism, mother-child interaction, and bronchial asthma. J Psychosom Res 1989; 33:469–476.

24. Schobinger R, Florin I, Zimmer C, Lindemann H, Winter, H. Childhood asthma: Paternal critical attitude and father-child interaction. J Psychosom Res 1992; 36:743–750.
25. Wamboldt FS, Wamboldt MZ, Gavin LA, Roessler TA, Brugman, SM. Parental expressed emotion and treatment outcome in adolescents hospitalized for severe, chronic asthma. J Psychosom Res 1995; 39:995–1005.
26. Gustafsson PA, Kjellman, NIM, Cederbald M. Family therapy in the treatment of severe childhood asthma. J Psychosom Res 1986; 30:369–374.
27. Lask B, Matthew D. Childhood asthma. A controlled trial of family psychotherapy. Arch Dis Child 1979; 54:116–119.
28. Liebman R, Minuchin S, Baker L. The use of structural family therapy in the treatment of intractable asthma. Am J Psychiatty 1974; 131:535–540.
29. Alexander AB. Systematic relaxation and flow rates in asthmatic children. J Psychosom Res 1972; 16:405–410.
30. Philipp RL, Willde GJ, Day JH. Suggestion and relaxation in asthmatics. J Psychosom Res 1972; 16:193–204.
31. DePiano FA, Salzberg HG. Clinical applications of hypnosis to three psychosomatic disorders 1979; 86:1223–1235.
32. Carswell F. Thirty deaths from asthma. Arch Dis Child 1985; 60:25–28.
33. Fritz GK, Rubenstein S, Lewiston NJ. Psychological factors in fatal childhood asthma. Am J Orthopsychiatry 1987; 57:253–257.
34. Kravis LP, Kolski GB. Unexpected death in childhood asthma: a review of 13 deaths in ambulatory patients. Am J Dis Child 1985; 139:558–563.
35. Rubenstein S, Hindi RD, Moss RB, Blessing-Moore, J, Lewiston NJ. Sudden death in adolescent asthma. Ann Allergy 1984; 53:311–318.
36. Sears MR, Rea HH, Fenwick J, Beaglehole R, Gilles AJD, Holst PE, O'Donnell TV, Rothwell RPG, Sutherland DC. Deaths from asthma in New Zealand. Arch Dis Child 1986; 61:6–10.
37. Strunk R, Mrazek D, Fuhrmann G, LaBrecque J. Physiologic and psychological characteristics associated with deaths due to asthma in childhood. JAMA 1985; 254:1193–1198.
38. Strunk RC. Asthma deaths in childhood: Identification of patients at risk and intervention. J Allergy Clin Immunol 1987; 80:472–477.
39. Wigal JK, Creer TL, Kotses H, Lewis P. A critique of 19 self-management programs for childhood asthma: Part I. Development and evaluation of the programs. Ped Asthma Allergy Immunol 1990; 4:17–39.
40. Bernard-Bonnin A, Stachenko S, Bonin D, Charette C, Rousseau E. Self-management teaching programs and morbidity of pediatric asthma: a meta analysis. J Allergy Clin Immunol 1995; 95:34–41.
41. Clark NM, Feldman CH, Evans DE, Levison MJ, Wasilewski Y, Mellins RB. The impact of health education on frequency and cost of health care use by low income children with asthma. J Allergy Clin Immunol 1986; 78:108–116.
42. Baum D, Creer TL. Medication non-compliance in children with asthma. J Asthma 1986; 23:49–59.
43. Chryssanthopoulos C, Laufer P, Torphy DE. Evaluation of compliance and combined drug therapy. J Asthma 1983; 20:35–38.

44. Epstein HH, Cluss PA. A behavioral medicine perspective on adherence to long-term medical regimens. J Consult Clin Psychol 1982; 50:960–971.
45. Whitelaw WA. Asthma deaths. Chest 1991; 99:1507–1510.
46. Nides MA, Tashkin DP, Simmons MS, Wise RA, Li VC, Rand CS. Improving inhaler adherence in a clinical trial through the use of the nebulizer chronolog. Chest 1993; 104:501–507.
47. Block J, Jennings PH, Harvey E, Simpson E. Interaction between allergic potential and psychopathology in asthma. Psychosom Med 1964; 26:307–320.
48. Mrazek D, Anderson I, Strunk R. Disturbed emotional development of severely asthmatic pre-school children. In: Stevenson JE, ed. Recent Research in Developmental Psychopathology. New York: Pergamon Press, 1985:81–94.
49. Carson DK, Schauer RW. Mothers of children with asthma: Perceptions of parenting stress and the mother-child relationship. Psychol Rep 1992; 71:1139–1148.
50. Wikran R, Faleide A, Blakar RM. Communication in the family of the asthmatic child. Acta Psychiatr Scand 1978; 57:11–26.
51. DiBlasio P, Molinari E, Peri G, Taverna A. Family competence and childhood asthma: a preliminary study. Family Sys Med 1990; 8:145–149.
52. Mrazek D. Asthma: Psychiatric considerations, evaluation, and management. In: Middleton, E, Reed CE, Ellis EF, Adkinson NF, Yuninger JW, eds. Allergy Principles and Practice. Vol 2. Washington, DC: C.V. Mosby, 1988:1176–1196.
53. Jessop DJ, Riessman CK, Stein REK. Chronic childhood illness and maternal mental health. J Dev Behav Pediatrics 1988; 9:147–156.
54. Kashani JH, Konig P, Shepperd JA, Wilfley D, Morris, DA. Psychopathology and self-concept in asthmatic children. J Pediatr Psychol 1988; 13:509–520.
55. Wamboldt MZ, Weintraub P, Settle J, Krafchick D. Psychiatric family history of severe asthmatic adolescents. Paper presented at the American Academy of Child and Adolescent Psychiatry, New York, 1994.
56. Klein DF. False suffocation alarms, spontaneous panics, and related conditions: an integrative hypothesis. Arch Gen Psychiatry 1993; 50:306–317.
57. Gorman JM, Liebowitz MR, Fyer AJ, Stine J. A neuroanatomical hypothesis for panic disorder. Am J Psychiatry 1989; 46:148–161.
58. Nutt D, Lawson, C. Panic attacks. A neurochemical overview of models and mechanisms. Br J Psychiatry 1992; 160:165–178.
59. Papp LA, Klein DF, Gorman JM. Carbon dioxide hypersensitivity, hyperventilation, and panic disorder. Am J Psychiatry 1993; 150:1149–1157.
60. Lebowitz MD, Thompson HC, Strunk RC. (1981). Subjective psychological symptoms in outpatient asthmatic adolescents. J Behav Med 1981; 4:439–449.
61. Woods SW, Charney DS, Delgado PL, Heninger GR. The effect of long-term imipramine treatment on carbon dioxide-induced anxiety in panic disorder patients. J Clin Psychiatry 1990; 51:505–507.
62. Karajgi B, Rifkin A, Doddi S, Killi, R. The prevalence of anxiety disorders in patients with chronic obstructive pulmonary disease. Am J Psychiatry 1990; 147:200–201.

63. Zandbergen J, Bright M, Pols H, Fernandez I, Loof C, Griez EJL. Higher lifetime prevalence of respiratory disease in panic disorders? Am J Psychiatry 1991; 148:1583–1585.
64. Goldstein R, Weissman M, Adams P, Horvath E, Lish J, Charney D, Woods S, Sobin C, Wickramaratne P. Psychiatric disorders in relatives of probands with panic disorder and/or major depression. Arch Gen Psychiatry 1994; 51(5):383–394.
65. Last C, Herse M, Kazdin A, Orvaschel H, Perrin S. Anxiety disorders in children and their families. Arch Gen Psychiatry 1991; 48:928–934.
66. Moreau D, Weissman M. Anxiety symptoms in nonpsychiatrically referred children and adolescents. In: Last C, ed. Anxiety Across the Lifespan: A Developmental Perspective. New York:Springer, 1993:37–62.
67. Gittelman-Klein R, Kein D. The relationship between separation anxiety and panic and agoraphobic attacks. Psychopathology 1984; 17:56–65.
68. Torgersen S. Childhood and family characteristics in panic and generalized anxiety disorders. Am J Psychiatry 1986; 143:630–632.
69. Pine DS, Weese-Mayer DE, Silvestri JM, Davies M, Whitaker AH, Klein DF. Anxiety and congenital central hypoventilation syndrome. Am J Psychiatry 1994; 151:864–870.
70. Biederman J, Milberger S, Faraone SV, Guite J, Warbuton R. Associations between childhood asthma and ADHD: issues of psychiatric comorbidity and familiality. J Am Acad Child Adolesc Psychiatry 1994; 33:842–848.
71. Wamboldt MZ, Wamboldt FS. Psychosocial aspects of severe asthma in children. In: Szefler SJ, Leung DYM, eds. Severe Asthma: Pathogenesis and Clinical Management. New York: Marcel Dekker, 1995:465–495.
72. Wamboldt M, Weintraub P, Krafchick D, Wamboldt F. Psychiatric family history in adolescents with severe asthma. J Am Acad Child Adolesc Psychiatry 1996; 35:1042–1049.
73. O'Byrne PM, Ramsdale EH, Hargreave FE. Isocapnic hyperventilation for measuring airway hyperresponsiveness in asthma and in chronic obstructive pulmonary disease. Am Rev Respir Dis 1991; 143:1444–1445.
74. Baron C, Marcotte J. Experience and reason—briefly recorded. Role of panic attacks in the intractability of asthma in children. Pediatrics 1994; 94:108–111.
75. Yellowlees PM, Kalucy RS. Psychobiological aspects of asthma and the consequent research implications. Chest 1990; 97:628–634.
76. Bennett, D. S. Depression among children with chronic medical problems. J Pediatr Psychol 1994; 19:149–169.
77. Miller BD, Wood BL. Psychophysiologic reactivity in asthmatic children: a cholinergically mediated confluence of pathways. J Am Acad Child Adolesc Psychiatry 1994; 33:1236–1245.
78. Klinnert MD, McCormic DR, Bryant NE. Adjustment, affect, and autonomic functioning in asthmatic children. Paper presented at the Society for Research in Child Development, Indianapolis, 1995.
79. Porges SW, Doussard-Roosvelt JA, Maiti AK. Vagal tone and the physiological regulation of emotion. Monogr Soc Res Child Dev 1994; 59:167–186.

80. Kobayashi S, Nishimura M, Yamamoto M, Akiyama Y, Kishi, F. Dyspnea sensation and chemical control of breathing in adult twins. Am Rev Respir Dis 1993; 147:1192–1198.
81. Pennebaker JW, Gonder-Frederick L, Cox DJ, Hoover CW. The perception of general vs. specific visceral activity and the regulation of health-related behavior. Adv Behav Med 1985; 1:169–198.
82. Cox DJ, Gordon-Frederick L, Pohl S et al. Reliability of symptom-blood glucose relationships among insulin-dependent diabetics. Psychosom Med 1986; 45:357–360.
83. Zechman FW, Wiley RL. Afferent inputs to breathing: respiratory sensation. In: Fishman AP, ed. The Respiratory System, Handbook of Physiology. Vol. II. Bethesda, MD: American Physiological Society, 1986.
84. Shahid SU, Goddard BA, Howell JBL. Detection and interaction of elastic and flow resistive respiratory loads in man. Clin Sci 1981; 61:339–343.
85. Suzuki S, Suzuki J, Ishii T, Akahori T, Okubo T. Relationship of respiratory effort sensation to expiratory muscle fatigue during expiratory threshold loading. Am Rev Respir Dis 1992; 145: 461–466.
86. Simon PM, Basner RC, Weinberger SE, Fencl V, Weiss JW, Schwartzstein RM. Oral mucosal stimulation modulates intensity of breathlessness induced in normal subjects. Am Rev Respir Dis 1991; 144:419–422.
87. Cosgrove J, Neuberger N, Levison H. The ventilatory response to carbon dioxide in asthmatic children, measured by the mouth-occlusion method (PM 100). Pediatrics 1976; 57:952–959.
88. Gottfried S, Altose M, Kelsen S, Cherniack N. Perception of changes in airflow resistance in obstructive pulmonary disorders. Am Rev Respir Dis 1981; 124:566–570.
89. Kelsen S, Fleegler B, Altose M. The respiratory neuromuscular response to hypoxia, hypercapnea, and obstruction in airflow in asthma. Am Rev Respir Dis 1979; 120:517–527.
90. Altose M, Kelsen S, Cherniack N. Respiratory responses to changes in airflow resistance in conscious man. Respir Physiol 1979; 36:249–260.
91. Isaza G, Posner J, Altose M, Kelsen S, Cherniack N. Airway occlusion pressure in awake and anaesthetized goats. Res Physiol 1976; 27:87–98.
92. Whitelaw W, Derenne J, Couture J, Milic Emili J. Adaptation of anaesthetized man to breathing through an inspiratory resistor. J Appl Physiol 1976; 41:285–291.
93. Turcotte H, Boulet LP. Perception of breathlessness during early and late asthmatic responses. Am Rev Respir Dis 1993; 148:514–518.
94. Steiner H, Higgs C, Fritz GK. Defense style and the perception of asthma. Psychosom Med 1987; 49:34–44.
95. Kinsman RA, Dirks JF, Jones NF. Psychomaintenance of chronic physical illness. In: Millon T, Green CJ, eds. Handbook of Clinical Health Psychology. New York: Plenum Publishing, 1982:435–465.
96. Hudgel DW, Cooperson DM, Kinsman RA. Recognition of added resistive loads in asthma: The importance of behavioral styles. Am Rev Respir Dis 1982; 26: 121–125.

97. Hudgel DW, Kinsman RA. Interactions among behavioral style, ventilatory drive, and load recognition. Am Rev Respir Dis 1983; 128:246–248.
98. Baron C, Lamarre A, Veilleux P, Ducharme G, Spier S, Lapierre JG. Psychomaintenance of childhood asthma: a study of 34 children. J Asthma 1985; 23:68 79.
99. Barsky AJ, Goodson JD, Lane RS, Cleary, PD. The amplification of somatic symptoms. Psychosom Med 1988; 50:510–519.
100. Simon PM, Schwartzstein RM, Weiss JW, Fencl V, Teghtsoonian M, Weinberger SE. Distinguishable types of dyspnea in patients with shortness of breath (see comments). Am Rev Respir Dis 1990; 142:1009–1014.
101. Carrieri VK, Kieckhefer G, Janson-Bjerklie S, Souza, J. The sensation of pulmonary dyspnea in school-age children. Nurs Res 1991; 40:81–85.
102. Bridges K, Goldberg, D. Somatic presentations of DSM-III disorders in primary care. J Psychosom Res 1985; 29:563–569.
103. Goldberg DP, Bridges K. Somatic presentations of psychiatric illness in primary care setting. J Psychosom Res 1988; 32:137–144.
104. Mellins RB. Asthma education: a national strategy. Am Rev Respir Dis 1989; 140:577–578.
105. Eichenhorn MS, Beauchamp RK, Harper PA, Ward JC. An assessment of three portable peak flow meters. Chest 1982; 82:306–309.
106. Klein RB, Fritz GK, Yeung A, McQuaid EL, Mansell A. Spirometric patterns in childhood asthma: peak flow compared to other indices. Pediatr Pulmonol 1995; 20:372–379.
107. Fritz GK, Spirito A, Yeung A, Klein R, Freedman E. A pictorial visual analog scale for rating severity of childhood asthma episodes. J Asthma 1994; 31:473–478.
108. McGrath PA. An assessmentof children's pain: a review of behavioral, physiological and direct scaling techniques. Pain 1987; 31:147–176.
109. Polgar G, Romadhat V. Pulmonary Functioning Testing in Children: Techniques and Standards. Philadelphia: W.B. Saunders, 1971.
110. Cox DJ, Gonder-Frederick LA, Carter W, Bennett-Johnson S, Rosenbloom A, Bradley C, Moses J. Symptoms and blood glucose levels in diabetics. JAMA 1985; 253:1558–1562.
111. Takishima T, Hilda W, Sasaki H, Suzuki S, Sasaki T. Direct-writing recorder of the dose-response curves of the airway to mehacholine. Chest 1981; 80:600–606.
112. Lebecque P, Desmond K, Swartebroeckx Y, Dubois P, Lulling J, Coates A. Measurement of respiratory system resistance by forced oscillation in normal children: a comparison with spirometric values. Pediatr Pulmonol 1991; 10: 117–122.
113. Mansell A, Levison H, Kruger K, Tripp TL. Measurement of respiratory resistance in children by forced oscillation. Am Rev Respir Dis 1972; 106:710–714.
114. Sly PD, Landau LL, Weymouth R. Home recording of peak flow rates and perception of asthma. AJDC 1985; 139:479–482.
115. Fritz GK, Klein RB, Overholser JC. Accuracy of symptom perception in childhood asthma. J Dev Behav Pediatr 1990; 11:69–72.

116. Fritz GK, McQuaid EL, Spirito A. Symptom perception in pediatric asthma: Relationship of functional morbidity and psychological factors. J Acad Adolescent Psychiatry 1996; 35:1033–1041.
117. McQuaid EL, Fritz GK, Yeung A, Mansell A Resistive-load detection in healthy school-aged children. Pediatr Pulmonology 1996; 22:357–363.
118. Zechman F, Burki N. Tracking procedure to assess load detection threshold. Chest 1976; 70(suppl):165–168.
119. Birkhead G, Attaway NJ, Strunk RC, Townsend MC, Teutsch, S. Investigation of a cluster of deaths of adolescents with asthma: evidence implicating inadequate treatment and poor patient adherence with medications. J Allergy Clin Immunol 1989; 84:484–491.
120. Weissman MM. The epidemiology of panic disorder and agoraphobia. In: Hales RE, Francis AJ, eds. Review of Psychiatry. Washington, DC: American Psychiatric Press, 1988:7:54–66.

9

Sensory Psychophysics

JOHN C. BAIRD

Dartmouth College
Hanover, New Hampshire

The human senses register environmental events and transfer relevant information to the brain, thus giving rise to sensory experience. These events are physical stimuli whose characteristics vary along quantifiable dimensions, such as chemical concentration (odor), luminance (vision), amplitude (hearing), and degree of respiratory load (breathing). The field consists of two major branches: "outer psychophysics" concerns the relationship between the external world of stimuli and the psychological world of sensory awareness; "inner psychophysics" concerns the relationship between physiological activity and sensory awareness. For the most part, laboratory investigations fall within the realm of outer psychophysics, whereas theoretical statements about the underlying causes of a subject's responses are often formulated in terms of inner psychophysics.

In the laboratory, a subject or patient estimates (or discriminates) the magnitude of physical stimuli and the experimenter uses quantitative tools to understand the relationship between perceived and stimulus intensity. Psychophysical laws are the quantitative relationships observed between dimensional variations of stimuli and corresponding changes in sensory experience, as reflected in the responses (1–3).

In recent years, the field has evolved along two paths. One approach treats phenomena and empirical laws as strictly due to sensory mechanisms (3). Contextual effects, such as stimulus range and subjects' preferences for certain numbers, are considered impediments whose influence must be neutralized in order to isolate the "true" sensory scale. Such approaches assign no special role to central, cognitive processes.

The alternative approach treats psychophysics in terms of cognitive variables, such as the effect of instructions and the impact of response biases (4,5). Sensory factors play a minor role in these formulations. The physiological concomitants of psychophysics are said to be centrally located in the brain, not in the peripheral sense organs themselves. In this chapter I emphasize the sensory side of outer psychophysics, but include some discussion of cognitive factors.

Major subdivisions of outer psychophysics distinguish between "local" and "global" processes (6,7). Global psychophysics refers to the methods and models involving functional relationships over a substantial range of stimulus intensities. Local psychophysics refers to those methods and models involving the perception of small stimulus differences.

The application of psychophysical approaches to respiratory variables is comparatively recent (8,9). Much of the research has a clinical motivation, and hence little research has been published that tests or extends the reach of psychophysical theory. Nonetheless, this work suggests strongly that respiratory psychophysics can be studied with essentially the same methods used for other sensory experiences, such as visual and auditory perception.

Experimenters in the field of respiratory psychophysics exhibit great ingenuity in fashioning stimulus conditions for normal "subjects" that induce sensory experiences to represent the experiences of patients with various breathing disorders, such as asthma and cardiopulmonary and lung disease. Resistive loads are imposed by having individuals breathe air through a connected series of porous materials to mimic the difficulties experienced by patients with obstructive lung disease. Elastic loads are imposed by having individuals breathe from containers of different sizes joined in parallel. Breathing through this circuit alters the pressure-volume relationships and mimics the difficulties experienced by patients with restrictive lung disease (8).

In using "local" methods, subjects decide whether a load is present or absent (detection tasks), or whether one load is more or less than another (discrimination tasks). In "global" methods a subject expresses the perceived magnitude (scaling tasks) of the sensation elicited by each of a series of different loads. The physical stimulus is measured in terms of the actual load imposed or as the peak inspiratory pressure, which indicates the tension induced in the respiratory muscles by the load (more details are given in Ref. 8).

I. Stevens Functions

Two assumptions underlie all psychophysical tasks. First, the human being is a measuring instrument whose operating characteristics are the same for all sensory inputs but whose sensitivity depends on the physical attribute. Second, subjects' overt judgments are the consequence of this measurement and can be ordered along psychological "scales." The four major scales are ratio, interval, ordinal, and nominal (10), but recent measurement theory claims that these types do not exhaust all the possibilities (11). One empirical aim of the field is to determine the scale type appropriate for representing subject judgments of stimulus intensity.

Sensory instruments usually process intensity information in a nonlinear manner. This is inferred from the nonlinear relation between stimulus intensity and the magnitude of the corresponding response. The most robust of empirical laws describing such relationships is the power function (3):

$$R = \lambda S^{\gamma} \tag{1}$$

where R is perceived magnitude, S is stimulus intensity, and λ and γ are free parameters. The logarithmic form of Eq. (1) yields a straight line in double logarithmic coordinates and is used to fit empirical data.

$$\log R = \gamma \log S + \log \lambda \tag{2}$$

Differences in a subject's sensitivity to values along different physical dimensions are indicated by the exponent (slope). An exponent of less than 1 means that a sense modality has a compressed response range (perceptual experience) relative to the dynamic range of physical intensity. An exponent of greater than 1 means that a modality has an expanded response range relative to the stimulus range, whereas an exponent of 1 means that the response ratios accurately mirror physical ratios.

A. Scaling

Methods

Most studies undertaken to test the power law instruct the subject to use numbers to reflect the ratios perceived among a set of 7–10 stimuli, spaced logarithmically and presented in a random sequence. In magnitude estimation, one stimulus may be designated as a standard and assigned a numerical value (modulus) by the experimenter, whereas in "free magnitude estimation" the subject chooses his or her own numbers, without a standard (2). Table 1 gives a representative list of exponents for a variety of stimulus attributes, including several for respiration. These values span a wide spectrum of laboratory findings.

Table 1 Representative Exponents of the Power Law

Attribute	Exponent	Stimulus condition
Brightness	0.33	5^0 target in dark
Smell	0.6	Heptane
Loudness	0.67	Sound pressure of 3000-Hz tone
Visual area	0.7	Projected square
Discomfort, warm	0.7	Whole body irradiation
Tactual hardness	0.8	Squeezing rubber
Respiration	0.8	Resistive loads (young adults)
Respiration	0.95	Elastic loads (young adults)
Vibration	0.95	Amplitude of 60 Hz on finger
Visual length	1	Projected line
Cold	1	Metal contact on arm
Respiration	1.05	Breath-hold discomfort
Pressure on palm	1.1	Static force on skin
Vocal effort	1.1	Vocal sound pressure
Duration	1.1	White noise stimuli
Respiration	1.2	Inspired lung volume
Lightness	1.2	Reflectance of gray papers
Taste	1.3	Sucrose
Warmth	1.3	Irradiation of skin, small area
Finger span	1.3	Thickness of blocks
Angular accel.	1.4	5-sec rotation
Heaviness	1.45	Lifted weights
Tactual roughness	1.5	Rubbing emery cloths
Muscle force	1.7	Static contractions
Electric shock	3.5	Current through fingers

Source: Refs. 2,3,8.

In magnitude production the experimenter states a number and the subject matches it by adjusting the intensity of a physical stimulus. In cross-modality matching, physical stimuli along one dimension (e.g., visual luminance) are matched to those along another (e.g., sound amplitude). In all of these methods the average estimates or settings collected from a group of subjects are the response values (R) in Eqs. (1) and (2). On a double logarithmic plot the points are summarized by a straight line (Eq. 2) with a slope of γ and a y-intercept of log λ. Figure 1 shows examples of data from cross-modality matching (12). The points represent the median force of handgrip (dynamometer) when subjects matched five to seven criterion intensities for the attributes noted. The positions of the functions are shifted along the abscissa to

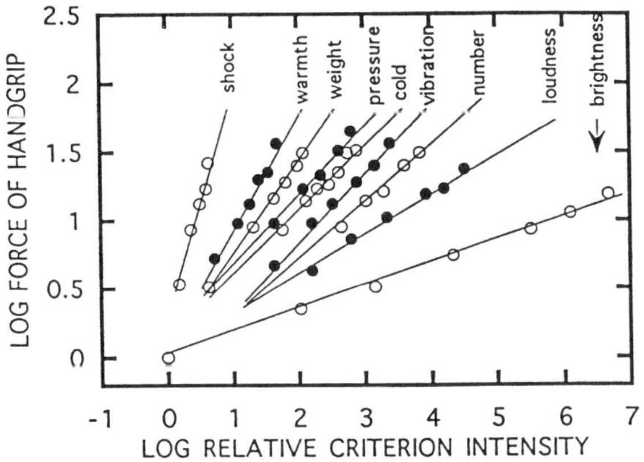

Figure 1 Median force of handgrip exerted on a dynanometer to match criterion intensities for various attributes (log-log coordinates). The positions of the Stevens functions are shifted along the abscissa. (Adapted from Ref. 12.)

aid in viewing. The slopes (exponents) range from those much greater than 1 (electric shock) to those much less than 1 (brightness). The lower exponents are typically associated with larger stimulus ranges.

Variations on the Power Law

Although the fits of the power law expressed by Eq. (1) are often acceptable, nonlinearities in the log-log plot sometimes occur for the weakest stimulus intensities. This nonlinearity appears as a downward concavity (2,13).

In order to secure a better fit, Stevens's law [Eq. (1)] is modified by subtracting a constant from each stimulus. That is:

$$R = \lambda(S - S_0)^{\gamma} \tag{3}$$

Inclusion of the constant (S_0) straightens out the curvature in the log-log plot, though its physiological meaning is uncertain. (It probably does not reflect the lower absolute threshold.)

B. Respiratory Research and the Power Law

Bakers and Tenney (14) were the first to apply global scaling techniques to respiratory sensations. Subjects used either an "estimation" or a "production" technique. In magnitude estimation, the subject produced a gradually increasing or gradually decreasing stimulus, and at various points the experimenter halted the change and the subject estimated the magnitude (in comparison with a standard). In magnitude production the subject produced stimuli to match numbers stated by the experimenter. The group exponents [Eq. (1)] were all greater than 1.0: pressure, 1.5; volume, 1.3; and ventilation, 1.9.

Exponents greater than 1.0 were also reported in later work by Stubbing et al. (15), who employed a magnitude production technique and determined exponents for the perceived magnitude of four respiratory variables: tidal volume, 1.14; inspiratory flow, 1.13; ventilation, 1.28; and respiratory frequency, 1.04.

Over the past two decades numerous clinical applications of such scaling methods have generated a reliable corpus of data indicating that the exponent of the power function for patients with chronic obstructive pulmonary disease (COPD) is reduced over what it is in normals. For example, Gottfried et al. (16) report that the exponent for normals and asthmatics is 0.93 for magnitude estimation of airflow resistance, whereas patients with COPD have lower exponents (0.64) when tested under the same conditions. The impairment of COPD patients is not due to a general inability to scale intensities. This is indicated by the fact that their estimates of lifted weights are similar to those given by normals. All three groups showed similar results when tested on this latter task.

Exponents have been compared for healthy young and old adults and patients with COPD. The key variable in this research appears to be the age of the subject, rather than the nature of their illness. Older (> 60 years) healthy subjects generate smaller exponents than those for younger subjects (< 31 years) for both resistive and elastic loads (17,18). When combined with data for patients with COPD (also showing a drop in the exponent), the possibility presents itself that the age of the subject is the critical variable determining the decline in the exponent. Ward and Stubbing (19) report that the exponents for perceived magnitude of resistive and elastic loads are higher in young adults than they are in patients, but that the latter do not differ in their performance from age-matched controls. The same conclusion is reached by Mahler and colleagues, who investigated the scaling performance of patients with either airway disease or interstitial lung disease (20,21).

C. Context and Method Effects

Stimulus Range

It is generally acknowledged that stimulus range affects the exponent of the power law: short stimulus ranges are associated with high exponents (4,22). This is true both for within-attribute and between-attribute variations. According to Poulton (4), the exponent is determined entirely by stimulus range, a conclusion at odds with the basic tenets of sensory psychophysics. If Poulton were correct, we would expect the between-attribute effects to be identical to the within-attribute effects. This is not the case. Varying the stimulus range of a single attribute does not change the exponent nearly as much as does a change in attribute type. Stimulus range may well be correlated with the exponent, but this stimulus variable probably does not determine exponent size (23,24).

Nevertheless, it may be worthwhile to reevaluate the data on respiratory psychophysics within the theoretical framework that treats the exponent as a function of stimulus range. The possible range of stimulus values employed in the studies of respiration is comparatively narrow (usually less than 1 log unit, base 10). Therefore, the possibility exists that the exponent observed in these studies depends, at least in part, on stimulus range. This hypothesis should be tested by determining the exponent of the power function with stimulus ranges that vary both within and across stimulus attributes (e.g., pressure, volume, and ventilation).

Number Preferences

The working assumption of global scaling is that subjects use numbers as if they were sampling from a continuous line. There are supposed to be no response biases, so the stimulus intensities are the sole determinants of the numbers given. One implication of this assumption is that numerical responses should be randomly distributed when given in the absence of stimulus presentation.

Contrary to this view, subjects give certain response values significantly more often than others when estimating physical stimuli (25) or when guessing numbers in the absence of a stimulus (26–28). If someone is requested to list 10 numbers from 1 to 100, the values tend to be integers reflecting a strong round-number bias. The choices are usually multiples of 5 or 10. This tendency occurs whenever subjects give a number in response to just about any query pertaining to quantity, such as: "How beautiful is this picture of my horse on a scale of 1 to 100?" or "How long does it take to drive from your office to your home?" Most individuals do not reply by saying things like "1.37 units of beauty" or "17 minutes." The answer is more likely to be an integer such as 1, 2, 5, 10, 20, 50, or 100.

Because subjects favor certain response values over others, it is possible to incorporate these biases into the output stage of psychophysical models. In fact, the power function and the exponent can be explained as partly due to number biases (1,29), but as with the case of range effects, it is not known whether such biases play a causative role or are merely a curious byproduct of the estimation methods.

Many of the procedures in respiratory psychophysics require a subject to "produce" a magnitude rather than give a numerical estimate. Therefore, one might not expect numerical preferences to interfere with the interpretation of results. Unfortunately, it may also be true that subjects have an inclination to produce certain respiratory states. One way to investigate this possibility would be to have subjects produce a large number of random values along a respiratory continuum and see whether values cluster together in particular regions along the continuum. In other words, subjects may show "preferences" for certain stimulus values, and these preferences would have to be accounted for in the interpretation of the power law and its exponent.

Category Methods

In category estimation, the subject's response options are the integers from 1 to *N*, where *N* usually does not exceed 10. The goal is to partition the stimulus range into equal-spaced response bins, each identified by an integer. Because of the limited number of responses, the total response range is truncated at the bottom and top of the scale. This constraint can be problematic because truncation builds in a dependency between the mean and standard deviation of the response distribution for each stimulus (1,30). That is to say, as the variance of a response distribution for a fixed stimulus increases, the mean of the distribution moves toward the center of the full-response scale. Thus, an experimenter might assume that one group perceives two stimuli as more alike than does another group, whereas the result is actually due to differences in response variability between the two groups. This constraint weakens as the number of categories is increased beyond 10. In fact, the functional relation between response and stimulus magnitude obtained by category and magnitude estimation methods is essentially the same when the number of categories is very large (31).

Borg (32) developed a hybrid category and magnitude scale for assessing the perceived effort expended by a patient when exercising. The goal of his scale is to provide a linear relation between a patient's expressed judgment of physical exertion and physiological heart rate. The argument is that the clinician can then interpret the patient's verbal response on this scale as a direct indicator of heart rate. For instance, patients may give subjective ratings of "how tired they are" as part of a test to evaluate fatigue from walking on a treadmill. So far, validation of a similar rating scale for the case of breathlessness has not appeared in the literature. One of the difficulties here is in finding a physiological response for calibrating the subjective scale—that is, what is the equivalent of heart rate for the case of breathlessness? (8).

On the other hand, the Borg scale has been used to secure reliable estimates and productions of breathlessness (dyspnea), a finding of substantial clinical significance (33). For example, Horowitz et al. (34) found that patients with cardiopulmonary disease could produce designated dyspnea ratings (from the Borg scale) through exercising on a cycle ergometer and that the ratings were reliably obtained over a 5–week time period. For many patients, subjective evaluations of dyspnea provides a more efficient and convenient way to monitor amount of physical exertion, in contrast to self-measures of heart rate, where it is necessary to halt ongoing exercise, find the pulse, etc.

Comparing Magnitude and Category Estimates

Early research indicated that the relation between category estimates and physical stimuli is logarithmic (35,36), while other work suggested that a power function is a better choice (37,38). If a power function is fit to each set of data, the category exponent is approximately half the size of the magnitude exponent (3,38). There seems to be a consensus nowadays that a power function fits the data from category estimation better than a logarithmic function.

Because the relation between magnitude estimates and physical intensity is a power function, the relation between category and magnitude estimates of the same stimulus series is also a power function. Category estimates are compressed relative to magnitude estimates (slope in a log-log plot of approximately 0.5).

Implicit Anchors

Robinson (39) conducted a study in which the experimenter called the standard "100" and assigned different examples of possible response magnitudes to the comparisons. One group was told to say "50" when a stimulus was 50% of the standard and "150" when it was 1.5 times the standard, whereas another group was told to say "25" when the stimulus was 25% of the standard and "750" when it was 7.5 times the standard. The exponent of Stevens's power law [Eq. (1)] was significantly less when the response examples spanned a narrow range as compared to when the examples spanned a wide range. In other words, exactly the same stimuli were assigned different responses, depending on the preexperimental examples.

Explicit Anchors

We once did a study to evaluate the importance of explicit standards on scaling behavior (40). Seven groups (and a control) were run with an identical series of visual line lengths. The experimental groups received different response anchors for the largest stimulus, while the smallest stimulus was always called "1." Subjects were told that the experimenter had a relationship in mind between the lines and the response scale. Power functions provided adequate fits to the mean estimates for each group (except for the two steepest functions), and the higher the explicit anchor, the higher the exponent. The conclusion is that different psychophysical relations are generated for exactly the same set of stimulus intensities, just by changing the value assigned to the maximum stimulus. It is not surprising that the functions change with the maximum response. The real significance is that a power function was appropriately "filled in" by the subjects and that the exponent was what one would predict based on the assigned maximum.

The implications of these findings for respiratory psychophysics are exactly the same as those for the psychophysics of other sensory modalities. One needs to exercise caution in the instructions to subjects, especially in regard to the sample values given with the aim of illustrating the method. Subjects seem to think these examples are supposed to be used in the experiment. Obviously, such a belief can distort the results. It is for this reason that many experimenters prefer to use a "free magnitude estimation" technique, in which the subject is encouraged to use any response option he or she chooses in representing the perceived magnitude of stimuli.

II. Ekman Functions

Ekman and Künnapas (41) proposed that the relation between the standard deviation (sd) and the mean (R) from global scaling methods was systematic in its course and depended on the attribute. They offered this proposal as a mathematical hypothesis, open to experimental test (42). Namely:

$$\mathrm{sd} = \kappa R^{\delta} + \xi \tag{4}$$

where the κ, δ, and ξ are free parameters. They then defined special cases holding for different attributes. For example, setting $\delta = 1$, we have a linear relation between response variability and the mean:

$$\mathrm{sd} = \kappa R + \xi \tag{5}$$

When $\xi = 0$, we have a relation of the form:

$$\mathrm{sd} = \kappa R \tag{6}$$

Equation (6) is referred to as Ekman's law (23,24), and the more general forms [Eqs. (4) and (5)] are called Ekman functions (1). The empirical data described by Ekman functions do not always follow Ekman's law.

A. Empirical Findings

Montgomery (36) published a thorough investigation of the variability arising in direct scaling methods. He found that the standard deviation was a nonlinear function of the mean for a wide assortment of experimental procedures

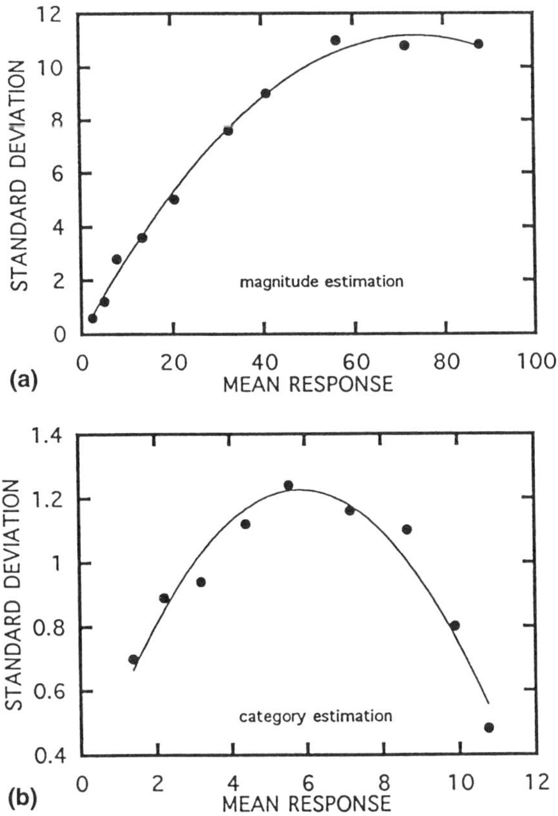

Figure 2 Standard deviation as a function of mean response (Ekman function) for (a) magnitude estimation and (b) category estimation. (Adapted from Ref. 36.)

(e.g., instructions to judge stimulus differences or ratios, stimulus range, and the number of response categories).

Figure 2 presents two Ekman functions (magnitude and category estimation) from Montgomery's research. For each data set, Eq. (4) provides a satisfactory description, where the weights of the parameters depend on the experimental conditions. The decreased variability at the ends of the stimulus scale for category estimation is caused by the truncation at the smallest and largest integer categories.

B. Relative Error

Another common measure of response variability is the relative error (standard deviation divided by the mean), which is often plotted as a function of the

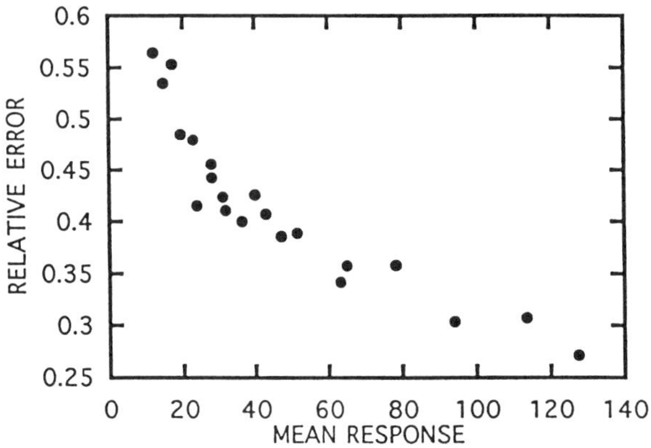

Figure 3 Relative error (standard deviation divided by the mean) as a function of mean response. Based on loudness estimates of a 1000-Hz tone. [Group averages (raw data provided by the authors) from Ref. 44.]

mean (43,44). Figure 3 presents an example of such a function for the mean of five subjects estimating the loudness of a 1000-Hz tone (44). The relative error declines with the mean, which is what must happen if the standard deviation is a negatively accelerated function of the mean (Fig. 2a).

The standard deviation of responses is seldom reported in studies of respiratory psychophysics, despite its potential value in helping the experimenter sort out the underlying physiological mechanisms responsible for the results. In particular, if the subject is not imposing limits on his or her response options, one expects to see the standard deviation increasing monotonically with the mean. If limits are introduced either by the subject or by the experimenter (perhaps inadvertently), response variability may be nonmonotonic with the mean. This is almost always the case with category estimation because of the truncation built into the ends of the response scale. It would be informative, for example, to determine whether the Borg scale (noted above) yields variability measures that increase monotonically with the mean response (characteristics of a ratio scale) or ones that increase nonmonotonically (characteristics of a category scale).

What is known is that the variability of Stevens exponents is substantial in both normal and patient populations. In a study by Mahler et al. (20) the range of exponents for added resistances in normals was approximately 0.3–0.9, and for patients with obstructive airway disease the range was approximately 0.25–1.2. Similar large interindividual variation in exponents are reported by the same authors for elastic loads.

III. Piéron Functions

Simple reaction time (response time) is how long it takes a person to detect a suprathreshold stimulus. If physical intensity is varied, a functional relation is described between reaction time and intensity (45).

Numerous studies show that reaction time to detect a stimulus decreases as physical intensity increases (for discrete stimuli). In the early part of this century, Piéron (46,47) suggested that the relation between reaction time and intensity is:

$$MRT = r_0 + \kappa S^{-\beta} \tag{7}$$

where *MRT* is mean reaction time, and r_0, κ, and β are free parameters. Equation (7) is referred to as Piéron's law (45), and any relation between mean reaction time and stimulus intensity is a Piéron function (1).

Reaction times are often transformed to simplify the estimation of parameters. Piéron's law can be rewritten by taking logarithms of both sides of Eq. (7):

$$\log (MRT - r_0) = \log \kappa - \beta \log S \tag{8}$$

Equation (8) has the same form as the logarithmic version of Stevens's law [Eq. (2)], except that the slope (β) of the reaction-time function is negative, and there is an additional constant, r_0. Figure 4 presents an example of reaction-time data for light intensity (48). Data are presented in both log-linear (a) and log-log (b) coordinates. The slope (2) of the function is – 0.33, which is the negative of the slope of Stevens's law for brightness of a 5^0 target in the dark (Table 1). Whereas the absolute value of the exponents in these two laws are not always identical, they each signify alternative sides of a common, underlying sensory code (1).

In my search of the literature, I uncovered no studies in which respiratory effort was the independent variable and reaction time the dependent variable. It would be informative to determine whether Piéron's law held for the application of respiratory loads, because if it did, we would be inclined to conclude that the physiological principles underlying respiration are comparable at the level of information processing to principles that apply to other sensory modalities.

IV. Absolute Identification

One task of global psychophysics is to model the ability of sensory systems to faithfully transmit information from the surrounding world. The experi-

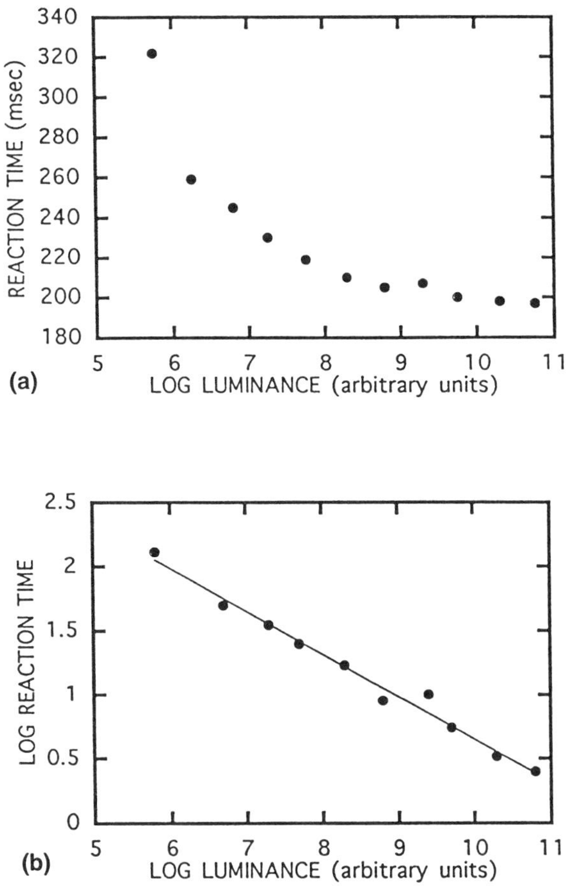

Figure 4 (a) Mean reaction time as a function of the logarithm of stimulus luminance and (b) the logarithm of mean reaction time as a function of the logarithm of stimulus luminance. (Adapted from Ref. 48.)

mental technique to determine the amount of information transmitted is "absolute identification" (2). In this procedure the subject tries to assign a unique response label to each member of a stimulus set. In most experiments, the labels are the integers 1 to N, where N is the total number of stimuli. The physical intensities are logarithmically spaced in equal steps. After each response is given, the subject is told the correct answer for that trial; that is, feedback is provided.

A. Channel Capacity

As the number of stimuli increase, identification performance keeps pace initially, but then levels off with larger stimulus sets. The subject begins to make errors of identification. According to information theory, a channel capacity is reached (2,49,50). Further increases in the number of stimuli beyond this point does not lead to any further improvement. The channel capacities typically found in the laboratory fall between 1.5 and 4 bits, depending on the attribute. A "bit" is the power to which 2 is raised to equal the number of stimulus alternatives. Generally speaking, attributes with low channel capacities have low Stevens exponents (1,51), though the connection is a weak one.

The experimenter computes information measures from the confusion matrix, whose entries are the number of times each response alternative is assigned to each stimulus (2). One such procedure expresses information transmitted (I_t) as a function of the average uncertainty of the response set [U(R)] and the average uncertainty of the response set conditional upon the stimulus set, $U(R|S)$:

$$I_t = U(R) - U(R|S) \tag{9}$$

The terms on the right side are nonmetric measures of response variance. Response uncertainty [$U(R)$] is the logarithm of the probability (relative frequency) of giving each possible response, summed over all the responses in the set. That is:

$$U(R) = \sum_{M}^{j=1} p(r_j) \log_2 p(r_j) \tag{10}$$

The second term [$U(R|S)$] represents the conditional response uncertainty. This is the average variability in assigning responses contingent on the stimulus:

$$U(R|S) = \sum_{N}^{j=1} p(s_j) \sum_{M}^{j=1} p(r_j \mid s_i) \log_2 p(r_j \mid s_i) \tag{11}$$

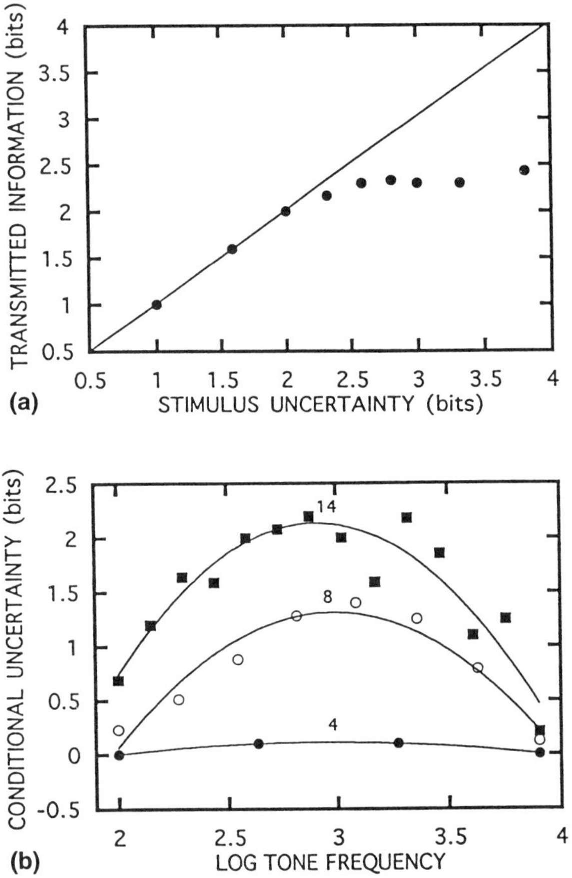

Figure 5 (a) Information transmitted as a function of stimulus uncertainty for auditory tones. The solid diagonal indicates perfect transmission. (b) Conditional response uncertainty as a function of the logarithm of tonal frequency (position in the stimulus series). Each curve represents data for a different stimulus set size. (Adapted from Ref. 52.)

This is an index of the subject's errors in assigning response labels.

Figure 5 shows the outcome of an early experiment by Pollack (52) in which the stimuli were tones, logarithmically spaced from 100 to 8000 Hz. In Figure 5a, the information transmitted is plotted as a function of stimulus uncertainty. Perfect matches between the stimulus uncertainty and the transmitted information are represented by the diagonal. The transmitted information begins to depart from the theoretical values around 1.5 bits of stimulus uncertainty. Beyond this point the transmitted information asymptotes at a maximum of approximately 2 bits. No matter how many stimuli are presented for judgment, the subject is only able to absolutely identify $2^2 = 4$ tones. It should be realized that the same subject is perfectly capable of distinguishing between any two such tones when they are presented as an isolated pair.

Figure 5b shows the conditional response uncertainty as a function of the logarithm of tone frequency. Set size is a stimulus parameter. The conditional variability is a function both of the position of a tone in the frequency series and of the number of tones in the total set. The bow-shaped functions are probably due to the fact that the response distributions are truncated at the bottom and top ends of the scale. A subject has fewer response options in the vicinity of the smallest and largest stimuli, because the end categories define the boundaries of the response scale.

Memory Effects

One possible reason the channel capacity is so low for unidimensional attributes is that the memory representation of each stimulus deteriorates over time. The subject in such tasks must recall previous occurrences of the stimulus along with the correct label (feedback given by the experimenter). Subjects may do poorly in this task because they cannot retrieve a degraded memory representation. In addition, stimuli occurring since the one to be recalled may interfere with memory retrieval. Siegel (53) offered an early test of these two hypotheses.

His stimuli were seven tones ranging from 500 to 560 Hz, and the parameter was the time between each presentation: 0, 0.5, 1, 3, or 6 sec. Both the retention interval and the number of intervening items influenced the experimental outcome. The percent correct decreased as retention interval increased, and performance also declined as the number of intervening items (0, 1, or 2) increased. In a later theoretical paper, Siegel and Siegel (54) explain their results by suggesting that attributes such as pure tones and lights are "ultra meaningless," and therefore difficult to recall. This type of stimulus is difficult to "chunk" into items that can be recovered as single entities. If stimulus chunking were possible, then when asked to do so, a subject could "unpack" a chunk to recover each of its constituents (50).

Range Effects

If the number of stimuli is held constant and the stimulus range is varied, information transmission at first increases, but eventually levels off at the channel capacity (55,56). This means that the total sensitivity to stimulus differences (information transmitted) is less for narrow than for broad stimulus ranges. This particular trend may be the result of the assimilation of the present response to those given earlier in the series (1). Such assimilation is more potent when the stimuli are closely bunched in intensity than when they are widely separated along the intensity dimension.

B. Clinical Implications

There apparently are no studies testing an individual's ability to absolutely identify the members of a set of respiratory loads. The clinical value of such data might be to give the physician some indication of the patient's ability to accurately recall and describe different types of respiratory episodes. If the channel capacity is relatively low for respiratory states, then one cannot expect to obtain reliable statements from the patient concerning recalled differences among states that were experienced some hours or days in the past.

V. Sequence Effects

In global scaling methods, responses depend on the stimulus presented, but also on the stimuli and responses from earlier trials in the series. The motivation to understand sequence effects comes from the promise of gaining a deeper understanding of subjects' judgment strategies when using such methods. In this way it may be possible to separate the contribution of sensory and cognitive factors to the eventual judgment.

A. The Triangle Pattern

Subjects in a method such as magnitude estimation are supposed to preserve ratios among stimulus intensities (over trials) that are reflected in response ratios. If subjects follow these instructions, and if the stimulus on the previous trial ($T-1$) is the critical referent, we anticipate a high correlation between responses on the current and previous trials. The size of the correlation should not depend on the magnitude of the difference between stimuli on successive trials. Contrary to this expectation, the empirical data are that the magnitude of sequential dependencies is a function of stimulus separation from one trial to the next.

These investigations are based on data from a small number of subjects, each of whom gives many (at least 100) responses (R) for each stimulus. Because each of N stimuli can occur on trials T and $T-1$, the possible sequences are represented by the $N \times N$ matrix of stimulus pairs. The correlation between log R_T and log R_{T-1} is computed for each cell of the matrix and averaged across stimulus pairs of the same log intensity separation.

Figure 6 presents results from a study by Green et al. (44). The correlation coefficient (log R_T; log R_{T-1}) is plotted against the rank stimulus

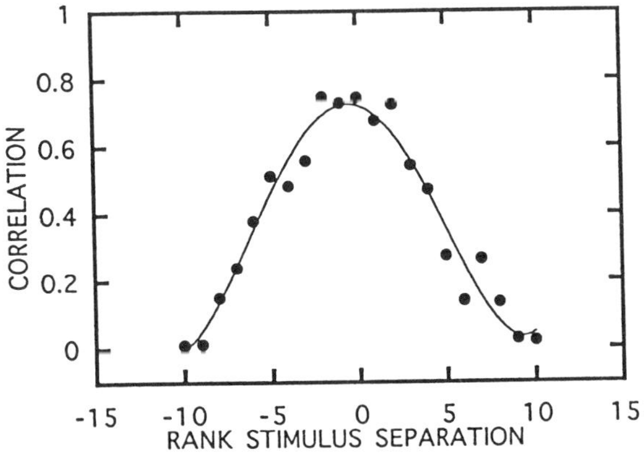

Figure 6 Correlation between responses on successive trials as a function of stimulus rank separation. Based on loudness estimates of a 1000-Hz tone. [Group averages (raw data provided by the authors) from Ref. 44.]

separation ($\log S_T - \log S_{T-1}$). The summary trend is referred to as the triangle pattern or the inverted V pattern (43,57). The correlation is substantial when successive stimuli are identical and approaches zero when successive stimuli are far apart in intensity. That is to say, the previous stimulus is a referent when it is close to the current stimulus intensity, but the utility of this referent declines with increasing stimulus separation.

B. Alternative Judgment Referents

Assimilation effects between the current and previous response are definitely influenced by instructions (58). These authors employed two types of instructions: "free" magnitude estimation, in which subjects were told to make all judgments in respect to a stimulus referent of their own choosing, and a "prior reference" condition, in which subjects were told to estimate the stimulus in respect to the one from the previous trial. Assimilation to the previous response was substantial when subjects used the prior stimulus/response pair as a referent, whereas it was weaker when the subject chose his or her own referent.

This result implies that alternative referents are available in global scaling tasks, suggesting that without explicit instructions, different subjects may employ different referents. This should increase response variability among a group of subjects giving estimates of the same stimuli. It is possible that the considerable variability observed in the respiratory psychophysics laboratory can be traced to the fact that different subjects employ different referents for their judgments. Experimenters must be careful to insure that subjects fully understand the task and that the requirements are unambiguously stated. On the other hand, it seems generally accepted that respiration depends on multidimensional stimulus conditions, and therefore, the opportunity is always there for the subject to rely on alternative stimulus referents (59,60).

VI. Detection Thresholds

The absolute lower threshold is the least intensity detected a certain percent of the time (e.g., 50%). In the early literature, the lower threshold is called the "stimulus threshold." The term implies that one detects the presence of real objects, such as thunder in the distance or an animal camouflaged by underbrush, rather than values along abstract stimulus dimensions, such as the amplitude of a 1000-Hz tone or the luminance of a spot of light. In the psychophysics laboratory, however, stimuli are not unique "objects," but rather values along continuous dimensions.

Campbell and colleagues determined detection thresholds for both resistive and elastic loads. In one study (61) healthy subjects breathed air through a circuit of negligible resistance, and occasionally an external elastance was added. The 50% detection level was 2.47 cm H_2O/L. In a second series of experiments (62) subjects detected the presence of added resistive loads, resulting in a detection threshold of 0.59 cm H_2O/L/sec.

Later experiments report somewhat smaller detection thresholds for elastic loads, but about the same range for resistive loads (9,63).

A. Theory of Signal Detectability

Modern approaches to signal detection often are done under the umbrella of the theory of signal detectability (TSD) (2,64,65). This theory and its associated psychophysical methods differ from its predecessors in claiming that cognitive variables influence the subject's predilection to detect and distinguish among stimuli. A measure of sensory discrimination (d') is obtained which is independent of response bias (β).

A simple example is when a subject attempts to distinguish between a stimulus and background noise by using a unique label for each of two possible events (e.g., "signal" and "noise"). On any given trial, either the signal or the noise is presented, and after the subject's response, the correct answer is given by the experimenter. The data are analyzed by assuming that both the signal and noise give rise to normal distributions of sensory effects with different means but identical standard deviations. The theory states that whether a subject says "signal" or "noise" on any given trial depends not only on the difference between the means of the two underlying distributions of sensory effects, but also on the probability of each of the two events, as well as on the payoffs and penalties the subject may receive as a consequence of giving a response ("signal" or "noise") contingent upon presentation of a stimulus (signal or noise).

By accepting these assumptions, one employs statistical techniques to arrive at independent measures of perceptual sensitivity and response bias. This is the chief advantage of the approach. In particular, the theory states that if μ_s and σ_s are the mean and standard deviation of the sensory distribution arising from repeated presentation of the same stimulus, and if μ_n, and σ_n are comparable parameters of the noise distribution, then d' is defined as

$$d' = \frac{\mu_s - \mu_n}{\sigma_n} \tag{12}$$

The measure of response bias (β) is a function of six variables:

1. The probability of signal, $p(s)$
2. The probability of noise, $p(n)$
3. The positive value for correctly identifying the signal, V_{hit}
4. The positive value for correctly identifying the noise—correct rejection, V_{cr}
5. The cost of saying "signal" to a noise presentation—false alarm, C_{fa}
6. The cost of saying "noise" to a signal presentation—miss, C_{miss}

These variables are combined to predict response bias:

$$\beta = \frac{p(n)}{p(s)} \times \frac{V_{cr} - C_{fa}}{V_{hit} - C_{miss}} \tag{13}$$

According to the theory, the subject employs a single β as a cutoff to decide whether to say "signal" or "noise" on any trial. If the perceived likelihood ratio (based on a single value from an internal evidence axis) of the signal to the noise (secured from the distributions stored in memory for signal and noise events) exceeds the cutoff, the subject says "signal." Otherwise, he or she says "noise." Accepting the tenets of the theory, it is possible to derive measures of perceptual sensitivity and response bias from the resulting stimulus-response matrix.

B. Respiratory Research and Detection

Several experiments in respiratory psychophysics have been conducted within the framework of TSD. Changes in both perceptual sensitivity and response bias have been brought under stimulus control. Narbed et al. (66) showed that strapping the chest at full expiration resulted in detection sensitivity decreases and response bias increases. Bonnel et al. (67) found that asthmatics demonstrated higher sensitivity both for inspiratory and expiratory load detection, when sensitivity was expressed in respect to airway resistance. No differences in response bias were obtained between normals and asthmatics.

More recently, Harver et al. (68) compared male and female subjects in detecting heartbeat, respiratory resistance, and light-tone matches. Men were more sensitive than women on the resistive load task but not on the light-tone task, suggesting that the gender differences in performance were task dependent. All subjects used a more conservative response criterion on the respiratory resistance task.

VII. Weber's Law

A. Discrimination

In discrimination tasks the stimuli are highly similar, and the subject indicates whether two events are the same or different. One purpose of discrimination tasks is to measure the just-noticeable-difference (JND) between stimuli and thereby find a functional relation between the JND and the stimulus intensity at which it is determined. This relationship is typically linear over the middle range of the stimulus continuum (Weber's law), but deviates from this pattern near the lower and upper absolute thresholds. The size of the JND across changes in physical intensity reflects the subject's sensitivity to stimulus differences, and therefore, the relative size of JNDs in respect to the intensity at which it is measured is an index of potential neurophysiological significance.

Perceptual sensitivity is indexed by the minimum stimulus change (ΔS) necessary for a subject to notice a difference. Over a century ago Weber (69) and then Fechner (70) proposed that the JND is a linear function of the stimulus intensity. Weber's law describes this relationship, where ΔS is the JND, S is the standard, and k is the Weber fraction.

$$\Delta S = kS \tag{14}$$

The value of k is inversely related to sensitivity. Table 2 lists typical Weber fractions for a variety of attributes. The values span the range from 0.02 for electric shock to 0.25 for resistive loads and lung volume (2,8,71).

Deviations at Threshold

Weber's law holds fairly well over the middle region of the stimulus continuum, but sensitivity is coarser near the lower, and sometimes near the upper, absolute threshold (72,73). The failure of Weber's law is seen by rearranging

Table 2 Representative Weber Fractions

Attribute	Weber fraction (ΔS/S)
Finger span	0.02
Saturation (red)	0.02
Electrical (skin)	0.03
Position of point (visual)	0.03
Length of lines (visual)	0.04
Area (visual)	0.06
Heaviness	0.07
Brightness	0.08
Loudness (1000 Hz, energy units)	0.10
Loudness (white noise, energy units)	0.10
Taste (salt)	0.14
Elastic loads	0.15
Taste (sweet)	0.17
Skin vibration (100–1100 Hz)	0.20
Smell (several substances)	0.24
Lung volume	0.25
Resistive loads	0.25

Source: Refs. 2,8.

Eq. (14) and then plotting the Weber fraction ($\Delta S/S$) as a function of stimulus intensity (S). Figure 7 gives two examples of such a plot for luminance (74) and sound intensity (75). Weber's law implies that the functions should be horizontal to the x-axis from the lower to the upper limits of the stimulus scale. In fact, the fraction is high near the lower threshold, decreases for the middle intensities, and then, for the case of luminance, shows a slight upward turn near the upper threshold.

Affine (Linear) Generalization

Other equations have been suggested to describe nonlinearities such as those in Figure 7 (2,76,77). One common practice is to add a constant to the simple form of Weber's law:

$$\Delta S = kS + c \tag{15}$$

This version of the Weber function offers a better description for many empirical data sets (e.g., Refs. 2,78). Equation (15) does not describe the upward turn of the Weber fraction sometimes observed at the upper limit of

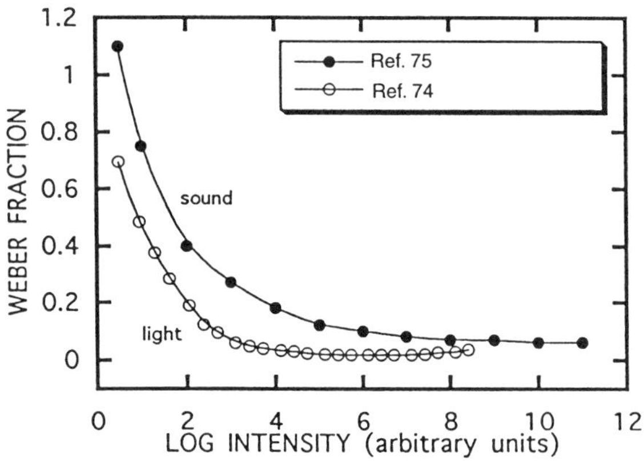

Figure 7 Weber fraction as a function of the logarithm of the standard intensity for sound and light. (From Refs. 74 and 75.)

sensory functioning. The common interpretation of this latter increase is that there are finite limits on the ability of any sense organ to track changes in intensity. A burned-out or damaged sense organ is unresponsive to stimulus change.

The Near Miss

In order to accommodate data that did not agree with Weber's law, Guilford (79) described the JND as a power function of stimulus intensity:

$$\Delta S = kS^{\alpha} \tag{16}$$

He gave empirical examples for several attributes, including visual line length and sound intensity. Because the exponent of Eq. (16) is often slightly less than 1, this formulation is referred to as the "near miss" to Weber's law (80). Jesteadt et al. (81) used this equation to summarize their findings for discrimination of loudness. They employed a two-interval forced-choice method and determined intensity discrimination of pulsed sinusoids for various frequencies and intensities. The exponent of Eq. (16) was 0.93, which agrees quite well with data reported by earlier investigators.

As is true for Ekman's law [Eq. (4)], a general formulation encompasses all the various Weber functions. That is,

$$\Delta S = kS^{\alpha} + c \tag{17}$$

Weber's law results when $\alpha = 1$ and $c = 0$, the affine version when $\alpha = 1$ and $c > 0$, and Guilford's power function when $\alpha > 0$ and $c = 0$.

B. Respiratory Research and Discrimination

A number of experiments have measured discrimination performance for the magnitude of resistive loads (8). The studies agree that respiratory discrimination is poor relative to the other senses (see Table 2), but that the JND is a linear function of the reference level over a range of values, in accord with Weber's law (62,82,83). As the reference level approaches the absolute lower threshold, however, the Weber fraction increases, thus indicating that sensitivity is even poorer in this region of the continuum (84). The latter data could be accommodated either by the affine version of Weber's law [Eq. (15)] or by the near miss formulation [Eq. (16)], where the exponent would be less than 1.

It has also been shown that the Weber fractions for asthma patients do not differ from normals (16,85), and patients with interstititial lung disease do not differ from normals with respect to their Weber fractions (63,86). On the other hand, Weber fractions are greater in patients with COPD than in normal controls (16,87).

C. Counting JNDs

An alternative measure of sensitivity applicable for any Weber function is the number of JNDs accumulated over the dynamic stimulus range of an attribute. The greater the number of JNDs per unit intensity, the greater the perceptual sensitivity. This measure has the advantage that it is not open to recent measurement concerns about using indices such as the Stevens exponent and the Weber fraction to imply differences in the sensitivity of different sense modalities (88). I have not come across any studies in respiratory psychophysics in which the number of JNDs over the dynamic range of a stimulus continuum is the perceptual sensitivity index.

D. Sequence Effects

Just as we saw with global methods, the methods of local psychophysics introduce "errors" into the assessment of discrimination. Two frequent types of errors are sequential in nature. The first is an *error of anticipation* (expectation) and the second is an *error of habituation* (89,90). The error of anticipation occurs with an adjustment method when the subject halts the change in the comparison at a magnitude less than the standard on ascending trials and greater than the standard on descending trials. The error of habituation occurs when the subject halts the change in the comparison when it is greater than the standard on ascending trials and less than the standard on descending trials. There really are no adequate models that elucidate the underlying causes of these errors (see, however, Refs. 1 and 91). They do not receive special attention in the literature on respiratory psychophysics, and it is not clear to what extent these errors actually occur in these experiments.

VIII. Multidimensional Psychophysics

The vast majority of psychophysical studies involve unidimensional stimuli, but some studies (especially with the attributes of odor and taste) are conducted in which stimuli vary along two or more dimensions.

A. Channel Capacity

The information transmitted [Eq. (9)] for multidimensional stimuli exceeds that for the unidimensional case. Pollack and Ficks (92) varied sounds along eight dimensions dichotomously (e.g., frequency, intensity, range, duration) and found that information transmission was seven bits out of a possible eight (2^8 alternatives). With overlearned stimuli (numerals, colors), the information transmitted can be as much as 15 bits (49,93).

The data indicate that the maximum information transmitted for a multidimensional stimulus is less than the sum of the transmitted information for each of the components. However, the transmitted information for a set of multidimensional stimuli depends on whether the components maintain their independence in the compound. The distinction here is between "integral" and "separable" stimuli. Integrality occurs when subjects perceive stimulus combinations as a new compound (94). Separability occurs when subjects still perceive individual components when they are presented as parts of a single compound.

Such issues are very important in olfactory research, where chemicals often do not retain their independence when sniffed in combination (95,96). When certain chemicals are mixed, the perceiver does not separate the unique contributions of the components, and consequently models are necessary to capture the rules by which percepts of individual elements are related to the composite percept aroused by the combination.

If stimulus independence is complete, the subject perceives the intensity of a dimension in the compound without interference from the other dimensions. Therefore, information transmission for the compound is the sum of the measures for each dimension when tested alone.

If dimensions are combined and the subject perceives an entirely new attribute, then information transmission for the compound is close to information transmission for the components tested in isolation.

B. Integration Psychophysics

Composition rules underlie the effects accruing from the presentation of objects comprised of two or more attributes (1,97–99). Suppose a set of stimuli is comprised of all pairs of values from two dimensions. The subject estimates on a linear scale either the total intensity of each pair or the intensity of one attribute in the compound. The resulting judgment is described as a weighted average of the contributions of the two components. In other instances, the strongest member of the pair dominates and the judgment of the compound is identical to the judgment of the strongest member of the pair (100).

C. Minkowski Metrics and Multidimensional Scaling

The Minkowski metric is a frequent model of dissimilarity ratings among stimuli that vary along several dimensions (2). The dissimilarities between pairs of values taken from separable dimensions are represented by the "city-block metric," whereas dissimilarities between pairs of values taken from integral dimensions are represented by the "Euclidean metric" (101,102). The general Minkowski metric is defined as

$$d_{ab} = \left(\sum_{N}^{k=1} | x_{ak} - x_{bk} |^{r} \right)^{1/r} \tag{18}$$

The perceived dissimilarity (d_{ab}) between two objects are the distances (x) along each of k dimensions. The power (r) specifies the metric that defines the coordinate location of an object. When $r = 1$, the city-block metric defines the interpoint distances; when $r = 2$, the Euclidean metric defines the interpoint distances. As r approaches infinity, Eq. (18) becomes the "supremum metric," where the distance between two objects is the largest difference along any one of their shared dimensions. These three cases are summarized as follows:

City-block $$d_{ab} = \sum_{N}^{k=1} | x_{ak} x_{bk} | \tag{19a}$$

Euclidean $$d_{ab} = \left(\sum_{N}^{k=1} | x_{ak} x_{bk} |^{2} \right)^{1/2} \tag{19b}$$

Supremum $$d_{ab} = \max_{N}^{k=1} | x_{ak} x_{bk} | \tag{19c}$$

D. Respiratory Research and Multidimensional Scaling

Studies have yet to be done on the use of multidimensional scaling techniques to determine the rules by which perception of individual respiratory variables combine to form a unified percept. Work that comes closest to this type of research was conducted by Killian and colleagues (103,104). They altered the breathing patterns of subjects and had them estimate flow-resistive and elastic loads added to breathing. Fixing tidal volume during the addition of elastic loads and flow rates caused an increase in the exponent of Stevens's power law (104), and the perceived magnitude of both resistive and elastic loads increased as inspiratory duration increased (103). In order to secure a complete picture of the interaction of these variables it would be necessary to employ a factorial design in which stimulus variables were tested in isolation, as well as in all possible pairs of magnitudes. This would permit the application of the Minkowski metrics in the manner outlined above.

A more recent study by Harver et al. (105) enlisted multidimensional scaling to determine the dimensionality and spacing of sensory experiences elicited by different breathing maneuvers. Dissimilarity judgments of the differences between maneuvers were described as distances in a one-dimensional Euclidean space. The single dimension appeared to be the degree of external loading, ranging from an infinite respiratory load to none at all. A second dimension, of less importance, seemed to implicate airflow direction (inspiration or expiration).

E. Vector Models

Another means to characterize the perception of a compound stimulus is by "vector models." The theoretical rationale here is that the perceived magnitude of a compound is captured by the vector addition of each of its components. For example, if the two component dimensions are a and b, then the perceived magnitude of the compound ($R_{a,b)}$ is described as the vector addition of the two individual percepts ($R_{a,}$ $R_{b)}$, then:

$$R_{ab} = (R_a^2 + R_b^2 + 2R_aR_b \cos \theta)^{1/2} \tag{20}$$

This model is very close in spirit to the Minkowski. For example, when $\theta = 90^\circ$ and therefore $\cos \theta = 0$, Eq. (20) becomes the Euclidean metric [Eq. (19b)]. When $\theta = 0^\circ$ and therefore $\cos \theta = 1$, Eq. (20) becomes the city-block metric [Eq. (19a)].

In sum, results for stimulus mixtures are often considered to depend on two processes: (1) a process in which each component is independent of the context (separable), and (2) another in which output depends on the interaction between the components. The relative influence of each of the two portions can vary from 0 to 100%, suggesting that compound stimuli should not be dichotomized into categories such as "separable" and "integral." Most compounds possess some features of both (106,107).

As noted above, these types of models have yet to be exploited by the field of respiratory psychophysics.

F. Cluster Analysis

A variety of clustering algorithms exist for grouping objects according to their similarity or dissimilarity (2,108). These algorithms suggest how subjects organize stimuli into meaningful categories and, as such, help us understand how subjects perceive relationships among stimuli.

A number of criteria are used to group items, but generally speaking there are two techniques: hierarchical and nonhierarchical. Most psychologists use the hierarchical methods. These start with a proximity matrix of perceived similarities or dissimilarities among the members of a stimulus set. Rules are specified for successive clustering of items. For example, in "agglomerative" techniques the two items that are closest together are grouped first, followed by the next closest items, etc. In the course of this computation procedure, the distance between an existing cluster and an item, or between two clusters, is determined according to a standard rule. For example, the distance between two clusters is taken as the closest distance between two items (one from each cluster), or as the farthest distance between two items. A tree structure, called a dendrogram, represents the clustering order, where the tips of the branches are stimuli and the nodes indicate the intersection of branches (2).

G. Respiratory Research and Cluster Analysis

Harver et al. (105) used hierarchical cluster analysis to reveal the perceived groupings among different types of respiratory experiences induced by different parameters, such as pressure, airflow, and lung volume displacement. The most similar elements were those associated with high and low inspiratory elastances. Subsequent clusters suggested the existence of a broad category involving kinds of external loading. Unimpeded inspiration was the maneuver most discrepant from the others.

At about the same time, Simon et al. (59) investigated the similarity of sensory experience elicited by such maneuvers as breath-holding, CO_2 inhalation, and breathing with a resistive load. Subjects selected descriptors of their sensations from a list of 19 options. The number of times each pair of descriptors was mentioned together was the measure of interitem similarity. Normal volunteers made clear distinctions among the maneuvers, as revealed by the clusters derived from the similarity matrix. The authors concluded that "breathlessness" involves multiple types of different sensations and, therefore, is unlikely to be explained by a single physiological mechanism.

This work was followed up by the authors in a study involving patients who had various medical disorders, including pulmonary vascular disease, congestive heart failure, asthma, and COPD (60). Based on their choice of descriptors for different maneuvers, patients reveal an ability to identify different sensations of breathlessness. Moreover, the patient's medical conditions were reliably related to particular clusterings of the descriptors.

More recently, Elliott et al. (109) conducted an extensive interview of patients with cardiopulmonary disease. Patients checked relevant descriptors from a list of 45, describing their feelings when experiencing breathing troubles during walking, running, or climbing stairs. The assessments were based on their memory for these experiences; that is, the patients did not actually perform any activities immediately before completing the questionnaire. Patients with chronic obstructive airways disease frequently checked items indicating "distress," asthma patients checked descriptors indicating "wheeziness," whereas restrictive patients checked "rapid breathing," and the cardiac group checked "a need to sigh." The aim of all such studies is the collection of clinical information to further our understanding of the etiology of disease, as well as to aid in prevention and treatment.

IX. Overview

The reliable connection between psychophysics and physiology indicates that the senses are instruments whose properties have predictable influences on the transformation of stimulus events into subjective experience. According to most theories of psychophysics, such measurements provide information about the relative sensitivity of the senses in response to external stimulation. The success of this viewpoint depends on the supposition that experimental outcomes are not markedly influenced by contextual and cognitive variables. This is a working assumption of respiratory psychophysics as well, although it must be admitted that the contextual variables discussed in this chapter have not received the same attention as stimulus variables with known ties to respiratory experience.

Classical (local) psychophysics represents one of psychology's major contributions to the understanding of which aspects of human perception must be explained by physiology. A multitude of studies examine the link between psychophysical results and sensory physiology. Experimentation confirms that psychophysical outcomes have correlates in the neurophysiology of vision (110), hearing (111–113), as well as other senses (e.g., Ref. 114). Many of the studies reviewed in this chapter draw meaningful links between behavioral data and possible neurophysiological mechanisms.

Global scaling also opens a window into the neurophysiological underpinnings of perception. This tradition furthers our understanding of sensory adaptation, the summation of neural events within sensory channels, and the functioning of the senses at low levels of stimulation (115–117). Psychophysical methods are not just relegated to the laboratory, but are frequently encountered in clinical settings. Historically, this has been especially true for vision and hearing, but more and more these methods are used to evaluate human performance in regard to other sensory modalities, such as those responsible for respiration.

The future prospects for inferring respiratory physiology from psychophysical results may depend on the expansion of response measures. In order to draw firm conclusions about sensory processes, it is necessary to have relationships between the mean response and other statistical measures, such as the variance, skewness, and kurtosis of the response distributions. Theoretical predictions about how these measures should relate to each other, as well as their implications for neural processing, are the main thrust of recent developments in theoretical psychophysics (1,118,119). Progress in respiratory psychophysics may depend on forging a closer link between this theoretical work and laboratory research.

One new spinoff of theoretical efforts is the realization that substantial data bases are required to reliably compute many statistical measures. Future studies might fruitfully employ small subject (patient) samples, rendering large numbers of responses for the same set of breathing maneuvers. The pattern emerging from such experiments will help us better identify the physiological mechanisms responsible for respiratory experience and, consequently, the most appropriate medical treatments for individuals suffering from respiratory illness.

References

1. Baird JC. Sensation and Judgment: Complementarity Theory of Psychophysics. Mahwah, NJ: Lawrence Erlbaum Associates, 1997.
2. Baird JC, Noma E. Fundamentals of Scaling and Psychophysics. New York: John Wiley & Sons, 1978.
3. Stevens SS. Psychophysics: Introduction to Its Perceptual, Neural and Social Prospects. New York: John Wiley & Sons, 1975.
4. Poulton EC. Bias in Quantifying Judgments. Hillsdale, NJ: Lawrence Erlbaum Associates, 1989.
5. Parducci A. Category ratings: still more contextual effects. In: Wegener B, ed. Social Attitudes and Psychophysical Measurement. Hillsdale, NJ: Lawrence Erlbaum Associates, 1982:262–282.
6. Luce RD, Green DM. A neural timing theory for response times and the psychophysics of intensity. Psychol Rev 1972; 79:14–57.

7. Luce RD, Krumhansl CL. Measurement, scaling, and psychophysics. In: Atkinson RC, Herrnstein RJ, Lindzey G, Luce RD, eds. Stevens' Handbook of Experimental Psychology. 2nd ed. Vol 1: Perception and Motivation. New York: John Wiley & Sons, 1986:3–74.
8. Harver A, Mahler DA. The symptom of dyspnea. In: Mahler DA, ed. Dyspnea. Mount Kisco, NY: Futura, 1990:1–53.
9. Zechman FW, Wiley RL. Afferent inputs to breathing: Respiratory sensation. In: Fishnan AP, ed. Handbook of Physiology. Bethesda, MD: American Physiological Society, 1986; 2:449–474.
10. Stevens SS. On the theory of scales of measurement. Science 1946; 103:677–680.
11. Luce RD, Krantz, DH, Suppes P, Tversky A. Foundations of Measurement: Representation, Axiomatization, and Invariance. New York: Academic Press, 1990:3.
12. Stevens JC, Mack JD, Stevens SS. Growth of sensation on seven continua as measured by force of handgrip. J Exp Psychol 1960; 59:60–67.
13. Mashhour M, Hosman J. On Stevens' "psychophysical law"—a validation study. Percept Psychophys 1968; 3:367–375.
14. Bakers JHCM, Tenney SM. The perception of some sensations associated with breathing. Respir Physiol 1970; 10:85–92.
15. Stubbing DG, Killian KJ, Campbell EJM. The quantification of respiratory sensations by normal subjects. Respir Physiol 1981; 44:251–260.
16. Gottfried SB, Altose MD, Kelsen SG, Cherniack NS. Perception of changes in airflow resistance in obstructive pulmonary disorders. Am Rev Respir Dis 1981; 124:566–570.
17. Tack M, Altose MD, Cherniack NS. Effect of aging on respiratory sensations produced by elastic loads. J Appl Physiol 1981; 50:344–350.
18. Tack M, Altose MD, Cherniack NS. Effect of aging on the perception of resistive ventilatory laods. Am Rev Respir Dis 1982; 126:463–467.
19. Ward ME, Stubbing DG. Effect of chronic lung disease on the perception of added inspiratory loads. Am Rev of Respir Dis 1985; 132:652–656.
20. Mahler DA, Rosiello RA, Harver A, Lentine T, McGovern JF, Daubenspeck JA. Comparison of clinical dyspnea ratings and psychophysical measurements of respiratory sensation in obstructive airway disease. Am Rev Respir Dis 1987; 135:1229–1233.
21. Mahler DA, Rosiello RA, Harver A, Lentine T, McGovern JF, Daubenspeck JA. Measurement of respiratory sensation in interstitial lung disease: evaluation of clinical dyspnea ratings and magnitude scaling. Chest 1989; 96:767–771.
22. Teghtsoonian R. Range effects in psychophysical scaling and a revision of Stevens' law. Am J Psychol 1973; 86:3–27.
23. Teghtsoonian R. On the exponents in Stevens' law and the constant in Ekman's law. Psychol Rev 1971; 78:71–80.
24. Teghtsoonian R. On facts and theories in psychophysics: Does Ekman's law exist? In: Moskowitz HR, Scharf B, Stevens JC, eds. Sensation and Measurement: Papers in Honor of SS Stevens. Dordrecht, Holland: D. Reidel, 1974:167–176.
25. Baird JC, Lewis C, Romer D. Relative frequencies of numerical responses in ratio estimation. Percept Psychophys 1970; 8:358–362.

26. Baird JC, Noma E. Psychophysical study of numbers: I. Generation of numerical responses. Psychol Res 1975; 37:281–297.
27. Noma E, Baird JC. Psychophysical study of numbers: II. Theoretical models of number generation. Psychol Res 1975; 38:81–95.
28. Ross BM, Engen T. Effects of round number preferences in a guessing task. J Exp Psychol 1959; 58:462–468.
29. Baird JC. Psychophysical study of numbers: V. Preferred state theory of matching functions. Psychol Res 1975; 38:189–207.
30. Baird JC, Harder K, Preis A. Annoyance and community noise: psychophysical model of dose-response relationships. J Envir Psychol (submitted).
31. Gibson RH, Tomko DL. The relation between category and magnitude estimates of tactile intensity. Percept Psychophys 1972; 12:135–138.
32. Borg G. Psychophysical bases of perceived exertion. Med Sci Sports Exercise 1982; 14:377–381.
33. Mahler DA, Harver A. Dyspnea. In: Fishman AP, ed. Pulmonary Rehabilitation. New York: Marcel Dekker, 1996:97–116.
34. Horowitz MB, Littenberg B, Mahler DA. Dyspnea ratings for prescribing exercise intensity in patients with COPD. Chest 1996; 109:1169–1175.
35. Helm CE, Messick S, Tucker LR. Psychological models for relating discrimination and magnitude estimation scales. Psychol Rev 1961; 68:167–177.
36. Montgomery H. Direct estimation: effect of methodological factors on scale type. Scand J Psychol 1975; 16:19–29.
37. Eisler H. On the problem of category scales in psychophysics. Scand J Psychol 1962; 3:81–87.
38. Foley HJ, Cross DV, Foley MA, Reeder R. Stimulus range, number of categories, and the "virtual" exponent. Percept Psychophys 1983; 34:505–512.
39. Robinson GH. Biasing power law exponents by magnitude estimation instructions. Percept Psychophys 1976; 19:80–84.
40. Baird JC, Kreindler M, Jones K. Generation of multiple ratio scales with a fixed stimulus attribute. Percept Psychophys 1971; 9:399–403.
41. Ekman G, Künnapas TM. Subjective dispersion and the Weber fraction. Reports from the Psychological Laboratory of the University of Stockholm, No. 41.
42. Olsson MJ, Harder K, Baird JC. What Ekman really said. Behav Brain Sc 1993; 16:157–158.
43. Baird JC, Green DM, Luce RD. Variability and sequential effects in cross modality matching of area and loudness. J Exper Psychol 1980; 6:277–289.
44. Green DM, Luce RD, Duncan JE. Variability and sequential effects in magnitude production and estimation of auditory intensity. Percept Psychophys 1977; 22: 450–456.
45. Luce RD. Response Times: Their Role in Inferring Elementary Mental Organization. New York: Oxford, 1986.
46. Piéron H. Recherches sur les lois de variation des temps de latence sensorielle en fonction des intensités excitatrices. L'Année Psychol 1914; 20:17–96.
47. Piéron H. Nouvelles recherches sur l'analyse du temps de latence sensorialle et sur la loi qui relie de temp a l'intensité d'excitation. L'Année Psychol 1920; 22:58–142.

48. Mansfield RJW. Latency functions in human vision. Vision Res 1973; 13:2219–2234.
49. Garner WR. Uncertainty and Structure as Psychological Concepts. New York: John Wiley & Sons, 1962.
50. Miller GA. The magical number seven, plus or minus two: some limits on capacity for processing information. Psychol Rev 1956; 63:81–97.
51. Baird JC. A cognitive theory of psychophysics: II. Fechner's law and Stevens' law. Scand J Psychol 1970; 11:89–102.
52. Pollack I. The information of elementary auditory displays. J Acoust Soc Am 1952; 24:745–749.
53. Siegel W. Memory effects in the method of absolute judgment. J Exp Psychol 1972; 94:121–131.
54. Siegel JA, Siegel W. Absolute judgment and paired-associate learning: kissing cousins or identical twins? Psychol Rev 1972; 79:300–316.
55. Braida LD, Durlach NI. Intensity perception. II. Resolution in one-interval paradigms. J Acoust Soc Am 1972; 51:483–502.
56. Luce RD, Nosofsky RM, Green DM, Smith AF. The bow and sequential effects in absolute identification. Percept Psychophys 1982; 32:397–408.
57. Luce RD, Baird JC, Green DM, Smith AF. Two classes of models for magnitude estimation. J Math Psychol 1980; 22:121–148.
58. DeCarlo LT, Cross DV. Sequential effects in magnitude scaling: models and theory. J Exp Psychol 1990; 119:375–396.
59. Simon PM, Schwartzstein, Weiss JW, Lahive K, Fencl V, Teghtsoonian M, Weinberger SE. Distinguishable sensations of breathlessness induced in normal volunteers. Am Rev Respir Dis 1989; 140:1021–1027.
60. Simon PM, Schwartzstein, Weiss JW, Lahive K, Fencl V, Teghtsoonian M, Weinberger SE. Distinguishable types of dyspnea in patients with shortness of breath. Am Rev Respir Dis 1990; 142:1009–1024.
61. Campbell EJM, Freedman S, Smith PS, Taylor ME. The ability of man to detect added elastic loads to breathing. Clin Sci 1961; 20:223–231.
62. Bennett FD, Jayson MIV, Rubenstein D, Campbell EJM. The ability of man to detect added non-elastic loads to breathing. Clin Sci 1962; 23:155–162.
63. Burki NK. Detection of added respiratory loads in patients with restrictive lung disease. Am Rev Respir Dis 1985; 132:1210–1213.
64. Green DM, Swets JA. Signal Detection Theory and Psychophysics. New York: Wiley, 1966/1974.
65. MacMillan N, Creelman CD. Detection Theory: A User's Guide. New York: Cambridge University Press, 1991.
66. Narbed PG, Marcer D, Howell JBL, et al. A signal detection theory analysis of the effects of chest cage restriction upon the detection of inspiratory resistive loads. Clin Sci 1983; 64:417–421.
67. Bonnel AM, Mathiot MJ, Crimaud C. Inspiratory and expiratory resistive load detection in normal and asthmatic subjects: a sensory decision theory analysis. Respiration 1985; 48:12–23.
68. Harver A, Katkin ES, Bloch E. Signal-detection outcomes on heartbeat and respiratory resistance detection tasks in male and female subjects. Psychophysiology 1993; 30:223–230.

69. Weber EH. Der Tastsinn und das Gemeinfühl. In: Wagner R, ed. Handwörterbuch der Physiologie. Braunschweig: Vieweg, 1846:481–588.
70. Fechner GT. Elements of psychophysics. Translated by Adler HE. New York: Holt, Rinehart and Winston, 1860/1966.
71. Baird JC. A cognitive theory of psychophysics: I. Information transmission, partitioning, and Weber's law. Scand J Psychol 1970; 11:35–46.
72. Ward LM, Davidson KP. Where the action is: Weber fractions as a function of sound pressure at low frequencies. J Acoustical Soc Am 1993; 94:2587–94.
73. Woodworth RS, Schlosberg H. Experimental Psychology. New York: Henry Holt, 1954.
74. König A, Brodhun E. Experimentelle Untersuchungen über die Psychophysische Fundamentalformel in Bezug auf den Gesichtsinn. Berlin: Akademie der Wissenschaften, 1889:614–644.
75. Riesz RR. Differential intensity sensitivity of the ear for pure tones. Phys Rev 1928; 31:867–875.
76. Falmagne JC. Psychophysical measurement and theory. In: Boff KR, Kaufman L, Thomas JP, eds. Handbook of Perception and Human Performance. New York: John Wiley & Sons, 1986.
77. Murray DJ. A perspective for viewing the history of psychophysics. Behav Brain Sci 1992; 16:115–137.
78. Miller GA. Sensitivity to changes in the intensity of white noise and its relation to loudness and masking. J Acoust Soc Am 1947; 19:609–619.
79. Guilford JP. A generalized psychophysical law. Psychol Rev 1932; 39:73–85.
80. McGill WJ, Goldberg JP. A study of the near-miss involving Weber's law and pure-tone intensity discrimination. Percept Psychophy 1968; 4:105–109.
81. Jesteadt W, Wier CC, Green DM. Intensity discrimination as a function of frequency and sensation level. J Acoust Soc Am 1977; 61:169–178.
82. Katz-Salamon M. The ability of human subjects to detect small changes in breathing volume. Acta Physiol Scand 1984; 120:43–51.
83. Wiley RL, Zechman FW. Perception of added airflow resistance in humans. Respir Physiol 1966; 2:73–87.
84. Stubbing DG, Killian KJ, Campbell EJM. Weber's law and resistive load detection. Am Rev Respir Dis 1983; 127:5–7.
85. Burki NK, Mitchell K, Chaudhary BA, et al. The ability of asthmatics to detect added resistive loads. Am Rev Respir Dis 1978; 117:71–75.
86. Chaudhary BA, Burki NK. The effects of airway anesthesia on detection of added inspiratory elastic loads. Am Rev Respir Dis 1980; 122:635–639.
87. Gottfried SB, Altose MD, Kelsen SG, Cherniack NS. The perception of changes in airflow resistance in normal subjects and patients with chronic airways obstruction. Chest 1978; 73(suppl): 286–288.
88. Narens L, Mausfeld R. On the relationship of the psychological and the physical in psychophysics. Psychol Rev 1992; 99:467–479.
89. Engen T. Psychophysics. I. Discrimination and detection. In: Kling JW and Riggs LA, eds. Woodworth and Schlosberg's Experimental Psychology. New York: Holt, Rinehart and Winston, 1972:11–46.

90. Gescheider GA. Psychophysics: Method, Theory, and Application. 2nd ed. Hillsdale, NJ: Lawrence Erlbaum Associates, 1985.
91. Hellström Å. The time-order error and its relatives: mirrors of cognitive processes in comparing. Psychol Bull 1985; 97:35–61.
92. Pollack I, Ficks L. Information of elementary multidimensional auditory displays. J Acoust Soc Am 1954; 26:155–158.
93. Anderson NS, Fitts PM. Amount of information gained during brief exposures of numerals and colors. J Exp Psychol 1958; 56:362–369.
94. Garner WR. The Processing of Information and Structure. Potomac, MD: Lawrence Erlbaum Associates, 1974.
95. Berglund B, Berglund U, Lindvall T. On the principle of odor interaction. Acta Psychol 1971; 35:255–268.
96. Berglund B, Olsson MJ. Odor-intensity interaction in binary and ternary mixtures. Percept Psychophys 1993; 53:475–482.
97. Anderson NH. Foundations of Information Integration Theory. New York: Academic Press, 1981.
98. Anderson NH. Methods of Information Integration Theory. New York: Academic Press, 1982.
99. Anderson NH. Integration psychophysics and cognition. In: Algom D, ed. Psychophysical Approaches to Cognition. Amsterdam: North-Holland, 1992:13–113.
100. McBride RL, Anderson NH. Integration psychophysics. In: McBride RL, MacFie HJH, eds. Psychological Basis of Sensory Evaluation. London: Elsevier, 1990:93–115.
101. Melara RD. The concept of perceptual similarity: from psychophysics to cognitive psychology. In: Algom D, ed. Psychophysical Approaches to Cognition. Amsterdam: North-Holland & Elsevier Science, 1992:303–388.
102. Shepard RN. Integrality versus separability of stimulus dimensions: from an early convergence of evidence to a proposed theoretical basis. In: Lockhead GR, Pomerantz JR, eds. The Perception of Structure. Washington, DC: American Psychological Association, 1991:53–71.
103. Killian KJ, Bucens DD, Campbell EJM. Effect of breathing patterns on the perceived magnitude of added loads to breathing. J Appl Physiol 1982; 52:578–584.
104. Killian KJ, Mahutte CK, Campbell EJM. Magnitude scaling of externally added loads to breathing. Am Rev Respir Dis 1981; 123:12–15.
105. Harver A, Baird JC, McGovern JF, Daubenspeck AJ. Grouping and multidimensional organization of respiratory sensations. Percept Psychophys 1988; 44:285–292.
106. Ashby FG, Maddox WT. A response time theory of separability and integrality in speeded classification. J Math Psychol 1994; 38:423–466.
107. Kingston J, Macmillan NA. Integrality of nasalization and F1 in vowels in isolation and before oral and nasal consonants: a detection-theoretic application of the Garner paradigm. J Acoust Soc Am 1995; 97:1261–1285.
108. Sneath PHA, Sokal RR. Numerical Taxonomy. San Francisco: W. H. Freeman, 1973.
109. Elliott MW, Adams L, Cockcroft A, Macrae KD, Murphy K, Guz A. The language of breathlessness: use of verbal descriptors by patients with cardiopulmonary disease. Am Rev Respir Dis 1991; 144:826–832.

110. Wandel B. Foundations of Vision. Sunderland, MA: Sinauer Associates, 1995.
111. Gulick WL, Gescheider GA, Frisina RD. Hearing: Physiological Acoustics, Neural Coding, and Psychoacoustics. New York: Oxford, 1989.
112. Green DM. An Introduction to Hearing. Hillsdale, NJ: Erlbaum Associates, 1976.
113. Luce RD. Sound and Hearing: A Conceptual Introduction. Hillsdale, NJ: Lawrence Erlbaum Associates, 1993.
114. Barlow HB, Mollon JD, eds. The Senses. Cambridge, England: Cambridge University Press, 1989.
115. Marks LE. Sensory Processes: The New Psychophysics. New York: Academic Press, 1974.
116. Marks LE, Algom D. Psychophysical scaling. In: Birnbaum MH, ed. Handbook of Perception and Cognition: Measurement, Judgment, and Decision. San Diego: Academic Press (in press).
117. Scharf B, Buus S. Audition I: Stimulus, physiology, thresholds. In: Boff KR, Kaufman L, Thomas JP, eds. Handbook of Perception and Human Performance. New York: John Wiley & Sons, 1986:14-1–14-71.
118. Link SW. The Wave Theory of Difference and Similarity. Hillsdale, NJ: Lawrence Erlbaum Associates, 1992.
119. Norwich KH. Information, Sensation, and Perception. New York: Academic Press, 1993.

Part Three

ASTHMA SELF-MANAGEMENT

10

Management of Asthma by Parents and Children

NOREEN M. CLARK

University of Michigan
Ann Arbor, Michigan

I. The Asthma Paradox

Much has been said in the past few years about the rising rates of asthma and the need to curb them. Dismal health status statistics and the pressure to contain medical care costs have led to scrutiny of clinical approaches to asthma care (1). Managed care organizations in the United States expect primary care pediatricians to handle most cases of asthma, and to refer only the most difficult cases to specialists is the goal. In new care systems, physicians can be penalized if their patients often require emergency care for asthma.

The backdrop for changes in asthma care delivery and financing has been described by Mellins (2) as a modern paradox. Our understanding of the pathogenesis and treatment of asthma has increased significantly in the recent past, yet morbidity and mortality from the disease have also increased worldwide. Understanding of the steps that can be taken to prevent severe asthma has increased, yet emergency visits and hospitalizations are on the rise (3).

These observations lead to some logical conclusions. Patients and families are not recognizing the symptoms of asthma. Physicians are not diagnosing the disease. Physicians are either not providing state-of-the-art care, or, if they are, patients are not adhering to the recommended therapies. It follows that the public needs to be educated to recognize asthma symptoms and seek informed medical care. The medical profession needs to be educated to provide state of the art treatment. And patients need to be educated to manage asthma effectively so that they can lead active and unrestricted lives.

Effective patient self-management appears to be linked to reducing death and disability and increasing functioning among people with asthma. Because pediatric asthma is the leading cause both of missed days from school and of emergency department visits (except accidents), children comprise a primary target for asthma interventions.

This chapter discusses effective patient education for self-management. It makes five points. Accounting for psychosocial factors is exceedingly important when educating families and children with asthma. A range of problems attends the efforts of patients and their caretakers to manage the disease. Studies are beginning to shed light on the psychosocial complexities influencing management by patients. Several model asthma-education programs have addressed social and behavioral influences and have yielded positive outcomes related to the health status and quality of life of children and their use of services. Finally, we have a long way to go before we fully realize the potential contribution behavioral research can make to controlling and preventing childhood asthma.

II. Social and Behavioral Factors in Asthma Mortality and Morbidity

Two authors have clearly illustrated the important influence of psychosocial factors on health care use and death from asthma (4,5). Janson et al. (5) studied patients seeking care in the emergency department of a large urban hospital. They found that the amount of distress perceived by patients during the asthma attack and the patients' appraisal of the social support available to help them in managing asthma were significant predictors of higher levels of service utilization.

Strunk (4) commenting on Weiss et al.'s (1) review of case studies of asthma deaths, posited that the final common pathway appeared to be a period of poor overall control including late arrival for emergency care. Psychosocial factors associated with deaths were disregard of symptoms, poor communication with the physician, and poor mobilization of family resources to seek care (6–9). Arriving late for care was associated with poor adherence to the medical regimen and accepting, almost without question, activity limitation due to symptoms (10–12). Social, cultural, and economic factors appeared to be associated with poor patient adherence and inadequate medical treatment (10).

Two types of asthma deaths seem evident. The first is associated with sudden onset of severe airway closure and death within an hour (13). Little is known about this type of death. The second type of mortality is characterized by Strunk as "slow onset late arrival" (4). He suggests that psychological, social, and economic features of the patient and family give rise to disregard of asthma symptoms and acceptance of disability. These in turn lead to poor use of medicines, late arrival for medical care, and death.

While in absolute numbers, deaths from asthma are comparatively rare events, they are especially unsettling because they are usually preventable. Strunk's analysis and observations by a large number of investigators (14–16) confirm the salient role of social and behavioral factors in asthma mortality as well as morbidity.

III. Some Current Findings Revealing Intrafamily Complexities in Asthma Self-Management

The careful examination of psychosocial and educational aspects of asthma management is in its formative stage. The work of the past two decades has been exceedingly fruitful and at the same time has underscored how little we understand the complex intra- and interpersonal dynamics that shape patient behavior. Findings from some recent studies regarding the family (perhaps the most significant of the social environments influencing behavior) serve to illustrate this point.

Gustafsson and colleagues (17) studied families of infants to examine a commonly held belief that family psychopathology gives rise to asthma. They found, on the contrary, that wheezing in infancy was significantly more likely to be the cause than result of family dysfunction. One infers from their work that education for parents of very young children should assist them in handling the pressures that asthma exerts on their interpersonal relationships.

Two studies by Schobinger and colleagues (18,19), one concerning mothers of asthma patients and one related to fathers, suggest that parent-child interactions are complex in families living with asthma. They studied both parents of children with the disease and parents of nonasthmatic children. Mothers and fathers of children with asthma were each significantly more likely to be overtly negative with their children than parents of nonasthmatic children. This type of parenting behavior no doubt affects a child's efforts to self-manage.

One group of investigators (20) has examined communication between pediatric asthma patients and their parents. They noted that it was in families where child and parent communicated more about asthma that the parent was a higher-level self-manager. These were also the families where children were more influential in the parent's decision making about asthma management, for example, whether or not to send the child to school, when to see a physician, and so on. The greater the information flow, the more likely the family was to be active in efforts to control the disease. The investigators also found that the mothers who were most likely to report that asthma was a significant disruption in their lives were also most likely to report that the child's father did not participate in asthma care (21), further evidence of the important interaction of asthma with interpersonal relationships.

For adolescents, the family is not the only source of significant interpersonal relationships. For these patients, peers constitute a powerful social influence. Brook and Shiloh (22) studied teenagers with asthma and found patients at this stage of development were actually more attitudinally disposed than other teens to smoke. Forero et al. (23) showed that, in fact, adolescents with asthma smoke just as much as other teenagers. Having asthma does not protect a child from strong pressures in the peer group social environment. Nor does it necessarily protect one from hazards at home. McIntosh and colleagues (24) found that after education and being provided with data regarding a child's cotenine level from passive smoking, most parents of children with asthma continued to smoke at home (although a number increased the frequency of smoking outside the house).

It is clear from many studies that health and illness beliefs are predictors of preventive and disease-management behavior (25). A family's interpersonal relationships are influenced to some extent by the beliefs the members hold about asthma and its effects. These beliefs, in part, enable or preclude realistic goal setting related to self-management. Our research group examined asthma beliefs of 290 low-income mothers and their 310 school-aged children with the disease. We found disturbing levels of misunderstanding about asthma among both parents and children. Some of their asthma beliefs are presented in Table 1. Further, a number of these mothers (45%), who were fearful about asthma medicines, also believed that they are addictive (39%), that they are not safe to use over a long time (36%), and/or that regular use reduces effectiveness (58%). In studying another group of parents (N = 259) who brought their children with asthma to the emergency department of a large urban hospital, we found that 15% believed that the asthma treatment received there was more effective than continuous care from a regular physician.

Of course, changing beliefs is not easy, and the education of pediatric patients and their caretakers is not simply a matter of providing information and a therapeutic plan. Parcel and colleagues (26) have discussed the importance of self-efficacy and outcome expectations in asthma management, that is, having confidence that one can carry out self-management tasks and believing that these actions make a significant difference in the course of the disease is fundamental to undertaking new behavior. To enable the family to

Table 1 Mothers' and Children's Beliefs About Asthma (N = 170)

Belief	Percent of mothers responding incorrectly	Percent of children responding incorrectly
Children with asthma can usually participate in physical activity	24	54
People frequently die from asthma	38	35
Even a mild asthma attack requires a visit to the hospital or clinic	42	55
Asthma usually leads to permanent lung damage	58	40
Even when a child with asthma feels okay, wheezing will occur with normal activity	80	39

develop effective management strategies, education must help them learn how to make difficult decisions, develop confidence, and take appropriate action. They must learn how to do this in light of complex intrapersonal factors such as their health beliefs and sense of self-efficacy, and in light of interpersonal factors, most certainly their relationships with family members and other significant actors in the social environment.

IV. Related Problems for Parents and Children in Managing Asthma

While quantitative data describing the problems families confront in managing asthma are not as plentiful as we would like, most clinicians are familiar with a range of issues that accompany patients into asthma treatment. For the purposes of this discussion, these issues will be categorized as those associated with the stage of development of the child, the child's social environment, and two facts about learning: (1) learning is a process considerably more complex than acquiring information and (2) assessment of successful learning varies according to who is doing the evaluating.

A. Stages of Development

It has been well accepted that a child's behavioral response to illness is in part a function of his or her developmental age (27). Children are able to acquire and process information according to the degree of their physical, psychological, and experiential maturity. A significant amount of discussion has centered on the point at which children are able to handle abstract concepts and enter into formal operational thinking. It is thought that this ability develops around 12 years of age. This does not mean that younger children cannot be taught to be very effective asthma managers, but their ability to fully comprehend information requires that it be presented in highly concrete, nontheoretical terms.

A given in pediatric asthma management is that the child is constantly changing physically, psychologically, and emotionally. Indeed the entire context for management is dynamic. The family constellation in which self-management is undertaken is continually changing as members, relationships, and external influences exert their effect. The exacerbating situations giving rise to symptoms can change according to individual and environmental factors. What is effective management in one set of circumstances may not work in another.

Not even the medical regimen remains constant. Asthma treatment varies greatly according to the age of the child. The physician makes therapeutic choices based on what is physiologically appropriate for a child and which forms of administration the child can tolerate. In the natural course of the disease the doctor changes the therapeutic plan as he or she tailors the regimen to the individual patient and tries new forms of therapy. The task for the family is to learn how to deal with constant change and to be able to make decisions and act appropriately given the "developmental" nature of management.

B. The Child's Social Environment

The problem of asthma self-management in children is complicated by the fact that the parent is the manager wholly or in part for the child until he or she reaches the age to undertake full responsibility. In addition, as Wilson-Pessano and Mellins (28) have suggested, sometimes there is confusion about the term self-management, i.e., for some it implies seeking alternatives to medical care. This is *not* how the term is used in asthma education and prevention. In chronic disease management, partnership with the physician is central to control. The physician brings to the partnership expertise in clinical care, whereas the patient brings an intimate personal knowledge of the illness. Both contributions are necessary to effect change. Together the doctor, parent, and child must determine what each will do to make the partnership work.

Negotiating a partnership isn't always easy. Patients and caretakers often become frustrated or even angry because of the burdens that asthma imposes. Episodes of symptoms can be frightening events. A study by our group of children's asthma attacks found that, during an attack, among low-income mothers of children with asthma, 62% felt frightened, 33% felt desperate, and 44% lacked the confidence to manage the attack. Learning how to cope with fear and anxiety is an important aspect of asthma patient education. By staying calm, parent and patient are better able to make the judgments and take the actions needed when a child is experiencing an asthma episode.

For many children the parent is not the only or even the central caretaker. Grandparents, aunts and uncles, siblings, and babysitters are often involved in asthma management. Each brings his or her own perceptions, beliefs, sense of confidence, and skills to the task. Further, for children in daycare or school, teachers, counselors, and principals play a significant role. Most families need to learn how to handle asthma at school. Effective self-management means informing significant school-based actors about the disease and equipping them to respond appropriately. Several studies have shown that asthma is not well accommodated by many school districts. Two showed that a majority of teachers were badly informed about the course of the disease (29), and another showed that teachers were very likely to prohibit physical exercise for children with asthma (30). In some areas, school policy prohibits children from using medications at school or allows use only when administered by a school official. Good asthma management requires that parent and child, assisted by the doctor, learn how to overcome obstacles such as these in the social environment.

C. The Learning Process

An important consideration in teaching and learning about asthma self-management is the definition of success. What is the goal of asthma patient education and what signals that the goal has been reached? As clinician and family bring different asthma expertise to disease control, so too do they emphasize different outcomes. The physician will necessarily be concerned with results of objective measures such as pulmonary function tests and need for medicines. The family will likely focus more on the quality of their lives (e.g., less disruption, full participation in sports, etc.). Juniper (31) has illustrated that measures assessing clinical outcome, functional health status, and quality of life do not always correlate well with each other. Each is a distinct construct, although somewhat related to the others. To achieve success, the greatest emphasis in therapy and counseling needs to be on the family's definition. The patient and parent are much more likely to be motivated to undertake the physician's recommendations when the target is their own interests and concerns.

As noted above, asthma management requires a high degree of judgment on the part of the patient. There is no absolute recipe for successful control. Sometimes asthma symptoms are predictable and sometimes they are not. A patient needs to be highly self-regulating, that is, have the ability to observe, judge, and act on the basis of subtle changes in symptoms or peak flow values

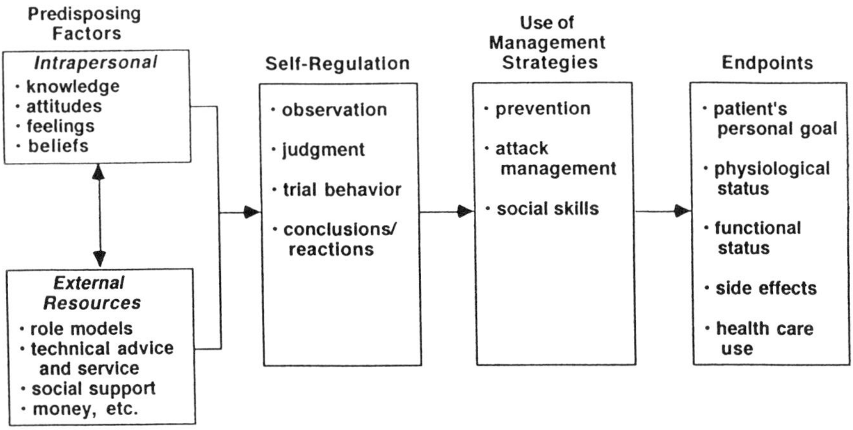

Figure 1 A model of patient management of asthma. (From Ref. 33.)

or functional status. It may be that developing skills of self-regulation is more important than learning asthma facts. A recent study (32) illustrated significant associations between parents' self-observations (e.g., using more ways to observe the seriousness of a child's symptoms) and self-efficacy (e.g., confidence that one can prevent symptoms) and using a greater number of asthma control strategies.

To conceptualize the goal of asthma management, and therefore asthma education, it helps to see management by the patient in a theoretical framework. Effective asthma education is based on a theoretical understanding of human behavior, that is, why people are motivated to do what they do. It moves beyond provision of clinical facts and information to skills, concepts, and data that help a family to decide and act within the complex context of change. Several theories of human behavior apply to asthma management. Figure 1 (33) is presented as an example of how theoretical constructs of self-regulation can be applied to asthma management.

The conceptual framework in Figure 1 is based on three ideas. First, several factors predispose or enable one to manage. Second, patient and family management is the conscious use of strategies to manipulate situations in order to reduce the impact of asthma on daily life. One learns which strategies work (or don't) through processes of self-regulation. Third, asthma management is not an end in itself, but the means to other ends. These ends include outcomes such as improved physiological status, reduction of symptoms and other clinical manifestations of disease, improved functioning, reduced side effects, appropriate health care use, and better quality of life as perceived by the patient.

Articles in the literature (32) have proposed that individuals draw from a base of asthma knowledge and, by virtue of their attitudes, beliefs, and feelings, make decisions and take actions. Important external factors that enhance or enable asthma management have also been proposed. Among other things, these include role models from whom management strategies can be learned, technical professional advice and service, social support, and money to obtain care and maintain an adequate physical environment. It also must be noted that while knowledge, attitudes, and feelings are the basis for action, these can also change as a result of behavior (34).

An individual uses his or her intrapersonal and external resources and, through processes of self-regulation, tries to reach a personal goal. An example would be a child who wants to play full-court basketball who believes he is susceptible to asthma problems, thinks medicines will help and so uses them preventively, works out a procedure for taking a breather when active, seeks moral support from his friends and coaches, and uses other strategies that enable him to reach his personal goal. He learns which strategies are effective through self-regulation. Self-regulation is particularly important in a disease such as asthma precisely because there is no sure-fire formula for an individual to use to control symptoms or deal with crucial interpersonal relationships. Patients and families must exercise a high degree of decision making in the absence of health professionals. The self-regulatory processes they engage in entail observing situations where asthma contributes to problems in reaching their goals, judging what types of actions might ameliorate the situation, using management strategies (i.e., trying out behavior), and drawing conclusions or reacting to the effects of the behavior. Two important reactions are that the behavior resulted in the desired effect, and that one can effectively carry out the behavior. Using strategies to prevent attacks, manage them effectively, and manage the interpersonal relationships needed to control the effects of the disease should lead to important outcomes. These likely include better quality of life for the patient including reaching his personal goal, as well as improved health status determined by the clinician and more appropriate use of health services.

Management is behavior, and to assess successful management, the behavior must be measured. There must be a determination of what the patient or family member actually does as opposed to what predisposes him to act—what he knows, believes, or feels. It is important to understand relationships among factors such as knowledge, feelings, attitudes, and behavior. But what makes the difference in outcome is that people actually behave in the desired way, not that they know how to behave appropriately or hold particular attitudes. Tests of knowledge, attitudes, and feelings have frequently been used in asthma research as proxies for behavior. However, knowledge tests have not proven very useful in predicting health behavior. The link between knowing general facts about disease, for example, pathophysiology, and behaving differently is not strong (35). The knowledge a patient needs is that which enables him to reach his personal goals. This information varies by individual, is very idiosyncratic, may not be what the health professional sees as important, or be amenable to standardized tests.

Similarly, attitudes are also not very useful as predictors of most health behavior. The link between general constructs of attitude and specific instances of health behavior is weak (35). Neither do feelings always predict behavior. As noted, feeling that one is able to carry out a management task has been shown to render people more likely to try the task (36,37), but even expressions of self-efficacy do not ensure that behavior will follow.

Health beliefs have consistently been associated with behavior, especially the beliefs that one is susceptible, that the consequences of disease could be serious, and that taking the recommended action produces greater benefit than cost (25). However, the variance in behavior accounted for by health beliefs is not strong enough for belief to serve as a proxy. Intention to act has repeatedly been shown in the health literature to be associated with actual behavior (e.g., intending to use relaxation exercises when facing an asthma episode), but intention cannot be considered the same as behavior (25).

There are two options in assessing patient-management behavior. First, the extent to which patients and their family members actually use asthma-management strategies can be measured. These measurements would include actions recommended by clinicians to prevent the onset of symptoms and manage episodes, and those uncovered in social and behavioral research that enable patients to communicate more effectively to family, health providers, and others and enable them to negotiate relationships. A second factor that can be measured is the behavioral processes that constitute self-regulation, i.e., patient judgment and subsequent action can be assessed, and such evaluation would include the extent to which patients are taking informed action, are more observant of situations precipitating asthma problems, try different ways to ameliorate these situations, evaluate the efficacy of the strategies tried, react by feeling more confident, and continue to refine management strategies through self-regulation.

As patient self-management is not an end in itself, the extent to which use of pharmacological and nonpharmacological management strategies actually predict better quality of life and other desired outcomes clearly is important in assessing interventions. The connection between management behavior and physiological measures, functional status, side effects, or health care use has not been examined very extensively in asthma research to date and continues to be an important area for study.

V. Evaluation of Pediatric Asthma Education Programs

Until recently, there were few models of systematic attempts beyond counseling provided by individual physicians to help patients improve their self-management skills. The past 15 years have given rise to several excellent models of pediatric asthma self-management education. These models have been based on theories of learning and behavior and have been carefully evaluated. There have been two generations of research. The first led to the development of clinic-based programs, the second to programs for school and home.

Krutzsch et al. (16) classified into three categories the self-management tasks that have been addressed in extant model programs: attack prevention, attack management, and social skills related to management. To prevent the onset of an attack, parent and child must recognize the early signs of asthma, act on these signs in appropriate ways, identify and avoid precipitants to wheezing, and take the prescribed medicines properly and on time. Attack management involves efforts to rest and stay calm, maintain adequate hydration, use medicines as prescribed, seek assistance as warranted, and follow other steps outlined by the physician. Social skills related to asthma management include communicating with the physician and handling problems in school, at home, and with friends.

A complement of evaluated studies based on these categories of skill is presented in Table 2 (24,26,30,38–52). In the main, the programs share three basic premises: the clinician and patient are partners in the management of asthma, the context for management is highly variable, and at-home management tasks must fit the family's lifestyle. At a general level, the programs cover the following topics: recognizing signs and symptoms of asthma, administering prescribed medicines correctly and managing side effects, avoiding panic, recognizing and responding to symptoms that require emergency care, reducing exposure to known triggers, normalizing the child's physical and social activities, and communicating effectively with clinicians.

As indicated in Table 2 (24,26,30,38–52), the programs have been developed using both group and individual formats and have been conducted in physicians' offices, outpatient clinics, emergency rooms, schools, camps, and homes. Several are conducted by skilled educators including physicians, nurses, and health educators. Some are self-administered through home study materials or computers. Programs are available for children in infancy through late adolescence.

Table 3 presents findings of the evaluations of these educational programs. Assessments have ranged in duration from 4 months to 3 years. Most have been clinical trials, that is, employed rigorous evaluation designs and

Table 2 Evaluated Pediatric Asthma Education Program

Year	Investigator (Ref.)	Who was targeted?		Child's age (yr)	Asthma educator(s)	Group/ individual format
		Parent	Child			
Programs in the clinical setting						
1981	Staudenmayer et al. (38)	√	√	8–17	Self	I
1981	Fireman et al. (39)		√	2–14	Nurse-Educator	G & I
1984	Lewis (40)	√	√	7–12	Physicians/Nurses	G
1984	Hindi-Alexander and Crapp (41)	√	√	6–14	Health educators	G
1985	McNabb et al. (42)	√	√	9–13	Health professionals	I
1985	LeBaron et al. (43)		√	6–17	Nurse-Educator	I
1986	Rubin et al. (44)		√	7–12	Computer	I
1986	Clark et al. (45)	√	√	4–17	Health educators	G
1994	Mesters et al. (46)	√	√	0–4	Physicians, nurses	I
1994	Wilson and Starr-Schneidkraut (47)		√	0–4	Health educator	G
Home-based programs						
1985	Rakos et al. (48)	√	√	7–12	Self	I
1986	Mitchell et al. (49)	√	√	2–14	Nurses	I
1988	Jenkinson et al. (50)	√	√	3–49	Self	I
1991	Hughes et al. (51)	√	√	6–16	Nurses	I
1994	McIntosh et al. (24)	√		6 months to 17 yr	Physicians, self	I
School-based programs						
1980	Parcel et al. (26)		√	5–11	Health educators	G
1987	Evans et al. (30)		√	8–11	Health educators	G
Camps						
1985	Robinson et al. (52)		√	10–14	Camp counselors	G

Table 3 Outcomes of Evaluated Pediatric Asthma Education

Study (Ref.)	*N*	Evaluation design	Follow-up period	Outcomes
Staudenmayer et al. (38)	40	Pre/Post	6 months	Change in direction of fewer school absences ($p = 0.09$) Reduced number of emergency physician visits for those with baseline use ($p = 0.03$) Reduced hospitalizations ($p = 0.03$)
Fireman et al. (39)	26	RCT	6–16 months	Decreased school absences ($p < 0.05$) Reduced ED use (no statistical analysis) Reduced hospitalization time (no statistical analysis)
Lewis et al. (40)	76	RCT	1 yr	Reduced ED visits ($p < 0.001$) Reduced number of hospital days ($p < 0.01$)
Hindi-Alexander and Crapp (41)	92	Pre/Post	1 yr	Reduced school absences ($p < 0.005$) Reduced ED use but not statistically significant No difference in hospitalizations
McNabb et al. (42)	14	Pre/Post	1 yr	Reduced ED use (no statistical analysis)
LeBaron et al. (43)	31	RCT	4 months	Increased knowledge about cromolyn (no statistical analysis) No change in pulmonary function or systems
Rubin et al. (44)	54	RCT	1 yr	No difference in school absences Improvement in ED use but not statistically significant
Clark et al. (45)	310	RCT	1 yr	Improved academic performance ($p < 0.05$) Increased self-management behavior: parent ($p = 0.0001$); child ($p = 0.05$) No difference in school absences Reduced ED use among those with baseline use ($p = 0.05$) Reduced hospitalizations among those with baseline use ($p = 0.05$)

Table 3 (Continued)

Study (Ref.)	N	Evaluation design	Follow-up period	Outcomes
Mesters et al. (46)	67	RCT	1 yr	Higher level of asthma self-efficacy ($p < 0.02$) Increased self-management behavior ($p < 0.01$) Decreased MD visits ($p < 0.01$)
Wilson and Starr-Schneidkraut (47)	80	RCT	1 yr	Increased symptom free days ($p = 0.004$) Fewer nights of parental sleep interruption
Rakos et al. (48)	43	RCT	1 yr	No significant difference in school absences No significant differences in ED use or emergency MD visits
Mitchell et al. (49)	368	RCT	6 months	Differences according to ethnicity of child No differences in hospital days
Jenkinson et al. (50)	177	RCT	3 and 12 months	Significant knowledge increases ($p = 0.002$) No other significant change
Hughes et al. (51)	95	RCT	1 yr	Reduced school absences ($p = 0.04$)
McIntosh et al. (24)	92	RCT	4–6 months	More treatment (35%) than control (17%) subjects reported smoking outside their homes at posttest (and their children's continine levels were lower) but not statistically significant
Parcel et al. (26)	104	Comparison group	1 yr	Improved ED use but not statistically significant Higher levels of health locus of control for treatment group ($p = 0.018$)
Evans et al. (30)	239	RCT	1 yr	Increased self-management ($p = 0.05$) Increased communication with parent ($p = 0.002$) Better academic grades ($p = 0.05$) Fewer days with symptoms ($p = 0.004$)
Robinson (52)	74	Pre/Post	3 yr	Reduced school absences (no statistical analysis) Better MDI technique (0.0005) Fewer hospital days (0.02)

methods (random controlled trials, RCT). Outcomes achieved through well-conceptualized and well-delivered patient education have been shown to include reductions in school absences and use of emergency department and hospital services, increases in patient self-efficacy and use of asthma-management strategies, less frequent wheezing, and improved academic performance.

VI. Role of the Clinician

In one way or another, each of these self management programs stresses the need for a regular systematic approach to care of patients with asthma, including the consistent use of medication as prescribed by the physician, the need to avoid exacerbating substances and conditions when these are known, and initiating early drug therapy when avoidance is not possible. These programs, together with effective communication with clinicians, facilitate the patient-physician partnership.

The growing appreciation that a variety of mediators involved in chronic airway inflammation play an important role in the genesis of the disease has redirected medical attention towards regimens that suppress or reverse airway inflammation (53). Self-management programs, by stressing adherence to these regimens, are a great help in the implementation of clinician recommendations. Leadership in the use, dissemination, refinement, and adaptation of the available patient-education programs rests with the clinical community and takes several forms. For example, practitioners can adapt existing programs for use in their private practices or encourage the adoption of programs by the institutions in which they collaborate. Further, practitioners can introduce the idea of self-management tasks, as embodied in the model programs, into their ongoing counseling of patients and family members. It is worth stressing the fact that self-management programs of the type presented in Tables 2 and 3 are an adjunct to, and do not replace, effective clinician communication and education.

Siminoff (54) has suggested that health professionals have four goals in communicating with patients.

1. *Cognitive outcomes*: the patient's recall and understanding of what the health professional has said and why the information is important
2. *Affective outcomes*: the reduction of patient anxiety and increasing the satisfaction of the health provider and patient with the encounter
3. *Behavioral outcomes*: the patient's adherence to the proposed regimen including self-management
4. *Clinical outcomes*: the actual improvement in the patient's clinical condition

Becker (55) has provided a synthesis of the research related to physician actions (as well as behavior of clinical support staff) that improve relationships with the patient and augur toward better adherence to the medical regimen. For example, understanding how the patient sees the problem helps the clinician to communicate more effectively (56,57). Several barriers to physician-patient interaction have also been identified by Becker, who described five specific patient concerns. Patients (1) often feel they are wasting the physician's valuable time, (2) omit details they deem unimportant, (3) are embarrassed to mention things they think will place them in an unfavorable light, (4) do not understand medical terms, and (5) may believe the physician has not really listened and therefore does not have the information needed to make a good treatment decision (58). Obviously, when patients do not fully inform physicians both diagnosis and therapy decisions are affected.

Certain behaviors on the part of the clinician increase the amount and quality of the information the patient provides. These behaviors (58) serve as useful guidelines for clinicians (although with some cultural subgroups of patients they may not always apply):

1. Be attentive to the patient (signaled by cues such as using eye contact, sitting rather than standing when conversing, moving closer to the patient, leaning slightly forward to attend to the discussion.
2. Elicit the patient's underlying concern about the condition.
3. Construct reassuring messages that alleviate fears (reduce fear as a distraction to enable the patient to focus on what is being said).
4. Address immediate concerns the family expresses (again enable patients to refocus their attention on the information being provided).
5. Engage the patient in interactive conversation through use of open-ended questions, simple language, analogies to teach important concepts (dialogue that is interactive produces richer diagnostic information).
6. Tailor the regimen by eliciting and addressing potential problems in the timing, dosage, or side effects of the medicine recommended.
7. Use appropriate nonverbal encouragement (e.g., a pat on the shoulder, nodding in agreement) and verbal praise when the patient reports using correct disease management strategies.
8. Elicit the patient's immediate objective related to control of asthma and reach agreement with the family on a short-term goal (i.e., a short-term objective that is important to the patient that both clinician and patient will strive to reach).
9. Review the long-term plan for the patient's treatment so the patient knows what to expect over time, knows the situations under which the physician will modify treatment, and knows the criteria for

judging the success of the treatment plan (e.g., the patient should sleep through the night, be symptom-free even when exercising, etc.).

10. Help the patient plan in advance for decision making about the asthma (e.g., using diary information or guidelines for handling potential problems, or exploring contingencies in managing attacks) (58).

Several behavioral theories support the notion that the relationship between patient and clinician is, in fact, a significant influence on the behavior of both. Social cognitive theory (34), for example, postulates that personal, social, and behavioral influences interact to cause behavior. Clinicians whose skills engender optimum responses from patients in the form of self-management are likely to be more motivated to continue education and counseling. Patients who feel their physicians communicate well and who see positive results from their own informed self-management will likely be more satisfied with the care they receive and more motivated to take further action to control disease. This reciprocal reinforcement, over time, may lead to stronger and stronger partnerships between clinician and patient. When patients receive effective clinician counseling and participate in asthma self-management education of the type evident in model programs, they and their physicians are in the best possible position to exert control over the disease.

VII. Next Steps

Evaluation of patient education in pediatric asthma has shown self-management to be a very promising route to disease control. The first and second generation of asthma programs have evolved excellent models for use in the clinical setting, home, and school. The focus of these programs on prevention and management of asthma episodes equips patients to contend with highly troublesome aspects of living with the disease.

A third generation of research into asthma self-management programs is needed, one that accounts more fully for the interpersonal relationships that help or hinder patients in managing optimally. New models to be derived would focus on the patient's social environment and on significant others who exert influence on management efforts. These models will be guided by studies of differences in children's beliefs and behaviors attributable to the social influences inherent in culture, economic status, and developmental stage. They will be based on sound social and behavioral theory related to social networks and interpersonal communication. They will be judged heavily on the extent to which they enhance the patient's perception of quality of life. While advances in asthma patient education have been significant in the past decade, we have some distance to travel before we realize the full potential of behavioral research to pediatric asthma prevention and control.

References

1. Weiss KB, Wagener DK. Changing patterns of asthma mortality: identifying target populations at high risk. JAMA 1990; 264:1683–1687.
2. Mellins RB. The role of health education in the management of asthma. In: Taussig L, Landau L, eds. Pediatric Respiratory Medicine. St. Louis, MO: Mosby, 1996.
3. Buist AS, Volmer WM. Reflections on the rise in asthma morbidity and mortality. JAMA 1990; 264:1719–1720.
4. Strunk RC. Death due to asthma: new insights into sudden unexpected deaths, but the focus remains on prevention. Am Rev Respir Dis 1993; 148(3):550–552.
5. Janson-Bjerklie S, Ferketich S, Benner P. Predicting the outcomes of living with asthma. Res Nurse Health 1993; 16(4):241–250.
6. Strunk RC, Mrazek DA, Fuhrmann GSW, LaBrecque JF. Physiologic and psychological characteristics associated with deaths due to asthma in childhood: a case-controlled study. JAMA 1985; 254:1193–1198.
7. Rea HH, Scragg R, Jackson R, Beaglehole R, Fenwick J, Sutherland DC. A case control study of deaths from asthma. Thorax 1986; 41:833–839.
8. Strunk RC. Deaths from asthma in childhood: Patterns before and after professional intervention. Pediatr Asthma Allergy Immunol 1989; 1:5–13.
9. Miller BD, Strunk RC. Circumstance surrounding the deaths of children due to asthma. AJDC 1989; 143:1294–1299.
10. Birkhead G, Attaway NJ, Strunk RC, Townsend MC, Teutsch S. Investigation of a cluster of deaths of adolescents from asthma: evidence implicating inadequate treatment and poor patient adherence with medications. J Allergy Clin Immunol 1989; 84:484–491.
11. Wissow LS, Gittelsohn AM, Szklo M, Starfield B, Mussman M. Poverty, race, and hospitalization for childhood asthma. AJPH 1988; 78:777–782.
12. Sears MR, Rea HH. Patients at risk for dying of asthma: New Zealand experience. J Allergy Clin Immunol 1987; 80:477–481.
13. Sur S, Crotty TB, Kephart GM, Hyma BA, Colby TV, Reed CE, Hunt LW, Gleich GJ. Sudden-onset fatal asthma: a distinct entity with few eosinophils and relatively more neutrophils in airway submucosa. Am Rev Respir Dis 1993; 148:713–719.
14. McNabb W, Wilson SR, Jacobs AM. Critical self-management competencies for children with asthma. J Pediatr Psychol 1986; 11:103–117.
15. Clark NM. Asthma self-management education: research and implications for clinical practice. Chest 1989; 95:1110–1113.
16. Krutzsch CB, Bellicha TC, Parker SR. Making childhood asthma management education happen in the community: translating health behavioral research into local programs. HEQ 1987; 14:357–373.

17. Gustafsson PA, Bjorksten B, Kjellman NI. Family dysfunction in asthma: a prospective study of illness development. J Pediatr 1994; 125(3):493–498.
18. Schobinger R, Florin I, Reichbauer M, Lindemann H, Zimmer C. Childhood asthma: mothers' affective attitude, mother-child interaction and children's compliance with medical requirements. J Psychosom Res 1993; 37(7):697–707.
19. Schobinger R, Florin I, Zimmer C, Lindemann H, Winter H. Childhood asthma: paternal critical attitude and father-child interaction. J Psychosom Res 1992; 36(8):743–750.
20. Clark NM, Levison MJ, Evans D, Wasilewski Y, Feldman CH, Mellins RB. Communication within low income families and management of asthma. Pat Educ Couns 1990; 15:191 210.
21. Wasilewski Y, Clark NM, Evans D, Feldman C, Kaplan D, Rips J, and Mellins RB. Effect of paternal social support on maternal disruption caused by childhood asthma. J Comm Health 1988; 13(1):33–42.
22. Brook U, Shiloh S. Attitudes of asthmatic and nonasthmatic adolescents toward cigarettes and smoking. Clin Pediatr 1993; 323(11):642–646.
23. Forero R, Bauman A, Young L, Larkin P. Asthma prevalence and management in Australian adolescents: results from three community surveys. J Adolesc Health 1992; 13(8):707–712.
24. McIntosh N, Clark NM, Howatt WF. Reducing tobacco smoke in the environment of the child with asthma: assisted minimal contact intervention. J Asthma 1994; 6(31):453–462.
25. Janz NK, Becker MH. The health belief model: a decade later. HEQ 1984; 11:1–47.
26. Parcel GS, Nader PR, Tiernan K. A health education program for children with asthma. J Dev Behav Pediatr 1980; 1(3):128–132.
27. Perrin EC, Garrity PS. There's a demon in your belly: children's understanding of illness. Pediatrics 1981; 667:841–849.
28. Wilson-Pessano SR, Mellins RB. Workshop on asthma self-management. J Allergy Clin Immunol 1987; 602:486–490.
29. Hill RA, Britton JR, Tattersfield AK. Management of asthma in schools. Arch Dis Child 1987; 62:414–415.
30. Evans D, Clark NM, Feldman CH, Rips J, Kaplan D, Levison MJ, Wasilewski Y, Levin B, Mellins RB. A school health education program for children with asthma aged 8–11 years. HEQ 1987; 14(3):367–279.
31. Juniper, E.F., Guyatt, G.H., Epstein, R.S., Ferrie, P.J., Jaeschke, R., Hiller, T.K. Evaluation of impairment of health related quality of life in asthma: development of a questionnaire for use in clinical trials. Respir Med 1991; 85:76–83.
32. Clark NM, Zimmerman BJ. Social cognitive view of self-regulated learning about health. Health Educ Res 1990; 5(3):371–379.
33. Clark NM, Starr-Schneidkraut NJ. Management of asthma by patients and families. Am J Respir Crit Care Med 1994; 149:S54–S66.
34. Bandura A. Social Foundations of Thought and Action: A Social Cognitive Theory. Englewood Cliffs, NJ: Prentice-Hall, Inc., 1986.

35. Becker MH. Patient adherence to prescribed therapies. Med Care 1985; 23(5): 539–555.
36. Clark N, Rosenstock I, Hassan H, Wasilewski Y, Evans D, Feldman C, Mellins R. The effect of health beliefs and feelings of self-efficacy on self-management behavior of children with a chronic disease. Pat Couns Educ 1988; 11(2):131–139.
37. O'Leary A. Self-efficacy and health. Behav Res Ther 1985; 23(4):437–451.
38. Staudenmayer H, Harris PS, Selner JC. Evaluation of a self-help education-exercise program for asthmatic children and their parents: six-month follow-up. J Asthma 1981; 18:1–5.
39. Fireman P, Friday GA, Gira C, Vierthaler WA, Michaels L. Teaching self-management skills to asthmatic children and their parents in an ambulatory care setting. Pediatrics 1981 68(3):341–348.
40. Lewis CE, Rachelsfsky G, Lewis MA, de la Sota A, Kaplan M. A randomized trial of A.C.T. (asthma care training) for kids. Pediatrics 1984; 74(4):478–486.
41. Hindi-Alexander MC, Cropp GJA. Evaluation of a family asthma program. J Allergy Clin Immunol 1984; 74:505–510.
42. McNabb WL, Wilson-Pessano SR, Hughes GW, Scamagas P. Self-management education of children with asthma: AIR WISE. AJPH 1985; 75(10):1219–1220.
43. LeBaron S, Zeltzer LK, Ratner P, Kniker WT. A controlled study of education for improving compliance with cromolyn sodium (Intel): the importance of physician-patient communication. Ann Allergy 1985; 55(6):811–818.
44. Rubin DH, Leventhal JM, Sadock RT, Letovsky E, Schottland P, Clemente I, McCarthy P. Educational intervention by computer in childhood asthma: a randomized clinical trial testing the use of a new teaching intervention in childhood asthma. Pediatrics 1986; 77(1):1–10.
45. Clark NM, Feldman CH, Evans D, Levison MJ, Wasilewski Y, Mellins RB. The impact of health education on frequency and cost of health care use by low income children with asthma. J Allergy Clin Immunol 1986; 78:108–115.
46. Mesters I, Meertens R, Kok G, Parcel GS. Effectiveness of a multidisciplinary education protocol in children with asthma (0–4 years) in primary health care. J Asthma 1994; 31(5):347–359.
47. Wilson SR, Starr-Schneidkraut N. State of the art in asthma education: the U.S. experience. Chest 1994; 106(4 Suppl):197S-205S.
48. Rakos RF, Grodek MV, Mack KK. The impact of a self-administered behavioral intervention program on pediatric asthma. J Psychosom Res 1985; 29(1):101–108.
49. Mitchell EA, Ferguson V, Norwood M. Asthma education by community child health nurses. Arch Dis Child 1986; 61(12):1184–1189.
50. Jenkinson D, Davison J, Jones S, Hawtin P. Comparison of effects of a self-management booklet and audiocassette for patients with asthma. BMJ 1988; 297(6643):267–270.
51. Hughes DM, McLeod M, Garner B, Goldbloom RB. Controlled trial of a home and ambulatory program for asthmatic children. Pediatrics 1991; 87(1):54–61.
52. Robinson LD. Evaluation of an asthma summer camp program. Chest 1985; 87(suppl):105s.

53. Barnes PJ. Inflammatory mediator receptors and asthma. Am Rev Respir Dis 1987; 135:26–31.
54. Siminoff LA. Cancer patient and physician communications: progress and continuing problems. Ann Behav Med 1989; 11(3):108 112.
55. Becker MH. Theoretical models of adherence and strategies for improving adherence. In: Schumaker SA, Schron EG, Ockene JK, eds. Handbook of Health Behavior Change. New York: Springer Publishing Company, 1990.
56. Inui TS, Yourtee EL, Williamson JW. Improved outcomes in hypertension after physician tutorials. Ann Intern Med 1976; 84:646–651.
57. Maiman LA, Becker MH, Kiptak GS, Nazarian LF, Rounds KA. Improving pediatricians' compliance-enhancing practices: a randomized trial. Am J Dis Child 1988; 142:773–779.
58. Clark NM, Nothwehr F, Gong M, Evans D, Maiman LA, Hurwitz ME, Roloff D, Mellins RB. Enhancing Physician-Patient Management of Chronic Disease: Elements of Partnership. Acad Med, October 1995; 70(11):957–959.

11

Adult Asthma Self-Management Programs

WILLIAM C. BAILEY, SUSAN L. DAVIES, and CONNIE L. KOHLER

University of Alabama at Birmingham
Birmingham, Alabama

I. Introduction

Asthma education is essential to help patients acquire adequate motivation and skill to manage their disease (1). The National Asthma Education Program (NAEP) expert panel on the management of asthma concluded in its guidelines that education is one of four critical components of asthma treatment, along with objective measures of lung function, pharmacological therapy, and environmental measures (1). This chapter provides a synopsis of the recommended, and often empirically tested, components of adult asthma self-management programs. The chapter is based on research in asthma education and self-management, expert panel opinion (1), and our own experience with developing, implementing, and evaluating adult asthma self-management programs (2,3). This chapter will cover objectives of self-management programs, recommended content to include in programs, the settings and formats for delivery of programs and key program evaluation issues. We will conclude with our recommendations regarding some practical considerations in implementing an adult asthma self-management program.

II. Objectives of Adult Asthma Self-Management Programs

A. Theoretical Foundations of Self-Management

There has been a recent renewal of interest in examining the theoretical foundations of asthma self-management and the education programs designed to promote these behaviors. Theories of empowerment and self-regulation have been recommended, among others (4,5). For example, the theory of empowerment asserts that individuals who have a sense of ownership or control over decisions that impact their health will be more successful in managing their condition (6). The theory of self-regulation describes processes of self-observation, self-judgment, and self-reaction, which can influence patients to use more strategies to control their asthma and lead to fewer emergency department and doctor's office visits (5).

While any number of theories of disease and human behavior may be compatible with the concept of patients' self-management of their chronic disease (7), there is a limited number of theories that provide a useful foundation for developing and evaluating programs to teach and promote self-management. One that has been used successfully is social cognitive theory, especially as outlined by Bandura (4). A self-management program developed within the framework of social cognitive theory will address a triad of interrelated factors, including (1) environmental factors, such as attack precipitants and social support, (2) cognitive factors such as knowledge and self-confidence to prevent and manage attacks, and (3) behavioral factors, such as trigger avoidance and medication compliance. Further, such a program will incorporate modeling techniques, self-monitoring, reinforcement and observational learning of skills and will aim to increase patients' confidence in their ability to self-manage their condition.

B. Program Objectives

In general, the self-management components identified in the literature are categorized into attack prevention, attack management, and social skills, although specific tasks within these categories may differ for adults and children (8–10). Prevention of asthma attacks has traditionally included recognizing early warning signs and acting on those signs to prevent an attack, using a peak flow meter to anticipate attacks, identifying and controlling triggers, and taking prescribed medications correctly and on schedule (11). In recent years, however, the recognition that asthma is basically an inflammatory disease has changed the nature of asthma management for both the clinician and the patient. Mainline asthma treatment with inhaled anti-inflammatory medicines is, in fact, preventive treatment, and the patient's adherence to this treatment is the single most important self-management and attack prevention activity. Given this new approach to asthma treatment, attack prevention will continue to require environmental control of triggers, while recognizing early warning signs and using a peak flow meter could be considered either secondary prevention of an attack (i.e., limiting the seriousness of the attack) or attack management.

The attack-management component also includes resting and staying calm in the face of an attack, taking medications as prescribed, using predetermined criteria to seek assistance when necessary, and following other instructions outlined by the clinician (11). The social skills component of asthma self-management entails those strategies that enable the patient to manage the interpersonal relationships affected by asthma and reduce the potential impact of asthma's disruption of daily life. Effective communication and negotiation strategies within the family, with health care personnel, at the workplace, and related to leisure pursuits are important social skills in managing this disease.

III. Content of Adult Asthma Self-Management Programs

Several documents describe important components of an asthma education program (1,11–13). Given the objectives of attack prevention, attack management, and social skills, asthma self-management program content optimally includes behavioral issues related to these objectives along with broader concerns related to family, environmental influences, and health care providers. The NAEP panel emphasizes the importance of information (e.g., the differences between bronchodilators and anti-inflammatories), skills training (e.g., using an MDI correctly), and self-confidence development as crucial elements in helping patients manage their condition. Specifically, the panel recommends inclusion of the key topics listed in Table 1. In addition to the expert recommendations regarding what should be taught, the application of social cognitive theory provides a theoretical rationale for the inclusion of key elements as described below.

A. Self-Monitoring

Self-monitoring provides patients with data useful in increasing self-control of their asthma. Self-monitoring refers to observation and recording of factors related to the behavior under consideration such as physiological processes, environmental stimuli, cognitive processes and behaviors (15). Categories for self-monitoring stressed in asthma self-management programs include monitoring physical signs and symptoms, peak flow monitoring, and keeping diaries of medication taking, attacks, precipitating events, and so forth.

Table 1 Recommended Content of an Asthma Self-Management Program

Definition of the three features that characterize asthma: temporary airflow obstruction, airway inflammation, and increased airway sensitivity
Signs and symptoms of asthma and use of a peak expiratory flow meter
Characteristic airway changes (inflammation, bronchospasm, and excessive mucus production) and the role of medications to reverse these physiological manifestations
How to avoid, eliminate, or control asthma triggers such as allergens and irritants, viral respiratory tract infections, and exercise
The five tenets of effective asthma treatment: (1) individualized, continuing care; (2) medications to prevent or relieve symptoms; (3) adverse effects of medication and how to reduce them; (4) preventive treatment in the absence of symptoms; and (5) early treatment of symptoms
Patient fears concerning medication related to long-term adverse effects, toxicity, addiction, and reduced effectiveness with continued use
Use of written guidelines in language that the patient can understand, including (1) specific instructions about medication dose and frequency, under normal conditions and when symptoms occur; (2) monitoring methods including body signs and symptoms and peak flow meters to detect airway narrowing; (3) criteria for initiating or modifying treatment when necessary; (4) steps to take in an acute episode; (5) specific criteria for seeking emergency care; and (6) observable signs that therapy is less than optimal
Use of written diaries to record medication use, peak flow rates, asthma symptoms, environmental exposures, and actions taken
Correct use of inhalers including demonstration and supervised practice
Criteria for premedicating to prevent onset of symptoms with cromolyn sodium or β_2 agonist agents
Criteria for detecting the onset of symptoms (e.g., a drop in predicted peak expiratory flow rate, cough, wheeze, chest tightness and shortness of breath) and initiating treatment
Indications for emergency care (e.g., cyanosis, failure of medications to control symptoms, significant drop in peak flow rate)
Optimal use of home peak flow rate monitoring in deciding to initiate or terminate treatment, to seek emergency care, or to consider additional treatment
Periodic evaluation of results of treatment plan to make therapy changes that better suit the patient's needs and capabilities (14)
Correct fears and misconceptions: (1) asthma is not caused by psychological factors; (2) most asthma fatalities are related to undertreatment; (3) asthma need not limit activity; and (4) asthma need not lead to permanent lung disability
Need for family education, understanding and support, especially in managing acute exacerbations
Acknowledge patient's negative feelings about asthma and refer to group self-management programs, support groups, or counseling when appropriate

Source: Ref. 10.

The value of self-monitoring is that by observing and recording disease-related factors and situations in which they occur, patients can identify cause-and-effect relationships between these factors and exacerbations of their asthma. Armed with this knowledge, they can then determine appropriate responses to these factors and develop cues to prompt these responses (15). For example, if patients use diaries to observe and record their medication-taking behaviors and link these behaviors to the status of their asthma, they will be able to analyze the situations in terms of cause and effect and come up with means of improving their own adherence. Similarly, by monitoring peak expiratory flow rate with a peak flow meter, patients will be able to link certain activities with a drop in peak flow and learn to avoid these activities as necessary.

That self-monitoring strategies are associated with improved morbidity outcomes has been supported empirically (6,16,17). For example, Kotses and colleagues (17) developed an intervention based on self-monitoring of symptoms, medications taken, peak expiratory flow rates, and asthma precipitants. They reasoned that keeping track of these variables provided individuals with information, which, when viewed in the context of recent asthma conditions, signaled patients to institute strategies contributing to asthma control. This type of training led to improved symptoms and fewer physician visits in the short term and to reduced frequency of attacks in the long term.

In the UAB Asthma Self-Management Program (2), patients are instructed to monitor their peak expiratory flow rate on a daily basis for 5 consecutive days when not experiencing breathing problems in order to establish an average peak flow rate. Patients can then monitor peak flow rates on a regular basis (daily to weekly) to detect drops. Drops in peak flow rate that are more than 10% below normal are self-treated as instructed by the patient's physician. Patients experiencing a severe drop in peak flow rates, or who do not respond to self-treatment, are instructed to contact their health care provider.

Alternatively, the NAEP Guidelines recommend that patients with moderate to severe asthma monitor their peak expiratory flow rate by using the peak flow meter upon waking in the morning and at bedtime before taking inhaled medications. By doing so, they will be able to recognize potential problems, such as a decrease from predicted or personal best rates, high variability in rates indicating airway hyperresponsiveness, evening dips below morning levels indicating a possible problem with dose or timing of medication, and identifying whether sudden episodes of breathlessness are due to actual obstruction or to panic or anxiety.

Self-monitoring of medication adherence is another important component of the UAB program and others. With the increased importance of anti-inflammatory medication regimens for attack prevention, medication adherence and monitoring are critical. Patients do not adhere to medication regimens for a number of reasons, including inaccessibility of medications or inability to pay for them, forgetting, social pressures, changing perceptions of the need for the medication (i.e., feeling better or worse), misconceptions about regular use (e.g., fear of dependence), and misunderstanding of clinician instructions. In the UAB program patients are asked to indicate the reasons for missed doses of each type of medication to provide a basis for solving adherence problems. This provides both patient and clinician with information about the environment and situations in which medication adherence failures occur. Armed with this information, they can develop strategies for addressing future problem situations.

B. Environmental Factors

According to social cognitive theory, environmental factors are reciprocal determinants of behavioral and cognitive factors, and so changes in the environment affect changes in behavior and vice versa (15). This "social cognitive" view of environment includes the patient's physical living environment, which is commonly thought of as requiring control for asthma management, but also includes a broader scope of influences such as the social environment and the availability of health care. The broad range of environmental factors are important for asthma self-management, especially with regard to attack prevention.

Controlling the Living Environment

A number of aspects of the asthma patient's living environment must be controlled for better asthma management. For example, every case of asthma has unique triggers, which instigate or exacerbate symptoms, and treatment measures taken without concurrent control of these triggers will be less effective. The UAB Asthma Self-Management Program suggests three behavioral options for dealing with triggers or other problems in the physical environment:

1. Avoid the trigger entirely. It is considered essential that patients actively identify the environmental problems exacerbating their asthma and take measures to eliminate these problems. The first line of defense is to completely remove irritants, allergens, and other triggers (e.g., house dust, animals, aerosols) from the living environment. The second line is to avoid contact by not entering homes with animals, having someone else do the housecleaning, etc.
2. Limit exposure to the trigger if complete avoidance is not possible. For example, patients who are in a place where someone starts smoking or enters wearing strong perfume must leave.
3. Take an extra dose of inhaled bronchodilator medication before trigger exposure if neither of the first two options is possible.

The Social Environment

Another important environmental factor is the asthma patient's social support. An environment of supportive others to provide encouragement for preventive behaviors and assistance in attack management is extremely important in a self-management program. This is one reason why improvement of social and communication skills is considered a main objective of asthma self-management programs (8–10). Patient treatment and activity can affect other family members, and the behavior of other family members can have a direct impact on patients. Like the physical environment, the social environment can be improved by the patient's behaviors. Educating the family and others who play a significant role in the patient's life can help create an environment conducive to favorable patient outcomes, especially in relation to adherence, positive attitude, and attack management (1). Many self-management programs offer either group classes or support groups. Support groups that include both patients and "support" persons provide an effective context in which the person without asthma can learn about special needs and concerns of asthma patients and see supportive behaviors modeled by others. Finally, developing an open and trusting relationship with health care staff will improve the social environment. In the UAB Program, for example, the health educator helps patients identify questions and issues for discussion with the health care team and encourages them to take this list of questions to their next clinic appointment.

C. Self-Medication Knowledge and Skills

Asthma self-management depends on both knowledge and skills regarding the medication regimen. Social cognitive theory holds that a person's behavior is influenced by the expected outcome of the behavior and how the person values that outcome. From this perspective it is very important that patients understand the different effects of the different types of medications they take. Patients are more likely to be adherent if the outcomes from taking the medication match their expectations. For example, they should know from which medications to expect immediate relief of symptoms and from which to expect more long-term control of symptoms. With increased emphasis on anti-inflammatory treatment, this is a critical point. If a patient has the expectation that use of an inhaled anti-inflammatory medication will improve breathing within the next 15 minutes, that patient may conclude that the medication does not work and discontinue its use. Given the value of anti-inflammatory treatment, it is essential that such misconceptions be corrected.

Most asthma medications are delivered by a metered dose inhaler (MDI), and incorrect technique in using MDIs has been a source of concern (18). Incorrect technique may often result in less than optimal doses of medicine reaching the patient's airways. While pharmaceutical companies have ad-

Table 2 UAB Inhaler Use Checklist

1.	Patient shakes the inhaler for 5 seconds
2.	Patient places the mouthpiece of the inhaler into one end of the spacer
3.	Patient positions the inhaler with the index finger on top of the medication canister with thumb supporting the bottom of the inhaler
4.	Patient places the spacer tube or mouthpiece between the teeth into the mouth
5.	Patient closes mouth around spacer or mouthpiece
6.	Patient exhales normally
7.	Patient presses down the top of the medication canister with the index finger to release the medication
8.	Patient inhales deeply and slowly while medication is being released
9.	Patient holds the medication within the lungs a minimum of 10 seconds before exhaling
10.	Patient waits a minimum of 2–3 minutes before repeating steps 1–9 for second puff

Source: Ref. 18.

dressed this problem by developing devices that are easier to use, even these require careful instruction when first introduced to the patient. Most asthma self-management programs include an inhaler use instruction component as recommended by the NAEP. Social cognitive theory emphasizes the importance of modeling and observational learning in acquiring new skills. Learners must attend to, retain, and produce the motor activities involved in a task, and they must be motivated to do so. Further, practice errors should be immediately corrected and correct performance should be reinforced (4). In the UAB Program, correct MDI use is demonstrated by a nurse or patient educator and followed by supervised practice by the patient (using placebo inhalers). An illustrated list of steps is included in the patient manual to help the patient remember the steps taught. Patients' technique is then periodically assessed at subsequent clinic visits to ensure that errors are caught and corrected, and that correct use is reinforced. Assessment of technique is guided by an inhaler-use checklist, such as the one in Table 2. This checklist was developed for use with bronchodilator medications delivered by a standard MDI. Different assessment instruments would be more appropriate for use with specialized inhaled steroid devices or with the newer breath-activated inhalers.

D. Enhancing Self-Efficacy

One important goal of self-management programs based on social cognitive theory is that patients develop positive expectations regarding their ability to perform the necessary tasks for asthma control. Positive expectations about the ability to behave in a certain way is referred to as self-efficacy. Although self-management programs do not "teach" self-efficacy as a topic area, they generally include components that enhance self-efficacy. For example, education and support groups will include a variety of patients with different levels of skill in managing their asthma. The most skilled patients become models for patients who have not developed self-efficacy, who may tell themselves, "There is another patient with asthma as bad as mine. If she can handle it, so can I." Another type of activity that can increase self-efficacy is a practice activity in which small successes are experienced. For example, a quick demonstration of correct inhaler use with no practice and feedback may leave the patient confused and overwhelmed. On the other hand, a step-by-step demonstration with practice and feedback at each step will foster small successes that can build patient self-confidence.

E. Matching Content to Clinic Population Needs

In many clinical settings, all of the components listed in Table 1 may not be feasible or appropriate to include. In such situations it is necessary to determine the most important areas to cover. One approach to identifying the most important aspects or components of an asthma-education program for a local clinic population is to conduct a needs assessment. This type of data collection should include information from patients, families, and providers. Needs assessment data are collected using standard data-collection methods, such as interviews, surveys, record reviews, and observations. For example, for the UAB Program we performed a needs assessment to identify the asthma self-management education needs among our clinic population (2). Data were collected through interviews with clinic and hospital asthma patients, pulmonary physicians, pulmonary nurses, and the spouses, parents, and children of asthma patients, and we found that many asthma patients lacked a clear understanding of (1) the function and proper use of asthma medications, (2) proper utilization of health care resources, (3) trigger identification and control, and (4) early detection and treatment of attacks. We also observed interactions among physicians, patients, and families and identified factors that hinder effective communication: (1) inadequate information sharing by patients, often as a result of fear or embarrassment, (2) rushing by physicians, (3) use of inappropriate terminology by physicians, and (4) failure of patients to remember what physicians said. The knowledge and communication factors identified through the needs assessment served to tailor the UAB Program to local clinic population needs.

Wilson (9) employed a critical incident research technique to determine what patients and providers were doing that contributed to good or poor

Table 3 Individual Patient Needs Survey

I know what asthma is and can recognize the signs and symptoms of asthma.

Strongly Agree	Agree	Undecided	Disagree	Strongly Disagree
1	2	3	4	5

I understand how each of my asthma medications work to improve my condition.

Strongly Agree	Agree	Undecided	Disagree	Strongly Disagree
1	2	3	4	5

I know what my asthma "triggers" are and how I can avoid them or reduce exposure to them.

Strongly Agree	Agree	Undecided	Disagree	Strongly Disagree
1	2	3	4	5

I am confident I can prevent asthma symptoms from occurring in most situations.

Strongly Agree	Agree	Undecided	Disagree	Strongly Disagree
1	2	3	4	5

I am confident I can manage asthma symptoms when they occur.

Strongly Agree	Agree	Undecided	Disagree	Strongly Disagree
1	2	3	4	5

I know how to monitor my airways with a peak flow meter to detect potential asthma episodes before they occur.

Strongly Agree	Agree	Undecided	Disagree	Strongly Disagree
1	2	3	4	5

Having asthma does not prevent me from doing activities that I would do if I did not have asthma.

Strongly Agree	Agree	Undecided	Disagree	Strongly Disagree
1	2	3	4	5

People around me (family, friends, coworkers, school personnel) are familiar with my asthma self-management plan.

Strongly Agree	Agree	Undecided	Disagree	Strongly Disagree
1	2	3	4	5

outcomes and what factors influenced their behavior. This technique required gathering reports of actual instances in which actions or omissions of actions by patients significantly affected the control of their disease. Examples of frequently reported ineffective behaviors included (1) failure to maintain continuity of medical care, (2) failure to bring symptoms to the attention of a physician, (3) not accepting the need for preventive medication, (4) allowing social circumstances to delay taking action, and (5) incorrect MDI use. This information was used by Wilson and colleagues to develop an asthma education program for their patients.

While needs assessment activities such as those described are useful in determining the important program components for a practice setting, an individual assessment of each patient's needs is also important. The individual patient's beliefs, educational level, understanding and experience of the disease, individual need for particular types of information, and ability to comprehend should all be assessed (8). Informal assessment techniques are conducted routinely by clinicians during patient encounters, such as checking the patient's understanding of the medication regimen and observing the patient's use of the metered dose inhaler. The information gathered in this way is invaluable in giving immediate feedback and correcting faulty understanding or incorrect techniques. Table 3 is a list of questions developed at UAB that target particular types of information needs. Use of this instrument provides the basis for an individually prescribed education plan. Areas of knowledge deficit can be particularly emphasized with individual patients while still employing the overall framework of the asthma self-management program.

IV. Adult Self-Management Program Settings and Formats

Self-management education is delivered in a variety of settings and formats. Settings include the hospital, the clinic, outside agencies such as the American Lung Association, or local community facilities such as the public library and even the telephone. Formats include large and small group instruction and support, one-to-one instruction and counseling, self-help or self-instructional materials (including written, audio, video, and computer materials), and combinations of these.

A. Settings for Asthma Self-Management Programs

In a 1993 review of large-scale evaluations of asthma self-management education, the authors described seven programs that took place primarily in the outpatient clinic setting (8). Other settings included schools (for programs aimed at children and parents) and a voluntary health organization. It appears from this and other programs described in the literature that the outpatient clinic is the most commonly used setting. For example, the UAB self-management program is delivered primarily in the clinic when the patient has a scheduled doctor appointment, while another program based in a West Coast health maintenance organization is delivered in the clinical setting, but not necessarily in conjunction with a health care visit (17). However, self-management education may be delivered in less traditional settings such as neighborhood community centers and over the telephone. The point of these nontraditional venues is to reach people with asthma who may not utilize the clinic on a regular basis.

B. Formats

Asthma self-management education is delivered using a number of different formats, and there is no empirical evidence to suggest the superiority of one format over another in terms of asthma outcomes. In a recent study of the effectiveness of individual teaching compared with small-group teaching in a clinic-based asthma education program, researchers concluded that both small-group education and individual education were associated with significant benefits (such as improved MDI technique and environmental control practices), but that the group program was easier to administer, better received by patients and educators, and more cost-effective (19). In another adult self-management program (17), the developers identified three format components that they felt were related to effectiveness: presentation of material by a highly credible individual, a small-group setting, and the inclusion of written material. While patients in this program experienced better asthma outcomes than those in a control group, the improved outcomes may be attributable to a number of factors in addition to the format, most notably the content and self-monitoring methods taught.

Printed materials are commonly used as a format for delivering self-management instruction. Many programs have their own handbooks, such as UAB's *Learn Asthma Control in Seven Days* and the *Patient's Guide to Living with Asthma* developed at Ohio University. The American Lung Association publishes a variety of materials designed to teach patients about their asthma. For example, *The Asthma Handbook* is a self-help manual that covers many of the topics included in person-to-person self-management education programs, such as physiological features of asthma, trigger control, medication use, and social support. Another source of printed materials is the NAEP. This program has developed a wide variety of materials for patients and clinicians. The advantage of the printed-material format is that the materials can be widely distributed and used by patients according to their own needs and learning styles. The disadvantage is the lack of social support and professional feedback featured in person-to-person programs. Other forms of self-help materials include audio tapes, videotapes, and computer software. Video and computer materials that show color and motion are especially useful in modeling skills such as metered dose inhaler and peak flow meter use. Audiovisual materials are often available from pharmaceutical companies that market asthma medications.

V. Evaluation Issues in Asthma Self-Management Programs

Effective programs rely on the feedback obtained from evaluation activities to remain effective. As asthma treatment changes with the advent of new medications and technologies, so must the self-management program. Thus, evaluation is optimally viewed as an integral part of the ongoing education program rather than an occasional or unnecessary activity. In recent years reports on a number of adult asthma self-management programs have been published. Evaluation research data gathered in these programs support their effectiveness in a number of areas such as improving asthma knowledge, attitudes and perceptions, decreasing health care utilization, improving adherence to medical regimens, improving asthma symptoms, and reducing the number of attacks (17).

Evaluation of adult asthma self-management programs generally includes two types of activities: evaluation of program processes and evaluation of program outcomes. The purpose of process evaluation is to monitor program activities to ensure they are in line with program objectives. For example, process evaluation would include gathering data on what type of instruction and materials are given to each patient, who attended instructional sessions, who (M.D.s, nurses) is providing what instruction, and so forth. The purpose of outcome evaluation is to assess the program's effectiveness by asking if patients are appropriately managing their asthma. Examples of patient outcomes that indicate effective self-management are frequently reported in published evaluation studies. These include improved knowledge and attitudes, morbidity, health and functional status, quality of well-being, and health care utilization. Very recent and future reports of program evaluation research will likely include the patient's need for "p.r.n." or "rescue" albuterol as a measure of effective attack prevention: patients who are appropriately using anti-inflammatory medications should need fewer doses of p.r.n. albuterol.

A significant issue in the evaluation and reporting of self-management program effectiveness is the lack of standardized measurement instruments to account for these outcomes (20). For example, health care utilization may be defined as clinic visits, emergency department visits, hospitalizations, telephone calls to physicians, or any combination of these. If standardized measures of health care utilization, functional status, and morbidity are developed, it will allow more valid comparisons of what works and does not work across programs.

A special case of outcome evaluation is a cost-effectiveness analysis. In an analysis of the efficacy and cost of the UAB Program, it was reported that the intervention cost was about $32 per patient. The cost of delivering brief intervention in the control group was estimated at about $3 per patient. When these costs were viewed as a function of increased medication adherence, the total cost for the control group ($243) was more than twice that for the intervention group ($96), arguing for the cost-effectiveness of providing more intense interventions (21).

VI. Practical Considerations

Providing a comprehensive, individually tailored asthma education program that focuses on the patient's most pertinent learning needs is not always feasible in a busy clinical practice. Practically speaking, a great deal of the education takes place outside the physician-patient encounter. A team approach to self-management has consistently been recommended in the literature (14). The self-management team consists of the patient, the clinical staff, and the patient's family, all of whom also comprise the teaching and learning team of an asthma-education program.

Even when the physician conducts brief assessment activities and encourages patients to be actively involved in learning about asthma, much of the assessment and instruction fall to other members of the asthma education team, such as a nurse or health educator. Practical steps to incorporate patient education on smoking cessation in the busy office routine can be modified for an asthma-education program as well (22). They include (1) prominent labels on charts of patients needing asthma education, (2) cards or forms inserted into charts to list the major learning needs (e.g., MDI technique, pet dander control, adherence to inhaled steroids) and to remind the physician to briefly address the needed topic, and (3) easily accessible self-help materials and systematic counseling and follow-up activities. A nurse or health educator often takes the lead in developing a feasible system for the clinic. The duties of this person include setting up the system, training the rest of the team in working within the established framework, and monitoring ongoing asthma education activities.

From a practical point of view, it appears there are times within the clinical encounter most convenient for teaching. When conducted by clinic staff other than the physician, patient education often takes advantage of the waiting time spent by patients, either in the outer office or in the exam room.

For example, an assessment instrument (Table 3) may be completed by the patient while in the waiting room, reviewed by the nurse, and inserted into the patient's chart. In addition to using waiting time, patients are often asked to return to the clinic for special asthma education sessions or for prescheduled classes.

VII. Summary

The appropriate characteristics of an adult asthma self-management program were outlined by the Expert Panel Guidelines of the NAEP and have been reflected and expanded upon by a number of self-management programs described in the research literature. A shift in emphasis over recent years from bronchodilator to anti-inflammatory treatment makes attack prevention through adherence to anti-inflammatory medications, modification and control of the physical environment, and trigger avoidance the most important objectives of self-management education. A well-developed behavioral theory, such as social cognitive theory, can provide a rational basis for deciding what element to include in a self-management program. In summary, an effective asthma self-management program will reflect current treatment practices and will include elements supported by theoretical, empirical, and experiential underpinnings.

Acknowledgment

This work was supported by the Division of Lung Diseases, National Heart Lung and Blood Institute Grant RO1 HL 31481, and the Department of Veterans Affairs Medical Center, Birmingham, Alabama.

References

1. National Asthma Education Program Expert Panel. Guidelines for the diagnosis and management of asthma. 1991; 91–3042:i–136.
2. Bailey WC, Richards JM, Brooks CM, Soong SJ, Windsor RA, Manzella BA. A randomized trial to improve self-management practices of adults with asthma. Arch Int Med 1990; 150:1664–1668.
3. Kohler CL, Dolce JJ, Manzella BA, Higgins D, Brooks, CM, Richards JM, Bailey WC. Use of focus group methodology to develop an asthma self-management program useful for community-based medical practices. Health Ed Q 1993; 20:421–429.
4. Bandura A. Self-efficacy: toward a unifying theory of behavior change. Psychol Rev 1977; 84:191– 215.
5. Clark NM, Evans D, Zirnmerman BJ, Levison MJ, Mellins RB. Patient and family management of asthma. Theory-based techniques for the clinician. J Asthma 1994; 31:427–435.

6. Green LW, Frankish CJ. Theories and principles of health education applied to asthma. Chest 1994; 106:5–9.
7. Bruhn JG. The application of theory in childhood asthma self-help programs. J Allergy Clin Immunol 1983; 72:561–577.
8. Clark NM, Gotsch A, Rosenstock IR. Patient, professional, and public education on behavioral aspects of asthma: a review of strategies for change and needed research. J Asthma 1993; 30:241–255.
9. Wilson SR. VA. Patient and physician behavior models related to asthma care. Med Care 1993; 31:MS49–MS60.
10. Krutzsch CB, Bellicha TC, Parker SR. Making childhood asthma management education happen in the community. Health Ed Q 1987; 14:357–373.
11. Clark NM, Starr-Schneidkraut NJ. Management of asthma by patients and families. Am J Respir Crit Care Med 1994; 149:S54–S66.
12. Bailey WC, Clark NM, Gotsch AR, Lemen RJ, O'Connor GT, Rosenstock IM. Asthma prevention. Chest 1992; 102:216s-231s.
13. Creer TL, Kotses H, Wigal JK. A second-generation model of asthma self-management. Ped Asthma Allergy Immunol 1992; 6:143–165.
14. National Asthma Education Program Expert Panel. Executive Summary: Guidelines for the diagnosis and management of asthma. 1991; 91–3042A:i-44.
15. Tobin DL, Reynolds RVC, Holroyd KA, Creer TL. Self-management and social learning theory. In: Holroyd KA, Creer TL, eds. Self-Management of Chronic Disease. Orlando: Academic Press, 1986:29–55.
16. Madsen J, Sallis JF, Rupp JW, Senn KL, Patterson TL, Atkins CJ, Nader PB. Relationship between self-monitoring of diet and exercise change and subsequent risk factor changes in children and adults. Patient Ed Counsel 1993; 21:61–69.
17. Kotses H, Bernstein L, Bernstein DI, Reynolds RVC, Korbee L, Wigal JK, Ganson E, Stout C, Creer TL. A self-management program for adult asthma. Part I: Development and evaluation. Allergy Clin Immunol 1995; 95:529–540.
18. Manzella BA, Brooks CM, Richards JM, Windsor RA, Soong S, Bailey WC. Assessing the use of metered dose inhalers by adults with asthma. J Asthma 1989; 26:223–230.
19. Wilson SR, Scamagas P, German DF, Hughes GW, Lulla S, Coss S, Chardon L, Thomas RG, Starr-Schneidkraut N, Stancavage FB, Arsham GM. A controlled trial of two forms of self-management education for adults with asthma. Am J Med 1993; 94:564–576.
20. Bailey WC, Wilson SR, Weiss KB, Windsor RA, Wolle JM. Measures for use in asthma clinical research. Am J Respir Crit Care Med 1994; 149:S1–S8.
21. Windsor RA, Bailey WC, Richards JM, Manzella B, Soong SJ, Brooks M. Evaluation of the efficacy and cost effectiveness of health education methods to increase medication adherence among adults with asthma. Am J Public Health 1990; 80:1519–1521.
22. Solberg LI, Maxwell PL, Kottke TE, Gepner GJ, Brekke ML. A systematic primary care office-based smoking cessation program. Family Pract 1990; 30:647–654.

12

Individualized Asthma Self-Management

HARRY KOTSES

Ohio University
Athens, Ohio

In recent years, a number of asthma self-management programs have been developed whose focus it is to train patients to avoid asthma when possible and to reduce asthma when necessary. Developers of the programs promoted the idea that the best care emerges from a setting wherein the patient and physician work together toward successful management of asthma. This theme has been echoed by a recent National Institute of Health report concerning the diagnosis and management of asthma, which advocates establishing a partnership between patient and clinician (1).

Asthma self-management programs have been successful. Children trained in asthma self-management demonstrated improvements in a number of quantities including knowledge of asthma, attitudes toward asthma, compliance to management regimes, cost of treatment, and morbidity indices. The conclusion from a general review of the pediatric programs was that self-management contributes to prevention or control of attacks, reduction of the cost of care, reduction of the impact of asthma on the lives of patients and their families, and an increase in the willingness of the patient's family to shoulder asthma-management responsibilities (2). The adult programs also have met with success. One recent program for adults (3) resulted in a decrease in asthma symptoms, an increase in asthma self-management behaviors, improvement in cognitive measures, and a decrease in frequency of physician visits. Improvement in these and other variables was demonstrated in earlier research (4–13).

Although it is widely accepted that asthma self-management can have a positive impact on the disease, self-management training is not offered to most asthma patients. The reasons for this have not been clearly determined, but two considerations stand out: first, asthma self-management training has not been available in a highly efficient form; and second, self-management training cannot be implemented conveniently by physicians within the context of present-day health care delivery systems.

Different elements of asthma self-management training contribute detrimentally to its efficiency and convenience. Its efficiency is compromised by the requirement of uniform instruction. Self-management programs consist of instruction intended for all patients, regardless of their individual needs. In some situations, teaching all individuals the same things is commendable because it ensures the establishment of a minimal level of competence throughout the target population. But in training patients to manage asthma, the goal is not to achieve competence in self-management, but to achieve competence in managing one's disease. Self-management procedures that may be useful to one patient may not be helpful to another. Teaching all asthma patients the same skills means that patients almost certainly will be taught skills not applicable to their asthma. The convenience of asthma self-management training, the second area of concern, is compromised in current programs because they are not compatible with the most widely used format for delivery of medical services: the office visit. Conducting the self-management training of today requires activities more common to education than to medicine. Most physicians are neither equipped nor prepared to provide formal educational services to their patients. This is an obstacle to the routine application of asthma self-management.

An approach to asthma self-management that may overcome both the inefficient and the inconvenient aspects of the current programs is to individualize the training. Individualization improves efficiency by reducing the amount of time needed to train the individual. The reduction of time needed for training results from limiting the training each individual receives to matters directly applicable to his or her asthma. Individualization increases convenience for physicians because individualized asthma self-management programs may be provided to a patient during an office visit. Programs that may be delivered during an office visit are much more likely to be implemented than programs requiring additional arrangements involving facilities and staff. A related benefit of individualization includes delivery of the program by medical personnel, individuals who are regarded by patients as highly credible. Research has shown that the effectiveness of a program increases as a function of the credibility of the individual presenting the program (14). In addition, individualization eliminates the need on the part of the patient to sift through a comprehensive program in an effort to identify useful material. All material in an individualized program would be applicable to the patient. Finally, physicians may be more attracted to individualized asthma self-management than to group programs because the individualized approach is consistent with medical practice: it involves both a diagnostic operation and treatment recommendations based on diagnosis. For these reasons, individualized asthma self-management may ultimately have a greater impact on asthma than programs based on the traditional group format.

Unlocking the potential of individualized asthma self-management, however, presupposes the development of individualized programs. This is not an easy task. Any serious effort to individualize asthma self-management must come to grips with the tremendous variability between asthma patients, a variability that has dampened the enthusiasm for developing individualized programs. As a result, few individualized programs have emerged. More importantly, a consensus about the way in which individualized programs should be developed has not been reached. But overcoming the developmental obstacles may have great implications for the management of chronic disease. In particular, the development of individualized asthma self-management programs may represent a blueprint for future self-management programs and health care in general.

In this chapter, I review the development of asthma self-management training focused more on the needs of the individual than on the needs of the group. In connection with this discussion, I also examine a partial solution to the problem of individualization, programs tailored for specific populations. Throughout this review, my chief objective is not to evaluate asthma self-management for effectiveness, but only to highlight those operations in each program that contribute to individualization of training. In a subsequent section of the chapter, I consider differences between patients in a number of self-management areas. Variability in these areas may serve as a guide to the development of full-fledged individualized programs of asthma self-management.

I. Individualized Asthma Self-Management

The goal of individualized asthma self-management is to tailor training to the needs of the individual patient. Two approaches to tailoring have developed. One, the obvious approach, is to tailor training to the needs of the individual patient. A difficulty of this approach is that the operations used to evaluate patient needs are either limited or cumbersome. The second approach to individualization is to tailor training to the needs of a specific group, for example, inner-city children. Such programs are successful to the extent that members of a group are similar in their asthma self-management needs. An advantage of tailoring to the group is that it obviates the necessity for evaluation of the self-management needs of each patient. A disadvantage of tailoring to the group is that the self-management needs of the members of the group are never entirely the same. Additional tailoring may be needed to provide each individual with the best available program.

A. Tailoring for the Individual Patient

Individualization is both one of the latest trends and one of the earliest features of behavioral efforts to manage asthma. Prior to the advent of formal programs of asthma self-management, individualized treatment was offered to asthma patients in the form of psychotherapy. Asthma education, though not a specific objective of psychotherapy, was provided on an informal basis and tailored to the needs of individual patients (15). Educating individuals about asthma during psychotherapy was taken one step further in comprehensive asthma-management programs that were developed at residential treatment centers (16). In these programs, psychotherapy was one of a number of procedures used to control asthma. Depending on the degree to which a psychosomatic orientation prevailed in a residential treatment program, the role of psychotherapy varied from that of adjunct treatment for those patients who needed it (17) to a central element of asthma control (18). Programs with a strong psychosomatic orientation placed a heavy emphasis on psychotherapy, while other programs relied on psychotherapy to a lesser degree. Regardless of the degree of emphasis on psychotherapy, counseling of individuals with asthma undoubtedly touched upon matters relating to control of the chronic disease. Because it was a vehicle for asthma education, psychotherapy is related to modern asthma self-management programs.

General Asthma Self-Management Tailoring

With the development of formal asthma self-management programs came a number of procedures for tailoring the programs to the individual patient. In most cases, tailoring operations were considered an integral part of a group program and investigators showed little inclination either to describe them fully or to evaluate them independently. Information upon which to base the tailoring decisions was obtained either from interviews of patients or from recordings of the experience of patients. Both procedures have their strengths and weaknesses.

Tailoring Based on Information from Patient Interviews

The most common tailoring procedure is based on information obtained by interviewing the patient. If conducted properly, the interview can provide information useful in tailoring a program. To be successful for tailoring, an interview must be designed with specific objectives in mind as to the kind of information needed from the patient. The information collected must relate to components of the asthma self-management program that can be modified in some way to accommodate the needs of the patient. A strength of the interviewing procedure is the ease with which information about individualization may be collected. Weaknesses include the inability to predict in advance the kind of information needed for effective tailoring and the difficulty some patients experience in providing accurate information about their asthma.

Interviewing the patient was the first procedure used in efforts to individualize asthma self-management programs. The interview was conducted either prior to the initiation of the training program or after the program was underway by a variety of personnel. In the community asthma management program in Minneapolis (19), a member of the program staff conducted an interview with each patient prior to the initiation of the program to identify individual problems. Information from the interviews was used to determine the portions of the program that would be emphasized when the program was given. A slightly different procedure for individualization was used in the Family Asthma Program, a pediatric program developed in Buffalo (20). In this case, the patient interviews were held after the program had begun and were conducted by the patient's physician. The tailoring decisions were made by the physician who interviewed the patient. Individual needs also were identified by interview in the You Can Control Asthma program (21). The interview, in this case, was a 10- to 20-minute group interview with several objectives in addition to identification of individual needs. It was conducted by a nurse toward the end of the presentation of the program.

The introduction of individual instruction to asthma self-management training provided ample opportunity for personalization. A one-to-one counseling session was a part of an adult asthma self-management program developed in Birmingham (22). The session, organized around instruction of self-management procedures, provided an opportunity for discussion of potential solutions to the patient's problems. Even more individual instruction was given in the Pittsburgh pediatric asthma self-management program (23). Patients in this program received 4 hours of individual instruction in addition to group classes. The individual instruction was provided by a nurse-educator, who was guided by a standard curriculum but who was free to make adjustments to the curriculum to satisfy the requirements of individual patients.

A different approach to individualization was taken in two asthma self-management programs developed at Kaiser Permanente. One of these programs was designed for use with pediatric patients (24), and one was developed for use with adults (25). Like the Pittsburgh and Birmingham programs, the Kaiser programs included interview-based assessment of individual needs and individual teaching. But unlike the Pittsburgh and Birmingham programs, the Kaiser programs did not adhere to a standard curriculum that was applied generally. Instead, they included procedures by which a unique curriculum was developed for each individual. The material from which individual programs were formulated was contained within group programs for the analogous age groups (25,26). The specific information each patient in an individualized program received was identified during a diagnostic interview. The information was then delivered to the patient in three to five meetings depending on the amount of information contained in the patient's program.

The Kaiser individualized programs are unique among programs derived from patient interviews in other ways. First, the outcome of the tailoring operation was made explicit in the Kaiser programs. This is in contrast to other interview-based programs in which the outcome of tailoring was merely suggested, never described. Second, the individualized Kaiser programs have undergone evaluation (25,27). Evaluation of interview-based individualization was not possible when it was one component of a complex program (22,23).

There is little doubt that information from patient interviews has been used successfully to tailor asthma self-management programs to the needs of individual patients. But the interview-derived programs have some deficiencies. With the exception of the Kaiser programs, neither the type of information obtained from the interview nor the way in which it affected the subsequent presentation of material in the program was reported. Although most asthma self-management programs that included interview-based tailoring were successful in controlling asthma, only in the Kaiser programs was success linked to the tailoring operation. But even the Kaiser programs did not encompass the full range of problems exhibited by asthma patients; they were limited to material comprising the Kaiser group programs.

Tailoring Based on Recordings of Patient Experience

Eliminating the patient interview as a tool for gathering information about tailoring eliminates a number of problems. Once the interview is removed, its validity is not an issue, and the ability of the patient to describe asthma is not a concern. The elimination of these problems, however, is purchased at considerable cost by the burdensome nature of the procedures used in place of the interview. Two procedures have been used in this role: a statistical procedure and training patients to tailor their own programs.

If enough information about an asthma patient is collected, it should be possible to identify events statistically related to the occurrence of asthma in the patient. The events identified in this way could be the basis of a patient's individualized program of asthma self-management. This was the strategy used in the development of statistically derived, individualized programs of asthma self-management (28). Patients in the program recorded both the occurrence of asthma and the presence of any one of a large number of potential asthma triggers during a 3-month baseline period. At the end of the 3-month period, analyses were conducted to identify those events associated with the occurrence of asthma in each patient. The events related statistically to a patient's asthma were the basis of self-management recommendations given to the patient in the form of a personalized report. In addition, the report was discussed in detail with the patient during a 90-minute conference. Only four patients participated in this program, but the programs differed considerably from one another in recommendations for avoidance of specific asthma triggers, medication compliance, and relaxation. A more recent version of the program featured fewer recording requirements and a shorter baseline period (29).

A novel solution to the tailoring problem is to train asthma patients to tailor their programs. This approach was taken in both the pediatric Living with Asthma program (30,31) and in an adult program (3,32,33) derived from the pediatric program. In both programs, patients collected information about each of their asthma attacks. The information included medication status, where and when the asthma occurred, the presence of early warning signs, a listing of the activities of the patient just prior to the occurrence of asthma, and steps the patient took to reduce asthma. In connection with the collection of this information, patients were taught to evaluate the information from a number of asthma attack reports and to make decisions about the management of their asthma based on this information. Homework assignments were made to facilitate the development of these skills, and discussions of the individual concerns of patients were held during group sessions to emphasise differences between patients. Both adult and pediatric patients mastered information gathering and decision-making skills. Evaluation of this component of the programs, however, was not made independently of evaluation of the entire programs.

It is clear that procedures other than the patient interview may be used to individualize asthma self-management training. Alternatives to the patient interview have the potential to yield a greater variety of tailored programs than the interview. On the negative side, alternatives to the patient interview require a considerable amount of work on the part of the patient. Whether patients are willing to perform this work in order to individualize their programs is not known. And whether such efforts can succeed outside the research setting is also unclear. Until the time when noninterview tailoring can be conducted with a minimum of effort, it may be relegated primarily to research applications. But development of novel approaches to tailoring, regardless of how impractical they might be today, hastens the day when a high degree of individualization is a standard feature of asthma self-management.

Limited Asthma Self-Management Tailoring

Tailoring of asthma self-management to accommodate the needs of individual patients usually has been approached through individualization of comprehensive asthma self-management plans. Tailoring with less ambitious objectives, however, may also be helpful. There is increasing interest in personalizing specific aspects of asthma management in an effort to maximize the effectiveness of a general self-management intervention. Self-management instruction, for example, may be tailored to the intellectual abilities of the patient (34). This is an obvious target of individualization and may be an essential element of self-management programs. But in addition to critical features such as intellectual ability, the tailoring of several other qualities may pay dividends in the management of asthma.

Medication requirement is a candidate for tailoring. An unusual approach to self-management instruction for children with asthma involved an interactive, asthma-specific computer game (35). The children were taught self-management skills by responding to game scenarios and receiving electronic feedback representing reward and punishment. Each child's medications and allergens were entered into the computer memory and became a part of the game. A more typical way of personalizing information concerning medication was reported recently in connection with a large British study of several features of asthma self-management (36). Patients were given enhanced education, which consisted of distribution of booklets containing information about asthma in addition to their usual asthma care. The booklets, delivered to patients at monthly intervals, were personalized according to patient medication requirement and to aspects of medical history that could conveniently be personalized. The latter included, when appropriate, antismoking messages and information requested by the patient.

Medication requirement also was personalized in a self-management program conducted in a Maori community in New Zealand (37). In addition to participating in several group sessions in which information related to asthma was discussed, patients in this program were given a card on which both instructions for self-assessment of asthma and guidelines for asthma management were recorded. The latter included information about when to contact a physician and when to seek emergency treatment. Medication information on the card was personalized, as was peak expiratory flow rate, an ojective measure used by patients for self-assessment.

Other components of asthma self-management that have been personalized include the formulation of personal emergency plans and the design of personal fitness programs (38). Personal emergency plans were derived from asthma diary information with input from the patient's physician, and personal fitness programs resulted from consideration of both the patient's condition and the limitations imposed on the patient by chronic disease.

Evaluation of the effectiveness of personalizing aspects of asthma self-management has not been a major concern of investigators. Up to the present, evaluations have been concerned not so much with the tailored features of a program but with the entire program. In this regard, positive changes followed application of several programs with personalized features (35,36). In other cases (34,37,38), research design precluded meaningful evaluation. Despite this, the research illustrated the use of a number of tailoring procedures that on the surface appear significant and that for the most part may be implemented easily. Reconsideration of these and other tailoring operations may contribute both to the effectiveness and to the acceptance of asthma self-management.

B. Tailoring for Populations

Tailoring for a population combines incongruous elements: it alludes to both group and individual interests. Blending the two tendencies results in a unique approach to development of asthma self-management programs. An overview of developmental methodologies illustrates how tailoring for populations is related both to group and to individual programs. Tailoring for a population lies on a dimension anchored at one end by individualized programs, and at the other by programs without clear population targets. Movement along the dimension is accompanied by variation in two qualities: effectiveness and applicability. Programs not designed with a particular population in mind enjoy fairly broad applicability, but their effectiveness in dealing with any one population is hampered. On the other hand, individualized programs may be extremely effective for use with an individual but are deficient when applied broadly. Targeting a specific population is a compromise between these two approaches to asthma self-management programs. As such it retains the strengths and weaknesses of both.

Although a framework for targets of asthma self-management programs is theoretically useful, it may be misleading in practice. It is difficult to identify a group program that is not tailored for a specific population; the distinction between programs along the individual-group dimension is one of degree. All group programs are developed with a particular population in mind. None is suitable for universal application. On the other end of the scale, it may not be helpful to think of individualized programs as unique. The overlap between individual programs may be such that in the aggregate they approach a group program.

Tailoring grows out of the process of program development. Prior to de-velopment, problems in some population are identified and translated into program objectives. Information about problems is obtained directly from members of a population or based on expert opinion. In the latter case, the population within which problems are identified is the population with which the expert is familiar. Because program objectives are based upon the problems of a specific group, the program is more applicable to that group than to others.

Programs for various populations differ less in asthma management than they do in delivery of asthma-management programs. Factors related to program delivery have been summarized (39) and include economic concerns, psychological factors, literacy, intellectual ability, language, and cultural considerations. A program for pediatric patients that includes significant parental involvement may not work well when parental involvement is restricted by economic factors; and one that assumes a level of literacy will fail if applied to a group whose abilities are limited. Considerations such as these cannot be a part of a general asthma self-management program because they would be of little or no value in many cases. On the other hand, inclusion of these elements is essential when the problems to which they relate are encountered.

The earliest asthma self-management programs were tailored for use with children (24,26,31,39). But additional tailoring occurred in those programs even though they were not necessarily intended to accommodate the needs of a more specific population. The tailoring proceeded from the needs-assessment operations conducted prior to the development of the programs. In developing the Open Airways program (40,41), a clinic-based pediatric asthma self-management program, investigators developed guidelines for the program by interviewing clinic patients and their caretakers. The children and their caretakers resided in the neighborhoods served by the clinics. The program was based on information derived from the interviews as well as information obtained from patient medical records. It was well suited to children who were members of the group from which it was derived and who obtained asthma care in a clinic. To offer the Open Airways program in a school setting rather than in the clinic (42), additional tailoring was conducted to accommodate the special needs of students. This too was based on interviews conducted prior to administration of the self-management programs. The special requirements of students included increased emphasis on seeking adequate medical care, on school attendance, and on participation in physical activities and decreased emphasis on participation of caretakers and on patient-physician communication.

Needs assessment was conducted differently in the development of other early programs. The Air Power program, a small-group program for pediatric asthma self-management, was derived not only from information obtained by interview from children with asthma, but also from information obtained by written report from physicians, nurses, teachers, and parents (43,44). The addition of expert input to needs assessment contributed to the applicability of the program; the program was appropriate for a wide range of populations. But it was not universally applicable in that it did not address the needs of some. Expert opinion also guided the pediatric Living with Asthma program (30). Although this small-group program included a mechanism for individualization, it too was not suitable for use with all children with asthma.

Subsequent programs were tailored for use with adult asthma patients (3,11,25). These programs, developed with techniques similar to those used to develop the earlier pediatric programs, were subject to limitations of applicability similar to those of the pediatric programs. Recognition of these limitations spurred interest in programs tailored for more specific populations. Work in this area, only recently begun, has led to development of asthma self-management programs for groups delineated by two factors: age and health risk status.

Despite their emphasis on age, the early self-management programs were not applicable to patients in some age categories. One patient category not included in the early programs was preschool children with asthma. Accordingly, a program for preschool children was developed recently (45,46). It was based on input from parents of preschool children with asthma. The information was gathered from recordings of group interviews and from responses to a questionnaire. Another group, young adults in college, was the focus of a program tailored for use in a college community (8). The program may have been superior to more general adult programs for use in a college community because it was developed with the college environment in mind. At the other end of the age continuum, interest exists in asthma self-management programs for elderly individuals (47), but programs in this area have yet to be reported.

Health risk status is a major consideration in the specialization of asthma self-management programs. Populations have been targeted because their members are either disproportionately affected by asthma or have little access to asthma care facilities. Among the earliest contributions to this area was a Spanish language version of the Open Airways program (40). More recently, the application of programs for children with special needs have been reported. One pediatric program was developed for Hispanic children (48), and several have focused on inner-city black children (49,50). Evaluations of these programs have been positive. Extensive efforts to develop programs for urban children from low-income families continue. One such program includes asthma education directed to the entire community, not only to patients within a health care setting (51). Beyone pediatric populations, there is interest in tailoring asthma self-management for women during pregnancy (52). Efforts to reach special populations may ultimately include programs for occupational asthma and even programs for individuals whose special needs grow out of lifestyle choices.

II. Guidelines for the Development of Individualized Programs

Recommendations for developing individualized asthma self-management programs are lacking. Beyond interviewing the patient to identify problems, little attention has been paid to a methodology of personalization. This section is concerned with discussion of issues that could be considered during development of an individualized asthma self-management program. The topics discussed represent important areas of self-management. Within each area, variations between patients have implications for development of individualized programs. The listing of areas is not meant to be comprehensive. The information is of interest only to the extent that it suggests avenues to personalization of asthma self-management.

A. Peak Flow/Air Flow Monitoring

Home monitoring of peak flow seems like a good idea. Because peak flow scores are related to the patency of the airways, they may be useful in managing asthma. Peak flow decrease, signaling asthma in an early stage, can signal the patient to initiate asthma-management procedures. Incorporating home peak flow monitoring in an asthma self-management program may improve the effectiveness of the program.

Despite its attractiveness, home peak flow monitoring is controversial. Some initial findings relating to home peak flow recordings were promising, but subsequent research has not supported its use. In an early study of two individuals, a threefold increase over base rate predictability of asthma resulted from use of peak flow scores to predict asthma (53). Refinement of analytical procedures and a larger sample of patients yielded a higher prediction rate (54). More recent work did not support the use of home peak flow monitoring: peak flow scores were related neither to changes in lung function (55) nor to improvement in asthma management (56).

Resolution of the disparate findings may require consideration of the individual patient. It is possible that asthma patients vary in their ability to use home peak flow instruments. Support for this was derived from studies of healthy individuals. In 13 healthy individuals, within-subject correlations of mini-Wright and standard Wright scores varied from −0.21 to 0.85 (57). A similar range of variation was reported for peak flow and total respiratory resistance scores in 10 healthy subjects (58); the relationship exhibited by different individuals, in this case, varied between −0.75 and 0.28. The differences in the degree of relationship occurred despite the fact that all subjects were trained to use the instruments. A similar distribution of ability may exist within the population of asthma patients trained to use a peak flow meter. This suggests one of two courses of action: the first is to limit home peak flow recording to patients whose scores are highly related to standard pulmonary measures; the second is to provide additional training in peak flow recording to patients whose scores are not highly related to standard pulmonary measures. An individualized program can accommodate either approach.

Ability to use a peak flow meter effectively would not be the only criterion on which to base inclusion of peak flow monitoring in a personalized program. Obviously, other factors such as age, asthma severity, general physical condition, and mental condition have a bearing on peak flow use. Careful consideration of these and other variables would be helpful in deciding whether peak flow monitoring should play a role in a patient's effort to control asthma.

An alternative to home peak flow monitoring is to augment sensitivity to changes in resistance to air flow. This is a new approach to early detection of asthma, and one that has not been evaluated to any great extent. It offers an advantage over home peak flow monitoring in that it does not require the use of an instrument. The idea behind improving sensitivity to changes in air flow resistance is to provide the patient the experience of breathing throughout a range of air flow resistance levels. This is done by adding resistive loads to a circuit through which the patient breathes. Asthma patients and healthy individuals who received training with added resistive loads exhibited improvement in their sensitivity to air flow resistance variation (59–61). Insofar as personalization is concerned, the research documented differences between individuals in baseline sensitivity to change air flow resistance. If training of this sort becomes a component of asthma self-management programs, it may be personalized according to baseline sensitivity.

B. Avoiding Asthma Triggers

Educating asthma patients on how to avoid asthma triggers is a basic element of asthma self-management. Identification of a patient's asthma triggers and avoidance of them would go a long way toward controlling asthma. But this is easier said than done. Asthma may be precipitated by many stimuli. Discussion of these stimuli during asthma self-management training usually involves their inclusion in a long list of potential triggers that is given to the patient. The list does not include all triggers. Most triggers mentioned do not apply to a particular patient.

The outcome of discussing asthma triggers with patients is variable. For some patients, the discussion will merely confirm what they already know about their triggers. These patients will likely be in the minority. A year-long study of asthma triggers in 17 patients revealed they responded to triggers of which they were unaware (62). Other patients may respond to discussions of asthma triggers by formulating and testing hypotheses about potential triggers. These are recommended courses of action. It is the remaining patients that are of concern: those who might respond to information about asthma triggers with either alarm or indifference. They may become alarmed after seeing a long list of asthma triggers because of a growing belief that they are surrounded by asthma stimuli, or they may respond with indifference because they quickly realize that the vast majority of the triggers discussed do not affect them. It is possible that the information about asthma triggers in self-management programs is of little benefit to these patients.

For maximum effectiveness in controlling asthma, information about asthma triggers and their avoidance should be personalized. This step has not been taken because effective procedures for personalizing information about triggers are not available. The procedures that have been used for this purpose have serious shortcomings. Interviewing the patient in an effort to gain information about triggers, the most common trigger identification procedure is easily conducted but may identify only the most salient triggers. Other procedures, such as long-term recordings of asthma and its precipitating factors, may identify more asthma triggers than the interview but require considerable effort and time (28). Continued research on diagnosing asthma triggers is needed. Personalization of training for asthma trigger avoidance will almost certainly be a consequence of the development of effective diagnostic procedures.

C. Knowledge of Asthma

One reason for the interest in asthma self-management is that knowledge of asthma among patients is deficient. Asthma patients lack knowledge about medication use, management of attacks, and physician utilization (63). In general, quality of self-management is poor (64). Application of asthma self-management programs improves knowledge of asthma among patients.

Teaching patients about asthma has been conducted mostly on a group basis, but it need not be approached in that way. Personalized instruction makes just as much sense in this area as it does in instruction about asthma triggers. Information from diagnostic instruments can guide formulation of educational objectives for each patient. Doing so might improve the quality of instruction by allowing the patient to concentrate on only those areas in which his or her knowledge is deficient.

Diagnostic instruments for assessing educational needs are available. One instrument was designed to record the responses of patients to a number of scenarios involving asthma self-management (65). The responses were graded and combined into a score relating to overall self-management ability. Another instrument was a questionnaire that evaluates knowledge and several other quantities (66). These and similar instruments may be useful in determining the type and amount of knowledge a patient has about aspects of asthma care and may help establish the content of instructional programs concerned with asthma knowledge.

D. Psychological Considerations: Behavior and Attitudes

Each individual displays unique behavior and holds a unique set of attitudes. Some of these quantities may be inconsistent with effective management of asthma. Asthma self-management programs that take account of disruptive behavior and attitudes will likely be more effective than programs that do not.

Among the behavior that can seriously undermine efforts to bring asthma under control is failure to comply with medical recommendations (67). Many, but not all, patients exhibit this detrimental behavior. For those patients who do not comply with medical recommendations, some sort of educational or behavioral intervention is in order. Deciding on an intervention of this sort may involve either the analysis and subsequent modification of factors that give rise to noncompliant behavior, or the application of a procedure effective in control of noncompliant behavior (68). Other behavior that may adversely affect asthma self-management includes any of a number of adjustment problems. In this, asthma patients are no different from anyone else who experiences a problem of adjustment. The health educator, however, must be sensitive to to adjustment problems and take steps to minimize their impact on the management of asthma. In some cases, this may be done within the context of an individual's self-management program (69). In other cases, adjustment problems may require referral for counseling. In any event, modifying behavior that negatively affects asthma management must be accomplished on an individualized basis.

Like behavior, attitudes may interfere with the management of asthma. The study of attitudes in the management of asthma is well established (70). Attitudes important for individuals undergoing asthma self-management training included perceived vulnerability to to disease, perceived quality of care, and perception of one's own ability to control asthma (71). Strongly held negative views in any one of these areas may require special attention during asthma self-management training. In particular, the attitude relating to perception of one's own ability to control asthma has received attention (66,72). The attitude, called asthma self-efficacy, has been used to evaluate asthma self-management programs (3,66), but it may also be used to signal the need for measures that increase a patient's feeling of control. Additional interventions based on diagnosis may be adopted to overcome other attitudes detrimental to the success of self-management training.

III. The Future of Asthma Self-Management

It's a good guess that the trend toward personalization of asthma self-management will continue. Asthma is treated one patient at a time, and self-management training will be conducted one patient at a time. The effective measures identified by research in asthma self-management eventually will become part of the armamentarium of the physician who treats asthma. There are signs of this already. A self-management program for use with individual patients has been developed (73). It includes instruction in self-management, peak flow monitoring, dairy recoding, medication use, and management plans—many of the effective features of asthma self-management. Like asthma self-management programs in general, it encourages formation of a partnership between physician and patient for the purpose of managing the patient's asthma. As more programs of this type are developed and adopted, the need for group approaches to self-management will diminish. The occurrence of this will signal maturity in asthma self-management training.

References

1. Expert Panel Report, National Asthma Education Program, National Heart, Lung, Blood Institute. Guidelines for the diagnosis and management of asthma. Pediatr Asthma Allergy Immunol 1991; 5:57–188.
2. Wigal JK, Creer TL, Kotses H, Lewis P. A critique of 19 self-management programs for childhood asthma. Pediatr Asthma Allergy Immunol 1990; 4:17–39.
3. Kotses. H, Burnstein IL, Burnstein DI, Reynolds RVC, Korbee L, Wigal JK, Ganson E, Stout C, Creer TL. A self-management program for adult asthma. Part I: Development and evaluation. J Allergy Clin Immunol 1995; 95:529–540.
4. Bauman AE, Craig AR, Dunsmore J, Browne G, Allen DH, Vandenberg R. Removing barriers to effective self-management of asthma. Patient Educ Counsel 1989; 14:217–226.
5. Ringsberg KC, Wiklund I, Wilhelmsen L. Education of adult patients at an "asthma school": effects on quality of life, knowledge and need for nursing. Eur Respir J 1990; 3:33–37.

6. Moldofsky H, Broder I, Davies A, Leznoff A. Videotape education program for adults with asthma. Can Med Assoc J 1979; 120:669–672.
7. Snyder SE, Winder JA, Creer TL. Development and evaluation of an adult asthma self-management program: Wheezers Anonymous. J Asthma 1987; 24:153–158.
8. Tehan N, Sloan BC, Walsh-Robart N, Chamberlain MD. Impact of asthma self-management education on the health behavior of young adults. J Adolesc Health Care 1989; 10:513–519.
9. Maiman LA, Green LW, Gibson G, Mackenzie EJ. Education for self-treatment by adult asthmatics. JAMA 1979; 241:1919–1922.
10. Bolton MB, Tilley BC, Kuder J, Reeves T, Schultz LR. The cost and effectiveness of an educational program for adults who have asthma. J Gen Intern Med 1991; 6:401–407.
11. Bailey WC, Richards JM Jr, Brooks CM, Soong S, Windsor RA, Manzella BA. A randomized trail to improve self-management practices of adults with asthma. Arch Intern Med 1990; 150:1664–1668.
12. Mulhauser I, Richter B, Kraut D, Weske G, Worth H, Berger M. Evaluation of a structured treatment and teaching program on asthma. J Intern Med 1991; 230:157–164.
13. Beasley M, Cushley M, Holgate ST. A self-management plan in the treatment of adult asthma. Thorax 1989; 44:200–244.
14. Cooper J, Croyle RT. Attitudes and attitude change. Annu Rev Psychol 1984; 35:395–426.
15. Hanley FW. Individualized hypnotherapy for asthma. Am J Clin Hypnosis 1974; 16:275–279.
16. Peshkin MM. Asthma in children: IX. Role of environment in the treatment of a selected group of cases: a plea for a "home" as a restoration measure. Am J Dis Child 1930; 39:774–781.
17. Feldman Ch, Davis WJ, Feldman BR, Grebin B. St. Mary's Hospital for Children: a model regional residential center for children with chronic asthma. J Asthma 1983; 20:17–25.
18. Steiner H, Fritz GK, Hilliard J, Lewiston NJ. A psychosomatic approach to childhood asthma. J Asthma 1982; 19:111–121.
19. Blumenthal MN, Cushing RT, Fashingbauer TJ. A community program for the management of bronchial asthma. Ann Allergy 1972; 30:391–398.
20. Hindi-Alexander M, Cropp GJA. Community and family program for children with asthma. Ann Allergy 1981; 46:143–148.
21. Taggert VS, Zuckerman AE, Sly RM, Steinmueller C, Newman G, O'Brien RW, Schneider S, Bellanti JA. You can control asthma: evaluation of an asthma education program for hospitalized inner-city children. Patient Educ Counsel 1991; 17:35–47.
22. Bailey WC, Richards JM, Brooks M, Soong S, Windsor RA, Manzella BA. A randomized trial to improve self-management practices of adults with asthma. Arch Intern Med 1990; 150:1664–1668.
23. Fireman P, Friday GA, Gira C, Virthaler WA, Michaels L. Teaching self-management skills to asthmatic children and their parents in an ambulatory care setting. Pediatrics 1981; 68:341–348.

24. National Heart, Lung, and Blood Institute. Air Wise: Self-Management of Asthma Through Individual Education. Bethesda, MD: National Institutes of Health, 1984; publication no. 85–2363.
25. Wilson SR, Scamagas P, German DF, Hughes GW, Lulla S, Coss S, Chardon L, Thomas RG, Starr-Schneidkraut N, Stancavage FB, Arsham GM. A controlled trial of two forms of self-management education for adults with asthma. Am J Med 1993; 94:564–576.
26. National Heart, Lung, and Blood Institute. Air Power: Self-Management of Asthma Through Group Education. Bethesda, MD: National Institutes of Health, 1984; publication no. 85–2362.
27. McNabb WL, Wilson-Pessano SR, Hughes GW, Scamagas P. Self-management education of children with asthma: AIR WISE. Am J Public Health 1985; 75: 1219–1220.
28. Kotses H, Stout C, Wigal JK, Carlson B, Creer TL, Lewis P. Individualized asthma self-management: a beginning. J Asthma 1991; 28:287–289.
29. Kotses H, Stout C, McConnaughy K, Winder J, Creer TL. Evaluation of individualized asthma self-management programs. J Asthma 1996; 33:113–118.
30. Creer TL, Backiel MA, Burns KL, Leung P, Marion RJ, Miklich DR, Morrill C, Taplin PS, Ullman S. Living with Asthma: I. Genesis and development of a self-management program for childhood asthma. J Asthma 1988; 25:335–362.
31. Creer TL, Backiel M, Ullman S, Leung P. Living with Asthma. Parts I & II. National Institutes of Health, publication no. 86–2364, 1986.
32. Creer TL, Kotses H, Reynolds RV. A Handbook for Asthma Self-Management: A Patient's Guide to Living with Asthma. Athens, Ohio: Ohio University Press, 1991.
33. Creer TL, Kotses H, Reynolds RV. A Handbook for Asthma Self-Management: A Leader's Guide to Living with Asthma. Athens, Ohio: Ohio University Press, 1991.
34. Richards W, Church JA, Roberts MJ, Newman LJ, Garon MR. A self-help program for childhood asthma in a residential treatment center. Clin Pediatr 1981; 20:453–457.
35. Rubin DH, Leventhal JM, Sadock RT, Letovsky E, Schottland P, Clemente I, McCarthy P. Educational intervention by computer in childhood asthma: a randomized clinical trial testing the use of a new teaching intervention in childhood asthma. Pediatrics 1986; 77:1–10.
36. Osman LM, Abdalla MI, Beattie JAG, Ross SJ, Russell IT, Friend JA, Legge JS, Douglas JG. Reducing hospital admissions through computer supported education for asthma patients. Br Med J 1994; 308:568–571.
37. D'Souza W, Crane J, Burgess H, Karu HT, Fox C, Harper M, Robson B, Howden-Chapman P, Crossland L, Woodman K, Pearce N, Pomare E, Beasley R. Community-based asthma care: trial of a "credit card" asthma self-management plan. Eur Respir J 1984; 7:1260–1265.
38. Byrne DM, Drury J, Mackay RC, Robinson S, Faranda C, MacAdam DB. Evaluation of the efficacy of an instructional programme in the self-management of patients with asthma. J Adv Nursing 1993; 18:637–646.

39. FitzGerald JM. Psychosocial barriers to asthma education. Chest 1994; 106: 260–263S.
40. National Heart, Lung, and Blood Institute. Open Airways/Respiro Abierto: Asthma Self-Management Program. Bethesda, MD: National Institutes of Health, 1984; publication no. 85–2635.
41. Clark NM, Feldman CH, Freudenberg N, Millman EJ, Wasilewski Y, Valle I. Developing education for children with asthma through study of self-management behavior. Health Educ Q 1980; 7:278–297.
42. Kaplan DL, Rips JL, Clark NM, Evans D, Wasilewski Y, Feldman CH. Transferring a clinic-based health education program for children with asthma to a school setting. J Sch Health 1986; 56:267–271.
43. McNabb WL, Wilson-Pessano SR, Jacobs AM. Critical self-management competencies for children with asthma. J Pediatr Psychol 1986; 11:103–117.
44. Wilson-Pessano SR, McNabb WL. The role of patient education in the management of childhood asthma. Prev Med 1985; 14:670–687.
45. Mesters I, Pieterse M, Meertens R. Pediatric asthma, a qualitative and quantitative approach to needs assessment. Patient Educ Couns 1991; 17:23–34.
46. Mesters I, Meertens R, Kok G, Parcel GS. Effectiveness of a multidisciplinary education protocol in children with asthma (0–4 years) in primary health care. J Asthma 1994; 31:347–359.
47. Chapman KR, Dales RE, Chan-Yeung M, Cote J. Future research in asthma education. Chest 1994; 106:270–273S.
48. Lewis MA, de la Sota A, Rachelefsky G, et al. ACT—Asthma control y tratamiento para ninos: a progress report. Health Ed Q 1987; 14:281–290.
49. Shields MC, Griffin KW, McNabb WL. The effect of a patient education program on emergency room use for inner-city children with asthma. Am J Public Health 1990; 80:36–38.
50. Taggart VS, Zuckerman AE, Lucas S, Acty-Lindsey A, Bellanti JA. Adapting a self-management education program for asthma for use in an outpatient clinic. Ann Allergy 1987; 58:173–178.
51. Fisher EB, Sussman LK, Arfkin C, Harrison D, Munro J, Sykes RK, Sykvia S, Strunk RC. Targeting high risk groups: Neighborhood organization of pediatric asthma management in the neighborhood asthma coalition. Chest 1994; 106:248–259S.
52. National Asthma Education Program. Executive Summary: Management of Asthma During Pregnancy. Bethesda, MD: National Institutes of Health, 1992; publication no. 93–3279A.
53. Taplin PS, Creer TL. A procedure for using peak expiratory flow rate data to increase the predictability of asthma episodes. J Asthma Res 1978; 16:15–20.
54. Harm DL. Kotses H, Creer TL. Improving the ability of peak expiratory flow rates to predict asthma. J Allergy Clin Immunol 1985; 76:688–694.
55. Sly PD, Cahill P, Willet K, Burton P. Accuracy of mini peak flow meters in indicating changes in children with asthma. Br Med J 1994; 308:572–574.
56. Drummond N, Abdella M, Beattie JAG, et al. Effectiveness of routine self-monitoring of peak flow in patients with asthma. Br Med J 1994; 308:564–567.

57. Kotses H, Harver A, Creer TL. An intraindividual comparison of standard and mini-Wright scores. Ann Allergy 1984; 52:419–422.
58. Wigal JK, Kotses H. Creer TL, Harm DL, Tinker TR. Total respiratory resistance and peak expiratory flow rate: an intrasubject comparison. J Asthma 1986. 23:11–13.
59. Stout C, Kotses H, Creer H. Improving recognition of respiratory sensations in healthy adults. Biofeedback Self-Reg 1993; 18:79–92.
60. Harver A. Effects of feedback on ability of asthmatic subjects to detect increases in flow-resistive components to breathing. Health Psychol 1994; 13:52–62.
61. Stout C, Kotses H, Creer TL. Improving perception of air flow obstruction in asthma patients. Psychosom Med (in press).
62. Stout C, Kotses H, Carlson BW, Creer TL. Predicting asthma in individual patients. J Asthma 1991; 28:41–47.
63. Avery CH, March J, Brook RH. An assessment of the adequacy of self-care by adult asthmatics. J Commun Heath 1980; 5:167–180.
64. Dekker FW, Kaptein AA, van der Waart MAC, Gill K, Quality of self-care of patients with asthma. J Asthma 1992; 29:203–208.
65. Taylor GH, Rea HH, McNaughton S, Smith L, Mulder J, Asher MI, Mitchell EA, Seelve E, Stewart AW. A tool for measuring the self-management competency of families. J Psychosom Res 1991; 35:483–491.
66. Wigal JK, Stout C, Brandon M, Winder JA, McConnaughy K, Creer TL, Kotses H. The knowledge, attitude, and self-efficacy asthma questionnaire. Chest 1994; 104:1144–1148.
67. Kinsman RA, Dirks JF, Dahlem NW. Noncompliance to prescribed as needed (PRN) medication use in asthma: Usage patterns and patients characteristics. J Psychsom Res 1980; 24:97–107.
68. Creer TL, Kotses H. Asthma. In: Ollendick TH, Hersen M, eds. Handbook of Child Psychopathology. New York: Plenum, 1989:341–357.
69. Perrin JM, MacLean WE, Gortmaker SL, Asher KN. Improving the psychological status of children with asthma: a randomized controlled trial. J Dev Behav Pediatr 1992; 13:241–247.
70. Kinsman RA, Dirks JF, Jones NF. Psychomaintenance of chronic physical illness: Clinical assessment of personality styles affecting medical management. In: Millon T, Green C, Meacher R, eds. Handbook of Clinical Health Psychology. New York: Plenum, 1982:435–466.
71. Richards JM, Dolce JJ, Windsor RA, Bailey WC, Brooks MC, Soong S. Patient characteristics relevant to effective self-management: scales for assessing attitudes of adults toward asthma. J Asthma 1989; 26:99–108.
72. Tobin DL, Wigal JK, Winder JA, Holroyd KA, Creer TL. The "Asthma Self-Efficacy Scale." Ann Allergy 1987: 59:273–277.
73. Plaut PF, Brennan CF. Asthma Charts and Forms for the Physician's Office. Amherst, MA: Pedipress, 1994.

13

Adherence

ANDREW G. WEINSTEIN

Jefferson Medical College
duPont Hospital for Children
Wilmington, Delaware

I. Introduction and Case Outline

This chapter addresses a critical question in the care of asthmatic individuals: How can health care providers improve adherence to medical recommendations to reduce the increasing pulmonary morbidity and mortality from asthma? It attempts to answer this question by presenting (1) specific asthma-management guidelines (pharmacological, environmental, educational, and immunotherapy) that need to be followed, (2) methods to assess adherence to these guidelines, and (3) specific approaches to promote compliance with these management principles.

In the past, patients who did not follow medical directives were defined as noncompliant. Compliance usually referred to the extent to which patients and families were obedient and faithful to health care providers' instructions, proscriptions, and prescriptions (1). It implied a passive role for the patient. The title of this chapter implies a more active, voluntary collaborative involvement in a process that provides a mutually acceptable preventive or therapeutic result (2). It implies choice and negotiation. From this chapter, it will become clear that asthma is a complex disease with multiple factors that produce symptoms and multiple therapeutic options that require patient/clinician cooperation. The basic therapeutic choices are avoidance of allergens, administration of medication, and immunotherapy when indicated. A case example will outline the clinical problem and potential strategies to promote adherence.

Case Outline

A 10-year-old girl is referred to an allergist because of a recent hospitalization for asthma. She lives with her mother during the week and stays with her father every other weekend. Allergy testing identifies pollen and cat sensitivity. Her mother smokes throughout the house, and the cat lives with her father. The physician recommends inhaled anti-inflammatory medicine with an inhaled bronchodilator as needed. For this asthmatic child, definitive therapy would include total elimination of smoking and the cat from the households. If presented as an ultimatum, most families would not change lifestyles. If the family were informed that reduction in irritants and allergens could reduce asthma symptoms, however, they might be agreeable to restrict smoking to one room, in which there is an air purifier, and keep the cat out of the patient's bedroom. If symptoms continue and the cat remains in the household, immunotherapy injections may be required. The parents agree to start immunotherapy injections because they both recognize that their daughter will continue to be around the smoke and the cat. The lifestyles will be partially altered with the mother limiting smoking and the father excluding the cat from the child's bedroom. In this case, pharmacological therapy and immunotherapy are required because the principal asthma triggers, smoke and cat allergen, will remain in the households, although in decreased concentration.

Adherence is not a dichotomous (yes/no) assessment. Blackwell suggests that patients and families may exhibit different adherence rates for different reasons for each component of a complex regimen (3). Returning to the case example, the physician prescribes two inhaled medications, beclomethasone (anti-inflammatory) three times a day and albuterol (bronchodilator) every 4 hours as needed. The mother administers the medication in the morning and at bedtime. The child frequently forgets to take the medications after school when she is not supervised. The father does not think the anti-inflammatory medication is effective and lets his daughter use the albuterol inhaler as needed. Both parents find time to take their daughter for immunotherapy injections because the physician's office is convenient to both households and there are evening and weekend hours. Prescription costs are not a factor as they are covered by both parents' health plans. In this family, adherence is highest for immunotherapy injections. The child receives anti-inflammatory therapy approximately 70% of the time if the child remembers. Need for bronchodilator therapy is a function of allergen/irritant exposure and location of the child. If the child spends more time at her father's house, for example during a summer vacation, anti-inflammatory medication stops altogether. Symptoms may be less, however, because there is no exposure to smoke during that time interval and the child is outside much of the time during the day. This family is not atypical of asthma-management problems that face health care providers. The complexities underscore the challenging task of not only how to conceptualize adherence, but also how to assess adherence.

The term "clinician" in this chapter refers to physicians as well as nurses who have responsibility in counseling asthmatic patients and their families. The term compliance will refer to subjects in research studies who are requested to follow specific directions and health care providers who are recommended to follow specific guidelines for treatment (i.e., use of anti-inflammatory medication, peak flow meters, etc.).

II. Epidemiology, Morbidity, and Mortality

The cost, prevalence, morbidity, and mortality from asthma in both children and adults have been increasing over the past 20 years. Weiss et al. analyzed the economic impact of asthma in the United States (4). Estimated costs in 1985 were $4.5 billion; this number increased by 38% in 1990 to $6.2 billion. In 1990, 60% of the direct costs resulted from hospitalization and emergency care. Indirect costs included days lost from work and school and premature deaths.

The National Health Interview Survey (NHIS) documented that the prevalence of asthma in the United States increased 29% from 1980 to 1987 (5). The prevalence rose from 31.2 to 40.1 per 1000 population. There was a higher incidence of asthma in both child and adult black populations when compared with white populations. The greatest increase in asthma prevalence occurred in individuals under 20 years of age, at a rate of 42%. Taylor and Newacheck examined NHIS data and reported that in 1987, 1.7 million children with asthma missed 10.1 million days from school, spent 7.3 million days restricted to bed, and required 1.9 million days of hospitalization (6).

The National Hospital Discharge Survey found that the hospital discharge rate for asthma as the first listed diagnosis increased 6% from 1980 to 1987 (7). Gergen and Weiss found that hospitalization increased 4.5% for children 16 years or younger from 1979 to 1987 (8). Asthma mortality increased 6.2% annually during the 1980s (9). Mortality was disproportionately higher for children, urban dwellers, blacks, and those in lower socioeconomic groups (9–11). Death rates are increasing fastest among children under the age of 14 years. Although the number of childhood and young adult asthma deaths is relatively small (481 in 1987), this represents a 63% increase since 1967, when this age group averaged 295 deaths per year (9).

The National Asthma Education Expert Panel Report has identified and published an effective asthma-management plan that, if followed, should significantly reduce morbidity and mortality from asthma (12). This plan includes the use of effective anti-inflammatory medication, home peak flow monitoring, and a written attack management plan developed by the physician for the asthmatic patient. The phrase "if followed" is synonymous with "if adhered to." Despite gains in understanding the pathogenesis and the treatment of asthma, poor adherence is the primary reason for the increase in morbidity, mortality, and costs for asthma. This cost increase reflects poor compliance by physicians who do not prescribe anti-inflammatory medications or peak flow meters (or fail to demonstrate their use), as well as patients who fail to obtain these treatment modalities or use them at all.

III. Pathophysiology

Asthma is an inflammatory disease of the airways in which there is intermittent and variable reversible airway obstruction (13). The degree of symptoms (cough, wheeze, dyspnea) and obstruction is a reflection of the amount of airway hyperresponsiveness. Hyperreactivity is defined by the contraction of large and small airways in response to chemicals and pharmacological agents inhaled in a pulmonary function laboratory (14). Asthma symptoms may develop in response to a variety of exposures that include allergens, irritants, viruses, cold air, and exercise. Airway inflammation and airway reactivity are closely related (15). The level of airway responsiveness correlates with the clinical severity of asthma and anti-inflammatory medication requirements (16). Therapeutic interventions that reduce bronchial inflammation appear to decrease airway hyperresponsiveness (17). The frequency and severity of symptoms is a function of the type and length of exposure to triggering agents and the type and amount of medications taken (13). Individuals with mild asthma will have intermittent symptoms requiring intermittent treatment compared with the moderately severe asthmatic who requires daily therapy. Those with intermittent symptoms may be at a disadvantage in adhering to a specific therapy that is used infrequently, easily misplaced, and requires the correct technique (18). Individual patients may incorrectly assess their expected response to an allergen or irritant and not be prepared for more severe symptoms. An asthmatic patient with cat sensitivity may have visited a home with a cat for one hour and left with mild congestion. The same individual who stayed longer may have more severe symptoms and may not be prepared to treat the more severe attack. Failure to anticipate and seek emergency care has been recognized as a risk factor of death from asthma (19).

IV. Strategies to Treat/Prevent Asthma

A. Goals of Therapy

The expert panel report has listed specific goals for asthma therapy (12). These goals include (1) maintaining normal activity levels, including exercise, (2) maintaining normal or highest level of pulmonary function, (3) preventing recurrent exacerbations of asthma, and (4) avoiding adverse effects from asthma medications. To reach these goals, clinicians need to consider the following general treatment principles: moderate to severe asthma is a chronic condition that requires continuous care; prevention (avoidance of allergens/irritants and continuous around-the-clock medication) is paramount; anti-inflammatory agents, used prophylactically, may decrease airway hyperresponsiveness and prevent irreversible obstruction; and anticipatory or early interventions in treating acute exacerbations of asthma reduce severe airway narrowing (12).

Asthma therapy has several integral components, including pharmacological therapy, patient education, environmental control, and immunotherapy (Figure 1). Each component will be examined to determine what is required

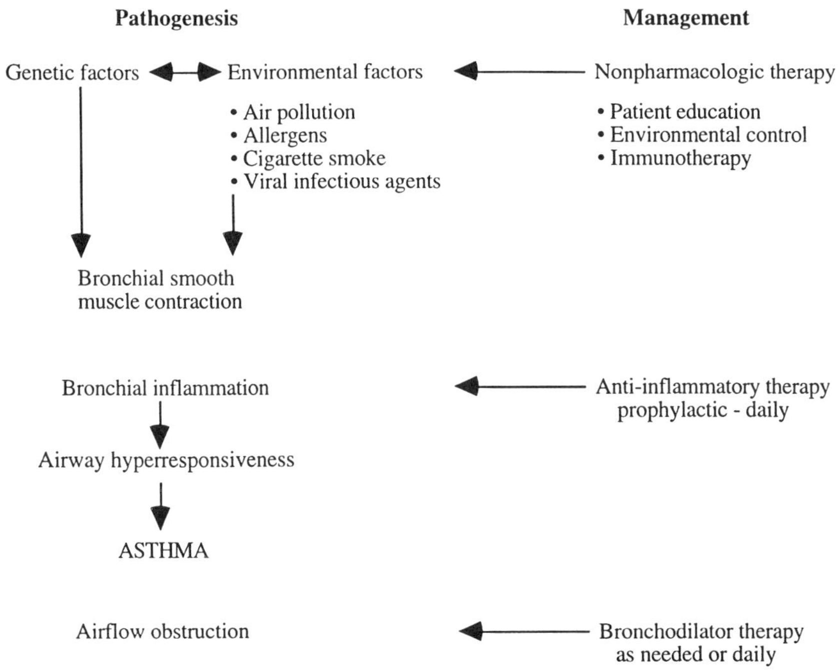

Figure 1 General approach to the understanding and treatment of asthma. (Adapted from Ref. 12.)

Table 1 Medication and Adherence Measures

Medication action	β_2 Agonist		Theophylline	Cromolyn	Nedocromil	Steroids		Leukotriene antagonists and synthesis inhibitors	Anticholinergic
	Inhaled	P.O.				Inhaled	P.O.		
Anti-inflammatory	—	—	+	+++	+++	++++	++++	+++	—
Bronchodilator	++++	+++	+++	—	—	—	—	+	
Biochemical measure	—	Blood RIA Experimental	Blood RIA Clinically available	Urine HPLC Experimental	Urine HPLC Experimental	—	Total eosinophil count	Blood RIA Experimental	—
Observation MDI technique	+			+	+	+			+
Medication measure	Weight	Pill count Liquid	Pill count	Weight	Weight	Weight	Pill count Liquid	Pill count	Weight
Medication monitors	NEB, Chron	MEMS	MEMS	NEB, Chron	NEB, Chron	NEB, Chron	MEMS	MEMS	NEB, Chron
Frequency of administration	q4 to q12 p.r.n. daily	q4 to q12 p.r.n. daily	qd/bid/tid daily	bid/qid daily	bid/qid daily	bid/qid daily	qd to tid q.o.d./p.r.n. daily	bid to qid	qid p.r.n. daily
Cost	++[a] ++++[b]	+	+	++++	+++	+++[c] ++++[d]	+	++++	+

Key: RIA = radioimmunoassay; HPLC = high pressure liquid chromatography; MEMS = medication event-monitoring system; NEB Chron = Nebulizer Chronolog.
[a] Albuterol.
[b] Salmeterol.
[c] Low dose.
[d] High dose.

to control asthma and how physicians and other health care workers can assist patients in adhering to these instructions.

B. Pharmacological Therapy

Pharmacological therapy is used to treat and prevent reversible air flow obstruction and airway hyperresponsiveness. Medications include bronchodilators and anti-inflammatory agents. Some drugs may act as both. Primary anti-inflammatory agents include corticosteroids, leukotriene antagonists and synthesis inhibitors, cromolyn sodium, and nedocromil sodium. Bronchodilators include beta-adrenergic agonists, methylxanthines, and anticholinergics (Table 1).

Anti-inflammatory Agents

Corticosteroids

Corticosteroids are the most effective anti-inflammatory drugs for the treatment of asthma. This class of medication decreases inflammation by (1) interfering with arachadonic acid metabolism and the synthesis of leukotrienes and prostaglandins, (2) preventing the directed migration and activation of inflammatory cells, and (3) increasing the responsiveness of beta-agonist receptors of the airway smooth muscle (20).

Corticosteroids are effective in decreasing/preventing the late-phase reaction associated with allergen triggers. Corticosteroids may be given parenterally, orally, or in aerosol form (21). Oral and parenteral forms are used primarily for the treatment of exacerbations of acute asthma. Oral corticosteroids may also be used in the day-to-day management of the severe asthmatic who has persistent asthma despite maximal use of all other treatment agents. Inhaled corticosteroids are safe and effective for the adult and child asthmatic when given in the recommended doses (22,23). These medications must be given preventively to be effective. Because of the importance of decreasing airway inflammation, inhaled corticosteroids are being used more frequently as a primary therapy of adult asthmatics (12). Adverse side effects associated with high-dose, short-term systemic therapy that discourage their use by asthmatic patients include increased appetite, fluid retention, weight gain, and mood alteration (20). Adherence with oral corticosteroids may be directly measured via a total eosinophil count, which should be markedly reduced or be zero if adherent. Adherence with inhaled corticosteroids may be monitored indirectly by the Nebulizer Chronolog (Medtrak, Auroa, Co) (24).

Leukotrine Antagonists and Synthesis Inhibitors

Leukotriene antagonists and synthesis inhibitors are the most recent anti-inflammatory medications for asthma. Leukotrienes are cytokines that result from the breakdown of a cellular membrane constituent, arachadonic acid, via the 5-lipoxygenase enzyme pathway. Leukotrienes have various anti-inflammatory biological activities associated with asthma: increased vascular permeability, enhanced mucus production, decreased mucociliary transport, and bronchoconstriction. Two general types of drugs have been developed with the goal of attenuating the leukotriene's effects on inflammation and asthma: leukotriene receptor blockers and leukotriene synthesis inhibitors (e.g., 5-lipoxygenase inhibitors) (25). Those drugs are indicated for prophylaxis of chronic asthma and not indicated for use in the reversal of bronchospasm in acute asthma attacks.

Leukotriene receptor antagonists have the ability to block the early and late phase responses to allergens and prevent chronic, as well as exercise- and cold-induced, asthma (25). Zafirlukast is a leukotriene antagonist that became available for adults and children 12 and over in the United States in 1996. It is taken in tablet form b.i.d. In placebo-controlled trials with Zafirlukast, there was a 30% reduction in rescue use of β_2 agonists, a 46% reduction in night waking, and a 26% improvement in morning asthma (26). In these early trials, the main adverse effects reported for leukotriene antagonists were headache, dry mouth, and somnolence. This drug should be taken 1 hr before or 2 hr after meals. Although further clinical studies are clearly necessary, leukotriene antagonists are safe and efficacious in the treatment of asthma.

Leukotriene synthesis inhibitors have been shown to decrease bronchial hyperresponsiveness, allergen-induced bronchoconstriction, and asthma secondary to aspirin and nonsteroidal antiinflammatory drugs (25). Zileuton is a 5-lipoxygenase inhibitor that became available for adults and children 12 or over in the United States in 1997. It is taken in a tablet form q.i.d. Zileuton has been shown to be effective in mild to moderate asthma, increasing FEV_1, as well as reducing asthma symptoms, β agonist, and inhaled corticosteroid use (27). Headache was the most common side effect associated with zileuton. Approximately 3% of individuals treated with zileuton may have reversible elevated liver function tests, thus requiring periodic monitoring. Patients who are receiving both zileuton and theophylline should receive lower doses of theophylline, since the clearance of theophylline is decreased. Zileuton may be taken with or without meals.

Leukotriene antagonists and synthesis inhibitiors can be directly monitored by radioimmunoassays, which are not commercially available.

Cromolyn Sodium/Nedocromil Sodium

These are the best aerosolized nonsteroidal anti-inflammatory drugs available (28). It is theorized that these agents stabilize and prevent mediator release from mast cells (29). These drugs are administered prophylactically and only by aerosolized form and may be given by the metered dose inhaler, roto-inhaler, or nebulizer. Both inhibit early- and late-phase allergen-induced narrowing and acute airway narrowing secondary to exercise, cold air, and sulfur dioxide. Nedocromil sodium may also decrease airway hyperresponsiveness (29). Cromolyn sodium is most effective in treating mild asthma. Biochemical monitoring of these drugs is primarily experimental and measured by analyzing urine samples by high-pressure liquid chromatography (HPLC). Indirect measure can be made with the Nebulizer Chronolog.

Bronchodilators

Beta-Adrenergic Agonists

Beta-adrenergic agonists are the primary bronchodilators used to reverse sudden onset of bronchospasm (30). The desirable pharmacological effects of these drugs result from their action on the β_2 adrenergic receptor, which primarily leads to smooth muscle relaxation and secondarily may modulate mediator release from mast cells and basophils (31). Although selective for β_2 receptors, these medications may stimulate β_1 receptors that may produce undesirable cardiovascular stimulation, anxiety, and tremor. Inhaled therapy is comparable to or better than oral therapy in producing bronchodilation and also generates fewer side effects (30). Aerosols to counter bronchospasm have a limited duration (4–6 hours). A 12-hour bronchodilator (Salmeterol), however, is available to maintain bronchodilation and is primarily for individuals who also receive inhaled corticosteroids (32).

β_2 agonists are the medication of choice for the treatment of acute exacerbation of asthma and for the prevention of exercise-induced asthma. These medications can be used either intermittently to control episodic narrowing or chronically to prevent persistent narrowing in more severely affected patients. Recent studies suggest that daily use of inhaled β_2 agonists may increase asthma symptoms either by increasing airway hyperresponsiveness or by developing tolerance or subsensitivity resulting from downregulation of β-adrenergic receptors (33). It is for these reasons that short-acting (every 4 hours) β_2 agonists are recommended for as-needed use at times of acute symptoms (12). Oral β_2 agonists may be administered two or three times daily to maintain bronchodilation. Oral β_2 agonists are more likely to cause side effects and have slower onset of action when compared with the inhaled group. Oral β_2 agonists can be monitored directly by radioimmunoassay, which is not commercially available (33). Inhaled β_2 agonists delivered by the metered dose inhaler can be monitored by the Nebulizer Chronolog.

Methylxanthines

Theophylline, a methylxanthine, is prescribed as a daily chronic therapy to reverse continuous bronchospasm (34). Recent studies have suggested that theophylline may have anti-inflammatory properties that attenuate late-phase reactions (35,36). Because theophylline is eliminated from the body rapidly, sustained release preparations have been developed to maintain serum levels in the therapeutic range (5–20 μg/ml) (37). Sustained-release preparations are absorbed slowly and are designed to eliminate peak-trough variations in blood levels. Periodic monitoring for toxicity is required to avoid higher blood levels outside the therapeutic range that could produce gastrointestinal, cardiac, and central nervous system symptoms (38). The dose of theophylline depends on the absorption distribution, metabolism, and elimination of the drug (38). Factors that affect absorption are sustained release delivery systems (37), as well as dietary carbohydrate and fat content (39). Because 90% of theophylline is metabolized in the liver, elimination is affected by liver disease (40), congestive heart failure (41), and medications that are cleared by the liver (42,43). Febrile illness may decrease clearance (44), whereas smoking may increase clearance (45). All of these factors need to be considered when interpreting the serum level, either for selecting the specific dosage or monitoring for adherence. Frequency of dosing varies depending on the age of the patient. Younger children (1–4 years) eliminate the drug most frequently, requiring three daily doses (46). Older children (4–17 years) require two daily doses (47). Healthy, nonsmoking adults require one or two doses daily (45). Direct measure of adherence can be made by readily available commercial assays (34).

Anticholinergics

Anticholinergics are the oldest form of bronchodilator therapy for asthma. Inhaled drugs block post ganglionic efferent vagal pathways and act by reducing vagal tone and decreasing bronchoconstriction caused by inhaled irritants (49). A newer form of anticholinergic, ipratropium bromide, has fewer atropinelike side effects (dry respiratory secretions, blurred vision, central nervous system stimulation) as compared with the older anticholinergics, and has been shown to be effective in treatment of acute exacerbations of asthma in children and adults (50). The benefit of its use in day-to-day management of asthma, however, has not been established. Regular use of anticholinergics as bronchodilators appears to be most effective in patients with chronic obstructive pulmonary disease and partially reversible air flow obstruction (51). Indirect measurement of adherence can be made by the Nebulizer Chronolog.

Aerosol Therapy

All aerosolized medications that are used to treat asthma are available as metered dose inhalers (MDI) (52). The major advantages of aerosol therapy are the direct delivery of medication to the site of need, a rapid onset of action, and the small quantities of drug required to produce beneficial effect. A major disadvantage using metered dose inhalers is that they are difficult for some patients to use and faulty technique leads to suboptimal therapeutic responses and poor control of asthma (53). The number of subjects who could not use inhalers correctly has varied from 12 to 89% with the average incidence of misuse being 38% (53). Problems occur in all age groups but are greatest in the young and elderly. The most common errors are failure to coordinate actuation of the metered dose inhaler on inhalation and inappropriately rapid inspiratory flow rates. Education and training in the correct use of MDI can improve technique (54). A significant number of physicians, nurses, and pharmacists, however, do not know how to assemble or use MDIs correctly (55).

To overcome the problems with coordination and enhance aerosol deposition in the lower airways, a number of inhalation aids have been added to conventional MDIs. These aids are called spacers or chambers, designed to facilitate delivery of the aerosol to the airways (56). Patients need to be instructed to use the MDI correctly with spacers, and their technique should be checked periodically. A 12-step systematic approach has been developed to assist patients in correctly using these devices. Spacers are effective in (1) reducing the irritant properties of the aerosol and tendency to cough and (2) reducing medication deposition in the mouth and oropharynx and the possibility of oral candidiasis when used to deliver corticosteroids (56). Spacers can increase the use of MDIs in young children and decrease the need for jet nebulizers, which are less portable and more time consuming. A breath-activated pressurized MDI has become available (57). This device eliminates problems of coordination of inhalation.

Metered-dose inhalers are pressurized devices that contain chloro-fluorocarbons as propellants. Environmental laws and treaties to reduce the atmospheric depletion of ozone are curbing the manufacturing of MDIs in favor of non–ozone-depleting alternative propellants (53). These dry-powder inhalers require an inhalation technique that is different from the MDI technique. They are breath-activated, nonpressurized, dry powder aerosols that can be delivered by the spinhaler, roto-haler, disk inhaler, and turbo-haler (58).

C. Peak Flow Meters

Asthmatic patients and their families are able to objectively monitor their lung function at home with hand-held portable peak flow meters (12). Objective measurements of air flow obstruction are desirable because subjective statements such as dyspnea or wheezing may be inaccurate (59). Poor perception of the severity of asthma on the part of the patient and physician has been cited as a major factor causing delay in treatment and possible mortality from asthma (60). The following uses of home peak expiratory flow rate monitoring have been reported for asthmatic patients: (1) daily monitoring to detect early stages of airway obstruction and initiate therapy before obstruction becomes serious, (2) monitoring the course of treatment using objective criteria to initiate or terminate steps in the treatment plan, (3) determining when emergency medical care is needed, and (4) facilitating communication between patient and clinician by providing an objective assessment of asthma severity and implement an asthma-management plan based on peak flow values (61,62). Peak flow meters can be used by adults and children over 5 years of age. The National Asthma Education Program Expert Panel Report has developed an eight-step systematic method of measuring and reporting peak flow values (12).

D. Patient Education

Health education by the clinician is a powerful tool for helping patients gain the motivation and skill to control their asthma. By establishing a partnership with the patient and/or family, the health care team has the opportunity to promote adherence to the treatment plan The National Asthma Education Program Expert Panel Report lists teaching guidelines that clinicians should present in a written format to their patients/families (12). Failure to do so would represent clinician noncompliance. The guidelines are listed in Table 2.

The pharmacy has a significant role in promoting patient adherence with asthma therapy. The American Society of Hospital Pharmacists developed guidelines in 1977 for pharmacists to follow when prescribing medications for patients (63). These guidelines became legal mandates with the passage of the Omnibus Budget Reconciliation Act of 1990 (OBRA '90). This law requires pharmacists to give oral and/or written instructions as presented in Table 3. Pharmacy practice laws in each state vary to the extent that they permit pharmacists to exercise their professional judgment in deciding which of the 14 items to discuss with the patient.

E. Environmental Control

Allergy has a significant role in the pathophysiology of asthma. Studies from the United States, Europe, Australia, and New Zealand have established that IgE-mediated sensitization to indoor allergens (dust mite, cockroach, cat, and dog) is a major risk factor for asthma and asthma exacerbation (64–69). Similarly, tobacco smoke inhaled directly or passively can induce the development of asthma or aggravate the asthmatic condition (70–73). These indoor allergens/irritants become increasingly important as individuals spend up to 95% of their time indoors (at least 50% at home) in temperature-controlled houses with low air-exchange rates, which promote mite growth and accumulation of animal dander allergens (74). Because persistent exposure to allergens and irritants induces airway inflammation and hyperreactivity, environmental con-

Table 2 Written Guidelines for Asthmatic Patients and Families

Written guidelines given by clinicians should include the following points:

Specific instructions about use of medications, including dose, frequency of administration, guidelines for changing dose or adding medications if appropriate, and adverse effects to report to the clinician.

Observable signs of asthma, such as sleep interruption and/or consistently low or highly variable expiratory flow rate (PEFR). Such signs should be discussed with the clinician.

How to monitor body signs or symptoms and/or PEFR to detect increasing airflow obstruction as early as possible; early signs of airflow obstruction vary according to the individual and should be identified for each patient.

Attack management

Criteria for initiating or modifying treatment: a drop in PEFR or early signs or symptoms.

List of steps to take in managing an acute asthma episode (i.e., removing the precipitating trigger, giving medication, avoiding strenuous physical activity, and keeping patient and family calm).

Specific criteria for seeking emergency medical care, including a pattern of declining PEFR; failure of medications at home to control worsening of symptoms; difficulty in breathing (wheeze may be absent), walking, and/or talking; intercostal retractions; blue fingernails or lips.

Source: Adapted from Ref. 12.

Table 3 Pharmacy Instructions to Asthmatic Patients and Families

1. Drug name
2. Intended use and expected action
3. Dosage: form and amount
4. Route of administration and schedule
5. Special directions for preparation
6. Special directions for administration
7. Precautions to be observed during administration
8. Common side effects that may be encountered, including their avoidance and action required if they occur
9. Techniques for self-monitoring of drug therapy
10. Proper storage
11. Potential drug-drug or drug-food interactions or other therapeutic contraindications
12. Prescription refill information
13. Action to be taken in the event of a missed dose
14. Information peculiar to a specific patient or drug

Source: Ref. 134.

Table 4 Allergen Avoidance Recommendations

Dust mites[a]
Use plastic, zippered casing for mattress, boxspring, and pillows
Wash bedding in hot water or hot dryer cycle (> 130°F) weekly
Remove carpeting and upholstered furniture from bedroom
Use chemical agents to kill dust mites or alter antigen
Animals with fur (dogs, cats, etc.)[b]
Remove from home. If removal is not possible:
Wash animal weekly
Remove carpets
Use HEPA air filter and vacuum cleaner

[a] Adapted from Refs. 202,203.
[b] Adapted from Refs. 204,205.

trol is a critical therapy to reduce or prevent asthma symptoms (75). House dust mites thrive in areas where there is a high concentration of human epithelial cells, such as bedding and carpeting. Dust mite populations grow best in high humidity (76). Recommendations to avoid dust mites are presented in Table 4. By eliminating mite exposure. one can reduce airway reactivity and medication required to control asthma.

Dogs and cats are found in 43% and 28% of American households, respectively (77). Dander and saliva are potent allergens, with dander being found in airborne samples. To eliminate exposure to animal dander, the animals should be removed from the house. Even after the pet is removed from the house and the house is cleaned, residual allergen can be found in the house months to years later (78). Most individuals sensitive to pets keep them. Recommendations to reduce animal allergen exposure are presented in Table 4. The major allergens produced by dust mite, cat, dog, cockroach, and several molds have been purified, leading to the development of enzyme-linked immunosorbent assays (ELISA), which can detect the presence and concentration of these allergens (79). Households of sensitive individuals can be monitored initially to determine the presence and concentration of the above allergens, and then again at a later time.

F. Immunotherapy

Allergy testing [skin testing, radioimmunosorbent test (RAST)] identifies potential asthma triggers that should be avoided (e.g., cat, dust mite) (80,81). When avoidance is not possible and appropriate medications are ineffective in preventing asthma symptoms, immunotherapy (allergy injection treatment) is considered as an additional modality (82). Immunotherapy has been shown to be effective in decreasing asthma symptoms as well as increasing the threshold of subsequent inhalation challenges to house dust mite, cat dander, grass pollen, and alternaria (83–86). This therapy has been shown to be effective in both children and adults. Patients receive injections weekly during the "build-up" phase. They continue to receive increasing doses of vaccine, until they reach the maintenance level in approximately 8–12 months (82). Injections may be given less frequently, every 2–4 weeks, during the 3- to 5-year period of immunotherapy. Injections may be discontinued after 2 years if there is no improvement. Patients may discontinue immunotherapy after 3–5 years, although the risk of relapse and return to a weekly schedule at a later date in the future need to be weighed against patient preference of continuing with monthly injections and minimal symptoms.

V. Measuring Adherence to Asthma Regimens

A. Direct Measures

Biochemical Assay

Biochemical analysis of blood, urine, or other bodily fluids is an objective, accurate method of determining if a patient has recently taken a specific medication (87). This analysis may be done by assaying the drug or its by-products. Unfortunately, most asthma medications delivered by MDI (cromolyn, β_2 agonists, corticosteroids, or anticholinergics) are poorly absorbed and therefore cannot be measured by commercial medical laboratories. Theophylline can be measured by radioimmunoassays in blood or saliva to permit monitoring for adherence (34). Saliva specimens are less intrusive and painful than blood specimens. The salivary assay is not as rapid or as accurate as a blood assay and is not commercially available (88). Cluss and colleagues have added a tracer substance (riboflavin) to theophylline capsules to permit a non- quantifiable (yes/no) measure of adherence when urine is examined under ultraviolet light (89). Radio-immunoassays are available for oral β_2 agonists and could be commercially available if there were significant clinical demands (33).

Biochemical analysis permits an objective and accurate measure of adherence; however, there are limitations to this approach. For example, there are many factors that confound the interpretation of theophylline blood levels. These factors include variations in theophylline absorption, distribution, metabolism, and elimination within the same patient (90); timing of sampling in relation to oral ingestion (91); time-released characteristics of the particular medication (37); and the metabolism of the drug in relation to other medications received (42,43), foods eaten (39), and smoking behavior (45). Repeated theophylline levels are intrusive and painful. Biochemical analysis can provide adherence information at the time of sampling, but it does not reliably predict adherence with the medication at other times. It is also possible that the patients may take monitored medication immediately prior to sampling so as to appear to be adherent (24,92,93). When biochemical monitoring becomes commercially available for leukotriene agents, physicians will need to refer to clinical pharmacokinetics and bioavailability data to assess adherence.

Observational Measures of MDI Technique by Medical Staff

Many asthma patients lack the fine motor skills to effectively use MDIs (53). The current technique includes (1) shaking the MDI prior to use, (2) holding the MDI in an inverted position, (3) activating the inhaler while inhaling slowly and deeply, and (4) breath holding after inhalation (12). The use of a spacer enhances medication delivery (56). Trained persons can measure the patient's use of MDI, and these results can be scored or quantified on observational checklists (54). Adherence with MDI technique can be measured within the medical office; however, the technique observed may not be the technique used at home.

B. Indirect Measures

Clinical judgment is the most common method health care providers use to assess patient adherence. It is based on the clinician's overall assessment of the clinical situation. The factors that affect this determination include the patient's personal characteristics (age, sex, race, and socioeconomic status), past medical history including prior adherence assessments, and the clinical condition at the time of the visit (92). Physicians frequently assume adherence if the patient is asymptomatic. Certain "adherent types" of patients or families are thought to be adherent even when symptomatic. Studies have shown that physicians greatly overestimate the degree to which patients and families adhere to a treatment plan (94,95). No cost and rapid determination also account for its wide use. The value of clinical judgment in clinical practice may be as a supplement to other more objective measures, because accuracy rate of clinical judgment by itself is thought to be no better than chance (96,97).

Self-Report/Asthma Diaries

These patient reports, presented at the time of an office visit, are the most common measure of adherence in a research setting. They are utilized because they are easy to obtain, quick to analyze, and inexpensive. They are useful in examining the relationship between disease exacerbation and adherence to asthma-management plans (93). It is possible to follow the use of multiple medications daily and/or as needed and correlate their use to environmental allergens/irritant exposure while observing peak flow values over time. Asthma diaries are most commonly used by clinical investigators, although the self-report instruments are also valuable in clinical practice. Diaries developed by Clark et al. (Asthma Management Index) (98) and Creer et al. (Asthma Problem Behavior Checklist) (99) have been used in a research setting to measure medication adherence. There are weaknesses in relying heavily on asthma diaries. Patients frequently overestimate self administration of medication and underreport nonadherence. When patients have completed self-reports at the same time that they have been monitored by other means (e.g., pill count or Nebulizer Chronolog), there have been wide variations between self-report and the other measures (24,100–103). These discrepancies may be a result of patient characteristics, memory, motivation, and desire to please physicians (93). Another weakness is that patient reports or diaries require training of patients and health care workers, who need to follow these patients over time.

Medication Measurement

A number of methods are used to measure the patient's remaining amount of medication upon the return to the office, which infers use during that time interval. Counting pills and measuring the weight of liquids or MDI canisters are examples of this procedure (92). Another method to indirectly measure how much medication "should have been used" is the accounting for prescriptions written and prescriptions obtained (87). These approaches are primarily used by clinical investigators. The advantages of medication measurement is that the data are simple to collect and are objective. Limitations include: (1) the patient may fail to return the medication, (2) the patient may deceive the investigator, and (3) there is no indication of the timing or accuracy of the dosing.

Electronic Medication Monitors

Electronic medication monitors are devices that utilize microprocessors to record and store the date and time that a medication container is opened or activated. The Nebulizer Chronolog encircles a commercial MDI and records the actuation of the canister (24). A microprocessor can be attached to the caps of medication containers to document the time of opening (104). These devices cannot confirm that the medication was taken, they only infer their administration. Research studies with the Nebulizer Chronolog have documented subjects emptying their canister at the time of their collection, unaware of being monitored (24). Electronic medication monitors such as these have documented the disparity between patient self-report and drug adherence and support general claims that patients consistently overestimate their adherence (87,102). Disadvantages of these devices are the expense and the frequency of malfunction (93).

Observation by a Family Member

One measure of adherence is the report of direct observation of the patient by another individual, such as a family member (105). The advantages of direct observation are that it is objective, quantifiable, and more valid and reliable than self-report. The disadvantages of direct observation are that the observation of family members may be obtrusive and can disrupt activity or interactions at home, "reactivity" may occur because someone is aware of being monitored and may change behavior, and it is difficult to observe and not to interfere with patients, particularly children (106). It is possible to place video monitoring cameras in proximity to medication administration areas to observe both technique and timing of administration (107). This approach may be helpful in selected patients who have persistent symptoms and when biochemical measurements are not feasible. An outline of procedures used to measure adherence to asthma therapy is given in Table 5.

In summary, no one method of measuring adherence is ideal. Multiple measures will give an investigator or clinician a sense of patient adherence. Clinicians may be able to identify nonadherence by asking the patient carefully worded questions. Use of closed-ended questions such as "Are you using your asthma inhalers?" are more likely to be answered in a dichotomous (yes/no) response than identifying irregular or incorrect use (108). Steele et al. found that physicians who used open-ended information-rich interviewing strategies were better able to identify actual adherence patterns than those using information-poor closed-ended questions (109). If nonadherence is identified by taking the medical history, clinicians do not need biochemical assays to corroborate this finding at the time of the office visit. If the patient states that he or she has adhered to the prescribed plan, then the clinician must rely on the adherence measures reviewed above to "guess" the patient's true adherence status.

VI. Adherence and Asthma Therapy

A. Patient Adherence

Adherence has been generally defined by Haynes as "the extent to which patient behavior (in terms of taking mediations, following diets, or executing lifestyle changes) coincides with medical or health advice" (1). Cluss and Epstein have expanded this definition to include knowledge of the correct name of the medication, attendance at follow-up appointments, and properly refilling prescriptions (110). Jerome and colleagues have listed a variety of common forms of nonadherent behavior including complete failure to follow a regimen, poor comprehension of dosing schedule, omission of doses, increasing or decreasing dosing schedule, discontinuing treatment earlier than prescribed, taking medications for the wrong reason, using outdated or discontinued medication, or taking more treatments than prescribed (111). Nonadherence with asthma therapy has been documented. Studies of asthmatic children prescribed theophylline have reported adherence with this medication by measuring theophylline blood levels (89,112–118). Adherence rates vary from 2% to 100% and reflect methodological differences in the patient population and definition of adherence.

Studies of children being treated for acute wheezing report theophylline adherence varying from 2 to 66% (114–117). Groups using lower theophylline levels (i.e., greater than or equal to 5 μg/ml) as a marker for adherence had greater adherence rates than those using higher levels (i.e., ≥ 10 μg/ml) as a definition for adherence. This finding was also true in outpatient studies (89,118). LeBaron et al. found that 83% of asthmatic children receiving cromolyn were adherent, as measured by urinary assays (119).

Adherence with MDIs can be compromised by improper technique as well as by not following the prescribed regimen. Several reports have documented that more than 50% of asthmatic children do not use their MDIs correctly (120–122). Manzella and colleagues developed a 10-item inhaler use checklist to assess a 10-step process in MDI use (54). They found that only 31 of 238 (13%) adult asthmatics performed all steps correctly. Compton reported that 72% of adult patients studied had poor technique (123). Underuse of MDI has been reported by both child and adult asthmatics. Coutts et al. studied asthmatic children prescribed beclomethesone (124). By measuring adherence with Nebulizer Chronolog, they found that 55% of asthmatic children were underusing this medication. Nides et al. studied MDI compliance with the Nebulizer Chronolog in adults prescribed ipratropium bromide (125). Subjects, on average, used this medication 1.65 actuations per day compared with the requested three actuations per day.

Adherence with medication has been a primary concern regarding optimal care of asthmatic patients. Maintaining adherence with immunotherapy (82) and environmental avoidance (77) (allergens and irritants), however, are also important in managing the patient and reducing medication requirements. Adherence rates for asthmatic individuals receiving immunotherapy in a university hospital setting have been reported to be 56% (126) in an adult clinic and 61% (127) in child clinics. Adherence rates for immunotherapy in all atopic patients (asthma, rhinitis, eczema) in a private practice setting have been found to be 83% (128). Adherence to dust mite and cat avoidance recommendations have not been measured but are presumed to be less than 10% of patients

Table 5 Measuring Adherence to Asthma Therapy

Measures	Description	Strength of strategy	Weakness of strategy	Cost
Biochemical	Analysis of medication in blood, saliva, urine, and biological tracer	Direct, objective quantifiable	Limited number of assays for asthma medication Results do not reflect adherence over time. Intrusive, painful Results vary by: timing of test; age of patient; absorption of drug; concommitant medication	Low to moderate
Direct observation	Use of trained staff to document appropriate inhaler use	Simple, objective, direct	Skills demonstrated in office may be different than at home Observer may become involved Difficult for some physicians	Low
Clinical judgement	Clinician's overall judgement of adherence	Fast, simple, inexpensive	Not valid, no standard criteria	Low
Self-report	Retrospective review of patient's progress	Fast, simple to administer, detailed information	Often overestimates adherence Subject to distortion or memory	Low
Asthma diaries	Daily record of medication use, peak flow values, days school/work missed	Detailed account of adherence and outcome data	Poor patient adherence over long term Subject to patient overestimate	Low
Medication measurement	Measuring: pill counts, liquid, canister weights, prescription refills	Objective, simple, validates other measures	Provides no information about daily patterns of medication usage Overestimates adherence Dependent on patient's cooperation to return medication	Low to moderate

Medication monitors	Electronic monitors that record date and time of medication use events (e.g., Nebulizer Chronolog)	Objective	Cost	High
		Records of date and time of actuation.	Records use of medication dispensing, not medication taken in	
	Medication event monitoring system	Minimal patient effort to provide long-term data		
Treatment outcome	Identify symptomatic patients who may be nonadherent	Identify patients who need to be adherent	Low validity	Low
		Data available to clinicians	Outcome influenced by factors other than adherence	
Assessment of immunotherapy	Indirect review of injection record and vaccine ordered	Quick, simple, quantifiable	May not account for patients switching to another physician and receiving immunotherapy there	Low
Assessment of environmental allergens/irritants:				
Environmental allergens	Direct measure of dust mite, cockroach, dog, cat by ELISA assay	Direct assessment by health care personnel, who also examine extent of allergen exposure	Reassessment of measurable allergen exposure may not reflect patient's efforts to reduce allergen	High
Cigarette smoke	Measure CO by breath analysis of patient	Direct	Intrusive	Moderate
		Identifies smoking		
	Home visit to determine if smoke in home environment	Identifies smoking in environment	Not quantitative	

Source: Adapted from Refs. 92,93.

and families treated by allergists in the United States (129). Butz and Rosenstein examined smoking behavior of parents with chronically ill children (130). They reported that 80% of parents with children with asthma and cystic fibrosis modified their smoking behavior after the diagnosis was made. Success in smoking cessation is difficult to achieve. Of the 17 million Americans who attempt to quit each year, only 1.3 million (7%) are successful (131).

B. Health Care Provider Compliance

Physician Compliance

The section on measuring patient adherence is based on the assumption that the health care provider is knowledgeable regarding current asthma-management principles and applies them appropriately. Physicians should be able to provide patients with correct medications, dosing schedules, and medication administration techniques, as well as peak flow monitoring and attack management plans. One might label physicians noncompliant if they are aware of the National Asthma Education Program Expert Panel Report: Guidelines for the Diagnosis and Management of Asthma (12) but are not following their recommendations. Surveys have been conducted estimating the prevalence of physicians prescribing inhaled corticosteroids and peak flow meters for asthma patients. In 1992, it was estimated that 55% of allergists/immunologists, 43% of pulmonologists, 25% of family practitioners, and 11% of pediatricians were prescribing inhaled corticosteroids (132). A survey of 66 primary care physicians attending continuing medical education credit asthma seminars at the University of Maryland in 1995 revealed that only 43% routinely use peak flow meters when evaluating patients with lower respiratory symptoms, and only 47% were familiar with the guidelines for the diagnosis and management of asthma (133).

The lack of compliance with or awareness of current asthma-management therapy underscores the need for continuing medical education to improve clinician adherence.

Pharmacist Compliance

Dispensing errors are one indicator of the quality of drug distribution tasks by pharmacists (134). Dispensing error rates of inpatient pharmacists have been documented to range from 3 to 12% (135,136). OBRA '90 requires pharmacists to counsel Medicaid patients orally and/or in writing on 14 categories (Table 2). Allen and colleagues studied randomly selected community pharmacies by having trained observers pose as patients and document pharmacists' actions (137). They requested pharmacists to fill prescriptions of three oral medications [Coumadin (DuPont Pharma, Wilmington, DE), Tegretol

(Ciba Geneva Pharmaceuticals, Summit, NJ), and a sustained release theophylline preparation, Theo-Dur (Key Pharmaceuticals, Kenilworth, NJ)]. These investigators found that these subjects were receiving an average of only 9 of 14 categories of counseling required by law. Analysis of 100 prescription orders dispensed detected 24 dispensing errors, 4 of which were clinically significant. One theophylline error was an underdose, which could have led to an increase in the signs and symptoms of asthma. The authors of this study suggested that "problems with the quality of community pharmacy medication counseling and dispensing require immediate attention." It is important to realize that these errors and deficiencies in counseling were for oral medications, which are the easiest to administer. Pharmacists dispense MDIs, spacers, and peak flow meters, which require significantly more time to demonstrate their use. One would expect even lower levels of compliance when dispensing these types of medications and devices.

VII. Factors Recognized as Influencing Adherence

Haynes has reviewed the medical literature regarding specific determinants that predict adherence behavior (138). He found that features of the disease (i.e., diagnosis, severity, type of symptoms, duration of symptoms, previous hospitalization, and clinical improvement) were not usually associated with adherence behavior. There are several exceptions, however: (1) patients with psychiatric diagnosis are believed to be less adherent than individuals with any other diagnosis (139); (2) patients with increasing severity of symptoms may demonstrate less adherence (140); and (3) developing a disability produced by a disease may increase adherence, most likely as a result of increased supervision that accompanies a disability (141) (Table 6). Haynes notes that the longer the elapsed time between referral and the actual appointment, the lower the likelihood that the appointment will be kept (142). Longer waiting time in the clinic and lack of specific appointment time were related to less likelihood of the patient returning to the clinic for another visit (143). The treatment process was related to adherence practices. Not surprisingly, parenteral drug administration was associated with compliance. The greater the duration of treatment, the number of medications and treatments prescribed, the complexity of treatment, the cost of treatment, and the frequency of pharmacy errors, the less likely the individual/family would adhere to the recommended plan. Haynes also emphasized the importance of communication, patient satisfaction, and the physician awareness of patients concerns in determining adherence rates (138). Physicians and other health care workers are more likely to experience nonadherence with asthma patients because asthma is a chronic and sometimes lifelong illness, because asthma is treated

Table 6 Important Factors Related to Adherence Patterns in General Medicine and Asthma

	Relationship to adherence	
	General medicine	Asthma (Ref.)
The Disease	–	
Mental illness, especially schizophrenia, paranoia, and personality disorder		
Psychological function	+	+ (103,147)
Symptoms/severity	–	– (124,148)
Disability	+	
The Referral Process		
Time from referral to appointment	–	
The Clinic		
Waiting time	–	
Individual appointment time	+	
The Treatment		
Parental drug administration	+	
Duration of treatment	–	– (148)
Type of medications	–	– (149)
Frequency of medications	–	– (151)
Complexity of medication	–	– (150)
Number of medications	–	– (54)
Cost	–	– (144)
Safety containers	–	– (93)
Erring and errant pharmacists	–	– (137)
The Doctor/Patient Relationship		
Communication	+	
Patient satisfaction	+	
Physician awareness of patient concern	+	

+ Related to adherence; – related to nonadherence.
Source: Adapted from Ref 138.

with multiple medications prescribed in varying frequency depending on monitored lung function and delivered in a complex vehicle and because asthma medications are expensive (144).

Psychological functioning of chronically ill children and their families has been identified as a factor contributing to poor adherence, particularly in adolescents (145,146). Christiaanse et al. studied 38 asthmatic children and found that both a child's psychological adjustment and family cohesiveness were positively related to adherence as measured by theophylline levels (147). Weinstein et al. studied 37 severe asthmatic children treated in an inpatient rehabilitation unit and followed for one year as outpatients (103). During the one-year follow-up period, families with adherent children had significantly higher psychological functioning than families of nonadherent children as measured by global assessment scales. This finding may reflect the family's ability to work effectively with the asthmatic child in following asthma-management guidelines at home.

Studies of both child and adult asthmatics support Haynes's observation that the more symptomatic patients may be nonadherent. Coutts et al. (125) monitored asthmatic children receiving inhaled corticosteroids with the Nebulizer Chronolog. They found that despite symptomatic asthma, there was underuse of inhaled steroids on 55% of study days. Mann et al. examined the relationship between patient adherence to a four-times-a-day beclomethasone schedule and periods of increased asthma severity in adult asthmatics (148). The patients had a mean beclomethasone adherence of 67%, as measured by Nebulizer Chronolog, and did not improve compliance either when peak flows were lower or when albuterol use increased because of symptoms.

Adherence rates may vary with types of medication prescribed. Kelloway et al. (149) studied adherence rates in a health-management organization where medical and pharmacy records could be evaluated. Individuals had higher mean adherence with theophylline (79%), followed by cromolyn (55%) and inhaled steroids (44%). These findings support the theory that asthmatic patients adhere best to simple oral regimens with a low frequency of administration. Low frequency of administration of oral medication for asthma is related to drug adherence. Tabachnik et al. evaluated adherence with theophylline in children given either two or four daily doses (150). Theophylline levels were higher and wheezing symptoms lower on the twice daily schedule. Tinkelman et al. studied asthmatic children receiving theophylline either two or four times a day and found that 75% of the children had better compliance on a two times a day regimen (151). With the advent of the oral leukotriene agents, one would expect improved adherence, especially when administered with less frequent dosing. With the development of commercially available assays, leukotriene antagonists and synthesis inhibitors will significantly assist clinicians in maintaining adherence and reducing morbidity.

VIII. Perspectives Regarding Adherence and Health Care Providers

A. Patient Perspective

Most patients are reluctant to take medications over a long period without careful consideration. They frequently do not use the correct dosage of medication. They may either underuse, overuse, or erratically use treatments (152). They may not follow a plan because of unpleasant side effects, confusion about dosing, or concerns about ingesting too much prescribed medicine (2) (Table 7). Individuals may reject part of a plan because of concerns about deterioration in their quality of life. This is a common finding in asthmatic individuals who are allergic to pets or dust mites or who smoke. Some individuals may not perceive that they have the illness that the physician diagnosed or do not feel that the medication is effective. The asymptomatic patient may discontinue a medication because he or she does not perceive that taking the medication makes any difference. Some individuals prefer to take the least amount of medication so as not to become too "dependent" on the medication and demonstrate their ability to "fight it." Patients observe how physicians adjust medications based on their clinical status. Patients recognize this pattern and alter therapy without physician consultation between office visits. This self-regulation may represent an attempt to assert some degree of control over a condition that may appear at times to be completely beyond control (153). Deaton has described "adaptive noncompliance" in a pediatric asthma practice in which children and families did not follow medication schedules because of differences in perspective with physicians (154). This situation occurs when

Table 7 Patient/Parent Variables Related to Treatment Nonadherence

Lack of comprehension of instructions (memory, intelligence, language barrier)
Inappropriate or conflicting health beliefs (prefers p.r.n. when symptomatic vs. daily preventitive)
Inappropriate expectations about treatment
Failure to recognize severity of symptoms or need for treatment
Apathy and pessimism
Peer pressure
Adolescence
Lack of family support
Inadequate psychological functioning of the individual or family
Lack of resources to afford health care or transportation to receive care
Prior history of nonadherence with other health problems

Source: Adapted from Ref. 153.

the patients have valid and appropriate reasons for not adhering, such as possible side effects or changes in lifestyle, that negatively impact the psychological functioning of the child and family. Patients with moderate to severe asthma may find that the small benefits of adhering to a complex medical and environmental plan may not justify the large cost or inconvenience and discomfort to the patient and family.

Adolescence is a difficult age to encourage adherence with any treatment plan (146). It is even more difficult with preventative ones such as the use of inhaled anti-inflammatory therapy for asthma. Some adolescents may deny disease severity and undertreat or ignore asthma symptoms (155). These individuals may be at risk for death and need to be managed carefully.

Patients with a disease make decisions to follow treatment based on their perception of its effectiveness and its suitability with their lifestyle. They decide to follow the plan if it gives them a positive functional result (i.e., improve school or work attendance or improved activity) (153). Social psychologists in the U.S. Public Health Service developed the Health Belief Model in the 1950s in an attempt to persuade citizens to receive immunization for polio (156). The theory argues that whether or not an individual will undertake the recommended health action is dependent upon that individual's perceptions of (1) the level of personal susceptibility to the particular illness or condition; (2) the degree of the consequences (physical or psychological) that may result from that illness, (3) the health action's potential benefits or efficacy in preventing or reducing symptoms, and (4) physical, psychological, financial, and other barriers or costs related to the health actions taken (157). The health belief, action, or stimulus must occur to trigger the appropriate behavior by making the individual consciously aware of his or her health beliefs without the health threat. The original model has been revised to evaluate parents' health beliefs regarding childhood illness including asthma (116). This maternal/child model was tested by a study of an inner-city pediatric emergency room population of child asthmatics who were prescribed theophylline as part of their asthma-treatment regimen. There was a significant correlation with theophylline level measures and mothers responses to the Health Belief Model instrument.

B. Health Care Provider Perspective

The relationship between the health care provider and patient affects patient adherence. Patients who perceive that the clinician is aware of their concerns and can communicate effectively are more likely to follow a treatment plan (158). This finding has also been found to be valid in interactions between patient and pharmacist (159). Svarstad has developed the Human Communication Model of treatment adherence (160), which suggests that adherence requires communication and recall of the regimen, as well as motivation to follow it. Because patients exhibit concerns and fears that undermine their willingness to follow a plan, physicians need to identify these concerns to promote adherence (153). Inadequate medication selection and instruction, failure to explain side effects and assess asthma-related health beliefs, or failure to evaluate the patient periodically may lead to nonadherence.

IX. Prescribing Strategies to Improve Adherence

Most individuals prefer not to take medication. If medication is needed, they want to take the least amount for the shortest period necessary. For patients with moderate to severe asthma, preventative anti-inflammatory medication is most effective in reducing symptoms. Factors that negatively influence adherence with medication include duration of treatment, frequency of medication administration, complexity of treatment, number of medications, and cost.

Physicians cannot predict the length of time an individual may need asthma therapy. Adults are more likely to require ongoing treatment than children, who may "outgrow" their asthma (13). Frequency of medication administration can be reduced. Once-daily sustained-release theophylline preparations have been developed to maintain pulmonary function over longer periods (161). Sustained-release inhaled β_2 agonists are available to be given twice daily (29). Inhaled corticosteroids are now available to be administered twice daily while maintaining potency and safety profiles (162,163). Physicians now have a combined medication, a β_2 agonist and an anticholinergic, in the same delivery system that "decreases" the number of medications to be taken (164).

Physicians can reduce the complexity of medication administration by selecting inhaled medications that do not require coordination. These medications include breath-activated MDI or a nonpressurized dry powder aerosol that negates the need for spacers (57,58). Some physicians have adopted a prescribing strategy to decrease the need for as-needed β_2 agonists by using inhaled corticosteroids as their "first-line drug" for asthma therapy (12). By giving this medication up to the maximal amount that will not affect adrenal function, the patient does not require frequent use of β_2 agonist or other concomitant therapies.

Cost may be a significant factor in selecting asthma therapy. By asking "Do you have a prescription plan?" the physician can identify the affordability of therapy. Patients who do not have the resources to pay for more expensive anti-inflammatory therapy or for newly introduced long-acting β_2 agonists may be prescribed less costly sustained release theophylline or oral and inhaled β_2 agonists. These less expensive medications can provide adequate lung function and maintain full activity for many asthmatic individuals.

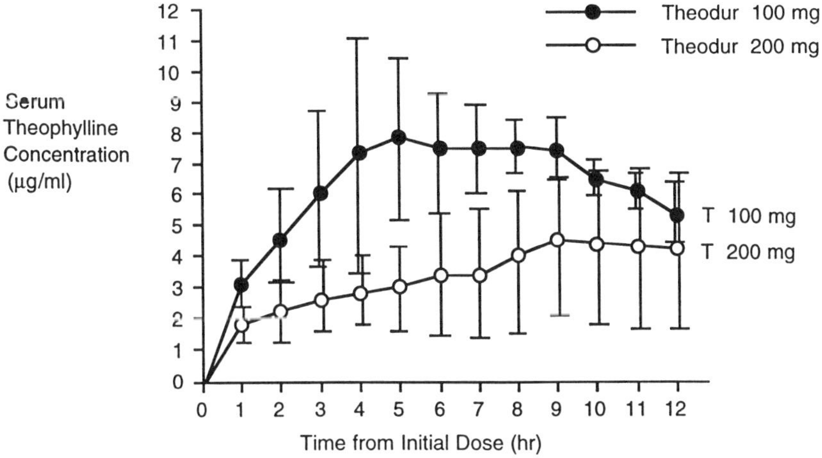

Figure 2 Prescribing strategy for "forgetful" asthmatics placed on sustained-release theophylline. Mean theophylline serum concentration of five subjects given an identical dose of 400 mg of theophylline in the form of Theodur 100 mg tablets and at a later time Theodur 200 mg tablets. (From Ref. 165)

Physicians need to take into account patient preference of therapy. Some individuals do not prefer to take inhaled medication, and their desire to take oral therapy should be respected and supported. Leukotriene antagonists and synthesis inhibitors could be considered for this patient population. Patients on sustained-release theophylline need to be adherent so they can maintain therapeutic theophylline levels and adequate pulmonary function. The pharmacokinetic properties of this medication are designed to give gradual timed release (34). Asthmatic patients starting sustained-release theophylline from 0 μg/ml may take 24–48 hours to achieve and maintain a therapeutic blood level. Weinstein studied the pharmacokinetic properties of the Theo-Dur 100 mg, 200 mg, and 300 mg sustained-release tablets and found that the 100 mg tablet had the fastest release characteristics (165). He suggested that the "forgetful asthmatic" might be prescribed multiples of the 100 mg tablet to achieve higher therapeutic levels sooner than with the other "slower" sustained-release preparations when restarting this type of medication (Fig. 2).

Clinicians may take advantage of an individual's preference to take the least amount of pharmacological therapy by encouraging elimination of environmental allergens and irritants. By decreasing exposure to asthma triggers, the patient can be expected to decrease medication need. Similarly, immunotherapy injections, when indicated, can be offered as a way of reducing pharmacological therapy over time.

The capability of drug monitoring should be an important factor when prescribing asthma therapy. Unfortunately, not many medications can be readily monitored. Only theophylline can be monitored by commercially available biochemical assay. Indirectly monitoring MDI use by the Nebulizer Chronolog is possible (24); however, its high cost limits its availability to practicing physicians. Oral β_2 agonists and leukotriene agents can be directly monitored; however, drug monitoring of is not available because the health care industry has not made this practice commercially viable. Health insurance plans, the pharmaceutical industry, and medical laboratories might want to consider the development of biochemical assays for all asthma medications. Such assays would permit clinicians to identify nonadherence and reduce unnecessary costs secondary to hospitalization and emergency care.

X. Educational Strategies to Improve Adherence

Asthma is a complex disease and requires education of the patient and family to allow successful management. Self-management programs that include both educational and behavioral components have been developed (166–173). The educational formats supply basic learning principles to promote adherence with asthma therapy. Neufield has summarized these basic learning principles as follows (174): learning proceeds more rapidly and efficiently if the learner understands clearly what he or she is expected to learn, learning is facilitated by rapid and complete individual feedback, learning is an individual process accomplished at individual rates by individual means, learning is more efficient and effective when the student perceives its relevance to him or her, and learning is enhanced when the student is motivated by these principles.

Green has made practical recommendations that are relevant to health care providers attempting to increase adherence by asthmatic patients (175):

1. Be brief, roughly one half of the statements made to a patient will be forgotten within five minutes (176).
2. It is important to be selective when presenting information, present the most important information first because this is best retained by the patient.
3. Present a written summary of the information at the sixth-grade level without medical jargon.
4. Present information in categories to make it more understandable (e.g., triggers, medications and side effects, asthma-management plan, and important telephone numbers).
5. Present written information that is specific to the individual, including customizing general instructions to the specific needs of each patient.

6. Supplement written instructions with oral ones.
7. Repeat instructions, as repetition permits better retention of information so that it will be recalled more rapidly.

Educational programs that combine written instructions with demonstration sessions have improved adherence with MDIs. One study evaluated the number of correct skills using MDI pre- and postintervention (177). Subjects receiving combined written and demonstration instruction had a 25% increase in skills as well as better retention at 1-year follow-up compared with those receiving written instructions alone.

Computer-assisted reinforcement of instruction has been shown to be effective in different aspects in asthma education. Rubin and colleagues have reported the benefit of computer programs to improve child asthmatic behaviors related to asthma management (169). Huss and colleagues demonstrated that adult asthmatics sensitive to dust mites can achieve higher adherence to dust mite environmental instruction by receiving both traditional and computer-assisted reinforcement than those receiving traditional education alone (178). Adherence was measured by subjects' self-rating and home visits by the investigator.

Acquisition of information does not necessarily lead to improved adherence. LeBaron and colleagues studied cromolyn adherence in a pediatric private practice setting (119). The educational intervention group had similar adherence to the control group as measured by urinary cromolyn levels. Although pediatric self-management programs have demonstrated an increase in the child and family's knowledge about asthma control, increased knowledge does not always lead to improved asthma-management behavior (179,180). Research on patient adherence suggests that psychological factors may be as or more important than knowledge in explaining adherence with medical regimens (181).

XI. Behavioral Strategies and Adherence

Behavioral strategies are those procedures that attempt to directly influence specific nonadherent behaviors through the use of techniques such as reminders, tailoring, contracting, graduated regimen implementation, monitoring, changing health beliefs, and reinforcement (181). When considering the use of behavioral strategies, it is helpful to consider the goal of the intervention. For the goal of preventing nonadherence, the target of the intervention is the entire population. The purpose is to shift the current or expected adherence distribution toward high adherence. For the goal of remediation, the target of the intervention is that group of individuals with identified adherence problems. Maintenance of compliance is recognized as difficult even when approximately one third of patients need no assistance in maintaining the regimen.

Reminders have a well-substantiated role in behavioral therapy; they help forgetful patients keep appointments and take medication properly (182). Use of reminders has been helpful in maintaining adherence both in asthmatic children followed in an asthma clinic (113) and asthmatic children followed as outpatients after inpatient asthma rehabilitation (103).

A. Tailoring

Tailoring refers to fitting the prescribed regimen and intervention strategies to specific patient characteristics. It is another effective behavioral method used to improve adherence. Basically, tailoring involves treatment of patients as unique individuals and may increase patient satisfaction and clinician-patient interaction (158). Tailoring has been used successfully in the pediatric asthmatic population (103,113).

B. Contracting

Contracting refers to a process specifying a set of rules regarding some behavior or interest and generating a commitment to adhere to them. Mahoney and Thorensen (183) list five guidelines for developing a contract: (1) the contract should be fair, (2) the terms of the contract should be very clear, (3) the contract should be generally positive, (4) procedures should be systematic and consistent, and (5) at least one other person should participate. Typically, these contracts take the form of "if-then rules." If X behavior occurs, then Y consequence follows. If X behavior does not occur, then Y consequence does not follow. Contingency contracts offer a number of advantages, including (1) the process of developing the contract involves the patient in planning of his or her regimen, (2) the contract provides a written outline of the behavioral expectation, (3) a formal public commitment is elicited, and (4) the contract provides incentive value through the establishment of rewards or attainment of a self established goal (184). Contracts have been useful in maintaining adherence for asthmatic children and their families (103,113,185). Examples of negotiating a contract have been published (184,186).

C. Graduated Regimen

Graduated regimen implementation or shaping refers to the process of introducing components of the regimen sequentially as the patient successfully masters prior steps in the sequence. The steps are graded according to difficulty for the patient. For example, adding one medication at a time and introducing medication that is the simplest to take followed by medication that is more difficult to administer would be simple forms of shaping (93). This approach assumes that a delay in reaching full medication dosage is not medically harmful. Shaping has been used successfully in helping asthmatic children use metered-dose inhalers (187) and compressor-driven nebulizers (188).

D. Monitoring

Self-Monitoring

Self-monitoring is a process of observing and recording one's own behavior. Asthma diaries, checklists, or charts kept by patients are an integral part of adherence research and self-management programs (107). Self-monitoring may be a useful intervention in that self-monitoring alters behavior. That is, the individual observes one of his or her behaviors, evaluates it, and then regulates it. Kotses suggests that asthma diaries serve as discriminative stimuli to prompt adherence behaviors in adult asthmatics (189).

Clinician Monitoring

Clinicians may monitor their patients in several ways. One way is the use of telephone contact by the medical staff to evaluate symptoms, peak flow values, and questions regarding care (184). Physicians and health care workers can monitor their patients through direct observation. Specter has noted that some children in inpatient rehabilitation programs refuse to take medicine by hiding it or pretending to take it (190). Health care providers can actually observe these individuals take their medication and document (with certain medications by blood levels) whether or not patients received their medication. Health care providers can also observe whether or not the patient is using metered-dose inhaler techniques correctly. By asking patients to bring all medications to the office, clinicians can see if all prescriptions have been filled or if other medications may be negatively interacting with asthma therapy (190). Weinstein observed families "role modeling" an asthma attack to determine if the specific sequences of attack management were being followed correctly (184,186).

Providing feedback to patients regarding medication adherence is an important clinical strategy. Drug monitoring with biochemical assays, such as theophylline, identifies adherence at one point in time. Weinstein and colleagues have used theophylline monitoring in conjunction with other behavioral strategies in two different populations of child asthmatics. Using the definition of theophylline adherence as theophylline levels greater than or equal to 5 μg/ml, 28 of 39 patients (72%) maintained mean theophylline levels greater than 5 μg/ml during a 6-month study of medication-dependent asthmatic children seen in an outpatient clinic (113). Weinstein and investigators also treated 37 severe asthmatic children in an inpatient rehabilitation program and followed them for one year as outpatients (103). Twenty-seven of the 37 patients (73%) were adherent during the 1-year study. By monitoring, one can identify nonadherent behavior and make interventions to promote adherence. Nides et al. used an indirect assessment of adherence with ipratroproium bromide delivered by MDI and monitored asthmatic adults with the Nebulizer Chronolog (125). Subjects using this electronic medication monitor had varying rates depending on their awareness of being monitored. Control subjects not aware of monitoring had on average 1.65 actuations per day compared with the recommended 3 per day. Subjects aware of the Nebulizer Chronolog's capability of monitoring had 1.95 actuations per day.

Health care providers can monitor indoor allergen levels to promote adherence with environmental recommendations. They can take advantage of ELISA assays to identify the presence of indoor allergens in homes of patients sensitive to dust mite, cockroach, cat, and dog (74). By obtaining vacuum samples of mattresses and rugs in the patient's home, health care providers can encourage adherence to specific environmental recommendations reviewed above. Follow-up samples, however, may not be helpful in assessing adherence with environmental recommendations because many dwellings in the United States have low air-exchange rates that retard clearance of indoor allergens (78). It may not reflect efforts taken by the individual family members.

Immunotherapy

Patients can be monitored for progress with immunotherapy injections. This may be done by reviewing the allergy injection record and observing the date of administration. Allergists may also monitor by recording the date of expiration on the allergy vaccine prepared and noting at a later time if the vaccine has been reordered by that date. Because immunotherapy injections may be given by primary care physicians, nurses, as well as allergists, all members of the health care team need to be involved in this process. Telephone reminders may be helpful for the "forgetful" patient/family member. For the individual who decides not to continue with immunotherapy injections, a review visit with the allergist during the time of year that the patient has been symptomatic might be valuable. At this consultation, the physician and patient can review the clinical course to determine if the patient may need and want to consider restarting immunotherapy injections. One may expect that the asthmatic individuals will be more likely to follow through with immunotherapy if the location is convenient, the times of administration are favorable, the process is quick, the patient experiences minimal untoward side effects, and the cost is affordable.

E. Reinforcement

Reinforcement is an essential component in each of the strategies mentioned above to improve adherence. Reinforcement refers to any consequences that increase the probability of the behavior being repeated. Positive reinforcement is the presentation of a desirable stimulus such as praise for a job well done, self-administered treat, or reward for successful performance. Negative reinforcement is the removal of an aversive stimulus for successful performance. Dunbar et al. suggests that clinicians' time and attention to the patient may be the most powerful available reinforcer (181). Studies by Korsch et al. suggest that the quantity of time a patient spends with the clinician is positively related to adherence (191). The patient's family may also be an important external reinforcer. Weinstein et al. negotiated contracts with childhood asthmatics and their families (144,186). A sample contract with family reinforcement reads as follows: "John has severe asthma and requires medication for him to breathe easily. If he reminds his father or mother when it is time to take his medication and if he is successful 90% of the time each week, he will receive $5.00 for video equipment. Signed: John, Mr. and Mrs. Brown, and Dr. Jones." By using the contract to include the family, the reinforcement is given by individuals who are not only most significant to the patient but also most readily available at the time the health behavior occurs.

F. Changing Health Beliefs

Individuals have health beliefs that influence medical decisions. These beliefs can be identified and potentially modified if found to be maladaptive. Weinstein and colleagues have used exercise challenges, with and without asthma medication, when asthmatic children and/or parents question the diagnosis of asthma or the effectiveness of asthma therapy (144). By exercising, an asthmatic patient who is off all asthma therapy will exhibit a decrease in peak flow values and may develop wheezing. By repeating the exercise challenge with pretreatment of an inhaled β_2 agonist and/or cromolyn, health care providers can demonstrate the effectiveness of asthma therapy to these individuals. Having the family attend the exercise challenge creates an opportunity at a subsequent counseling session to reevaluate health beliefs. In the context of the Health Belief Model, the exercise challenge serves as a "cue to action" or stimulus that makes the child or family aware of the health threat of asthma and the benefits of medication.

XII. Maintenance Interventions

Achieving and maintaining adherence over a long time period is a difficult task for health care providers. Adherence-improving strategies that have been tested in the areas of weight management and hypertension have not shown evidence of continuing effects when the intervention is withdrawn (192,193). With this observation in mind, investigators in the area of child and adult asthma management have developed self-management programs to enable the patient and family to manage asthma efficiently and effectively in conjunction with their physician. Wigal et al. (194) evaluated 19 management programs for children. They concluded that when asthma skills were acquired, the children and parents were more able to prevent and manage attacks, reduce asthma costs, improve the family and child adaptation to illness, and accept greater responsibility for asthma management. Adult asthmatic self-management programs have been developed and found to be effective in reducing asthma morbidity (195,196).

Taitel et al. developed and demonstrated the cost-effectiveness of an adult self-management program (197). Creer suggests that the key element of a self-management program is that self-regulation is necessary for medication adherence to occur (106). The processes and skills necessary for self-regulation include goal selection, information collection, information processing and evaluation, decision making, action, and self-reaction (198). To apply this process of self-regulation to asthma, a clinical example is presented in Table 8. In conjunction with the physician, one goal selected may be no emergency room visits. The patient would collect peak flow values that would be compared with personal best values. These values would then be related to an asthma-management plan that dictates medications to use in relation to peak flow measures. A specific target peak flow number is selected at which point the patient starts inhaled β_2 agonist. After 30 minutes, the patient evaluates for wheezing and obtains another peak flow value that will start the process again.

The successful outcomes of the self-management programs mentioned above were presumed to be related to adherence to medical instructions documented by patient self-report. Several studies did evaluate patients correct use of MDIs by investigators (pre- and postintervention) (170,196,197). Weinstein et al. developed a self-management program for severe asthmatic children that was delivered in an inpatient setting (199). Patients and families were admitted to the rehabilitation program only if the families agreed to participate in

Table 8 Clinical Application of Self-Regulation and Asthma

Goal selection	No emergency visits
Information collection	Peak flow measures
Information processing/evaluation	Record peak flow measure in relation to personal best values
Decision making	In context of peak flow measures in Attack Management Plan (see Table 2)
Action	Take appropriate medication and implement environmental avoidance (see Attack Management Plan, Table 2)
Self-reaction	Repeat peak flow

self-management meetings. These individuals were followed over a 4-year period. As a result of the rehabilitation intervention, there was a significant and sustained reduction in hospitalization, emergency care, corticosteroid bursts, and medical costs in spite of decreased frequency of rehabilitative out-patient visits (200) (Table 9). The success of this program was related to adherence with medication (103,199). Multiple interventions (educational, behavioral, and home environmental visit) were used to generate this sustained result. An out-patient asthma rehabilitation program has been developed with comparable outcomes.

XIII. Summary

No one single intervention has been shown to be useful on its own to maintain long-term adherence. Clinicians must rely on multiple strategies delivered by different members of the health care team. Monitoring of the patient's medication, self-monitoring by the patient, contracts with incentive rewards and involvement by the family, peer group discussions, counseling by health educator or nurse, and mechanical or electronic devices to assist in remembering when to take medication are some of the interventions that need to be considered. Meichenbaum and Turk recommend that health care providers need to consider the patient's perspective and expectations when developing a treatment plan (2). The considerations of patients include their expectations of the office visit; their health belief system regarding the cause of illness, susceptibility of the patient, and the effectiveness of the treatment; their perceptions of the goals, benefits, and risks of treatment; their existing health knowledge and skills to implement therapy; their degree of adaptation to the disease; their sense of hopelessness or lack of confidence to implement treatment; and their family support . Haynes encourages a collaborative clinician/patient relationship in which patients should be involved in the planning and decisions of treatment (201). The clinician should (1) ask the patient to repeat what has been done, (2) keep the prescription as simple as possible, (3) give clear instructions on the exact treatment regimen, preferably in writing, (4) call the patient when an appointment is missed, (5) prescribe a self-care regimen that takes into account the patient's daily schedule, (6) emphasize the importance of adherence at each visit, (7) schedule the frequency of visits to the patient's adherence needs, (8) acknowledge the patient's efforts to adhere at each visit, and (9) involve the patient's spouse/partner or family member.

Development of safe and effective asthma medications that are affordable, readily monitored by clinicians, requiring simple and infrequent administration, and having minimal adverse effects would facilitate adherence to asthma therapy.

Table 9 Asthma Cost and Resource Use Pre- and Postinpatient Rehabilitation

	1 Year Prior, Median (5–95 Pctl.), n = 59	1st Year Post, Median (5–95 Pctl.), n = 59	2nd Year Post, Median (5–95 Pctl.), n = 58	3rd Year Post, Median (5–95 Pctl.), n = 57	4th Year Post, Median (5–95 Pctl.), n = 57
Total cost (dollars)	10,240 (1,874–27,964)	4,036 (1,696–13,450)*	2,801 (940–8,625)*	2,316 (450–8,425)*	1,936 (340–7,311)*
Inpatient days	7 (0–25)	0 (0–6)*	0 (0–4)*	0 (0–4)*	0 (0–4)*
Emergency care	4 (1–11)	0 (0–5)*	0 (0–5)**	0 (0–2)*	0 (0–2)*
Corticosteroid bursts	2 (0–9)	2 (0–8)	1 (0–5)*	1 (0–4)*	1 (0–5)*
Physician visits	8 (3–24)	2 (0–11)	1.5 (0–8)	2 (0–8)	2 (0–8)
Rehabilitation follow-up					
Physician visits	—	9 (1–20)	4 (0–14)	3 (0–11)	1 (0–12)
Psychology visits	—	1 (3–24)	0 (0–5)	0 (0–2)	0 (0–0)

* $p < 0.001$, ** $p < 0.01$

Source: Adapted from Ref. 200.

Note In Proof: Drug Monitoring with Metered-Dose Inhalers

SmartMist (Aradigm, Hayward, CA) is a commercially available, portable monitor for adherence with metered-dose inhalers propelled by chlorofluorocarbons. It has three functions. It is a microprocessor-controlled device that records the time and date of activation. It activates delivery from a standard MDI at the optimal combination of inspiratory flow rate and inspired volume. It is also a pulmonary function device that maintains a record of each FEV1 and peak expiratory flow rate and the time of the maneuver. It is currently designed for MDIs manufactured by Schering (Proventil, Vanceril) and GlaxoWellcome (Ventolin, Serevent, Beclovent, Flovent). Data can be downloaded into a personal computer. It is anticipated that physicians will select SmartMist for patients with persistent asthma where nonadherence to treatment may be suspected but not documented. See Wighton et al. (206) for more information.

Acknowledgments

I would like to thank Dr. Alan Spitzer for reviewing this manuscript and Flossie Cushing for her assistance in typing this manuscript.

References

1. Haynes RB. Introduction. In: Haynes RB, Taylor DW, Sackett DI, eds. Compliance in Healthcare. Baltimore: Johns Hopkins University Press, 1979:1–10.
2. Meichenbaum D, Turk DC. Facilitating Treatment Adherence: A Practitioner's Guidebook. New York: Plenum Press, 1987.
3. Blackwell B. Treatment adherence: a contemporary point of view. Psychosomatics 1979; 20:27–35.
4. Weiss KB, Gergen PH, Hodgson TA. An economic evaluation of asthma in the United States. N Engl J Med 1992; 326:862–866.
5. NCHS National Health Interview Survey (machine readable public use data tape). Hyattsville, MD: U.S. Department of Health and Human Services, PHS, CDC, 1970–1987.
6. Taylor WR, Newacheck PW. The impact of childhood asthma on health. Pediatrics 1992; 90:657–662.
7. NCHS National Hospital Discharge Survey (machine readable public use data files). Hyattsville, MD: U.S. Department of Health and Human Services, PHS, CDC, 1970–1987.
8. Gergen PJ, Weiss KB. Changing patterns of asthma hospitalization among children: 1979–1987. JAMA 1990; 264:1688.

9. Weiss KB, Wagener DK. Changing patterns of asthma mortality: identifying populations at high risk. JAMA 1990; 264:1683
10. Evans RE III, Mullally DI, Wilson RW, et al. National trends in the morbidity and mortality of asthma in the U.S. Prevalence, hospitalization, and death from asthma over two decades: 1965–1984. Chest 1987; 91:65S.
11. Buist AS, Vollmer WM. Reflections on the rise in asthma morbidity and mortality. JAMA 1990; 264:1719.
12. National Asthma Education Program Expert Panel Report: Guidelines for the Diagnosis and Management of Asthma. U.S. Department of Health and Human Services, National Institutes of Health Publication No. 91:3042, August 1991.
13. Busse WW, Reed CE. Asthma: definition and pathogenesis. In: Middleton E, Reed CE, Ellis ER, et al., eds. Allergy: Principles and Practice. 3rd ed. St. Louis: CV Mosby, 1988; p 969.
14. Boushey HA, Holtzman MJ, Sheller JR, Nadel JA. Bronchial hyperreactivity. Am Rev Respir Dis 1980; 121:389–414.
15. Cockroft DW. Airway hyperresponsiveness: therapeutic implications. Ann Allergy 1987; 59:405–414.
16. Bleecker ER. Airways reactivity and asthma: significance and treatment. J Allergy Clin Immunol 1985; 75:21–24.
17. Holgate ST, Beasley R, Twentymen OP. The pathogenesis and significance of bronchial hyperresponsiveness in airways disease. Clin Sci 1987; 73:561–572.
18. Renne CM, Creer TL. Asthmatic children and their families. In: Walraich ML, Routh DK, eds. Advances in Developmental and Behavioral Pediatrics. Greenwich, CT: Jai Press, Inc., 1985:41–81.
19. Sheffer AL, Buist AS. Proceedings of the Asthma Mortality Task Force. J Allergy Clin Immunol 1987; 80:361–514.
20. Schleimer RP. Glucocorticosteroids: the mechanism of action and use in allergic diseases. In: Middleton E Jr, Reed CE, Ellis EF, Adkinson NF Jr, Yunginger JW, Busse WW, eds. Allergy Principles and Practice. 4th ed. St. Louis: CV Mosby, 1993:893–925.
21. Toogood JH. How to use steroids in asthma therapy. J Respir Dis 1982; 3:15–22.
22. Kerrebijn KF, van Ess-Zandvilet EM, Neijens HJ. Effect of long-term treatment with inhaled corticosteroids and beta-agonists on the bronchial responsiveness in children with asthma. J Allergy Clin Immunol 1987; 79:653–659.
23. Juniper EF, Kline PA, Vanzielehgem MA, Ramsdale EH, O'Bryrne PM, Hargreave FE. Effect of long-term treatment with an inhaled corticosteroid on airway hyperresponsiveness and clinical asthma in nonsteroid-dependent asthmatics. Am Rev Respir Dis 1990; 142:632–636.
24. Spector SL, Kinsman R, Mawhinney H, Siegel SC, Rachelefsky GS, Katz RM, Rohr AS. Compliance of patients with asthma with an experimental aerosolized medication: implications for controlled clinical trials. J Allergy Clin Immunol 1986; 77:65–70.
25. Holgate ST, Bradding BM, Sampson AP. Leukotriene antagonists and synthesis inhibitors: new directions in asthma therapy. J Allergy Clin Immunol 1996; 98:1–13.

26. Spector S, Smith LJ, Glass M. Accolate Asthma Trialists Group. Effects of 6 weeks of therapy with oral doses of ICI 204, 219. A leukotriene D4 receptor antagonist in subjects with bronchial asthma. Am J Respir Crit Care Med 1994; 150:618–623.
27. Israel E, Cohn J, Dube L, Drazen J. Zileuton Clinical Trial Study Group. Effect of treatment with zileuton, a 5-lipoxygenase inhibitor, in patients with asthma: A randomized controlled trial. JAMA 1996; 275:931–936.
28. Bernstein IL. Cromolyn sodium. Chest 1985; 87:685–735.
29. Foreman JC, Pearce FL. Cromolyn and nedocromil. In: Middleton E Jr, Reed CE, Ellis EF, Adkinson NF Jr, Yunginger JW, Busse WW, eds. Allergy Principles and Practice. 4th ed. St. Louis: CV Mosby, 1993:926–940.
30. Nelson HS. Adrenergic therapy of bronchial asthma. J Allergy Clin Immunol 1986; 77:771–785.
31. Webb-Johnson DC, Andrews JL. Bronchodilator therapy (two parts). N Engl J Med 1977; 297:476–482.
32. Ullman A, Svedmyr N. Salmeterol: a new long-acting inhaled beta-2agonist: comparison with salbutamol in adult asthmatic patients. Thorax 1988; 43:674–679.
33. Fraser CM, Nelson HS, Middleton E Jr. Adrenergic agents. In: Middleton E Jr, Reed CE, Ellis EF, Adkinson NF Jr, Yunginger JW, Busse WW, eds. Allergy Principles and Practice. 4th ed. St. Louis: CV Mosby, 1993:778–815.
34. Weinberger M, Hendeles L. Theophylline. In: Middleton E Jr, Reed CE, Ellis EF, Adkinson NF Jr, Yunginger JW, Busse WW, eds. Allergy Principles and Practice. 4th ed. St. Louis: CV Mosby, 1993:816–855.
35. Crescioli S, Spinazzi A, Piebani M, et al. Theophylline inhibits early and late asthmatic reactions induced by allergens in asthmatic subjects. Ann Allergy 1991; 66:245–251.
36. Barnes PJ, Pauwels RA. Theophylline in the management of asthma: time for reappraisal. Eur Respir J 1994; 7:579–591.
37. Weinberger M, Hendeles L, Bighley L. The relation of product formulation to absorption of theophylline. N Engl J Med 1978; 299:852–857.
38. Paloucek RP, Todvold KA. Evaluation of theophylline over dose and toxicities. Ann Emerg Med 1988; 17:135–142.
39. Kappas A, Alvares AP, Anderson KE, et al. Effect of charcoal beef on antipyrine and theophylline metabolism. Clin Pharmacol Ther 1978; 23:445–451.
40. Staib AH, Chupann D, Lisner R, et al. Pharmacokinetics and metabolism of theophylline in patients with liver disease. Int J Clin Pharmacol Ther Toxicol 1980; 18:500–506.
41. Piafsky KM, Sitar DS, Rangno RE, et al. Theophylline kinetics in acute pulmonary edema. Clin Pharmacol Ther 1977; 21:310–316.
42. Campbell MA, Plachetka JR, Kackson JE, et al. Cimetadine decreases theophylline clearance. Ann Intern Med 1981; 95:672–679.
43. Prince RA, Wing DS, Weinberger MM, et al. Effect of erythromycin on theophylline kinetics. J Allergy Clin Immunol 1981; 68:427–431.
44. Chang KC, Bell TD, Lauer BA, et al. Altered theophylline pharmacokinetics during acute respiratory viral illness. Lancet 1978; 1:1132–1135.

45. Jenne J, Nagasawa H, McHugh R, et al. Decreased theophylline half-life in cigarette smokers. Life Sci 1975; 17:195–201.
46. Loughnan PM, Sitar DS, Ogilvie RI, et al. Pharmacokinetic analysis of the disposition of intravenous theophylline in young children with asthma. J Pediatr 1976; 88:874–882.
47. Ginchansky E, Weinberger M. Relationship of theophylline clearance to oral dosage in children with chronic asthma. J Pediatr 1977; 91:655–663.
48. Hendeles L, Weinberger M, and Bighley L. Disposition of theophylline after a single intravenous infusion of aminophylline. Am Rev Respir Dis 1978; 118:97–104.
49. Gross NJ and Gold WM. Anticholinergic drugs. In: Middleton E, Reed CE, Ellis ER, et al., eds. Allergy: Principles and Practice. 3rd ed. St. Louis: CV Mosby, 1988:941–962.
50. Gross NJ. Ipratropium bromide. N Engl J Med 1988; 319:486–494.
51. Tashkin DP, Ashutosh K, Bleecker ER, et al. Comparison of the anticholinergic bronchodilator ipratropium bromide with metaproternol in chronic obstructive pulmonary disease. Am J Med 1986; 81:81–89.
52. Newhouse MT, Dolovich MB. Control of asthma by aerosols. N Engl J Med 1986; 315:870–874.
53. McFadden ER Jr. Improper patient techniques with metered dose inhalers. Clinical consequences and solutions to misuse. J Allergy Clin Immunol 1995; 96:278–283.
54. Manzella AB, Brooks CM, Richards JM Jr, et al. Assessing the use of metered dose inhalers by adults with asthma. J Asthma 1989; 26:223–230.
55. Guidry GG, Brown WD, Stogner SW, George RB. Incorrect use of metered-dose inhalers by medical personnel. Chest 1992; 101:31–33.
56. Crompton GK. Inhalation devices. Eur J Respir Dis 1982; 63:489–492.
57. Crompton GK, Duncan S. Clinical assessment of a new breath-activated inhaler. Practitioner 1989; 233:268–277.
58. Crompton GK. New inhalation devices. Eur Respir J 1988; 1:679–680.
59. Banner AS, Shad RS, Addington WW. Rapid prediction of need for hospitalization in acute asthma. JAMA 1976; 235:1337–1338.
60. Nowak RM, Pensler MI, Sarkar DD, et al. Comparison of peak expiratory flow and FEV 1: Admission criteria for acute bronchial asthma. Ann Emerg Med 1982; 11:64–69.
61. Taplin PS, Creer TL. A procedure for using peak expiratory flow rate data to increase the predictability of asthma episodes. J Asthma Res 1978; 16:15–19.
62. van As A. The accuracy of peak expiratory flow meters (editorial). Chest 1982; 82:263.
63. Statement on Pharmacist-Conducted Patient Counselling. Approved by the American Society of Hospital Pharmacists Board of Directors at its meeting of November 1975 and by ASHP House of Delegates, April 1976.
64. Call RS, Smith TF, Morris E, Chapman MD, Platts-Mills TAE. Risk factors for asthma in inner city children. J Pediatr 1992; 121:862–866.
65. Gelber LE, Seltzer LH, Bonzoukis JK, Pollart SM, Chapman, MD, Platts-Mills TAE. Sensitization and exposure to indoor allergens as risk factors for asthma among patients presenting to hospital. Am Rev Respir Dis 1993; 147:573–578.

66. Lau S. Falkenhorst G, Weber A, et al. High mite-allergen exposure increases the risk of sensitization in atopic children and young adults. J Allergy Clin Immunol 1989; 84:718–725.
67. Peak JK, Tovey E, Mellis CM, Leeder SR, Woolcock AJ. Importance of house dust mite and alternaria allergy—an epidemiological study of children living in two climatic areas of northern New South Wales. Clin Exp Allergy 1993; 23:812–820.
68. Sears MR, Hervison GP, Holdaway MD, Hewitt J, Flannery EM, Silva PA. The relative risks of sensitivity to grass pollen, house dust mite, and cat dander in the development of childhood asthma. Clin Exp Allergy 1989; 19:419–424.
69. Sporik R, Holgate ST, Platts-Mills TAE, Cogswell J. Exposure to house dust mite allergen (Der p I) and the development of asthma in childhood: a prospective study. N Engl J Med 1990; 323:502–507.
70. Colley JRT, Holland WW, Corkhill RT. Influence of passive smoking and parental phlegm on pneumonia and bronchitis in early childhood. Lancet 1974; 2:1031–1034.
71. Fergusson DM, Horwood LJ, Shannon FT, et al. Parental smoking and lower respiratory illness in the first three years of life. J Epidemiol Commun Health 1981; 35:180–184.
72. Weiss ST, Tager IB, Speizer FE, et al. Persistent wheeze: its relationship to respiratory illness, cigarette smoking, and level of pulmonary function in a population sample of children. Am Rev Respir Dis 1980; 122:697–707.
73. Liard R, Perdrizet S, Reinert P. Wheezy bronchitis in infants and parents' smoking habits. Lancet 1982; 1:334–335.
74. Chapman MD, Heymann PW, Sporik RP, Platts-Mills TAE. Monitoring allergy exposure in asthma: new treatment strategies. Allergy 1995; 50:29–33.
75. Ehnert B, Lau S, Weber A, Wahn U. Reduction of mite allergen exposure and bronchial hyper-reactivity. J Allergy Clin Immunol 1991; 87:320.
76. Platts-Mills TAE, Thomas WR, Aalberse RC, Vervloet D, Chapman MD, Co-Chairmen. Dustmite allergens and asthma: report of a 2nd international workshop. J Allergy Clin Immunol 1992; 89:1046–1060.
77. Platts-Mills TAE, Solomon WR. Aerobiology of inhallant allergens. In: Middleton E Jr, Reed CE, Ellis EF, Adkinson NF Jr, Yunginger JW, Busse WW, eds. Allergy Principles and Practice. 4th ed. St. Louis: CV Mosby, 1993:469–528.
78. Wood RA, Chapman MD, Adkinson NF Jr, Eggleston PA. The effect of cat removal on allergen content in household-dust samples. J Allergy Clin Immunol 1989; 83:730–734.
79. Chapman MD. Monoclonal antibodies as structural probes for mite, cat, and cockroach allergens. In: Elshami S, Merett T, eds. Allergens and Molecular Biology. Oxford: Pergamon Press, 1988:281–295.
80. Bousquet J, Michel FB. In vivo methods for the study of allergy. In: Middleton E Jr, Reed CE, Ellis EF, Adkinson NF Jr, Yunginger JW, Busse WW, eds. Allergy Principles and Practice. 4th ed. St. Louis: CV Mosby, 1993:573–594.
81. Hamburger HA, Katzman JA. Methods in laboratory immunology. In: Middleton E Jr, Reed CE, Ellis EF, Adkinson NF Jr, Yunginger JW, Busse WW, eds. Allergy Principles and Practice. 4th ed. St. Louis: CV Mosby, 1993:554–572.

82. Van Metre TE Jr, Adkinson NF Jr. Immunotherapy for aeroallergen disease. In: Middleton E Jr, Reed CE, Ellis EF, Adkinson NF Jr, Yunginger JW, Busse WW, eds. Allergy Principles and Practice. 4th ed. St. Louis: CV Mosby, 1993:1489–1509.
83. Warner JO, Price JF, Soothill JF, et al. Controlled trial of hyposensitization to dermatophgoides pterronyssinus in children with asthma. Lancet 1978; 2:912.
84. Ohman JL Jr, Findlay SR, Leiterman KM. Immunotherapy in cat-induced asthma: double-blind trial with evaluation of in vivo and in vitro responses. J Allergy Clin Immunol 1984; 74:230–239.
85. Bousquet J, Maasch HJ, Hejjaoui A. Double-blind, placebo-controlled immunotherapy with mixed grass-pollen allergoids. III. Efficacy and safety of unfractionated and high-molecular-weight preparations in rhinoconjunctivitis and asthma. J Allergy Clin Immunol 1989; 84:546–556.
86. Horst M, Hejjaoui A, Horst V, Michel F-B, Bousquet J. Double-blind, placebo-controlled rush immunotherapy with a standardized alternaria extract. J Allergy Clin Immunol 1990; 85:460–472.
87. Gordis L. Conceptual and methodologic problems in measuring patient compliance. In: Haynes RB, Taylor DW, Sackett DL, eds. Compliance in Healthcare. Baltimore: Johns Hopkins University Press, 1979:23–48.
88. Hendeles L, Burkey S, Bighley L, et al. Unpredictability of theophylline saliva measurements in chronic obstructive pulmonary disease. J Allergy Clin Immunol 1977; 60:335.
89. Cluss PA, Epstein LH, Galvis SA, et al. Effect of compliance for chronic asthmatic children. J Consult Clin Psychol 1984; 52:909–910.
90. Dederich RA, Szefler SJ, Green ER. Intrasubject variation in sustained release theophylline absorption. J Allergy Clin Immunol 1981; 67:465–471.
91. Hendeles L, Weinberger M. Theophylline product and dosing interval selection for chronic asthma. J Allergy Clin Immunol 1985; 76:285–291.
92. Rand CS, Wise RA. Measuring adherence to asthma medication regimens. Am J Respir Crit Care Med 1994; 149:569–576.
93. Creer TL, Levstek D. Adherence to asthma regimens. In: Gochman DS, (ed.): Handbook of Health Behavior Research. New York: Plenum.
94. Davis MS. Variations in patients' compliance with doctors' orders: analysis of congruence between survey responses and results of empirical investigations. J Med Educ 1966; 41:1037– 1048.
95. Kasl SV. Issues in patient adherence to health care regimens. J Human Stress 1975; 1:5–17.
96. Mushlin AI, Appel FA. Diagnosing potential non-compliance: physician's ability in a behavioral dimension of medical care. Arch Intern Med 1977; 137:318–321.
97. Francis V, Korsch BM, Morris MJ. Gaps in doctor-patient communications. N Engl J Med 1969; 280:535–540.
98. Clark NM, Feldman CH, Evans D, Duzey O, et al. Managing better: children, parents and asthma. Patient Educ Couns 1986; 8:27–38.
99. Creer TL, Wigal JK, Tobin DL, Kotses H, Snyder SE, Winder JA. The revised asthma problem behavior checklist. J Asthma 1989; 26:17–29.
100. Rand CS, Wise RA, Nides M, Simmons MS, Bleecker ER, Jusek JW, Li VC, Tashkin DP. MDI adherence in a clinical trial. Am Rev Respir Dis 1992; 146: 1559–1564.

101. Mawhinney H, Spector SL, Kinsman RA, Siegel SC, Rachelefsky GS, Katz RM, Rohr AS. Compliance in clinical trials of two nonbronchodilator, antiasthma medications. Ann Allergy 1991; 66:294–299.
102. Gong H Jr, Simmons MS, Clark VA, Tashkin DP. Metered-dose inhaler usage in subjects with asthma. Comparison of nebulizer chronolog and daily diary recordings. Allergy Clin Immunol 1988; 82:5–10.
103. Weinstein AG, Faust D. Maintaining Theophylline compliance in severely asthmatic children. The role of psychological functioning of the child and family. Ann Allergy Asthma Immunol 1997 (in press).
104. Becker LA, Glanz K, Sobel E, Mossey J, Zinn SL, Knott KA. A randomized trial of special packaging of antihypertensive medication. J Family Pract 1986; 22:357–361.
105. Rapoff MA, Christophersen ER. Improving compliance in pediatric practice. Pediatr Clin North Am 1982; 29:339–357.
106. Creer TL. Medication compliance and childhood asthma. In: Krasnegor NA, Epstein S, Johnson SB, Yaffe SJ, eds. Developmental Aspects of Health Compliance Behavior. Hillsdale, NJ: Lawrence Erlbaum Associates, 1993:303–333.
107. Weinstein, AG. Compliance and asthmatic children: Workshop Annual Meeting of the American Academy of Allergy Asthma and Immunology 1995, New York.
108. Rand CS, Wise RA. Adherence with asthma therapy. In: Szefler SJ, Leuang DY, eds. Severe Asthma: Pathogenesis and Clinical Management. New York: Marcel Dekker Inc., 1996:435–464.
109. Steele DJ, Jackson TC, Gutmann MC. Have you been taking your pill? The adherence-monitoring sequence in the medical interview. J Family Pract 1990; 30:294–299.
110. Cluss PA, Epstein LH. The measurement of medical compliance in the treatment of diseases. In: Karoly P, (ed.): Measurement Strategies in Health Psychology. New York: John Wiley & Sons, 1985:140–161.
111. Jerome A, Wigal JK, Creer TL. A review of medication compliance in children with asthma. Pediatr Asthma Allergy Immunol 1987; 1:193–210.
112. Eney RD, Goldstein EO. Compliance of chronic asthmatics with oral administration of theophylline as a one measured by serum and salivary levels. Pediatrics 1976; 57:513–517.
113. Weinstein AG, Cuskey W. Theophylline compliance in asthmatic chilren. Ann Allergy 1985; 54:19–24.
114. Miller KA. Theophylline compliance in adolescent patients with chronic asthma. J Adolesc Health Care 1982; 3:177–179.
115. Sublett JL, Pollard SJ, Kadlec GJ, Karibo JM. Noncompliance in asthmatic children: a study of theophylline levels in a pediatric emergency room population. Ann Allergy 1979; 43:95–97.
116. Radius SM, Becker MH, Rosenstock IM, et al. Factors influencing mothers' compliance with medication regimens for asthmatic children. J Asthma Res 1978; 15:133–149.
117. Wood PR, Casey R, Kolski GB, McCormick MC. Compliance with oral theophylline therapy in asthmatic children. Ann Allergy 1985; 54:400–404.
118. Baum D, Creer TL. Medication compliance with children and asthma. J Asthma 1986; 12:49–59.

119. LeBaron S, Zeltzer LK, Raner P, Kniker WT. A controlled study of education for improving compliance with cromolyn sodiun (Intel): the importance of physician-patient communication. Ann Allergy 1985; 55:811–818.
120. Baciewicz AM, Kyllonen KS. Aerosol inhaler technique in children with asthma. Am J Hosp Pharm 1989; 46:2510–2511.
121. Lee HS. Proper aerosol inhalation technique for delivery of asthmatic medications. Clin Pediatr 1983; 22:440–443.
122. Pederson S. Ensuring compliance in children. Eur Respir J 1992; 5:143–145.
123. Compton GK. Problems patients have using pressurized aerosol inhalers. Eur J Respir Dis 1982; 63:101–104.
124. Coutts JAP, Gibson NA, Paton JY. Measuring compliance with inhaled medication in asthma. Arch Dis Child 1992; 67:65–70.
125. Nides, MA, Tashkin DP, Simmons MS, Wise RA, Li VC, Rand CS. Improving inhaler adherence in a clinical trial through the use of the nebulizer chronolog. Chest 1993; 104:501–507.
126. Cohn JR, Pizzi A. Determinants of patient compliance with allergen immunotherapy. J Allergy Clin Immunol 1993; 1:734–737.
127. Lower T, Henry J, Mandik L, et al. Compliance with allergen immunotherapy. Ann Allergy 1993; 70:480–482.
128. Tinkelman D, Smith F, Cole WQ, Silk HJ. Compliance with an allergen immunotherapy regimen. Annal Allergy 1995; 74:241–246.
129. Platts-Mills TAE. Effective environmental control methods: what works and what doesn't. Plenary session American College of Allergy, Asthma, Immunology, Annual Meeting November 13, 1995, Dallas, TX.
130. Butz AM, Rosenstein EJ. Passive smoking among children with chronic respiratory disease. J Asthma 1992; 29:265–272.
131. Fiore MC, Novotny TE, Pierce JP, et al. Methods used to quit smoking in the United States. JAMA 1990; 263:2760–2765.
132. National Disease Therapeutic Index 1992, Rhone Poulenc Rorer, Fort Washington, PA.
133. Lavy R. Respiratory examination and asthma guideline survey. J Allergy Clin Immunol 1995; 95:225A.
134. McKenney JM. The clinical pharmacy and compliance. In: Haynes RB, Taylor DW, Sackett DL, eds. Compliance in Health Care. Baltimore: Johns Hopkins University Press, 1979:260–277.
135. Allan BL, Barker KN. Fundamentals of medication error research. Am J Hosp Pharm 1990; 47:665–671.
136. Kistner UA, Keith MR, Sergeant KA, et al. Accuracy in dispensing in a high volume hospital based unpatient pharmacy. Am J Hosp Pharm 1994; 51:2793–2797.
137. Allen EL, Barker KN, Malloy MJ, Heller WM. Dispensing errors in counselling in community practice. Am Pharm 1995; NS 35:25–33.
138. Haynes RB. Determinants of compliance: the disease and the mechanics of treatment. In: Haynes RB, Taylor DW, Sackett DL, eds. Compliance in Healthcare. Baltimore: Johns Hopkins University Press, 1979:49–62.
139. Barnum JF. Outlook for treating patients with self-destructive habits. Ann Intern Med 1974; 81:387–393.

140. Kanzler M, Jaffe J, Zeidenberg P. Long and short term effectiveness of a large-scale proprietary smoking cessation program—a four year follow-up of Smokenders participants. J Clin Psychol 1976; 32:661–669.
141. Donabedian A, Rosenfeld LS. Follow-up study of chronically ill patients discharged from hospital. J Chron Dis 1964; 17:847–862.
142. Finnerty FA, Shaw LW, Himmelsbach CK. Hypertension in the inner city, ii). Detection and follow-up. Circulation 1973; 47:76–78.
143. Alpert JJ. Broken appointments. Pediatrics 1964; 34:127–132.
144. Weinstein AG. Clinical management strategies to maintain drug compliance in asthmatic children. Ann Allergy 1995; 74:304–310.
145. Jay S, Litt IF, Durant RH. Compliance with therapeutic regimens. J Adolesc Health Care 1984; 5:124–136.
146. Varni JW, Wallander JL. Adherence to health-related regimens in pediatric chronic disorders. Clin Psych Rev 1984; 4:585–596.
147. Christiaanse ME, Lavigne JV, Lerner CV. Psychosocial aspects of compliance in children and adolescents with asthma. J Dev Behav Pediatr 1989; 10:75–80.
148. Mann MC, Eliasson O, Patel K, ZuWallack RL. An evaluation of severity-modulated compliance with q.i.d. dosing of inhaled beclomethasone. Chest 1992; 102:1342–1346.
149. Kelloway JS, Wyatt RA, Adlis SA. Comparison of patients' compliance with prescribed oral and inhaled asthma medications. Arch Intern Med 1994; 154: 1349–1352.
150. Tabachnik E, Scott P, Correia J, et al. Sustained release theophylline: a signifiacnt advance in the treatment of childhood asthma. J Pediatr 1982; 100:489–494.
151. Tinkelman DG, Vanderpool GE, Carroll MS, et al. Compliance differences following administration of theophylline at six and twelve hour intervals. Ann Allergy 1980; 44:283–288.
152. Kinsman RA, Dirks JF, Dahlem NW. Noncompliance to prescribed as-needed (PRN) medication use in asthma: usage patterns and patient characteristics. J Psychosom Res 1980; 24:97–107.
153. Turk DC, Meichenbaum D. Adherence to self-care regimens. In: Sweet JJ, Rosensky RH, Tovian SM, eds. Handbook of Clinical Psychology in Medical Settings. New York: Plenum Press, 1991:249–266.
154. Deaton AV. Adaptive noncompliance in pediatric asthma: the parent as expert. J Pediatr Psychol 1985; 10:1–14.
155. Strunk RC, Mrazek DA, Fuhrmann GS, LaBregue JF. Physiologic and psychological characteristics associated with deaths due to asthma in childhood. JAMA 1985; 254:1193–1200.
156. Rosenstock IM. Why people use health services. Milbak Mem Fun Q 1966; 254:1193– 1200.
157. Becker MH, Maiman LA, Kirscht JP, et al. Patient perceptions and compliance: recent studies of the health belief model. In: Haynes RB, Taylor DW, Sackett DI, eds. Compliance in Healthcare. Baltimore: Johns Hopkins University Press, 1979:78–109.
158. Hulka BS. Patient-clinician interaction and compliance. In: Haynes RB, Taylor DW, Sackett DI, eds. Compliance in Healthcare. Baltimore: Johns Hopkins University Press, 1979:63–77.

159. Ludy J, Gagnon J, Caiola S. The patient-pharmacist interaction in two ambulatory settings—its relationship to patient satisfaction and drug misuse. Drug Intell Clin Pharm 1977; 11:81–89.
160. Svarstad BL. Physician-patient communication and patient conformity with medical advice. In: Mechanic D, ed. The Growth of Bureaucratic Medicine: An Inquiry Into the Dynamics of Patient Behavior and the Organization of Medical Care. New York: Wiley, 1976.
161. Gonzales MA, Kisicki J, Straughn AB. Pharmacokinetic comparison of a once-daily and twice-daily theophylline delivery system. Clin Ther 1994; 16:686–692.
162. Tukiainen H, Vaara, Terho E, et al. Comparison of twice-daily administration and four-times daily administration of beclomethasone diproprionate in patients with severe chronic bronchial asthma. Eur J Clin Pharmacol 1986; 30:319–322.
163. Malo JL, Cartier A, Chezzzo H, et al. Comparison of four-times-a-day and twice-a-day dosing regimens in subjects requiring 1200 ug or less of budesonide to control mild to moderate asthma. Respir Med 1995; 89:537–543.
164. Bone R. In chronic obstructive pulmonary disease, a combination of ipratroprium and albuterol is more effective than either agent alone. An 85 day multi-center clinical trial. Chest 1994; 105:1411–1419.
165. Weinstein AG. Sustained-release theophylline for the forgetful asthmatic. Ann Allergy 1985; 55:563–568.
166. Lewis C, Rachelefsky G, Lewis MA. ACT for kids. In: Self-Management Educational Programs for Childhood Asthma. Washington, DC: National Institute of Allergy and Infectious Diseases, 1981:21–52.
167. Wilson-Pessano SR, McNabb WL. The role of patient education in the management of childhood asthma. Prev Med 1985; 14:670–687.
168. McNabb WL, Wilson-Pessano SR, Hughes GW, Scamagas P. Self-management education of children with asthma: AIR WISE. Am J Pub Health 1985; 75:1219–1220.
169. Rubin DH, Leventhal JM, Sadock RT, Letovsky E, Schottland P, Clemente I, McCarthy P. Educational intervention by computer in childhood asthma: a randomized clinical trial testing the use of a new teaching intervention in childhood asthma. Pediatrics 1986; 77:1–10.
170. Fireman P, Firday GA, Gira C, Vierthaler WA, Michaels L. Teaching self-management skills to asthmatic children and their parents in an ambulatory care setting. Pediatrics 1981; 68:341–348.
171. Hindi-Alexnader M, Cropp GJA. Evaluation of a family asthma program. J Allergy Clin Immunol 1984; 74:505–510.
172. Creer TL, Backial MA, Burns KL, Leung P, Marion RJ, Miklich DR, Morrill C, Taplin PS, Ullman S. Living with asthma. I. Genesis and development of a self-management program for childhood asthma. J Asthma 1988; 25:335–362.
173. Clark NM, Feldman CH, Freudenberg N, Millman EJ, Wasilewski Y, Valle I. Developing education for children with asthma through study of self-management behavior. Health Ed Quart 1980; 7:278–297.
174. Neufeld VR. Patient education: a critique. In: Sackett DS, Haynes RB, eds. Compliance With Therapeutic Regimens. Baltimore: Johns Hopkins University Press, 1976:83–92.

175. Green LW. Educational strategies to improve compliance with therapeutic and preventative regimens: the recent evidence. In: Haynes RB, Taylor DW, Sackett DL, eds. Compliance in Healthcare. Baltimore: Johns Hopkins University Press, 1979:121–143.
176. Ley P, Bradshaw PW, Eaves D, Walkter CM. A method for increasing patients' recall of information prestented by doctors. Psychol Med 1973; 3:217–220.
177. Wilson SR, Scamagas P, German DF, et al. A controlled trial of two forms of self-management education for adults with asthma. Am J Med 1993; 94:56–63.
178. Huss K, Salerno M, Huss RW. Computer-assisted reinforcement of instruction: effects on adherence in adult atopic asthmatics. Res Nursing Health 1991; 14: 259–267.
179. Howland J, Bauchner H, Adair R. The impact of pediatric education on morbidity: assessing the evidence. Chest 1988; 94:964–969.
180. Rubin DH, Bauman LH, Lauby JL. The relationship between knowledge and reported behavior in childhood asthma. Dev Behav Pediatr 1989; 10:307–312.
181. Dunbar JM, Marshall GD, Hovell MF. Behavioral strategies for improving compliance. In: Haynes RB, Taylor DW, Sackett DI, eds. Compliance in Healthcare. Baltimore: Johns Hopkins University Press, 1979:157–178.
182. Nazarian L, Mechaber J, Charney E, Coulter M. Effect of a mailed appointment reminder on appointment keeping. Pediatrics 1974; 53:349–352.
183. Mahoney MJ, Thoreson CE. Self-Control: Power to the Person. Monterey, CA: Brooks-Cole, 1974.
184. Weinstein AG. Behavioral strategies and theophylline compliance in asthmatic children. Ann Allergy 1985; 55:16–21.
185. Creer TL. Asthma Therapy: A Behavioral Health-Care System for Respiratory Disorders. New York: Springer Publishing Co, 1979.
186. Weinstein AG. Drug compliance in asthmatic children: prevention of hospitalization and emergency care (video). 1994.
187. Marion, RJ, Creer TL, Burns KL. Training asthmatic children to use a nebulizer correctly. J Asthma 1983; 20:183–188.
188. Renne CM, Creer TL. Training children with asthma to use inhalation therapy equipment. J Appl Behav Anal 1976; 9:1–11.
189. Kotses H, Bernstein IL, Bernstein DI, Reynolds RVC, Korbee L, Wigal JK, Ganson E, Stout C, Creer TL. A self-management program for adult asthma. Part I. Development and evaluation. J Allergy Clin Immunol 1995; 95:529–540.
190. Spector SL. Is your asthmatic patient really complying? Ann Allergy 1985; 55:552–556.
191. Korsch BM, Negrete VF. Doctor-patient communication. Sci Am 1972; 227:66–74.
192. Harris MB,d Bruner CG. A comparison of a self-control and a contract procedure for weight control. Behav Res Ther 1971; 9:347–354.
193. Haynes RB, Sackett DL, Gibson ES, et al. Improvement of medication compliance in uncontrolled hypertension. Lancet 1976; 1:1265–1268.
194. Wigal JK, Creer TL, Kotses H, Lewis PD. A critique of 19 self-management programs for childhood asthma. Part I. The development and evaluation of the programs. Pediatr Asthma Allergy Immunol 1990; 4:17–39.

195. Bailey WC, Richards JM, Brooks CM, Soong S, Windsor RA, Manzella BA. A randomized trial to improve self-management practices of adults with asthma. Arch Inter Med 1990; 150:1664–1668.
196. Windsor RA, Bailey WC, Richards JR Jr, Manzella B, Soong S, Brooks M. Evaluation of the efficacy and cost effectiveness of health education methods in increased medication adherence among adults with asthma. Am J Public Health 1990; 80:1519–1521.
197. Taitel MS, Kotses H, Bernstein IL, Bernstein DI, Creer TL. A self-management program for adult asthma Part II: Cost-benefit analysis. J Allergy Clin Immunol 1995; 95:672696.
198. Creer TL, Holroyd KA. Self-management. In: Baum A, McManus C, Newman J, Weinman J, West R, eds. Cambridge Handbook of Psychology, Health, and Medicine. Cambridge: Cambridge University Press.
199. Weinstein AG, Faust DS, McKee L, Padman R. Outcome of short term hospitalization for children. J Allerg Clin Immunol 1992; 90:66–75.
200. Weinstein AG, McKee L, Stapleford J, Faust D. An economic evaluation of short-term inpatient rehabilitation for severe asthmatic children. J Allergy Clin Immunol 1996; 98:264–723.
201. Haynes RB, Wang E, DaMota Gomes M. A critique of interventions to improve compliance with prescribed medications. Patient Educ Counselling 1987; 10:155–166.
202. MacDonald L, Tovey E. The role of water temperature and laundry procedures in reducing house dust mite populations and allergen content of bedding. J Allergy Clin Immunol 1992; 90:559–608.
203. Owen S, Morganstern M, Hepworth J, Woodcock A. Control of house dust mite antigen in bedding. Lancet 1990; 335:396–397.
204. DeBlay F, Chapman MD, Platts-Mills TAE. Airborne cat allergen (Fel d I): environmental control with the cat in situ. Am Rev Respir Dis 1991; 143:1334–1339.
205. Woodfolk JA, Luczynska CM, DeBlay F, Chapman MD, Platts-Mills TAE. The effect of vacuum cleaners on the concentration and particle size distribution of airborne cat allergen. J Allergy Clin Immunol 1993; 91:829–837.
206. Wighton TG, Friedli KJ, Slatore CG, Okikawa J, Jacinowski D, Rubsamen R, Shapiro GS. Patient medication and lunch function measurement compliance using an electronic asthma management system. J Allergy Clin Immunol 1997; 99(No. 1, Part 2): SA71.

14

History and Evaluation

THOMAS L. CREER and DEIRDRE A. LEVSTEK

Ohio University
Athens, Ohio

RUSS V. C. REYNOLDS

Optimum Behavioral Health, Inc.
Memphis, Tennessee

I. Introduction

While asthma and self-management each have separate histories, they share a striking similarity. McFadden and Stevens (1) traced the history of asthma from the early Christian era when Aretaeus and Galen wrote about the disorder of modern times. It was pointed out that little progress was made regarding asthma until 35–40 years ago. Despite the explosion of knowledge regarding asthma since that time, McFadden and Stevens concluded that our understanding of the disorder still echoes that of Floyer who, in 1698, lamented that neither the true nature of asthma nor its cure was known. The history of self-management, on the other hand, can be traced back to the admonishment from the inscription at the Delphic oracle to "know thyself," somewhere between 650 and 550 B.C. (2), to modern-day philosophers and behavioral scientists. Many of the latter scientists (e.g., Ref. 3) have described impressive success in teaching individuals to control aspects of their behavior. As we are a nation where one-third of the citizens are overweight, one-fourth smoke, and one in 10 abuse alcohol (4), it is apparent that we face an almost insurmountable challenge in regulating our own behavior. Despite exhortations by politicians that we assume responsibility for our own actions, achieving this goal will require more than empty rhetoric by orators who, in many cases, have yet to demonstrate that they can control their own behavior. Thus, there is a similarity between asthma and what is referred to as self-management, self-control, or self-regulation: we have learned much in the past half century, but we have only started the journey.

A. History of Asthma Self-Management

Despite what has often seemed a hopeless task, other chapters in this volume have discussed the progress made in developing and assessing self-management programs for asthma primarily in the past two decades. The evolution of the programs has been steady and cumulative. This fact is borne out not only by the programs reviewed by Clark (Chapter 10) for children and by Bailey et al. (Chapter 11) for adults, but by the variety of approaches taken in creating and assessing self-management programs (Chapters 10–12). The development of asthma self-management programs has not only benefited patients with the disorder, but has won acceptance as an integral component in the overall management and control of asthma (5–7). Putting the findings into perspective, a period of 10 years or less from the appearance of outcome data concerning a procedure to the widespread integration of the technique into a treatment regimen is remarkable given the time line that usually marks progress in the control of physical disorders. The result should not only bring satisfaction to the dedicated scientists and clinicians who developed and evaluated pioneer asthma self-management programs, but offer a harbinger of promise for the refinement and expansion of asthma self-management.

One frequently asked question is: Why has so much progress been made in creating and evaluating self-management programs for asthma? Self-management approaches are more sophisticated and better synthesized into treatment regimens for asthma than is the case for comparable chronic diseases and disorders (8,9). There are a number of reasons for the fabric created by asthma self-management. Several threads were the natural outgrowth of research that became prominent 25 or 30 years ago in biological science, particularly immunology, such as the discovery of IgE by Ishizawka and Ishizawka (10) in 1968; other strands of the fabric, however, must be considered as coincidental events that often result in significant advances in science (e.g., increasing prevalence and costs of asthma). Still other fibers in the tapestry of asthma self-management were the culmination of a unique fusion of medical and behavioral expertise to create a vision of what might be possible in the control of asthma. Chapters in this volume have reviewed many issues that have contributed to our understanding of comprehensive asthma treatment. These have included investigations of the influences of psychological and behavioral factors in asthma self-management (Chapters 2, 4, 6, and 8), applications of health-related quality of life as an outcome

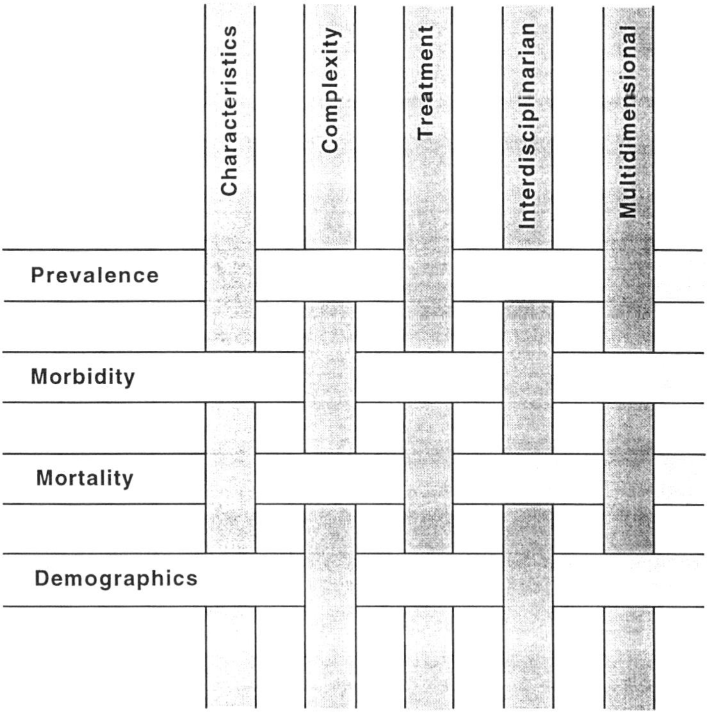

Figure 1 Fabric of asthma self-management.

variable in asthma (Chapter 5), and explorations of trends of asthma self-management as they influence individuals (Chapter 12) and populations of asthma patients (Chapters 10 and 11).

Advances in the self-management of asthma are increasingly inseparable from the variegated fabric created by our knowledge of asthma. This is shown in Figure 1. The richness and complexity of an asthma self-management program can be appreciated only within the context of our growing understanding of the disorder at a multifaceted level that includes physiological, molecular/biological, environmental, genetic, and psychosocial/behavioral variables (11). We will review the strands of our growing knowledge of asthma and its management in some detail.

Prevalence

One thread in the tapestry of asthma knowledge concerns the prevalence of asthma. As described in Chapter 1, there is ample evidence that because of either the increased incidence of asthma or improved diagnoses and reporting, the disorder is becoming more common. The growing prevalence of asthma has demanded that the combined armature of medical and behavioral sciences be synthesized to produce effective self-management approaches to control the disorder.

Morbidity

Morbidity data reflects the qualitative and quantitative states or conditions affected by asthma. Several indices of morbidity were discussed in Chapter 1. As was also noted, it will only continue to increase as we end one century and start another.

Mortality

Another fiber in the pattern of asthma self-management concerns deaths from asthma. The increase in deaths from asthma, summarized in Chapter 1, is nothing less than tragic. That anyone should die from the disorder is bad enough, but, as so aptly put by Kaliner (12), "No asthmatic should die, and if properly treated very, very few asthmatics would die."

Demographics

The strand of fabric concerning demographics of populations and population segments prone to asthma has only begun to emerge in the past two decades. As noted by Taitel et al. (Chapter 1), there has recently been considerable attention paid to populations, particularly minority groups, who experience higher rates of asthma. Currently, a major attempt is being made to combat asthma in populations where the disease has its greatest impact—minority groups in the inner city. This reflects the trend, noted by Marwick (13), not only to tackle asthma not only by looking at the numbers, but by using the demographics for the disorder. It can be anticipated that ongoing attention will also be directed towards reviewing both the genetics of asthma and worksite stimuli that contribute to the disorder.

B. Characteristics of Asthma

The characteristics of asthma defined the content of asthma self-management programs from their infancy. The characteristics are, at all levels, a primary factor that demands the comprehensive asthma education included in most self-management programs. The three characteristics of asthma—the intermittency, variability, and reversibility of attacks—were significant factors in stimulating research on asthma self-management. An impetus for such programs, however, was the increasing recognition by both medical and behavioral scientists that any control over a patient's asthma would disappear unless the patient played a major role in the management of his or her asthma. This was particularly evident at many residential and medical centers for pediatric asthma, which, after bringing a child's asthma under control, returned the youngster to an uncertain future in the very environment where he or she had previously experienced severe asthma. It was known that the ability to promote stimulus generalization across desperate environments, such as a hospital and a tenement in an inner city, was limited. Therefore, the decision was to promote response generalization by teaching children to perform self-management or self-control skills across settings and in different contexts (13,14). The approach was to make patients allies with their physicians in the management of asthma; the framework used was based upon 10 tenets common both to self-management in general (15) and to asthma self-management in particular (16).

1. Only patients know all the behaviors they should perform. Unless they experience attacks in the presence of medical personnel, patients and those around them must perform whatever skills are necessary to help abort an asthmatic episode.
2. Only patients know what and when they should perform in a given way. This factor is especially important given the intermittent and variable nature of asthma; patients are apt to be the only ones who know what they should do to help manage an episode, such as using an inhaler correctly.
3. Only patients can reward themselves for many of the behaviors they perform. The ability they have to control an unexpected attack is intrinsically rewarding to patients.
4. Patients are more motivated when they perceive they have some personal control over a situation. If patients have had past successes in helping manage their asthma attacks, they will be more motivated to perform similar self-management steps in the future.
5. Patients are more apt to comply to a change plan they helped develop. The intermittent and variable characteristics of asthma mean that patients must be prepared to act on an action plan they and their physicians have jointly developed. This is why goal selection assumes a major role in asthma self-management.
6. Pursuing and attaining the goals patients help establish is a strong source of motivation. Patient commitment is a necessary component to the suc- cess of any self-management program, including one designed for asthma.
7. Self-attributions following self-management reinforce a patient's independent actions. If patients believe their actions helped control asthma, they will be more likely to perform whatever actions are required to help manage future episodes.

8. Self-management is ethically and socially valued more by patients than passive approaches. Any activity that included goal setting by patients with their physicians to establish steps they can use to help control their asthma falls into this category.
9. Only patients can control their own behavior and achieve goals they helped select in helping to manage asthma. This point is often misunderstood, particularly by medical personnel.
10. Self-management facilitates generalization. Given the intermittent, variable, and reversible nature of asthma, there is no more important reason for self-management than the promotion and maintenance of response generalization by patients. Patients must learn not only to perform asthma-management skills in different settings and with attacks of varying intensity, but over time. Given what may be the intermittent pattern of a patient's episodes, the generalization prompted by a patient performing self-management skills is the only way attacks can be consistently controlled.

II. Complexity of Asthma

The emerging complexity of asthma formed another fiber in the tapestry of asthma self-management. Twenty years ago, the varied and complicated nature of the pattern was cogently summarized by the NIAID Report (17). The discovery of IgE, the immunoglobulin related to allergic responses, led to a burst of optimism that asthma might soon be conquered. This was followed by a few halcyon years that gradually dissolved with the knowledge that the discovery of IgE, as important as it was, served mainly to stimulate research into what continues to evolve into the complex enigma of asthma. Research, in particular, was directed towards the immunological basis of the disorder. By 1979, there was increased recognition of the role of mediators released from the basophil and mast cells during asthma attacks, as well as the role played by inflammation in the disorder. It was also recognized that a number of new pharmaceutical advances might occur on the basis of increased knowledge emerging from studies of the immunological and biological basis of asthma. Implications of the complexity of asthma were also beginning to be recognized by those concerned with self-management of the disorder.

In the period since the NIAID Report (17), impressive progress has occurred in investigating the underlying biological and immunological bases of asthma. All one needs to do is to compare any definition of asthma from that period, which only described characteristics of the disorder, with definitions proposed by the National Asthma Education Program (5) and the International Consensus Report (6). The growing complexity of the disorder, however, is

Table 1 Pathological Changes and Mast Cell Mediators

Pathological changes	Mediators
Bronchial smooth-muscle contraction	Histamine (H-1 response)
	Leukotrienes C-4, D-4, E-4
	Prostaglandins and thromboxane A-2
	Bradykinin
	Platelet activating factor
Mucosal edema	Histamine (H-1 response)
	Leukotrienes C-4, D-4, E-4
	Prostaglandins E-2
	Bradykinin
	Platelet activating factor
	Cymase
	Reactive oxygen species
Mucosal inflammation	Inflammatory factors and cytokines
	Eosinophil and neutrophil chemotactic factors
	Leukotriene B-4
	Platelet activating factor
Mucosal secretion	Histamine (H-2)
	Prostaglandins E-2
	Hydroxyeicosatetraenoic acids
	Leukotriene B-4
Desquamation	Reactive oxygen species
	Proteolytic enzymes
	Inflammatory factors and cytokines

Source: Adapted from Ref. 18.

captured best by briefly summarizing one aspect of this knowledge: pathological changes and mediators thought to be responsible for asthma. Advances in the identification of these factors and their role in asthma, reviewed by Kaliner and Lemanske (18), are presented in Table 1. As will be noted, information about of the pathogenesis of asthma has had a major influence on the way the disorder is treated. This change, in turn, will affect which skills are taught to patients in asthma self-management.

A. Treatment of Asthma

The treatment of asthma, primitive as it sometimes is, has always been a fiber in the tapestry of asthma self-management. The NIAID Report (17) noted that treatment of asthma relied upon the use of β-adrenergic agents, theophylline, cromolyn sodium, and corticosteroids. Concern was expressed regarding the continued use of ephedrine; mention was made about the role inhaled steroids might play in the future control of asthma. Although many of these medications, particularly theophylline, sometimes required that doses be taken every 6 hours to be effective, nothing was noted about the importance of medication compliance. Educational needs of physicians were stressed, and it was recommended that a multidisciplined approach be taken to the overall treatment of asthma, particularly in children. A concluding chapter of the NIAID Report issued a plea for formal educational programs or what today would be referred to as self-management. A number of outcome variables were listed; most of these were included in the pioneer programs in self-management, particularly those developed for childhood asthma (19).

In the time since the NIAID Report (17), major changes have occurred in the treatment and management of asthma. How asthma should be treated has been the basis of a number of reports, including the National Asthma Education Program (5), the International Consensus Report (6), the Global Asthma Report (7), and Practice Parameters for the Diagnosis and Treatment of Asthma (20). These reports emphasized that the effective management of asthma depended upon four integral components: (1) objective measurement of lung function, (2) environmental measures to control precipitants of asthma, (3) patient education, and, in particular, (4) the pharmacological treatment of asthma. Much of what is described concerning treatment of asthma, particularly chronic asthma, was incorporated into the National Asthma Education Report (5). This included (1) treatment of the underlying pathology of asthma with maintenance or preventative drugs, (2) tailoring of treatment programs for individual patients, (3) controlling asthma triggers, (4) using step-care pharmacological therapy, and (5) continual monitoring of patients. Changes in medication regimens were cogently described by Kaliner and Lemanske (18):

> Over the past 5 years, there has been a dramatic change in the pharmacological approach to asthma treatment. This change is based on the division of available agents into those that act symptomatically (reversing the cause of airflow obstruction but not the underlying event) and those that act specifically (reducing the underlying process and eventually affecting airflow obstruction). Depending upon the severity of the patient's asthma, symptomatic treatment is initiated to control the patient's airflow obstruction while specific treatments are given time to act. In other words, all patients who have more than occasional symptoms are treated with specific agents, along with a limited course of symptomatic drugs as well. The plan is to reduce the symptomatic agents as soon as the patient responds to the specific approaches.

The complexity of asthma and its treatment has proved a significant factor in the evolution of asthma self-management programs. Changes in the programs were detailed in Chapters 10–12. It is not the intention of this chapter, even if possible, to present a compendium of both current and future changes in asthma self-management sparked both by the complexity of asthma and by knowledge of the disorder. These are nicely summarized in the National Asthma Education Project (5), the International Consensus Report (6), the Global Asthma Report (70), and Practice Parameters for the Diagnosis and Treatment of Asthma (20). However, based upon the information gleaned from these reports, it can be expected that greater attention will be paid in future self-management programs to: (1) risk assessment, (2) prediction and prevention of attacks, including the performance of environmental modification and improved medication compliance, (3) the sequential performance of steps required to manage acute episodes, (4) refined assessment of patient self-reaction, and (5) the generalization and maintenance of asthma self-management skills over time and settings.

B. Interdisciplinarian Approach

One of the most salient fibers in the fabric of asthma self-management is often overlooked: the transition of asthma management from a multidisciplinarian model to an interdisciplinarian approach. A multidisciplinarian approach to the management of asthma was recommended in the NIAID Task Force Report (17). However, Melvin (21) and Varni (22) clearly delineated the distinction between multidisciplinarian and interdisciplinarian approaches. Multidisciplinary refers to activities that involve professionals from a number of disciplines who approach the patient primarily through an uncoordinated discipline-specific fashion. The approach, noted Varni (22), requires that the each professional only know the skills specific to his or her own discipline. In contrast, the interdisciplinarian approach requires that physicians, nurses, physical therapists, social workers, health educators, psychologists, and other allied health care professionals have a working knowledge of the other team members' skills and specialties. Thus, concluded Varni, "the interdisciplinary approach is synergistic, integrating the knowledge and skills from the various disciplines into a coordinated plan for patient care" (22).

The interdisciplinarian model provided the optimal format for the development and evaluation of an asthma self-management program. The model would not have been introduced or successful, however, if not for both the vision and the willingness of a number of physicians who, two decades ago, encouraged and participated in interdisciplinarian asthma self-management programs. While running the risk of omitting people, these physicians included Claude Lenfant at the National Heart, Lung, and Blood Institute, Elliott Middleton and Hyman Chai in the early 1970s and Pat Leung and John Winder in the late 1970s at the National Asthma Center; Bob Mellins and Charles

Feldman at Columbia University; Elliott Middleton, Elliott Ellis, and Jerry Cropp at SUNY/Buffalo; Bill Bailey at UAB; Gary Rachelefsky and Charles Lewis at UCLA; Joanne Blessing-Moore at Stanford University; Phil Fireman and Gil Friday at the University of Pittsburgh; and Stuart Rusnak in Honolulu.

C. Multidimensional Educational and Behavioral Model

Purcell and Weiss (23) pointed out that, historically, asthma traveled a winding path from being considered a symptom of the nervous system, to being an immunological disorder, to being "a symptom whose interpretation depends on the particular spectacles through which one is peering." The NIAID Report (17) stressed the biological and immunological basis of asthma. A similar position was suggested by both the National Asthma Education Project (5) and the International Consensus Report (6), with a particular emphasis upon asthma as a disorder that featured airway obstruction, airway inflammation, and airway hyperresponsiveness. Both of the latter reports, however, stressed the need for patient education as an integral ingredient of the overall management of asthma.

What is often overlooked is the multidimensional education and behavioral model that has emerged with respect to asthma in the past two decades. The model occurred, in part, because of the demise of two prominent theories of behavioral acquisition. The first theory was the psychoanalytic position of French and Alexander (24), which suggested that asthma represented a repressed cry for a patient's mother. While no empirical evidence was ever put forward to support this position, it contributed to the perception that patients with asthma were psychologically abnormal. The position is patently untrue, although it took several decades for the idea to be dispelled by behavioral scientists. Facts prevailed, however, and today the dominant view of patients with asthma, including children, is that they are normal people in an abnormal situation (14,25–27). Asthma does present problems, but, overall, asthma patients are influenced by the same types and variety of environmental developmental, cognitive, and contextual variables as other people.

The second event was the weakening of support for theories of learning that were developed to permit the description, control, and prediction of behavior within a simple set of principles. The development of these theories, once regarded with high optimism, was replaced by strong disillusionment by the end of the 1970s. From a behavioral viewpoint, self-control or self-management was recently described as primarily control of behavior by delayed events (28). Considerable debate was generated by this premise, including arguments by Premack and Premack (29), who suggested that "neither 'self' nor 'decision'—major prerequisites of self-control—are properly explicated." Unfortunately, maintained Premack and Premack, "self" remains intractable, and although "decision" has attracted interesting treatments, final accounts are lacking. Other models for analyzing behaviors, including the health belief model (30), were also developed; however, judging from the literature, the health belief model was not influential in the creation and development of asthma self-management programs. Social learning theory (31,32) appeared within a few months after the deadline for proposals on self-management programs for childhood asthma issued by the National Heart, Lung, and Blood Institute. All this led to the serendipitous result that the developed programs were not wedded to any particular theoretical orientation, other than the insistence that outcomes be valid and reliably assessed. The result was programs that were pragmatic, dynamic, empirically based, and interdisciplinarian in nature. The emphasis was on processes of self-management, particularly on teaching patients skills required at each stage of self-management. These characteristics are the underpinnings of most of the programs reviewed in Chapters 10–12.

The National Heart, Lung, and Blood Institute and the National Institute of Allergy and Infectious Diseases sponsored a workshop on the self-management of childhood asthma in 1983. At the meeting, Thoresen and Kirmil-Gray (33) presented a cognitive social learning model for self-management. A component of the model was particularly relevant to the self-management of asthma: reciprocal determinism. The concept of reciprocal determinism refers to the constant interaction of personal, environmental, and behavioral determinants. Influences of determinants were multidirectional where an event acts and reacts, often modifying the future probability of the prior event itself as well as that of other events. The model presumes that social and environmental, physiological, cognitive, and behavioral sources will be studied in concert. Whether an existing asthma self-management program was based on social learning theory or not, all programs had elements of reciprocal determinism inherently built into them. For example, all were concerned about environmental stimuli, the physiological reaction of asthma often triggered by the stimuli, the detection of changes in breathing that signaled an asthma attack, and the behaviors taken to alleviate the episode. The model has certain disadvantages, particularly complexity, but the concepts most clearly fit the reality of asthma experienced by patients, medical, and behavioral scientists. This was especially true considering the levels of empirical assessment and explanations sought in early self-management programs.

Another factor reintroduced by social learning theory was the distinction between learning and performance (32). In most cases, learning consists of changes in behavior that occur concurrently with practice. Most of us learned how to read by repeating words spoken and pointed out to us in a book. Tolman (34), the person who established many of the basic principles now subsumed under the rubric of social learning theory, distinguished between learning and performance. Learning refers to changes produced by practice, and performance refers to the translation of learning into behavior. The distinction is chiefly in terms of temporal considerations (35). While the distinction is not relevant in most learning situations, it is with respect to asthma self-management. While patients can be taught what to do when they have an attack, the intermittent nature of asthma means a period of time may elapse before patients have an attack and must perform the skills they have been taught. The learning-performance distinction spawns a number of methodological problems, particularly those related to assessment and study design. In the case of the former, we can assess knowledge of skills after instruction. How do we assess the later performance of the patient? What dependent measures are necessary to distinguish between learning and performance? In the case of the design of a study, how can it be designed to fit the reality of what patients actually do? This point will be amplified in the discussion section.

III. Processes of Self-Management

Processes of self-management have been described by others, particularly Ford (36), Kanfer and Schefft (15), Creer and Christian (37), and Karoly (38). Despite different terminology, there is considerable commonality among asthma self-management programs with respect to processes; the differences that exist often result from the way particular processes are described, not from differences in conceptualization (4,39,40). Processes relevant to the self-management of asthma include: (1) goal selection, (2) information collection, (3) information processing and evaluation, (4) decision making, (5) action, and (6) self-reaction (Fig. 2). As noted, all of the processes are embedded within context. This means that how the processes are performed is influenced by the context within which they occur. For example, context influences how information is collected or action is taken, e.g., whether such processes occur at home or work. Different contexts influence different processes.

A. Goal Selection

Goal selection occurs only after systematic preparation, including patient education. When patients have relevant information necessary to perform self-management skills, they should meet with their physicians to establish reasonable and specific goals. As noted by Weinstein (41; see also Chapter 13), the establishment of these goals requires negotiation between patients and their physicians. Through such negotiation, physicians can establish the goals they want pursued and patients can help determine how the goals can be achieved. The give-and-take that characterizes negotiation, contracting, and

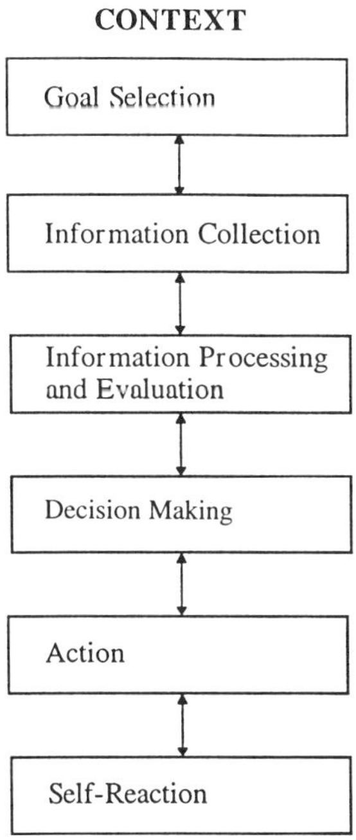

Figure 2 Processes relevant to the self-management of asthma.

goal selection processes has three positive consequences (36,38): (1) it establishes preferences about desired outcomes of self-management, (2) it strengthens the commitment of patients to perform goal-relevant self-management skills, and (3) it establishes expectations on the part of patients that augment their effort and performance.

Goal selection has been the most overlooked process in asthma self-management. It is ignored although it is the only process where both physicians and medical personnel are involved in establishing preferences and priorities for a patient's management. It is, in short, the only time when there is what some physicians and medical personnel refer to as "co-management" of asthma with patients (6). After goal selection, patients are on their own. They will or will not perform the self-management skills they have been taught; all medical personnel can do is track the performance of patients and, if necessary, renegotiate goals on the basis of the obtained data. For this reason, greater attention needs to be paid to goal selection by physicians and patients, such as recommendations suggested by Weinstein (Chapter 13), as long as a patient performs asthma self-management skills.

B. Information Collection

Self-monitoring, or the self-observation and recording of data, is the basis of information collection. In pursuing selected goals, individuals must attend to and record information that bears upon the pursuit of these goals (40). The overall conclusion is that self-monitoring is a necessary, but not sufficient, condition for the success of self-management programs (42).

Several suggestions have been offered to improve the monitoring of asthma-related behaviors (43). First, patients should attend only to phenomena that have been operationally defined. There are signs that this step has been considered in self-management programs for asthma. Second, an objective measure, such as a peak flow meter, should be included as a monitoring instrument. These devices have also been included in many asthma self-management programs. Finally, whenever information is gathered by patients over a period, it is best to have them observe and record data only during specified windows of time. Otherwise, unless the patients exhibit what Kirschenbaum (44) referred to as a healthy obsession, patients are not going to be vigilant in constantly collecting data on themselves and their asthma. It is noteworthy that these suggestions were followed in many of the programs developed for the self-management of asthma (Chapters 10–12).

There remains the temptation to ask patients to gather too much data about themselves and their asthma. A prime example is offered by the guidelines for asthma management proposed by the National Asthma Education Project (5). The numerous suggestions they offer make sense when considered individually; if combined together, however, they would overwhelm the average patient. At the same time, attending to and gathering too much information forces patients to center their lives around their asthma. For this reason, the minimal number of essential elements of asthma self-management should be established and followed (40).

C. Information Processing and Evaluation

After they gather data on themselves, patients must learn to process and evaluate the information they have collected. Creer and Holroyd (40) pointed out that five distinct steps are involved.

First, patients must be able to ferret out any changes that occur in the information they observe, record, and process about themselves. The peak flow meter can assist in detecting changes, but, as noted by Fritz and Wamboldt (Chapter 8) and Creer et al. (Chapter 6), there are weaknesses in using the instrument. Subjective symptoms of asthma can also be unreliable, although many patients believe their subjective feelings provide better indications of asthma attacks than peak flow values. Patients especially skilled at detecting changes in their breathing use a variety of information, including private symptoms and public PEF values, to determine if changes have occurred. Integrating valid and reliable aspects from both objective and subjective systems seems the most appropriate way to collect information. Considerably more information is provided on this topic in Chapters 7–9.

Second, standards must be established to permit individuals to evaluate the data they collect about themselves and their asthma. In recent years, the personal best peak flow values have replaced predicted peak flow rates as the preferred standard for comparing PEF values (5,6). The zone approach defines fixed PEFR ranges, using a traffic signal analogy of green, yellow, and red to indicate asthma severity at each level. The zones in approach have been developed according to expected PEFR ranges based upon arbitrary norm data rather than upon individualized patterns of PEFR data and symptom fluctuations. As noted in Chapter 6, the use of the system lacks empirical support. As the zone system has limited applicability in assessing the performance of groups of patients, standards must be tailored for individual patients to permit them to make accurate evaluations about their breathing.

Third, patients must learn to make judgments about the data they have processed and evaluated. Making judgments may be easier for individual patients using objective data, such as PEF rates. Patients can match any deviations from personalized standards and determine if action should be taken. Matching-to-standard may be a more difficult task with respect to subjective symptoms, although many patients claim this not to be the case (Chapter 6). Individuals must acquire and refine their skills at matching their respiratory reactions to what, at the outset of their performance, may be fuzzy and ambiguous standards.

Fourth, individuals should evaluate any changes in terms of the antecedent conditions that may have altered their breathing, the behaviors they can perform to change their breathing, and the potential consequences of their action. An analysis of antecedents, behaviors, and consequences within which attacks occur provides considerable information about attacks. The result should be better decisions and more effective action.

Finally, contextual factors must be considered in processing and evaluating information about asthma. A multidirectional interaction always exists between context, asthma, and elements of self-management. In addition, contextual variables, including setting events, establishing stimuli, and establishing operations, may influence the course of asthma self-management (4,43).

D. Decision Making

Decision making is a critical function in self-management in that after patients collect, process, evaluate, and judge data on themselves and their asthma, they are expected to make appropriate decisions based upon the information

(40). Despite an increasing amount of data available regarding medical decision making, there is limited information as to how asthma patients make decisions. Yet, decision making is at the core of successful asthma self-management. Two studies illustrate the extremes that can be found in decision making in asthma. First, Creer (45) compared decision-making skills of medical personnel with asthmatic children and their families. Results indicated that both physicians and patients skilled at asthma management used similar decision-making strategies or heuristics. Both groups used advanced cognitive strategies, coupled with remarkable flexibility, both in evaluating processed information and in making appropriate decisions based upon their evaluation. The conclusion from this study is that patients can become highly skilled at making effective and efficient decisions regarding their asthma. A second study shows the other extreme. Gibson and colleagues (46) examined autonomy, reflected in the preferences of subjects for information seeking and decision making, with respect to asthma. It was found that patients expressed strong preferences for acquiring information regarding their asthma. They did not want to make their own decisions about the management of attacks, however. As the severity of the asthma exacerbation increased, the desire to make decisions decreased. Older subjects expressed less desire for decision making than younger subjects. The results obtained by Gibson et al. (46) not only have implications for the success of asthma self-management programs and action plans, but suggest another reason for the high drop-out rate among patients involved in asthma self-management programs.

E. Action

Action includes all the steps performed by patients to control their asthma. Self-instruction, or what some refer to as verbal rules (47), underlies any action taken by individuals. Self-instruction includes the prompting, directing, and maintenance of self-management skills. How successful patients are in asthma self-management is a function, to a major extent, of the instruction they provide to themselves to initiate and maintain whatever action is necessary to halt an attack.

Self-instruction is important in two other ways (4, 43). First, establishing control over asthma requires that patients perform, in a stepwise manner, the treatment strategies they have agreed on with their physicians. Individuals must use self-management skills to reach the goals they and their physicians jointly set at the outset. Second, self-instruction prompts other strategies and techniques for managing or coping with a disorder. Strategies summarized by Karoly (38), including allocating greater attention to a problem, effort mobilization, planning and problem solving, verbal self-cuing, facilitative cognitive sets of expectations, stimulus control, and mental thought or cognitive control.

Other strategies might include relaxation, self-desensitization, skill rehearsal, modeling, linking or unlinking of behavioral chains, and self-reinforcement (4,43).

F. Self-Reaction

Bandura (32) defined self-reaction as the attention individuals direct towards evaluating their performance. On the basis of their evaluation, patients can establish realistic expectations about their performance, as well as evaluate whether they need more training and expertise. Individuals should also develop realistic expectations about the limits of self-management in attaining goals that they may select.

Locus of control, or the degree to which we perceive events as under our control or the control of others, is a part of self-reaction. One of the initial studies of asthma self-management resulted in the development of an instrument to assess locus of control in children with asthma (48).

Another key concept in self-management is self-efficacy. Self-efficacy is the belief of individuals that they can adequately perform specific skills in a given situation (49). Knowing asthma self-management skills is not enough to ensure that the skills will be used appropriately; individuals must also believe they are capable of performing these skills to reach whatever goal they have helped establish. Self-referent thought is credited as mediating the relationships between selected goals and action. As noted by Karoly (38), a belief in domain-specific efficacy, in contrast to beliefs individuals have regarding the consequences of their performance, "is the self-referent, generative capability that stands out as a singularly powerful self-motivating force." Self-efficacy arises, in part, from performance accomplishments, a situation that enhances the role self-efficacy plays in helping to guide and regulate action. Self-efficacy has been shown to be a significant component in a number of asthma self-management programs (50,51).

IV. Conclusions and Recommendations

This volume detailed the strides made in asthma self-management. While we are only at the start of the journey—a point that will become more clear as additional research is conducted—impressive progress has been made. The fact that asthma self-management has, in a relatively brief period of time, become a significant component in the treatment and management of asthma is impressive. Continued development and investigation will only enhance the value of self-management techniques in the control of asthma.

Five related questions frequently arise about asthma self-management. The first question is: Why is self-management different from any other intervention procedure? An answer is often mumbled about the stages of self-management or the processes of the procedure. In essence, stages of self-management are identical to those taken to train most health care providers, including physicians. Consider the training of physicians: The first phase occurs when budding medical doctors attend medical school. This comprises the *acquisition stage* of their training. By the end of the stage, the candidate has demonstrated that he or she has acquired knowledge about medicine. The second stage occurs when physicians apply the knowledge acquired in medical school to patients during internships and residencies. This constitutes the *performance stage* of training. Finally, physicians go out into the world as practicing doctors. Learning does not stop, however, as physicians are expected to attend meetings, continuing education workshops, and read current literature to stay abreast of developments in their field. This final stage is the *maintenance stage* of training.

The same three stages involved in training physicians comprise asthma self-management. In stage 1, patients acquire knowledge about asthma and how self-management can be used to help control the disorder. By the end of the stage, patients should know the self-management skills they can perform to help manage their condition. In stage 2, patients must translate the knowledge they have acquired into performance. In short, they must demonstrate competency in performing self-management steps to help control their asthma. It is also in this stage where initial breakdown of self-management performance may begin to occur. Procedures, described as relapse prevention or management techniques (53), have been developed specifically to deal with any deterioration of performance. In stage 3, patients must continue to perform self-management skills to control attacks. This is the maintenance stage of asthma self-management. Maintenance is not always necessary if there is an abatement in a patient's asthma, a factor that often occurs in childhood asthma (54). However, children and many adults who continue experiencing asthma throughout their lives, cannot terminate the efforts to help manage their disorder. It is here that the relapse prevention and management techniques, suggested by a number of authors (42,53), become so important.

Self-management programs for asthma have included both acquisition and performance phases, although the distinction between the two stages is often blurred. Little has been done with respect to the maintenance of self-management. A conclusion is that asthma self-management be considered within a dynamic three-stage (acquisition-performance-maintenance) model. The model must be dynamic, much like the model of physician education. It would require that specific measurement procedures be tailored to assess what occurs in each of these stages. This will require more work on our part, but it will permit us to answer such questions as to whether changes in a program are due to the translation of knowledge into performance, the development

Table 2 Steps in Developing Effective Treatments

1. *Conceptualization of the disorder*. Conceptualization refers to delineating key areas related to the development, onset, and exacerbation of a disorder.
2. *Research on processes related to disorder*. Research conducted to examine the relation of processes proposed as critical to the disorder.
3. *Conceptualization of treatment*. Conceptualization of treatment focuses on how specific procedures relate to other processes implicated in the disorder and to desired treatment outcomes.
4. *Specification of treatment*. Concrete operationalism of interventions, preferably in treatment manual form, so that the integrity of treatment can be evaluated, outcome data can be codified, and treatments can be replicable.
5. *Tests of treatment process*. Studies are conducted to identify whether or not the intervention techniques, methods, and procedures actually affect those processes critical to the model.
6. *Tests of treatment outcome*. Studies are conducted to evaluate the impact of treatment (e.g., open studies, single-case studies, clinical trials, combined design trials) and can provide evidence that change is produced. Several types of studies (e.g., dismantling, parametric, comparative outcome) are relevant.
7. *Test of boundary conditions and moderators*. These studies include an examination of child, adult, parent, family, and contextual factors that interact with treatment.

Source: Adapted from Ref. 55.

and subsequent breakdown of self-efficacy, etc. The goal would be to achieve a state such as that which exists in medicine, where acquisition, performance, and maintenance of skills can be readily assessed and, if necessary, altered.

A second question is: Can we improve the design and implementation of asthma self-management programs? Kazdin (55) outlined seven stages of developing effective treatments. An adaptation of these steps is presented in Table 2. With asthma self-management, it is clear that considerable progress has been made in steps 1 through 4. Progress with steps 5 through 7 is less clear. In particular, our understanding of the process or mediating variables associated with behavior change and maintenance of these changes remain limited. For example, why do we often see behavior change during stage one—acquisition—in the absence of improvements in self-efficacy? There are many process-outcome variable relationships in asthma self-management and, more broadly, in health behavior change that remain to be deciphered. Most asthma self-management research has focused on high strength or broad-spectrum interventions. We have begun to show that self-management programs that provide comprehensive asthma education and training in behavioral change

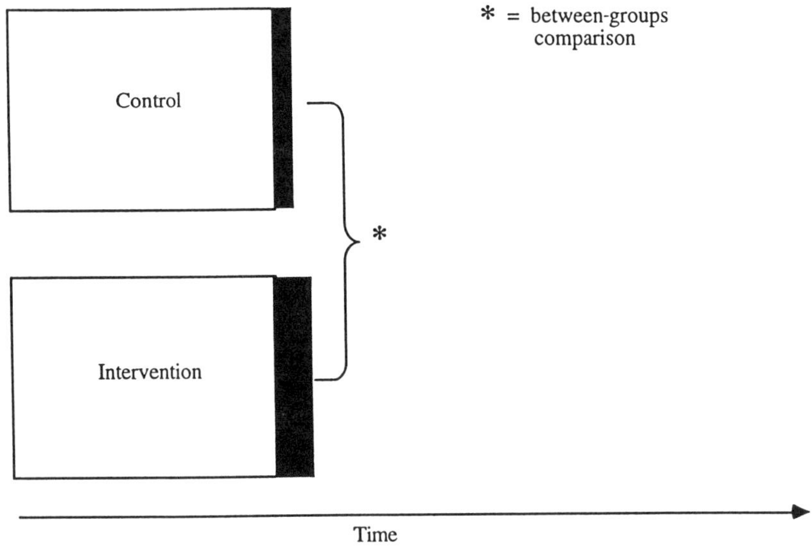

Figure 3 Clinical trial.

skills, such as self-monitoring, goal setting, and self-reaction, produce significant reductions in asthma morbidity (56). Considerable within-group process research will be necessary to fully understand the psychological variables that mediate positive outcome.

Our understanding of treatment outcomes is limited. We do not know the long-term durability of self-management outcomes or the full impact of complex interventions. The impact of quality of life has been difficult to measure, although newer instruments in this area show promise in helping us tease out the impact asthma self-management has on the life of patients (57). Finally, we have little understanding of the boundary conditions and moderators that influence self-management outcomes. For example, what environmental, cognitive, behavioral, and physiological variables interact with contextual factors to produce inadequate or even negative responses to self-management interventions?

A third question is: Are traditional clinical trial models appropriate for evaluating asthma self-management? The answer is no. The typical model used in clinical trial research is shown in Figure 3. As depicted, this consists of patients being randomly assigned to two groups: a control group and an intervention group. The comparison is then made between findings obtained in each of the two groups. The design is useful in assessing the effects of an independent variable, such as an asthma medication, in comparison to those

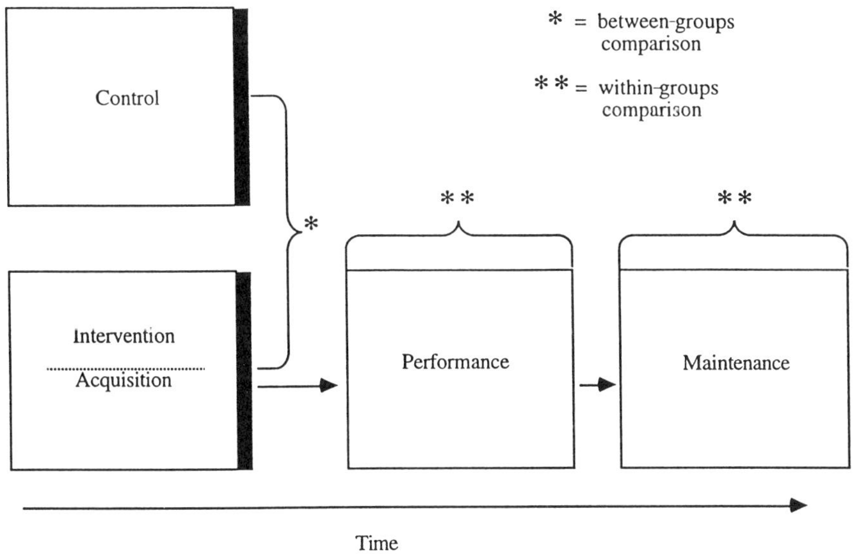

Figure 4 Self-management trial.

who do not receive the medication. Figure 4 depicts what occurs in asthma self-management. It starts out like a clinical trial: patients are randomly assigned to a control or intervention group during the acquisition phase. Comparison between the two groups is warranted at this point, as it would be in comparing the untreated control groups to the teaching both of physicians and asthma self-management skills.

In stage two—performance—patients continue to be compared to an untreated control group. While such comparisons do not violate the null hypothesis—the premise in inferential statistics that populations in the two samples do not differ from one another—in reality the groups are inherently different. Consider what occurs in medicine. In stage two, physicians have the opportunity in internships and residencies to perform the skills they acquired in medical school if they wish to do so. What if a comparison were made between physicians who complete internships and residencies to those physicians who elect not to continue such training, e.g., those who go directly into research? Can a legitimate comparison be made between the two groups of physicians in the performance of clinical skills taught in medical school? While it can be argued that the null hypothesis has not been violated, the results would be highly predictable: physicians who complete internships and residencies will perform clinical skills they were taught in medical school significantly better than physicians who elect not to complete such training.

The same rationale applies to patients who perform self-management skills versus those taught the skills but who fail to perform them. Are the two groups equal? According to statistical procedures, the answer would be yes. Indeed, the assumption of a meta-analysis of a self-management program for pediatric asthma was that the programs were identical to simple clinical trials (58). They are not, however; such a comparison defies logic. Patients who actually practice asthma self-management skills will perform them significantly better than those who do not. The same rationale is also relevant in comparing the performance of both physicians or patients in the third or maintenance stage. The only solution to this problem is, after the acquisition stage of asthma self-management, to use within-group designs where both key process and outcome variables can be evaluated across the stages of performance and maintenance. Combination designs permit the degree of linear relationships between variables to be determined and, in addition, has the advantage of providing both within-group comparisons and complex comparisons across variables. This sophisticated approach not only is more relevant but more ethical than simple between-group designs (54,56). In asthma self-management, how can you ethically withhold education about asthma and how to manage the disorder for a year or more from a control group of patients? You cannot.

The discussion seems forthright, but it is deceiving. Process variables in asthma self-management, including acquisition and performance factors, are usually analyzed in a simplistic linear fashion. Consequently, results are ambiguous and unclear; true relationships cannot be delineated between variables at each phase of self-management. Most of the process and outcome variables of interest to self-management investigators have unique and mutable rates of change. They attain maximum and minimum values at different points along the self-management continuum. Teasing out these unique processes and relationships requires that assessment be repeated and, at times, continuous throughout a self-management investigation. For example, while knowledge of asthma reaches asymptote early during the acquisition phase, it is a necessary precursor for the later execution of self-management skills during the performance phase. Morbidity data, such as hospitalization and emergency room use, costs of asthma, and respiratory problems, decrease gradually over time and often do not bottom out until a year after self-management training. Quality-of-life variables show gradual improvement over the course of a self-management study; psychological variables, on the other hand, show an array of changes with depression scores gradually decreasing and self-efficacy and positive attitudes towards asthma and its management gradually increasing over the course of a study.

The fourth question is: How do we mainstream asthma self-management into managed health care? For a number of years, asthma has been the leading reason for hospitalization among pediatric patients (59). For this reason, managed care organizations, particularly health maintenance organizations (HMOs), have focused on reducing hospitalization for asthma as a way to reduce overall costs. There have been two weaknesses to strategies taken by these organizations, however. First, many have eschewed the expertise of asthma specialists and relied upon asthma treatment by primary care physicians. However, as Marwick (13) pointed out, only one of every 10 physicians surveyed were aware of or familiar with the guidelines issued by the National Asthma Education Project (5) or that the National Heart, Lung, and Blood Institute had issued such guidelines. Lack of adequate medical management has been blamed in two reports for the rise of deaths due to asthma (12,60). Physicians, particularly primary care doctors who care for asthma, must know these guidelines if there is to be a reduction in asthma mortality. These physicians must recognize that asthma is not a broken leg that only requires setting; rather, asthma, like other chronic conditions, is a situation characterized by physicians, uncertain that their treatment strategies will work, who are managing patients who are uncertain that the tactics used to control their asthma will work. Only considerable expertise at managing asthma, by both physicians and patients, will demonstrate that asthma can be managed over the long run.

A second problem is the time required to teach patients self-management skills and to assess the performance of these skills. Kotses and colleagues (56) demonstrated how the performance of self-management skills could sharply curtail the costs of asthma (61). These gains are being maintained for up to 7 years in an ongoing follow-up of these patients (D.A. Levstek, unpublished, 1996). HMOs and other managed care units, however, expect immediate benefits from asthma-education programs; they are often unwilling to delay larger long-term reductions in asthma costs in favor of what they perceive as short-term benefits of brief educational programs. Perceptions of short-term reductions in health care costs are illusionary, however: there are few short-term benefits from brief educational, particularly self-management, programs. For example, a quiz can be given to determine the knowledge of patients with asthma. Following this assessment, a brief asthma-education program can be provided. The knowledge quiz could then be readministered. Significant changes would likely be found in the patients knowledge about asthma. Would this increase in knowledge translate to the performance of self-management skills by patients, especially given the intermittency of asthma? A few patients might remember what to do, but the majority would not. They would be as unskilled at treating asthma as they were before the brief intervention. There is a good analogy to the situation with asthma: suppose a group of 15-year-olds were shown a film on how to drive a car a year before they were eligible for their drivers' license. If they were tested before and after the film, most would show an improvement in their knowledge of driving an automobile. What would happen, however, if they were made to drive an automobile on a busy freeway 6 months or a year after they had seen the film? This, after all, could represent the period between when patients learn self-management skills and when they have to perform them. There may be a few novice drivers who could successfully navigate the freeway because they have had other experiences related to driving a car. Most budding drivers, however, would be like most asthma patients exposed only to a brief educational program: they simply would not know what to do. Under these circumstances, they would not only be a danger to themselves, but to others. Only through a strategy that combines acquisition, performance, and maintenance of self-management skills can a program be developed and implemented to help control asthma.

The final question is the most complex: How do you get patients to perform asthma self-management skills as long as required to control their asthma? If anyone had the answer to this question, particularly anyone who has conducted an asthma self-management program, they could retire by selling their answer. *The aim is to have patients voluntarily perform whatever skills they have been taught for as long as necessary to help manage their asthma. The number of factors that can impinge upon the patient in performing these skills is infinitesimal; the complexity of the interaction of these variables dwarfs any other topic in asthma, including the immunology and biological basis of the disorder.* Many people who have worked with asthma for a period know patients who adhere to every medical instruction they have received. Indeed, we once received data on a person who had faithfully taken a spirometric reading on himself and, on the basis of his data, adjusted his medications every 6 hours for 8 consecutive years. His vigilance to physician instructions was perfect. However, at the same time, the attention paid to his asthma was at the center of the patient's life. Do medical and behavioral scientists want asthma to be at the center of their patients' lives? Few will answer yes. At the other extreme, there are patients who do not follow the advice of their physician and rely instead upon emergency room service. These patients are more difficult to deal with, even though asthma is certainly not at the center of their lives. Striking the right balance between competent medical care and the patient's performance to control asthma is the goal of asthma self-management. This book has described the pioneering efforts to achieve asthma control through the development of self-management programs. The research provides a firm foundation for both current investigators and those who follow them to build upon.

References

1. McFadden ER Jr, Stevens JB. A history of asthma. In: Middleton E Jr, Reed CE, Ellis EF, eds. Allergy: Principles and Practice. 2nd ed. St Louis: CV Mosby, 1983:805–809.

2. Bartlett J, Kaplan J, eds. Bartlett's Familiar Quotations. 16th Ed. Boston: Little, Brown and Company, 1992.
3. Prochaska JO, Norcross JC, DiClemente CC. Changing for Good. New York, Morrow, 1994.
4. Creer TL. Psychology of Adjustment: An Applied Approach. Upper Saddle River NJ: Prentice-Hall, 1997.
5. National Asthma Education Program, Executive Summary: Guidelines for the Diagnosis and Management of Asthma. Washington, DC: National Institutes of Health, NIH Publication No. 91–3042, 1991.
6. International Consensus Report on Diagnosis and Treatment of Asthma. Bethesda, MD: National Institutes of Health, Publication No. 92–3091, 1992.
7. Global Initiative for Asthma. Bethesda, MD: National Institutes of Health, Publication No. 95–3659, 1995.
8. Holroyd KA, Creer TL, eds. Self-Management of Chronic Disease. Handbook of Clinical Interventions and Research. Orlando, FL: Academic Press, 1986.
9. Roberts MC, ed. Handbook of Pediatric Psychology. 2nd ed. New York: Guilford Press, 1995.
10. Ishizawka K, Ishizawka T. Identification of IgE antibodies as a carrier of reaginic activity. J Immunol 1967; 99:1189–1198.
11. Reed CE, Townley RG. Asthma: Classification and pathogenesis. In: Middleton E Jr., Reed CE, Ellis EF, eds. Allergy: Principles and Practice. 2nd ed. St. Louis: CV Mosby, 1983:811–831.
12. Altman LK. Rise in asthma deaths is tied to ignorance of many physicians. New York Times, May 4, 1993:B8.
13. Marwick C. Inner-city asthma control campaign under way. JAMA 1995; 274: 1004.
14. Creer TL. Asthma Therapy: A Behavioral Health Care System for Respiratory Disorders. New York: Springer Publishing, 1979.
15. Kanfer FH, Schefft BK. Guiding the Process of Therapeutic Change. Champaign IL: Research Press, 1988.
16. Creer TL, Kotses H, Reynolds RVC. A Handbook for Asthma Self-Management: A Leader's Guide to Living with Asthma. Athens, OH, Ohio University Press, 1991.
17. Asthma and Other Allergic Diseases. NIAID Task Force Report. Bethesda, MD: National Institutes of Health, NIH Publication No. 79–387, 1979.
18. Kaliner M, Lemanske R. Rhinitis and asthma. JAMA 1992; 268:2807–2829.
19. Wigal JK, Creer TL, Kotses H, Lewis PD. A critique of 19 self-management programs for childhood asthma. Part I. The development and evaluation of the programs. Pediatr Asthma Allergy Immunol 1990; 4:17–39.
20. Spector SL, Nicklas RA. Practice parameters for the diagnosis and treatment of asthma. J Allergy Clin Immunol 1995; 96:707–870.
21. Melvin JL. Interdisciplinary and multidisciplinary activities and the ACRM. Arch Phys Med Rehab 1980; 61:379–380.
22. Varni JW. Clinical Behavioral Pediatrics. New York: Pergamon Press, 1983.
23. Purcell K, Weiss JH. Asthma. In: Costello CC, ed. Symptoms of Psychopathology. New York: Wiley, 1970:507–623.

24. French TM, Alexander F. Psychogenic Factors in Bronchial Asthma. Psychosom Med Monogr 1941; 4:1–92.
25. Desquin BW, Holt IJ, McCarthy SM. Comprehensive care of the child with a chronic condition. Part I. Understanding chronic conditions in childhood. Cur Prob Pediatrics 1994; 24:199–218.
26. Perrin JM, MacLean WE. Children with chronic illness. Peds Clin North Amer 1988; 35:1325–1337.
27. Russo DC, Varni JW. Behavioral Pediatrics: Research and Practice. New York: Plenum Press, 1982.
28. Rachlin H. Self-control: beyond commitment. Behav Brain Sci 1995:18:109–121.
29. Premack D, Premack AJ. Why self-control is both difficult and difficult to explicate. Behav Brain Sci 1995; 18:140–141.
30. Becker MH. The health belief model and sick role behavior. Health Ed Monogr 1974; 2:409–419.
31. Bandura A. Social Learning Theory. Englewoods Cliffs, NJ: Prentice-Hall, 1977.
32. Bandura A. Social Foundations of Thought and Action. A Social Cognitive Theory. Englewood Cliffs, NJ: Prentice-Hall, 1986.
33. Thoresen CE, Kirmil-Gray K. Self-management psychology and the treatment of childhood asthma. J Allergy Clin Immunol 1983; 72:596–606.
34. Tolman EC. Principles of performance. Psychol Rev 1955; 62:315–325.
35. Kimble GA. Hilgard and Marquis' Conditioning and Learning. New York: Appleton-Century-Crofts, 1961.
36. Ford DH. Humans as self-constructing living systems: a developmental perspective on behavior and personality. Hillsdale, NJ: Erlbaum, 1987.
37. Creer TL, Christian WP. Chronically-Ill and Handicapped Children: Their Management and Rehabilitation. Champaign, IL: Research Press, 1976.
38. Karoly P. Mechanisms of self-regulation: a systems view. Ann Rev Psychol 1993; 44:23–52.
39. Creer TL. Asthma. In: Ollendick TH, Herson M, eds. Handbook of Child Psychopathology. 3rd ed. New York: Plenum.
40. Creer TL, Holroyd KA. In: Baum A, McManus C, Newman S, Weinman J, West R, eds. Cambridge Handbook of Psychology, Health, and Medicine. Cambridge: Cambridge University Press.
41. Weinstein A. Clinical management strategies to maintain drug compliance in asthmatic children. Ann Allergy Asthma Immunol 1995; 74:304–310.
42. Kirschenbaum DS. Self-regulatory failure: a review with clinical implications. Clin Psychol Rev 1987; 7:77–104.
43. Creer TL, Bender BG. Asthma. In: Gatchel RJ, Blanchard EB, eds. Psychophysiological Disorders: Research and Clinical Applications. Washington, American Psychological Association, 1993:151–204.
44. Kirschenbaum DS. Weight Loss Through Persistence: Making Science Work for You. Oakland, CA: New Harbinger, 1994.
45. Creer TL. Strategies for judgment and decision-making in the management of childhood asthma. Ped Asthma Allergy Immunol 1990; 4:253–264.

46. Gibson PG, Talbot PI, Toneguzzi RC. Population Medicine Group 93C. Self-management, autonomy, and quality of life in asthma. Chest 1995; 107:1103–1108.
47. Martin G, Pear J. Behavior Modification. 5th ed. Upper Saddle River, NJ: Prentice-Hall, 1996.
48. Parcel GS, Meyer MP. Development of an instrument to measure children's health locus of control. Health Ed Monogr 1978; 6:149–159.
49. Bandura A. Self-efficacy: toward a unifying theory of behavioral change. Psychol Bull 1977; 84:191–215.
50. Clark NM, Rosenstock IM, Hassan H, Evans D, Wasilewski Y, Feldman C, Mellins RB. The effect of health beliefs and feelings of self-efficacy on self-management behavior of children with a chronic disorder. Pat Ed Counseling 1988; 11:131–139.
51. Schlosser M, Havermans G. A self-efficacy scale for children and adolescents with asthma: construction and validity. J Asthma 1992:29:99–108.
52. Tobin DL, Wigal JK, Winder JA, Holroyd KA, Creer TL. A self-efficacy scale for asthma. Ann Allergy 1987; 59:273–277.
53. Marlatt GA, Gordon JR. Relapse Prevention: Maintenance Strategies for Addictive Behavior Change. New York: Guilford Press, 1985.
54. Creer TL, Backial M, Burns KL, Leung P, Marion RJ, Miklich DR, Morrill C, Taplin PS, Ullman S. Living with asthma. Part I. Genesis and development of a self-management program for childhood asthma. J Asthma 1988; 25:335–362.
55. Kazdin AE. Treatment of conduct disorder: Progress and directions in psychotherapy research. Dev Psychopathol 1993; 5:277–310.
56. Kotses H, Bernstein IL, Bernstein DI, Reynolds RVC, Korbee L, Wigal JK, Ganson E, Stout C, Creer TL. A self-management program for adult asthma. Part I. Development and evaluation. J Allergy Clin Immunol 1995; 95:529–540.
57. McSweeney AJ, Creer TL. Health-related quality of life assessment in medical care. Disease-a-Month 1995; 41:1–72.
58. Bernard-Bonnin A, Stachenko S, Bonin D, Charette C, Rousseau E. Self-management teaching programs and morbidity of pediatric asthma: a meta-analysis. J Allergy Clin Immunol 1995; 95:34–41.
59. Reed CE. New therapeutic approaches to asthma. J Allergy Clin Immunol 1986; 77:537–543.
60. Creer TL. Psychological factors and death from asthma: creation and critique of a myth. J Asthma 1986; 23:261–169.
61. Taital MS, Kotses H, Bernstein IL, Bernstein DI, Creer TL. A self-management program for adult asthma. Part II. Cost-benefit analysis. J Allergy Clin Immunol; 1995; 95:672–676.

15

Modification of Respiratory Perceptions
A Self-Management Perspective

ANDREW HARVER

University of North Carolina at Charlotte
Charlotte, North Carolina

EDWARD S. KATKIN

State University of New York at Stony Brook
Stony Brook, New York

I. Introduction

The development and implementation of asthma self-management and asthma education programs during the past several decades encompass a variety of efforts to increase patient awareness of behaviors, beliefs, and symptoms to control asthma. Most programs seek to improve knowledge about asthma in patients (1–4), but increases in knowledge are not necessarily associated with changes in behavior that relate to improved asthma morbidity and mortality (Chapter 4). Behaviorally oriented programs additionally emphasize the self-monitoring and recording of asthma-related variables such as peak expiratory flow rates (5) (Chapter 6). Home monitoring of peak expiratory flow in asthma patients has been endorsed uniformly by both national and international assemblies, and summaries of the National Asthma Education and Prevention Program have recommended daily peak flow monitoring for patients with moderate to severe persistent asthma (6–8). Peak flow records provide an objective way to verify decreasing levels of lung function and to document wide diurnal variations in lung function that are related to worsening asthma.

Peak flow monitoring seems especially relevant for patients who are poor perceivers either of asthma symptoms or of the seriousness of symptoms. Since the widely cited observations of Rubinfeld and Pain in 1976 (9), an increasing number of clinical investigations concerned with the relationship between symptoms and lung function in patients with asthma have confirmed that patients vary widely in their ability to perceive airway obstruction, and that some asthmatics seem more aware of symptoms than others despite similar degrees of airway obstruction (10–13). A developing literature has heightened suspicions that poor perception of asthma is likely related, in part, to more frequent and severe attacks of asthma, including life-threatening episodes (14–18).

Blunted perception of declines in lung function may lead both to underestimation of the seriousness of acute asthma and to undertreatment, but the measurement of asthma perception is ill-defined. At least three operational definitions for reduced perceptual sensitivity to increases in airway resistance are evident in the clinical literature: a lack of symptoms for a given level of airflow obstruction, variability in the intensity of symptoms for a given level of airflow obstruction, and inconsistency in the correspondence between repeated measures of symptoms and lung function (Chapter 7). Although the relevance of symptoms in the assessment and diagnosis of asthma is well appreciated (wheeze, dyspnea, cough, sputum), individual differences in the perception of symptoms among individuals either with or without asthma have not been studied systematically (19). On the other hand, it is well established that the absence of symptoms does not preclude the presence of airflow obstruction and that significant improvement in symptoms does not signal complete reversal of acute airflow limitation (20–24).

Despite keen interest by clinicians and researchers in the ability of patients to assess acute increases in airflow limitation accurately, little experimental work to date relates directly to the problem of symptom discrimination in asthma (10,12,14,25). Symptom discrimination—the awareness and appropriate interpretation of increased resistance to breathing—represents the first stage in the detection and treatment of an asthmatic episode (26). Failure to discriminate the onset of an asthma attack reduces the likelihood that learned skills designed to manage or otherwise control the episode will be performed. In this chapter, we consider efforts to affect perceived increases in the resistance to breathing relevant to the patient with asthma. First, we examine historical perspectives on the self-regulation and perception of physiological sensations. Next, we review interrelationships among sensations, symptoms, and illness, and then consider the perception of symptoms in patients with asthma. Finally, we summarize strategies to improve detection of airflow increases and attempts to modify accuracy of respiratory perceptions. Our primary purpose is to develop a framework to portray evolving concerns of reduced perception of airflow limitation in asthma self-management.

II. The Biofeedback Tradition

Although the experimental study of sensations arising from internal organs historically attracted little attention from either physiologists or psychologists, there has been accelerated interest in the neurophysiological and psychological bases and mechanisms of visceral sensation in recent decades (27,28). In psychology, research and theory on the perception and control of autonomic nervous system responses are linked to the development of biofeedback techniques, with strong emphasis on the use of augmented visceral sensory feedback as a mediator of acquired visceral self-control (29). The relationship between visceral perception and visceral self-control was stated clearly by Brener (30), who asserted that persons who are trained to discriminate a visceral response should display improvement in their ability to control that response as well as in their ability to identify specific occurrences of it. These and related hypotheses prompted behavioral scientists to develop valid and reliable measures of individual differences in visceral self-perception (27).

A. Conditioning of Physiological Responses

The biofeedback tradition derived from demonstrations that physiological responses can be brought under stimulus control through either classical or instrumental conditioning procedures. Classical conditioning experiments represent, perhaps, the first coherent contribution to the psychophysiology of interoception by providing substantive evidence for (1) the existence of a highly organized afferent system that underlies transfer of visceral signals to the central nervous system and (2) specific information about the projection and termination of visceral afferent fibers in the central nervous system (27,28).

Until very recently, Soviet physiologists appeared to have monopolized the study of visceral afferents. Between 1900 and 1961, nearly 5000 experiments concerned with visceral afferent processes were reported by Soviet scientists and reviewed extensively by Bykov (31), Chernigovskiy (32), and Razran (33). Many experiments reviewed in these volumes share a surprisingly similar paradigm (see Ref. 33). The conditional stimulus or the unconditional stimulus, or both, are delivered to the mucosa of a specific viscus. A typical stimulus used in many experiments is the distension of the lumen of an organ by a fine-walled rubber balloon inflated with either air or water. The nature of these studies dictates that most of the data are collected on animals, but reflex responses to visceral stimulation and conditional visceral responses have also been demonstrated in humans. To accomplish this, stimuli are applied through fistulas in patients, or through swallowed balloons. A great many reflex responses have been examined, including both visceral-visceral and visceral-somatic connections. Examples include conditional skeletal and visceral activity elicited by uterine contractions, stomach distensions, and gall bladder stimulations; conditional urinary secretions, heart rate accelerations, and vascular responses; and, in humans, conditional gastric secretions, vasoconstriction, vasodilation, hyperventilation, and bladder fullness (31,33,34).

While the voluminous research on the classical or Pavlovian conditioning of autonomic responses evolved, virtually no comparable work was evident on the instrumental conditioning of such responses (35). As recently as 1961, influential texts reported that "for autonomically mediated behavior, the evidence points unequivocally to the conclusion that such responses can be modified by classical, but not instrumental, training methods" (see Ref. 36, p. 100). Within the subsequent decade, however, these beliefs were tempered by a series of influential papers supporting the view that autonomic responses could be modified by instrumental conditioning methods (37–39). Efforts to demonstrate conditioned cardiovascular responses have always dominated the field, but limited efforts to alter respiratory behaviors through instrumental or operant conditioning were evident during this period (28,40).

B. Conditioning of Respiratory Responses

Studies demonstrating classical conditioning of respiratory responses in both animals and humans have been reviewed by Freedman in 1951 (41) and more recently by Ley (42). Many of these studies share the same procedures and involve pairing an auditory (conditioned) stimulus with a shock (unconditioned) stimulus to produce conditioned increases in respiratory frequency or variations in the pattern of breathing such as inspiratory gasps. In both reviews, connections were fashioned between the plasticity of respiratory behavior and psychosomatic complaints (41) or psychophysiological disorders such as panic (42). Contemporary research in the self-regulation and self-management of asthma, however, is far removed from both classical conditioning and instrumental learning models proposed to account for the psychogenic origin and maintenance of asthma and asthma-like responses (43–46). One case study (asthma elicited by an artificial rose) and one theoretical model (equating asthma to a cry for help) are sufficient to evoke the psychodynamic hold placed on asthma in earlier decades (43,46).

The application of biofeedback to the treatment of asthma represents a distinct phase of research and theory on the perception and control of respiratory responses since it serves to link past demonstrations that physiological responses can be brought under stimulus control through either classical or instrumental conditioning procedures with current efforts to affect perceived increases in the resistance to breathing. Biofeedback training has been shown to improve asthma-related measures, but additional testing is warranted to determine its clinical utility (47–50). Biofeedback training for asthma has involved both direct and indirect attempts to alter lung function through instrumental, or operant, procedures (see Ref. 50 for review). In direct approaches, feedback is provided to improve a specific measure of lung function. The effectiveness of feedback, based on either forced expiratory volume in 1 second (FEV_1) or peak expiratory flow rate (PEFR), is inconclusive since the studies performed to date have not provided evidence that lung function was reliably altered by a biofeedback treatment; on the other hand, feedback based on total respiratory resistance (R_t) has provided evidence of conditioned R_t responding in asthmatics (47,50,51). In indirect approaches, training is provided to improve lung function by reducing levels of skeletal muscle tension. The control of asthma obtained through general relaxation is not enhanced by the addition of biofeedback training for general relaxation, but specific reductions in facial muscle tension lead to improvements in lung function and to better attitudes toward asthma (50) (Chapter 3).

III. Sensations, Symptoms, and Illness

An area of consensus among clinicians and researchers is that obstructed or hindered breathing in acute asthma is a sensory experience similar to those initiated by external stimuli. But asthma patients present a constellation of complaints that result not only from active processing of primary respiratory signals (e.g., difficulty in breathing, chest tightness) but also from multiple collateral signals (e.g., fatigue, panic, "pins and needles"). In the process, physiological sensations undergo a process of elaboration in becoming symptoms (52). The visibility of symptoms and their frequency of occurrence, as well as the representation of their causes and consequences, guide the search for verification of disease and steer health-seeking behavior (53–56). Illness refers to the interaction between disease processes and the cognitions, beliefs, and behaviors that reflect a patient's understanding of, and responses to, disease (53, 57–59).

A. Interoception: Visceral Afferents and Sensations

Sechenov (1866), the founder of Russian physiology, and Pavlov (1911) hypothesized that sensory nerves—Pavlov's "internal analyzers"—permeated all internal tissues (32). But the system used to characterize sensory receptors by anatomical location originated with Sherrington, who organized the distribution of receptor organs into two "receptive fields": the "deep field" and the "surface field." Sherrington's deep field consists of proprioceptors mediating sensations arising from muscles and joints, such as position and movement; the relationship of body segments to one another; and forces on the tendons and joints. His surface field is composed of the external (exteroceptive field) and internal (interoceptive field) surfaces of the body. The external field is open to the environment, and its exteroceptors mediate sensations that impinge on the surface of the body such as sound and pressure on the skin (i.e., the classic and cutaneous senses). The internal field consists of all the internal bodily receptors (i.e., interoceptors) that reside in the walls, coverings, and attachments of the viscera, including all thoracic, abdominal, and pelvic organs, as well as blood vessels, bone, and endocrine glands (31,32).

Until the late 1800s, philosophers and physiologists tended to ignore sensations arising from internal structures because feelings, regarded as the avenue by which the mind learns about things, must normally have an object, whereas feelings such as cramps, tickle, and hunger appear to have no object (60). The field of sensory psychology, which emerged from the synthesis of sensory physiology and the epistemological ideology of the philosophers in the middle of the nineteenth century, defined sensation as the basic issue for study—including sensations elicited from within the body. Some of the first experiments designed to examine the ability of individuals to perceive visceral sensations were conducted by Boring in 1915 (61,62). In those experiments, he examined the sensitivity of humans to pressure, temperature, and chemicals (e.g., mustard and alcohol) applied to portions of the alimentary canal and stomach. At about the same time, the coincidence between hunger pangs and stomach contractions was described by Cannon and Washburn (as reviewed in Ref. 60), and the association between pain sensations and stimulation of the pleural membranes was described by Capps (63).

Visceral afferents serve at least three functions. The primary function is homeostasis and the normal functioning of organs and systems during wakefulness and sleep. A second function is to inform higher centers about conditions that necessitate voluntary responses, such as hunger and thirst, as well as to provide warning signals in emergency conditions. Finally, visceral afferents may serve an intermediate function by influencing behaviors without giving rise to specific sensations as might occur in emotional states. We can therefore envision a continuum of signals arising from visceral events ranging from glucose activation in the liver and gastric secretions to synthetic sensations such as hunger, thirst, and shortness of breath to emergency states such as vomiting and angina. Along this continuum, a psychophysiology of visceral sensation must contend with many serious obstacles including the strong homeostatic mechanisms that keep all functions in check, the relatively small cortical space devoted to visceral afferents, and the overshadowing of visceral senses by cutaneous ones, as in referred pain (27,28).

Virtually all motivational and behavioral theories within psychology have assumed that awareness of internal states affects behavior in some way. Perhaps the most influential theoretical discussion of the role of visceral sensibility for behavior was generated at the turn of the century by William James (64). James suggested that when certain stimuli excited the viscera, the perception or "feeling" of the elicited bodily changes defined an emotion. The "object-simply-apprehended" became the "object-emotionally-felt." Preoccupations with the role of visceral sensation in social behavior, including human emotion, and struggles to develop adequate methods to assess individual differences in the perception of physiological sensations continue to characterize a great deal of contemporary research in psychology (27,28,65). Interest in the perception of physical sensations has evolved, in parallel, to include consideration of the processes and individual difference variables influencing the perception of symptoms (66,67).

B. Sensations and Symptoms

Acute bronchoconstriction always involves increases in the resistance to breathing that may be documented with tests of lung function. Both chronic and acute increases in the resistance to breathing are accompanied by less reliable signs and symptoms of asthma including wheeze, cough, sputum production, and shortness of breath endorsed by patients as an increased "effort" or "work" of breathing and as chest "tightness" or "constriction" (68,69). In general, patterns of asthma symptomatology do not vary among patients with respect to age, age at onset of asthma, or illness duration, but females tend to report more symptoms than males (70–72).

The detection of physiological change and the perception of symptoms are uniquely subjective processes, apparent only to the affected person. Pennebaker and Hoover (67,73) have elaborated a distinction between sensory-based "detection" and information-based "perception" processes. In their analysis, detection is highly dependent on sensory information arising from stimulation of physiological receptors; perception, on the other hand, results from the active processing of multiple sources of information (73,74). The perception of symptoms is, in part, an attribution process that is influenced by demographic variables, individual differences, and sociocultural factors (67). Symptom attribution includes processes involving the ways in which persons identify and evaluate symptoms and make interpretations about their causes and consequences. The information-processing models sparked by the "cognitive revolution" in psychology emphasize that individuals are active and selective processors of information, and operate in the environment by selectively organizing ongoing experience into subjectively meaningful patterns (75). People process information about symptoms according to their preconceived notions about the relatedness among physiological, psychological, and environmental quantities (76) and form "disease prototypes"—mental models, or representations—that serve as standards against which to match and to evaluate symptoms (77). Schema theory is the primary theoretical model used to assess how cognition, including thinking about the causes and effects of symptoms (58,74), affects behavior (75). Individual cognitive models, or schemata, are constructed to give meaning to behavior; they reflect understanding of the relatedness among physiological, cognitive, social, and environmental quantities.

C. Illness Behavior

Mechanic (53) has defined illness behavior as the way in which people monitor their bodies, define and interpret their symptoms, take remedial action, and utilize various sources of help including the health-care system. A sequence of decisions, each with a particular temporal component, can be imagined to account for processes involved in health-care judgments: appraisal delay, illness delay, and utilization delay (78). Appraisal delay refers to the time it takes to determine that symptoms indicate disease and illness delay is the time between the appraisal or self-examination processes and the decision to seek medical care. Utilization delay refers to the time between the decision to seek medical care and entry into the health-care system.

An analogous set of processes were described by Dracup and colleagues (79) to account for the delay in seeking treatment for heart-attack symptoms. The delay between the onset of symptoms and emergency treatment is more than 4 hours in up to 44% of acute myocardial infarction (AMI) patients. Three phases of action were outlined for AMI patients: a symptom-recognition phase, a prehospital action phase (travel to hospital), and a hospital action phase (from arrival to definitive care). Additional analyses by Dracup and colleagues revealed that symptom-recognition delays were shorter in patients considered hemodynamically unstable, in those with larger infarcts, and in those who referenced sensations accurately to the cardiac system. Delays were longer for females and for older individuals but were unaffected by either education levels or knowledge about AMI symptoms. In their analyses, Dracup and colleagues adopted a health belief model, in part to explain the response of an individual to the signs and symptoms of AMI. In that model, a person's beliefs about a health threat, and about whether a specific action will lead to a reduction of the health threat, are used to predict individual behavior.

Theoretical analyses are not available to explain the response of an individual to the signs and symptoms of acute asthma, but delays in seeking medical care for acute asthma may be common. A survey of 50 patients attending a hospital clinic at least four times a year indicated that 72% of patients delayed seeking help in the preceding 12 months during an attack of asthma lasting at least 48 hours (80). In 10 severe cases requiring admission to a respiratory intensive care unit, the mean duration of symptoms prior to hospital admission was 24 hours; five cases reported having had symptoms for over 48 hours (17). Progress in efforts to determine the value of symptom discrimination in asthma self-management will depend, therefore, not only on the capability of patients to distinguish airflow increases accurately but also on (1) the sequence of decisions that affect health-care judgments in acute asthma and (2) the delays individuals exhibit in response to decreasing levels of lung function and to worsening asthma. Data comparable to those provided by a British Thoracic Society subcommittee will contribute to these efforts (81). The subcommittee reviewed the management and treatment of 90 patients who died from asthma in the year prior to the fatal episode, a month before the episode, and at the time of the fatal attack. Information about the duration of the fatal attack and the timing of requests for medical services in patients during the last phase of the fatal attack, for example, was obtained from medical records and from interviews conducted with the general medical practitioner and a relative.

IV. Perception of Asthma

Patients with asthma are characterized by hypersensitivity to allergens, chemicals, air temperature, dust, and other stimuli that result in widespread narrowing of the airways and, consequently, in increases in the resistance to breathing and in abnormalities in the distribution of air into the lungs (82). Smooth muscle contraction, edema with or without inflammatory cellular infiltration, and decreases in the clearance of secretions serve to increase the resistance to airflow through the tracheobronchial tree. The inflammatory response to allergens, infections, exercise, and other stimuli develops over minutes, hours, or days; produces variable degrees of airflow limitation both within and between individuals; and results in uneven and nonuniform degrees of obstruction throughout the lungs that can remit spontaneously or reverse with treatment (8).

A. Increased Resistance to Breathing

One index of symptom awareness in asthma patients includes the relationship (i.e., correlation) between perceived symptoms and abnormalities in lung function. Symptom awareness in patients has been calculated in two general ways: (1) single pairs of observations are obtained from patients exhibiting a range of symptom scores and lung-function values, and a correlation is computed to determine the magnitude of correspondence between the two quantities (i.e., between-subjects analysis), and (2) repeated measures of symptoms and lung function are obtained from the same patient and used to compute a single, intraindividual correlation (i.e., within-subjects analysis). The overall interindividual relationship between symptoms and lung function in patients with asthma is surprisingly high. In one study, the correlation coefficient between the intensity of symptoms at the height of induced bronchoconstriction and predicted FEV_1 values in 61 patients was 0.97 (9); in another, a significant overall relationship (r = 0.88) was observed in 45 patients with asthma between Borg ratings of breathlessness and reductions in FEV_1 during methacholine challenge (11). Reliable correlations were also evident between scores acquired with a 100-mm visual analog scale (VAS) and percent predicted FEV_1 (r = –0.532) in 40 children ranging in age from 9 to 17 years during routine outpatient review (83), and between VAS dyspnea ratings and percent predicted PEFR (r = –0.72) recorded in 40 stable patients ranging in age from 16 to 60 years (84).

On the other hand, correlations between pairs of observations obtained in the same individual vary over a wide range. For example, individual correlation coefficients computed between VAS ratings and PEFR obtained twice daily during hospitalization for acute asthma in 12 patients ranged from –0.17 to –0.79 (84), and VAS ratings and PEFR ranged from –0.16 to 0.86 in 37 children at summer camp ranging in age from 7 to 15 years (72). In a study involving 11 general practices in and around Bristol, U.K., both VAS ratings of asthma severity and PEFR were recorded up to four times daily for 14 days in 255 asthma patients whose ages ranged between 17 and 76 years (13). In that investigation, 60% of participants (n = 152) were considered "poor discriminators" of asthma because the intraindividual correlation coefficient calculated between VAS scores and percent predicted PEFR was statistically unreliable.

We have advocated previously the use of intrasubject, rather than intersubject, correlations to interpret relationships among measures of pulmonary function, and provided both conceptual and computational justifications for our recommendation (85). Individuals with high scores on one measure tend to have high scores on a second measure, whereas those with low scores on one tend to have low scores on the second. The similarity of a given individual's scores on two measures exaggerates the correlation between them. The contribution of this artifact increases as the population range increases. When the population range is great, a high interindividual correlation may not reflect the extent to which an individual's score on one measure predicts his or her score on a second measure. These same caveats apply to the interpretation of interindividual relationships observed between perceived symptoms and abnormalities in lung function in patients with asthma.

B. Symptom Discrimination

Symptom discrimination refers to the awareness and appropriate interpretation of the changes in internal state that precede or predict the onset of an asthmatic episode. It represents the first stage in the detection and treatment of an asthmatic episode. Failures by patients to detect variations in internal state, including increases in the resistance to breathing that accompany acute bronchoconstriction, delay the initiation of behaviors designed to manage or otherwise control acute bronchoconstriction (26,86). Creer (26) enumerated reasons that patients fail to recognize increases in airflow resistance or correctly discriminate the onset of an asthmatic episode (Table 1). In his analysis, Creer emphasized that a lack of correlation among physiological, cognitive, behavioral, and environmental events leads to failures in symptom discrimination.

The clinical significance of reduced perception of airflow obstruction in asthma self-management remains to be determined; experimental work that relates directly to symptom discrimination in asthma has only started. One recent study (16) demonstrated that most patients with near-fatal asthma have a blunted perception of dyspnea during resistive loading and diminished chemoresponsiveness, but reduced perception of breathlessness was not a significant predictor of risk for a near-fatal asthma attack in one cohort of 45

Table 1 Classification of Errors in Symptom Discrimination

Lack of knowledge about:
Precipitants of attacks
Precipitant of a given attack
Misinterpretation or misperception of environmental/physical stimuli as precipitants
Failure to detect symptoms:
Physiological failure
Cognitive failure
Contextual failure
Misidentification of physiological signs:
Identification of a sign as a signal of an attack when it is not
Misidentification of respiratory change as not signaling an attack when it does
Misidentification of normal respiratory change as a signal for asthma
Failure of interpretation:
Failure to understand meaning of physiological sign
Failure to use physiological sign
Failure to initiate treatment:
Lack of knowledge as to what to do
Failure to seek assistance

Source: Ref. 26.

patients (87). Those who have experienced a near-fatal episode of asthma exhibit greater variability in daily peak flow scores, are less compliant with prescribed treatment, and manifest fewer adaptive personality traits compared to those who have never experienced such an episode (15). Recurrent risk factors for fatal asthma that likely relate to the undertreatment of asthma include: age, airway lability, delays in seeking medical attention for symptoms, denial, duration of asthma, persistent airflow obstruction, poor compliance, previous hospital admission, previous occurrence of life-threatening asthma, reduced hypoxic responses, and severe obstruction without awareness (e.g., 10,11,14–18,87). Casual review of these factors is illustrative of the variety of sensory factors, perceptual processes, and illness behaviors that pertain to greater asthma morbidity and mortality.

V. Changing Respiratory Perceptions

We emphasized previously that symptom perception is influenced by demographic variables, individual differences, and sociocultural factors (66,67; see Ref. 88 for review). Cioffi (88) appraised the issue of whether it is more adaptive to distract attention away from an unpleasant sensation or to focus attention on it. On the one hand, active distraction from noxious physical sensations or from one's own reaction to them often facilitates tolerance of and adaptation to a physical stressor. For example, distraction underlies many psychological interventions for pain management. On the other hand, the assumption that increases in somatic awareness (i.e., sensory monitoring) lead to appropriate self-regulatory behaviors by improving knowledge of physiological status (88) suggests that it may be adaptive to focus attention on the distress. In this sense, a developing literature in asthma self-management relates to two related hypotheses: (1) asthma patients can improve their ability to detect airflow increases and to assess respiratory perceptions accurately and (2) accurate awareness of acute increases in airflow obstruction accompanying an episode of asthma motivates adaptive illness-reducing behaviors. Adequate testing of the second hypothesis depends on confirmation of the first.

A. Peak Flow Training

The need for repeated, objective measurement of airflow obstruction in asthma is widely accepted. Summaries of the National Asthma Education Program, for example, have recommended that all patients "undergo objective measurements of airflow in the physician's office during each visit, and those patients with moderate-to-severe asthma, who therefore would be utilizing daily pharmacologic therapy, should have daily peak flow assessment" (8). Daily peak flow assessment at home serves to identify decreasing levels of lung function and wide diurnal variations in lung function that are related to worsening asthma. Barnes (14) recommended that patients with poor symptom perception should be provided with written plans if peak flow levels drop by more than 30%. Such action plans, however, have received little controlled evaluation (89). Moreover, the management of asthma based on peak flow scores occurring within particular ranges, or zones, is not without controversy (Chapters 6 and 8).

The history and development of home peak flow monitoring have been described well by Creer and colleagues (Chapter 6). Many portable meters are available to check peak flow (Figure 1), and the recent proliferation of peak flow monitors has prompted a range of concerns, many related to the validity and reliability of peak flow measurement. A number of investigators have evaluated the performance characteristics of various peak flow meters available commercially as well as the reliability of peak flow scores obtained with different meters within the same person and between persons (90–94). The available devices provide repeatable results, but inaccuracies evident in some meters lead to errors in estimates of peak flow variability, to reductions in the correspondence between peak flow scores obtained with different meters, and to doubts about the ability of portable meters to detect clinically meaningful reductions in lung function (91,92,95,96). In response to these observations, most researchers have recommended that the same device always be used by a patient at home and at the clinic to enable meaningful interpretation and consideration of the scores (91,94–97).

Peak flow monitoring can impact favorably on the management and treatment of asthma. But no consensus is evident on who benefits from monitoring, or whether monitoring contributes reliably to better control of asthma (95,98–101). Several studies demonstrate that few patients currently check peak flow routinely. In a New Zealand study, 16% of 352 patients reported measuring peak flow daily (99), and in a study of 255 patients treated in a general practice in the United Kingdom only 10% reported using a peak flow meter regularly (13). Interviews conducted with patients who own peak flow meters suggest limited perceived value in their use: although 70% of 50 parents found peak flow meters very useful for their children, only 20% of parents suggested that peak flow monitoring helped them to detect worsening asthma (100). Relatedly, only 25% of 191 peak flow owners mentioned the use of peak flow in an evolving attack of asthma, and 75% of 352 patients preferred making asthma-management decisions based on symptoms rather than on peak flow (99). These self-report data, plus problems in patient compliance and instrument decay, serve to support more skeptical views that the costs of peak flow monitoring exceed its presumed benefits (102–104).

Peak flow meters have been used not only as tools to assist in the identification of asthma episodes but also to improve the prediction and

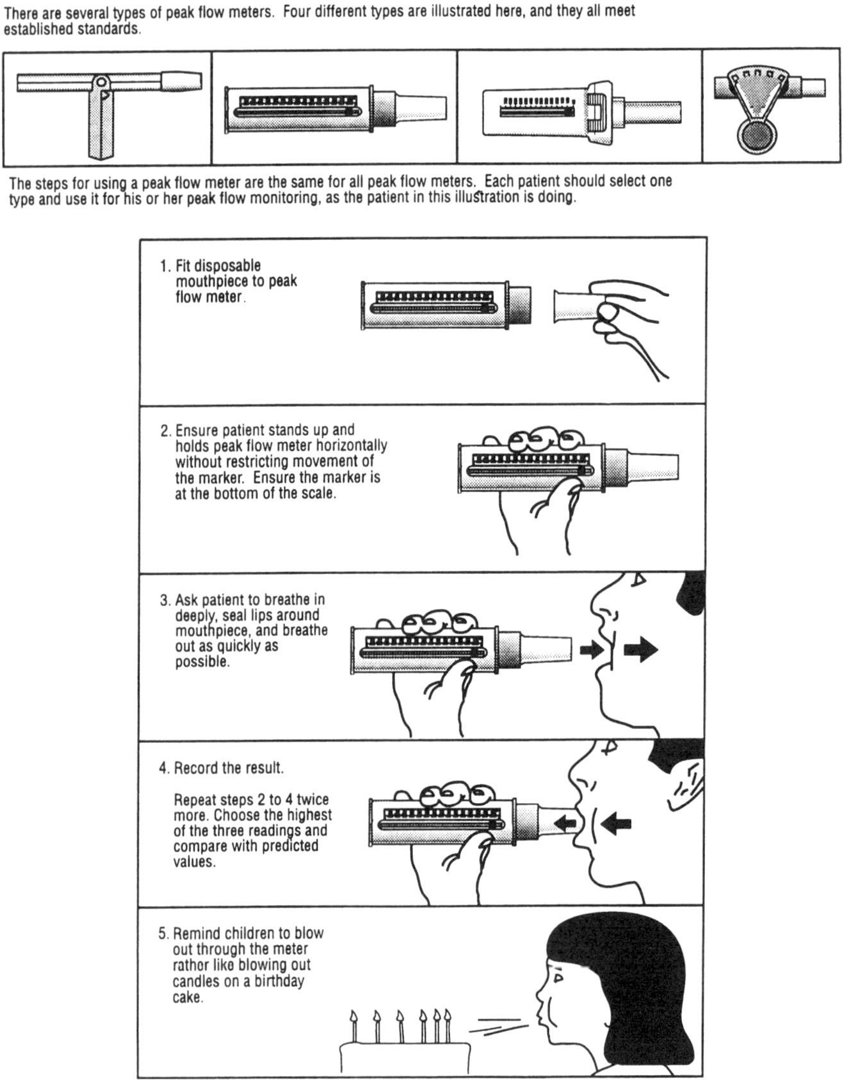

Figure 1 How to teach and check peak flow. (From Asthma Training Center, Warwickshire, England. Reprinted from Ref. 7.)

perception of worsening asthma. Airway lability, or wide diurnal variations in lung function, has long been recognized as a marker of asthma severity. Increased peak flow variability is a distinguishing feature of those patients susceptible to a worsening of their condition within a few days or weeks (105); therefore, determining the most adequate index of airway lability is an ongoing objective (106,107). In a series of papers, Creer and colleagues (108–110) examined the ability of peak flow data to predict the likelihood of asthma within a 12-hour period. Taplin and Creer (110) found an approximately threefold increase in the predictability of asthma episodes for two subjects, and Harm and colleagues (108) found an approximately fivefold increase in the predictability of asthma episodes in 25 subjects. Most recently, peak flow readings, exercise, time of day, and the presence of an attack during the previous 24 hours were found to improve—to varying degrees—the predictability of asthma in 10 subjects (109). Together, these studies demonstrate that improvement in the prediction of asthma attacks is gained from repeated, objective measurement of airflow obstruction by patients at home.

The highly variable association between subjective and objective measures of asthma has led investigators to consider the extent to which patient perceptions of lung function improve as a function of structured peak flow monitoring experiences (111–115). Five studies have been conducted to determine whether improvements in the relationship between estimated and actual peak flow scores (112–115), or between symptoms and actual peak flow scores (111–113), follow daily assessment and recording of these variables. The five studies share strikingly similar designs, including a lack of control or comparison conditions, to evaluate the effects of assessment and recording procedures on patient perceptions. Patients recorded peak flow values at least twice each day for 1 month in three studies (112,114,115), for three months in a fourth study (111), and for between 5 and 21 days in a fifth (113). In four studies (112–115) an estimated peak flow was recorded before actual peak flow measurement, and in three studies (111–113) a symptom score was recorded before actual peak flow measurement.

Intraindividual correlations between estimated and actual peak flow ranged from 0.17 to 0.96 in one study (112), and intraindividual correlations between actual peak flow and symptoms ranged from –0.23 to 0.92 in a second (113). In two other studies (111,115), perception of peak flow did not improve as a function of the training procedure. Silverman and colleagues (113,114), on the other hand, reported more favorable effects of peak flow training in two studies. In contrast to the other studies, intraindividual correlations between estimated and actual peak flow, in 18 patients in one study and in 10 patients in a second study, ranged from 0.97 to 0.99. Taken together, all these studies provide evidence that structured peak flow monitoring experiences lead, at best, to mixed confidence in their ability to affect perception of airflow obstruction in asthma patients. On the other hand, rigorous clinical testing involving adequate control groups is needed to evaluate more fully the impact of peak flow training on asthma-symptom perception.

B. Discrimination Training

An alternative approach to enhancing perceptual sensitivity to acute increases in airflow resistance involves improving the ability of individuals to discriminate directly between the presence or absence of actual airflow increases, or between levels of airflow increases. Providing immediate feedback (e.g., knowledge of results) in this context might facilitate the effectiveness of discrimination training. In other physiological systems, perception of internal state improves as a function of feedback experience. For example, subjects provided knowledge of results in a two-alternative forced-choice paradigm exhibited significant improvements in sensitivity to heart rate (116) and to both systolic and diastolic blood pressure (117). In a series of single-subject experiments, significant reductions in error rates between estimated and actual blood glucose levels were observed when feedback for actual blood glucose levels was provided to patients (118). In the earliest demonstration of the effects of discrimination training in asthma, Renne and his colleagues (119) trained children with asthma to seek treatment for worsening asthma at increasingly higher percent-of-predicted normal peak flow levels.

Effects of feedback on the ability of patients with asthma to detect increases in the flow-resistive component to breathing were examined by Harver (120) in subjects who breathed through external devices (i.e., resistive loads) that increased the resistance to breathing (Chapter 7). In two multisession experiments, the threshold resistance to breathing was determined in an initial session and the effects of feedback were assessed in subsequent sessions. The threshold for increases in respiratory resistance was used as a premeasure of subjects' ability to detect flow-resistive loads and was determined using the method of constant stimuli (i.e., the stimulus intensity detected in 50% of the trials). Signal-detection methodology was used to assess the effects of feedback (121). In signal detection experiments, subjects participate in a series of trials: in half the trials a signal is present and in the other half the signal is absent. Subsequent analyses of hit, miss, and false-alarm rates provide separate assessments of sensitivity and response bias (Chapter 7).

In Experiment 1 (120), detection ability of asthma patients who received feedback about the correctness of their responses in a signal-detection task (experimental subjects) was compared with that of asthma patients who did not receive feedback (control subjects). Experimental subjects determined the presence of flow-resistive loads one-half the magnitude of threshold levels

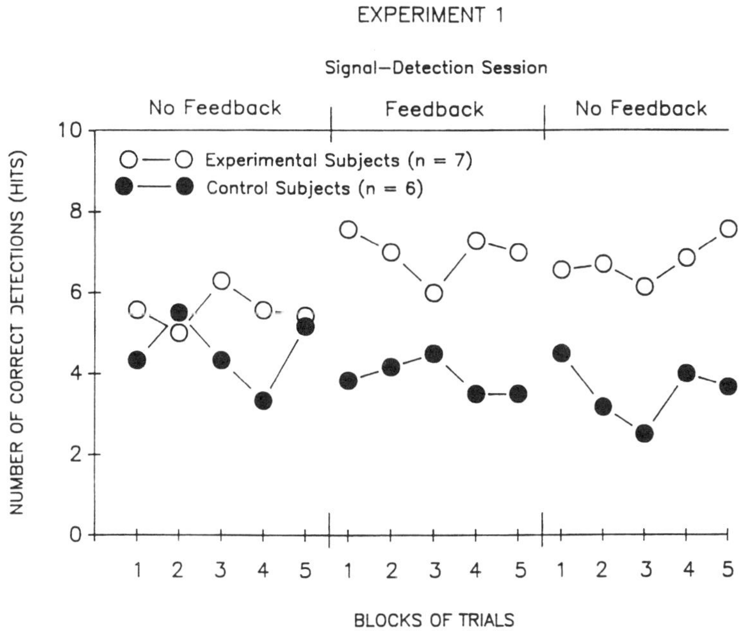

Figure 2 Mean number of correct detections (hits) of an added resistive load one-half the magnitude of threshold levels, in each of 5 blocks of 10 signal and 10 nonsignal trials, in each of three signal-detection sessions held about a week apart. Subjects in the experimental group received feedback about the correctness of their responses after each trial only in the second (feedback) signal-detection session. (From Ref. 120; Copyright © 1994 by the American Psychological Association. Reprinted with permission.)

(M = 1.4 $cmH_2O/L/s$) more accurately than control subjects (Figure 2). Significant differences between groups were observed both in a session in which feedback was provided continuously and in a subsequent session about a week later in which knowledge of results was withheld. In Experiment 2, effects of feedback in a signal-detection task were compared between asthmatic and nonasthmatic subjects. The results from Experiment 2 demonstrated that perception of flow-resistive loads in a signal-detection paradigm was comparable for subjects with and without asthma, and replicated the effects of feedback in discrimination training that had been obtained in the first experiment for asthmatic patients.

In two related studies, Stout and colleagues (122,123) determined the ability of adults with or without asthma to discriminate between two increas-

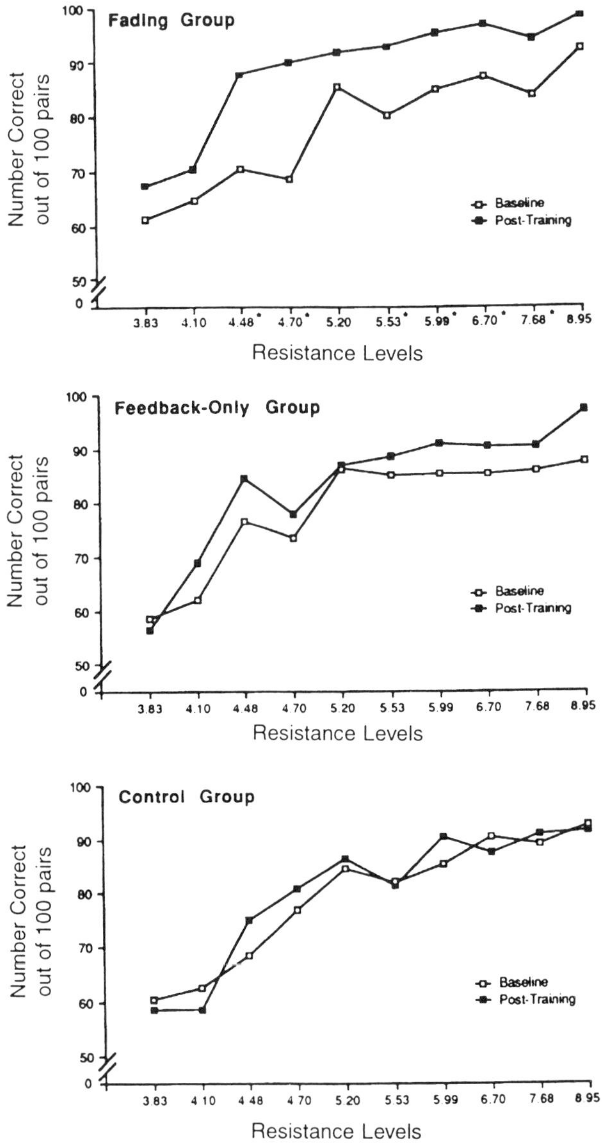

Figure 3 Number of correct responses made by subjects in each of three groups during both baseline and posttraining for each resistive load level. Asterisks along the abscissa indicate statistically significant ($p < .05$) improvements. (From Ref. 122.)

ingly similar levels of added resistance. Stout used a fading procedure that involves presenting standard and comparison loads to subjects in order of increasing difficulty (i.e., from most dissimilar to most similar) on successive breaths. In the first study, with healthy subjects, the standard load was set to 3.41 $cmH_2O/L/s$, and 10 comparison loads were set to range from one that subjects could discriminate from the standard 50% of the time to one that subjects could discriminate from the standard 90% of the time. Healthy individuals who participated in the fading procedure and were informed about the accuracy of their responses improved more in their discrimination of added loads than individuals in either a feedback-only or a control group (122). Such improvements were observed in subjects trained to discriminate between added loads presented either on inspiration (Figure 3) or, in a separate experiment, on expiration.

In the second study (123), asthma patients were tested to see if they could also improve in their ability to detect differences between added loads after training that incorporated both fading and feedback procedures. The standard load was set to 3.41 $cmH_2O/L/s$, and the 10 comparison loads ranged from 3.83 to 7.67 $cmH_2O/L/s$. Each resistive load was applied for two breaths, and resistive load pairs were applied during four consecutive breaths. Patients in the fading condition compared the standard load with the comparison loads in serial order, from the comparison load with the highest resistive value to the load with the lowest resistive value, and were informed by the experimenter as to the accuracy of their responses. As in the study with healthy individuals, asthma patients in the fading-plus-feedback group improved more in their ability to discriminate external resistive loads than patients in either the feedback-only or the control group.

These experiments demonstrate that young adults, with or without asthma, improve in their ability to discriminate between the presence or absence of airflow increases, or between levels of airflow increases, in training tasks involving feedback or fading-plus-feedback. But many uncertainties surround the application of load-discrimination training in asthma. Stout and colleagues (123) emphasized a critical issue: do the improvements observed in these studies generalize to improvements in the ability of patients to detect intrinsic increases in the resistance to breathing as occurs in asthma? Intrinsic loads may be perceived differently than external loads. It is well recognized, for example, that mechanical loads may not fully duplicate the pattern of sensory stimulation produced by lung disease, and intrinsic loads are frequently associated with complex and localized stressors not produced by external loads such as inflammation, vascular changes, and hormonal and electrolyte disturbances (Chapter 7). Finally, it is not known whether improved awareness or perception of acute increases in airflow obstruction that accompany an episode of asthma motivates adaptive illness behaviors.

VI. Summary and Conclusions

We have reviewed a history of interest in the perception of increased resistance to breathing that has its origins in the 100-year-old study of visceral afferents and its contemporary expression in the uncertain clinical significance of reduced perception of airflow obstruction in asthma self-management. Along the way, we have reviewed: applications of biofeedback to the treatment of asthma and to the perception and control of respiratory responses; developments in the study of symptoms, including the distinction between "detection" and information-based "perception" processes prevalent in interactions between health and behavior; and ideas that somatic awareness and interpretation can provide information about actual physiological status and lead to more appropriate self-regulatory behaviors. We intimated that progress in efforts to determine the value of symptom discrimination in asthma self-management will depend not only on the capability of patients to distinguish airflow increases accurately, but also on the sequence of decisions that affect health-care judgments in acute asthma and the delays individuals exhibit in response to decreasing levels of lung function and to worsening asthma.

Despite growing concerns that some proportion of asthma patients are poor perceivers either of asthma symptoms or of the seriousness of symptoms, little experimental work to date relates directly to the problem of symptom discrimination in asthma. Peak flow training is a familiar and convenient strategy to improve patient perceptions, but whether asthma patients improve in their ability to detect airflow increases or assess respiratory symptoms more accurately as a result of structured peak flow monitoring experiences is largely unresolved. Doubts about the ability of portable meters to detect clinically meaningful reductions in lung function, and more skeptical views that the costs of peak flow monitoring exceed its benefits, increase the likelihood that only results obtained in randomized control trials will kindle interest in peak flow training for symptom discrimination. Several recent discrimination studies have demonstrated that young adults either with or without asthma improve in their ability to discriminate between the presence or absence of actual airflow increases in training tasks involving feedback or fading-plus-feedback. But whether the improvements observed in these studies generalize to improvements in the ability of patients to detect intrinsic increases in the resistance to breathing as occurs in asthma remains to be answered. Finally, whether training for perception of increases in the resistance to breathing as occurs in acute asthma serves to motivate adaptive illness-reducing behaviors in patients requires concerted research efforts.

References

1. Boulet LP, Chapman KR. Asthma education: the Canadian experience. Chest 1994; 106: 206S–210S.
2. Boulet LP, Chapman KR, Green LW, FitzGerald JM. Asthma education. Chest 1994; 106:184S–196S.
3. Parker SR, Mellins RB, Sogn DD. Asthma education: a national strategy. Am Rev Respir Dis 1989; 140:848–853.
4. Wilson SR, Starr-Schneidkraut N. State of the art in asthma education: the US experience. Chest 1994; 106:197S–205S.
5. Kotses H, Lewis P, Creer TL. Environmental control of asthma self-management. J Asthma 1990; 27:375–384.
6. Highlights of the Expert Panel Report 2: Guidelines for the Diagnosis and Management of Asthma. Bethesda, MD: National Heart, Lung, and Blood Institute, US Department of Health and Human Services Publication (NIH)97-4051A, 1997.
7. Global Initiative for Asthma. Global Strategy for Asthma Management and Prevention NHLBI/WHO Report. Bethesda, MD: National Heart, Lung, and Blood Institute, US Department of Health and Human Services Publication (NIH)95-3659, 1995.
8. Sheffer AL, Taggart VS. IIIB. The National Asthma Education Program: expert panel report guidelines for the diagnosis and management of asthma. Medical Care 1993; 31:MS20–MS28.
9. Rubinfeld AF, Pain MCF. Perception of asthma. Lancet 1976; i:882–884.
10. Barnes PJ. Poorly perceived asthma [editorial]. Thorax 1992; 47:408–409.
11. Burdon JGW, Juniper EF, Killian KJ, Hargreave FE, Campbell EJM. The perception of breathlessness in asthma. Am Rev Respir Dis 1982; 126:825–828.
12. Higgs CMB, Laszlo G. Perception of asthma. In: Adams L, Guz A, eds. Respiratory Sensation. New York: Marcel Dekker, 1996:285–309.
13. Kendrick AH, Higgs CMB, Whitfield MJ, Laszlo G. Accuracy of perception of severity of asthma: patients treated in general practice. Br Med J 1993; 307:422–424.
14. Barnes PJ. Blunted perception and death from asthma [editorial]. N Engl J Med 1994; 330:1383–1384.
15. Boulet LP, Deschesnes F, Turcotte H, Gignac F. Near-fatal asthma: clinical and physiologic features, perception of bronchoconstriction, and psychologic profile. J Allergy Clin Immunol 1991; 88:838–846.
16. Kikuchi Y, Okabe S, Tamura G, Hida W, Homma M, Shirato K, Takishima T. Chemosensitivity and perception of dyspnea in patients with a history of near-fatal asthma. N Engl J Med 1994; 330:1329–1334.
17. Molfino NA, Nannini LJ, Martelli AN, Slutsky AS. Respiratory arrest in near-fatal asthma. N Engl J Med 1991; 324:285–288.
18. Molfino NA, Slutsky AS. Near-fatal asthma. Eur Respir J 1994; 7:981–990.

19. Woolcock AJ. How does inflammation cause symptoms? Am J Respir Crit Care Med 1996; 153:S21–S22.
20. Adelroth E, Hargreave FE, Ramsdale EH. Do physicians need objective measurements to diagnose asthma? Am Rev Respir Dis 1986; 134:704–707.
21. Baumann UA, Haerdi E, Keller R. Relations between clinical signs and lung function in bronchial asthma: how is acute bronchial obstruction reflected in dyspnoea and wheezing? Respiration 1986; 50:294–300.
22. Brand PLP, Rijcken B, Schouten JP, Koter GH, Weiss ST, Postma DS. Perception of airway obstruction in a random population sample: relationship to airway hyperresponsiveness in the absence of respiratory symptoms. Am Rev Respir Dis 1992; 146:396–401.
23. Dales RE, Ernst P, Hanley JA, Battista RN, Becklake MR. Prediction of airway reactivity from responses to a standardized respiratory symptom questionnaire. Am Rev Respir Dis 1987; 135:817–821.
24. McFadden ER Jr, Kiser R, DeGroot WJ. Acute bronchial asthma: relations between clinical and physiologic manifestations. N Engl J Med 1973; 288:221–225.
25. Fritz GK, Yeung A, Taitel MS. Symptom perception and self-management in childhood asthma. Curr Opin Pediatr 1994; 6:423–427.
26. Creer TL. Response: self-management psychology and the treatment of childhood asthma. J Allergy Clin Immunol 1983; 72:607–610.
27. Reed SD, Harver A, Katkin ES. Interoception. In: Cacioppo JT, Tassinary LG, eds. Principles of Psychophysiology: Physical, Social, and Inferential Elements. Cambridge: Cambridge University Press, 1990:253–291.
28. Vaitl D. Interoception. Biolog Psychol 1996; 42:1–27.
29. Obrist PA, Black AH, Brener J, DiCara LV, eds. Cardiovascular Psychophysiology: Current Issues in Response Mechanisms, Biofeedback and Methodology. Chicago: Aldine, 1974.
30. Brener J. Sensory and perceptual determinants of voluntary visceral control. In: Schwartz GE, Beatty J, eds. Biofeedback: Theory and Research. New York: Academic Press, 1977:29–66.
31. Bykov KM. The Cerebral Cortex and the Internal Organs. New York: Chemical Publishing, 1957. (Original work published 1954.)
32. Chernigovskiy VN (Onischenko G, trans). Interoceptors. Washington, DC: American Psychological Association, 1967. (Original work published 1960.)
33. Razran G. The observable unconscious and the inferable conscious in current Soviet psychophysiology: interoceptive conditioning, semantic conditioning, and the orienting reflex. Psycholog Rev 1961; 68:81–147.
34. Newman PP. Visceral Afferent Functions of the Nervous System. London: Edward Arnold, 1974.
35. Kimmel HD. Instrumental conditioning of autonomically mediated responses in human beings. Am Psychologist 1974; 29:325–335.
36. Kimble GA. Hilgard and Marquis' Conditioning and Learning. 2nd ed. New York: Appleton-Century, 1961.
37. Katkin ES, Murray EN. Instrumental conditioning of autonomically mediated behavior: theoretical and methodological issues. Psycholog Bull 1968; 70:52–68.

38. Kimmel HD. Instrumental conditioning of autonomically mediated behavior. Psycholog Bull 1967; 67:337–345.
39. Miller NE. Learning of visceral and glandular responses. Science 1969; 163:434–445.
40. Miller NE. Biofeedback and visceral learning. Annu Rev Psychol 1978; 29:373–404.
41. Freedman B. Conditioning of respiration and its psychosomatic implications. J Nerv Ment Dis 1951; 113:1–19.
42. Ley R. The modification of breathing behavior: Pavlovian and operant control in emotion and cognition. Behav Modification. In press.
43. Dekker E, Pelser HE, Groen J. Conditioning as a cause of asthmatic attacks. J Psychosom Res 1957; 2:97–108.
44. Franks CM, Leigh D. The theoretical and experimental application of a conditioning model to a consideration of bronchial asthma in man. J Psychosom Res 1959; 4:88–98.
45. Kotses H, Hindi-Alexander M, Creer TL. A reinterpretation of psychologically induced airway changes. J Asthma 1989; 26:53–63.
46. Turnbull JW. Asthma conceived as a learned response. J Psychosom Res 1962; 6:59–70.
47. Mass R, Dahme B, Richter R. Clinical evaluation of a respiratory resistance biofeedback training. Biofeedback Self-Reg 1993; 18:211–223.
48. Kotses H, Harver A, Segreto J, Glaus KD, Creer TL, Young GA. Long-term effects of biofeedback-induced facial relaxation on measures of asthma severity in children. Biofeedback Self-Reg 1991; 16:1–21.
49. Lehrer P, Sargunaraj D, Hochron S. Psychological approaches to the treatment of asthma. J Consulting Clin Psychol 1992; 60:639–643.
50. Kotses H, Glaus KD. Applications of biofeedback to the treatment of asthma: a critical review. Biofeedback Self-Reg 1981; 6:573–593.
51. Janson-Bjerklie S, Clarke E. The effects of biofeedback training on bronchial diameter in asthma. Heart & Lung 1982; 11:200–207.
52. Leventhal H. Symptom reporting: a focus on process. In: McHugh S, Vallis TM, eds. Illness Behavior: A Multidisciplinary Model. New York: Plenum Press, 1986:219–237.
53. Mechanic D. The concept of illness behaviour: culture, situation, and personal predisposition. Psycholog Med 1986; 16: 1–7.
54. Pennebaker JW, Epstein D. Implicit psychophysiology: effects of common beliefs and idiosyncratic physiological responses on symptom reporting. J Personality 1983; 51:468–496.
55. Lacroix JM. Assessing illness schemata in patient populations. In: Skelton JA, Croyle RT, eds. Mental Representation in Health and Illness. New York: Springer-Verlag, 1991:193–219.
56. Mechanic D. Illness behavior: an overview. In: McHugh S, Vallis TM, eds. Illness Behavior: A Multidisciplinary Model. New York: Plenum Press, 1986:101–109.
57. McHugh S, Vallis M. Illness behavior: operationalization of the biopsychosocial model. In: McHugh S, Vallis TM, eds. Illness Behavior: A Multidisciplinary Model. New York: Plenum, 1986:1–31.

58. Skelton JA, Pennebaker JW. The psychology of physical symptoms and sensations. In: Sanders GS, Suls J, eds. Social Psychology of Health and Illness. Hillsdale, NJ: Lawrence Erlbaum Associates, 1982:99–128.
59. Cott A. The disease-illness distinction: a model for effective and practical integration of behavioural and medical sciences. In: McHugh S, Vallis TM, eds. Illness Behavior: A Multidisciplinary Model. New York: Plenum Press 1986:71–99.
60. Boring EG. Sensation and Perception in the History of Experimental Psychology. New York: Appleton-Century-Crofts, 1942.
61. Boring EG. The sensation of the alimentary canal. Am J Psychol 1915; 26:1–57.
62. Boring EG. The thermal sensitivity of the stomach. Am J Psychol 1915; 26:485–494.
63. Capps JA. An experimental study of the pain sense in the pleural membranes. Arch Intern Med 1911; 8:717–733.
64. James W. What is an emotion? Mind 1884; 9:188–205.
65. Öhman A. The psychophysiology of emotion: an evolutionary-cognitive perspective. Adv Psychophysiol 1987; 2:79–127.
66. Harver A, Mahler DA. Dyspnea: sensation, symptom, and illness. In: Mahler DA, ed. Dyspnea. New York: Marcel Dekker, 1998:1–34.
67. Pennebaker JW. The psychology of physical symptoms. New York: Springer-Verlag, 1982.
68. Simon PM, Schwartzstein RM, Weiss JW, Fencl V, Teghtsoonian M, Weinberger SE. Distinguishable types of dyspnea in patients with shortness of breath. Am Rev Respir Dis 1990; 142:1009–1014.
69. Mahler DA, Harver A, Lentine T, Scott JA, Beck K, Schwartzstein RM. Descriptors of breathlessness in cardiorespiratory diseases. Am J Respir Crit Care Med 1996; 154:1357–1363.
70. Kinsman RA, O'Banion K, Resnikoff P, Luparello TJ, Spector SL. Subjective symptoms of acute asthma within a heterogeneous sample of asthmatics. J Allergy Clin Immunol 1973; 52:284–296.
71. Lebowitz MD, Thompson HC, Strunk RC. Subjective psychological symptoms in outpatient asthmatic adolescents. J Behav Med 1981; 4:439–449.
72. Fritz GK, Klein RB, Overholser JC. Accuracy of symptom perception in childhood asthma. Developmental Behav Pediatr 1990; 11:69–72.
73. Pennebaker JW, Hoover CW. Visceral perception versus visceral detection: disentangling methods and assumptions. Biofeedback Self-Reg 1984; 9:339–352.
74. Pennebaker JW, Brittingham G L. Environmental and sensory cues affecting the perception of physical symptoms. In: Baum A, Singer JE, eds. Advances in Environmental Psychology. Vol 4. Environment and Health. Hillsdale, NJ: Lawrence Erlbaum Associates, 1982:115–136.
75. Pace TM. Schema theory: a framework for research and practice in psychotherapy. J Cogn Psychother 1988; 2:147–163.
76. Bishop GD, Converse SA. Illness representations: a prototype approach. Health Psychol 1986; 5:95–114.
77. Bishop GD. Understanding the understanding of illness: lay disease representations. In: Skelton JA, Croyle RT, eds. Mental Representation in Health and Illness. New York: Springer-Verlag, 1991:32–59.

78. Safer MA, Tharps QJ, Jackson TC, Leventhal H. Determinants of three stages of delay in seeking care at a medical clinic. Medical Care 1979; 17:11–29.
79. Dracup K, Moser DK, Eisenberg M, Meischke H, et al. Causes of delay in seeking treatment for heart attack symptoms. Soc Sci Med 1995; 40:379–392.
80. Ellis ME, Friend JAR. How well do asthma clinic patients understand their asthma? Br J Dis Chest 1985; 79:43–48.
81. British Thoracic Society. Deaths from asthma in two regions of England. Br Med J 1982; 285:1251–1255.
82. American Thoracic Society. Guidelines for the evaluation of impairment/disability in patients with asthma. Am Rev Respir Dis 1993; 147:1056–1061.
83. Couriel JM, Demis T, Olinsky A. The perception of asthma. Austral Paediatr J 1986; 22:45–47.
84. Dhand R, Kalra S, Malik SK. Use of analogue scales for assessment of the severity of asthma. Respiration 1988; 54:255–262.
85. Kotses H, Harver A, Harm DL, Wigal JK, Creer TL. More on intrasubject and intersubject correlation: a reply to Popa and Singleton. J Asthma 1987; 24:319–322.
86. Creer TL, Kotses H, Reynolds RVC. Living with asthma: Part II. Beyond CARIH. J Asthma 1989; 26:31–51.
87. Ruffin RE, Latimer KM, Schembri DA. Longitudinal study of near fatal asthma. Chest 1991; 99:77–83.
88. Cioffi D. Beyond attentional strategies: a cognitive-perceptual model of somatic interpretation. Psycholog Bull 1991; 109:25–41.
89. Gibbons PG, Wlodarczyk J, Hensley MJ, Murree-Allen K, Olson LG, Saltos N. Using quality-control analysis of peak expiratory flow recordings to guide therapy for asthma. Ann Intern Med 1995; 123:488–492.
90. Hankinson JL, Das MK. Frequency response of portable PEF meters. Am J Respir Care Crit Care Med 1995; 152:702–706.
91. Harm DL, Kotses H, Creer TL. Portable peak flow meters: intrasubject comparisons. J Asthma 1984; 21:9–13.
92. Miller MR, Dickinson SA, Hitchings DJ. The accuracy of portable peak flow meters. Thorax 1992; 47:904–909.
93. Pedersen OF, Miller MR, Sigsgaard T, Tidley M, Harding RM. Portable peak flow meters: physical characteristics, influence of temperature, altitude, and humidity. Eur Respir J 1994; 7:991–997.
94. Simmons M, Wynegar T, Hess D. Evaluation of the agreement between portable peak flow meters and a calibrated pneumotachometer. Respir Care 1993; 38:916–922.
95. Clark NM, Evans D, Mellins RB. Patient use of peak flow monitoring. Am Rev Respir Dis 1992; 145:722–725.
96. Sly PD, Cahill P, Willet K, Burton P. Accuracy of mini peak flow meters in indicating changes in lung function in children with asthma. Br Med J 1994; 308:572–574.
97. Li JTC. Home peak expiratory flow rate monitoring in patients with asthma. Mayo Clinic Proceedings 1995; 70:649–656.

98. Cross D, Nelson HS. The role of the peak flow meter in the diagnosis and management of asthma. J Allergy Clin Immunol 1991; 87:120–128.
99. Garrett J, Mitchell E, Fenwick JM, Rea H, Taylor G. Peak expiratory flow meters (PEFMs): who uses them and how does education affect the pattern of utilisation? Austral NZ J Med 1994; 24:521–529.
100. Lloyd BW, Ali MH. How useful do parents find home peak flow monitoring for children with asthma? Br Med J 1992; 305:1128–1129.
101. Ruffin RE, Pierce RJ. Peak flow monitoring: which asthmatics, when, and how? [editorial]. Austral NZ J Med 1994; 24:519–520.
102. Charlton I, Charlton G, Broomfield J, Mullee MA. Evaluation of peak flow and symptoms only self management plans for control of asthma in general practice. Br J Med 1990; 301:1355–1359.
103. Jones KP, Mullee MA, Middleton M, Chapman E, Holgate ST, and the British Thoracic Society Research Committee. Peak flow based asthma self-management: a randomised controlled study in general practice. Thorax 1995; 50:851–857.
104. Malo JL, l'Archevêque J, Trudeau C, d'Aquino C, Cartier A. Should we monitor peak expiratory flow rates or record symptoms with a simple diary in the management of asthma? J Allergy Clin Immunol 1993; 91:702–709.
105. Bellia V, Cibella, Coppola P, Greco V, Insalaco G, Milone F, Oddo S, Peralta. Variability of peak expiratory flow rate as a prognostic index in asymptomatic asthma. Respiration 1984; 46:328–333.
106. Enright PL, Sherrill DL, Lebowitz MD. Ambulatory monitoring of peak expiratory flow: reproducibility and quality control. Chest 1995; 107:657–661.
107. Reddel HK, Salome CM, Peat JK, Woolcock AJ. Which index of peak expiratory flow is most useful in the management of stable asthma? Am J Respir Crit Care Med 1995; 151:1320–1325.
108. Harm DL, Kotses H, Creer TL. Improving the ability of peak expiratory flow rate to predict asthma. J Allergy Clin Immunol 1985; 76:688–694.
109. Pinzone HA, Carlson BW, Kotses H, Creer TL. Prediction of asthma episodes in children using peak expiratory flow rates, medication compliance, and exercise data. Ann Allergy 1991; 67:481–486.
110. Taplin PS, Creer TL. A procedure for using peak expiratory flow rate data to increase the predictability of asthmatic episodes. J Asthma Res 1978; 16:15–19.
111. Reeder KP, Dolce JJ, Duke L, Raczynski JM, Bailey WC. Peak flow meters: are they monitoring tools or training devices? J Asthma 1990; 27:219–227.
112. Schandry R, Leopold C, Vogt M. Symptom reporting in asthma patients and insulin-dependent patients. Biolog Psychol 1996; 42:231–244.
113. Silverman BA, Mayer D, Sabinsky R, Williams-Akita A, Feldman J, Schneider AT, Chiaramonte LT. Training perception of air flow obstruction in asthmatics. Ann Allergy 1987; 59:350–354.
114. Silverman B, James C, Misra S, Schneider AT, Chiaramonte LT. Training perception of acute airflow obstruction. Ann Allergy 1990; 64:373–375.
115. Sly PD, Landau LI, Weymouth R. Home recording of peak expiratory flow rates and perception of asthma. Am J Dis Childhood 1985; 139:479–482.

116. Ashton R, White KD, Hodgson G. Sensitivity to heart rate: a psychophysical study. Psychophysiology 1979; 16:463–466.
117. Greenstadt L, Shapiro D, Whitehead R. Blood pressure discrimination. Psychophysiology 1986; 23:500–509.
118. Gross AM, Wojnilower DA, Levin RB, Dale J, Richardson P, Davidson PC. Discrimination of blood glucose levels in insulin-dependent diabetics. Behav Modification 1983; 7:369–382.
119. Renne CM, Nau E, Dietiker KE, Lyon R. Latency in seeking asthma treatment as a function of achieving successively higher flow rate criteria. Presented at 10th Annual Convention of the Association for the Advancement of Behavior Therapy, New York, Dec 1976.
120. Harver A. Effects of feedback on the ability of asthmatic subjects to detect increases in the flow-resistive component to breathing. Health Psychol 1994; 13:52–62.
121. Grossberg JM, Grant BF. Clinical psychophysics: applications of ratio scaling and signal detection methods to research on pain, fear, drugs, and medical decision making. Psycholog Bull 1978; 85:1154–1176.
122. Stout C, Kotses H, Creer TL. Improving recognition of respiratory sensations in healthy adults. Biofeedback Self-Reg 1993; 18:79–92.
123. Stout C, Kotses H, Creer TL. Improving perception of air flow obstruction in asthma patients. Psychosom Med 1997; 59:201–206.

INDEX

A

Q

R

S

T

V

W

Z